AF361874

Advanced Technology Related to Radar Signal, Imaging, and Radar Cross-Section Measurement

Advanced Technology Related to Radar Signal, Imaging, and Radar Cross-Section Measurement

Special Issue Editors

Hirokazu Kobayashi
Toshifumi Moriyama

MDPI • Basel • Beijing • Wuhan • Barcelona • Belgrade • Manchester • Tokyo • Cluj • Tianjin

Special Issue Editors

Hirokazu Kobayashi
Osaka Institute of Technology
Japan

Toshifumi Moriyama
Nagasaki University
Japan

Editorial Office
MDPI
St. Alban-Anlage 66
4052 Basel, Switzerland

This is a reprint of articles from the Special Issue published online in the open access journal *Electronics* (ISSN 2079-9292) (available at: https://www.mdpi.com/journal/electronics/special_issues/radar_related).

For citation purposes, cite each article independently as indicated on the article page online and as indicated below:

LastName, A.A.; LastName, B.B.; LastName, C.C. Article Title. *Journal Name* **Year**, *Article Number*, Page Range.

ISBN 978-3-03936-142-7 (Hbk)
ISBN 978-3-03936-143-4 (PDF)

Contents

About the Special Issue Editors

Hirokazu Kobayashi (Prof., Dr.) was born in Hokkaido, Japan. He received a B.E. and M.E. degree from Shizuoka University, Japan, in 1978 and 1980, respectively, and a Doctorate Eng. from University of Tsukuba, Japan, in 2000. He joined Fujitsu Ltd., Japan, in 1980. He served as a researcher and manager on developments for micro/millimeter wave wide band antennas, active phased array radar systems, and theoretical investigations of electromagnetic scattering and antennas. In 2010, he joined Niigata University, Japan, as a professor. Since 2013 he has been a professor at Osaka Institute of Technology, Japan.

His current research interests include high-frequency scattering field theoretical analysis; computation of radar cross-section (RCS) for large objects; electromagnetic scattering analysis using physical optics; physical and geometrical theory of diffraction; RCS near- to far-field transformation technology for electrical large objects, such as airplanes, based on microwave radar imaging theory; radar SAR and ISAR imaging; theoretical radiation analysis such as conformal array with curved patch antenna; and recognition analysis by micro-Doppler information.

Dr. Kobayashi has published around two hundred papers including international and domestic conference papers and three books. He is a Senior Member of the Institute of Electrical and Electronic Engineers (IEEE) and was a Chair of the Technical Committee on Space, Aeronautical and Navigational Electronics (SANE) of the Institute of Electronics, Information and Communication Engineers (IEICE) from 2015 to 2016. He has been Editor-in-Chief of Microwave and Wireless Communications of the MDPI journal *Electronics* since 2019.

Toshifumi Moriyama (Dr.) received his B.E., M.E., and D.E. degrees in Information Engineering from Niigata University, Japan, in 1994, 1995, and 1998, respectively. In his thesis research, he engaged with radar polarimetry and the polarimetric radar sensing of buried objects. He was with Fujitsu System Integration Laboratories, Ltd., Japan, from 1998 to 2003, the National Institute of Information and Communications Technology, Japan (NICT) from 2003 to 2005, and the Earth Observation Research and Application Center (EORC), Japan Aerospace and Exploration Agency (JAXA) in 2006, respectively.

He is now an associate professor at Nagasaki University. He is involved in research activities in collaboration with ELEDIA Research Center, and he is the Director of the ELEDIA@UniNAGA. His interests are inverse scattering, radar polarimetry, microwave remote sensing, and wireless sensor networks. Currently, he is the Vice Chair of IEEE Aerospace Electronics System Society Japan Chapter and the IEEE Antennae and Propagation Society Fukuoka Chapter.

Preface to "Advanced Technology Related to Radar Signal, Imaging, and Radar Cross-Section Measurement"

The existence of the electromagnetic wave was first established by Maxwell and the measurement by Hertz. Since then, radar application technology has seen great progress. In particular, against the background of the rapid development of hardware and software signal processing in the 1980s, the conventional single function for distance measurement has evolved to advanced imaging and identification processing.

A radar system is electromagnetic wave timing equipment that generates waves by itself and radiates a beam from an antenna toward a target. The echo scattered by the target is received by the antenna again and the signal is detected. It is said that this was first confirmed by Hülsmeyer in Germany. A radar system is made of many elemental and hardware/software technologies. Recent applications expanded to short-distance radars, with functions such as security, nondestructive observation, and aerial monitoring, as well as long-distance radars, with functions such as remote-sensing, surveillance, and weather observation. Furthermore, short-distance radar technology is essential for car sensors in autonomous driving systems. In these various applications, the key technologies supporting the radars are fundamentally the signal, the image, and data processing. This works to detect a target more explicitly and includes synthetic aperture imaging (SAR) and inverse-SAR (ISAR), polarimetry, compressive sensing, multiple-input multiple-output (MIMO) processing, and radar beam scanning.

The radar-related technology is mainly processed within time and frequency domains but, at the same time, it is a multi-dimensional integrated system including the spatial domain for transmitting and receiving electromagnetic waves. Due to the enormous technological advancements of the pioneers actively discussed in this book, research and development in multi-dimensional undeveloped areas is expected to continue. This book contains state-of-the-art work that should guide your research.

Hirokazu Kobayashi, Toshifumi Moriyama

Special Issue Editors

Editorial

Special Issue on "Advanced Technology Related to Radar Signal, Imaging, and Radar Cross-Section Measurement"

Hirokazu Kobayashi [1],* and Toshifumi Moriyama [2]

[1] Department of Electronics and Information Systems Engineering, Osaka Institute of Technology, Osaka 535-8585, Japan

[2] Graduate School of Engineering, Nagasaki University, Nagasaki 852-8521, Japan; t-moriya@nagasaki-u.ac.jp

* Correspondence: hirokazu.kobayashi@oit.ac.jp or h.kobayashi@kobayashi-jim.com

Received: 14 April 2020; Accepted: 14 April 2020; Published: 16 April 2020

---

1. Introduction

A radar system is made of many elemental and hard/software technologies. Recent applications are expanding to short-distance radars, such as security, nondestructive observation, and aerial monitoring, as well as long-distance radars, such as remote-sensing, surveillance, and weather observation. Further, short-distance radar technology is essential for car sensors in autonomous driving systems. In these various applications, the key technologies supporting radars are essentially the signal, image, and data processing in order to detect a target more explicitly, which includes synthetic aperture imaging (SAR) and inverse-SAR (ISAR), polarimetry, compressive sensing, multiple-input multiple-output (MIMO) processing, and radar beam scanning, in a broad sense. On the other hand, radar cross-section (RCS) evaluation and electromagnetic modeling technologies of radar targets are also important for the development of future smart radars.

2. The Present Issue

This Special Issue [1] focuses on the state-of-the-art investigations on various important radar technologies for future applications. We received many paper submissions for this Special Issue. After a very careful peer-review process, a total of 32 papers were accepted. These works include SAR/ISAR [2–9], polarimetry [10–12], MIMO [13,14], direction of arrival (DOA)/direction of departure (DOD) [13–15], sparse sensing [5,14,16], ground-penetrating radar (GPR) [17–19], through-wall radar [20,21], coherent integration [22,23], clutter suppression [24,25], and meta-materials, among others [26–31]. All of these accepted papers are the latest research results and are expected to be further advanced, applied, and diverted. For example, [32] introduces the analytical approach for the development of radio frequency microelectromechanical switches, and [33] explains the comprehensive SAR approach for identifying the scattering mechanisms of radar backscatter caused by vegetated terrains. This Special Issue's editors hope that these papers attract much attention in the research and development of radar technology.

3. Future

As with many applications, radar-related technology is, of course, also deeply linked to the evolution of computer cluster technology. For example, deep learning technology, which is used for the classification of radar targets, has been significantly advanced and will be applied to many applications we covered in our Special Issue. Although a high-capacity storage and high-speed processor are required, these technologies will create new areas of radar application. Furthermore, the electromagnetic scattering behavior of millimeter waves is different from that of microwaves, and it can be a new research target. The expectation and analysis, or modeling and simulation of hardware and software signals will inevitably require knowledge of electromagnetic field theory. Consequently, the development and research of hardware and software in the radar field will be more active in the future.

The radar-related technology is mainly processed within the time and frequency domains, but at the same time, is a multi-dimensional integrated system including the spatial domain for transmitting and receiving electromagnetic waves. Based on the enormous technological assets of the pioneers, as actively discussed in this Special Issue, research and development in multi-dimensional undeveloped areas are expected to continue.

Author Contributions: H.K. and T.M. worked together in the whole editorial process of the special issue, "Advanced Technology Related to Radar Signal, Imaging, and Radar Cross-Section Measurement" published by journal *Electronics*. H.K. drafted this editorial summary. H.K. and T.M. reviewed, edited and finalized the manuscript. All authors have read and agreed to the published version of the manuscript.

Acknowledgments: First of all, we would like to appreciate not only all the researchers who submitted articles to this Special Issue for their excellent contributions but also all the reviewers who contributed to the evaluations of scientific merit and quality of the manuscripts. We would like to acknowledge the editorial board of the *Electronics* journal, who invited us to guest edit this Special Issue. We are also grateful to the *Electronics* editorial office staff who worked thoroughly to maintain the rigorous peer-review schedule and timely publication.

Conflicts of Interest: The authors declare no conflict of interest.

References

1. Kobayashi, H.; Moriyama, T. *Electronics* Special Issues Advanced Technology Related to Radar Signal, Imaging, and Radar Cross-Section. Available online: https://www.mdpi.com/journal/electronics/special_issues/radar_related (accessed on 21 June 2018).
2. Lv, Y.; Wu, Y.; Wang, H.; Qiu, L.; Jiang, J.; Sun, Y. An Inverse Synthetic Aperture Ladar Imaging Algorithm of Maneuvering Target Based on Integral Cubic Phase Function-Fractional Fourier Transform Electronics. *Electronics* **2018**, *7*, 148. [CrossRef]
3. Álvarez López, Y.; García Fernández, M.; Grau, R.; Las-Heras, F. A Synthetic Aperture Radar (SAR)-Based Technique for Microwave Imaging and Material Characterization. *Electronics* **2018**, *7*, 373. [CrossRef]
4. Wang, Z.; Liu, M.; Lv, K. Retrieval of Three-Dimensional Surface Deformation Using an Improved Differential SAR Tomography System. *Electronics* **2019**, *8*, 174. [CrossRef]
5. Li, G.; Ye, W.; Lao, G.; Kong, S.; Yan, D. Narrowband Interference Separation for Synthetic Aperture Radar via Sensing Matrix Optimization-Based Block Sparse Bayesian Learning. *Electronics* **2019**, *8*, 458. [CrossRef]
6. Shi, L.; Guo, B.; Han, N.; Ma, J.; Zhu, X.; Shang, C. Bistatic-ISAR Linear Geometry Distortion Alleviation of Space Targets. *Electronics* **2019**, *8*, 560. [CrossRef]
7. Qian, Y.; Zhu, D. Focusing of Ultrahigh Resolution Spaceborne Spotlight SAR on Curved Orbit. *Electronics* **2019**, *8*, 628. [CrossRef]
8. Zang, B.; Zhu, M.; Zhou, X.; Zhong, L. Application of S-Transform in ISAR Imaging. *Electronics* **2019**, *8*, 676. [CrossRef]
9. Li, G.; Lu, Q.; Lao, G.; Ye, W. Wideband Noise Interference Suppression for Sparsity-Based SAR Imaging Based on Dechirping and Double Subspace Extraction. *Electronics* **2019**, *8*, 1019. [CrossRef]
10. Wang, F.; Pang, C.; Li, Y.; Wang, X. Designing Constant Modulus Sequences with Good Correlation and Doppler Properties for Simultaneous Polarimetric Radar. *Electronics* **2018**, *7*, 153. [CrossRef]
11. Wu, H.; Pang, B.; Dai, D.; Wu, J.; Wang, X. Unmanned Aerial Vehicle Recognition Based on Clustering by Fast Search and Find of Density Peaks (CFSFDP) with Polarimetric Decomposition. *Electronics* **2018**, *7*, 364. [CrossRef]
12. Han, M.; Dou, W. Atomic Norm-Based DOA Estimation with Dual-Polarized Radar. *Electronics* **2019**, *8*, 1056. [CrossRef]
13. Chen, Z.; He, X.; Cao, Z.; Jin, Y.; Li, J. Position Estimation of Automatic-Guided Vehicle Based on MIMO Antenna Array. *Electronics* **2018**, *7*, 193. [CrossRef]
14. Chen, P.; Cao, Z.; Chen, Z.; Yu, C. Sparse DOD/DOA Estimation in a Bistatic MIMO Radar with Mutual Coupling Effect. *Electronics* **2018**, *7*, 341. [CrossRef]

15. Wu, G.; Zhang, M.; Guo, F.; Xiao, X. Direct Position Determination of Coherent Pulse Trains Based on Doppler and Doppler Rate. *Electronics* **2018**, *7*, 262. [CrossRef]
16. Ma, X.; Hu, S.; Liu, S.; Fang, J.; Xu, S. Remote Sensing Image Fusion Based on Sparse Representation and Guided Filtering. *Electronics* **2019**, *8*, 303. [CrossRef]
17. Jia, Z.; Liu, S.; Zhang, L.; Hu, B.; Zhang, J. Weak Signal Extraction from Lunar Penetrating Radar Channel 1 Data Based on Local Correlation. *Electronics* **2019**, *8*, 573. [CrossRef]
18. Wang, Q.; Shen, Y. Calculation and Interpretation of Ground Penetrating Radar for Temperature and Relative Water Content of Seasonal Permafrost in Qinghai-Tibet Platea. *Electronics* **2019**, *8*, 731. [CrossRef]
19. Qin, H.; Xie, X.; Tang, Y. Evaluation of a Straight-Ray Forward Model for Bayesian Inversion of Crosshole Ground Penetrating Radar Data. *Electronics* **2019**, *8*, 630. [CrossRef]
20. Jia, Y.; Song, R.; Chen, S.; Wang, G.; Guo, Y.; Zhong, X.; Cui, G. Multipath Ghost Suppression Based on Generative Adversarial Nets in Through-Wall Radar Imaging. *Electronics* **2019**, *8*, 626. [CrossRef]
21. Bivalkar, M.; Singh, D.; Kobayashi, H. Entropy-Based Low-Rank Approximation for Contrast Dielectric Target Detection with Through Wall Imaging System. *Electronics* **2019**, *8*, 634. [CrossRef]
22. Jin, K.; Lai, T.; Wang, Y.; Li, G.; Zhao, Y. Coherent Integration for Radar High-Speed Maneuvering Target Based on Frequency-Domain Second-Order Phase Difference. *Electronics* **2019**, *8*, 287. [CrossRef]
23. Pan, J.; Hu, P.; Zhu, Q.; Bao, Q.; Chen, Z. Feasibility Study of Passive Bistatic Radar Based on Phased Array Radar Signals. *Electronics* **2019**, *8*, 728. [CrossRef]
24. Wu, Q.; Zhao, F.; Wang, J.; Liu, X.; Xiao, S. Improved ISRJ-Based Radar Target Echo Cancellation Using Frequency Shifting Modulation. *Electronics* **2019**, *8*, 46. [CrossRef]
25. Zhu, Q.; Li, T.; Pan, J.; Bao, Q. PBR Clutter Suppression Algorithm Based on Dilation Morphology of Non-Uniform Grid. *Electronics* **2019**, *8*, 708. [CrossRef]
26. Ji, L.; Hu, X.; Wang, M. Saliency Preprocessing Locality-Constrained Linear Coding for Remote Sensing Scene Classification. *Electronics* **2018**, *7*, 169. [CrossRef]
27. Reggiannini, M.; Bedini, L. Multi-Sensor Satellite Data Processing for Marine Traffic Understanding. *Electronics* **2019**, *8*, 152. [CrossRef]
28. Wang, H.; Qiu, H.; Zhi, P.; Wang, L.; Chen, W.; Akhtar, R.; Zahoor Raja, M.A. Study of Algorithms for Wind Direction Retrieval from X-Band Marine Radar Images. *Electronics* **2019**, *8*, 764. [CrossRef]
29. Hannan, S.; Islam, M.; Hoque, A.; Singh, M.; Almutairi, A. Design of a Novel Double Negative Metamaterial Absorber Atom for Ku and K Band Applications. *Electronics* **2019**, *8*, 853. [CrossRef]
30. Labun, J.; Kurdel, P.; Češkovic, M.; Nekrasov, A.; Gamec, J. Low Altitude Measurement Accuracy Improvement of the Airborne FMCW Radio Altimeters. *Electronics* **2019**, *8*, 888. [CrossRef]
31. Zhao, F.; Liu, X.; Xu, Z.; Liu, Y.; Ai, X. Micro-Motion Feature Extraction of a Rotating Target Based on Interrupted Transmitting and Receiving Pulse Signal in an Anechoic Chamber. *Electronics* **2019**, *8*, 1028. [CrossRef]
32. Lysenko, I.; Tkachenko, A.; Sherova, E.; Nikitin, A. Analytical Approach in the Development of RF MEMS Switches. *Electronics* **2018**, *7*, 415. [CrossRef]
33. Arii, M.; Yamada, H.; Kojima, S.; Ohki, M. Review of the Comprehensive SAR Approach to Identify Scattering Mechanisms of Radar Backscatter from Vegetated Terrain. *Electronics* **2019**, *8*, 1098. [CrossRef]

Review

Analytical Approach in the Development of RF MEMS Switches

Igor E. Lysenko [1], Alexey V. Tkachenko [1,*], Elena V. Sherova [2] and Alexander V. Nikitin [1]

[1]   Department of Electronic Apparatuses Design, Southern Federal University, Taganrog 347929, Russia; ielysenko@sfedu.ru (I.E.L.); designcenter61@gmail.com (A.V.N.)

[2]   Taganrog Scientific Research Institute of Communication, Taganrog 347913, Russia; Lenochka_polai@mail.ru

*   Correspondence: msqk@mail.ru; Tel.: +7-928-605-9722

Received: 22 November 2018; Accepted: 6 December 2018; Published: 10 December 2018

Abstract: Currently, the technology of microelectromechanical systems is widely used in the development of high-frequency and ultrahigh-frequency devices. The most important requirements for modern and advanced devices of the ultra-high-frequency range are the reduction of weight and size characteristics, power consumption with an increase in their functionality, operating frequency and level of integration. Radio frequency microelectromechanical switches are developed using the technology of the manufacture of CMOS-integrated circuits. Integrated radio frequency control circuits require low control voltages, the high ratio of losses to the isolation in the open and closed condition, high performance and reliability. This review is devoted to the analytical approach based on the knowledge of materials, basic performance indices and mechanisms of failure, which can be used in the development of radio-frequency microelectromechanical switches.

Keywords: RF MEMS; switch; analytical approach; low control voltage; high switching speed; high reliability

---

1. Introduction

Radio frequency (RF) electronics of high-frequency (HF) and ultra-high-frequency (UHF) microwave electronics have always needed switches that make low losses in the open state and provide high isolation in a closed state with high permissible signal power and low control power. The first switches fully meet the requirements of steel electromagnetic relays. After many years of improvement modern RF relays provide low on signal loss and high isolation in the off state, but they have such disadvantages as large overall dimensions, high cost and limited resource (from hundreds of thousands to tens of millions of cycles).

The switch on PIN-diodes and field-effect transistors based on GaAs have a number of advantages over the RF relay, consisting in high speed operation, small size and weight, as well as low-power control. However, they introduce higher losses into the microwave transmission line and, therefore, cannot replace electromechanical switches in a number of applications. Further research led to the development of RF switches based on microelectromechanical systems (MEMS), which combined some of the advantages of semiconductor and relay devices.

The comparative characteristics of semiconductor switches are based on PIN-diodes and field-effect transistors, MEMS switches and electromechanical switches of the UHF signal [1].

Micromechanical microwave switches were first demonstrated by Petersen as cantilever beams using electrostatic actuation [2].

According to the forecast of MEMS devices market development, published in 2015 in [3], the following areas of MEMS technologies are noted for high growth: pressure sensors, inertial measurement systems, microfluidic devices and systems. Nevertheless, RF MEMS devices have shown a significant growth from 2014 to 2020, which have great market potential and have been declared as one of the main applications of MEMS technologies.

The RF MEMS segment consisting of filters based on volume acoustic waves and switches boost the income level of the entire MEMS area as a whole with estimated revenues in 2022 at 10 billion dollars [4].

Driven by the complexities associated with the move to 5G and the higher number of bands it brings, there is an increasing demand for RF filters in 4G/5G, making RF MEMS the largest-growing MEMS segment. This market will probably soar from US $2.3B in 2017 to US $15B in 2023. Excluding RF MEMS market will grow at 9% over 2018–2023. With RF MEMS the CAGR is 17.5% [4].

With the miniaturization of RF chips and the growing need to improve their functions, the trend towards integration of RF MEMS and CMOS devices for RF applications will only increase.

RF MEMS switches utilize various actuation designs including electromagnetic [5–8], magnetostatic [9], electrostatic [10], thermal-electric [11] and various structural designs including a rotating transmission line [12], surface micromachined cantilevers [13–17], multiplesupported or membrane based designs [18,19], bulkmicromachined or wafer bonded designs [20–22], diamond cantilever and contact [23], polysilicon switch [24], mercury micro-drop contact [25,26] and bistable microrelays [27,28]. All switches designs have their advantages and disadvantages. The micromachined electrostatic switches or relays are the most widely studied devices to date.

RF MEMS switches with electrostatic activation mechanism are characterized by their own power consumption of the order of several mW, which is considerably lower in comparison with other type of RF MEMS switches. Therefore, most of the developed RF MEMS switches have an electrostatic activation mechanism. At the technological level of electrostatic control it does not require the deposition of specific materials, for example, with piezoelectric or ferromagnetic properties which simplifies the manufacturing technology and, therefore, reduces the cost of finished devices.

However, research conducted in the field of RF MEMS switches is largely limited only by the RF characteristics of these devices, although taking into account the interdisciplinary nature of MEMS technologies, it is necessary to take into account the relationship of electrical parameters with the mechanical parameters of MEMS structures. For example, residual stresses in deposited metal layers of RF MEMS switches have a much greater impact on the performance and life expectancy of the device than expected. In addition, the vast majority of RF MEMS switches require large values of control voltage (40–100 V for pull-down voltage and 15–30 V for hold-down voltage, respectively), along with high switching time, which significantly reduces the scope of their application, excluding wireless communication devices, which involve the use of low-voltage power supplies given the growing demand for RF MEMS switches in this area [3,4,29]. In [30] is also shown that the life expectancy of MEMS switches with an electrostatic activation mechanism strongly depends on the magnitude of the applied voltage.

In addition, achieving a high switching speed remains a major limitation; few studies have been devoted to improving the switching speed, except that proposed in [31], by miniaturizing the MEMS switch. In [32], the authors demonstrate the switching time at the level of nanoseconds through the use of the membrane dielectricas as a structural material. It is shown in [33] that the addition of curved sides to miniature elastic suspension elements increases the stiffness coefficient, causing a further increase in the resonance frequency, which leads to a low switching time. Consequently, improving the switching speed has a negative effect on the control voltage. Increasing the stiffness of the elastic suspension elements will inevitably lead to an increase in the value of the control voltage.

In general, our review presents new design solutions and an analytical approach to the development of RF MEMS switches. The analytical approach is based on the knowledge of materials, key performance indices and failure mechanisms of RF MEMS switches, and parametric analysis is based on a mathematical model.

2. Analytical Approach to the Development of RF MEMS Switches of Electrostatic Type

The analytical approach in the development of RF MEMS switches has the following structure, shown in Figure 1.

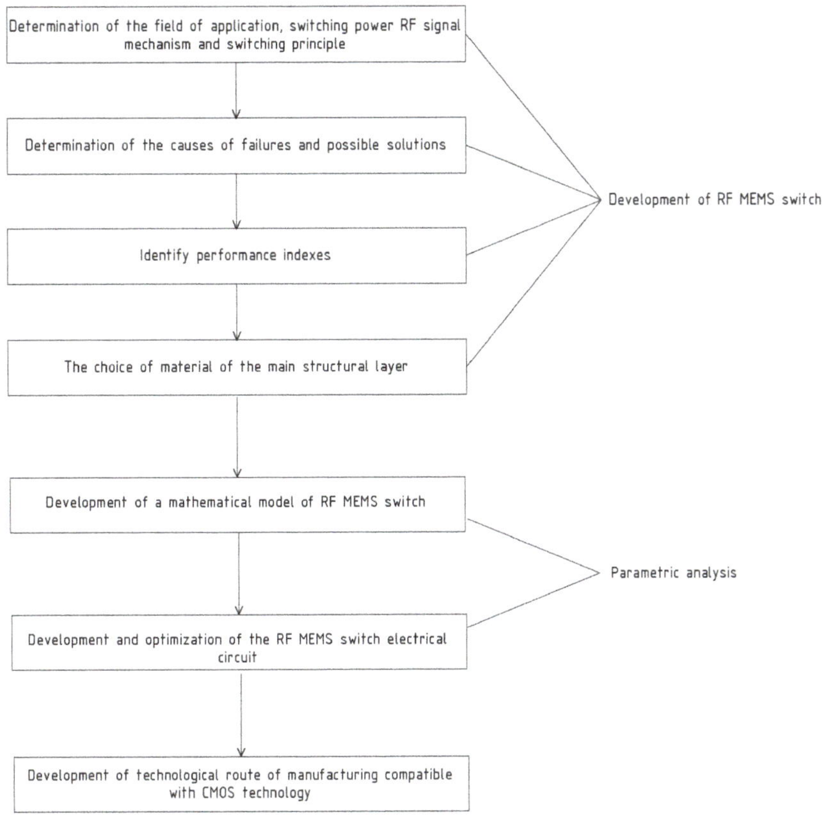

Figure 1. Structure of the analytical approach.

2.1. The Influence of Switching Characteristics of Switch

RF MEMS switches differ in both the activation mechanism and the principle of switching—resistive and capacitive.

The main criterion in the development of RF MEMS resistive type switches (see Figure 2) is the contact area. Large contact areas have less resistance and, therefore, lower contact temperature. The contact area is determined by the applied force and hardness of the material, as well as its ability to form a surface layer with high-resistance. It is also necessary to take into account the adhesive force in the contact—the cantilever or the switch membrane must be sufficiently strong to overcome the adhesive force after removal of the control voltage [1].

Typically, switches with a resistive switching principle are used in a wide frequency range from 0 to 50 GHz, contact resistance of resistive switch based on gold, as a rule, lies in the range of 0.15–0.4 Ω, contact force on the site is from 80 to 500 μN. However, a feature of resistive switches is the need to supply a sufficiently high control voltage—from 60 to 80 V [13,16,17,34].

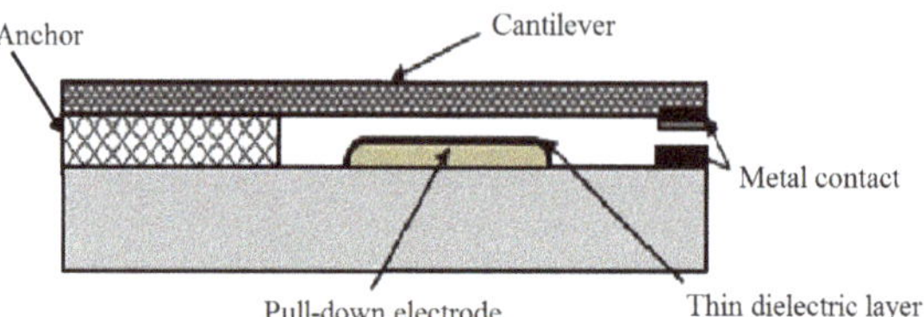

Figure 2. Schematic view of the resistive RF MEMS switch.

RF MEMS switches with capacitive switching principle (see Figure 3) operate by changing the capacitance between the waveguide and the grounded electrode. Typically, switches with capacitive switching principle are used as shunt switches [35] with operating frequencies from 10 to 100 GHz, losses 0.2 dB, isolation from 15 to 35 dB for operating frequencies [19,36,37].

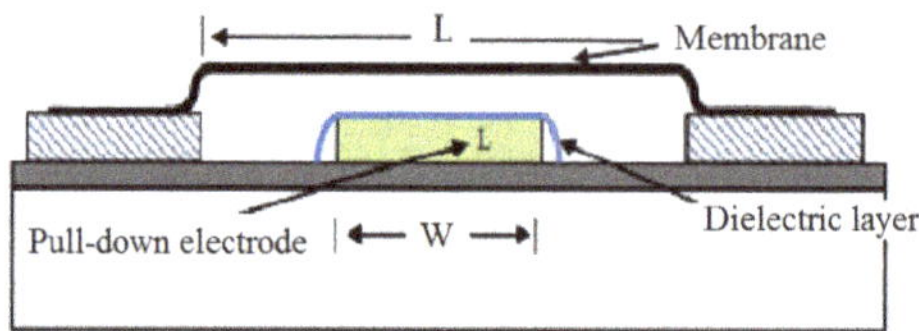

Figure 3. Schematic view of the capacitive RF MEMS switch.

The main advantage of RF MEMS capacitive type switches is the ability to develop switches with low control voltage, since there is no need to make a significant effort to create a contact. However, RF MEMS devices of this type are sensitive to surface roughness and internal stresses in the membrane [38] due to the appearance of an additional air gap, which significantly affects the ratio of capacities in the ON and OFF states. The power handling of RF MEMS switches of the capacitive type does not exceed 7 W, since the current density can exceed the critical value due to the high-resistance of the membrane fastening compared to the switches of resistive type [1].

For the manufacture of high-power handling RF MEMS switches (100 mW and above) more preferred type of contact is the resistive switching principle. This is due to the need to use membranes with a rigid mount for the capacitive switching principle due to the effects of arbitrary operation and high current density, which eliminates their advantages over switches with a resistive switching principle.

For the manufacture of switches of medium and low-power handling (up to 100 mW), switch with capacitive switching principle has an important advantage; the ability to manufacture a switch with low-voltage operation, which allows the use of a single power supply circuit for semiconductor devices and to control the switches.

2.2. Causes of Failure of RF MEMS Switches

For RF MEMS switches with capacitive switching principle at low-power handling (1 mW and below) the main failure mechanism is the electric charge of the dielectric film. The solution to this problem is the use of control voltages with variable polarity [39,40]. For medium-power handling switches (10–100 mW) the main failure mechanisms are related to the electric charge of the dielectric film and failure due to high current density. Problems of failure due to high current density are usually solved by increasing the thickness of the membrane [41,42]. For high-power handling switches the main mechanisms are random tripping and high current density. Random triggering occurs at a voltage in the RF transmission line comparable to the pull-down or return voltage. The solution to this problem is to increase the stiffness of the switch design and control the membrane using separate electrodes [43,44].

For RF MEMS switches with a resistive switching principle at low-power handling (less than 1 mW), the main failure mechanisms are erosion, contact hardening and the formation of dielectric films on the contact. The solution to this problem lies in the selection of materials with the best contact characteristics [45,46]. Medium-power handling switches (10–100 mW) fail due to high contact current density and contact microwelding. This problem is solved by the selection of materials with better contact characteristics [47–50]. For high-power handling switches (100 mW or more), the main failure mechanism is temperature rise, high current density and microwelding of contacts. This problem is solved by constructive work on cooling the contact area, the selection of materials contact [35,51,52].

2.3. Selection of Structural Material of RF MEMS Switch with Capacitive Switching Principle

Due to the fact that the stiction of the membrane to the pull-down electrode as a consequence of the electric charge of the dielectric film is one of the main problems of the reliability of RF MEMS switches with capacitive switching principle, significant efforts of researchers in this area have been aimed at reducing or completely eliminating this problem. In this regard, the properties of materials play a very important role in preventing stiction.

The choice of structural material depends on the hardness, resistance and complexity of the process. The properties of the materials significantly depend on the conditions and deposition technology of the material. For example, the specific resistance of the deposited metal film is almost twice that of the original material. Pure Au provides the lowest contact resistance and is inert to the formation of oxides, but it has been empirically found that the predominant number of failures are associated with pure Au due to the point destruction of the contact area under repeated efforts. Hence, pure Au is not suitable as a structural material of RF MEMS switches that require a long lifespan. Solid metals, such as tungsten or molybdenum, are capable of processing sufficiently high-frequency signal power and do not exhibit any stiction problems. However, they are more sensitive to oxidation and require a relatively high initial contact force. Thus, tungsten and molybdenum are not suitable as structural material. From the obtained results, it was found that the evaporated aluminum is one of the suitable candidates as a contact material [53].

Such materials as gold, aluminum, platinum, molybdenum, copper, and nickel can be used to make an RF MEMS switch with capacitive switching principle [54–57].

When choosing the appropriate material, we should address three main performance indexes of the RF MEMS switches with capacitive switching principle: pull-down voltage, the level of RF losses, thermal residual stress. The key material properties are Young's modulus, Poisson's ratio, thermal expansion coefficient and thermal conductivity.

The pull-down voltage is determined by Equation (1) [1]:

$$V_{pull\text{-}down} = \sqrt{\frac{8k}{27\epsilon_0 A} g_0^3}$$

(1)

where k—the coefficient of stiffness of elastic suspension elements; g_0—the value of the air gap between the electrodes; A—the area of electrostatic interaction.

The coefficient of stiffness k depends on the Young's modulus, thermal residual stress and Poisson's ratio of the material of the membrane and the elastic suspension elements. Consequently, the pull-down voltage can be optimized by selecting a material with the corresponding above properties.

The second performance index is the RF loss level, which can be reduced by selecting a material with good conductivity. The power dissipated by RF is determined by Equation (2) [56]:

$$P_{loss} = I^2 R$$

(2)

where I, R—the current and the resistance of the membrane.

At the RF signal of high-power (from 100 mW and above), the membrane of the switch design experiences self-heating, which leads to a change in the thermal residual stress, which is determined by Equation (3) [56]:

$$\Delta\sigma = E\Delta\alpha P_{loss}R_{TH} \tag{3}$$

where P_{loss}—the level of power dissipated by RF; R_{TH}—the thermal resistance.

Thus, as a result of the analysis, as well as on the basis of material selection diagrams [58], it was found that the most suitable candidate as the main structural material is aluminum after gold and copper.

The use of structures with elastic suspension elements in the form of a meander or thin beams allows us to obtain low values of the stiffness coefficient and control voltage, which is often a compromise between reliability and switching speed.

However, capacitive RF MEMS switches with a low coefficient of rigidity of the elastic suspension elements and low values of the control voltage are subject to deformation, curvature or deformation of the edges of the elastic elements, which prevents the formation of an air gap. The solution to this problem is to use the sandwich structure as a material of elastic suspension elements and the membrane. It embodies two materials of a given geometry and scale, as shown in Figure 4, formed in order to provide high strength and bending rigidity at low mass.

The separated surfaces on both sides of the base material increase the moment of inertia of the structure and form a structure that is well resistant to various kinds of deformations and warping [59]. Another major advantage of sandwich structure is that the balanced structure minimizes the thermal-stress-induced upward warpage in the release process [60,61].

As shown in Figure 4, surfaces with a thickness of t take up most of the load, so they must be tough and durable. In this case, it is proposed to use TiN the advantages of which are high hardness, good adhesion and plasticity, high chemical resistance, high wear resistance, resistance to the formation of oxides and high temperature.

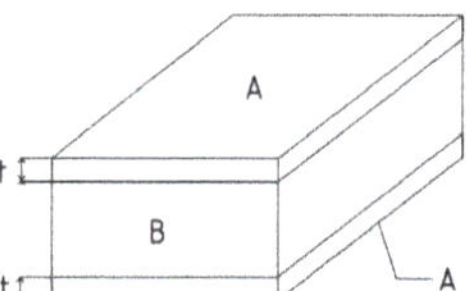

Figure 4. Schematic view of sandwich structure.

In addition, TiN layers prevent the formation of warps, deformations of the edges of elastic suspension elements, which may be present after deposition in the process of manufacturing the structure. In other words, the TiN layers balance the residual stresses in the Al layer that have arisen after the technological processes. Also, TiN is characterized by high-resolution in the process of photolithography, due to good light adsorption.

In the end, the composite material is proposed to use the sandwich structure TiN/Al/TiN, as a material of elastic suspension elements and the membrane.

2.4. Material Properties of the Sandwich Structure

The various composite material properties, such as Young's modulus (E), Poisson's ratio (v), thermal expansion coefficient (α), thermal conductivity (K), electrical resistivity (ρ) and density can be determined by Equation (4) [62]:

$$P_{eff} = \frac{\Sigma P_n A_n}{\Sigma A_n} \tag{4}$$

where P_n—the property of material; A_n—the cross-sectional area.

2.5. Mathematical Model of the RF MEMS Switch with Capacitive Switching Principle

The applied bias voltage is generally separated with respect to the RF signal. One way or another, the potential difference is applied between the membrane connected to the ground lines of the coplanar waveguide and the signal line. Under these conditions, an electrostatic force acts on the design of the switch, which is balanced by the elastic force, depending on the stiffness coefficient k of the elastic suspension elements. In theory, the balance exists until the membrane drops 1/3 of the initial air gap. The membrane then collapses onto the pull-down electrode and a lower voltage value is required to hold it in this position.

The dynamics of the switch are also affected by the presence of a medium (usually air or nitrogen during sealing to eliminate the influence of humidity), which introduces its own friction and causes damping and change in the switching speed [63–65]. Currently, there are several models that take into account the detailed effect of damping, including the presence of holes in the membrane [1,66–68]. In addition, the damping changes the natural oscillation frequency of the membrane.

Another contribution to the movement of the membrane is made by the contact forces of the membrane with respect to the plane of the coplanar waveguide. They are due to the interaction between the two surfaces and the local redistribution of charges. So we should take into account the impact of the forces Van der Waals forces with the effect of the approximation [63,65].

Both last components are important in the case when the membrane is close to the fixed electrode or is in contact with the dielectric film.

However, a phenomenological approach is usually applied, which takes into account the influence of specific most important parameters useful to describe the required mechanical and electrical response.

Equation (5) of the balance of forces acting on the design of the switch has the form:

$$m\frac{\partial^2 z}{\partial t^2} = \Sigma Forces \tag{5}$$

where z—the displacement of the membrane; $m = \rho A t$—the mass of the membrane determined by the material density ρ, area A and thickness t.

Taking into account the given physical quantities, Equation (5) can be rewritten in Equation (6) in the following form:

$$m\ddot{z} = F_e + F_s + F_m + F_d + F_c \tag{6}$$

where $F_e = \frac{1}{2}\frac{\partial C}{\partial z}V^2$—the electrostatic force as a consequence of the applied potential difference with the capacitance change along the direction of motion z; $F_m = -k\left[z - (t_d + g)\right]$—the mechanical force due to the rigidity of the elastic suspension elements, the opposite direction of the electrostatic force F_e; $F_s = -k_s\left[z - (t_d + g)\right]^3$—the nonlinear tensile force of the elastic suspension element [1]; $F_d = -\alpha\dot{z}$—the damping force due to the action of the medium, which depends on the speed of the membrane $\dot{z}$ and the damping parameter α, which in turn is associated with the geometric parameters of the membrane and the viscosity of the medium; F_c—contact forces, which can be divided into Van der Waals and surface forces; the first acts as an attractive and the second as a repulsive, with a possible equilibrium position at a given distance from the fixed electrode [63].

The total capacity according to Figure 3 is calculated using Equation (7):

$$C(z) = \frac{\epsilon_0 \epsilon_r A}{t_d + \epsilon_r(z - t_d)}, z \in [t_d, t_d + g] \tag{7}$$

where $\epsilon_0 = 8.85 \times 10^{-12}$ F·m^{-1}—the dielectric constant of vacuum; ϵ_r—the relative permittivity of the dielectric material film.

The derivative of $C(z)$ is used to determine the electrostatic force, which is given by Equation (8):

$$\frac{\partial C}{\partial z} = -\frac{\epsilon_0 \epsilon_r}{\left[t_d + \epsilon_r (z - t_d) \right]^2} A \tag{8}$$

Thus, Equation (6) can be represented in Equation (9):

$$m\ddot{z} + k\left[z - (t_d + g) \right] + k_s \left[z - (t_d + g) \right]^3 + \alpha \dot{z} = -\frac{1}{2} \frac{\epsilon_0 \epsilon_r A}{\left[t_d + \epsilon_r (z - d) \right]^2} V^2 \tag{9}$$

2.6. Coefficient of Stiffness of Elastic Suspension Elements

Various variants of elastic elements of membrane fastening are known, which are used to reduce the coefficient of elasticity. The most frequently used variants of elastic elements of the membrane suspension are shown in Figure 5.

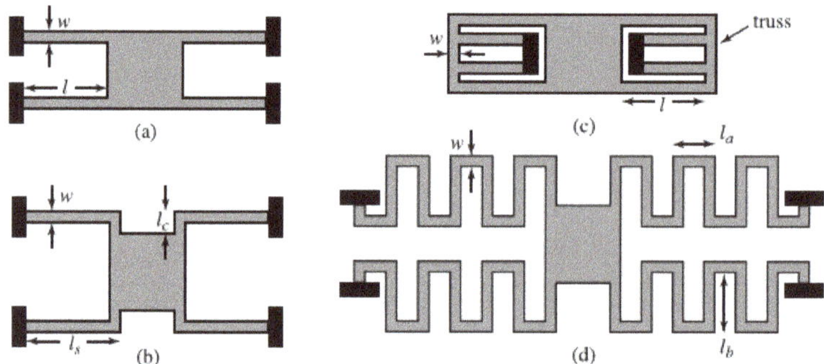

Figure 5. Various support beams used to reduce the spring constant. (**a**) Fixed–fixed flexures. (**b**) Crab-leg flexures. (**c**) Folded flexures. (**d**) Serpentine flexures.

The resulting value of the stiffness coefficient k of the elastic elements fastening and membrane is determined by Equation (10):

$$k = k' + k'' \tag{10}$$

where k'—the effective stiffness coefficient of the elastic suspension; k''—the coefficient of residual stresses in elastic suspension defined by Equation (11):

$$k'' = 4\left[8\sigma(1 - v)w\left(\frac{t}{l}\right) \right] \tag{11}$$

where σ—the tensile strength; v - the Poisson's ratio; w, t, l—the width, the thickness, the length of elastic suspension.

The effective stiffness coefficient k' depends on the type of elastic suspension fastening:

(a)   Fixed-fixed flexure [69]:

$$k' = 4Ew\left(\frac{t}{l}\right)^3 \tag{12}$$

(b) Crab-leg flexure [70]:

$$k' = \frac{4Ew\left(\frac{t}{l_c}\right)^3}{1 + \frac{l_s}{l_c}\left[\left(\frac{l_s}{l_c}\right)^2 + 12\left(\frac{1+v}{1+(\frac{w}{t})^2}\right)\right]} \approx 4Ew\left(\frac{t}{l_s}\right)^3 \tag{13}$$

for $l_s \gg l_c$.

(c) Folded flexure [71]:

$$k' \approx 2Ew\left(\frac{t}{l}\right)^3 \tag{14}$$

for very stiff truss.

(d) Serpentine flexure [71]:

$$k' \approx \frac{48GJ}{l_a^2\left(\frac{GJ}{EI_x}l_a + l_b\right)n^3} \tag{15}$$

for $n \gg \frac{3l_b}{\frac{GJ}{EI_x}l_a + l_b}$. Where n—the number of meanders in the serpentine flexure; $G = \frac{E}{2(1+v)}$—the torsion modulus; $I_x = \frac{wt^3}{12}$—the moment of inertia; $J = \frac{1}{3}t^3w\left(1 - \frac{192}{\pi^5}\frac{t}{w}\sum_{i=1,n}^{\infty}\frac{1}{i^5}\tanh(\frac{i\pi w}{2t})\right)$—the torsion constant. For the case where $l_a \gg l_b$, $k' \approx 4Ew\left(\frac{t}{(nl_a)^3}\right)$.

3. Approbation of the Analytically Approach in the Development of RF MEMS Switches

Schematic view of the developed design of the integrated RF MEMS switch is shown in Figure 6a–c.

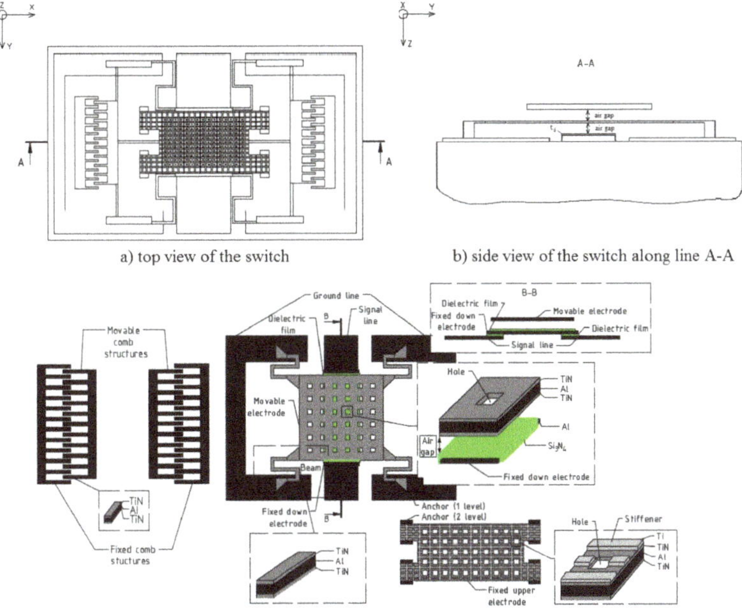

Figure 6. Schematic view of the developed design of the integrated RF MEMS switch.

The available design of an integrated RF MEMS switch contains substrate with a coplanar transmission line located on it, consisting of two grounding lines and a transmission line. Membrane is made in the form of a plate with perforation, located above the dielectric film with air gap between them. The dielectric film is deposited on the surface of the fixed down electrode, which integrated in the transmission line. Membrane is fixed on the anchor areas (1 level) by means of elastic suspensions consisting of elastic beams in the form of a meander. The fixed upper electrode is made in the form of a plate with perforation used to pull-up the membrane. Movable comb structures and two side electrodes are made in the form of comb structures designed to attract the movable comb structures in the case of action on the construction switch of the external acceleration in the negative or positive direction of the X, Y, Z axes, vibrations or stiction of the membrane in the down or up position.

The switch operates as a variable capacitor with two states depending on the position of the membrane. In the neutral position of the membrane, the capacitance between the membrane and the transmission line is small and the signal flows freely to the output of the transmission line. When a control voltage is applied between the membrane and the fixed down electrode, the charge redistribution occurs, which leads to the appearance of an electrostatic force between them, independent of the polarity of the applied voltage. The electrostatic force causes the membrane to fall to a stationary one and since the structure is bent, tensile forces appear in it seeking to return it to its original position. When the applied control voltage reaches the threshold value, the tensile forces cease to balance the electrostatic forces and the membrane drops sharply to the fixed down electrode. In the down position of the membrane the capacitance value increases sharply and the signal coming to the input of the transmission line shunts to the ground lines.

Materials and topological dimensions of the elements of the integrated RF MEMS switch design are presented in Tables 1 and 2. Table 3 shows the material properties of the sandwich structure.

Table 1. Materials of integrated RF MEMS switch.

Structural Element	Material
Coplanar waveguide membrane	Al TiN/Al/TiN
Fixed down electrode	Al
Fixed upper electrode	TiN/Al/TiN
Second electrostatic drive	TiN/Al/TiN
Anchor area	Al
Dielectric layer	Si_3N_4
Elastic suspension elements	TiN/Al/TiN
Substrate	SiO_2-well in *Si*-substrate SOI technology

Table 2. The topological dimensions of the structure elements of integrated RF MEMS switch.

Structural Element	Dimension, μm
Signal line	$600 \times 150 \times 1$
Ground line	$600 \times 100 \times 1$
membrane	$300 \times 300 \times 2$
Hole in the membrane	$5 \times 5 \times 2$
Fixed down electrode	$400 \times 150 \times 1$
Fixed upper electrode	$400 \times 300 \times 2$
Elastic suspension elements	$50 \times 2 \times 2$
Dielectric film	$300 \times 150 \times 0.15$
Air gap	0.8
Anchor area	$50 \times 10 \times 0.8$
Substrate	$600 \times 600 \times 100$

Table 3. The material properties of the sandwich structure

E, GPa	v	$\alpha, 10^{-6}(°C^{-1})$	K, W/m·K	$\rho, \Omega\cdot m$	Density, kg·m³
57.5	0.3175	16.7	196.55	6.32×10^{-8}	4×10^3

In the proposed design of the integrated RF MEMS switch serpentine flexures are used (see Figure 7). Equation (16) presented in [72] was used for the calculation the effective stiffness coefficient k' of the elastic suspension:

$$k' = \left[\left(\frac{8N^3a^3 + 2Nb^3}{3EI_x} \right) + \left(\frac{abN(3b + (2N+1)(4N+1))a}{3GJ} \right) - \left(\frac{Na^2 \left[\frac{2Na}{EI_x} + \frac{(2N+1)b}{GJ} \right]^2}{2(\frac{a}{EI_x} + \frac{b}{GJ})} \right) - \frac{Nb^2}{2} \left(\frac{a}{GJ} + \frac{b}{EI_x} \right) \right]^{-1}$$

(16)

where t—the thickness of meander; a, b—the geometrical parameters of the meander; E—the Young's modulus; v—the Poisson's ratio; I_x—the moment of inertia about the axis x; I_z—the moment of inertia about the axis z; J—the torsion constant defined by the following expression: $J = 0.413I_p$; I_p—the polar moment of inertia: $I_p = I_x + I_z$; N—the number of meanders in the elastic mount.

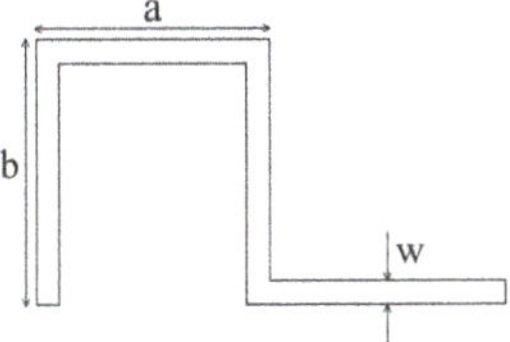

Figure 7. View of the elastic suspension element.

3.1. Study of the Dynamic Characteristics

In the electrostatic drive with parallel plates, the device is triggered sharply, which makes some uncertainty. The tensile forces cannot balance the electrostatic forces for a long time and the switch closes as soon as the threshold voltage is reached. The membrane type capacitive microstructure can be modeled as an elastic beam, which is affected by the electrostatic force and the damping force on the film (see Figure 8), which is described by Equation (17):

$$m\frac{\partial^2 x}{\partial t^2} + b\frac{\partial x}{\partial t} + kx = F_e$$

(17)

At low pressures, the damping term in (17) can be neglected and the system is inertia-dominated.

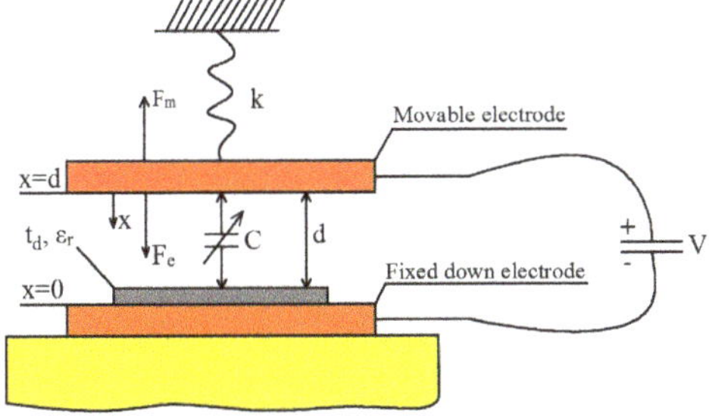

Figure 8. One-dimensional model of the integrated RF MEMS switch.

The electrostatic force acting on the membrane is determined by Equation (18):

$$F_e = \frac{1}{2}\epsilon_0 A \frac{V^2}{x^2} \tag{18}$$

where A—the plate area; d—the distance between the plates; V—the applied voltage. The distance x at different voltages is determined from the balance of forces $F_m = F_e$, using Equation (19):

$$x^3 - dx^2 + \frac{\epsilon_0 A V^2}{2k} = 0 \tag{19}$$

By solving Equation (19) in inverse form, it is possible to obtain Equation (20) to determine the value of the threshold voltage:

$$V(x) = \sqrt{\frac{2k}{\epsilon_0 A} x^2 (d - x)} \tag{20}$$

At a distance of $x = x_{pull-down} = 2d/3$, the $V(x)$ reaches its maximum value $V_{pull\text{-}down}$ and the membrane falls on the fixed down electrode. $V_{pull\text{-}down}$ determined by Equation (21):

$$V_{pull\text{-}down} = \sqrt{\frac{8k}{27\epsilon_0 A} d^3} \tag{21}$$

For the MEMS switch to stay in the down-state position the electrostatic force must be larger than the mechanical restoring force $F_m = k(g_0 - g)$ and this is achieved for a hold-down voltage. The hold-down voltage is determined by the following Equation (22):

$$V_{hold\text{-}down} = \sqrt{\frac{2k}{\epsilon\epsilon_0 A}(d - x)(x + \frac{t_d}{\epsilon_r})^2} \tag{22}$$

The voltage at which the membrane is raised by the elastic forces is determined by Equation (23):

$$V_{return} = \sqrt{\frac{2t_d^2 k d}{\epsilon_0 \epsilon_r A}} \tag{23}$$

The switching time of RF MEMS switches with electrostatic activation mechanism strongly depends on the applied voltage. In most cases, this voltage is 1.3–1.4 of the pull-down voltage in order to obtain a small switching time. A very high switching voltage leads to an increase in electrostatic force, which in turn adversely affects reliability.

The resonant frequency of any system modelled as spring is given by Equation (24):

$$\omega = \sqrt{k/m} \tag{24}$$

The frequency of operation of the integrated RF MEMS switch can be defined using Equation (25):

$$f = \frac{1}{2\pi}\sqrt{\frac{k}{m}} \tag{25}$$

The switching time in down state of the integrated RF MEMS switch can be determined using Equation (26):

$$t_{pull\text{-}down} = 3.67 \frac{V_{pull\text{-}down}}{V_s}\sqrt{\frac{m}{k}} \tag{26}$$

Analytical Equation (27) for the determining the switching time in initial state can be derived by using the energy method ($V_s = 0$) [73].

$$t_{pull\text{-}up} = \frac{1}{4f} \tag{27}$$

3.2. The Calculation Capacity of the Membrane Considering Perforated Holes

Capacitance of parallel plate thickness t, as shown in Figure 9 is defined by Equation (28) presented in [74]:

$$C = \frac{\varepsilon_0 w}{g_0}\left[1 + \frac{2g_0}{\pi w}\ln\left(\frac{\pi w}{g_0}\right) + \frac{2g_0}{\pi w}\ln\left(1 + \frac{2t}{g_0} + 2\sqrt{\left(\frac{t}{g_0} + \frac{t^2}{g_0^2}\right)}\right)\right] \tag{28}$$

Each of the perforated square holes of the membrane can be defined as a separate parallel flat capacitor. There are three components of the container: (a) the capacitance of the membrane; (b) the capacitance in the area of the membrane thickness edging; (c) the capacitance in the area of the hole edging.

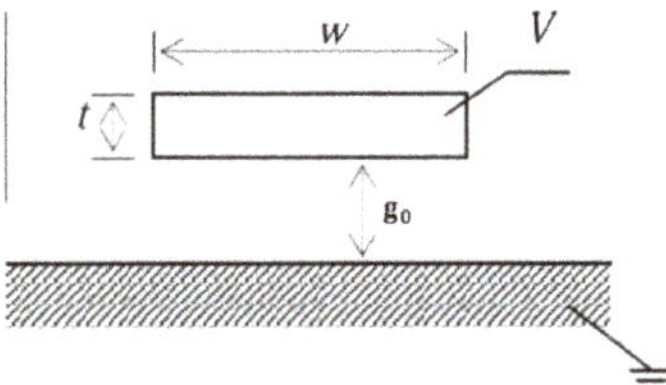

Figure 9. A parallel plate with thickness t.

In this case, the total capacitance in the up state is calculated using Equation (29) [75]:

$$C_{total} = (a) - (b) + (c)$$

$$(a) = \frac{\varepsilon_0 w W}{\left(g_0 + \frac{t_d}{\varepsilon_r}\right)} + \frac{2\varepsilon_0 W}{\pi}\left[\ln\left(\frac{\pi w}{\left(g_0 + \frac{t_d}{\varepsilon_r}\right)}\right) + \ln\left(1 + \frac{2t_b}{\left(g_0 + \frac{t_d}{\varepsilon_r}\right)} + \right.\right.$$

$$\left.\left. +2\sqrt{\left(\frac{t_b}{\left(g_0 + \frac{t_d}{\varepsilon_r}\right)} + \frac{t_b^2}{\left(g_0 + \frac{t_d}{\varepsilon_r}\right)^2}\right)}\right)\right];$$

$$(b) = n_l n_w \frac{\varepsilon_0 w_h^2}{\left(g_0 + \frac{t_d}{\varepsilon_r}\right)};$$

$$(c) = \frac{2n_l n_w \varepsilon_0 w_h}{\pi}\left[\ln\left(\frac{\pi w}{\left(g_0 + \frac{t_d}{\varepsilon_r}\right)}\right) + \ln\left(1 + \frac{2t_b}{\left(g_0 + \frac{t_d}{\varepsilon_r}\right)} + 2\sqrt{\left(\frac{t_b}{\left(g_0 + \frac{t_d}{\varepsilon_r}\right)} + \frac{t_b^2}{\left(g_0 + \frac{t_d}{\varepsilon_r}\right)^2}\right)}\right)\right]. \tag{29}$$

where w—the signal line width; W—the beam width; t_d—the dielectric film thickness; t_b—the beam thickness; n_l—the number of holes along the length; n_w—the number of holes along the width; w_h—the hole area.

To find the down state capacitance air height $g_0 = 0$ will be considered in Equation (29).

3.3. The Results of the Study of the Dynamic Characteristics and Discussion of the Results

Figure 10 shows the dependence of the membrane displacement on the applied bias voltage.

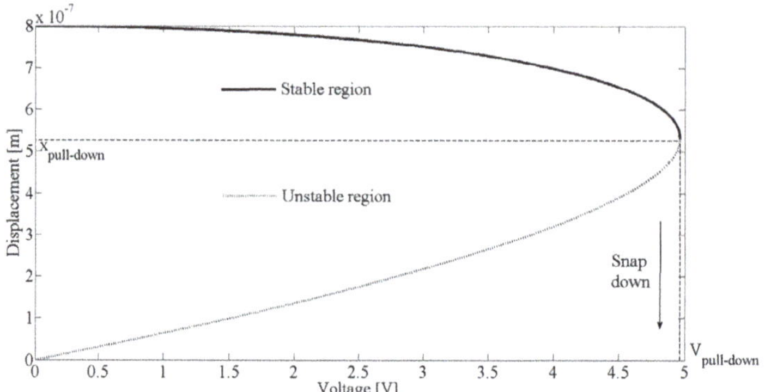

Figure 10. Displacement of the membrane from the applied bias voltage.

At $V(x) < V_{pull\text{-}down}$ the membrane is in a stable region. At $V(x) > V_{pull\text{-}down}$ the membrane collapses on the fixed down electrode, which corresponds to the unstable part in the Figure 10. The calculated value of the pull-down voltage is 5 V.

The dependence of the stiffness coefficient k of the elastic fastening elements of the membrane on the number of meanders N is shown in Figure 11. Figure 12 shows the dependence of the value of the pull-down voltage on the stiffness coefficient k of the elastic suspension elements.

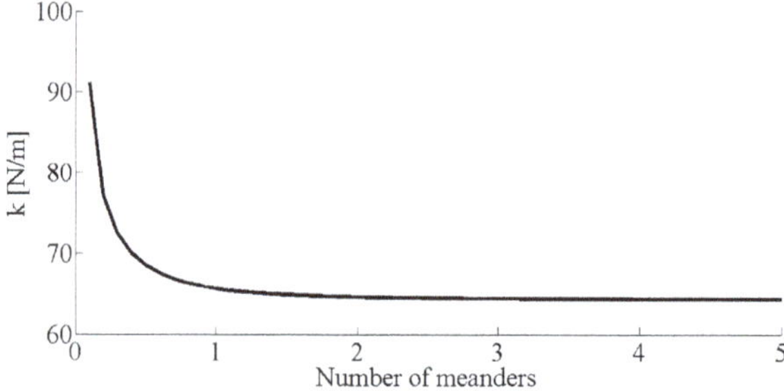

Figure 11. Graph of dependence of the coefficient of elasticity on the number of meanders.

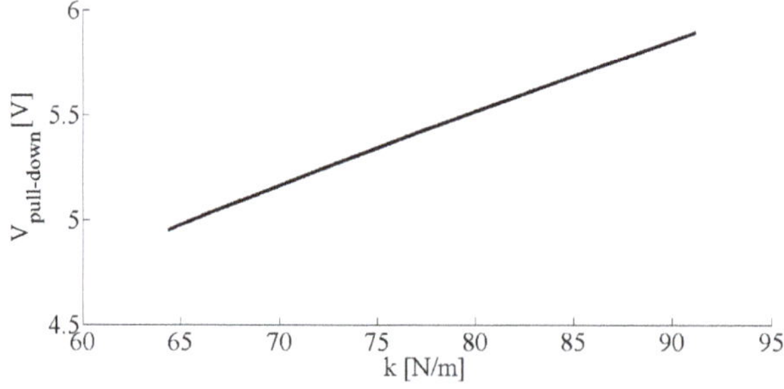

Figure 12. The dependence of the pull-down voltage on the stiffness coefficient.

As can be seen from Equation (21), $V_{pull\text{-}down}$ can be reduced in several ways. For example, when the distance between the membrane and fixed electrode is reduced, both the pull-down voltage and the ratio of the capacitances of the control variable capacitor are reduced, which greatly affects the ratio of the insertion loss and the isolation of the RF MEMS signal of the switch.

The increase in the area of the membrane entails an increase in its own losses, as well as an increase in residual stresses.

Thus, there remains a third option, which involves reducing the stiffness coefficient of the elastic elements of the membrane. Mounting with geometry in the form of a meander allows us to flexibly solve the problem in different frequency ranges and for different pull-down voltage.

Figure 13 shows the dependence of the $t_{pull\text{-}down}$ from different switching voltage (V_s) values.

Switching time analysis is critical to the design of RF MEMS switches, as achieving a short switching time from ON to OFF is a major challenge. It should be noted that the time of pull-down is slightly longer than the time of return to the initial state even in the case of a significantly greater potential difference than the $V_{pull\text{-}down}$.

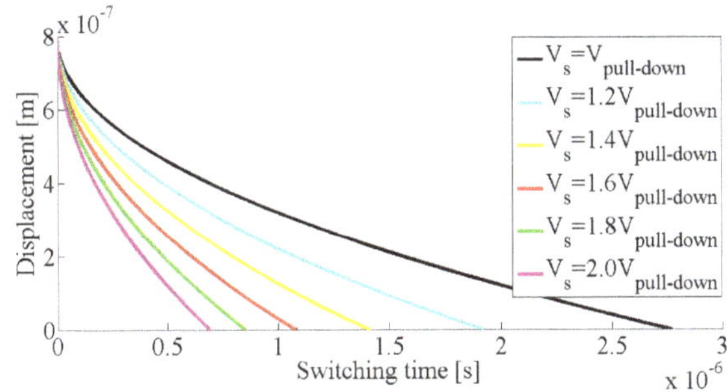

Figure 13. The switching time simulation.

Figures 14 and 15 present graphs of the dependence of the total capacitance of the membrane taking into account the perforation in the down- and initial state on the thickness of the dielectric film deposited on the surface of the fixed down electrode.

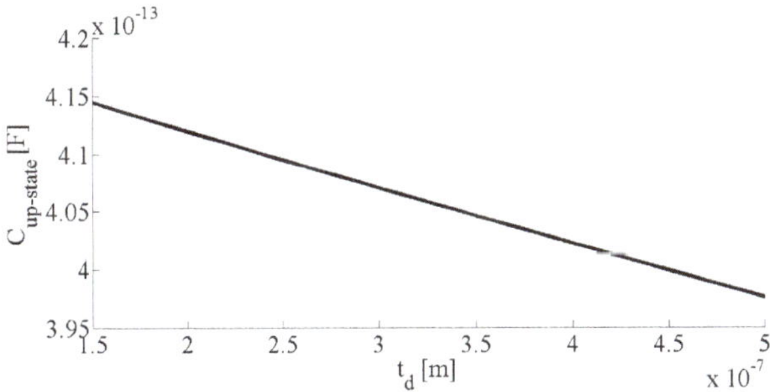

Figure 14. Capacitance in the up-state.

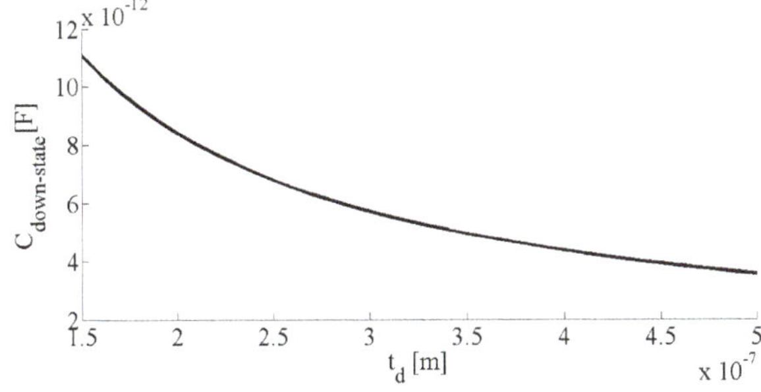

Figure 15. Capacitance in the down-state.

Table 4 presents the calculated values of the coefficient of stiffness of elastic suspension elements, pull- and hold-down voltage, return voltage, switching time in the down and initial states.

Table 4. Calculated values of the coefficient of stiffness, pull- and hold-down voltage, return voltage, switching time.

N	k, N/m	$V_{pull\text{-}down}$, V	$V_{hold\text{-}down}$, V	V_{return}, V	$t_{pull\text{-}down}$, µs	$t_{pull\text{-}up}$, µs
1	65.65	5	3.8	0.89	9.3	2.1

3.4. Development and Modeling of Equivalent Circuit

For this integrated RF MEMS switch, the electrical impedance is determined by Equation (30):

$$Z = R_s + jwL + \frac{1}{jwC} \tag{30}$$

For a given LC circuit the resonance frequency is determined by Equation (31):

$$f_0 = \frac{1}{2\pi} \frac{1}{\sqrt{LC}} \tag{31}$$

Thus, the resistance of the switch can be represented by Equation (32):

$$\begin{cases} \frac{1}{jwC}, f << f_0 \\ R_s, f = f_0 \\ jwL, f >> f_0 \end{cases} \tag{32}$$

The elastic fastening of the membrane plays an important role for the RF parameters of the switch, since they have inductance and resistance, affecting the passage of the RF signal.

To calculate the RF characteristics of the proposed design of an integrated RF MEMS switch LC isolation elements ($C_1, C_2, C_3, C_4, C_5, C_6$) for the control voltage from the RF signal were introduced into the existing electrical circuit which is implemented directly on the topology—capacitances located on the signal line (C_1, C_2) and membrane anchors (C_3, C_4, C_5, C_6) separating the constant control voltage from the RF signal. Inductance preventing the passage of the RF signal in power supply is located outside the circuit.

To calculate the inductance of the elastic fastening element (L_1, L_2, L_3, L_4) were used Equation (33) [76].

$$L = 0.002l(\ln(\frac{2l}{w+t}) + 0.50049 + (\frac{w+t}{3l})) \tag{33}$$

where l, w, t—the length, width, thickness of the conductor.

When calculating the switch, the inductance resistance was calculated by Equation (34):

$$R = \rho \frac{L}{S} \tag{34}$$

where ρ—the resistivity of multilayer structure; L—the length of the conductor; S—the area of the conductor.

As a result of the calculations, the parameters of the electrical circuit of the proposed integrated RF MEMS switch were obtained, as shown in Table 5.

Table 5. Calculated parameters of the electrical circuit.

Parameter	Value
L	0.0375 nH
R	0.0341 $\Omega \cdot m$
C_1, C_2	1.1 pF
C_3, C_4, C_5, C_6	22.12 pF
$C_{up\text{-}state}$	41.4 pF
$C_{down\text{-}state}$	0.011 pF

A electrical circuit of a integrated RF MEMS switch with LC isolation control signal is shown in Figure 16. The results of simulation of RF signal transmission are shown in Figure 17. Optimization and calculation of RF parameters of integrated RF MEMS switch is carried into effect in the Advanced Design System (ADS) CAD.

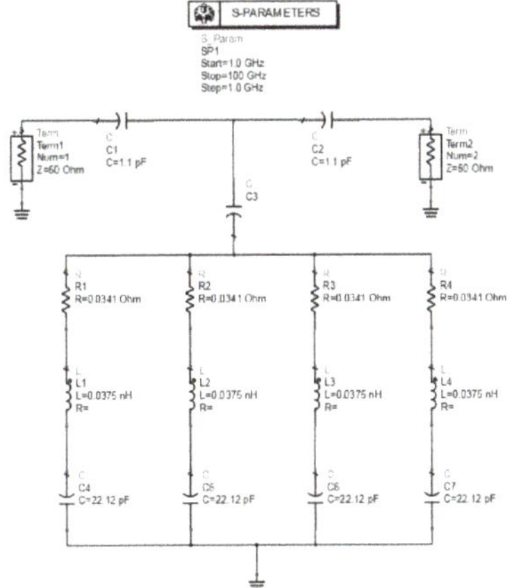

Figure 16. The electrical circuit of the integrated RF MEMS switch with LC isolation control signal.

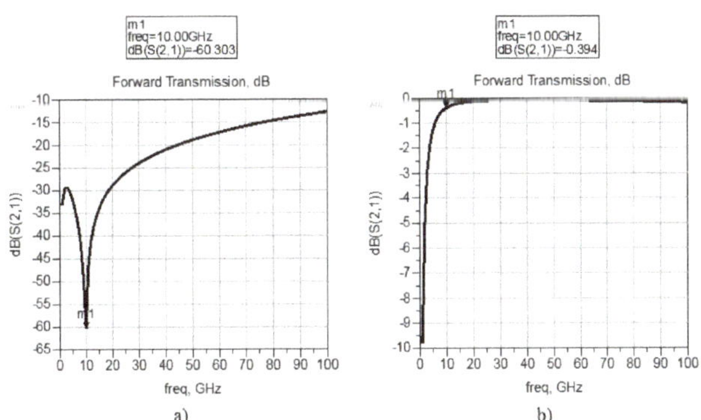

Figure 17. Passing RF signal through the switch: (**a**) in down state; (**b**) in up state.

3.5. Technological Route of Manufacturing of the Developed Design of a RF MEMS Switch

The technological process of manufacturing an integrated RF MEMS switch is shown in Figure 18. The process is characterized by a low deposition temperature of layers (not more than 350 °C) in order to prevent damage to the CMOS IC.

At the beginning of the CMOS IC, it is manufactured using a standard CMOS process with formation of through holes. Then, the lower structural elements of the switch are formed, such as coplanar waveguide, fixed down electrode, anchor areas. The sacrificial layer is the photosensitive polyamide at various stages of formation of structural elements of the switch. Then, the upper structural elements such as the membrane, elastic suspension elements, the pair of side electrodes, the fixed upper electrode are formed. After forming the walls and cover for the capsule sealing, the sacrificial layers of polyamide are removed through holes in the capsule and fixed upper electrode membrane using dry etching. Finally, the holes in the capsule are sealed. The layer of sealing prevents switch failures that can be caused by contamination or high humidity. Consequently, the integrated RF MEMS switch shown in Figure 6 is manufactured using CMOS IC technology.

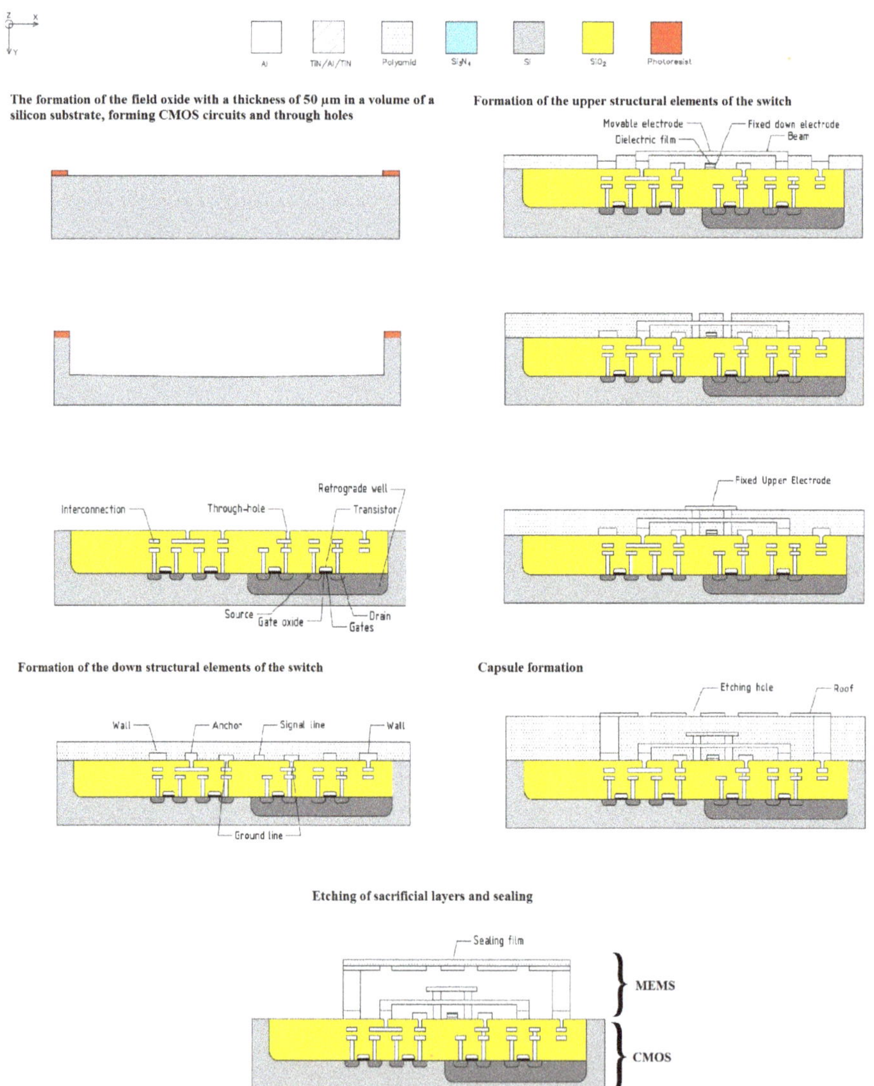

Figure 18. Technological route of integrated RF MEMS switch manufacturing.

4. Conclusions

RF MEMS switches are of classic MEMS devices for RF and microwave applications. In these devices, mechanical motion is used to close or open the RF transmission line (for a given frequency band). In other words, the total input impedance of the RF transmission line is controlled by the mechanical motion of the moving element of the RF MEMS switch design.

The field of application of RF MEMS switches is rigidly connected with their parameters. Solid-state analogs of RF MEMS switches have a good combination of low-power consumption, high switching power, low insertion losses and high isolation value in the closed state. However, electromechanical parameters such as control voltage and switching time are characterized by extremely high values, which complicates their integration into RF control circuits and reduces the scope of application. Electromechanical parameters are interrelated and the improvement of some can lead to the deterioration of others. It is possible to distinguish several main directions of improvement of electromechanical parameters: replacement of traditionally used materials on new perspective, and development of new original designs and technological decisions.

The analytical approach to the development of RF MEMS switches allows us to develop designs with the best ratio of parameters for specific applications, which is demonstrated by the example of the developed design of an integrated RF MEMS switch with a capacitive switching principle with low control voltages, high switching speed and suitable for using in the X frequency range (8–12 GHz) in terrestrial and satellite radio communication devices.

Author Contributions: All authors contributed to the present paper with the same effort in finding available literature resources, as well as writing the paper.

Funding: This research received no external funding.

Acknowledgments: The results were obtained using Equipment of the student design bureau "Elements and devices of inertial navigation systems of robotics" of the Institute of Nanotechnologies, Electronics and Electronic Equipment Engineering, Southern Federal University (Taganrog, Russia).

Conflicts of Interest: The authors declare no conflict of interest.

References

1. Rebeiz, G.M. *RF MEMS: Theory, Design, and Technology*; John Wiley & Sons: New York, NY, USA, 2004.
2. Petersen, K.E. Micromechanical membrane switches on silicon. *JBM J. Res. Dev.* **1971**, *23*, 376–385. [CrossRef]
3. Robin, L. *Status of MEMS Industry*; Tech. Rep.; Yole Developpement: Phoenix, AZ, USA, 2015; p. 7.
4. Robin, L. *Status of the MEMS Industry*; Tech. Rep.; Yole Developpement: Phoenix, AZ, USA, 2017; p. 2.
5. Hosaka, H.; Kuwano, H.; Yanagisawa, K. Electromagnetic microrelays: Concepts and fundamental characteristic. *Sens. Actuators A* **1994**, *40*, 41–47. [CrossRef]
6. Taylor, W.P.; Allen, M.G.; Dauwalter, C.R. Batch fabricated electromagnetic microrelays. In Proceedings of the 45th Relay Conference, Lake Buena Vista, FL, USA, 23 July 1997; pp. 8.1–8.6.
7. Taylor, W.P.; Allen, M.G. Integrated magnetic microrelays: Normally open, normally closed, and multi-pole devices. In Proceedings of the International Solid State Sensors and Actuators Conference (Transducers '97), Chicago, IL, USA, 19 June 1997; pp. 1149–1152.
8. Tilmans, H.A.; Fullin, E.; Ziad, H.; Van de Peer, M.D.; Kesters, J.; Van Geffen, E.; Bergqvist, J.; Pantus, M.; Beyne, E.; Baert, K.; et al. A fully-packaged electromagnetic microrelay. In Proceedings of the Technical Digest, IEEE International MEMS 99 Conference, Twelfth IEEE International Conference on Micro Electro Mechanical Systems, Orlando, FL, USA, 21 January 1999; pp. 25–30.
9. Wright, J.A.; Tai, Y.-C. Magnetostatic MEMS relays for the miniaturization of brushless DC motor controllers. In Proceedings of the Technical Digest, Twelfth IEEE International Conference on Micro Electro Mechanical Systems, Orlando, FL, USA, 21 January 1999; pp. 594–599.
10. Gretillat, M.A.; Gretillat, F.; Rooij, N.F. Micromechanical relay with electrostatic actuation and metallic contacts. *J. Micromech. Microeng.* **1999**, *9*, 324–331. [CrossRef]
11. Sanders, C. *MCNC Thermally Actuated Microrelays*; News Release; MEMS Technology Application Center MCNC: Durham, NC, USA, 30 November 1998.

12. Larson, L.E.; Hackett, R.H.; Lohr, R.F. Microactuators for GaAs-based microwave integrated circuits. In Proceedings of the 1991 International Conference on Solid-State Sensors and Actuators, Digest of Technical Papers, San Francisco, CA, USA, 24–27 June 1991; pp. 743–746.

13. Yao, J.J.; Chang, M.F. A surface micromachined miniature switch for telecommunications applications with signal frequencies from DC up to 4 GHz. In Proceedings of the International Solid-State Sensors and Actuators Conference, Stockholm, Sweden, 25–29 June 1995; pp. 384–387.

14. Schiele, I.; Huber, J.; Evers, C.; Hillerich, B.; Kozlowski, F. Micromechanical relay with electrostatic actuation. In Proceedings of the International Solid State Sensors and Actuators Conference, Chicago, IL, USA, 19 June 1997; pp. 1165–1168.

15. De Los Santos, H.J.; Kao, Y.-H.; Caigoy, A.L.; Ditmars, E.D. Microwave and mechanical considerations in the design of MEM switches for aerospace applications. *Proc. IEEE Aerosp. Conf.* **1997**, *3*, 235–254.

16. Hyman, D.; Lam, J.; Warneke, B.; Schmitz, A.; Hsu, T.Y.; Brown, J.; Schaffner, J.; Walston, A.; Loo, R.Y.; Mehregany, M.; et al. Surface-micromachined RF MEMS switches on GaAs substrates. *Int. J. RF Microw. CAE* **1999**, *9*, 348–361. [CrossRef]

17. Hyman, D.; Schmitz, A.; Warneke, B.; Hsu, T.Y.; Lam, J.; Brown, J.; Schaffner, J.; Walston, A.; Loo, R.Y.; Tangonan, G.L.; et al. GaAs-compatible surface-micromachined RF MEMS switches. *Electron. Lett.* **1999**, *35*, 224–226. [CrossRef]

18. Sovero, E.A.; Mihailovich, R.; Deakin, D.S.; Higgins, J.A.; Yao, J.J.; DeNatale, J.F.; Hong, J.H. Monolithic GaAs PHEMT MMICs integrated with high performance MEMS microrelays. In Proceedings of the 1999 SBMO/IEEE MTT-S International Microwave and Optoelectronics Conference, Rio de Janeiro, Brazil, 9–12 August 1999; pp. 257–260.

19. Muldavin, J.B.; Rebeiz, G.M. 30 GHz tuned MEMS switches. In Proceedings of the 1999 IEEE MTT-S International Microwave Symposium Digest, Anaheim, CA, USA, 13–19 June 1999; pp. 1511–1514.

20. Sakata, M.; Komura, Y.; Seki, T.; Kobayashi, K.; Sano, K.; Horiike, S. Micromachined relay which utilizes single crystal silicon electrostatic actuator. In Proceedings of the Technical Digest, IEEE International MEMS 99 Conference, Twelfth IEEE International Conference on Micro Electro Mechanical Systems, Orlando, FL, USA, 21 January 1999; pp. 21–24.

21. Hiltmann, K.M.; Schmidt, B.; Sandmaier, H.; Lang, W. Development of micromachined switches with increased reliability. In Proceedings of the International Solid State Sensors and Actuators Conference, Chicago, IL, USA, 19 June 1997; pp. 1157–1160.

22. Drake, J.; Jerman, H.; Lutze, B.; Stuber, M. An electrostatically actuated micro-relay. In Proceedings of the International Solid-State Sensors and Actuators Conference, Stockholm, Sweden, 25–29 June 1995; pp. 380–383.

23. Admschik, M.; Ertl, S.; Schmid, P.; Gluche, P.; Floter, A.; Kohn, E. Electrostatic diamond micro switch. In Proceedings of the 10th International Conference on Solid-State Sensors and Actuators, Sendai, Japan, 7–10 June 1999; pp. 1284–1287.

24. Gretillat, M.A.; Thiebaud, P.; Linder, C.; Rooij, N.F. Integrated circuit compatible polysilicon microrelays. *J. Micromech. Microeng.* **1995**, *5*, 156–160. [CrossRef]

25. Simon, J.; Saffer, S.; Kim, C.-J. A micromechanical relay with a thermally-driven mercury micro-drop. In Proceedings of the Ninth International Workshop on Micro Electromechanical Systems, San Diego, CA, USA, 11–15 February 1996; pp. 515–520.

26. Saffer, S.; Simon, J.; Kim, C.-J. Mercury-contact switching with gap-closing microcantilever. *Proc. SPIE* **1996**, *2882*, 204–208.

27. Sun, X.-Q.; Farmer, K.R.; Carr, W.N. A bistable microrelay based on two-segment multimorph cantilever actuators. In Proceedings of the Eleventh Annual International Workshop on Micro Electro Mechanical Systems. An Investigation of Micro Structures, Sensors, Actuators, Machines and Systems, Heidelberg, Germany, 25–29 January 1998; pp. 154–159.

28. Kruglick, E.J.J.; Pister, K.S.J. Bistable MEMS relays and contact characterization. In Proceedings of the Solid State Sensor and Actuator Workshop, Hilton Head Island, SC, USA, 8–11 June 1998; pp. 333–337.

29. Robin, L. *Status of MEMS Industry*; Tech. Rep.; Yole Developpement: Phoenix, AZ, USA, 2013; pp. 7–8.

30. Senturia, S.D. *Microsystem Design*, 1st ed.; Springer: New York, NY, USA, 2001.

31. Mercier, D.; Caekenberghe, K.; Rebeiz, G.M. Miniature RF MEMS Switched Capacitors. In Proceedings of the IEEE MTT-S International Microwave Symposium Digest, Long Beach, CA, USA, 17 June 2005; pp. 745–748.

32. Mercier, D.; Charvet, P.L.; Berruyer, P.; Zanchi, C.L.; Lapierre, L.O.; Vendier, O.; Cazaux, J.L.; Blondy, P. A DC to 100 GHz high performance ohmic shunt switch. In Proceedings of the 2004 IEEE MTT-S International Microwave Symposium Digest, Fort Worth, TX, USA, 6–11 June 2004; pp. 1931–1934.

33. Lacroix, B.; Pothier, A.; Crunteanu, A.; Cibert, C.; Dumas-Bouchiat, F.; Champeaux, C.; Catherinot, A.; Blondy, P. Sub-microsecond RF MEMS switched capacitors. *IEEE Trans. Microw. Theory Tech.* **2007**, *55*, 1314–1321. [CrossRef]

34. Majumder, S.; Lampen, J.; Morrison, R.; Maciel, J. A Packaged, High-Lifetime Ohmic MEMS RF Switch. *IEEE MTT-S Int. Microw. Symp. Dig.* **2003**, *3*, 1935–1938.

35. Yan, X.; McGruer, N.E.; Adams, G.G.; Majumder, S. Thermal Characteristics of Microswitch Contacts. In Proceedings of the National Association of Relay Manufacturer's (NARM) 49th Annual International Relay Conference, Oak Brook, IL, USA, 23–25 April 2001.

36. Randy, J.; Richards; Hector, J.; De Santos, H.J. MEMS for RF/Microwave Wireless Applications: The Next Wave. *Microw. J.* **2001**, *44*, 20.

37. Goldsmith, C.; Lin, T.-H.; Powers, B.; Wu, W.-R.; Norvell, B. Micromechanical membrane switches for microwave applications. In Proceedings of the 1995 IEEE MTT-S International Microwave Symposium, Orlando, FL, USA, 16–20 May 1995; pp. 91–94.

38. Jaafar, H.; Sidek, O.; Miskam, A.; Korakkottil, S. Design and Simulation of Microelectromechanical System Capacitive Shunt Switches. *Am. J. Eng. Appl. Sci.* **2009**, *2*, 655–660.

39. Ya, M.L.; Soin, N.; Nordin, A.N. Theoretical and simulated investigation of dielectric charging effect on a capacitive RF-MEMS switch. In Proceedings of the 2016 IEEE International Conference on Semiconductor Electronics (ICSE), Kuala Lumpur, Malaysia, 17–19 August 2016; pp. 17–20.

40. San, H.S.; Deng, Z.Q.; Yu, Y.X.; Li, G.; Chen, X.Y. Study on dielectric charging in low-stress silicon nitride with the MIS structure for reliable MEMS applications. *J. Micromech. Microeng.* **2011**, *21*, 125019. [CrossRef]

41. Zaghloul, U.; Coccetti, F.; Papaioannou, G.J.; Pons, P.; Plana, R. A novel low cost failure analysis technique for dielectric charging phenomenon in electrostatically actuated MEMS devices. In Proceedings of the 48th Annual IEEE International Reliability Physics Symposium (IRPS), Anaheim, CA, USA, 2–6 May 2010; pp. 237–245.

42. Zaghloul, U.; Papaioannouc, G.; Coccettia, F.; Ponsa, P.; Planaa, R. Dielectric charging in silicon nitride films for MEMS capacitive switches: Effect of film thickness and deposition conditions. *Microelectron. Reliab.* **2009**, *49*, 1309–1314. [CrossRef]

43. Li, M.; Zhao, J.; You, Z.; Zhao, G. Design and fabrication of a low insertion loss capacitive RF MEMS switch with novel micro-structures for actuation. *Solid State Electron.* **2016**, *127*, 32–37. [CrossRef]

44. Chu, C.; Liao, X.; Yan, H. Ka-band RF MEMS capacitive switch with low loss, high isolation, long-term reliability and high-power handling based on GaAs MMIC technology. *IET Microw. Antenna Propag.* **2017**, *11*, 942–948. [CrossRef]

45. Kwon, II., Choi, D.J.; Park, J.H.; Lee, H.C. Contact materials and reliability for high-power RF-MEMS switches. In Proceedings of the International Conference on MICRO Electro Mechanical Systems, Hyogo, Japan, 21–25 January 2007; pp. 231–234.

46. Coutu, R.A.; Kladitis, P.E.; Leedy, K.D.; Crane, R.L. Selecting metal alloy electric contact materials for MEMS switches. *J. Micromech. Microeng.* **2004**, *14*, 1157. [CrossRef]

47. Coutu, R.A.; Reid, J.R.; Cortez, R.; Strawser, R.E.; Kladitis, P.E. Microswitches with sputtered Au, AuPd, Au-on-AuPt, and AuPtCu alloy electric contacts. *IEEE Trans. Compon. Packag. Technol.* **2006**, *29*, 341–349. [CrossRef]

48. Broue, A.; Dhennin, J.; Courtade, F.; Dieppedale, C.; Pons, P.; Lafontan, X.; Plana, R. Characterization of Au/Au, Au/Ru and Ru/Ru ohmic contacts in MEMS switches improved by a novel methodology. *J. Micro/Nanolithogr. MEMS MOEMS* **2010**, *9*, 041102-1–041102-8

49. Czaplewski, D.A.; Nordquist, C.D.; Dyck, C.W.; Patrizi, G.A.; Kraus, G.M.; Cowan, W.D. Lifetime limitations of ohmic contacting RF MEMS switches with Au, Pt and Ir contact materials due to accumulation of 'friction polymer' on the contacts. *J. Micromech. Microeng.* **2012**, *22*, 105005. [CrossRef]

50. Liu, B.; Lv, Z.; He, X.; Liu, M.; Hao, Y.; Li, Z. Improving performance of the metal-to-metal contact RF MEMS switch with a Pt-Au microspring contact design. *J. Micromech. Microeng.* **2011**, *21*, 065038. [CrossRef]

51. Patel, C.D.; Rebeiz, G.M. RF MEMS metal-contact switches with mn-contact and restoring forces and low process sensitivity. *IEEE Trans. Microw. Theory. Tech.* **2011**, *59*, 1230–1237. [CrossRef]

52. Patel, C.D.; Rebeiz, G.M. A high-reliability high-linearity high-power RF MEMS metal-contact switch for DC-40-GHz applications. *IEEE Trans. Microw. Theory* **2012**, *60*, 3096–3112. [CrossRef]
53. Tang, M.; Agarwal, A.; Li, J.; Zhang, Q.X.; Win, P.; Huang, J.M.; Liu, A.Q. An approach of lateral RF MEMS switch for high performance. In Proceedings of the Symposium on Design, Test, Integration and Packaging of MEMS/MOEMS, Cannes, France, 7 May 2003.
54. Puyal, V.; Dragomirescu, D.; Villeneuve, C.; Ruan, J.; Pons, P.; Plana, R. Frequency scalable model for MEMS capacitive shunt switches at millimeter-wave frequencies. *IEEE Trans. Microw. Theory Tech.* **2009**, *57*, 2824–2833. [CrossRef]
55. Ekkels, P.; Rottenberg, X.; Puers, R.; Tilmans, H.A.C. Evaluation of platinum as a structural thin film material for RF-MEMS devices. *J. Micromech. Microeng.* **2009**, *19*, 065010–065018. [CrossRef]
56. Palego, C.; Deng, J.; Peng, Z.; Halder, S.; Hwang, J.C.M.; Forehand, D.I.; Scarbrough, D.; Goldsmith, C.L.; Johnston, I.; Sampath, S.K.; et al. Robustness of RF MEMS capacitive switches with molybdenum membranes. *IEEE Trans. Microw. Theory Tech.* **2009**, *57*, 3262–3269. [CrossRef]
57. Alam, A.H.M.Z.; Islam, M.R.; Khan, S.; Mohd Sahar, N.B.; Zamani, N.B. Effects of MEMS material on designing a multiband reconfigurable antenna. *Iran. J. Electr. Comput. Eng.* **2009**, *8*, 112–118.
58. Parate, O.; Gupta, N. Material selection for electrostatic microactuators using Ashby approach. *Mater. Des.* **2011**, *32*, 1577–1581. [CrossRef]
59. Kuwabara, K.; Norio, S.; Hiroki, M.; Junichi, K.; Toshikazu, K.; Katsuyuki, M.; Hiromu, I. RF-MEMS Switch Structure for Low-Voltage Actuation and High-Density Integration. In Proceedings of the Extended Abstracts of the 2007 International Conference on Solid State Devices and Materials, Ibaraki, Japan, 18–21 September 2007; pp. 90–91.
60. Lei, L.M.; Shun-Meen, K.; Tien-Yu, T.L.; Lianjun, L. A Mechanical Approach to Overcome RF MEMS Switch Stiction Problem. In Proceedings of the 53rd Electronic Components and Technology Conference, New Orleans, LA, USA, 27–30 May 2003; pp. 377–384.
61. Haixia, Z.; Yilong, H.; Zhiyong, X.; Dongmei, L.; Finch, N.; Marchetti, J.; Keating, D.; Narasimha, V. Design of A Novel Bulk Micro-machined RF MEMS Switch. *Int. J. Nonlinear Sci. Numer. Simul.* **2002**, *3*, 369–374.
62. Jun, S.C.; Huang, X.M.H.; Manolidis, M.; Zorman, C.A.; Mehregany, M.; Hone, J. Electrothermal tuning of Al-SiC nanomechanical resonators. *Nanotechnology* **2006**, *17*, 1506–1511. [CrossRef]
63. Al-Dahleh, R.; Mansour, R.R. High-capacitance-ratio warped-beam capacitive MEMS switch designs. *J. Microelectromech. Syst.* **2010**, *19*, 538–547. [CrossRef]
64. Chang, C.; Chang, P. Innovative micromachined microwave switch with very low insertion loss. *Sens. Actuators A Phys.* **2010**, *79*, 71–75. [CrossRef]
65. Bozler, C.; Drangmeister, R.; Duffy, S.; Gouker, M.; Knecht, J.; Kushner, L.; Parr, R.; Rabe, S.; Travis, L. MEMS microswitch arrays for recofigurable distributed microwave components. In Proceedings of the IEEE Antennas and Propagation Society International Symposium, Salt Lake City, UT, USA, 16–21 July 2000; pp. 587–591.
66. Entesari, K.; Rebeiz, G.M. A differential 4-bit 6.5-10-GHz RFMEMS tunable filter. *IEEE Trans. Microw. Theory Tech.* **2005**, *53*, 1103–1110. [CrossRef]
67. Demirel, K.; Yazgan, E.; Demir, S.; Akin, T. A new temperature-tolerant RF MEMS switch structure design and fabrication for Ka-Band applications. *J. Microelectromech. Syst.* **2016**, *25*, 60–68. [CrossRef]
68. Fernandez-Bolanos, M.; Tsamados, D.; Dainesi, P.; Ionescu, A.M. Reliability of RF MEMS capacitive switches and distributed MEMS phase shifters using AlN dielectric. In Proceedings of the IEEE 22nd International Conference on Micro Electro Mechanical Systems, Sorrento, Italy, 25–29 January 2009; pp. 638–641.
69. Roark, R.J.; Young, W.C. *Formulas for Stress and Strain*, 6th ed.; McGraw-Hill: New York, NY, USA, 1989.
70. Yun, W. A Surface Micromachined Accelerometer with Integrated CMOS Detection Circuitry. Ph.D. Thesis, University of California at Berkeley, Berkeley, CA, USA, 1992.
71. Fedder, G.K. Simulation of Microelectromechanical Systems. Ph.D. Thesis, University of California at Berkeley, Berkeley, CA, USA, 1994.
72. Peroulis, D.; Pacheco, S.P.; Sarabandi, K.; Katehi, L.P.B. Electromechanical considerations in developing low-voltage RF MEMS switches. *IEEE Trans. Microw. Theory Tech.* **2003**, *51*, 259–270. [CrossRef]
73. Shekhar, S.; Vinoy, K.J.; Ananthasuresh, G.K. Switching and Release Time Analysis of Electrostatically Actuated Capacitive RF MEMS Switches. *Sens. Transducers J.* **2011**, *130*, 77–90.

74. Palmer, H.B. Capacitance of a parallel-plate capacitor by the Schwartz-Christoffel transformation. *Trans. AIEE* **1927**, *56*, 363.
75. Koushik, G.; Mithlesh, K.; Saurabh, A.; Srimanta, B. A modified capacitance model of RF MEMS shunt switch incorporating fringing field effects of perforated beam. *Solid-State Electron.* **2015**, *114*, 35–42.
76. Stojanovic, G.; Zivanovt, L.; Damjanovic, M. Compact Form of Expressions for Inductance Calculation of Meander Inductors. *Serb. J. Electr. Eng.* **2004**, *1*, 57–68. [CrossRef]

Review

Review of the Comprehensive SAR Approach to Identify Scattering Mechanisms of Radar Backscatter from Vegetated Terrain

Motofumi Arii [1,*]**, Hiroyoshi Yamada** [2]**, Shoichiro Kojima** [3] **and Masato Ohki** [4]

[1]   Mitsubishi Space Software Co., Ltd. Kamakura 247-0065, Japan
[2]   Department of Information Engineering, Niigata University, Niigata 950-2181, Japan;
     yamada@ie.niigata-u.ac.jp
[3]   Applied Electromagnetic Research Institute, National Institute of Information and Communications
     Technology, Tokyo 184-8795, Japan; skojima@nict.go.jp
[4]   Earth Observation Research Center, Japan Aerospace Exploration Agency, Tsukuba 305-8505, Japan;
     ohki.masato@jaxa.jp
*   Correspondence: Arii.Motofumi@mss.co.jp

Received: 28 June 2019; Accepted: 27 September 2019; Published: 29 September 2019

Abstract: In a field of polarimetric synthetic aperture radar (SAR) remote sensing, various kinds of polarimetric decomposition techniques have been proposed. However, poor validations prevent them from operational applications. A true composition ratio of scattering mechanisms within a radar backscatter plays a key role. To overcome the issue, a novel comprehensive SAR approach to accurately identify a contribution of each scattering mechanism has been introduced. This is based on multiparametric SAR observation combined with a numerical model simulation. In this article, a comprehensive SAR approach is concisely reviewed to accelerate the research in this field. First, popular model-based polarimetric decompositions are introduced and their limitations are shown. Then, a behavior of scattering mechanisms is analyzed by the discrete scatterer model with some results using real multiparametric SAR data. A comprehensive SAR approach must be essential to realize an operational use of polarimetric SAR data.

Keywords: comprehensive SAR; multiparametric SAR observation; discrete scatterer model

---

1. Introduction

One of the main motivations of polarimetric synthetic aperture radar (SAR) remote sensing is to invert physical information from radar signal from vegetated terrain. Its flexible operability and an obvious sensitivity to a vegetation structure make polarimetric SAR an attractive monitoring tool [1–11]. A full polarimetric SAR capability such as Advanced Land Observing Satellite-2 (ALOS-2), developed by the Japan Aerospace Exploration Agency, has been universal in these days, and so various polarimetric decomposition techniques have been proposed [11–20]. Nonetheless, the decomposition techniques still have never been recognized as a solid practical application even after 30 years.

The backscattering cross section from vegetated terrain consists of various scattering mechanisms [21]. Simple forest model in Figure 1 generates a number of scattering mechanisms such as volume scattering from canopy or trunks, double-bounce scattering between canopy and ground or between trunk and ground, and surface scattering. In addition, the scattering mechanisms passing through a volume layer of leaves, twigs or trunks are attenuated [22,23], and of course, there are a large number of multiple scatterings within a layer. The variation makes an inversion of physical parameters seriously complicated. Suppose that a more backscattering cross section would be observed in a forest after rain. It could be explained by a mixture of various hypotheses such as a growth of vegetation, increased vegetation water content, increased soil moisture under vegetation, varied vegetation structure, and so forth. Each hypothesis must be tightly connected to a different set of scattering mechanisms in Figure 1 so that it could be difficult to directly retrieve physical parameters such as soil moisture, vegetation water content, and biomass without knowing a portion of scattering mechanisms.

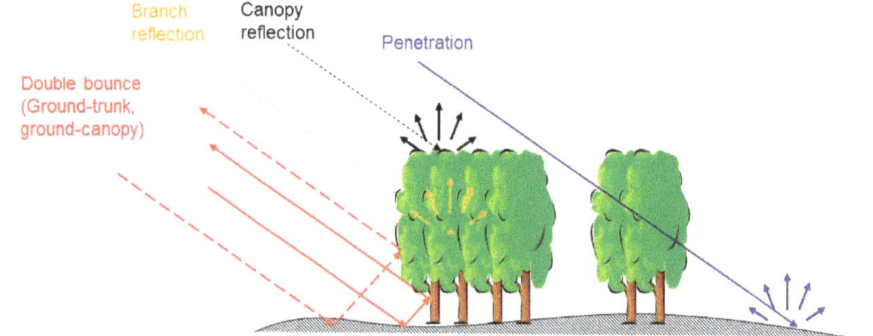

Reprinted, with permission, from Arii, M. Soil moisture retrieval under vegetation using polarimetric radar. Ph.D. Thesis, California Institute of Technology, 2009.

Figure 1. Backscattering from a vegetated terrain [21].

Main motivation to introduce full polarimetric SAR is to understand radar backscatter based on scattering mechanisms, and this leads to a potential problem. Unfortunately, there is no way to directly measure a composition ratio of scattering mechanisms within radar backscatter so that only a limited validation has been conducted to polarimetric decomposition results. This could prevent the polarimetric decomposition, even polarimetric SAR data, from being widely established.

To overcome the issue, a forward model, which theoretically calculates backscatter from randomly vegetated terrain, has been introduced [15,21,24–26]. The model simply provides us a contribution of each scattering mechanism in terms of specific observation condition such as the incidence angle and polarization so that one can quantitatively compare a set of experimental data with the numerical simulation results. The radar backscatter could be precisely characterized.

Forward models generally have a high degree of freedom so that many parameters must be reasonably determined. For this purpose, a multiparametric SAR observation, where various observation conditions such as frequency, polarization and incidence angle are gradually varied in a wide range of each value during a short span of time, has been proposed [27–36]. Simulation parameters can be found by fitting the model to the series of data set obtained by the multiparametric SAR observation. This comprehensive SAR approach can provide reliable model parameters.

In this article, essential elements for the comprehensive SAR approach are concisely reviewed to accelerate the research in this field. First, popular model-based polarimetric decompositions are introduced and their limitations are shown. Then, forward models are reviewed with real SAR data based on a multiparametric SAR observation. Note that the main part of this review article is based on our previous works [18,19,21,32,34,36,37].

2. Model-Based Polarimetric Decompositions and Their Limitation

Most polarimetric decomposition techniques proposed for nearly three decades could be categorized into two groups. An eigenvector-based decomposition was introduced by Cloude in the context of radar imaging [12] and the other researchers follow [13,16]. This method always provides a unique solution mathematically, whereas some approximations must be required to interpret the results in terms of established physical scattering mechanisms as described in [13].

Model-based polarimetric decomposition, as proposed by Freeman and Durden [14], is another group. There are three well-known physical scattering mechanisms: A volume scattering from canopy, a double-bounce scattering between tree and surface and a backscattering from the ground. The authors proposed that the measured covariance matrix can be expressed by a linear sum of them. Due to its simplicity, this method has been widely used. Our original purpose to introduce the polarimetric decomposition technique was to understand the behavior of physical scattering mechanisms so that the model-based polarimetric decomposition would be reviewed in the rest of this article.

The Freeman and Durden algorithm can decompose an observed covariance matrix of C to three elemental scattering mechanisms as:

$$\langle [C] \rangle = f_v \langle [C_v] \rangle + f_d [C_{db}] + f_s [C_s], \tag{1}$$

where f_v, f_d and f_s are coefficients of volume, double-bounce and surface scattering, respectively, whereas C, C_v, C_{db} and C_s are covariance matrices of total, volume, double-bounce and surface scatterings, respectively. Note that $<$ and $>$ indicates the ensemble average of adjacent pixels. Now the algorithm in [14] was applied to real SAR data of Black Forest in Germany, obtained by the National Aeronautics and Space Administration/Jet Propulsion Laboratory (NASA/JPL) Airborne Synthetic Aperture Radar (AIRSAR) system in the summer of 1991, shown in Figure 2.

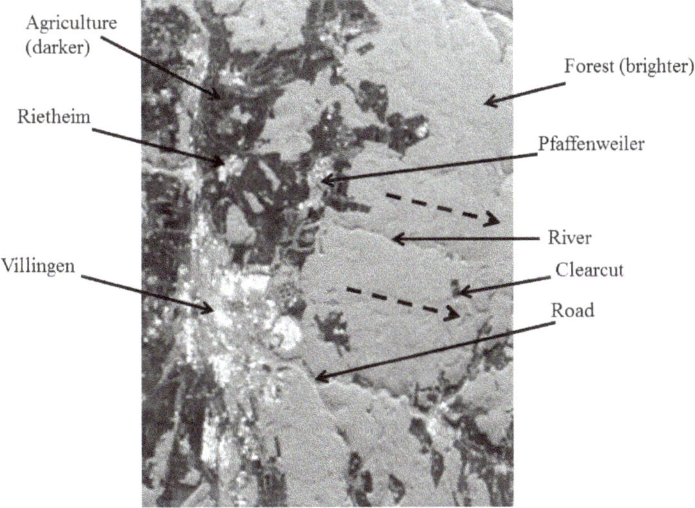

© Courtesy NASA/JPL-Caltech.
Reprinted, with permission, from Arii, M. Soil moisture retrieval under vegetation using polarimetric radar. Ph.D. Thesis, California Institute of Technology, 2009.

Figure 2. L-band image of the Black Forest in Germany obtained by the NASA/JPL AIRSAR system in the summer of 1991. The solid arrows indicate the name of city or area type. The dotted lines specify the direction of topographic change. The terrain slopes upward in the direction of the arrows [21].

The observations at C-band (6 cm), L-band (24 cm) and P-band (68 cm) were conducted, and a portion of three scattering mechanisms is shown in Figure 3.

The double-bounce component clearly discriminates urban areas such as Villingen and Rietheim for all wavelengths. A good penetration of canopy layer was obviously shown by the longer wavelengths in the agricultural area, and double-bounce scattering appears instead of the volume component.

C-Band   L-Band   P-Band

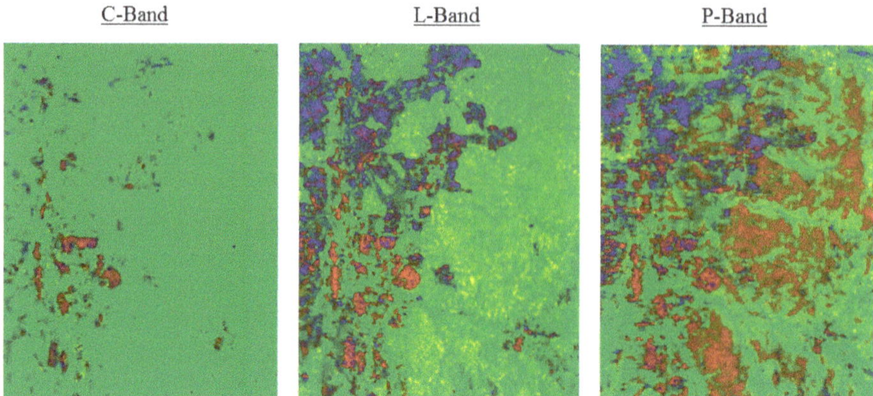

Reprinted, with permission, from Arii, M. Soil moisture retrieval under vegetation using polarimetric radar. Ph.D. Thesis, California Institute of Technology, 2009.

Figure 3. Results of the Freeman algorithm applied to three different wavelength images in Freiburg obtained by AIRSAR. From left to right, C-band (5 cm), L-band (24 cm) and P-band (68 cm) images are displayed. Green, red and blue are assigned to volume scattering, double-bounce scattering and ground scattering, respectively [21].

The Freeman and Durden algorithm sometimes breaks a conservation law of energy so that negative power could be estimated for a term on the right-hand side of Equation (1) as discussed in [18]. Since the algorithm by Freeman assumes that all of cross-polarization term is automatically assigned to the volume component, it may underestimate the contribution from other scattering mechanisms. Therefore, van Zyl et al. introduced Equation (2) to investigate a pixel having negative power [18]:

$$[C_{remainder}] = \langle [C] \rangle - f_v \langle [C_v] \rangle. \tag{2}$$

The negative power in Equation (2) can be easily obtained through eigenvalue analysis [18], and a result is shown in Figure 4 where pixels having negative power are indicated by black color. Unfortunately, most of the negative power happened in the forested area.

Original Image   Result

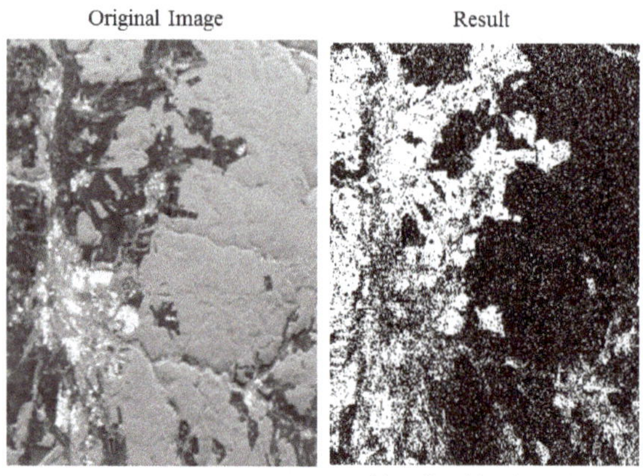

© Courtesy NASA/JPL-Caltech.   White: all positive eigenvalues
Black: at least one negative eigenvalue

Reprinted, with permission, from Arii, M. Soil moisture retrieval under vegetation using polarimetric radar. Ph.D. Thesis, California Institute of Technology, 2009.

Figure 4. Pixels with negative eigenvalue are displayed using the L-band Black Forest image. The left image is the total power image at L-band, and the right image is the result of the validation test [21].

The negative power distributed at most of the forested area was physically unacceptable. For this violation of physics, the nonnegative eigenvalue decomposition (NNED) model [18] has been proposed as follows:

$$\langle[C]\rangle = f_v\langle[C_v]\rangle + f_d[C_{db}] + f_s[C_s] + [C_{reminder}]. \tag{3}$$

By combining the remainder term with the eigenvalue decomposition technique [18], the three-component model guarantees that any term on the right-hand side of Equation (3) would be positive semidefinite. The result by the NNED is shown in Figure 5.

C-Band   L-Band   P-Band

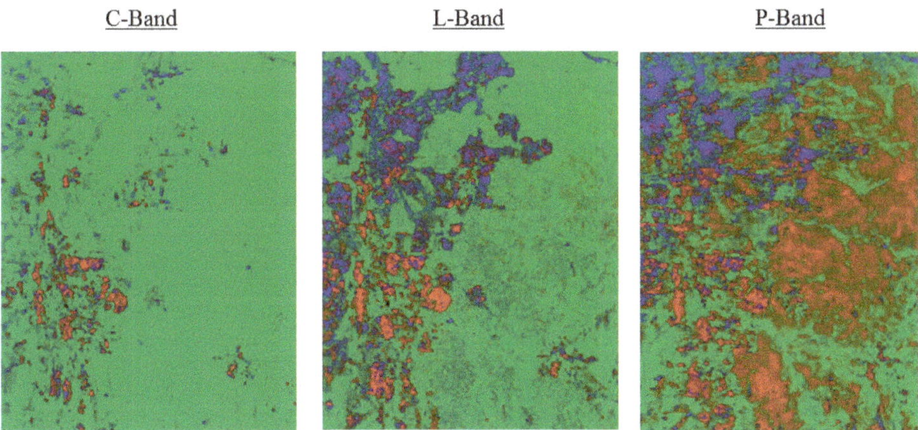

Reprinted, with permission, from Arii, M. Soil moisture retrieval under vegetation using polarimetric radar. Ph.D. Thesis, California Institute of Technology, 2009.

Figure 5. Decomposition results using nonnegative eigenvalue decomposition (NNED) are shown. The original images with three different frequencies are the same as in Figure 3. Color assignments are the same as Figure 3 as well [21].

It can be easily seen that the volume scattering component was distinctly reduced whereas surface scattering and double-bounce scattering were exaggerated in the images at L- and P-bands, respectively. Some faint red was recognized in the middle of the Black Forest at L-band. Any difference between the two techniques could not be seen in the agricultural and urban areas. The volume scattering component was also still recognized around the river halfway down the image at the P-band.

As mentioned in [19,37], the decomposition techniques shown above were operated on a pixel-to-pixel basis, either supposing the same volume scattering component for an entire image or utilizing a limited number of scattering models (three in the case of Yamaguchi et al. [15]) to select for each pixel. Randomly distributed dipoles are generally assumed to calculate a volume scattering component, where probability distribution function (pdf) of their orientation angle is specified. For example, Freeman and Durden algorithm utilizes a volume scattering component as follows:

$$\langle[C_v]\rangle = \begin{bmatrix} 3 & 0 & 1 \\ 0 & 2 & 0 \\ 1 & 0 & 3 \end{bmatrix}. \tag{4}$$

This assumes uniformly distributed thin cylinders so that the model disagrees to a pixel having methodically distributed thin cylinders with distinct mean orientation angle. To conquer the limitation, Arii et al. completely generalizes the volume component models:

$$\langle[C_v(\theta_0,\sigma)]\rangle = [C_\alpha] + p(\sigma)\big[C_\beta(2\theta_0)\big] + q(\sigma)\big[C_\gamma(4\theta_0)\big],$$

$$[C_\alpha] = \tfrac{1}{8}\begin{bmatrix} 3 & 0 & 1 \\ 0 & 2 & 0 \\ 1 & 0 & 3 \end{bmatrix},$$

$$\big[C_\beta(2\theta_0)\big] = \tfrac{1}{8}\begin{bmatrix} -2cos2\theta_0 & \sqrt{2}sin2\theta_0 & 0 \\ \sqrt{2}sin2\theta_0 & 0 & \sqrt{2}sin2\theta_0 \\ 0 & \sqrt{2}sin2\theta_0 & 2cos2\theta_0 \end{bmatrix},$$

$$\big[C_\gamma(4\theta_0)\big] = \tfrac{1}{8}\begin{bmatrix} cos4\theta_0 & -\sqrt{2}sin4\theta_0 & -cos4\theta_0 \\ -\sqrt{2}sin4\theta_0 & -2cos4\theta_0 & \sqrt{2}sin4\theta_0 \\ -cos4\theta_0 & \sqrt{2}sin4\theta_0 & cos4\theta_0 \end{bmatrix},$$

$$(5)$$

where θ_0 and σ are the mean orientation angle and pdf of the randomly distributed dipoles, respectively. Both p and q are predetermined functions in terms of σ [37]. One of the most significant merits of this model was that any volume component could be expressed by only two parameters, θ_0 and σ, which were related to physical vegetation as shown in Figure 6.

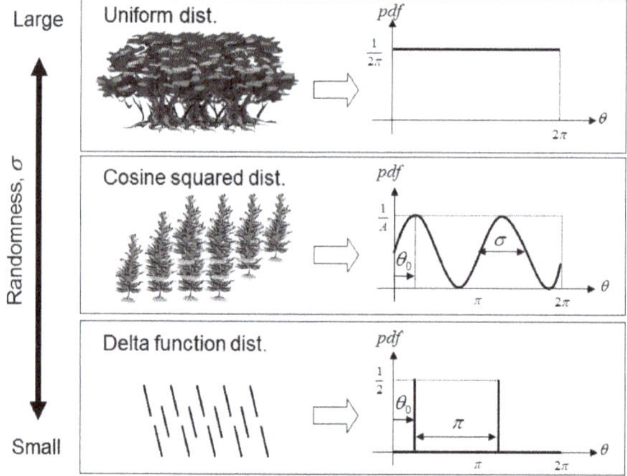

Figure 6. Various probability distribution functions (pdfs) in terms of randomness. There are two extreme cases: Uniform distribution with the highest variance and delta function distribution with the lowest variance. Cosine squared distribution sits in between these two. The randomness is defined by σ as shown in the cosine squared distribution [32].

Adaptive NNED (ANNED) technique has been realized by introducing the generalized volume scattering component as follows:

$$\langle[C]\rangle = f_v\langle[C_v(\theta_0,\sigma)]\rangle + f_d[C_{db}] + f_s[C_s] + [C_{reminder}]. \tag{6}$$

To fix the additional parameters θ_0 and σ, the remainder term is again utilized to avoid negative power [19]. A result by the adaptive model for the Black Forest data is shown in Figure 7.

C-Band   L-Band   P-Band

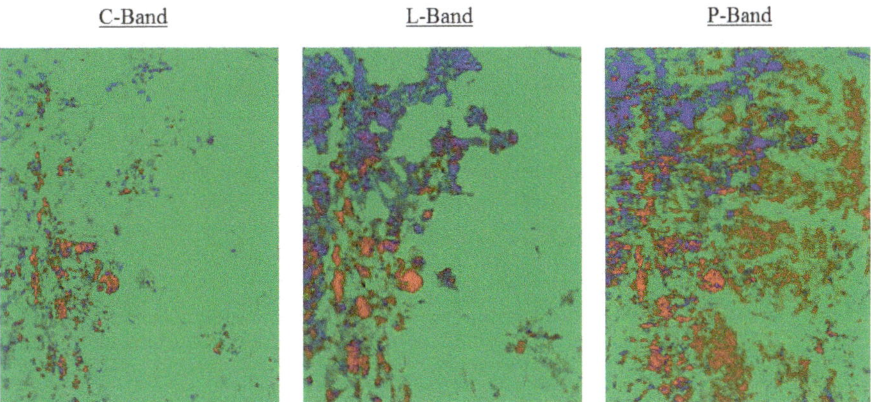

Reprinted, with permission, from Arii, M. Soil moisture retrieval under vegetation using polarimetric radar. Ph.D. Thesis, California Institute of Technology, 2009.

Figure 7. Results of the ANNED algorithm applied to three different wavelength images in Freiburg obtained by AIRSAR. Green, red and blue are assigned to the volume, double bounce and ground components [21].

Much more of the volume scattering contribution can be recognized in the forested area from the ANNED results at L-band whereas a similar tendency can be seen around the river halfway down the image at the P-band. The remainder term of Equation (6) was mapped to further study the applicability of ANNED in Figure 8 at the L-band. By supposing two different distributions: Uniform and cosine squared having no orientation angle for the volume scattering model, the results of NNED and ANNED were compared. The model tells that the smaller value in pixel in Figure 8 would be considered as the better fit to the selected parameters of the model.

ANNED   NNED' (cos2)   NNED' (uniform)

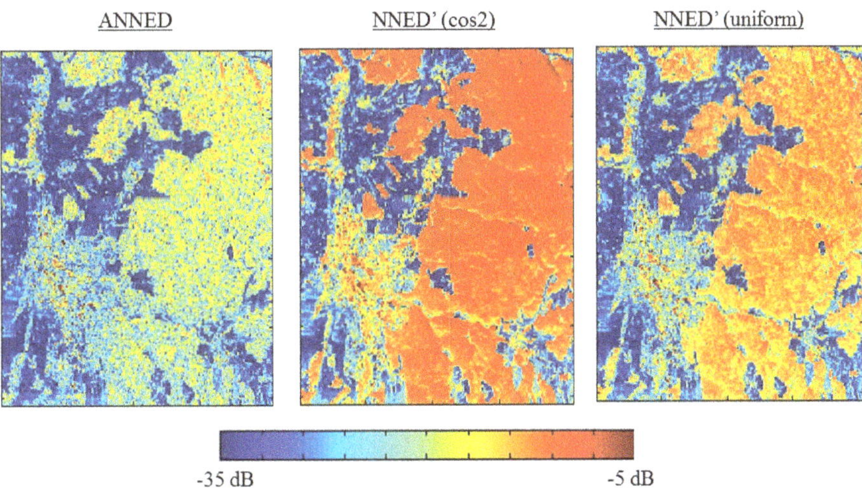

Reprinted, with permission, from Arii, M. Soil moisture retrieval under vegetation using polarimetric radar. Ph.D. Thesis, California Institute of Technology, 2009.

Figure 8. $C_{remainder}$ of ANNED (left) for the L-band Black Forest image compared with those of NNED' (dash) using two distributions: cosine squared (center) and uniform (right) distribution. Note that the cosine squared distribution has a zero orientation angle [21]. Note that NNED' is an extended version of NNED so that reflection symmetry [38] is never required.

This shows that the best fit parameter set could be reasonably found by ANNED especially in the forested area. Therefore, it should be concluded that an excellent applicability to the various vegetated surfaces were realized by ANNED. The results obtained by ANNED from multipolarization and multifrequency SAR data were qualitatively validated.

In [32], Arii et al. have first applied the model-based polarimetric decomposition technique to real multi-incidence angle and multi-polarimetric SAR (MIMP SAR) data sets from rice paddies at the X-band obtained by Pi-SAR2 developed by the National Institute of Information and Communications Technology. Contribution of each scattering mechanism derived by the Freeman and Durden algorithm is shown in Figure 9.

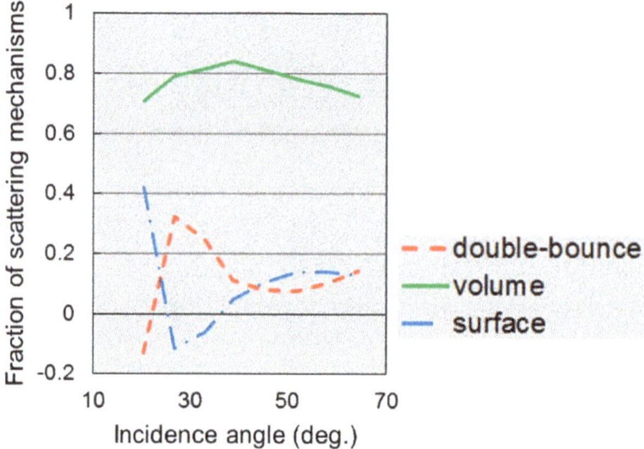

© 2019 IEEE. Reprinted, with permission, from Arii, M.; Yamada, H.; Kobayashi, T.; Kojima, S.; Umehara, T.; Komatsu, T.; Nishimura, T. Theoretical characterization of X-band multi-incidence angle and multi-polarimetric SAR data from rice paddies at late vegetative stage. IEEE Trans. Geosci. Remote Sens., 2017, 55, 2706-2715.

Figure 9. Freeman and Durden's three-component scattering decomposition result for block D [32].

It was obvious that the only dominant scattering mechanism over an entire range of incidence angles was the volume scattering. As previously mentioned, this is caused by all of HV, where vertically polarized wave is transmitted and horizontally polarized wave is received, being blindly considered as the volume scattering element. Therefore, as shown in the figure, negative power might occur if significant HV was obtained. Moreover, interaction between rice crops and ground surfaces and surface scattering alternate in terms of incidence angles. Only with a 10-degree difference of incidence angles that the second most dominant scattering was exchanged. A phase of the correlation between HH, where horizontally polarized wave is transmitted and received, and VV, where vertically polarized wave is transmitted and received, could be a potential reason because the phase plays a key role to estimate the amount of a double-bounce scattering and surface scattering by most of the model-based polarimetric decomposition techniques such as the three-component scattering decomposition [14,15,18,19]. The algorithm may not properly work if a dominant scattering mechanism to each co-polarization becomes different.

It could be an issue that the analysis above cannot be deepened any more without knowing the true fraction of each scattering mechanism. In addition, there were several critical oversights in the polarimetric decomposition models. First, attenuation by the volume layer was never taken into account. As shown in [21,29], attenuation may dictate backscatter from vegetated terrain when the amount of scatterers in a volume layer exceeds a certain amount. Second, a volume scattering component was estimated by assuming a thin cylinder, that is, dipoles, even though, natural scatterers are generally more like cylinders with a certain thickness. With the current decomposition models, a number of unknown parameters were more than those of polarimetric SAR observations. To make

SAR remote sensing an effective tool to monitor vegetated surfaces, an accurate fraction of scattering mechanisms to compare with the observation becomes essential.

3. Comprehensive SAR Approach

There is no way to directly measure the individual scattering mechanism within radar backscatter from vegetated terrain. To obtain an accurate fraction of scattering mechanisms, a forward model numerically calculating a contribution of each scattering mechanism was introduced. The model usually has a high degree of freedom so that many parameters have to be predetermined. The authors also have proposed multiparametric SAR observation that monitors a specific target by gradually varying the observation conditions so that the model parameters could be accurately estimated by comparing with a set of observation data. In this section, both technical elements, that is, a forward model and multiparametric SAR observation, were explained as essential elements of a comprehensive SAR approach.

3.1. Forward Model

To describe radar backscatter from vegetated surfaces, there are two models widely used. One is a radiative transfer model consisting of the layer structure, and the other is a discrete scatterer model (DSM) consisting of randomly distributed scatterers.

Chandrasekhar [39] originally introduced the radiative transfer theory. Then the concept was successfully applied to radar scattering from vegetated surfaces by Ulaby et al. [40–43]. This model is called Michigan microwave canopy scattering (MIMICS), in which a three-layer structure is considered in a typical forest, that is, the canopy layer, trunk layer and ground layer, with border condition. Differential equations can be formed to model the net intensity for each of the upward and downward directions within a layer by assuming the conservation law of energy for the infinitely thin slab. The backscattering cross section can be obtained by integrating the thin slab over an entire layer structure to the height direction under border conditions. Multiple scattering considered in the MIMICS causes higher accuracy than the DSM whereas the MIMICS may not provide sufficient physical insight due to its complexity. The other models based on radiative transfer are also discussed in [44,45].

Durden et al. proposed the DSM in [25] assuming the model for vegetated terrain as shown in Figure 10.

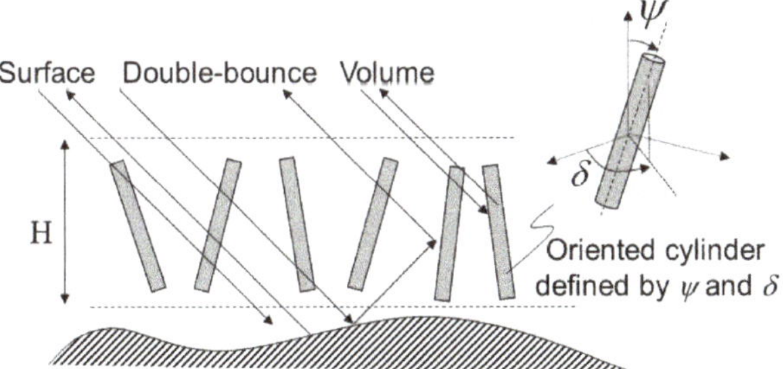

Figure 10. Various scattering mechanisms, volume, double-bounce and surface scatterings, from a cloud of oriented cylinders on the ground. An orientation of each cylinder is defined by ψ (elevation) and δ (azimuth) [32].

The model was filled up with discrete scatterers, and their multiple scatterings were ignored. Since the scatterers inside the canopy were assumed to be sparsely distributed, the scattered wave could be attenuated well through multiple scatterings [21,22]. Based on this assumption, the only three distinct scattering mechanisms such as volume scattering, interaction between the ground and volume scatterers and surface scattering describe the model.

Randomly oriented thin cylinders on the dielectric surfaces were assumed by the DSM as described in Figure 10. Surface scattering [46], double-bounce scattering, and volume scattering [47–49] were incoherently integrated by considering attenuations given by an optical theorem caused by randomly distributed elemental scatterers [22]. Randomly oriented thin cylinders were expressed not only by the cylinder distribution but also by the mean orientation angle and randomness as described in Figure 6. The randomness, σ, had a range of 0 (Delta function distribution: Methodical distribution) and 0.91 (uniform distribution) as a standard deviation, whereas the mean orientation angles, θ_0, appeared at every 180°. Note that a cosine squared distribution corresponds to the randomness of 0.56 as a medium randomness. In addition to the use as a validation tool of the polarimetric decomposition technique, the model could be directly utilized to invert physical parameters such as soil moisture as shown in [21,24,35].

To demonstrate a data analysis based on the DSM, rice paddies at late vegetative stage are modeled for example [32,34,36]. Three scatterers such as a grain, a stem and a leaf have been taken into account as elements of rice paddies at late vegetative stage. Leaves and stems are assumed to be in the same layer, whereas grains are on the top layer. The two-layer model with three types of distributed elemental scatterers is conceptually expressed in Figure 11.

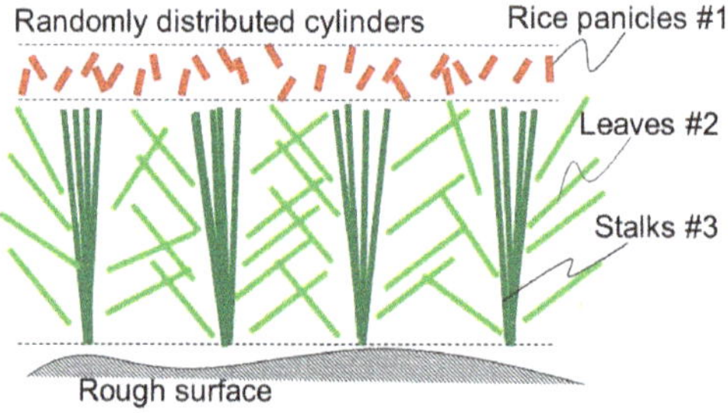

© 2019 IEEE. Reprinted, with permission, from Arii, M.; Yamada, H.; Kojima, S.; Ohki, M. Sensitivity analysis of multifrequency MIMP SAR data from rice paddies. IEEE Trans. Geosci. Remote Sens., 2019, 57, 3543-3551.

Figure 11. Schematic model for the discrete scatterer model (DSM). Randomly distributed cylinders stand on the dielectric rough surfaces [36].

An incoherent model cannot express a coherent effect such as radio interferometry. The effect could be significant on backscatter from densely distributed rice plants, where each scatterer is much smaller than the wavelengths. In this case, in-situ measured data should not be directly used as a set of input parameters, because backscatter from each element could be considerably weak so that incoherently summed power from all of the plants still stays weak. To explain certain scattering power by the incoherent model, a radio interferometry is realized by clustering the densely distributed thin scatterers at a scale of wavelength. The ears of rice plants at this stage form a slender cluster, which can be simply expressed as a virtual cylinder. The same concept is applied to leaves and stalks, bunches of which could be described as virtual cylinders, as shown in Figure 12.

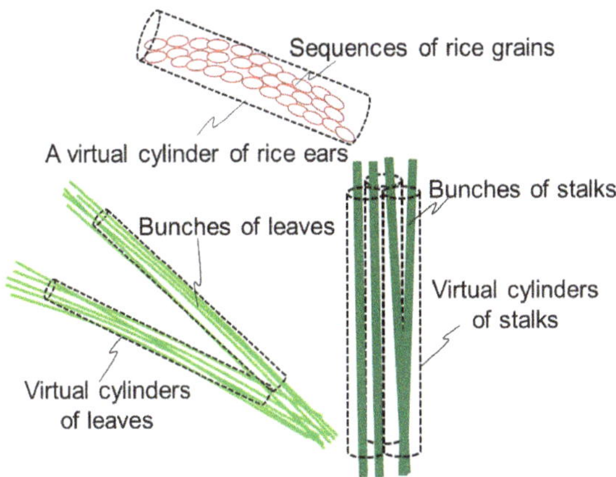

© 2019 IEEE. Reprinted, with permission, from Arii, M.; Yamada, H.; Kojima, S.; Ohki, M. Sensitivity analysis of multifrequency MIMP SAR data from rice paddies. IEEE Trans. Geosci. Remote Sens., 2019, 57, 3543-3551.

Figure 12. Cylinder models of sequences of rice ears, bunches of leaves and bunches of stalks [36].

The model allows calculating backscattering cross section at different bands so that multifrequency data analysis can be conducted. In this example [36], two blocks, A and B, of rice paddies were selected, and each block is simulated at the X- and L-bands. Table 1 shows model input parameter sets for XA, XB, LA and LB, which are determined by experimentally obtained SAR data (X-band on August 21, 2014 and L-band on August 3, 2016) as described in the Section 3.2. Note that the volumetric water content of each element is related to the dielectric constant by the Dual-dispersion model proposed in [50], where dielectric constants corresponding to 5% and 35% at X-band, and 12%, 15%, and 31% at the L-band of volumetric water contents are 3.1 and 17.2 at X-band, and 8.7, 10.7 and 21.0, respectively. The volumetric water contents in Table 1 may imply dryer vegetation condition than those shown in [51–53]. Since the dielectric constant at the center of a trunk is much higher than that on the surface as experimentally shown in [54], backscatter from rice paddy could be characterized as dryer condition than its true vegetation water content if the incident wave would not reach to a center of a grain, a stalk or a leaf.

Table 1. Model input parameters [36].

Parameter	Value XA	Value XB	Value LA	Value LB
Incidence angle (deg.)	25–65	25–65	24–55	24–55
Wavelength (cm)	3	3	24	24
Stem	-	-	-	-
Volumetric water content (%)	5	5	12	12
Radius (mm)	1.0	1.0	16	16
Layer height (m)	0.80	0.80	0.60	0.60
Mean orientation angle (deg.)	0.0	0.0	2.0	2.0
Distribution	0.40	0.40	0.30	0.30
Density (m^{-3})	400	400	150	150
Length (cm)	80	80	42	42
Leaf	-	-	-	-
Volumetric water content (%)	5	5	15	15
Radius (mm)	1.0	1.0	3.0	3.0
Layer height (m)	0.80	0.80	0.60	0.60
Mean orientation angle (deg.)	43	43	80	80
Distribution	0.42	0.42	0.30	0.30
Density (m^{-3})	1200	1200	700	700
Length (cm)	35	35	90	90

Table 1. *Cont.*

Parameter	Value XA	Value XB	Value LA	Value LB
Grain	-	-	-	-
Volumetric water content (%)	35	35	31	31
Radius (mm)	2.6	2.6	16.8	17.3
Layer height (m)	0.20	0.20	0.30	0.30
Mean orientation angle (deg.)	76	76	0.0	0.0
Distribution	0.38	0.38	0.09	0.088
Density (m^{-3})	500	500	430	430
Length (cm)	10.0	10.0	4.5	4.5
Volumetric soil moisture (%)	6	6	31	31
Surface roughness (mm)	0.2	0.2	1.0	1.0
Correlation length (mm)	5	5	20	20

Backscattering cross sections in terms of polarization, incidence angle and frequency are numerically calculated by the DSM as shown in Figure 13.

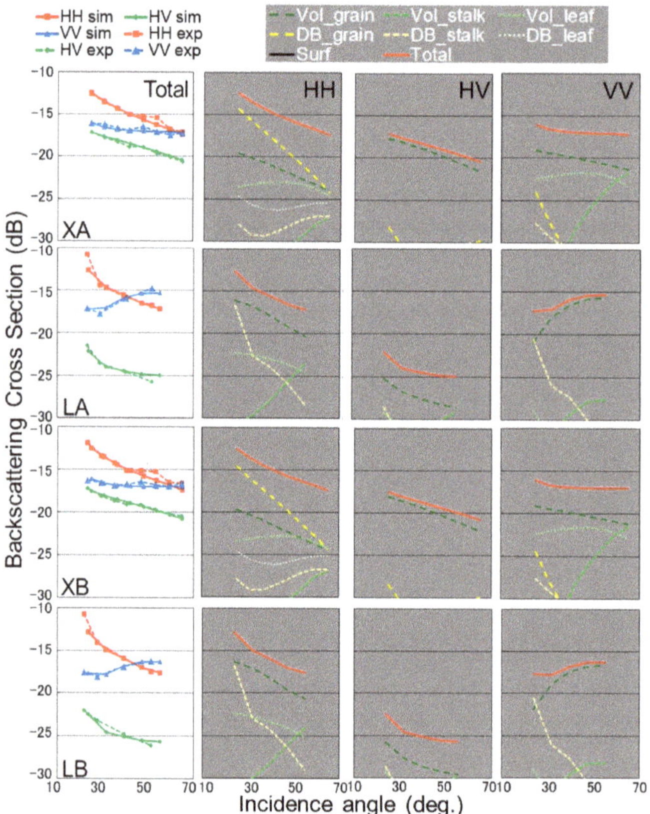

Figure 13. Comparison of the experimental data and DSM simulation results on blocks A and B at the X- and L-bands (left). Contribution of each scattering mechanism is also shown on the right-hand side with respect to polarization [36].

Based on the DSM modeling described in [32,34,36], each scattering mechanism is contributed to the total power as also shown on the right-hand side in Figure 13. The most dominant for all polarizations except HH is the volume scattering from rice panicles at the X-band, where interaction between grains and the surfaces is dominant. Another interaction between stalks and the ground surfaces is also dominant for HH and VV only at small incidence angles at the L-band. The most complicated curve is formed by VV at the X-band because the total power is pushed up by the volume scatterings from grains, leaves and stalks in terms of incidence angles one after another. Although numerical simulations by another set of input parameters assigning dominant vegetation water content to stem or leaf were conducted, strong attenuation of upper layer with rice panicles prevents from finding a set of input parameters to reasonably explain the observations. Nonetheless, it must be noted that the simulation results may not be a unique solution so that the interpretation could be varied by using a different forward model or a different set of input parameters.

Radius is the grain parameter with the most significant gap between the X- and L-bands, as shown in Table 1. In [49], scattering from a cylinder can be theoretically characterized by a wave number, k, and a cylinder radius, r, so that the kr must be conserved at different wavelengths as long as a received power stays at same level. This happened in the observation on HH and VV in Figure 13. The radii of virtual cylinders of grains at the L- and X-bands are 17.3 mm and 2.6 mm so that the ratio (17.3:2.6 mm) is approximately comparable to the ratio of the wavelength of the L-band to that of the X-band (24:3 cm). Therefore, the ratio of the radius to the wavelength is almost conserved. Based on the analysis, the effective radius of a virtual cylinder of rice grains at the late vegetative stage would vary due to the wavelength in use.

Clearly, the visible breakdown of the individual scattering mechanism in terms of polarization, frequency, and the incidence angle could deepen our understanding. The DSM could be an essential candidate of tools to provide a visible fraction among scattering mechanisms.

3.2. Multiparametric SAR Observation

The simulation results in the previous section show that a distinct sensitivity of the backscattering cross section from vegetated terrain to a various observation parameter is indicated. On the other hand, the DSM has a high degree of freedom so that it consists of many physical parameters as shown in Table 1. It is crucial to find a set of simulation parameters to reasonably describe the behavior of each scattering mechanism for an entire range of the observation condition.

To fix the model parameters, a great advantage potentially exists on MIMP-SAR data [32,34]. Although only a single point must be interpreted by a single incidence angle and single-polarimetric SAR observation, the MIMP SAR observation can provide multiple continuous curves in terms of incidence angles on each polarization. A few decades ago, it was extremely difficult to realize the MIMP SAR observation by a spaceborne SAR [55,56]. However, the concept becomes realistic by a recent advancement of SAR technology, where the ALOS-2, for example, can observe a specific area by the full polarimetric mode in a wider range of incidence angles between 20° and 40° [57].

Nonetheless, collecting MIMP SAR data by only a single satellite is still not very feasible at the present stage because it requires a number of dates in which plant growth potentially affects the data [58,59]. Dominant scattering mechanisms could also be misled by a sparsely sampled incidence angle [60–62]. Multi-incidence angle data can be obtained in another way by assuming uniform vegetation within an entire swath [20,63]. Although it is an efficient way, the assumption could be strict for natural vegetation with widely variable biological parameters. Hence, a direct multi-incidence angle SAR observation by an airborne SAR [32,34,36] was adapted to prevent from all the potential error sources above. Sufficiently small intervals are realized by the repeat path observations so that a time decorrelation can be minimized. The obtained MIMP SAR data are precisely characterized by a theoretical model based on a scattering mechanism. This multiparametric SAR approach improves the polarimetric decompositions to emphasize the sensitivity of backscattering cross section, and optimizes an observation condition for the limited spaceborne SAR monitoring.

A multi-incidence angle observation has been realized by multiple flight paths as proposed by Arii et al. [32,34,36] as shown in Figure 14. The incidence angle is varied to a specific point within the overlapped area by gradually shifting each orbit. For instance, eight observations in every 20 minutes were repeated from noon to 3:00 PM (local time) on August 21, 2014 as described in [32]. Therefore, any external factor such as a meteorological change and vegetation growth could be minimized. The incidence angle at the center of each scene in this study was varied from 24° to 65° by every 6° [27,28,32,34,36].

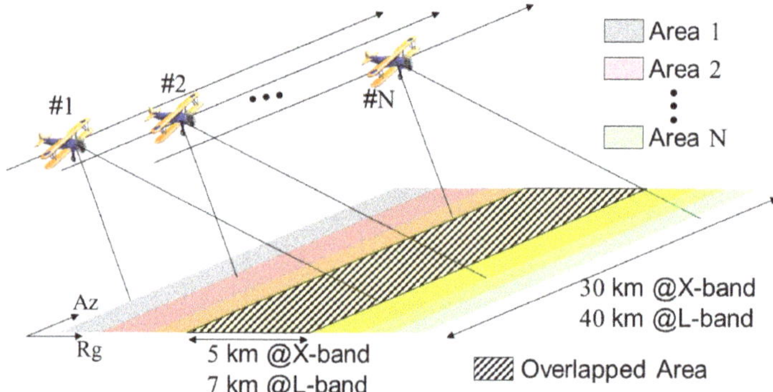

© 2019 IEEE. Reprinted, with permission, from Arii, M.; Yamada, H.; Kojima, S.; Ohki, M. Sensitivity analysis of multifrequency MIMP SAR data from rice paddies. IEEE Trans. Geosci. Remote Sens., 2019, 57, 3543-3551.

Figure 14. Concept of multi-incidence angle observation by an airborne synthetic aperture radar (SAR). Each orbit is gradually varied so that a specific point within the overlapped area has *N* sets of fully polarimetric data in terms of the incidence angle [36].

Currently, a full-polarimetric observation has been common for airborne SAR [64–69] so that the observation can be easily extended to an MIMP SAR observation. In addition, multifrequency MIMP SAR can be realized by simply repeating a set of MIMP SAR flights at a different band. Once multiparametric SAR data are collected, one has to explore simulation parameters as shown in Table 1, which hold for an entire range of observation condition. This concept can be called the comprehensive SAR approach.

Model fitting results are overlapped in Figure 13, and they show excellent agreement so that the selected parameters could be a reliable candidate to describe the backscatter from the paddy fields. This means that the simulated curves of each scattering mechanism shown on the right-hand side of Figure 13 could be also reliable.

As seen above, a history of each scattering mechanism in terms of observation conditions must be essential to quantitative validation of polarimetric decomposition results so that a sensitivity to a particular physical parameter is improved. In addition, a direct inversion from the backscattering cross section could be also feasible [21,24], which could be suitable for spaceborne SAR with a limited chance and observation condition.

4. Conclusions

In this article, the importance of the comprehensive SAR approach and its essential elements was thoroughly reviewed to broaden the use of polarimetric SAR remote sensing.

Popular model-based polarimetric decompositions were first introduced, and their limitations were also discussed with concrete examples, where only qualitative validation has been conducted. A fraction among scattering mechanisms of the backscattering cross section from vegetated surfaces must be essential for quantitative validation. To conquer the problem, a comprehensive SAR approach consisting of two key technologies was thoroughly reviewed.

One element is a forward model. A radiative transfer model and DSM widely used were concisely explained with pros and cons. As an example of the DSM, numerical simulations of backscatter from rice paddies at late vegetative stage were conducted under various observation conditions such as frequency, incidence angle and polarization, and a distinct sensitivity of backscatter to the physical parameters was shown.

The other is the multiparametric SAR observation where a number of SAR observations are conducted by gradually varying observation conditions. A data set of comprehensive SAR observations plays an important role to find reliable DSM parameters holding in a wide range of the observation conditions. Once the model parameters are fixed, a behavior of individual scattering mechanism could be accurately analyzed.

A comprehensive SAR approach consisting of a forward model and multiparametric SAR observation must be essential to move forward a field of polarimetric SAR remote sensing for future operational use.

Author Contributions: Conceptualization, methodology, simulation, M.A.; data analysis, M.A. and H.Y.; validation, H.Y., S.K. and M.O.

Funding: This review paper received no external funding.

Conflicts of Interest: The authors declare no conflict of interest.

References

1. Le Toan, T.; Laur, H.; Mougin, E.; Lopes, A. Multitemporal and dual-polarization observations of agricultural vegetation covers by X-band SAR images. *IEEE Trans. Geosci. Remote Sens.* **1989**, *27*, 709–718. [CrossRef]
2. Zebker, H.A.; van Zyl, J.J.; Durden, S.L.; Norikane, L. Calibrated imaging radar polarimetry: Technique, examples, and applications. *IEEE Trans. Geosci. Remote Sens.* **1991**, *29*, 942–961. [CrossRef]
3. Dubois, P.C.; van Zyl, J.; Engman, T. Measuring Soil Moisture with Imaging Radars. *IEEE Trans. Geosci. Remote Sens.* **1995**, *33*, 916–926. [CrossRef]
4. Dobson, M.C.; Pierce, L.E.; Ulaby, F.T. Knowledge-based land-cover classification using ERS-1/JERS-1 SAR composites. *IEEE Trans. Geosci. Remote Sens.* **1996**, *34*, 83–99. [CrossRef]
5. Le Toan, T.; Ribbes, F.; Wang, L.-F.; Floury, N.; Ding, K.-H.; Kong, J.-A.; Fujita, M.; Kurosu, T. Rice crop mapping and monitoring using ERS-1 data based on experiment and modeling results. *IEEE Trans. Geosci. Remote Sens.* **1997**, *35*, 41–56. [CrossRef]
6. Cloude, S.R.; Pottier, E. An entropy based classification scheme for land applications of polarimetric SAR. *IEEE Trans. Geosci. Remote Sens.* **1997**, *35*, 68–78. [CrossRef]
7. Shi, J.; Wang, J.; Hsu, A.Y.; O'Neil, P.E.; Engman, E.T. Estimation of bare surface soil moisture and surface roughness parameter using L-band SAR image data. *IEEE Trans. Geosci. Remote Sens.* **1997**, *35*, 1254–1266.
8. Pierce, L.E.; Bergen, K.M.; Dobson, M.C.; Ulaby, F.T. Multitemporal land-cover classification using SIR-C/X-SAR imagery. *Remote Sens. Environ.* **1998**, *63*, 24–39. [CrossRef]
9. Hajnsek, I.; Pottier, E.; Cloude, S.R. Inversion of surface parameters from Polarimetric SAR. *IEEE Trans. Geosci. Remote Sens.* **2003**, *41*, 724–744. [CrossRef]
10. Koay, J.-Y.; Tan, C.-P.; Lim, K.-S.; Bakar, S.; Ewe, H.-T.; Chuah, H.-T.; Kong, J.-A. Paddy fields as electrically dense media: Theoretical modeling and measurement comparisons. *IEEE Trans. Geosci. Remote Sens.* **2007**, *45*, 2837–2849. [CrossRef]
11. Lucas, R.; Armston, J.; Fairfax, R.; Fensham, R.; Accad, A.; Carreiras, J.; Bunting, P.; Clewley, D.; Bray, S.; Metcalfe, D.; et al. An evaluation of the ALOS PALSAR L-band backscatter above ground biomass relationship Queensland, Australia: Impacts of surface moisture condition and vegetation structure. *IEEE Trans. Geosci. Remote Sens.* **2010**, *3*, 576–593. [CrossRef]
12. Cloude, S.R. Uniqueness of target decomposition theorems in radar polarimetry. In *Direct and Inverse Methods in Radar Polarimetry*; Part 1, NATO-ARW; Boerner, W.M., Brand, H., Cram, L.A., Holm, W.A., Stein, D.E., Wiesbeck, W., Keydel, W., Giuli, D., Gjessing, D.T., Molinet, F.A., Eds.; Kluwer: Norwell, MA, USA, 1992; pp. 267–296.

13. Van Zyl, J.J. Application of Cloude's target decomposition theorem to Polarimetric imaging radar data. In *Radar Polarimetry*; SPIE-I748; SPIE: Bellingham, WA, USA, 1992; pp. 184–212.

14. Freeman, A.; Durden, S.L. A three-component scattering model for Polarimetric SAR data. *IEEE Trans. Geosci. Remote Sens.* **1998**, *36*, 963–973. [CrossRef]

15. Yamaguchi, Y.; Moriyama, T.; Ishido, M.; Yamada, H. Four-component scattering model for polarimetric SAR image decomposition. *IEEE Trans. Geosci. Remote Sens.* **2005**, *43*, 1699–1706. [CrossRef]

16. Touzi, R. Target scattering decomposition in terms of roll invariant target parameters. *IEEE Trans. Geosci. Remote Sens.* **2007**, *45*, 73–84. [CrossRef]

17. Freeman, A. Fitting a two-component scattering model to polarimetric SAR data from forests. *IEEE Trans. Geosci. Remote Sens.* **2007**, *45*, 2583–2592. [CrossRef]

18. Van Zyl, J.J.; Arii, M.; Kim, Y. Model based decomposition of polarimetric SAR covariance matrices constrained for non-negative eigenvalues. *IEEE Trans. Geosci. Remote Sens.* **2011**, *49*, 3452–3459. [CrossRef]

19. Arii, M.; van Zyl, J.J.; Kim, Y. Adaptive model-based decomposition of polarimetric SAR covariance matrices. *IEEE Trans. Geosci. Remote Sens.* **2011**, *49*, 1104–1113. [CrossRef]

20. Jagdhuber, T.; Hajnsek, I.; Bronstert, A.; Papathanassiou, K.P. Soil moisture estimation under low vegetation cover using a multi-angular polarimetric decomposition. *IEEE Trans. Geosci. Remote Sens.* **2013**, *51*, 2201–2215. [CrossRef]

21. Arii, M. Soil Moisture Retrieval under Vegetation Using Polarimetric Radar. Ph.D. Thesis, California Institute of Technology, Pasadena, CA, USA, 2009.

22. Ishimaru, A. *Wave Propagation and Scattering in Random Media*; IEEE Press: New York, NY, USA, 1978.

23. Bohren, C.F.; Huffman, D.R. *Absorption and Scattering of Light by Small Particles*; John Wiley & Sons: New York, NY, USA, 1983.

24. Van Zyl, J.J.; Kim, Y. *Synthetic Aperture Radar Polarimetry*; Wiley: Hoboken, NJ, USA, 2011.

25. Durden, S.L.; van Zyl, J.J.; Zebker, H.A. Modeling and observations of the radar polarization signatures of forested areas. *IEEE Trans. Geosci. Remote Sens.* **1989**, *27*, 290–301. [CrossRef]

26. Durden, S.L.; Klein, J.D.; Zebker, H.A. Polarimetric radar measurements of a forested area near Mt. Shasta. *IEEE Trans. Geosci. Remote Sens.* **1991**, *29*, 444–450. [CrossRef]

27. Arii, M.; Kitta, H.; Watanabe, T.; Yamada, H. Theoretical study of backscatter from rice paddy using discrete scatterer model. In Proceedings of the Asia-Pacific Conference on Synthetic Aperture Radar (APSAR), Tsukuba, Japan, 23–27 September 2013; pp. 27–30.

28. Arii, M.; Komatsu, T.; Nishimura, T.; Watanabe, T.; Yamada, H.; Kobayashi, T. Theoretical characterization of X-band full polarimetric SAR data from rice paddies in terms of incidence angles. In Proceedings of the IEICE Technical Report, Niigata, Japan, 29 August 2014; Volume 114, pp. 67–72.

29. Arii, M.; Watanabe, T.; Yamada, H. Sensitivity study of radar backscatter from boreal forest using discrete scatter model. In Proceedings of the IEEE International Geoscience and Remote Sensing Symposium, Munich, Germany, 22–27 July 2012; pp. 1425–1428.

30. Watanabe, T.; Yamada, H.; Arii, M.; Park, S.-E.; Yamaguchi, Y. Model experiment of permittivity retrieval method for forested area by using Brewster's angle. In Proceedings of the IEEE International Geoscience and Remote Sensing Symposium, Munich, Germany, 22–27 July 2012; pp. 1477–1480.

31. Watanabe, T.; Yamada, H.; Arii, M.; Sato, R.; Park, S.-E.; Yamaguchi, Y. Study on moisture effects on polarimetric radar backscatter from forested terrain. *IEICE Trans. Commun.* **2014**, *97*, 2074–2082. [CrossRef]

32. Arii, M.; Yamada, H.; Kobayashi, T.; Kojima, S.; Umehara, T.; Komatsu, T.; Nishimura, T. Theoretical characterization of X-band multi-incidence angle and multi-polarimetric SAR data from rice paddies at late vegetative stage. *IEEE Trans. Geosci. Remote Sens.* **2017**, *55*, 2706–2715. [CrossRef]

33. Arii, M.; Yamada, H. Rice paddy monitoring by L-band MIMP SAR Approach. In Proceedings of the IEEE International Geoscience and Remote Sensing Symposium (IGARSS), Fort Worth, TX, USA, 23–28 July 2017; pp. 2442–2445.

34. Arii, M.; Yamada, H.; Ohki, M. Characterization of L-band MIMP SAR data from rice paddies at late vegetative stage. *IEEE Trans. Geosci. Remote Sens.* **2018**, *56*, 3852–3860. [CrossRef]

35. Kim, S.; Arii, M.; Jackson, T. Modeling L-band synthetic aperture radar data through dielectric changes in soil moisture and vegetation over shrublands. *IEEE J. Sel. Top. Appl. Earth Observ. Remote Sens.* **2017**, *10*, 4753–4762. [CrossRef]

36. Arii, M.; Yamada, H.; Kojima, S.; Ohki, M. Sensitivity analysis of multifrequency MIMP SAR data from rice paddies. *IEEE Trans. Geosci. Remote Sens.* **2019**, *57*, 3543–3551. [CrossRef]
37. Arii, M.; van Zyl, J.J.; Kim, Y. A general characterization for polarimetric scattering from vegetation canopies. *IEEE Trans. Geosci. Remote Sens.* **2010**, *48*, 3349–3357. [CrossRef]
38. Borgeaud, M.; Shin, R.T.; Kong, J.A. Theoretical models for polarimetric radar clutter. *J. EM Waves Appl.* **1987**, *1*, 73–91. [CrossRef]
39. Chandrasekhar, S. *Radiative Transfer*; Dover: Mineola, NY, USA, 1960.
40. Ulaby, F.T.; Sarabandi, K.; Mcdonald, K.C.; Whitt, N.W.; Dobson, M.C. *Michigan Microwave Canopy Scattering Model (MIMICS)*; Rep. 022486-T-1; Radiation Laboratory, University of Michigan: Ann Arbor, MI, USA, 1988.
41. Ulaby, F.; Sarabandi, K.; McDonald, K.; Whitt, M.; Dobson, C. Michigan microwave canopy scattering model. *Int. J. Remote Sens.* **1990**, *11*, 1223–1253. [CrossRef]
42. MacDonald, K.C.; Dobson, M.C.; Ulaby, F.T. Using MIMICS to model multiangle and multitemporal backscatter from a walnut orchard. *IEEE Trans. Geosci. Remote Sens.* **1990**, *28*, 477–491. [CrossRef]
43. Dobson, M.C.; Ulaby, F.T. Active microwave soil moisture research. *IEEE Trans. Geosci. Remote Sens.* **1986**, *GE-24*, 23–36. [CrossRef]
44. Tsang, L.; Kong, J.A.; Shin, R. *Theory of Microwave Remote Sensing*; Wiley: New York, NY, USA, 1985.
45. Tsang, L.; Kong, J.A.; Ding, K.H. *Scattering of Electromagnetic Waves*; Vol. 1: Theory and Applications; Wiley: Hoboken, NJ, USA, 2000.
46. Rice, S.O. Reflection of electromagnetic waves from slightly rough surfaces. *Commun. Pure Appl. Math.* **1951**, *4*, 351–378. [CrossRef]
47. Van de Hulst, H.C. *Light Scattering by Small Particles*; Dover: Mineola, NY, USA, 1957.
48. Ruck, G.T.; Barrick, D.E.; Stuart, W.D.; Krichbaum, C.K. *Radar Cross Section Handbook*; Plenum Press: New York, NY, USA, 1970.
49. Ulaby, F.T.; Elachi, C. *Radar Polarimetry for Geoscience Applications*; Artech House: Norwood, MA, USA, 1990; pp. 92–101.
50. Ulaby, F.T.; El-Rayes, M.A. Microwave dielectric spectrum of vegetation-Part II: Dual-dispersion model. *IEEE Trans. Geosci. Remote Sens.* **1987**, *GE-25*, 550–557. [CrossRef]
51. Ribbes, F.; Le Toan, F. Rice field mapping and monitoring with RADARSAT data. *Int. J. Remote Sens.* **1999**, *20*, 745–765. [CrossRef]
52. Oh, Y.; Hong, S.; Kim, Y.; Hong, J.; Kim, Y. Polarimetric backscattering coefficients of flooded rice fields at L- and C-bands: Measurements, modeling, and data analysis. *IEEE Trans. Geosci. Remote Sens.* **2009**, *47*, 2714–2721.
53. Jia, M.; Tong, L.; Chen, Y. Multifrequency and multitemporal ground-based scatterometers measurements on rice fields. In Proceedings of the IEEE International Geoscience and Remote Sensing Symposium, Munich, Germany, 22–27 July 2012; pp. 642–645.
54. El-Rayes, M.A.; Ulaby, F.T. Microwave dielectric spectrum of vegetation-part I: Experimental observations. *IEEE TGRS* **1987**, *GE-25*, 541–549. [CrossRef]
55. Cimino, J.; Casey, A.D.; Rabassa, J.; Wall, S.D. Multiple incidence angle SIR-B experiment over Argentina: Mapping of forest units. *IEEE Trans. Geosci. Remote Sens.* **1986**, *24*, 498–509. [CrossRef]
56. Ahmed, A.; Richards, J.A. Multiple incidence angle SIR-B forest observations. *IEEE Trans. Geosci. Remote Sens.* **1989**, *27*, 586–591. [CrossRef]
57. Arii, M.; Nishimura, T. Landslide monitoring using ALOS-1 and 2 data. In Proceedings of the IEEE International Geoscience and Remote Sensing Symposium (IGARSS), Milan, Italy, 26–31 July 2015; pp. 4129–4132.
58. Shao, Y.; Fan, X.; Liu, H.; Xiao, J.; Ross, S.; Brisco, B.; Brown, R.; Staples, G. Rice monitoring and production estimation using multitemporal RADARSAT. *Remote Sens. Environ.* **2001**, *76*, 310–325. [CrossRef]
59. Lopez-Sanchez, J.M.; Vicente-Guijalba, F.; Ballester-Berman, J.D.; Cloude, S.R. Influence of incidence angle on the coherent copolar polarimetric response of rice at X-band. *IEEE Geosci. Remote Sens. Lett.* **2014**, *12*, 249–253. [CrossRef]
60. Inoue, Y.; Kurosu, T.; Maeno, H.; Uratsuka, S.; Kozu, T.; Dabrowska-Zielinska, K.; Qi, J. Season-long daily measurements of multifrequency (Ka, Ku, X, C, L) and full-polarization backscatter signatures over paddy rice field and their relationship with biological variables. *Remote Sens. Environ.* **2002**, *81*, 194–204. [CrossRef]

61. Lopez-Sanchez, J.M.; Vicente-Guijalba, F.; Ballester-Berman, J.D.; Cloude, S.R. Retrieving rice phenology with X-band co-polar data: A multi-incidence multi-year experiment. In Proceedings of the IEEE International Geoscience and Remote Sensing Symposium (IGARSS), Milan, Italy, 26–31 July 2015; pp. 3977–3980.

62. Cable, J.W.; Kovacs, J.M.; Jiao, X.; Shang, J. Agricultural monitoring in northeastern Ontario, Canada, using multi-temporal polarimetric RADARSAT-2 data. *Remote Sens.* **2014**, *6*, 2343–2371. [CrossRef]

63. Wang, H.; Ouchi, K. Accuracy of the K-distribution regression model for forest biomass estimation by high-resolution polarimetric SAR: Comparison of model estimation and field data. *IEEE Trans. Geosci. Remote Sens.* **2008**, *46*, 1058–1064. [CrossRef]

64. Van Zyl, J.J.; Carande, R.; Lou, Y.; Miller, T.; Wheeler, K. The NASA/JPL three-frequency polarimetric airsar system. In Proceedings of the IEEE International Geoscience and Remote Sensing Symposium, Houston, TX, USA, 26–29 May 1992; Volume I, pp. 649–651.

65. Nadai, A.; Uratsuka, S.; Umehara, T.; Matsuoka, T.; Kobayashi, T.; Satake, M. Development of X-band airborne polarimetric and interferometric SAR with sub-meter spatial resolution. In Proceedings of the IEEE International Geoscience and Remote Sensing Symposium, Cape Town, South Africa, 12–17 July 2009; pp. 913–916.

66. Shimada, M.; Kawano, N.; Watanabe, M.; Motooka, T.; Ohki, M. Calibration and validation of the Pi-SAR-L2. In Proceedings of the Asia-Pacific Conference on Synthetic Aperture Radar (APSAR), Tsukuba, Japan, 23–27 September 2013; pp. 194–197.

67. Shimada, M.; Watanabe, M.; Motooka, T.; Kankaku, Y. PALSAR-2 polarimetric performance and the simulation study using the PI-SAR-L2. In Proceedings of the IEEE International Geoscience and Remote Sensing Symposium, Melbourne, Australia, 21–26 July 2013; pp. 2309–2312.

68. Gebert, N.; Almeida, F.Q.; Krieger, G. Airborne demonstration of multichannel SAR imaging. *IEEE Geosci. Remote Sens. Lett.* **2011**, *8*, 963–967. [CrossRef]

69. Alexander, G.F.; Chapman, B.D.; Hawkins, P.; Hensley, S.; Jones, C.E.; Michel, T.R.; Muellerschoen, R.J. UAVSAR polarimetric calibration. *IEEE Trans. Geosci. Remote Sens.* **2015**, *53*, 3481–3491.

 *electronics*

Article

An Inverse Synthetic Aperture Ladar Imaging Algorithm of Maneuvering Target Based on Integral Cubic Phase Function-Fractional Fourier Transform

Yakun Lv [1,*], Yanhong Wu [1], Hongyan Wang [2], Lei Qiu [1], Jiawei Jiang [1] and Yang Sun [3]

[1] Department of Electronic and Optical Engineering, Space Engineering University, Beijing 101416, China; mail2wyh@163.com (Y.W.); qiulei2013@nudt.edu.cn (L.Q.); herojjw@163.com (J.J.)
[2] School of Space Information, Space Engineering University, Beijing 101416, China; yhgnaw@163.com
[3] Science and Technology on Complex Electronic System Simulation Laboratory, Beijing 101416, China; fireflypd@buaa.edu.cn
* Correspondence: qiul06@mails.tsinghua.edu.cn; Tel.: +86-010-6636-5117

Received: 11 July 2018; Accepted: 13 August 2018; Published: 15 August 2018

Abstract: When imaging maneuvering targets with inverse synthetic aperture ladar (ISAL), dispersion and Doppler frequency time-variation exist in the range and cross-range echo signal, respectively. To solve this problem, an ISAL imaging algorithm based on integral cubic phase function-fractional Fourier transform (ICPF-FRFT) is proposed in this paper. The accurate ISAL echo signal model is established for a space maneuvering target that quickly approximates the uniform acceleration motion. On this basis, the chirp rate of the echo signal is quickly estimated by using the ICPF algorithm, which uses the non-uniform fast Fourier transform (NUFFT) method for fast calculations. At the best rotation angle, the range compression is realized by FRFT and the range dispersion is eliminated. After motion compensation, separation imaging of strong and weak scattering points is realized by using ICPF-FRFT and CLEAN technique and the azimuth defocusing problem is solved. The effectiveness of the proposed method is verified by a simulation experiment of an aircraft scattering point model and real data.

Keywords: inverse synthetic aperture ladar (ISAL); maneuvering target; integral cubic phase function (ICPF); fractional Fourier transform (FRFT); non-uniform fast Fourier transform (NUFFT); CLEAN technique

---

1. Introduction

With the development of radar detection technology, high-precision target imaging has become an important aspect of the detection task. Inverse synthetic aperture ladar (ISAL) combines coherent laser technology and inverse synthetic aperture technology, overcoming the limitations of the actual aperture and diffraction. ISAL also overcomes the shortfalls of traditional microwave imaging radars that cannot provide enough range resolution for remote target and small target imaging and solves the problem experienced by traditional laser imaging radar, which cannot perform the high-resolution imaging of a moving target [1]. ISAL is the only optical means by which centimeter-level resolution can be obtained at a range of thousands of kilometers [2]. Therefore, ISAL imaging can fulfil the requirement for high precision imaging and quasi real-time imaging for target surveillance.

ISAL imaging is similar to the traditional inverse synthetic aperture radar (ISAR) imaging principle but due to the use of laser as a radiation source, ISAL has an ultra-high carrier frequency, ultra-large bandwidth and extremely short wavelength. Compared with ISAR, ISAL has higher resolution, smaller imaging angle and shorter imaging time [3]. Research on ISAL has mainly focused on principal analysis and algorithm simulation. Some close-range field tests have been reported [4–8].

Research on space target imaging using the space-based platform ISAL is still in its infancy, so the research in this area is important.

ISAL uses an ultra-high carrier frequency and ultra-large bandwidth signal and the existing range compression method disperses and distorts the echo range profile. In addition, the ISAL azimuth Doppler of the maneuvering target is time variant. If the conventional azimuth image method is used, the imaging quality will be seriously defocused. Two kinds of methods are currently available for imaging the maneuvering target in the microwave section ISAR: parameter estimation and time-frequency analysis. The parameter estimation method models the signal echo into a multi-component signal model, then uses the signal estimation method to estimate the maneuvering parameters of the signal and then compensates the signal after the estimation. The commonly used estimation methods include maximum likelihood (ML) estimation [9], high-order ambiguity function (HAF) [10], discrete chirp Fourier transform (DCFT) [11] and a variety of improved algorithms [12]. The disadvantage with these methods is that it is generally required to search for parameters and the computational complexity is considerable for high precision estimation. The time-frequency analysis method involves the instantaneous Doppler, which obtains the signal using the time frequency analysis method and then uses range instantaneous Doppler (RID) for imaging. The common RID methods include short time Fourier transform (STFT) [13], wavelet transform, S transform, Wigner-Ville distribution (WVD), the smoothed pseudo WVD (SPWVD) [14] and the adaptive time-frequency decomposition method [15,16]. The disadvantage of these methods is that the time-frequency resolution is low and cross terms exist, which affects the imaging quality. Given the high-resolution capability of ISAL, a large amount of data is generated. Therefore, finding a fast and efficient imaging method suitable for ISAL imaging is necessary.

In this study, to solve the above problems, we first analyzed the exact echo model of the target. Secondly, given the problem of the one-dimensional (1D) range dispersion of the echo signal, the frequency modulation rate of the echo signal was quickly estimated using the ICPF algorithm and then used the modulation frequency to calculate the best rotation angle in the FRFT domain. At this angle, the FRFT [17] method was used to achieve the range compression and eliminate the range dispersion. Then, to address the problem of azimuth defocus, ICPF combined with the CLEAN technique was proposed to estimate the frequency modulation rate of the strong and weak scattering points and then the azimuth compression imaging was realized using FRFT. Finally, the effectiveness of the method was verified by a simulation experiment of the plane scattering point model.

2. ISAL Signal Echo Model

The three-dimensional (3D) imaging geometry of the maneuvering target is shown in Figure 1a. Where the coordinate origin O is the target turntable center, point $P(x_p, y_p, z_p)$ is any scattering point on the target and r_p is the scattering point P position vector starting from O. ω is the rotational angular velocity vector of the target three-dimensional motion. The ISAL imaging projection plane Γ is determined by the vector ω and the radar line-of-sight direction (LOS) unit vector **R**, ω can be decomposed into the radial rotational component, ω_r along the LOS and the rotational component, ω_e perpendicular to the LOS. ω_r cannot cause the radial movement of the scattering point, that is, it will not cause the phase change of the echo and ω_e will cause the scattering point to move radially, resulting in Doppler frequency variation, which can achieve high-resolution ISAL imaging of the target, ω_e is called effective rotation component. The three-dimensional motion velocity of the target can be decomposed into a component v in the Γ plane and a component perpendicular to the Γ plane and the vertical component does not affect the imaging of the target, so this component can be ignored. For the parallel component v, it can be decomposed into the radial component v_r along the LOS and the component v_e perpendicular to the LOS. v_r causes the Doppler shift of the target echo which cause phase change, while v_e does not generate Doppler shift.

After the above analysis, the effective component in the three-dimensional (3D) imaging geometry can be projected onto the imaging plane Γ to obtain a two-dimensional (2D) turntable imaging geometry as shown in Figure 1b [18]. In two-dimensional (2D) imaging geometry, only the relative motion between radar and target is considered. Where O is the reference point and XOY is the rectangular coordinate system fixed on the target, the Y axis is the direction of the radar LOS. The target moves along the Y axis with the speed of v_r and rotates around the O point at the angular velocity of ω_e. Suppose that at time t (t is full-time and satisfies the equation $t = t_k + t_m$, where t_k is the range fast time and t_m is the azimuth slow time, $m = 1, 2, \ldots, M$), the range between the target geometry center and the radar is $R_0(t)$. The rotation angle of the target relative to the radar is $\theta(t)$. Then, the range $R_P(t)$ between any point $P(x, y)$ in the target and the radar is:

$$R_P(t) = \sqrt{R_0(t)^2 + r_p^2 - 2R_0 r_p \cos[\theta(t) + \frac{\pi}{2}]} \approx R_0(t) + x\sin\theta(t) + y\cos\theta(t) \tag{1}$$

Considering the inertia of the conventional target motion and the short imaging time of ISAL, which is less than 1 s, maneuvering target motion can only approximate to the second-order motion component. In other words, $R_0(t)$ and $\theta(t)$ can be approximated as the second-order function of t^2:

$$R_0(t) = R_0 + v_0 t + \frac{1}{2!}at^2 \tag{2}$$

$$\theta(t) = \theta_0 + \omega t + \frac{1}{2!}\Omega t^2 \tag{3}$$

where R_0 is the initial range, v_0 is the initial radial velocity, a is the radial acceleration, θ_0 is the initial rotation angle, ω is the rotation angular velocity and Ω is the rotation angular acceleration.

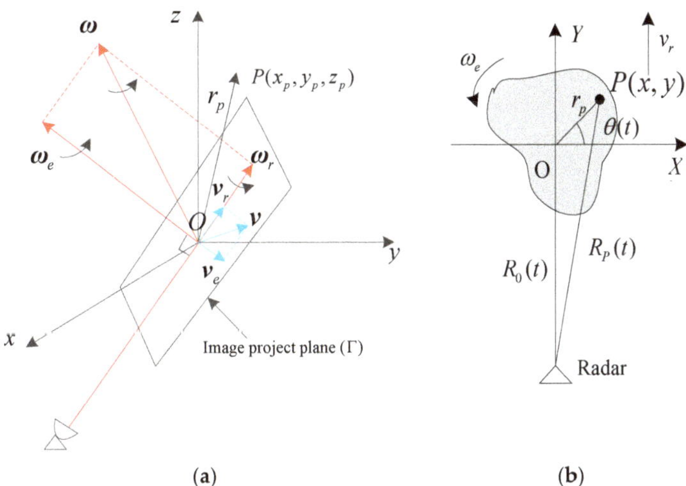

Figure 1. Inverse synthetic aperture ladar (ISAL) turntable imaging geometry; (**a**) three-dimensional (3D) imaging geometry; (**b**) two-dimensional (2D) imaging geometry.

Because wavelength of ISAL is on the micron scale, it is sensitive to the motion of the target, so the effect of the fast time t_k on the radial range cannot be ignored [19]. When the pulse duration is short, the influence of t_k on rotation components can be neglected. So, Equations (2) and (3) can be resolved as:

$$\begin{aligned}
R_0(t) &= R_0 + v_0(t_m + t_k) + \frac{1}{2}a(t_m + t_k)^2 \\
&= R_0 + v_0 t_m + \frac{1}{2}at_m^2 + (v_0 + at_m)t_k + \frac{1}{2}at_k^2 \\
&= R_0(t_m) + v(t_m)t_k + \frac{1}{2}at_k^2
\end{aligned} \tag{4}$$

$$\theta(t) \approx \theta(t_m) = \theta_0 + \omega t_m + \frac{1}{2}\Omega t_m^2 \tag{5}$$

where $R_0(t_m)$ and $v(t_m)$ represent the radial range and velocity varies with the azimuth time t_m, respectively.

According to the above equation, Equation (1) can be rewritten as:

$$R_P(t) \approx R_0(t_m) + R_d(t_m) + v(t_m)t_k + \frac{1}{2}at_k^2 \tag{6}$$

where $R_d(t_m) = x\sin\theta(t_m) + y\cos\theta(t_m)$.

ISAL usually uses ultra-bandwidth Linear Frequency Modulated (LFM) signals to achieve high range resolution, the expression of the transmitted signal can be written as:

$$s_t(t_k, t_m) = rect\left(\frac{t_k}{T_p}\right)\exp\left(j2\pi f_c t + j\pi k t_k^2\right) \tag{7}$$

$$rect\left(\frac{t_k}{T_p}\right) = \begin{cases} 1, & |t_k| \le T_p/2 \\ 0, & |t_k| > T_p/2 \end{cases} \tag{8}$$

where T_p is the width of the pulse, f_c is the carrier frequency and k is the chirp-rate.

Considering $c \gg v_r + at$, we ignore the effect of the target speed of the irradiating and receiving radar signals. Suppose the radar receives the echo signal of point P after time delay $\tau = 2R_P(t)/c$, the radar receiving signal is:

$$s_r(t_k, t_m) = \sigma rect\left(\frac{t_k - \tau}{T_p}\right)\exp\left(j2\pi f_c(t_k - \tau) + j\pi k(t_k - \tau)^2\right) \tag{9}$$

To reduce the data rate, ISAL often uses optical heterodyne coherent detection [20] to handle the echo signals. Suppose the reference delay of the coherent pulses is $\tau_{ref} = 2R_{ref}(t_m)/c$, the reference signal is:

$$s_{ref}(t_k, t_m) = rect\left(\frac{t_k - \tau_{ref}}{T_p}\right)\exp\left(j2\pi f_c(t - \tau_{ref}) + j\pi k(t_k - \tau_{ref})^2\right) \tag{10}$$

Therefore, the output signal after optical heterodyne detection is:

$$\begin{aligned} s_o(t_k, t_m) &= s_r(t_k, t_m) \cdot s^*_{ref}(t_k, t_m) \\ &= \sigma_p rect\left(\frac{t_k - \tau}{T_p}\right)\exp\left[j2\pi f_c(\tau_{ref} - \tau)\right] \cdot \exp\left[-j\pi k(\tau_{ref}^2 - \tau^2)\right] \cdot \exp\left[j2\pi k(\tau_{ref} - \tau)t_k\right] \end{aligned} \tag{11}$$

We substitute Equations (4) and (6) into Equation (11) to obtain a polynomial function about time t_k:

$$s_o(t_k, t_m) = \sigma_p rect\left(\frac{t_k - \tau}{T_p}\right) \cdot \exp\left[j2\pi\left(P_0 + P_1 t_k + P_2 t_k^2 + P_3 t_k^3 + P_4 t_k^4\right)\right] \tag{12}$$

$$P_0 = 2k\frac{\Delta R_{mP}(\Delta R_{mP} + 2R_{ref}(t_m))}{c^2} - 2f_c\frac{\Delta R_{mP}}{c} \tag{13}$$

$$P_1 = v(t_m)\left(4k\frac{R_0(t_m) + R_d(t_m)}{c^2} - \frac{2f_c}{c}\right) - \frac{2k}{c}\Delta R_{mP} \tag{14}$$

$$P_2 = -\frac{af_c}{c} + 2ak\frac{R_0(t_m) + R_d(t_m)}{c^2} + 2k\frac{v(t_m)^2}{c^2} - 2k\frac{v(t_m)}{c} \tag{15}$$

$$P_3 = 2ak_r\frac{v(t_m)}{c^2} - \frac{ak_r}{c} \tag{16}$$

$$P_4 = \frac{k_r a^2}{2c^2} \tag{17}$$

where $\Delta R_{mP} = R_0(t_m) + R_d(t_m) - R_{ref}(t_m)$. When $R_{ref}(t_m)$ is accurately estimated, it is approximately equal to $R_0(t_m)$, then $\Delta R_{mP} = R_d(t_m)$, which is only related to the scattering point on the target in azimuth time. In practice, considering the impact of the target velocity on τ and τ_{ref}, the envelope in the Equation (12) will cause a contraction in time. But the impact does not affect the analysis of the range dispersion, which is ignored here [1]. P_0 is the phase term related only to the azimuth time t_m, in which $2f_c\Delta R_{mP}/c$ is a necessary term for azimuth compression and has no influence on range compression. P_1 is the first-order phase term. The first item in P_1 is mainly affected by the high carrier signal frequency, which produces the signal in the pulse Doppler. For the uniformly accelerated moving target, the Doppler coupling time shift with azimuth time change is produced, resulting in a range move. The second item contains the range information $2k\Delta R_{mP}/c$, which is key to attaining range compression. P_2 is the chirp-rate phase term, mainly influenced by the ultra-high carrier frequency and large bandwidth. It is the root cause of the division and broadening of the peaks of the range. The range dispersion effect occurs if the conventional DFT is used for the compression processing of the range direction. From the expression P_2, the chirp-rate term of all scattering points in the single pulse echo is the same fixed value, whereas for different pulse echoes, the chirp-rate rate varies with slow time t_m. Therefore, processing the pulse echo sequence one at a time is necessary. P_3 and P_4 are the high-order phase terms. Because in a pulse period $c \gg av(t_m)$, $T_p^2 \ll c/2k_rT_p$ and $T_p^3a^2 \ll c^2/2k_rT_p$, the influence of the P_3 and P_4 on the intra-pulse Doppler spectrum broadening can be ignored.

According to the above analysis, in the imaging time, the second-order polynomial approximation can appropriately reflect the motion state of the maneuvering target and meet the imaging needs. The effect of the third- and fourth-order terms can be ignored. The range echo signals after heterodyne detection can be approximated to multicomponent LFM signals with the same frequency modulation slope:

$$s_0(t_k, t_m) = \sigma_p rect(\frac{t_k - \tau}{T_p}) \cdot \exp\left[j2\pi\left(P_0 + P_1 t_k + P_2 t_k^2\right)\right] \tag{18}$$

3. Range Imaging Based on ICPF-FRFT

FRFT is a kind of generalized Fourier transform that better focuses LFM signals [21]. The FRFT of the signal $s(t)$ is defined as:

$$S_\alpha(u) = F^p[s(t)] = \int_{-\infty}^{\infty} s(t)K_\alpha(t, u)dt \tag{19}$$

where p is the order of the FRFT, which can be any real number and α is the rotation angle; $\alpha = p\pi/2$. When $\alpha = \pi/2$, FRFT becomes a traditional Fourier transform. $K_\alpha(t, u)$ is the transformation operator, the expression is:

$$K_\alpha(t, u) = \begin{cases} \sqrt{1 - j\cot\alpha}\exp\left(j\pi\left((t^2 + u^2)\cot\alpha - 2ut\csc\alpha\right)\right), \\ \quad \alpha \neq n\pi; \\ \delta(t - u), \quad \alpha = 2n\pi \\ \delta(t + u), \quad \alpha = (2n \pm 1)\pi \\ \quad \alpha \neq n\pi; \end{cases} \tag{20}$$

FRFT is equivalent to projecting the signal on the frequency axis after the counterclockwise rotation α of the signal in the time-frequency plane. When the u axis of the FRFT is rotated to the time-frequency ridge of the signal, the amplitude of the signal projection to the fractional frequency u axis is maximized and the rotation angle at this time is called the best angle α_k of rotation. Therefore, the projection of FRFT at the best angle of rotation can be used for range imaging and the imaging principle is shown in Figure 2.

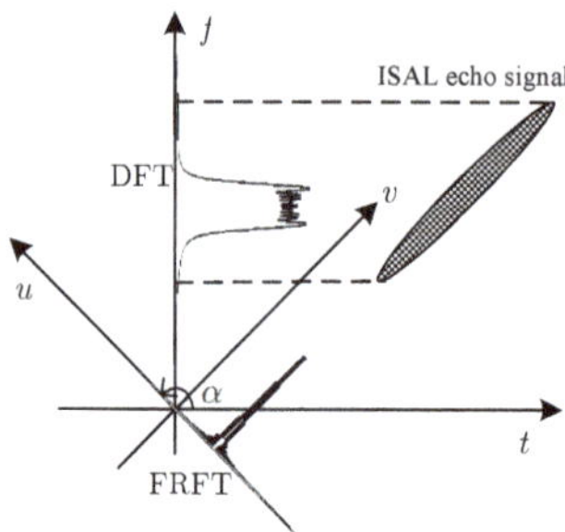

Figure 2. Schematic diagram of range compression via Discrete Fourier transform (DFT) and fractional Fourier transform (FRFT).

FRFT requires the peak search method in the two-dimensional (2D) plane (α, u) to obtain the optimal rotation angle. Therefore, the effect of range image compression depends on the value of α_k and its precision is easily influenced by the resolution of the search angle. The computation requirement is considerable in the high precision search. So, accuracy and computation are difficult to achieve. Paper and colleagues [22–25] proposed that ICPF can quickly estimate the chirp rate of LFM signals. The method only requires a 1D search and has good anti-noise performance and high estimation accuracy without being affected by subjective factors such as search resolution. Therefore, in this paper, ICPF was firstly used to estimate the modulation frequency of the optical heterodyne output signal and then the optimal rotation angle and the order of the FRFT were calculated. Finally, the range compression imaging was completed by FRFT at the optimal rotation angle. The ICPF definition of signal $x(t)$ is as follows:

$$ICPF(\mu) = \frac{1}{T} \int\limits_{-\infty}^{\infty} \left| \int\limits_{-\infty}^{\infty} x(t+\tau)x(t-\tau)\exp(-j\mu\tau^2)d\tau \right|^2 dt \tag{21}$$

From the definition, ICPF is a kind of transformation that detects the chirp-rate of the signal, which can concentrate the signal energy on the chirp-rate of the signal, in line with the energy distribution of the linear frequency modulation signal. Since ICPF needs to calculate the τ^2 of the signal, using FFT for fast calculations is not possible. Therefore, we used the non-uniform fast Fourier transform (NUFFT) [26,27] to overcome the rigorous data sampling requirements of the FFT and to improve the algorithm's calculation speed. The NUFFT is defined as:

$$\hat{z}_k = \sum_{l=1}^{M} z_l \exp(-j2\pi x_l/N), k = -N/2, \cdots, N/2+1 \tag{22}$$

where z_l is non-uniform sampling time and x_l is the corresponding non-uniform sampling position. Here, interpolation time domain non-uniform sampling data z_l is replaced by an interpolation index term to achieve fast non-uniform Fourier transform.

Suppose that $\hat{\varphi}(x) = \sqrt{\frac{2}{\pi}} \frac{\sinh(\alpha\sqrt{K^2-x^2})}{\sqrt{K^2-x^2}}, -K \leq x \leq K$, $\hat{\varphi}(\xi) = \begin{cases} I_0 K\sqrt{\alpha^2 - \xi^2}, & |\xi| \leq \alpha \\ 0, & |\xi| > \alpha \end{cases}$, where, K is the length of interpolation kernel function. According to P. O'Shea [27], the exponential function can be expanded as shown in Equation (23):

$$\exp(-jx\xi) = \frac{1}{\sqrt{2\pi}\varphi(\xi)} \sum_{m \in Z} \hat{\varphi}(x-m)\exp(-jm\xi) \tag{23}$$

where $|\xi| \leq \pi/c$, c is the oversampling factor. Suppose $x = cx_1$, $\xi = 2\pi k/(cN)$, $|k| \leq N/2$, $\alpha = \pi(2 - 1/c) - 0.01$, $\varphi_k = \varphi(2\pi k/(cN))$ and $\hat{\phi}_{lm} = \frac{1}{2\pi}\hat{\phi}(cx_l - (\mu_l + m))$. Substituting Equation (23) into Equation (22) yields a uniform frequency output:

$$\hat{z}_k = \frac{1}{\varphi_k}\sum_{j=-cN/2}^{cN/2-1} u_j \exp(-j2\pi kj/(cN)) \tag{24}$$

where

$$u_j = \sum_{l=1}^{M}\sum_{m\in Z} z_l\hat{\phi}_{l,j+cmN-\mu_l}, j = -cN/2, \cdots , cN/2 - 1 \tag{25}$$

The specific NUFFT implementation process is shown in Figure 3. First, the intermediate parameters $\mu_l, \varphi_k, \hat{\phi}_{lm}$ are calculated from the input non-uniform sampling data z_l and the corresponding position x_l. Then, the intermediate variable u_j is calculated according to Equations (25). Finally, the corresponding frequency output value $\hat{z}_k$ is calculated using Equation (24) with fast FFT.

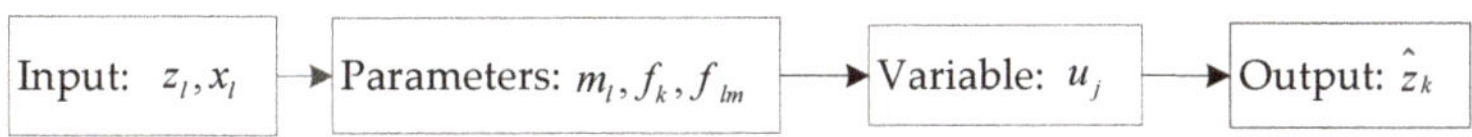

Figure 3. Non-uniform fast Fourier transform (NUFFT) diagram.

According to the above NUFFT principle, the non-uniform Fourier transform in the proposed ICPF is quickly implemented with NUFFT to reduce the computational complexity of the algorithm, so Equation (26) can be written as:

$$ICPF(\mu) = FFT_t[NUFFT_{\tau^2}(ICPF(t, \mu_{\tau^2}))] \tag{26}$$

where $NUFFT_{\tau^2}$ indicates the NUFFT operation on the variable τ^2 and FFT_t indicates FFT operation on the variable t.

Assuming N is the sampling point of a single pulse echo and M is the number of DFT search points, the complexity of the ICPF-based DFT direct calculation is $O(N^2M)$, whereas the complexity of the non-uniform Fourier transform calculation method is $O(N\log_2(N))$ [28]. Assuming that K is the number of scanning points in the FRFT transform domain α, which is determined by the step size and range of α, the complexity of the discrete FRFT is $O(NK\log_2(N))$. If we want an accurate α_k, K is usually large. The transformation and 2D search need to be coordinated [21,24], so the proposed NUFFT-based ICPF-FRFT algorithm does not need to perform any parameter search and has high anti-noise performance. These features enable the implementation of the ISAL imaging algorithm in real time.

As a result, the ICPF transform of the optical heterodyne output signal results in spikes only at its chirp-rate slope. The chirp-rate at the peak is μ_k:

$$\mu_k = \underset{\mu}{\mathrm{argmax}}|ICPF(\mu)| \tag{27}$$

Calculate the rotation angle α_k and FRFT order p_k corresponding to the tuning frequency, which are the best FRFT rotation angle and order, respectively. When using discrete FRFT calculations, the signal parameters need to be dimension normalized [29]. The relationship between the chirp-rate and the rotation angle is provided in Equation (28) and the FRFT order is provided in Equation (29):

$$\mu_k = -\cot(\alpha_k) \times \frac{f_s^2}{N} \tag{28}$$

$$p_k = \frac{2\alpha_k}{\pi} = \frac{2}{\pi} arc\cot(-\frac{\mu_k f_s^2}{N}) \tag{29}$$

The optical heterodyne signal in Equation (12) is subjected to p_k order FRFT:

$$
\begin{aligned}
S_o(u, t_m) &= F^{p_k}[s_o(t_k, t_m)] \\
&= \sigma_p A(u) \cdot \exp(j2\pi P_0) \int_{-\infty}^{\infty} rect(\frac{t_k - \tau}{T_p}) \cdot \exp[j2\pi(P_1 - u\csc\alpha)t_k] \cdot \exp\left[j\pi(2P_2 + \cot\alpha)t_k^2\right] dt_k \big|_{\alpha = \alpha_k} \\
&= \sigma_p T_p A(u) \exp(j2\pi P_0) \cdot \sin c\left[T_p(\frac{u}{\sin\alpha_k} - P_1)\right]
\end{aligned}
\tag{30}
$$

where $A(u) = \sqrt{1 + j2P_2} \exp(-j2\pi P_2 u^2)$. From the result, the peak value of the signal is obtained at $u = P_1 \sin\alpha_k$. That is, the range compression of the echo signal is achieved by one FRFT and the phase $\exp(j2\pi P_0)$ of the azimuth compression is retained.

4. Azimuth Imaging Based on ICPF-FRFT

4.1. Feature of Azimuth Echo Signal

In order to facilitate analysis, we assumed that the radar echo has completed range compression and motion compensation, so the echo signal can be converted into a turntable model with centroid as the reference point and the azimuth echo signal at the point can be expressed as:

$$s_p(t_m) = \sigma \exp\left[-j\frac{4\pi}{\lambda}\left(x_p \sin\theta(t_m) + y_p \cos\theta(t_m)\right)\right] \tag{31}$$

where σ is the amplitude of the signal after the motion compensation.

When the target is maneuvering, $\theta(t)$ can be expanded into a function of time t according to Taylor [30] due to the inertia of space targets. For a space target with certain inertia, the ISAL imaging time is shorter and the cumulative rotation angle required by the imaging is smaller, so the motion of the target and radar can approximate to the second-order component, meaning it approximates the uniform acceleration motion.

$$\theta(t) = \theta_0 + \omega t + \frac{1}{2!}\Omega t^2 \tag{32}$$

where θ_0 is the initial rotation angle, ω is the rotation angular velocity and Ω is the rotation angular acceleration.

As the ISAL wavelength is in the order of μm, to achieve the imaging resolution of mm magnitudes, the required rotation accumulation angle is in the order of mrad, so the following small angle approximation conditions are satisfied:

$$
\begin{cases}
\sin\theta(t) = \sin\left(\theta_0 + \omega t + \frac{1}{2}\Omega t^2\right) \approx \theta_0 + \omega t + \frac{1}{2}\Omega t^2 \\
\cos\theta(t) = \cos\left(\theta_0 + \omega t + \frac{1}{2}\Omega t^2\right) \approx 1 - \left(\theta_0 + \omega t + \frac{1}{2}\Omega t^2\right)^2
\end{cases}
\tag{33}
$$

According to Equations (29) and (31), the P point azimuth echo can be approximated to a linear frequency modulated signal.

$$s_p(t) = \sigma \exp\left[-j\frac{4\pi}{\lambda}\left(\phi_0 + f_a t + \frac{1}{2}k_a t^2\right)\right] \tag{34}$$

where

$$\phi_0 = x_p \theta_0 + y_p\left(1 - \frac{1}{2}\theta_0^2\right) \tag{35}$$

$$f_a = x_p \omega - y_p \theta_0 \omega \tag{36}$$

$$k_a = \frac{1}{2}\left(x_p \Omega - y_p \omega^2 - y_p \theta_0 \Omega\right) \tag{37}$$

In practice, multiple scattering points with different intensities are distributed in the same range cell, so the azimuth echo becomes a multicomponent LFM signal with a different LFM rate:

$$\hat{s}(t_m) = \sum_{i=1}^{I} \sigma_i \exp\left(-j\frac{4\pi}{\lambda}\left(\phi_{0i} + f_{ai}t_m + \frac{1}{2}k_{ai}t_m^2\right)\right) \tag{38}$$

where K is the number of scattered points and ϕ_{0i}, k_{ai} and f_{ai} satisfy Equations (35)–(37), respectively.

4.2. Azimuth Compression Based on ICPF-FRFT

ICPF can effectively suppress the cross and pseudo peaks caused by the interference of multicomponent signals, so the ICPF-FRFT can be used for imaging azimuth signals but the strong signal components affect the detection of the weak signal components. Therefore, we combined the CLEAN technique with ICPF-FRFT to estimate the strong to weak signals. The frequency modulation slope of the signal was calculated and then FRFT was used to image the signal components of different frequency modulation slopes. The imaging procedure for the azimuth compression based on ICPF-FRFT is shown in Figure 4.

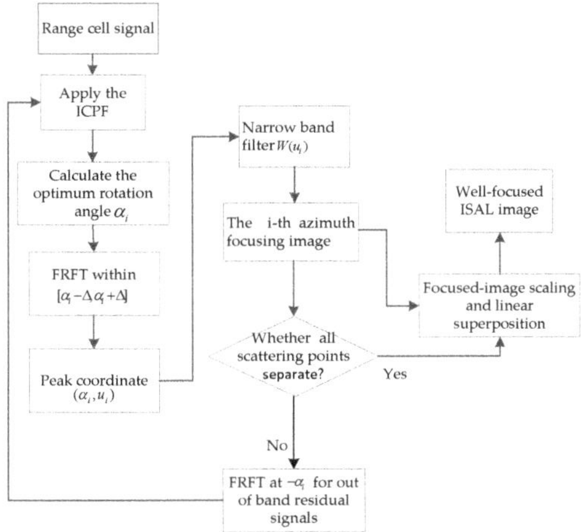

Figure 4. The imaging procedure for the azimuth compression.

The concrete steps are as follows:

Step 1: Calculate the ICPF of a range cell of the echo signal and estimate the frequency modulation rate of the strongest signal component:

$$\hat{s}(t_m) = \hat{s}(\phi_{0i}, f_{ai}, k_{ai}, t_m) + \sum_{l=1,l\neq i}^{I} \hat{s}(\phi_{0l}, f_{al}, k_{al}, t_m) \tag{39}$$

$$k_{ai} = \underset{k_a}{\mathrm{argmax}}|ICPF[\hat{s}(t_m)]| = ICPF[\hat{s}(\phi_{0i}, f_{ai}, k_{ai}, t_m)] \tag{40}$$

where $\hat{s}(\phi_{0i}, f_{ai}, k_{ai}, t_m)$ is the ith signal component.

Step 2: Calculate the best rotation angle α_i and the corresponding order p_{ki}, according to Equations (28) and (29), respectively. Then, calculate FRFT of the range cell signal in the range

$[\alpha_i - \Delta, \alpha_i + \Delta]$ where Δ is the calculation error of α_i. Search the peak to obtain the corresponding position u_i:

$$\hat{S}_\alpha(u) = F^{p_{ki}}[\hat{s}(\phi_{0i}, f_{ai}, k_{ai}, t_m)] + \sum_{l=1, l \neq i}^{I} F^{p_{ki}}[\hat{s}(\phi_{0l}, f_{al}, k_{al}, t_m)] = \hat{S}^i_{\alpha i}(u) + \sum_{l=1, l \neq i}^{I} \hat{S}_{\alpha,l}(u) \tag{41}$$

$$\{u_i\} = \underset{u}{\mathrm{argmax}}\left[|\hat{S}_{\alpha_i}(u)|\right] \tag{42}$$

Step 3: Separate the peak point by using the CLEAN technique to construct a narrowband filter $W(u_i)$ centered on u_i. Filter the strongest component and the peak value $\hat{S}^i_{\alpha i}(u)$ is considered the azimuth focusing image of the ith component.

$$\hat{S}^i_{\alpha_i}(u) = \hat{S}_{\alpha_i}(u)W(u_i) \tag{43}$$

$$\hat{S}_{\alpha,i}(u) = A(u)\sigma_i \exp(j2\pi\phi_{0i}) \sin c[T_p(\frac{u}{\sin \alpha} - f_{ai})] \tag{44}$$

where $A(u) = \sqrt{1 - j\cot \alpha}\, \exp(j\pi u^2 \cot \alpha)$.

Step 4: Transform the rest of the signal to the time domain using FRFT with a rotation angle of $-\alpha_i$.

$$\hat{s}_{i+1}(t_m) = \int_{-\infty}^{+\infty} \hat{S}_{\alpha_i}(u)(1 - W(u_i))K_{-\alpha_i}(t_m, u)du \tag{45}$$

Step 5: Repeat the above steps until all the scattered points in the current range cell are separated. This separation can be judged by when the residual signal component energy E of the ith range cell is less than a certain energy threshold E_H, which is usually 5% of the original signal [31,32].

Step 6: The target image is obtained by scaling the scattered images $u' = u / \sin \alpha$ and stacking them linearly.

Step 7: The 2D ISAL images can be obtained by using the above methods according to the sequence numbers of the range cells.

5. Imaging Procedure

For a maneuvering target with approximately uniformly accelerated motion, the ISAL imaging algorithm flow is shown in Figure 5.

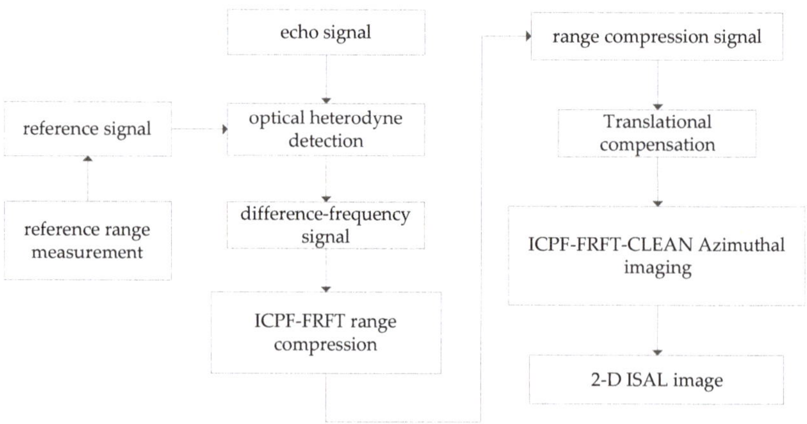

Figure 5. ISAL imaging algorithm flow for maneuvering target.

The detailed procedure is as follows: (1) Input the received echo signal; (2) Construct the reference signal based on the reference range; (3) Perform the optical heterodyne detection of the echo signal and the reference signal, with the output being the differential frequency signal; (4) Complete range compression with the ICPF-FRFT outlined in Section 3; (5) Translational compensation for the signal after range compression; (6) Azimuthal compression with the ICPF-FRFT-CLEAN outlined in Section 4.2; (7) Output the 2D ISAL image.

6. Experimental Results of Simulation and Real Data

In order to verify the effectiveness of the proposed algorithm, simulation and real data experiments were completed. Some other imaging algorithms were considered for comparison.

6.1. Experimental Simulation Results

The ladar and target simulation parameters used in the simulation experiment are provided in Table 1, which refers to the scattering point model in Papers [1,11]. The simulation model shown in Figure 6 is a plane model that contains 52 scattering points. The ladar parameters are typical parameters that can be realized and the range resolution was 0.001 m. We assumed that the positional relationship between the ladar and the target was as shown in Figure 1. The target motion parameters were set as the turntable motion parameters as shown in Table 1.

Table 1. Simulation parameters.

Radar Parameter	Value	Target Parameter	Value
Wavelength (μm)	1.55	Initial range (km)	100
bandwidth (Ghz)	150	Initial velocity (m/s)	100
Pulse width (μs)	100	Velocity acceleration (m/s^2)	30
PRF ()	3.3	Angular velocity (rad/s)	0.005
Range sampling number	256	Angular acceleration (rad/s^2)	0.01
Pulse number	512	Angular acceleration rate (rad/s^3)	0.006
Processing time (s)	0.155	-	-

Figure 7 is the smooth pseudo Wigner distribution (SPWVD) time frequency graph of the 128th pulse echo. The ISAL single echo signal is a multicomponent LFM signal with the same modulation frequency, which confirms the analysis of the echo signal in Section 2. Therefore, the compression of all scattering points can be accomplished through one compression of a range. Figure 8 shows the 128th pulse range compression result using the DFT method. From the display results, the direct use of DFT compression results in a serious dispersion effect of the range image and the scattering points of the adjacent resolution units form. Serious mutual interference occurs for scattering points of adjacent resolution units. Figure 9 shows the 128th pulse range compression result using the ICPF-FRFT method proposed in this paper. As seen in the figure, the results show that a better compression effect was achieved and the range dispersion was eliminated.

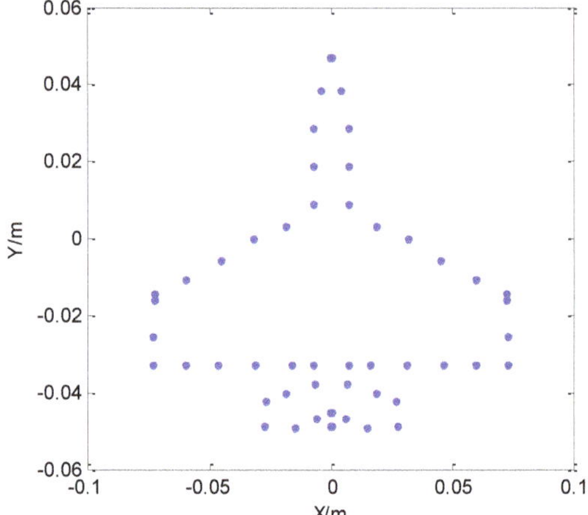

Figure 6. Simulation model of aircraft.

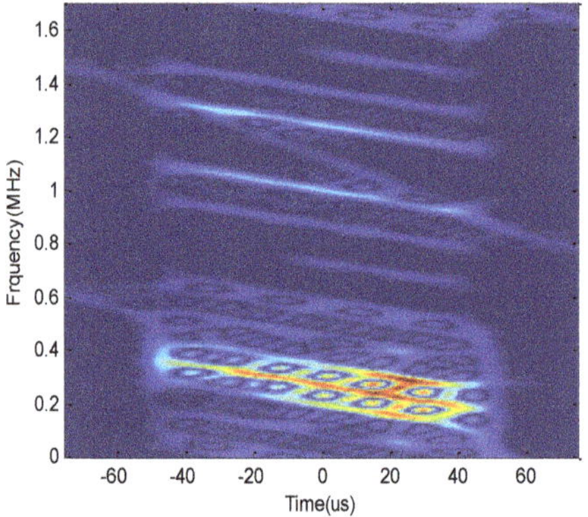

Figure 7. The 128th pulse's smooth pseudo Wigner distribution (SPWVD).

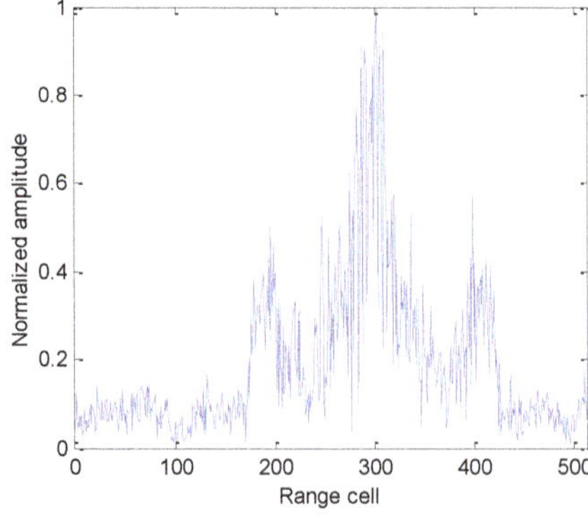

Figure 8. The 128th pulse's range compression via Discrete Fourier transform (DFT).

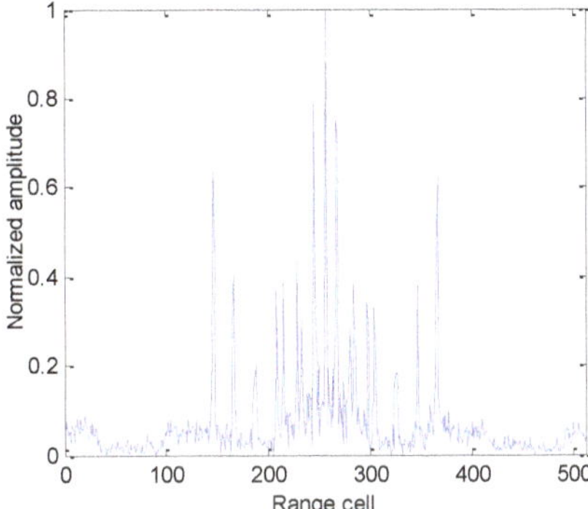

Figure 9. The 128th pulse's range compression via integral cubic phase function-fractional Fourier transform (ICPF-FRFT).

Figure 10 provides the SPWVD time frequency graph of the 79th (the lower edge of the aircraft) and 128th (the aircraft range center) range cell. From the diagram, due to the short ISAL imaging time, the target azimuth echo signals can be approximated as a multi-component LFM signal, even if there is a third order rotational component (angle acceleration rate) in the target. This demonstrates a slope with different slopes on the time frequency graph, which is the same as the previous theoretical analysis. So, dividing the scattering imaging into scattering points on different range cells was necessary.

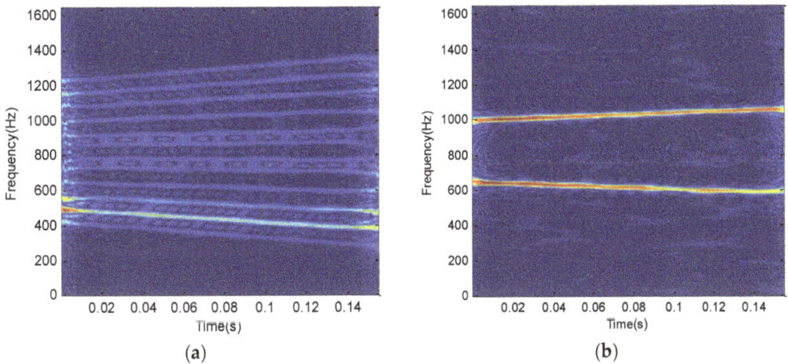

Figure 10. Range cell's SPWVD: (**a**) the 79th range cell and (**b**) the 128th range cell.

Figure 11a shows the result of traditional DFT azimuthal compression, which highlights that the scattered points from the center of the azimuth were seriously defocused—a poor imaging result. For comparison, we also provide three instantaneous Doppler (RID) imaging results based on STFT, WVD and SPWVD, as shown in Figure 11b–d, respectively. The image results used the 24th frame, which was t = 0.116 s. From the results of STFT in Figure 11b, the time-frequency resolution is affected by the window function and the azimuth defocus was severe. From Figure 11c, WVD can improve upon the time-frequency resolution but since the azimuth echo is a multi-component signal, the imaging result produces a cross term, so the imaging results are poorly readable and cannot identify the target. The SPWVD provides windowing and smoothing of WVD, so it weakens the cross terms but the time-frequency resolution also decreases. From Figure 11d, we can verify that the SPWVD has no cross-scattering point compared to the result of Figure 11c but the resolution is reduced. From the four results in Figure 11, the direct imaging range-Doppler (RD) algorithm and the three range-instantaneous Doppler (RID) imaging methods cannot achieve better imaging results and the azimuth defocus still exists.

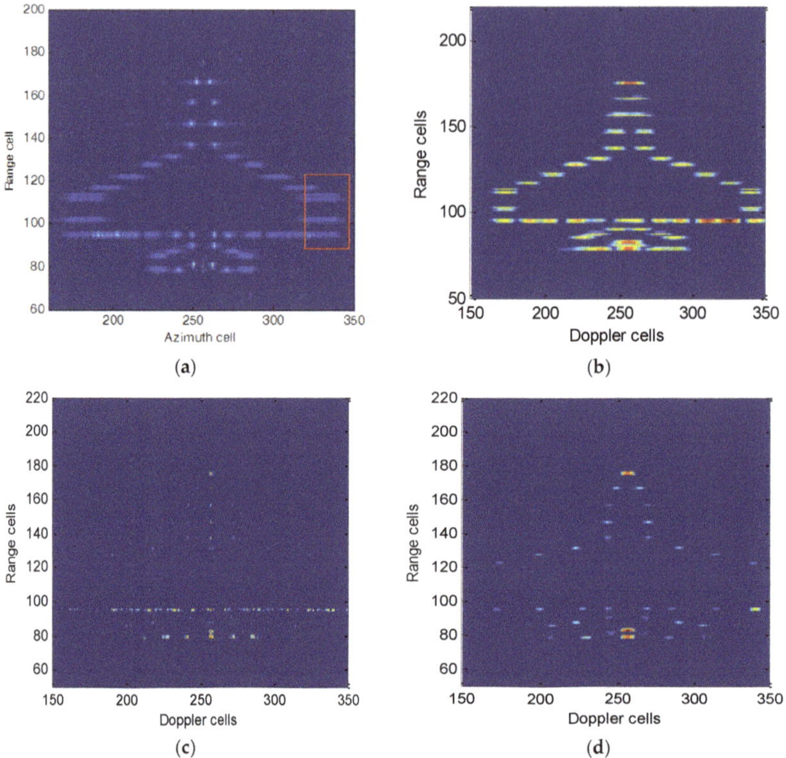

Figure 11. Azimuth compression via classical methods: (**a**) direct DFT, (**b**) RID image based on STFT, (**c**) RID image based on WVD and (**d**) RID image based on SPWVD.

Figure 12 is the result of azimuth compression using ICPF-FRFT. In Figure 12a, the CLEAN technique is not used, whereas it is applied in Figure 12b. Figure 12a shows that the imaging results of a single scattering point on a certain range cell are good. With multiple scattering points, the strong signal suppresses the separation of the weak signal due to the different signal intensities. Most of the signal is strong and is the signal of its side lobe component, and the weak signal is missing. Figure 12b shows that when the CLEAN technique was used to separate the scattering points on different range cells from strong to weak separation imaging, a better focusing effect was achieved. The diagram demonstrates that this method can effectively separate the scattering points of two range cells from each other, proving the effectiveness of the square method and further illustrates that the ISAL can be applied in real situations. High precision (millimeter level) imaging of space targets is now being performed.

Figure 13a–c are schematic diagrams of the separation of the first three peaks of the 95 range cell on the lower wing of the aircraft. The left image is the search process of the FRFT peak in the ICPF estimation error, the right side is the Clean processing for the peak point, the red frame is a narrow band filter and the frame content is the transverse focus image. Through this process, the azimuth scattering points with different intensities were separated and appropriately imaged.

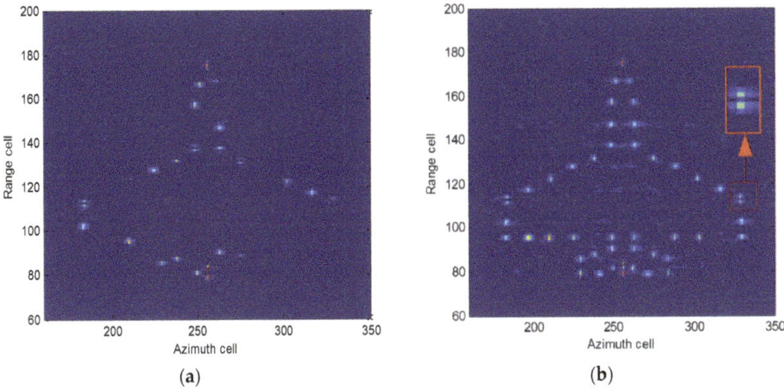

Figure 12. Azimuth compression via ICPF-FRFT: (**a**) without the CLEAN technique and (**b**) with the CLEAN technique.

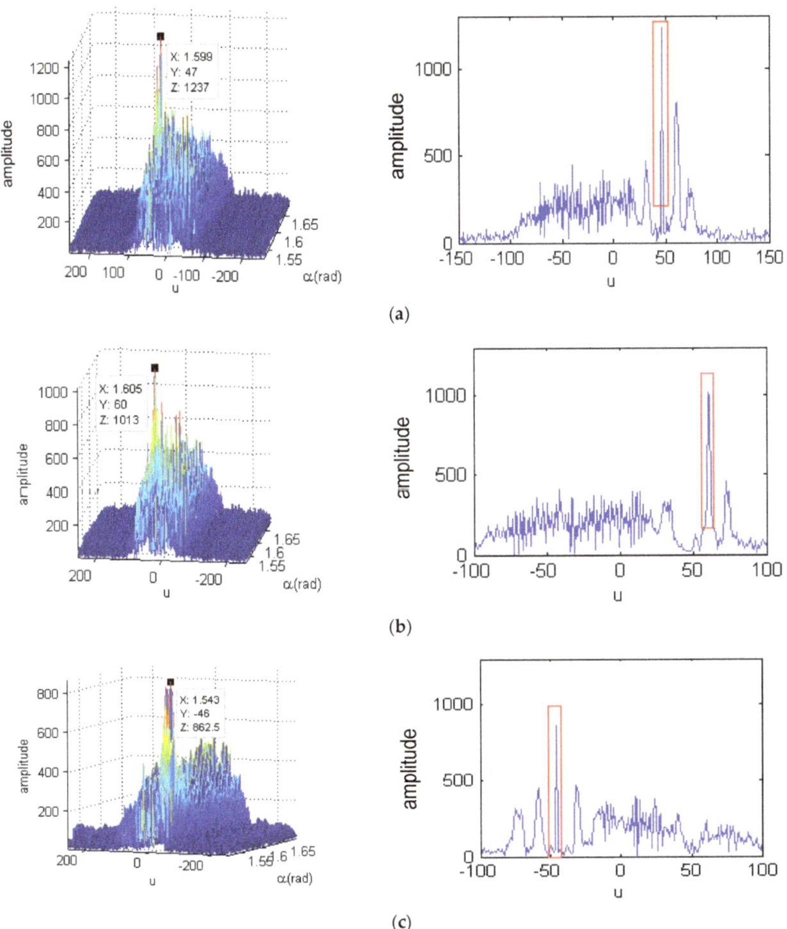

Figure 13. Separation imaging of the 95th range cell: (**a**) peak one, (**b**) peak two and (**c**) peak three.

In order to quantitatively evaluate the effectiveness of the proposed ICPF-FRFT algorithm, image entropy, contrast and running time were used to illustrate the imaging quality of the algorithm. Suppose the acquired ISAL image is $f(n,k)$, where n and k are the range and azimuth number of the sampling unit, respectively. The definition of image entropy is:

$$E = -\sum_{k=1}^{K}\sum_{n=1}^{N} \frac{|f(n,k)|^2}{F} \ln \frac{|f(n,k)|^2}{F} \tag{46}$$

$$F = \sum_{k=1}^{K}\sum_{n=1}^{N} |f(n,k)|^2 \tag{47}$$

where F is the total energy of the ISAL image. The image entropy is small when the image is well-focused. Conversely, a large image entropy indicates that the compensation effect is worse.

The definition of image contrast is:

$$C = \frac{\sqrt{E\left\{[|f(n,k)| - E(|f(n,k)|)]^2\right\}}}{E(|f(n,k)|)} \tag{48}$$

where $E(\cdot)$ represents the average operation. The image contrast is large when the image is well-focused. Conversely, a small image entropy indicates that the compensation effect is worse.

The results of the proposed ICPF-FRFT algorithm compared with the RD algorithm, FRFT algorithm, the ICPF-FRFT algorithm without CLEAN technique and three RID imaging methods based on STFT, WVD and SPWVD (Table 2). From the table, the algorithm proposed in this paper has smaller image entropy and a larger image contrast than the other algorithms, which shows that the image quality of the algorithm proposed in this paper is better. Notably, although all the indexes of the CLEAN technique are better, the loss of the scattering points cannot correctly reflect the distribution of the target scattering point, so the imaging quality was not the best. When the FRFT imaging algorithm with a small step size is performed directly, the result can reach an image entropy and contrast close to that of the proposed ICPF-FRFT algorithm paper but considerable computation time is required, indicating that the proposed algorithm is more efficient.

Table 2. Comparison of simulation aircraft imaging results.

Imaging Algorithm	RD	STFT	WVD	SPWVD	FRFT	No-CLEAN	ICPF-FRFT
Image entropy	8.0546	9.2598	7.6822	6.0802	5.7540	4.1374	5.2629
Contrast ratio	1.6678	1.3632	3.4347	8.4177	10.8059	29.6294	13.0397
Running time (s)	0.2066	3.3042	4.9585	215.8427	77.0294	2.3453	3.8554

6.2. Experimental Results of Real Data

Since no ISAL data have been published to date, the research on ISAL at this stage is mainly based on simulation data to verify algorithms. However, considering the problem of azimuth Doppler time-varying when imaging a maneuvering target in ISAR, ISAR is consistent with ISAL imaging in the pursuit of azimuth focusing. Therefore, the ISAL azimuth imaging algorithm based on ICPF-FRFT is also suitable for imaging ISAR maneuvering targets but the radar signal bandwidth in ISAR is much smaller than ISAL, so the spread over the range can be ignored. To further validate the effectiveness of the algorithm, the publicly available Boeing B727 ISAR aircraft data from Victor C. Chen of the U.S. Naval Research Laboratory (Washington, DC, USA) was used for experimental verification [33]. The data included 256 continuous pulses with a carrier frequency of 9 GHz, a bandwidth of 150 MHz and a pulse repetition rate of 20 kHz. Range compression and the motion compensation for the data were completed.

The imaging results using the imaging method proposed in this paper and other comparison methods described in the previous section are shown in Figure 14. The evaluation indexes of each imaging result are shown in Table 3. As can be seen from Figure 14a, the azimuth defocusing that occurred when using the RD algorithm was severe. From Figure 14b–d, the RID imaging results are related to the time-frequency method used, in which the time-frequency resolution of STFT was the worst and WVD had the highest time-frequency resolution but the cross-term was the most serious and SPWVD was somewhere in between. The result of the notable time-frequency imaging method shows that as the azimuth Doppler dynamically changes, the results displayed at different azimuths are different. In addition, some weak scattering point energy loss occurs, as shown in the wing part of the figure. Some scattering points are missing. It can be seen from Figure 14e–g that all three imaging methods can effectively improve the azimuth focusing effect but the direct FRFT requires a long computation time to achieve the same focusing effect as ICPF-FRFT. However, for the No-CLEAN technique, although all the indicators are superior, this algorithm only focuses on the strong scattering point, resulting in a lack of partial scattering points. Considering the minimum entropy, contrast and running time, the ICPF-FRFT algorithm is optimal, which is consistent with the results of the previous simulation analysis.

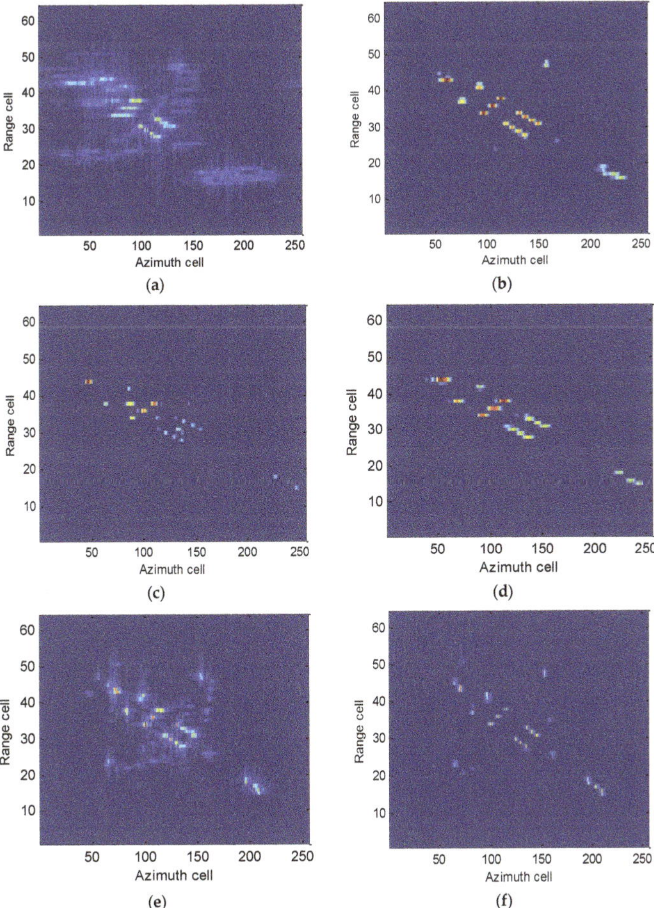

Figure 14. *Cont.*

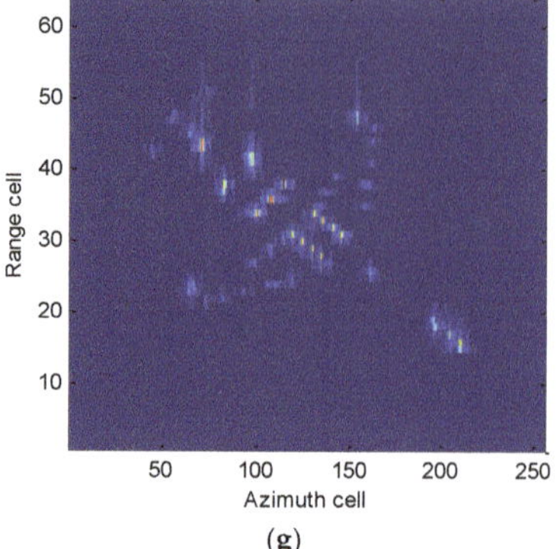

(g)

Figure 14. Inverse synthetic aperture radar (ISAR) imaging results: (**a**) RD algorithm, (**b**) STFT algorithm, (**c**) WVD, (**d**) SPWVD, (**e**) FRFT, (**f**) No-CLEAN technique and (**g**) ICPF-FRFT algorithm. The simulated echo pulses of Boeing-727 shown in Figure 14 are available online at http://www.mdpi.com/2072-4292/10/4/593/s1.

Table 3. Comparison results of the Boeing-727 images.

Imaging Algorithm	RD	STFT	WVD	SPWVD	FRFT	No-CLEAN	ICPF-FRFT
Image entropy	6.6428	6.1032	5.7609	4.8711	5.5242	3.9852	4.2940
Contrast ratio	2.3968	2.8949	3.6200	6.4915	4.8906	12.7516	7.5289
Running time (s)	0.0014	0.2350	0.4022	9.6358	24.0325	1.0412	1.2939

7. Conclusions

ISAL can meet the high precision and quasi real-time imaging requirements for targets. However, due to the use of ultra-high carrier frequency and large bandwidth signals, the ISAL radar's target echo signal produces distortion and a 1D dispersion profile. In addition, when the target moves, the radar signal echo direction changes to Doppler. To address these issues, an ISAL imaging algorithm based on ICPF-FRFT was proposed for space maneuvering targets, which was able to quickly image uniformly accelerated motion. The algorithm first uses the ICPF algorithm based on NUFFT computing to quickly estimate the frequency modulation rate of the echo signal and then uses FRFT to compress the range image at the best rotation angle and eliminate the range dispersion. After motion compensation, the ICPF-FRFT and CLEAN technique are used again to separate the strong and weak scattering points and solve the azimuth defocusing problem. The validity of the method was verified with a simulation experiment of an aircraft scatter point model and Boeing-727 data.

Author Contributions: Y.L. and L.Q. conceived and designed the method; Y.W. guided the students to complete the research; Y.L. performed the simulation and experiment tests; H.W., J.J. and Y.S. helped in the simulation and experiment tests; and Y.L. wrote the paper.

Funding: The authors are grateful for the financial support received from the State 863 Project of China (No. 2014AA7113016) and the Research Project of the State Key Laboratory of Complex Electromagnetic Environment Effects on Electronics and Information Systems (No. 2017Z0203B).

Conflicts of Interest: The authors declare no conflict of interest.

References

1. He, J.; Zhang, Q.; Yang, X.Y.; Luo, Y.; Zhu, X.P. High resolution imaging algorithm for inverse synthetic aperture imaging LADAR. *Syst. Eng. Electron.* **2011**, *33*, 1750–1755.

2. Liu, L.R. Synthetic aperture laser imaging radar (I): Defocused and phase biased telescope for reception antenna. *Acta Opt. Sin.* **2008**, *28*, 997–1000.

3. Beck, S.M.; Buck, J.R.; Buell, W.F.; Dickinson, R.P.; Kozlowski, D.A.; Marechal, N.J.; Wright, T.J. Synthetic-aperture imaging laser radar: Laboratory demonstration and signal processing. *Appl. Opt.* **2005**, *44*, 7621–7629. [CrossRef] [PubMed]

4. Crouch, S.; Barber, Z.W. Laboratory demonstrations of interferometric and spotlight synthetic aperture ladar techniques. *Opt. Express* **2012**, *20*, 24237–24246. [CrossRef] [PubMed]

5. Turbide, S.; Marchese, L.; Terroux, M.; Bergeron, A. Synthetic aperture ladar concept for infrastructure monitoring. In *Electro-Optical Remote Sensing, Photonic Technologies and Applications VIII; and Military Applications in Hyperspectral Imaging and High Spatial Resolution Sensing II*; International Society for Optics and Photonics: Bellingham, WA, USA, 2014.

6. Barber, Z.W.; Dahl, J.R. Synthetic aperture ladar imaging demonstrations and information at very low return levels. *Appl. Opt.* **2014**, *53*, 5531–5537. [CrossRef] [PubMed]

7. Trahan, R.; Nemati, B.; Zhou, H.; Shao, M.; Hann, I.; Schulze, W. Low-CNR inverse synthetic aperture LADAR imaging demonstration with atmospheric turbulence. In *Long-Range Imaging*; International Society for Optics and Photonics: Bellingham, WA, USA, 2016.

8. Luan, Z.; Sun, J.; Zhou, Y.; Wang, L.; Yang, M.; Liu, L. Down-looking synthetic aperture imaging ladar demonstrator and its experiments over 1.2 km outdoor. *Chin. Opt. Lett.* **2014**, *12*, 111101. [CrossRef]

9. Ikram, M.Z.; Abed-Meraim, K.; Hua, Y. Estimating the parameters of chirp signals: An iterative approach. *IEEE Trans. Signal Process.* **2002**, *46*, 3436–3441. [CrossRef]

10. Barbarossa, S.; Scaglione, A.; Giannakis, G.B. Product high-order ambiguity function for multicomponent polynomial-phase signal modeling. *IEEE Trans. Signal Process.* **1998**, *46*, 691–708. [CrossRef]

11. Xia, X.G. Discrete chirp-Fourier transform and its application to chirp rate estimation. *IEEE Trans. Signal Process.* **2000**, *48*, 3122–3133.

12. Djurović, I.; Simeunović, M.; Wang, P. Cubic phase function: A simple solution to polynomial phase signal analysis. *Signal Process.* **2017**, *135*, 48–66. [CrossRef]

13. Chen, V.C.; Miceli, W.J. Time-varying spectral analysis for radar imaging of maneuvering targets. *IET Proc. Radar Sonar Navig.* **1998**, *145*, 262–268. [CrossRef]

14. Xing, M.D.; Wu, R.B.; Li, Y.C.; Bao, Z. New ISAR imaging algorithm based on modified Wigner Ville distribution. *IET Proc. Radar Sonar Navig.* **2009**, *3*, 70–80. [CrossRef]

15. Trintinalia, L.; Ling, H. Joint time-frequency ISAR using adaptive processing. *IEEE Trans. Antennas Propag.* **1997**, *45*, 221–227. [CrossRef]

16. Lao, G.; Yin, C.; Ye, W.; Sun, Y.; Li, G. A frequency domain extraction based adaptive joint time frequency decomposition method of the maneuvering target radar echo. *Remote Sens.* **2018**, *10*, 266. [CrossRef]

17. Tao, R.; Deng, B.; Wang, Y. Research progress of the fractional Fourier transform in signal processing. *Sci. China Inf. Sci.* **2006**, *49*, 1–25. [CrossRef]

18. Wang, B.; Xu, S.; Wu, W.; Hu, P.; Chen, Z. Adaptive ISAR imaging of maneuvering targets based on a modified Fourier transform. *Sensors* **2018**, *18*, 1370. [CrossRef] [PubMed]

19. Attia, E.H. Data-adaptive motion compensation for synthetic aperture LADAR. In Proceedings of the 2004 IEEE Aerospace Conference, Big Sky, MT, USA, 6–13 March 2004.

20. Shapiro, J.H.; Capron, B.A.; Harney, R.C. Imaging and target detection with a heterodyne-reception optical radar. *Appl. Opt.* **1981**, *20*, 3292–3313. [CrossRef] [PubMed]

21. Song, J.; Liu, Y.; Zhu, X. Parameters estimation of LFM signals by interpolation based on FRFT. *Syst. Eng. Electron.* **2011**, *33*, 2188–2193.

22. O'Shea, P. A new technique for instantaneous frequency rate estimation. *IEEE Signal Process. Lett.* **2002**, *9*, 251–252. [CrossRef]

23. Wang, P.; Li, H.; Djurović, I.; Himed, B. Integrated cubic phase function for linear FM signal analysis. *IEEE Trans. Aerosp. Electron. Syst.* **2010**, *46*, 963–977. [CrossRef]

24. Su, J.; Tao, H.; Xie, J.; Sun, Y.; Li, G. Imaging and Doppler parameter estimation for maneuvering target using axis mapping based coherently integrated cubic phase function. *Digital Signal Process.* **2018**, *10*, 266. [CrossRef]

25. Li, H.; Qin, Y.L.; Li, Y.P.; Wang, H.Q.; Li, X. Analysis of multi-component LFM signals by the integrated quadratic phase function. *J. Electron. Inf. Technol.* **2012**, *28*, 926–931.

26. Liu, Q.H.; Nguyen, N. An accurate algorithm for nonuniform fast Fourier transforms (NUFFT's). *IEEE Microw. Guided Wave Lett.* **2002**, *8*, 18–20. [CrossRef]

27. O'Shea, P. Improving polynomial phase parameter estimation by using nonuniformly spaced signal sample methods. *IEEE Trans. Signal Process.* **2012**, *7*, 3405–3414. [CrossRef]

28. Xing, M.; Wu, R.; Lan, J.; Bao, Z. Migration through resolution cell compensation in ISAR imaging. *IEEE Geosci. Remote Sens. Lett.* **2004**, *1*, 141–144. [CrossRef]

29. Simeunović, M.; Djurović, I. Non-uniform sampled cubic phase function. *Signal Process.* **2014**, *101*, 99–103. [CrossRef]

30. Wang, C.; Wang, Y.; Li, S.B. Inverse synthetic aperture radar imaging of ship targets with complex motion based on match Fourier transform for cubic chirps model. *IET Radar Sonar Navig.* **2013**, *7*, 994–1003. [CrossRef]

31. Wang, Y.; Lin, Y. ISAR imaging of non-uniformly rotating target via range-instantaneous Doppler derivatives algorithm. *IEEE J. Sel. Top. Appl. Earth Obs. Remote Sens.* **2014**, *7*, 167–176. [CrossRef]

32. Zheng, J.; Liu, H.; Liu, Z.; Liu, Q.H. ISAR imaging of ship targets based on an integrated cubic phase bilinear autocorrelation function. *Sensors* **2017**, *17*, 498. [CrossRef] [PubMed]

33. B727S.mat. Available online: http://www.mdpi.com/2072-4292/10/4/593/s1 (accessed on 12 April 2018).

Article

A Synthetic Aperture Radar (SAR)-Based Technique for Microwave Imaging and Material Characterization

Yuri Álvarez López [1],*, María García Fernández [1], Raphael Grau [2] and Fernando Las-Heras [1]

[1]   Área de Teoría de la Señal y Comunicaciones, Universidad de Oviedo, 33203 Gijón (Asturias), Spain;
       garciafmaria@uniovi.es (M.G.F.); flasheras@uniovi.es (F.L.-H.)
[2]   Faculty of Computer Science, Hochschule Mannheim, 68163 Mannheim, Germany;
       raphael.grau@stud.hs-mannheim.de
*   Correspondence: alvarezyuri@uniovi.es; Tel.: +34-985-182-281

Received: 30 October 2018; Accepted: 29 November 2018; Published: 2 December 2018

Abstract: This contribution presents a simple and fast Synthetic Aperture Radar (SAR)-based technique for microwave imaging and material characterization from microwave measurements acquired in tomographic systems. SAR backpropagation is one of the simplest and fastest techniques for microwave imaging. However, in the case of heterogeneous objects and media, a priori information about the constitutive parameters (conductivity, permittivity) is needed for an accurate imaging. In some cases, a first guess of the constitutive parameters can be extracted from an uncorrected SAR image, and then the estimated parameters can be introduced in a second step to correct the SAR image. The main advantage of this methodology is that there is little or no need for a priori information about the object to be imaged. Besides, calculation time is not significantly increased with respect to conventional SAR, thus allowing real-time imaging capabilities. The methodology has been validated by means of measurements acquired in a cylindrical setup.

Keywords: Synthetic Aperture Radar (SAR); microwave imaging; constitutive parameters; conductivity; permittivity; tomography

---

1. Introduction

Electromagnetic imaging is one of the most widespread techniques for nondestructive testing thanks to the capability of the electromagnetic waves to penetrate through different media [1]. The different responses of these media depending on the working frequency band (terahertz [2], millimeter waves [3], etc.) have resulted in a wide variety of electromagnetic imaging systems, not only in terms of hardware, but also concerning processing techniques.

Electromagnetic inverse scattering and imaging techniques are able to provide the geometry of the object/area under inspection, the constitutive parameters (permittivity, conductivity), or both. The capability of recovering these parameters with a certain degree of accuracy depends not only on the setup/hardware of the imaging system and the working frequency band(s), but also on the post-processing algorithms. Factors such as the dynamic range or processing time have to be taken into account when selecting an imaging system that best fits the requirements for a particular nondestructive testing application. As an example, detecting 15–20 cm size metallic targets buried 30 cm in dry sand [4], imaging of targets behind a 10 cm thick wall [5], or locating tumors in breast tissue [6] require different microwave imaging hardware and methods.

In general, inverse scattering and imaging techniques can be classified into two main groups: on the one hand, those based on scattered field backpropagation and, on the other hand, model-based imaging techniques.

In the first group, standard Synthetic Aperture Radar (SAR) imaging, also known as backpropagation or range migration techniques [4–8], are the most common techniques for radar applications, due to their simplicity, which makes these techniques computationally efficient thanks to the use of the Fast Fourier Transform (FFT). Their main limitation is the amount of spatial and frequency bandwidth required for accurate imaging. Nevertheless, improvements in microwave and radiofrequency hardware have made affordable the development of ultrawideband systems for imaging applications.

The second group includes model-based techniques that require setting an electromagnetic model of the scenario-under-test. Then, a cost function relating the measured scattered field and the calculated one for the model is defined. Global minimum of the cost function corresponds to the best fit between the true and the modelled target/object-under-test (OUT). Equivalent currents [9,10], level-set [11], linear sampling method [12], local optimization strategies [13], and global optimization based on evolutionary algorithms [3,14] fall within this second group. As opposed to backpropagation techniques, the strength of model-based inverse scattering lies on the little amount of information needed, being capable of reconstructing the profile accurately using a single frequency and few field-of-views. The price to pay is a high calculation time, mainly due to the iterative nature of algorithms.

Hybrid backpropagation and model-based techniques have been also considered in order to obtain accurate imaging results [15]. In these cases, the former is used to provide a first guess of the profile of the target/OUT for the latter method.

In the area of nondestructive testing for detection of targets/objects embedded in a surrounding opaque medium (e.g., detection of tumors in breast tissue [6]), the aforementioned techniques require a priori information about the problem, which varies depending on the inverse scattering or imaging technique to be applied. For example, those based on multilayered Green's Function formulation need an initial guess of the constitutive parameters of the surrounding medium [16,17]. Inverse scattering techniques based on cost function minimization [3,14] require a set of first guess solutions and the definition of the search space boundaries.

Sometimes having an accurate estimate of the constitutive parameters of the surrounding medium can be difficult, such as in Ground Penetrating Radar (GPR) and Through-The-Wall Imaging (TTWI) applications, where ground and wall composition is not homogeneous, and conductivity and permittivity can be affected by moisture levels. In these cases, additional measurements (and hardware) are required for a proper estimation of these parameters, mostly reflectometry [18] and transmission/reflection-based techniques [19,20]. Besides, these constitutive parameters can be also the unknown of the inverse scattering problem, as in security screening systems for detecting weapons and explosives.

Aiming to reduce the need for additional measurements to characterize the constitutive parameters, SAR-based techniques have proven to be successful in recovering the geometry and also getting an estimate of the conductivity and permittivity of the OUT and/or the surrounding medium [21,22]. The theoretical fundament is the different velocity of the electromagnetic waves when passing through different media, so that the reflectivity of the imaged targets is displaced backwards with respect to their expected position. In order to detect this shifting, a reference background is needed (the human body surface in the case of [21], a reference metallic plate in [22,23]).

This contribution extends the SAR-based imaging techniques presented in [21–23] to provide a better recovery of geometry and constitutive parameters, making a more efficient use of the imaging information. More precisely, the proposed methodology takes advantage of the reflections at the interfaces between different media to obtain an estimate of the permittivity and the conductivity, avoiding the need of a reference target or a background medium.

2. Methodology

2.1. Synthetic Aperture Radar imaging

The basics of SAR processing are presented in this section. For the sake of simplicity, a two-dimensional (2D) scenario in the XY plane is considered. A 3D scenario with translation symmetry along z-axis could be assumed as well without loss of generality. Given the scattered field $E_{scatt}(f,r,\varphi)$ measured at the position (r,φ) over a certain bandwidth $BW = [f_1\ f_2]$, the reflectivity $\rho(x', y')$ evaluated at the position (x', y') of the scenario-under-test is defined in Equation (1), assuming an homogeneous propagation medium:

$$\rho(x', y') = \Sigma_{f=f1:f2}\ E_{scatt}(f,r,\varphi)\exp(j\ k_{medium}\ R) \tag{1}$$

where k_{medium} is the wavenumber defined as $k_{medium} = 2\pi f / v_{prop,medium}$, f being the frequency and $v_{prop,medium}$ the propagation velocity of the electromagnetic wave in a particular medium. R is the Euclidean distance between (r,φ) and (x', y'). In case the medium is vacuum or air, $k_0 = 2\pi f / c$. The center of the scenario-under-test (e.g., the center of a rotary platform in the case of a tomographic imaging system) is defined as the origin of the coordinate system. A monostatic or quasi-monostatic setup is considered, so that the transmitting and receiving antennas are placed at (r,φ). Image resolution in the radial (range) direction, Δr, is given by Equation (2):

$$\Delta r = 0.5\ v_{prop,medium} / (f_2 - f_1) \tag{2}$$

The problem can be even reduced to a one-dimensional case in the range direction, for those points satisfying $x' = r'\cos(\varphi)$, $y' = r'\sin(\varphi)$, so $R = r - r'$.

Let us consider now the imaging scenario depicted in Figure 1, where the OUT is a cylindrical wax candle of diameter d_{OUT}. The axis of the OUT is aligned with the axis of the rotary table, so that the distance between the Tx/Rx antennas and the candle surface is $r - d_{OUT}/2$ for any rotation angle φ. The medium surrounding the wax candle is air ($k_{medium} = k_0$). Thus, the reflectivity at any point $r' \in [d_{OUT}/2, r]$ is given by Equation (3). The range $[d_{OUT}/2, r]$ will be denoted as Region #1.

$$\rho(r',\varphi) = \Sigma_{f=f1:f2}\ E_{scatt}(f,r,\varphi)\exp(j\ k_0\ (r - r')),\ r' \in [d_{OUT}/2, r] \tag{3}$$

Next, the reflectivity for a point *inside* the wax candle has to be calculated taken into account the different propagation velocity inside the wax, $v_{prop,OUT} = c/(\varepsilon_{r,OUT})^{1/2}$. Reflectivity in the interval $r' \in [-d_{OUT}/2, d_{OUT}/2]$ (Region #2) is then calculated as indicated in Equation (4):

$$\rho(r',\varphi) = \Sigma_{f=f1:f2}\ E_{scatt}(f,r,\varphi)\exp(j\ k_0\ (r - d_{OUT}/2))\ \exp(j\ k_{OUT}\ (d_{OUT}/2 - r')),\ r' \in [-d_{OUT}/2, d_{OUT}/2] \tag{4}$$

And finally, for those points behind the wax candle (Region #3), the reflectivity is given by Equation (5), where the interval within the wax candle is taken into account:

$$\rho(r',\varphi) = \Sigma_{f=f1:f2}\ E_{scatt}(f,r,\varphi)\exp(j\ k_0\ (r - d_{OUT} - r'))\exp(j\ k_{OUT}\ d_{OUT}),\ r' < -d_{OUT}/2 \tag{5}$$

Now, let us assume that neither the position nor the constitutive parameters of the OUT (the wax candle) are known. In this case, one could make use of Equation (3) to evaluate the reflectivity at any point r'. If free-space propagation is considered, a first peak of the reflectivity should appear at the interface between the air and the OUT (denoted as interface #1 in Figure 1). Similarly, a reflectivity peak should appear at any position r' where there is an interface between two media with different constitutive parameters. But, as free-space propagation is assumed for evaluating the reflectivity at any position (Equation (1), $k_{medium} = k_0$), reflectivity peaks will be shifted backwards, as illustrated in Figure 1 (dashed blue line). If the position of the interfaces and the permittivity values of the different media were known, Equations (3)–(5) could be used, resulting in a proper recovery of the reflectivity (Figure 1, solid red line).

Inaccurate recovery of the permittivity may result in inaccurate location of embedded targets in homogeneous media (e.g., tumors in breast tissue [6], or landmines buried in the ground [4]). Furthermore, depending on the imaging setup, free-space SAR approach could result in the concealed targets to be imaged *outside* the object where they are embedded, as it will be shown in a latter example.

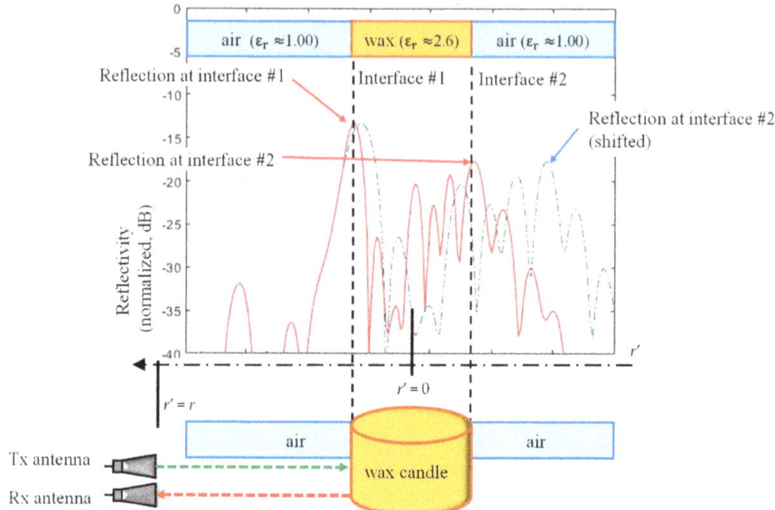

Figure 1. Illustration of the reflectivity delay due to the consideration of propagation in free-space, and comparison with corrected reflectivity when considering true permittivity (ε_r).

2.2. Constitutive Parameters Estimation and Range Correction

Under the assumption that the outer profile/geometry of the OUT is known, it is possible to recover the conductivity and permittivity of the OUT from the shifted reflectivity peaks. For the sake of simplicity, let us consider again the example presented in Figure 1 (a wax candle of diameter d_{OUT}).

If the permittivity of the wax is not known, then the reflectivity calculated according to Equation (1) for all r' with $k_{\text{medium}} = k_0$ corresponds to the dashed blue line in Figure 2. From the theoretical analysis presented in Section 2.1, it is known that the reflectivity peak corresponding to the interface #2 (rear side of the candle) has to be shifted, as k_0 instead of k_{OUT} was used to calculate the reflectivity. Although its exact position cannot be estimated a priori, a search region can be defined taking into account the size of the OUT. For this example, it can be expected the reflectivity peak corresponding to the interface #2 to appear at $r' < -d_{\text{OUT}}$. The distance between the shifted reflectivity peak of interface #2 and the reflectivity peak of interface #1 is denoted as d_{echo} (Figure 2).

Next, the relationship between the delay (or phase shift) considering free-space propagation (Equation (1), $k_{\text{medium}} = k_0$) and propagation through the OUT considering a permittivity estimate $\varepsilon_{r,\text{est}}$, yields Equation (6). An explanation about how to obtain this equation is given in [21,22]:

$$\varepsilon_{r,\text{est}} = (d_{\text{echo}}/d_{\text{OUT}})^2 \tag{6}$$

Note that, for the scenario considered to illustrate this methodology, no additional reference targets are required for recovering the permittivity. In this case, the shifted reflection at the interface #2 corresponds to the OUT-air interface.

Once the permittivity is estimated, Equations (3)–(5) can be applied to recover the reflectivity with the corrected propagation velocity within the OUT, so that the reflectivity peak of the interface #2 will appear at the correct position, that is, without shifting (solid red line in Figure 2).

In addition to the permittivity, an estimate of the conductivity (σ_{OUT}) can be obtained as well, by measuring the difference on the reflectivity levels at interfaces #1 and #2. The attenuation constant α (measured in Np/m) is given by Equation (7) [22]:

$$\alpha = \ln(|\rho(r'_{\text{interface \#1}})| / |\rho(r'_{\text{interface \#2}})|)/d_{\text{OUT}} \tag{7}$$

α is related to the conductivity according to Equation (8) [22]:

$$\sigma \approx \mathrm{Im}\{((\varepsilon_r)^{1/2} + j\alpha c/(2\pi f_c))^2\}, f_c = (f_1 + f_2)/2 \tag{8}$$

A summary of the methodology described in this section is illustrated in Figure 3.

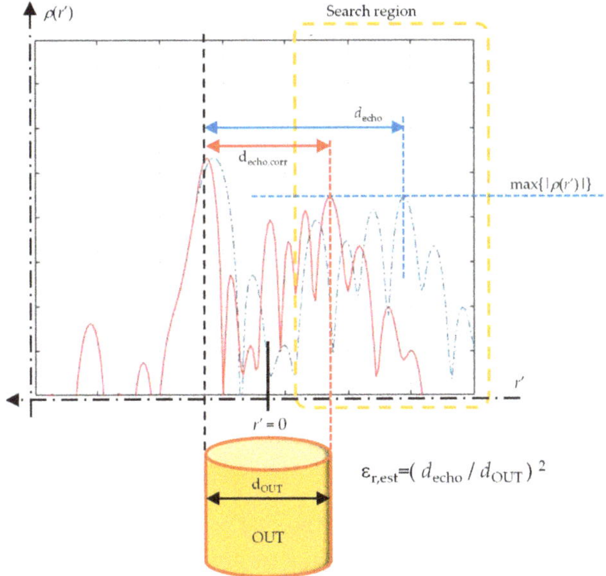

Figure 2. Methodology to estimate the permittivity of the object-under-test (OUT) ($\varepsilon_{r,est}$) from scattered field measurements given the thickness of the OUT (d_{OUT}).

3. Results

Validation of the proposed methodology for fast and simple estimation of constitutive parameters from SAR images is conducted in this section.

3.1. Measurement Setup

A 3D tomographic measurement setup is proposed, consisting of a rotary platform where the OUT is placed, and an XYZ positioner [24]. The Tx/Rx probe antennas (Standard Gain Horn, SGH [25]) are mounted in a quasi-monostatic configuration on a vertical slider of the XYZ positioner. Vertical (z-axis) motion is allowed along 27 cm, in 1 cm steps, while the OUT can be rotated 360°, with 1° step. With these parameters, the entire measurement of the OUT takes around 1 h and 30 min. Alignment of the Tx/Rx antennas with respect to the center of the rotary platform was conducted using a laser leveler. The distance from the rotation axis of the rotary platform to the aperture plane of the Tx/Rx antenna is 89 cm.

Tx/Rx antennas are connected to a Microwave Vector Network Analyzer (VNA) [26], as shown in the scheme of Figure 4 and in the picture of Figure 3. In order to set a reference phase for SAR imaging, calibration is done at the end of the cables connecting the VNA and the SGH antennas. A frequency band from $f_1 = 12$ GHz to $f_2 = 18$ GHz is selected as a trade-off between resolution and penetration of the electromagnetic waves in the targets to be tested. This bandwidth yields $\Delta r' = 2.5$ cm resolution in range.

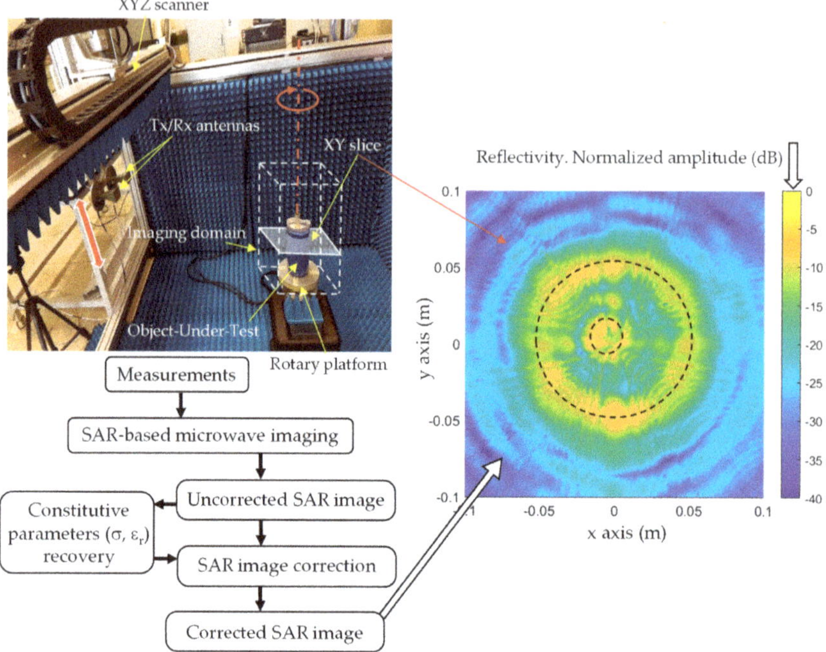

Figure 3. Picture of the measurement setup and flowchart of the Synthetic Aperture Radar (SAR)-based technique for microwave imaging and constitutive parameters characterization.

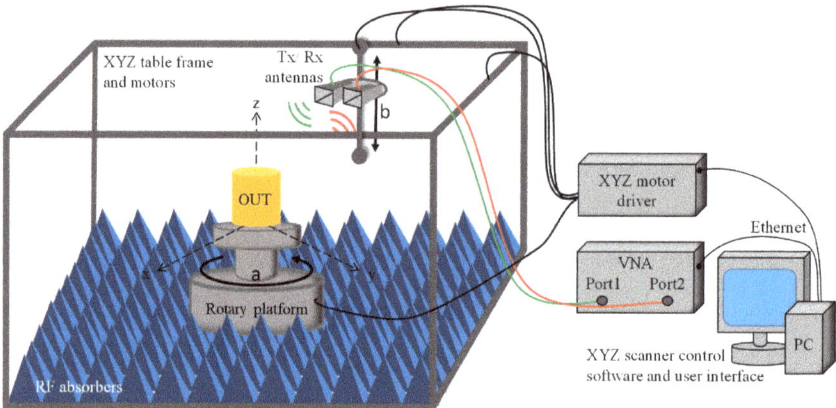

Figure 4. Scheme of the measurement for dielectric objects imaging using a rotary platform and vertical slider. Tx and Rx antennas are placed according to a quasi-monostatic configuration with respect to the OUT. Full (360°) angular rotation (*a*) is allowed. Vertical motion range is *b* = 27 cm.

SGH aperture size is 5.6×4.4 cm [25], so $D_{SGH} = (5.6 \times 4.4)^{1/2} = 5$ cm. The far field distance is $R_{FF} = 2(D_{SGH})^2/\lambda = 29.4$ cm at $f = 18$ GHz. As the distance between the Tx/Rx antennas and the center of the rotary platform is 89 cm, then, the OUT is placed in the far field region of the antennas. Besides, for this SGH, -3 dB antenna beamwidth (θ_{-3dB}) ranges from 25.5° at 18 GHz, to 37° at 12 GHz. That means that, at the distance of 89 cm, the θ_{-3dB} is wider than 42.5 cm in the working frequency band, thus fully covering the area where the OUT is placed.

As the OUT is placed in the far field region of the Tx/Rx antennas, and it is fully illuminated by the antenna beams, spherical wave propagation can be assumed in the imaging domain. Note that if the OUT were placed in the near field region of the Tx/Rx antennas, the spherical wave model could result in poorer imaging results, requiring accurate characterization of the near field in the imaging domain.

3.2. Wax Candle

The first OUT selected for testing the proposed methodology for accurate SAR imaging and constitutive parameters retrieval was a wax candle, with 40 cm length and 10 cm diameter, as depicted in Figure 5. The fact of having both rotation and translation symmetry around vertical (z-) axis motivated the choice of this OUT.

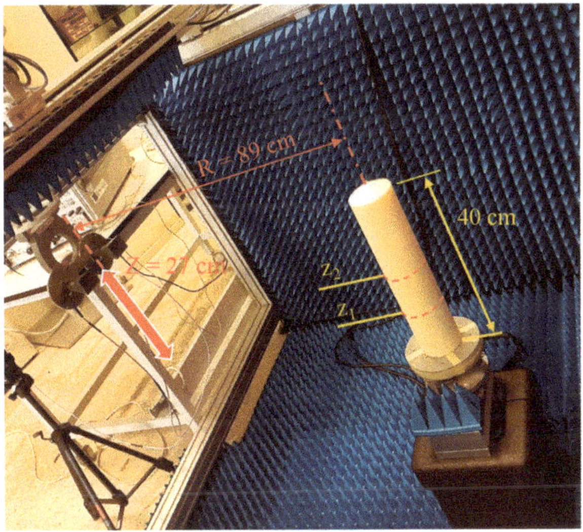

Figure 5. Picture of the wax candle placed on the rotary platform.

Once the scattered field for each Tx/Rx position and rotation angle was measured, it was processed according to the flowchart depicted in Figure 3. The recovered reflectivity of the OUT for each rotation angle φ in the range $r' = [0.7, 1.15]$ at two different XY planes (or slices) z_1 and z_2 is shown in Figure 6. Range r' is defined from the position of the Tx/Rx antennas, being the center of rotation (rotation axis in Figure 6) located at $r' = 0.89$ m. As the constitutive parameters of the wax are not known, reflectivity is calculated using Equation (1) ($k_{medium} = k_0$). The reflection at the air-wax interface (#1) can be clearly visible in Figure 6, having a mean value of $|\rho_{(interface\ \#1)}| \approx -10$ dB $= 0.32$. Note that the wax candle was not perfectly centered at the rotation axis, so the reflectivity peak of the air-wax interface fluctuates between $r' = [0.83, 0.85]$ m. As the wax diameter is $d_{OUT} = 10$ cm, the reflectivity peak of the wax-air interface (#2) can be expected to be found at $r' > [0.83 + d_{OUT}, 0.85 + d_{OUT}]$ m. For each rotation angle (φ) the maximum of the reflectivity in the range $r' = [0.95, 1.15]$ is registered. As observed in Figure 6, the reflectivity peak of the interface #2 ranges between $r' = [0.97, 1.01]$ m, with an average amplitude of $|\rho_{(interface\ \#1)}| \approx -15$ dB $= 0.18$. Finally, d_{echo} is calculated as the distance between the first and second reflectivity peaks. As the OUT has a cylindrical shape, d_{echo} can be calculated individually for each rotation angle, then averaging the result, yielding $d_{echo} = 0.15$ m.

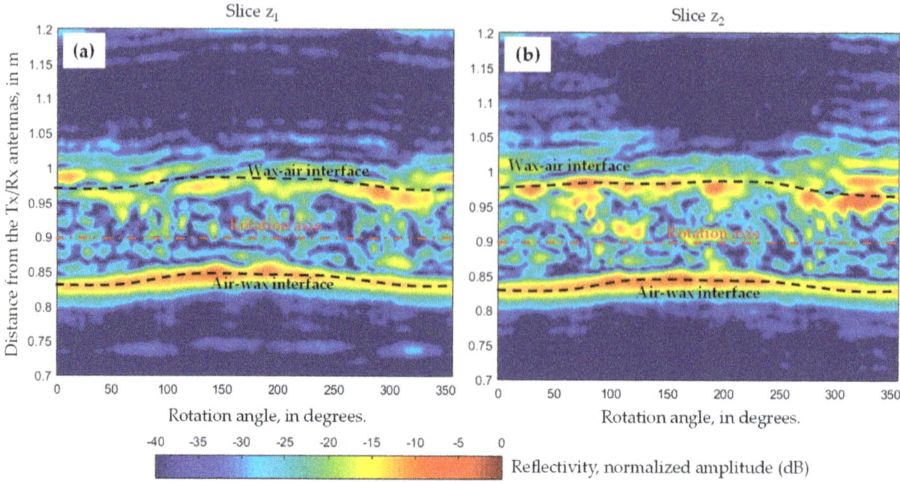

Figure 6. Reflectivity calculated for each observation angle as a function of the distance from the Tx/Rx antennas, for two different XY slices ((**a**) $z_1 = -12$ cm and (**b**) $z_2 = -5$ cm with respect to the top of the wax candle). Air-wax (front reflection) and wax-air (rear reflection) interfaces are noticed.

Now, by applying Equations (6)–(8), an estimate of the permittivity and the conductivity for the wax candle can be calculated (Equations (9)–(11)):

$$\varepsilon_{r,est} = (d_{echo}/d_{OUT})^2 = (0.15 \text{ m}/0.1 \text{ m})^2 = 2.3 \tag{9}$$

$$\alpha = \ln(|\rho(r'_{interface\ \#1})|\,/\,|\rho(r'_{interface\ \#2})|)/\,d_{OUT} = \ln(0.32/0.18)/0.1 = 5.76 \text{ Np/m} = 50 \text{ dB/m} \tag{10}$$

$$\sigma \approx \mathrm{Im}\{((\varepsilon_{r,est})^{1/2} + j\alpha\,c/(2\pi f_c))^2\} = 0.06 \text{ S/m, with } f_c = (f_1 + f_2)/2 = 15 \text{ GHz} \tag{11}$$

As listed in Table 1, these values are in agreement with the expected ones for paraffin (wax), as discussed in [27] (Figure 3, parameter x = 0) and in [3] (f = 9.4 GHz: $\varepsilon_{r,est}$ = 2.17, σ_{est} = 0.03 S/m), where an integral equation-based technique was used to recover these constitutive parameters.

Table 1. Constitutive parameters of the media considered in the examples. Comparison with other techniques at microwave frequencies.

Material	Frequency (GHz)	Permittivity (ε_r)	Conductivity (σ) (S/m)	Method	Reference
Wax (paraffin)	12–18	2.3 ± 0.2	0.06 ± 0.02	Backpropagation SAR	This contribution
Wax (paraffin)	9–15	2.2	0.35	X-ray powder diffraction analysis	[27]
Wax (paraffin)	9.4	2.17	0.03	Model-based monochromatic inverse scattering	[3]
Sand	12–18	2.5 ± 0.2	0.08 ± 0.02	Backpropagation SAR	This contribution
Sand	3–6	[2.7, 3.5]	[0.27, 0.4]	Backpropagation SAR, with reference target	[22]
Sand	Up to 10	2.4 ± 0.2	0.02 ± 0.005	Coaxial probe	[28]

Apart from the constitutive parameters, the goal of the proposed methodology is to recover the geometry of the OUT. For this purpose, the (r',φ) representation of the reflectivity has to be converted into cartesian coordinates. If the rotation axis of the rotary table is defined as z-axis, then, for each slice, the reflectivity in cartesian coordinates is given by Equation (12):

$$\rho(x', y', z') = \rho\,((r' - R)\cos(\varphi),\,(r' - R)\sin(\varphi),\,z'),\ \text{with } R = 89 \text{ cm.} \tag{12}$$

Figure 7a,b corresponds, respectively, to the reflectivity depicted in Figure 6a,b, after applying Equation (12). While the profile of the wax candle can be noticed ($R = 5$ cm), several echoes outside the wax contour are observed as well. These correspond to the uncorrected position of the wax-air interface (#2), which is imaged further than its true range distance.

SAR images can be corrected by applying Equations (3)–(5), as $\varepsilon_{r,wax}$ has been already estimated. Resulting reflectivity images in cartesian coordinates are depicted in Figure 7c,d, where it can be verified that air-wax (#1) and wax-air (#2) reflections are imaged on the contour of the wax candle.

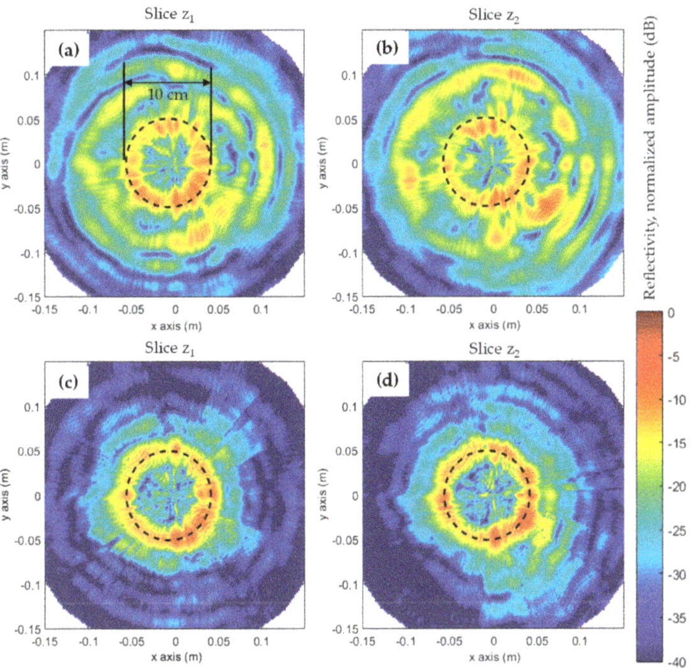

Figure 7. Polar representation of the reflectivity for two different XY slices ($z_1 = -12$ cm and $z_2 = -5$ cm with respect to the top of the wax candle). (**a,b**) Without dielectric delay correction. (**c,d**) After dielectric delay correction, considering $\varepsilon_{r,cat} = 2.3$. Dashed line represents the true contour of the wax candle.

Concerning calculation time, the number of measurement points for each slice is 360. For each rotation angle, SAR along r' axis is calculated in ~5 ms (7 ms in the case of the corrected SAR) using a conventional laptop with no parallelization of the SAR code. Thus, the calculation time to obtain the SAR image on each slice is 18 s for uncorrected SAR and 25 s for corrected SAR. As the estimation of the conductivity and permittivity values requires less than 2 s, the overall calculation time for each slice is approximately 45 s. It must be remarked that the methodology is fully parallelizable, so that the calculation time can be decreased proportionally to the number of processors used.

It is worth mentioning that all the required information for estimating the conductivity and the permittivity, and thus being able to correct the SAR image as proved in Figure 7, is obtained just from the representation of the reflectivity assuming free-space propagation condition depicted in Figure 6.

Finally, corrected reflectivity images for different XY slices can be stacked to obtain a 3D representation of the OUT. For this example, the reconstructed geometry of the wax candle is shown in Figure 8. Graphics post-processing techniques could be applied to convert reflectivity isosurfaces into a 3D geometry model.

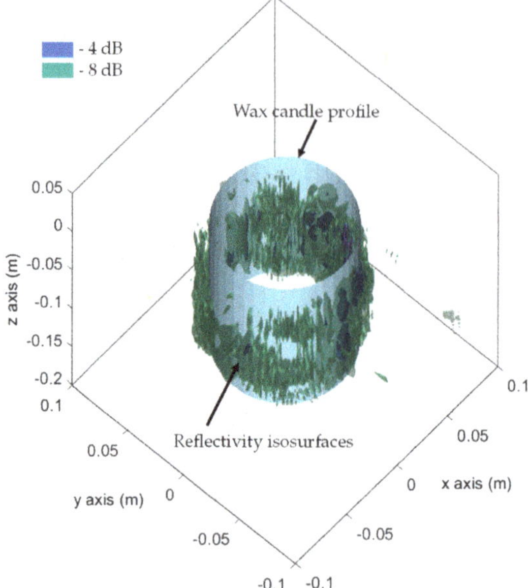

Figure 8. 3D representation of the recovered reflectivity of the wax candle, after dielectric delay correction.

3.3. Plastic Bottle Filled with Sand

In order to remark on the consequences of not considering the permittivity of the OUT for SAR imaging, the second OUT consists of a $d_{OUT} = 12$ cm diameter plastic bottle filled with sand, with two metallic plates concealed on it, as depicted in Figure 9a. The two metallic plates are placed approximately symmetrical with respect to the center of the bottle. As in the previous example, the OUT is placed on top of the rotary table (Figure 9b).

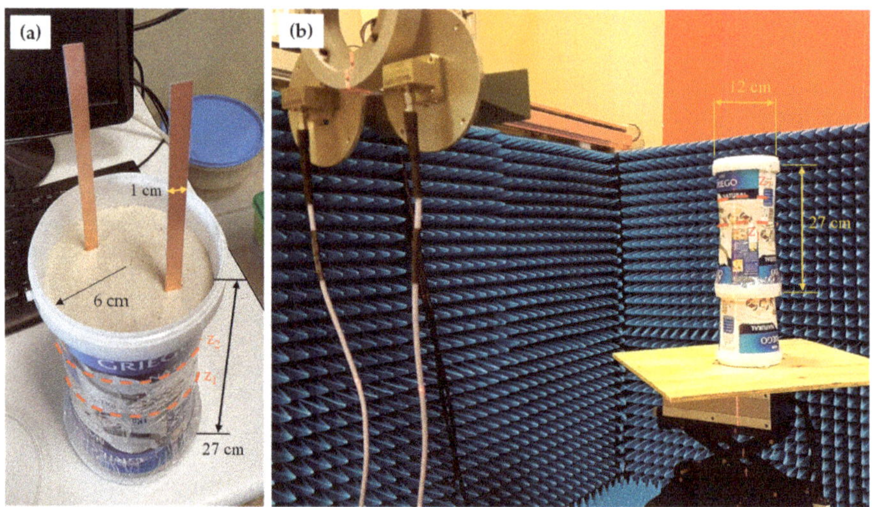

Figure 9. (**a**) Picture of the plastic box filled with sand, with two metallic plates embedded. (**b**) Picture of the OUT placed on the rotary platform.

First, SAR algorithm assuming free-space propagation conditions (Equation (1)) is applied. Reflectivity for slice $z_1 = -13$ cm as a function of the rotation angle and the distance from the Tx/Rx is depicted in Figure 10. As in the first example, the air-sand (#1) and the sand-air (#2) interfaces can be observed, together with the placement of the two metallic plates inside the plastic bottle. Note that the two metallic plates are facing the Tx/Rx antennas twice during the 360° acquisition. Thus, two main echoes of the same metallic plates (front and rear) appear in the reflectivity image, both shifted backwards proportionally to the distance between the air-sand interface and the metallic plate placement. The rear echo is more noticeable, as it is further from the air-sand interface (which partially masks the front reflection of the metallic plates).

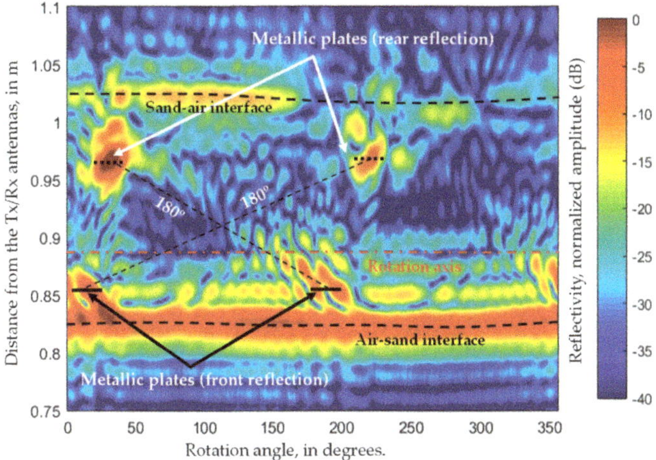

Figure 10. Reflectivity calculated for each observation angle as a function of the distance from the Tx/Rx antennas, for a XY slice ($z_1 = -13$ cm with respect to the top of the plastic box). Air-sand (front reflection) and sand-air (rear reflection) interfaces are noticed, as well as the stronger reflection in the metallic plates.

If this reflectivity image is converted into cartesian coordinates by applying Equation (12), the image depicted in Figure 11a is obtained. Not only the sand-air interface (#2) is shifted backwards, but also the rear reflection of the metallic plates is imaged *outside* the sand box. The reason is that the displacement of the rear reflection due to free-space propagation conditions is greater than the distance from the metallic plates to the plastic bottle. If another slice is chosen ($z_2 = -7$ cm), the same effect can be observed (Figure 11b). Thus, there is a clear need for using an estimate of the permittivity of the sand in order to recover a correct reflectivity image of the OUT. As in the previous example, conductivity and permittivity can be estimated from the uncorrected reflectivity depicted in Figure 10.

An analysis of Figure 10 allows estimating the air-sand interface (#1), located at $r' = 0.83$ m on average, and the sand-air interface (#2), placed at $r' = 1.02$ m. Thus, as $d_{echo} = 19$ cm and $d_{OUT} = 12$ cm, the relative permittivity estimated using Equation (6) is $\varepsilon_{r,est} = 2.5$.

For the conductivity, values of the reflectivity within the angular range $\varphi = [60°, 150°]$ can be considered, yielding $|\rho(\text{interface \#1})| \approx -5\,\text{dB} = 0.56$, $|\rho(\text{interface \#2})| \approx -13\,\text{dB} = 0.22$. From Equation (7), the attenuation is $\alpha = 7.8\,\text{Np/m} = 67.6\,\text{dB/m}$, and, finally, the conductivity is (Equation (8)) $\sigma_{est} = 0.08\,\text{S/m}$.

As a reference, typical values for sand with a moisture content below 1% is $\varepsilon_{r,est} \approx 2.4$ and $\sigma = 0.02$ at 10 GHz (Figure 2 of [28]). In this example, the same sand as in [22] was used, where values of ε textsubscriptr,est $\approx [2.7, 3.5]$ and $\sigma = [0.27, 0.40]$ were estimated in the 3 to 6 GHz frequency band. As shown in [28], the value of these constitutive parameters tends to decay with frequency. A summary of the recovered constitutive parameters is shown in Table 1.

Once the constitutive parameters have been estimated, the corrected SAR image can be computed. SAR images corresponding to slices $z_1 = -13$ cm and $z_1 = -7$ cm are depicted in Figure 11c,d. Not only the contour of the plastic bottle is correctly imaged, but also the two metallic plates are found inside the area filled with sand. Note that, as observed in Figure 9, the plastic bottle is not a perfect cylinder. In the case of slice z_1 the surface of the bottle is almost parallel to the Tx/Rx horn antennas aperture, whereas in the case of slice z_2 the surface of the bottle is slightly tilted, so that reflection on the air-sand interface will not be fully reflected back to the Tx/Rx antennas.

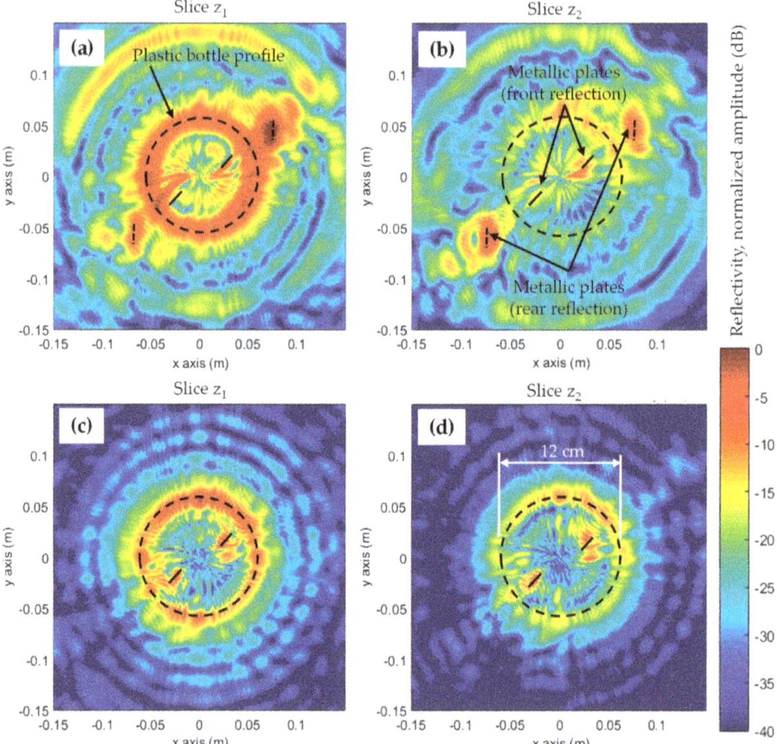

Figure 11. Polar representation of the reflectivity for two different XY slices ($z_1 = -13$ cm and $z_2 = -7$ cm with respect to the top of the plastic bottle). (**a,b**) Without dielectric delay correction. (**c,d**) After dielectric delay correction, considering $\varepsilon_{r,est} = 2.5$. Dashed line represents the true contour of the plastic bottle. Solid line represents the true position of the metallic plates. In (**a,b**), the straight dashed lines indicate the location of the imaged metallic plates, which appear outside the plastic box contour due to the dielectric delay.

4. Discussion

Results presented in Section 3 confirm the effectiveness of the proposed methodology to recover the constitutive parameters of the OUT using the imaged reflectivity assuming free-space propagation conditions. Then, the estimated reflectivity value is introduced into a modified SAR technique that takes into account the different media that compose the microwave imaging scenario, so that a corrected SAR image is produced.

With respect to similar techniques where conductivity and permittivity were retrieved from reflectivity images [21,22], the main novelty is that all the information is extracted from the SAR image of the OUT, avoiding the need of placing external references such as a metallic plate (buried or acting as background). In the proposed methodology, the challenge is the development of processing algorithms capable of extracting the information from the uncorrected SAR image, taking advantage of the a priori information about the OUT geometry. In the case of the examples presented in this contribution, rotation symmetry around vertical (z-) axis made this processing easy, as the reflectivity could be represented in cylindrical coordinates (r',φ). For those targets with more complex geometry, pattern recognition algorithms could be used to identify and extract the position of the interfaces between media.

Recovered permittivity and conductivity values are summarized in Table 1. In the case of the permittivity, the estimated values are within the range provided by other methods based on different hardware and processing techniques. Conductivity values are more dependent on the exact composition of the medium (e.g., moisture content), so larger dispersion can be expected.

Calculation time is also a key issue for the development of inverse scattering and imaging systems. Backpropagation SAR-based techniques are by far faster than model-based methods. For the examples presented in this contribution, recovery of the final SAR image required around 45 s for each XY slice using a non-parallelized software code. Although this is not real-time imaging, it must be remarked that (i) measurement time was around 180 s per slice (360 acquisition points), and (ii) the proposed SAR-based technique is fully parallelizable. If the code is run on a 4-core processor (available in most conventional computers nowadays), calculation time would be reduced to less than 12 s per slice. The use of a Graphics Processing Unit (GPU) could result in 60–80 times speedup, as explained in [9], thus enabling real-time imaging.

5. Conclusions

A simple, fast method for microwave imaging using a SAR-based technique has been presented. The proposed methodology is capable of providing an estimate of the permittivity and conductivity of the OUT from a SAR image retrieved under free-space propagation conditions, and then, correcting the SAR image by introducing the estimated permittivity value (or, in other words, introducing the correct propagation velocity at each medium on the imaging problem). Results showed the effects of inaccurate SAR imaging, and the capability of the proposed methodology to provide accurate microwave images of the targets under test.

Author Contributions: Conceptualization, Y.Á.L.; Methodology, Y.Á.L., M.G.F. and F.L.-H.; Software, Y.Á.L. and R.G.; Measurements and Validation, Y.Á.L., M.G.F. and R.G.; Data Curation, Y.Á.L., M.G.F. and R.G.; Writing-Original Draft Preparation, Y.Á.L. and R.G.; Writing-Review & Editing, M.G.F. and F.L.-H.; Supervision, F.L.-H.

Funding: This research was partially supported by the "Ministerio de Economía y Competitividad" of Spain/FEDER under grant FPU15/06341; by the Principado de Asturias under project GRUPINN-18-000191, and by the European Union under the framework of the Erasmus+ mobility agreement between the University of Oviedo (E OVIEDO 01) and Hochschule Mannheim (D MANNHEI 03).

Conflicts of Interest: The authors declare no conflict of interest.

References

1. Deng, Y.; Liu, X. Electromagnetic Imaging Methods for Nondestructive Evaluation Methods. *Sensors* **2011**, *11*, 11774–11808. [CrossRef] [PubMed]
2. Cooper, K.B.; Dengler, R.J.; Llombart, N.; Thomas, B.; Chattopadhyay, G.; Siegel, P.H. THz imaging radar for standoff personnel screening. *IEEE Trans. Terahertz Sci. Technol.* **2011**, *1*, 169–182. [CrossRef]
3. Álvarez, Y.; García-Fernández, M.; Poli, L.; García-González, C.; Rocca, P.; Massa, A.; Las-Heras, F. Inverse Scattering for Monochromatic Phaseless Measurements. *IEEE Trans. Instrum. Meas.* **2017**, *66*, 45–60. [CrossRef]
4. García-Fernández, M.; Álvarez-López, Y.; Arboleya-Arboleya, A.; González-Valdés, B.; Rodríguez-Vaqueiro, Y.; Las-Heras Andrés, F.; Pino García, A. Synthetic Aperture Radar Imaging System for Landmine Detection Using a Ground Penetrating Radar on Board a Unmanned Aerial Vehicle. *IEEE Access* **2018**, *6*, 45100–45112. [CrossRef]

5. Narayanan, R.M.; Gebhardt, E.T.; Broderick, S.P. Through-Wall Single and Multiple Target Imaging Using MIMO Radar. *Electronics* **2017**, *6*, 70. [CrossRef]

6. Elahi, M.A.; O'Loughlin, D.; Lavoie, B.R.; Glavin, M.; Jones, E.; Fear, E.C.; O'Halloran, M. Evaluation of Image Reconstruction Algorithms for Confocal Microwave Imaging: Application to Patient Data. *Sensors* **2018**, *18*, 1678. [CrossRef] [PubMed]

7. Zhuge, X.; Yarovoy, A.G. A sparse aperture MIMO-SAR-based UWB imaging system for concealed weapon detection. *IEEE Trans. Geosci. Remote Sens.* **2011**, *49*, 509–518. [CrossRef]

8. Soumekh, M. Bistatic synthetic aperture radar inversion with application in dynamic object imaging. *IEEE Trans. Signal Process.* **1991**, *39*, 2044–2055. [CrossRef]

9. Lopez-Portugues, M.; Alvarez-Lopez, Y.; Lopez-Fernandez, J.A.; Garcia-Gonzalez, C.; Ayestaran, R.G.; Las-Heras Andres, F. A multi-GPU sources reconstruction method for imaging applications. *Prog. Electromagn. Res.* **2013**, *136*, 703–724. [CrossRef]

10. Lin, C.Y.; Kiang, Y.W. Inverse scattering for conductors by the equivalent source method. *IEEE Trans. Antennas Propag.* **1996**, *44*, 310–316. [CrossRef]

11. Woten, D.; Hajihashemi, M.R.; Hassan, A.M.; El-Shenawee, M. Experimental microwave validation of level set reconstruction algorithm. *IEEE Trans. Antennas Propag.* **2010**, *58*, 230–233. [CrossRef]

12. Eskandari, A.R.; Naser-Moghaddasi, M.; Eskandari, M. Reconstruction of Shape and Position for Scattering Objects by Linear Sampling Method. *Int. J. Soft Comput. Eng.* **2012**, *2*, 2231–2307.

13. Caorsi, S.; Donelli, M.; Massa, A. Detection, location, and imaging of multiple scatterers by means of the iterative multiscaling method. *IEEE Trans. Microw. Theory Tech.* **2004**, *52*, 1217–1228. [CrossRef]

14. Rocca, P.; Benedetti, M.; Donelli, M.; Franceschini, D.; Massa, A. Evolutionary optimization as applied to inverse scattering problems. *Inverse Probl.* **2009**, *25*, 123003. [CrossRef]

15. Gonzalez-Valdes, B.; Alvarez, Y.; Martinez-Lorenzo, J.A.; Las-Heras, F.; Rappaport, C.M. On the Combination of SAR and Model Based Techniques for High-Resolution Real-Time Two-Dimensional Reconstruction. *IEEE Trans. Antennas Propag.* **2014**, *62*, 5180–5189. [CrossRef]

16. Fallahpour, M.; Case, J.T.; Ghasr, M.T.; Zoughi, R. Piecewise and Wiener filter-based SAR techniques for monostatic microwave imaging of layered structures. *IEEE Trans. Antennas Propag.* **2014**, *62*, 282–294. [CrossRef]

17. Laviada, J.; Wu, B.; Ghars, M.T.; Zoughi, R. Nondestructive Evaluation of Microwave-Penetrable Pipes by Synthetic Aperture Imaging Enhanced by Full-Wave Field Propagation Model. *IEEE Trans. Instrument. Meas.* **2018**, 1–8, in press. [CrossRef]

18. Robinson, D.A.; Jones, S.B.; Wraith, J.M.; Or, D.; Friedman, S.P. A Review of Advances in Dielectric and Electrical Conductivity Measurement in Soils Using Time Domain Reflectometry. *Vadose Zone J.* **2003**, *2*, 444–475. [CrossRef]

19. Garret, J.D.; Fear, E.C. Average Dielectric Property Analysis of Complex Breast Tissue with Microwave Transmission Measurements. *Sensors* **2015**, *15*, 1199–1216. [CrossRef] [PubMed]

20. Costa, F.; Borgese, M.; Degiorgi, M.; Monorchio, A. Electromagnetic Characterisation of Materials by Using Transmission/Reflection (T/R) Devices. *Electronics* **2017**, *6*, 95. [CrossRef]

21. Gonzalez-Valdes, B.; Alvarez-Lopez, Y.; Martinez-Lorenzo, J.A.; Las Heras, F.; Rappaport, C.M. SAR processing for profile reconstruction and characterization of dielectric objects on the human body surface. *Prog. Electromagn. Res.* **2013**, *138*, 269–282. [CrossRef]

22. Álvarez-López, Y.; García-Fernández, M.; Arboleya, A.; González-Valdés, B.; Rodríguez-Vaqueiro, Y.; Las-Heras, F.; Pino García, A. SAR-based technique for soil permittivity estimation. *Int. J. Remote Sens.* **2017**, *38*, 5168–5185. [CrossRef]

23. López-Rodríguez, P.; Escot-Bocanegra, D.; Poyatos-Martínez, D.; Weinmann, F. Comparison of Metal-Backed Free-Space and Open-Ended Coaxial Probe Techniques for the Dielectric Characterization of Aeronautical Composites. *Sensors* **2016**, *16*, 967. [CrossRef] [PubMed]

24. Arboleya, A. Novel XYZ Scanner-Based Radiation and Scattering Measurement Techniques for Antenna Diagnostics and Imaging Applications. Ph.D. Thesis, Universidad de Oviedo, Oviedo, Spain, 2016. Available online: http://digibuo.uniovi.es/dspace/bitstream/10651/40222/1/TD_AnaArboleya.pdf (accessed on 12 October 2018).

25. Bell Electronics. Narda 639 Standard Gain Horn, 12.4 to 18 GHz. Available online: https://www.bellnw.com/manufacturer/Narda/639.htm (accessed on 5 October 2018).

26. Keysight. N5247A PNA-X Microwave Network Analyzer. Available online: https://www.keysight.com/en/pdx-x201825-pn-N5247A/pna-x-microwave-network-analyzer-67-ghz?cc=EN&lc=eng (accessed on 5 October 2018).

27. Bayrajdar, H. Complex permittivity, complex permeability and microwave absorption properties of ferrite-paraffin polymer composites. *J. Magn. Magn. Mater.* **2011**, *323*, 1882–1885. [CrossRef]

28. Abdelgwad, A.H.; Said, T.M. Measured dielectric Permittivity of Contaminated sandy soils at Microwave Frequency. *J. Microw. Optoelectron. Electromagn. Appl.* **2016**, *15*, 115–122. [CrossRef]

Article

Retrieval of Three-Dimensional Surface Deformation Using an Improved Differential SAR Tomography System

Zhigui Wang, Mei Liu * and Kunfeng Lv

School of Electronics and Information Engineering, Harbin Institute of Technology, Harbin 150001, China; wangzhiguihit@outlook.com (Z.W.); wanshicun@foxmail.com (K.L.)
* Correspondence: liumei@hit.edu.cn; Tel.: +86-0451-8641-8052

Received: 13 December 2018; Accepted: 31 January 2019; Published: 2 February 2019

Abstract: Conventional differential synthetic aperture radar tomography (D-TomoSAR) can only capture the scatterers' one-dimensional (1-D) deformation information along the line of sight (LOS) of the synthetic aperture radar (SAR), which means that it cannot retrieve the three-dimensional (3-D) movements of the ground surface. To retrieve the 3-D deformation displacements, several methods have been proposed; the performance is limited due to the insufficient sensitivity for retrieving the North-South motion component. In this paper, an improved D-TomoSAR model is established by introducing the scatterers' 3-D deformation parameters in slant range, azimuth, and elevation directions into the traditional D-TomoSAR model. The improved D-TomoSAR can be regarded as a multi-component two-dimensional (2-D) polynomial phase signal (PPS). Then, an effective algorithm is proposed to retrieve the 3-D deformation parameters of the ground surface by the 2-D product high-order ambiguity function (PHAF) with the relax (RELAX) algorithm. The estimation performance is investigated and compared with the traditional algorithm. Simulations and experimental results with semi-real data verify the effectiveness of the proposed signal model and algorithm.

Keywords: synthetic aperture radar; differential SAR tomography; squinted SAR; 3-D deformation; 2-D PPS

---

1. Introduction

Differential synthetic aperture radar (SAR) tomography (D-TomoSAR) [1,2] is a kind of multi-baseline SAR processing framework that allows the joint resolution capability of multiple scatterers' velocities and elevations in the same range-azimuth cell through a two-dimensional (2-D) baseline-time spectral estimation. It is favored over similar differential SAR interferometry (D-InSAR) technologies [3], such as persistent scatterer interferometry (PSI) [4,5] and small baseline subset (SBAS) [6], because D-TomoSAR can overcome the layover phenomenon in the forest or urban areas. At present, D-TomoSAR is widely used in mapping and monitoring the infrastructure deformation and ground subsidence caused by the over-pumping of groundwater or mining.

One of the disadvantages of the D-TomoSAR technique is that only one-dimensional (1-D) deformation along the radar line of sight (LOS) can be captured, which cannot fully describe the actual deformation information of ground. The extraction of three-dimensional (3-D) (East-West, North-South, Up-Down) deformation information plays an important role in monitoring seismic and volcanic activities and the urban surface subsidence. Thus, it is significant to decompose the observed 1-D LOS deformation information into the 3-D deformation parameters.

Several methods of 3-D deformation retrieval based on D-InSAR and D-TomoSAR technology have already been reported. These methods use a combination of multiple LOS deformation observations to retrieve the three deformation components [7,8], while the equal-precision 3-D deformation estimation cannot be obtained. In particular, the estimation error in the North-South direction is much greater than that in the other two directions. This is heightened by the fact that most of the spaceborne SARs in orbit fly along the near-pole orbits [8], namely, the azimuth direction of SAR imaging is almost parallel to the North-South direction. To improve the accuracy of North-South deformation retrieval from LOS measurements, several methods have been proposed, such as a combination of ascending/descending SAR acquisitions and the multi-aperture interferometry (MAI) algorithm [9] or offset-tracking [10] technology. However, these methods are only suitable for retrieving the large deformation caused by earthquakes or volcanic activities and are unable to be applied in D-TomoSAR directly because they are developed from D-InSAR. Combining the D-InSAR with global positioning system (GPS) data [11] is another approach to retrieving the 3-D deformation maps. Nevertheless, this method strongly depends on the external GPS data and cannot be widely used.

Results of existing methods for retrieving the 3-D deformation show that the higher the diversity of the geometric configuration (the satellite heading angle and incidence angle of antenna) of the combined SAR data is, the more accurate the North-South retrieval results will be [12]. This provides an effective way to improve the accuracy of North-South deformation retrieval. A common method is to combine the SAR data acquired from different incidence angles, e.g., combining SAR data acquired from the ascending and descending tracks with right- and left-looking configurations [13]. However, extreme attitude control is required for this method to change the incidence angle of the antenna, which is difficult for satellites with large antennas. Another method is to change the heading angle of the satellite and make it deviate from the North-South direction, which can fundamentally solve the problem of low precision of deformation estimation in the north component [14,15], whereas the satellite heading angle is directly related to the orbital inclination of the satellite, and the non-polar orbit will limit the imaging coverage in the latitudinal direction.

At present, the SuperSAR [12,16] and BiDiSAR [17] are proposed to improve the accuracy of the deformation retrieval in the North-South direction by increasing the squint angle of SAR imaging in D-InSAR. In this paper, an improved D-TomoSAR signal model is proposed by introducing the squint angle of SAR imaging into the traditional D-TomoSAR model to achieve a higher accuracy for North-South deformation estimation. In the proposed model, the relationship between the phase term of the improved D-TomoSAR and the 3-D deformation velocities in the slant range, azimuth, and elevation directions are established through imaging geometry. As a result, the elevation and 3-D deformation parameters of scatterers can be retrieved by estimating the coefficients of a 2-D polynomial phase signal (PPS). Analyses of numerical simulations and the semi-real data experiments indicate that a comparable effect to change the satellite heading angle can be achieved by increasing the squint angle of SAR imaging.

This paper is organized as follows. The improved D-TomoSAR signal model is proposed in Section 2. In Section 3, the 2-D product high-order ambiguity function (PHAF) with the relax (RELAX) algorithm is adopted to estimate the 3-D deformation components, and an analysis is carried out to evaluate the performance of the proposed model. Section 4 presents experiments on simulated and semi-real data to demonstrate the precision and efficiency of the proposed algorithm. A general conclusion is presented in Section 5.

2. System Model

2.1. Review of D-TomoSAR

The observation geometry of conventional D-TomoSAR is shown in Figure 1, where X represents the ground-range direction, Y is the azimuth direction, Z is the vertical height direction, r is the reference slant range direction along the LOS of the radar, and s is the elevation direction orthogonal to the slant range-azimuth plane. Consider the processing of SAR acquisitions from N passes with M single looking complex (SLC) SAR images acquired simultaneously at each pass by M similar spaceborne SAR. Let $t_n, n = 1, 2, \cdots, N$, be the acquisition time of the phase centers in the nth pass, and let t_{mn}, $m = 1, 2, \cdots, M, n = 1, 2, \cdots, N$, be the imaging time of the mth SAR radar at the nth pass, thus, we have $t_{11} = \ldots t_{M1} = t_1, \ldots, t_{1N} = \ldots t_{MN} = t_N$. The registration of the $M \times N$ SAR images to a reference master image is firstly performed, and the correction of atmospheric disturbance is carried out to eliminate the measurement noise in the SAR images. The focused complex-valued measurement of each calibrated range-azimuth pixels in the $M \times N$ SAR images can be arranged in the $M \times N$ baseline-time plane to form a data matrix $\mathbf{G}$. Each element g_{mn} in this data matrix denotes the focused measurement of the range-azimuth pixel (r', y') in SAR images acquired from the mth SAR radar at time t_n, and it is the integral of the focused signal of scatterers distributed along the elevation direction s. As a result, g_{mn} can be written as:

$$
\begin{aligned}
g_{mn}(r', y') &= \int \int f(r - r', y - y') dr dy \int \gamma(r, y, s) \exp\left\{-j\tfrac{4\pi}{\lambda} R_{mn}[r, s, \Delta d(s, t_n)]\right\} ds \\
m &= 1, 2, \cdots M, n = 1, 2, \cdots N
\end{aligned}
\tag{1}
$$

where $\gamma(\cdot)$ is the superposition of the backscattering function along elevation s in the pixel (r', y'), λ is the wavelength, $R_{mn}[r, s, \Delta d(s, t_n)]$ represents the distance between the scatterer located at coordinates of (r, s) on the ground and the mth satellite at time t_n, $\Delta d(s, t_n)$ is the scatterers' displacement in the LOS direction, $f(r', y')$ is the 2-D point spread function (PSF) of the focused SAR image.

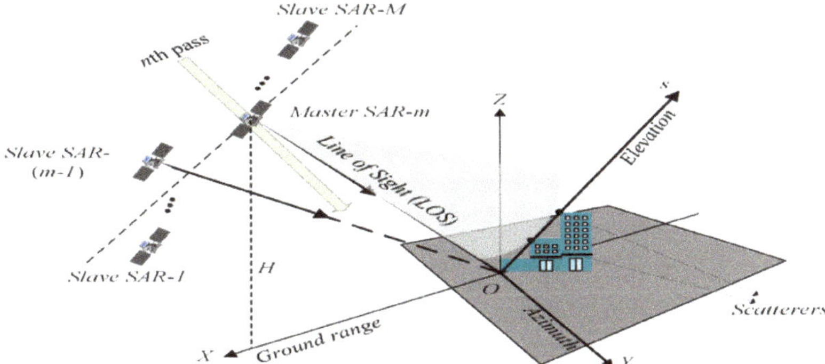

Figure 1. System geometry of differential synthetic aperture radar tomography (D-TomoSAR).

For simplicity, we assume the 2-D PSF to approximate an ideal 2-D Dirac function, and a deramping operation [18] is performed to compensate the phase quadratic distortion of the received data. Finally, the focused measurement of each pixel (r', y'), can be rewritten as:

$$
\begin{aligned}
g_{mn} &= \int \gamma(s) \exp\left\{-j\tfrac{4\pi}{\lambda}[R_{mn}(s, \Delta d(s, t_n)) - R_{mn}(0)]\right\} ds \\
m &= 1, 2, \cdots M, n = 1, 2, \cdots N
\end{aligned}
\tag{2}
$$

Note that only the LOS deformation can be observed by the conventional D-TomoSAR system, and at least three LOS observations from different acquisition geometries are required to decompose the LOS displacement into the 3-D deformation components [8]. However, the deformation component in the North-South direction cannot be reliably resolved because the current SAR satellites operate in the near-polar orbits. To improve the accuracy of the North-South deformation retrieval, the usage of squint imaging mode is helpful, which was verified in the D-InSAR [12,16]. In this context, an improved D-TomoSAR signal model is proposed by introducing the 3-D deformation components of scatterers into the traditional D-TomoSAR system through the SAR squint imaging mode in this paper to achieve an improvement for the accuracy of deformation retrieval in the North-South direction.

2.2. Signal Model of Improved D-TomoSAR

As shown in Figure 2, P_1 and P_m are the antenna phase center of the first and mth radar, respectively, and the corresponding imaging time for the same region is t_1 and t_n. In this paper, the acquisition acquired from radar P_1 at time t_1 is defined as the master image in the registration operation. The green plane is the 2-D imaging plane of the radar P_1, and the point P'_m is the position of projecting the radar P_m onto the imaging plane of radar P_1. B is the spatial baseline between P_1 and P_m. $B_{mn,r}$ is the horizontal baseline along the slant range direction, $B_{mn,y}$ is the horizontal baseline along the azimuth direction, and $B_{mn,s}$ is the orthogonal baseline along the elevation direction. The incidence angle of the beam center of P_1 is α, and θ_{sq} denotes the squint angle. $R_{11}(0)$ is the distance between P_1 at time t_1 and the reference point T_0, whose elevation is zero. Accordingly, $R_{mn}(s)$ is the distance between P_m and the scatterer with an elevation s at time t_n. $R_{mn}(0)$ is the reference range for the deramping operation, and it can be expressed as the distance between P_m and the reference point T_0. The scatterer is assumed to occur in a 3-D displacement during the imaging time from t_1 to t_n, where the deformation components in the slant range, azimuth, and elevation direction are Δr, Δy, and Δs, respectively.

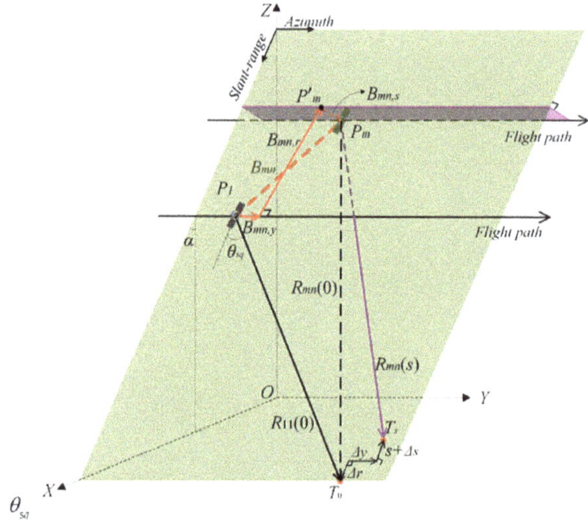

Figure 2. Squinted synthetic aperture radar (SAR) data acquisition geometry for the improved D-TomoSAR.

Referring to Figure 2, $R_{mn}(s)$ and the reference range $R_{mn}(0)$ for the deramping operation can be expressed as:

$$\begin{cases} \vec{R}_{mn}(s) = \vec{R}_{11}(0) - \Delta\vec{r} + \Delta\vec{y} - \vec{B}_{mn,y} - \vec{B}_{mn,r} - \vec{B}_{mn,s} + \vec{s} + \Delta\vec{s} \\ \vec{R}_{mn}(0) = \vec{R}_{11}(0) - \vec{B}_{mn,y} - \vec{B}_{mn,r} - \vec{B}_{mn,s} \end{cases} \tag{3}$$

As mentioned in (2), $R_{mn}(s) - R_{mn}(0)$ is given by:

$$R_{mn}(s) - R_{mn}(0) = \left\langle \vec{R}_{mn}(s), \vec{R}_{mn}(s) \right\rangle^{1/2} - \left\langle \vec{R}_{mn}(0), \vec{R}_{mn}(0) \right\rangle^{1/2} \tag{4}$$

where $\langle \cdot \rangle$ represents the inner product of the vectors. According to the Fresnel approximation, the difference in slant range can be rewritten as:

$$
\begin{aligned}
R_{mn}(s) - R_{mn}(0) \approx & \frac{s^2}{2R_{11}(0)} - \frac{s}{R_{11}(0)}B_{mn,s} - \frac{\Delta s}{R_{11}(0)}B_{mn,s} + \frac{\Delta r^2+\Delta y^2+\Delta s^2}{2R_{11}(0)} \\
& - \frac{\left(\cos\left(\theta_{sq}\right)R_{11}(0)+B_{mn,r}\right)\Delta r - \left(\sin\left(\theta_{sq}\right)R_{11}(0)-B_{mn,y}\right)\Delta y - s\Delta s}{R_{11}(0)}
\end{aligned}
\tag{5}
$$

In this paper, we assume that the acquisitions were registered to the master image with subpixel accuracy. Hence, the horizontal baseline along the azimuth direction $B_{mn,y}$ is nearly zero. In addition, the acquisitions acquired from different orbits were also corrected to the reference image with a constant slant range distance to eliminate the effect of the spatial decorrelation [19], by which the horizontal baseline along the slant range direction $B_{mn,r}$ is also approximately equal to zero. On the other hand, the first term in (5) is a constant and will not affect the amplitude of the image [20]. As a result, the first term in (5) can be incorporated in the backscattering function $\gamma(\cdot)$. Finally, the focused measurement in the improved D-TomoSAR model can be expressed as:

$$
\begin{aligned}
g_{mn} = \quad & \int \gamma(s)ds \exp\left\{ j\frac{4\pi}{\lambda}\frac{s}{r}B_{mn,s} \right\} \exp\left\{ j\frac{4\pi}{\lambda}\frac{\Delta s}{r}B_{mn,s} \right\} \\
\times \quad & \exp\left\{ j\frac{4\pi}{\lambda}\left[\cos\left(\theta_{sq}\right)\Delta r - \sin\left(\theta_{sq}\right)\Delta y - \frac{s\Delta s}{r} \right] \right\} \exp\left\{ -j\frac{2\pi}{\lambda}\left[\frac{\Delta r^2+\Delta y^2+\Delta s^2}{r} \right] \right\} \\
m = \quad & 1,2,\cdots M, n = 1,2,\cdots N
\end{aligned}
\tag{6}
$$

where $r = R_{11}(0)$. Assuming that the 3-D deformation model of the scatterer follows the linear model, we have:

$$
\Delta r = v_r t_n, \quad \Delta y = v_y t_n, \quad \Delta s = v_s t_n
\tag{7}
$$

where v_r, v_y, and v_s are the deformation velocities of the scatterer along the slant range, azimuth, and elevation directions, respectively. Finally, (6) can be rewritten as:

$$
\begin{aligned}
g_{mn} = \quad & \int_{\delta s}\int_{\delta v_r}\int_{\delta v_y}\int_{\delta v_s} \gamma(s,v_r,v_y,v_s)dsdv_rdv_ydv_s \exp\left\{ j\frac{4\pi}{\lambda}\frac{s}{r}B_{mn,s} \right\} \exp\left\{ j\frac{4\pi}{\lambda}\frac{v_s}{r}B_{mn,s}t_n \right\} \\
\times \quad & \exp\left\{ j\frac{4\pi}{\lambda}\left[\cos\left(\theta_{sq}\right)v_r - \sin\left(\theta_{sq}\right)v_y - \frac{sv_s}{r} \right]t_n \right\} \exp\left\{ -j\frac{2\pi}{\lambda}\left(\frac{v_r^2+v_y^2+v_s^2}{r} \right)t_n^2 \right\} \\
m = \quad & 1,2,\cdots M, n = 1,2,\cdots N
\end{aligned}
\tag{8}
$$

where δs is the elevation extent of the scatterers, and δv_r, δv_y, and δv_s are the range of possible velocities along the slant range, azimuth, and elevation, respectively.

In the conventional D-TomoSAR, the scatterers in each range-azimuth resolution cell are assumed to have only the linear displacement along the LOS direction, and the elevation and deformation parameters of the scatterers are independent of each other. As a result, the retrieval of the elevation and deformation parameters of scatterers fits well into the framework of 2-D spectral estimation. However, the elevation of each scatterer is coupled with its deformation parameter in the elevation direction in the improved D-TomoSAR signal model, as shown in (8). This means that the 2-D spectral estimation cannot be used to estimate the elevation and the 3-D deformation parameters for the improved D-TomoSAR.

In addition, no more than four scatterers are assumed to locate in the same range-azimuth resolution cell in most of the typical ground scenes [21]. Therefore, the backscattering coefficient $\gamma(\cdot)$ in (8) can be assumed to be sparse in the object domain. Let K be the number of the scatterers lying in the same range-azimuth resolution cell, and the backscattering coefficient can be written as [22]:

$$
\gamma\left(s,v_r,v_y,v_s\right) = \sum_{k=1}^{K}\gamma\left(s_k,v_{k,r},v_{k,y},v_{k,s}\right)\delta(s-s_k)\delta(v_r-v_{k,r})\delta\left(v_y-v_{k,y}\right)\delta(v_s-v_{k,s})
\tag{9}
$$

where s_k, $v_{k,r}$, $v_{k,y}$, and $v_{k,s}$ are the elevation and 3-D deformation components of the kth scatterer, respectively. The coefficient $\gamma\left(s_k, v_{k,r}, v_{k,y}, v_{k,s}\right)$ represents the complex amplitude of the kth scatterer. We assume that the PSF of the scatterers in the elevation and 3-D deformation direction is the ideal Dirac function in this paper. Substituting (9) into (8), we yield:

$$
\begin{aligned}
g_{mn}(B_{mn,s}, t_n) &= \sum_{k=1}^{K} \gamma\left(s_k, v_{k,r}, v_{k,y}, v_{k,s}\right) \exp\left\{j\frac{4\pi}{\lambda}\frac{s_k}{r}B_{mn,s}\right\} \exp\left\{j\frac{4\pi}{\lambda}\frac{v_{k,s}}{r}B_{mn,s}t_n\right\} \\
&\times \exp\left\{j\frac{4\pi}{\lambda}\left[\cos\left(\theta_{sq}\right)v_{k,r} - \sin\left(\theta_{sq}\right)v_{k,y} - \frac{s_k v_{k,s}}{r}\right]t_n\right\} \\
&\times \exp\left\{-j\frac{2\pi}{\lambda}\left(\frac{v_{k,r}^2 + v_{k,y}^2 + v_{k,s}^2}{r}\right)t_n^2\right\} + e_{mn} \\
m &= 1, 2, \cdots M, n = 1, 2, \cdots N
\end{aligned}
\tag{10}
$$

where e_{mn} is the residual error signal after atmospheric phase compensation. Inspecting the phase terms in (10), the improved D-TomoSAR model is found to be the multi-component second order 2-D PPS.

Let $s(\mu, v)$ be a general discrete-time multi-component second order 2-D PPS, which can be expressed as:

$$
s(\mu, v) = \sum_{k=1}^{K} A_k \exp[j2\pi\Phi_k(\mu, v)]
\tag{11}
$$

where:

$$
\Phi_k(\mu, v) = \alpha_{0,0}^{(k)} + \alpha_{1,0}^{(k)}\mu + \alpha_{0,1}^{(k)}v + \alpha_{2,0}^{(k)}\mu^2 + \alpha_{1,1}^{(k)}\mu v + \alpha_{0,2}^{(k)}v^2
\tag{12}
$$

where $\mu = 1, \ldots, N$ and $v = 1, \ldots, M$, K is the number of the components, and A_k is the amplitude of the kth component. Comparing (10) with (11) and (12), we have:

$$
\begin{cases}
A_k = \gamma\left(s_k, v_{k,r}, v_{k,y}, v_{k,s}\right) \\
\mu = B_{mn,s}, v = t_n \\
\alpha_{0,0}^{(k)} = 0, \alpha_{2,0}^{(k)} = 0 \\
\alpha_{1,0}^{(k)} = \frac{2s_k}{\lambda r}, \alpha_{1,1}^{(k)} = \frac{2v_{k,s}}{\lambda r} \\
\alpha_{0,1}^{(k)} = \frac{2\cos\left(\theta_{sq}\right)v_{k,r} - 2\sin\left(\theta_{sq}\right)v_{k,y}}{\lambda} - \frac{2s_k v_{k,s}}{\lambda r} \\
\alpha_{0,2}^{(k)} = -\frac{v_{k,r}^2 + v_{k,y}^2 + v_{k,s}^2}{\lambda r}
\end{cases}
\tag{13}
$$

Equation (13) reveals that the elevation and deformation parameters of scatterers can be retrieved by estimating the coefficients of the 2-D PPS shown in (11). Given that $\alpha_{0,1}^{(k)}$ contains two unknowns—$v_{k,r}$ and $v_{k,y}$—at least two sets of combined SAR acquisitions with different squint angles are required to solve the three deformation components of scatterers. In addition, the larger the diversity of the squint angles of the different combined acquisitions is, the higher the achieved precision would be for the 3-D retrieval [12].

3. 3-D Deformation Retrieval for the Improved D-TomoSAR

As can be seen from the previous section, the improved D-TomoSAR signal model can be equivalent to a multi-component 2-D PPS, and its coefficients correspond to the elevation and deformation parameters of the scatterers. However, different from the general 2-D PPS with uniform sampling, the equivalent 2-D PPS is typically non-uniform and very sparse in the spatial and temporal baseline plane. This is because the acquisitions from the D-TomoSAR system are obtained by a multiple-pass of SAR systems with nonuniformly spaced orbits, and only a limited number of SARs or channels are used to acquire the data for each pass. As a result, the traditional estimation algorithm for general 2-D PPS is unable to be applied to the improved D-TomoSAR directly.

Considering the above-mentioned, the 2-D product high-order ambiguity function (2-D PHAF) with the RELAX algorithm were proposed to achieve the estimations of the components in the 2-D PPS under the nonuniform and sparse sampled conditions [23]. In this section, the 2-D PHAF with the RELAX algorithm is firstly used to acquire the estimation of the coefficients in (10). Then, two sets of combined SAR acquisitions with different squint angles are adopted to retrieve the 3-D deformation velocities.

3.1. Review of the 2-D PHAF with RELAX Algorithm

For each single component $s^{(k)}(\mu, v), k = 1, 2, \ldots, K$ in (11), its 2-D high-order instantaneous moment (HIM) of second order is defined in [24]:

$$\begin{cases} s_1^{(k)}(\mu, v) = s^{(k)}(\mu, v) \\ s_2^{(k)}(\mu, v; \tau_1, \theta_1) = s_1^{(k)*}(\mu, v) \cdot s_1^{(k)}(\mu + \tau_1, v + \theta_1) \end{cases} \tag{14}$$

where τ_1 and θ_1 are the lags. Then, the 2-D high-order ambiguity function (HAF) is defined as the 2-D Fourier transform of the 2-D HIM, thus, the 2-D HAF of $s_2^{(k)}(\mu, v; \tau_1, \theta_1)$ is:

$$S_2^{(k)}(f, g; \tau_1, \theta_1) = \sum_{\mu=0}^{N-\tau_1} \sum_{v=0}^{M-\theta_1} s_2^{(k)}(\mu, v; \tau_1, \theta_1) \exp(-j2\pi(f\mu + gv)) \tag{15}$$

As mentioned in [24], the second order 2-D HIM of $s^{(k)}(\mu, v)$ has a linear phase for a second order 2-D PPS. Thus, the phase coefficients of the single component 2-D PPS can be estimated by searching the location of the peak of the 2-D HAF. According to (11), (12), (14), and (15), the peak's coordinates of the 2-D HAF of $s_2^{(k)}(\mu, v; \tau_1, \theta_1)$ can be given by:

$$\begin{cases} f^{(k)}(\tau_1, \theta_1) = 2\tau_1 \alpha_{2,0}^{(k)} + \theta_1 \alpha_{1,1}^{(k)} \\ g^{(k)}(\tau_1, \theta_1) = \tau_1 \alpha_{1,1}^{(k)} + 2\theta_1 \alpha_{0,2}^{(k)} \end{cases} \tag{16}$$

Inspecting (16), it is an equations group with two measurements—$f^{(k)}(\tau_1, \theta_1)$ and $g^{(k)}(\tau_1, \theta_1)$—and three unknowns—$\alpha_{2,0}^{(k)}$, $\alpha_{1,1}^{(k)}$, and $\alpha_{0,2}^{(k)}$. Thus, at least two sets of lags are required to estimate these unknown parameters. Once the second order coefficients are estimated, the second order phase term in $s^{(k)}(\mu, v)$ can be removed by a conjugate multiplication. After that, the original 2-D PPS degenerates into a one-order 2-D PPS whose phase coefficients can be achieved by the 2-D Fourier transform.

However, the pervious procedure is only applicable for the single component 2-D PPS. For the multi-component 2-D PPS, the cross terms caused by computing the 2-D HIM significantly affect the estimation results. In this case, the 2-D PHAF can be used to enhance the peaks of the auto terms and weaken the cross terms by multiplying the 2-D HAF with different sets of lags. Choosing L sets of the two lags as $\left(\tau_1^{(l)} = \tau_1, \theta_1^{(l)} = 0\right)$ or $\left(\tau_1^{(l)} = 0, \theta_1^{(l)} = \theta_1\right)$, where $l = 1, 2, \cdots, L$, the second order 2-D PHAF with two lags can be given by:

$$\begin{cases} PHAF_2(f, g, \boldsymbol{\tau}_1, 0) = \prod_{l=1}^{L} S_2\left(\dfrac{\tau_1^{(l)}}{\tau_1^{(1)}} f, \dfrac{\tau_1^{(l)}}{\tau_1^{(1)}} g; \tau_1^{(l)}, 0\right) \\ PHAF_2(f, g, 0, \boldsymbol{\theta}_1) = \prod_{l=1}^{L} S_2\left(\dfrac{\theta_1^{(l)}}{\theta_1^{(1)}} f, \dfrac{\theta_1^{(l)}}{\theta_1^{(1)}} g; 0, \theta_1^{(l)}\right) \end{cases} \tag{17}$$

where $\boldsymbol{\tau}_1 = \left[\tau_1^{(1)}, \cdots, \tau_1^{(L)}\right]$ and $\boldsymbol{\theta}_1 = \left[\theta_1^{(1)}, \cdots, \theta_1^{(L)}\right]$ are the two sets of lags vectors. $\tau_1^{(l)} / \tau_1^{(1)}$ and $\theta_1^{(l)} / \theta_1^{(1)}$ are the scaling factors to align the peaks of the auto terms. Let $\left(f_1^{(k)}, g_1^{(k)}\right)$ and $\left(f_2^{(k)}, g_2^{(k)}\right)$,

$k = 1, 2, \cdots, K$ be the locations of the peaks of $PHAF_2(f, g, \tau_1, 0)$ and $PHAF_2(f, g, 0, \theta_1)$, respectively. The second order phase coefficients of each component in (11) can be estimated as follows:

$$\hat{a}_{2,0}^{(k)} = \frac{f_1^{(k)}}{2\tau_1^{(1)}}, \quad \hat{a}_{1,1}^{(k)} = \frac{g_1^{(k)}}{2\tau_1^{(1)}} + \frac{f_2^{(k)}}{2\theta_1^{(1)}}, \quad \hat{a}_{0,2}^{(k)} = \frac{g_2^{(k)}}{2\theta_1^{(1)}} \tag{18}$$

After estimating the second order coefficients of the kth component, we multiply the phase factor $\exp\left[-j2\pi\left(\hat{a}_{2,0}^{(k)}\mu^2 + \hat{a}_{1,1}^{(k)}\mu v + \hat{a}_{0,2}^{(k)}v^2\right)\right]$ by the original signal in (11), then the kth component becomes a 2-D PPS of order one while the other components remain 2-D PPS of order two. Therefore, the 2-D Fourier transform can be used to estimate the first order coefficients of the kth component. Repeat the above procedure to achieve the estimates of the phase coefficients of all components.

The key problem of the above algorithm is calculating the spectrum of the 2-D HIM for searching the locations of the peaks of 2-D PHAF. However, the operation of (15) requires the uniform sampling of the signals with a sampling rate that satisfies the Nyquist theorem, thus, the original 2-D PHAF algorithm is unable to be applied to the D-TomoSAR model directly. In this context, the RELAX algorithm [23] is preferred to solve the above problem, and it is able to achieve the peak's location of 2-D PHAF in low signal noise ratio (SNR).

3.2. 3-D Deformation Motion Retrieval

Using the 2-D PHAF with the RELAX algorithm, the elevation and deformation parameters of multiple scatterers located in the same slant range-azimuth resolution cell can be estimated, and the deformation parameters are given from (13) as follows:

$$\begin{cases} \omega_{ry} = \cos(\theta_{sq})v_{k,r} - \sin(\theta_{sq})v_{k,y} \\ \omega_s = v_{k,s} \end{cases} \tag{19}$$

where:

$$\omega_{ry} = \frac{\lambda a_{0,1}^{(k)}}{2} + \frac{\lambda^2 r a_{1,0}^{(k)} a_{1,1}^{(k)}}{4} \tag{20}$$

For one set of combined acquisitions with a non-zero squint angle, we have one measurement ω_{ry} and two unknowns $v_{k,r}$ and $v_{k,y}$. To retrieve these two unknown deformation components, at least two sets of combined SAR data with different squint angles are required. However, it should be noted that the slant range and azimuth direction are related to the SAR imaging geometry defined by the satellite heading and antenna incidence angle. Thus, the master images used in the two sets of combined data must be acquired by the same imaging geometry. This can be achieved by correcting the imaging geometry of master images into a reference geometry parameter through the registration operation. Assuming that the squint angles used in the two sets of combined acquisitions are θ_{sq1} and θ_{sq2}, respectively, (19) can be rewritten in the matrix form as:

$$\mathbf{Y} = \mathbf{H}\mathbf{V}_{rys} \tag{21}$$

where $\mathbf{Y} = \left[\omega_{ry1}, \omega_{ry2}, \omega_{s1} \text{ or } \omega_{s2}\right]^T$, $\mathbf{H} = \begin{bmatrix} \cos(\theta_{sq1}) & -\sin(\theta_{sq1}) & 0 \\ \cos(\theta_{sq2}) & -\sin(\theta_{sq2}) & 0 \\ 0 & 0 & 1 \end{bmatrix}$, and $\mathbf{V}_{rys} = \left[v_{k,r}, v_{k,y}, v_{k,s}\right]^T$.

Once the scatterers' 3-D deformation components along the slant range, azimuth, and elevation directions are estimated, the observed 3-D deformation components can be mapped to the original 3-D deformation parameters of v_N, v_E, and v_U in North-South, East-West, and Up-Down directions. Assuming that the antenna incidence angle is α and the satellite heading angle is β, as shown in Figure 3, the explicit projection relationship between the observed 3-D deformation parameters $\left(v_{k,r}, v_{k,y}, v_{k,s}\right)$ and the original deformation components can be written as:

$$\begin{cases} v_r = -v_U \cos(\alpha) + v_E \sin(\alpha) \cos(\beta) - v_N \sin(\alpha) \sin(\beta) \\ v_y = v_E \sin(\beta) + v_N \cos(\beta) \\ v_s = v_U \sin(\alpha) + v_E \cos(\alpha) \cos(\beta) - v_N \cos(\alpha) \sin(\beta) \end{cases} \tag{22}$$

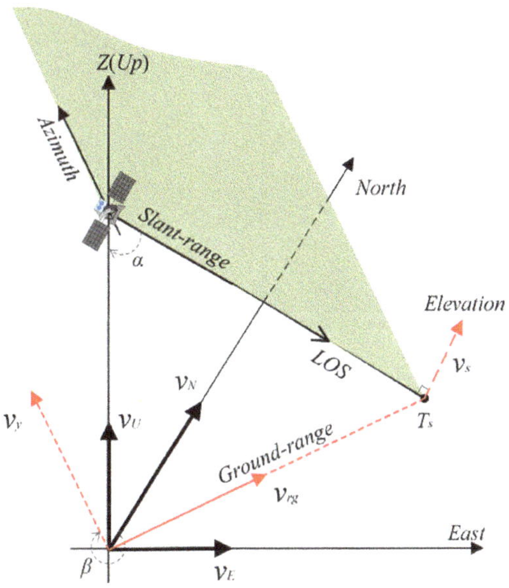

Figure 3. Diagram for the original three dimensional (3-D) deformation components v_N, v_E, and v_U in North-South, East-West, and Up-Down directions, and the displacement components v_{rg}, v_y, and v_s in ground-range, azimuth, and elevation directions.

Similarly, (22) can be rewritten in the matrix form as:

$$\mathbf{V}_{rys} = \mathbf{I}\mathbf{V}_{UEN} \tag{23}$$

where $\mathbf{V}_{NEU} = [v_U, v_E, v_N]^T$, and $\mathbf{I} = \begin{bmatrix} -\cos(\alpha) & \sin(\alpha)\cos(\beta) & -\sin(\alpha)\sin(\beta) \\ 0 & \sin(\beta) & \cos(\beta) \\ \sin(\alpha) & \cos(\alpha)\cos(\beta) & -\cos(\alpha)\sin(\beta) \end{bmatrix}$.

Combining (21) with (23), we have:

$$\mathbf{Y} = \mathbf{HIV}_{UEN} \tag{24}$$

Subsequently, the original 3-D deformation components in North-South, East-West, and Up-Down directions can be retrieved by weighted least squares as follows:

$$\mathbf{V}_{UEN} = \left[(\mathbf{HI})^T Q_{rsy}^{-1}(\mathbf{HI})\right]^{-1} (\mathbf{HI})^T Q_{rsy}^{-1} Y \tag{25}$$

where Q_{rys} is the covariance matrix caused by the residual atmospheric phase error of the measurements. Assuming the deformation parameters ω_{ry1} and ω_{ry2} acquired by the two sets of combined SAR data are independent, the non-diagonal elements in Q_{rys} are zeros, and Q_{rys} is only determined by its diagonal elements, i.e.,:

$$Q_{rys} = diag\left(\sigma^2_{\tilde{\omega}_{ry1}}, \sigma^2_{\tilde{\omega}_{ry2}}, \sigma^2_{\tilde{\omega}_{s1}} \text{ or } \sigma^2_{\tilde{\omega}_{s2}}\right) \tag{26}$$

where $\sigma^2_{\tilde{\omega}_{ry1}}, \sigma^2_{\tilde{\omega}_{ry2}}, \sigma^2_{\tilde{\omega}_{s1}}$ or $\sigma^2_{\tilde{\omega}_{s2}}$ are the estimation variance of the deformation parameters obtained by (19), respectively.

The results in (25) show that the accuracy of the 3-D deformation retrieval depends on the imaging geometry of the combined data. Here, we introduce the concept of position dilution of precision (PDOP) in the global navigation satellite system (GNSS) to assess the accuracy of the 3-D deformation retrieval [8,25]. According to the definition of PDOP, the estimation covariance matrix of 3-D deformation retrieval is:

$$Q_{UEN} = \left[(\mathbf{HI})^T Q_{rys}^{-1}(\mathbf{HI})\right]^{-1} = \begin{bmatrix} \sigma^2_U & \sigma_{UE} & \sigma_{UN} \\ \sigma_{EU} & \sigma^2_E & \sigma_{EN} \\ \sigma_{NU} & \sigma_{NE} & \sigma^2_N \end{bmatrix} \tag{27}$$

The matrix Q_{UEN} is a square symmetric matrix, and its diagonal elements denote the estimation error variance for the 3-D deformation components. The smaller the value the diagonal element is, the higher the precision of the estimation result for the corresponding deformation component will be.

3.3. Performance of 3-D Deformation Estimation

In this sub-section, a comparison of 3-D deformation retrieval performance between the proposed algorithm in this paper and the traditional motion decomposition method in [8] is carried out. In [8], the 3-D deformation components are estimated by decomposing the multiple LOS displacement observed from different viewing geometries. As shown in Figure 3, the relation between LOS deformation measurement v_{LOS} observed by the traditional D-TomoSAR and the 3-D displacement components can be written as:

$$v_{LOS} = -v_U \cos(\alpha) + v_E \sin(\alpha)\cos(\beta) - v_N \sin(\alpha)\sin(\beta) \tag{28}$$

The results in (28) show that at least three LOS measurements from different acquisition geometries are required to achieve the three deformation components v_U, v_E, and v_N. Assuming $\mathbf{b}_{m \times 1}$ is the deformation vector contains m ($m \geq 3$) LOS observations, $\mathbf{X}_{3 \times 1}$ is the 3-D deformation vector consisting of the three deformation components in North-South, East-West, and Up-Down direction, and $\mathbf{\Phi}_{m \times 3}$ is the coefficient matrix corresponding to the projection vectors. As a result, the relationship between the multiple LOS measurements and the 3-D deformation components can be expressed as:

$$\mathbf{b}_{m \times 1} = \mathbf{\Phi}_{m \times 3}\mathbf{X}_{3 \times 1} \tag{29}$$

Accordingly, the estimation covariance matrix of 3-D deformation retrieval from (29) can be written as:

$$Q'_{UEN} = \left[\mathbf{\Phi}^T_{m \times 3}Q_{Los}^{-1}\mathbf{\Phi}_{m \times 3}\right]^{-1} = \begin{bmatrix} \sigma^2_U & \sigma_{UE} & \sigma_{UN} \\ \sigma_{EU} & \sigma^2_E & \sigma_{EN} \\ \sigma_{NU} & \sigma_{NE} & \sigma^2_N \end{bmatrix} \tag{30}$$

where Q_{LOS} represents the covariance matrix of the LOS measurements.

To analyze the performance of the proposed and the traditional methods objectively, three cases of geometric configurations of acquisitions were considered. For the first two cases, three stacks of combined data were adopted, and the acquisitions were all obtained from stripmap mode of SAR with a squint angle of zero. In Case I, the first two stacks were both acquired from the ascending orbit, while the third stack was acquired from the descending orbit. In Case II, we changed the satellite heading angle of the third stack to keep it away from the near-polar orbit to evaluate the effect of the heading angle for the 3-D deformation retrieval. Finally, the 3-D deformation parameters were retrieved by the traditional method from three LOS deformation measurements, as shown in (28). For comparisons, in Case III, we considered two stacks of combined data with different squint angles to retrieve the 3-D deformation parameters with the proposed method.

It should be noted that the estimation covariance matrix (27) and (30) are related to the measurement covariance matrix Q_{rys} and Q_{LOS}, which are dependent on the SNR of acquisitions and the variance of the temporal baseline distribution. In order to evaluate the effect of variation of geometry for the 3-D deformation retrieval, we set the elements in the matrix Q_{rys} and Q_{LOS} to be $\sigma^2 = 1$ cm/year [8]. The geometric configurations of combined data are detailed in Table 1. Accordingly, the standard deviations of the estimation of 3-D deformation components in these three cases can be calculated by (27) and (30), respectively, as shown in Table 1.

Table 1. Acquisition parameters of each scenarios and the performance of 3-D retrieval.

Case	Heading Angle (deg)	Incidence Angle (deg)	Squint Angle (deg)	Standard Deviations of 3-D Retrieval (cm/year)
I	Ascending: 350 Ascending: 352 Descending: 187	Right-looking: 40 Right-looking: 51 Right-looking: 37	0 0 0	σ_U : 2.183 σ_E : 1.701 σ_N: 18.282
II	Ascending: 350 Ascending: 352 250	Right-looking: 40 Right-looking: 51 Right-looking: 37	0 0 0	σ_U : 0.615 σ_E : 0.452 σ_N: 1.049
III	Ascending: 350 Ascending: 350	Right-looking: 40 Right-looking: 40	5 20	σ_U : 0.123 σ_E : 0.112 σ_N: 0.534

Results of Case I in Table 1 show that the accuracy of the deformation component in North-South direction was far lower than that of the other two directions due to the influence of the near-polar orbits. Therefore, the traditional 3-D deformation retrieval mode cannot reliably estimate the North-South motion parameter. Compared with Case I, the third set of data in Case II was acquired from the non-polar orbit. The experimental results show that the accuracy of the North-South deformation component could be improved by changing the satellite heading angle to make it deviate from the near-polar orbit. In Case III, two sets of combined data with different squint angles were adopted to retrieve the 3-D deformation components by using the proposed method. We can draw a conclusion that the existence of squint angles of combined data can improve the sensitivity to the North-South component retrieval and achieve a comparable effect as changing the satellite heading angle from the experimental results of Case III.

Although changing the satellite heading angle can improve the accuracy of the North-South deformation component, the satellite with non-polar orbit limits the latitudinal imaging coverage, as shown in the results of Cases II and III. On the contrary, the proposed method only uses the squint imaging mode of SAR with no changes to the existing satellite orbit design, which provides an effective way to enhance the North-South deformation component. To further verify the effectiveness of the proposed method, a detailed comparison analysis between Case II and Case III is presented. Firstly, we considered a scenario with three stacks of acquisitions. The geometric configurations of first two stacks were same as Case II, while the third stack had a varying satellite heading angle in the range of 187 degrees to 260 degrees. Figure 4a depicts the 3-D deformation precision as a function of the satellite heading angle. The results indicate that the accuracy of the North-South deformation component could be significantly improved by changing the heading angle of the satellite to make it deviate from the near-polar orbit. Secondly, we evaluated the effect of squinted acquisitions by using the geometric configurations in Case III, where the squint angle of first stack was fixed to 5 degrees, and another one changed from 6 degrees to 40 degrees. In this case, the 3-D deformation precisions are shown in Figure 4b. Comparing Figure 4a with Figure 4b, we find that a squint angle yields an effect comparable to changing the satellite heading angle.

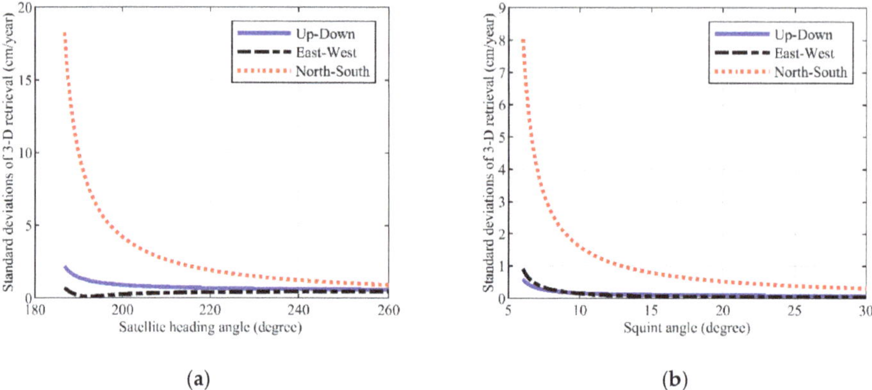

Figure 4. Impact of changing the satellite heading angle and using the acquisitions with different squint angles for the 3-D deformation retrieval. (**a**) and (**b**) are the diagrams of the standard deviations of 3-D retrieval varying with the satellite heading angle and squint angle, respectively.

3.4. Some Considerations

3.4.1. Resolution Analysis

Here, the analysis of the resolution for the proposed algorithm is presented. According to [26], the resolution of the elevation estimation of a scatterer is determined by the overall orthogonal baseline length. Therefore, the Rayleigh resolution of the elevation estimation is:

$$\rho_s = \frac{\lambda r}{2B_\perp} \tag{31}$$

where $B_\perp$ is the overall orthogonal baseline length.

Similarly, the resolution of the 3-D deformation velocity is determined by the span of time baseline, and the resolution of 3-D velocity along the slant range, azimuth, and elevation directions are expressed as follows:

$$\begin{cases} \rho_{v_r} = \dfrac{\lambda}{2T\cos\left(\theta_{sq}\right)} \\ \rho_{v_y} = \dfrac{\lambda}{2T\sin\left(\theta_{sq}\right)} \\ \rho_{v_s} = \dfrac{\lambda r}{2B_\perp T} \end{cases} \tag{32}$$

where T is the span of time baseline.

It can be seen that the resolution of deformation velocity in slant range and azimuth directions are related to the squint angle of the SAR imaging.

3.4.2. Discussion about the Nonlinear Deformation

In (7), a linear deformation model of scatterers was assumed in this paper. However, in reality, the deformation can be nonlinear. Therefore, it is necessary to analyze the performance of the proposed algorithm in the nonlinear deformation case.

Generally, the most common types of nonlinear motion of scatterers are accelerating motion and periodic motion [27]. Among them, the accelerating motion may be caused by the over-exploitation of groundwater or minerals, while the thermal dilation effect of scatterers caused by the seasonal temperature variations usually presents as a periodic movement characteristic. Therefore, the following analyses are divided into two cases: (a) the deformation of the scatterer contains linear and accelerating motions; (b) the deformation of the scatterer contains linear and periodic motions. In addition, we also give a brief analysis of the mixed deformation, which contains both of the above kinds of motions.

(a) Deformation contains linear and accelerating motion:

In this case, the 3-D deformation model described in (7) can be rewritten as:

$$\Delta r = v_r t_n + \frac{1}{2} a_r t_n^2, \ \Delta y = v_y t_n + \frac{1}{2} a_y t_n^2, \ \Delta s = v_s t_n + \frac{1}{2} a_s t_n^2 \tag{33}$$

where a_r, a_y, and a_s are the deformation accelerations of the scatterers along the slant range, azimuth, and elevation directions, respectively.

Substituting (33) into (6) and using the discretization operation, we have:

$$
\begin{aligned}
g_{mn}&\left(B_{mn,s}, t_n\right)\\
&= \sum_{k=1}^{K} \gamma\left(s_k, v_{k,r}, v_{k,y}, v_{k,s}\right) \exp\left\{ j \frac{4\pi}{\lambda} \frac{s_k}{r} B_{mn,s} \right\} \exp\left\{ j \frac{4\pi}{\lambda} \frac{v_{k,s}}{r} B_{mn,s} t_n \right\} \exp\left\{ j \frac{2\pi}{\lambda} \frac{a_{k,s}}{r} B_{mn,s} t_n^2 \right\}\\
&\quad \times \exp\left\{ j \frac{4\pi}{\lambda} \left[\cos(\theta_{sq}) v_{k,r} - \sin(\theta_{sq}) v_{k,y} - \frac{s_k v_{k,s}}{r} \right] t_n \right\} \exp\left\{ -j \frac{2\pi}{\lambda} \left(\frac{v_{k,r}^2 + v_{k,y}^2 + v_{k,s}^2}{r} \right) t_n^2 \right\}\\
&\quad \times \exp\left\{ j \frac{2\pi}{\lambda} \left[\cos(\theta_{sq}) a_{k,r} - \sin(\theta_{sq}) a_{k,y} - \frac{s_k a_{k,s}}{r} \right] t_n^2 \right\} \exp\left\{ -j \frac{\pi}{2\lambda} \left(\frac{a_{k,r}^2 + a_{k,y}^2 + a_{k,s}^2}{r} \right) t_n^4 \right\} + e_{mn}\\
&\quad m = 1, 2, \cdots M, n = 1, 2, \cdots N
\end{aligned}
\tag{34}
$$

Similarly, inspecting the phase terms in (34), the signal model can be regarded as a multi-component fourth order 2-D PPS. Accordingly, the relationship between the phase coefficients in (34) and those of the general fourth order 2-D PPS can be expressed as:

$$
\begin{cases}
A_k = \gamma\left(s_k, v_{k,r}, v_{k,y}, v_{k,s}\right), \mu = B_{mn,s}, v = t_n\\
\alpha_{0,0}^{(k)} = 0, \alpha_{2,0}^{(k)} = 0\\
\alpha_{1,0}^{(k)} = \frac{2 s_k}{\lambda r}, \alpha_{1,1}^{(k)} = \frac{2 v_{k,s}}{\lambda r}, \alpha_{1,2}^{(k)} = \frac{a_{k,s}}{\lambda r}\\
\alpha_{0,1}^{(k)} = \frac{2 \cos(\theta_{sq}) v_{k,r} - 2 \sin(\theta_{sq}) v_{k,y}}{\lambda} - \frac{2 s_k v_{k,s}}{\lambda r}\\
\alpha_{0,2}^{(k)} = -\frac{v_{k,r}^2 + v_{k,y}^2 + v_{k,s}^2}{\lambda r} + \frac{\cos(\theta_{sq}) a_{k,r} - \sin(\theta_{sq}) a_{k,y}}{\lambda} - \frac{s_k a_{k,s}}{\lambda r}\\
\alpha_{0,4}^{(k)} = -\frac{a_{k,r}^2 + a_{k,y}^2 + a_{k,s}^2}{4 \lambda r}
\end{cases}
\tag{35}
$$

Since the original 2-D PHAF algorithm is suitable for solving the high-order 2-D PPS, the phase coefficients in (34) can still be estimated by using the proposed algorithm in this paper. Once the estimations of phase coefficients in (34) are achieved, the elevation, 3-D deformation velocity and 3-D deformation acceleration of scatterer can be estimated from the following equations:

$$
\begin{cases}
s_k = \lambda r \alpha_{1,0}^{(k)} / 2, \ v_{k,s} = \lambda r \alpha_{1,1}^{(k)} / 2\\
\cos(\theta_{sq}) v_{k,r} - \sin(\theta_{sq}) v_{k,y} = \frac{\lambda \alpha_{0,1}^{(k)}}{2} + \frac{s_k v_{k,s}}{r}
\end{cases}
\tag{36}
$$

$$
\begin{cases}
a_{k,s} = \lambda r \alpha_{1,2}^{(k)}\\
\cos(\theta_{sq}) a_{k,r} - \sin(\theta_{sq}) a_{k,y} = \lambda \alpha_{0,2}^{(k)} + \frac{v_{k,r}^2 + v_{k,y}^2 + v_{k,s}^2}{r} + \frac{s_k a_{k,s}}{r}
\end{cases}
\tag{37}
$$

According to (36) and (37), at least two sets of SAR acquisition with different squint angles are required to solve the 3-D deformation velocities and accelerations of scatterers. However, it should be noted that in the process of phase coefficients estimation using the PHAF-based algorithm, the phase differentiation technique is employed to reduce the order of the PPS. Thus, the estimation errors of the highest-order coefficient affect the estimation of low-order coefficients, which is a so-called error propagation phenomena. As a result, it is necessary to improve the SNR of the SAR data to retrieve the 3-D deformation parameters of scatterers accurately.

(b) Deformation contains linear and periodic motion:

Previous research has shown that with the decrease of wavelength, SAR sensors are more sensitive to small surface displacements. Especially in urban areas, buildings with steel structures such as roofs, bridges, and tunnels not only have linear deformation displacement but are also affected by the nonlinear seasonal deformation caused by the thermal dilation effects [28].

Generally, there are two methods to retrieving the nonlinear seasonal deformation component. One is to use the temperature distribution of the monitoring area at the imaging time to form a synthetic aperture to carry out the five-dimensional (5-D) imaging. However, this method is not an optimal strategy because sometimes the temperature data of the monitoring area at the acquisition instants are difficult obtain, and the real temperature corresponding to the different structures also depends on the solar irradiation and the materials of the area. An alternative method is to use a sinusoidal function to simulate the thermal dilation effects caused by the seasonal temperature variations [27,29,30], which was proven to be effective under the condition of missing the temperature of the monitoring area. Therefore, in this part of the analysis, the periodic motion of the scatterer is described by a sinusoidal function rather than the former method. In this case, the 3-D deformation model described in (7) can be rewritten as:

$$\Delta r = v_r t_n + \beta_r \sin(2\pi f t_n), \; \Delta y = v_y t_n + \beta_y \sin(2\pi f t_n), \; \Delta s = v_s t_n + \beta_s \sin(2\pi f t_n) \tag{38}$$

where β_r, β_y, and β_s are the amplitude of the periodic motion of the scatterers along the slant range, azimuth, and elevation directions, respectively. f is the seasonal frequency. It should be noted that the period of the seasonal temperature variations is one year, thus, $f = 1$.

Similarly, substituting (38) into (6) and using the discretization operation, we have:

$$
\begin{aligned}
g_{mn}&(B_{mn,s}, t_n) \\
&= \sum_{k=1}^{K} \gamma\left(s_k, v_{k,r}, v_{k,y}, v_{k,s}\right) \exp\left\{ j\tfrac{4\pi}{\lambda}\tfrac{s_k}{r} B_{mn,s} \right\} \exp\left\{ j\tfrac{4\pi}{\lambda}\tfrac{v_{k,s}}{r} B_{mn,s} t_n \right\} \exp\left\{ j\tfrac{4\pi}{\lambda}\tfrac{\beta_s}{r} B_{mn,s} \sin(2\pi t_n) \right\} \\
&\times \exp\left\{ j\tfrac{4\pi}{\lambda}\left[\cos(\theta_{sq}) v_{k,r} - \sin(\theta_{sq}) v_{k,y} - \tfrac{s_k v_{k,s}}{r} \right] t_n \right\} \exp\left\{ -j\tfrac{2\pi}{\lambda}\left(\tfrac{v_{k,r}^2 + v_{k,y}^2 + v_{k,s}^2}{r} \right) t_n^2 \right\} \\
&\times \exp\left\{ j\tfrac{4\pi}{\lambda}\left[\cos(\theta_{sq}) \beta_r - \sin(\theta_{sq}) \beta_y - \tfrac{s_k \beta_s}{r} \right] \sin(2\pi t_n) \right\} \\
&\times \exp\left\{ -j\tfrac{2\pi}{\lambda}\left(\tfrac{\beta_{k,r}^2 + \beta_{k,y}^2 + \beta_{k,s}^2}{r} \right) \sin^2(2\pi t_n) \right\} + e_{mn} \\
m &= 1, 2, \cdots M, n = 1, 2, \cdots N
\end{aligned}
\tag{39}
$$

The signal shown in (39) is a multi-component 2-D hybrid sinusoidal frequency modulated (FM) and PPS (2-D hybrid sinusoidal FM-PPS), while the algorithm proposed in this paper is only suitable for the pure 2-D PPS. Therefore, in this case, the amplitude of the periodic motion of the scatterer is unable to be estimated. Moreover, owing to the existence of the periodic motion of the scatterer, the estimation error of linear deformation is increased.

In conclusion, for the scatterer with linear and accelerating motion, the proposed algorithm can be used to estimate the 3-D velocity and 3-D acceleration of the deformation. However, in order to achieve the accurate estimation results, a high SNR of the SAR acquisitions is required. For the scatterer with linear and periodic motion, only the linear deformation components with large estimation errors can be obtained by using the proposed algorithm, and the amplitude of the periodic motion fails to be achieved. Furthermore, when the scatterer contains the above three kinds of deformation, the estimation errors of 3-D deformation velocities and 3-D accelerations are further increased.

4. 3-D Deformation Retrieval Simulation

In this section, simulation experiments are carried out to verify the effectiveness of the proposed model and algorithm. In the following experiments, two sets of combined acquisitions were acquired to retrieve the 3-D deformation by using the pursuit monostatic mode of a bistatic SAR system through repeat-passes [31]. In order to be more realistic, the system parameters of TanDEM-X are introduced here for the simulation of bistatic SAR system in this paper, and the system parameters for each SAR sensor are shown in Table 2. The experiments in this section were composed of two parts; the first part was the numerical simulation experiment for point targets, and the second one was the validation performance for the scene target using the semi-real data.

Table 2. Simulation parameters of SAR systems for 3-D deformation retrieval.

Description	Satellite 1	Satellite 2
a: Semimajor axis (km)	6870.14	6870.14
i: Inclination (deg)	97.42	97.42
e: Eccentricity	0.003	0.003
ω: Argument of perigee (deg)	60	59.2833
Ω: Ascending node (deg)	-60	-59.9995
M: Mean anomaly (deg)	254.1	253.4
Beam direction	Right-look	Right-look
Incidence angle (deg)	40	40
Squint angle (deg)	5	21.3
Heading angle (deg)	350	350
Carrier frequency (GHz)	9.65	
Pulse duration (µs)	2	
Bandwidth (MHz)	100	
Sampling frequency (MHz)	110	

4.1. Numerical Simulation for Point Target

In this part, the point targets simulation was performed to verify the effectiveness of the proposed improved D-TomoSAR model, and the accuracy of the 3-D deformation retrieval was analyzed. To this end, we assumed that there were a total of three scatterers located in a same slant range-azimuth resolution cell along the elevation direction. The scatterers' elevations and the 3-D deformation velocities in East-West, North-South, and Up-Down directions are listed in Table 3. Accordingly, the corresponding scatterers' 3-D deformation velocities in slant range, azimuth, and elevation directions under the case of squint imaging mode can be calculated, which is also listed in Table 3. The Gaussian random noise with a mean value of zero and a standard deviation of 1 cm/year was added to the 3-D deformation velocities of each scatterer for realistic purpose. Previous results show that the combined data for the 3-D deformation retrieval needed to be acquired by the 2-D imaging of SAR. However, due to the existence of the squint angle, the conventional focusing algorithm for the side-looking SAR could not be applied to the squint mode SAR imaging directly. At this point, the algorithm in [32] was adopted to achieve the 2-D SAR imaging. This algorithm could still provide a stable 2-D focusing performance with a squint angle of 65 degrees, which meets the requirement of the 2-D imaging with squint mode in this paper. To approach the real imaging environment, the signal received by the SAR system was added by Gaussian noise with SNR = 5 dB in this experiment.

Table 3. Elevation and deformation parameters of scatterers.

Scatterer	Elevation (m)	3-D Deformation (v_U, v_E, v_N) (cm/year)	3-D deformation (v_r, v_y, v_s) (cm/year)
1	-40	$(2.647, -0.454, 2.208)$	$(-2.3, 1.4, -2.2)$
2	-15	$(2.051, -3.639, 0.313)$	$(-1.5, -2.1, -3.3)$
3	25	$(-1.425, -0.711, -2.667)$	$(1.2, -2.5, 1.4)$

To retrieve the 3-D deformation parameters of scatterers, two SAR sensors in Table 2 were used through 30 repeat-passes to achieve two stacks of combined SAR acquisitions with different squint angles. Subsequently, the scatterers' elevation s and the deformation parameters ω_{ry} and ω_s in (19) could be estimated by the 2-D PHAF with the RELAX algorithm for each combined SAR acquisition. The estimated results are shown in Figure 5a,b. Finally, the estimation of 3-D deformation components in the slant range, azimuth, and elevation directions could be obtained by using the weighted least squares method to solve (21), and the retrieved results are summarized in Table 4. The experimental results show that the estimations of the 3-D deformation in three directions were very close to the real values. The estimation error was less than 0.5 cm/year in the slant range and the elevation direction and was no more than 1 cm/year in the azimuth direction. Although the accuracy of deformation estimation in azimuth direction was inferior to the other two directions, the proposed method still achieved a great improvement in accuracy of retrieval for the azimuth direction deformation compared with the traditional method. This proves the effectiveness of the proposed improved D-TomoSAR model, which provides a feasible solution to the realization of estimations of the 3-D deformation. Furthermore, once the 3-D deformation components along the slant range, azimuth, and elevation direction were estimated, the corresponding deformation parameters in North-South, East-West, and Up-Down directions could be calculated by (25), as shown in Table 5.

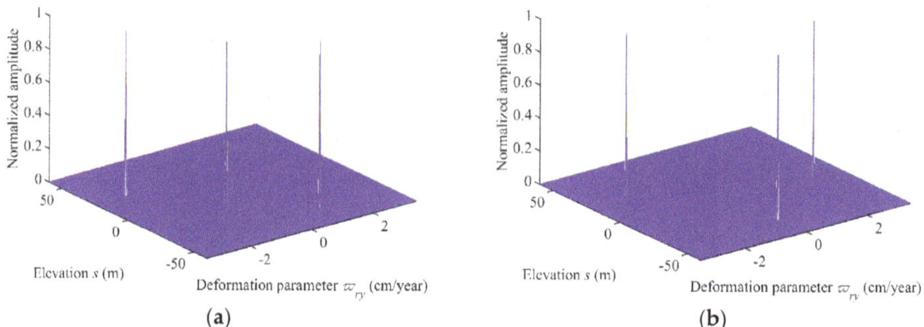

Figure 5. Estimation of elevation and deformation parameters of two stacks of combined acquisitions. (**a**) The result of first stack with a squint angle of 5 degrees. (**b**) The result of second stack with a squint angle of 21.3 degrees.

Table 4. Estimation of elevation and 3-D deformation velocity of three scatterers in slant range, azimuth, and elevation directions.

Scatterer	Estimation Value		Estimation Error	
	Elevation (m)	3-D Deformation (v_r, v_y, v_s) (cm/year)	Elevation (m)	3-D Deformation (v_r, v_y, v_s) (cm/year)
1	-40.05	$(-1.83, 2.03, -1.91)$	0.05	$(-0.47, -0.63, -0.29)$
2	-15.15	$(-1.12, -1.43, -3.73)$	0.15	$(-0.38, -0.67, 0.43)$
3	25.18	$(1.52, -1.61, 1.06)$	-0.18	$(-0.32, -0.89, 0.34)$

Table 5. Estimation of elevation and 3-D deformation velocity of three scatterers in East-West, North-South, and Up-Down directions.

Scatterer	Estimation Value	Estimation Error
	3-D Deformation (v_U, v_E, v_N) (cm/year)	3-D Deformation (v_U, v_E, v_N) (cm/year)
1	$(2.134, 0.107, 2.559)$	$(0.513, -0.561, -0.351)$
2	$(1.751, -3.584, 1.141)$	$(0.299, -0.056, -0.827)$
3	$(-1.681, -0.437, -1.735)$	$(0.256, -0.273, -0.933)$

Furthermore, in order to illustrate the advantages of the proposed algorithm, the above estimation results were compared with the motion decomposition method [8]. In the following comparative simulation, three sets of SAR acquisitions were used to retrieve the 3-D deformation components. The parameters of SAR satellites shown in Table 1 of [8] were adopted for this experiment, as shown in Table 6.

Table 6. Parameters of each satellite in [8].

Satellite	Incidence Angle (deg)	Heading Angle (deg)	Track Type
1	41.9	350.3	Ascending
2	51.1	352	Ascending
3	36.1	190.6	Descending

First, three sets of SAR acquisitions for the D-TomoSAR processing were obtained by the three satellites in Table 6 through 30 repeat-passes. Then, the sparse reconstruction algorithm was used to estimate the LOS deformation velocities for each set of SAR acquisitions. As a result, three LOS deformation observations from different acquisition geometries were obtained, and the reconstructed elevations and deformation velocities for the three sets of SAR acquisitions are shown in Figure 6. The estimations of deformation velocity along LOS are listed in Table 7. Subsequently, the L1-norm minimization algorithm in [8] was used to decompose the LOS observations to achieve the 3-D deformation components, and the estimated results are summarized in Table 8.

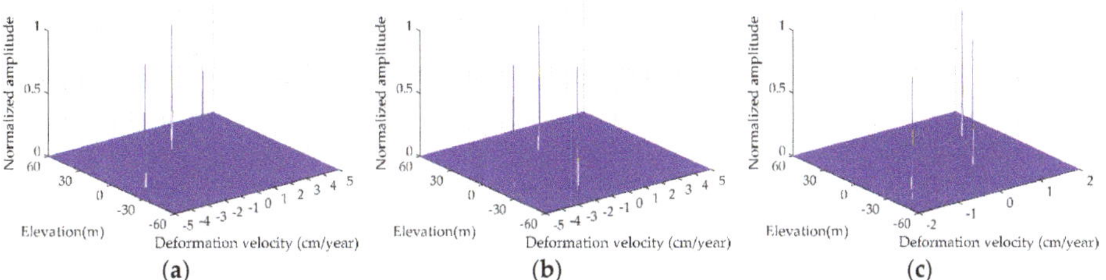

Figure 6. Estimation of elevation and deformation parameters along line of sight (LOS). (a–c) correspond to the estimated results of Satellite 1, Satellite 2, and Satellite 3, respectively.

Table 7. The results of the LOS deformation estimations.

Satellite	Satellite 1	Satellite 2	Satellite 3
	Estimations of LOS Deformation (cm/year)		
1	−2.119	−1.684	−1.539
2	−4.012	−4.059	0.494
3	0.494	0.159	1.178

Table 8. Results of 3-D deformation estimations using the motion decomposition [8].

Scatterer	Estimation Value	Estimation Error
	3-D Deformation (v_U, v_E, v_N) (cm/year)	3-D Deformation (v_U, v_E, v_N) (cm/year)
1	(5.273, −0.809, 20.787)	(−2.626, 0.356, −18.578)
2	(3.773, −3.876, 11.978)	(−1.723, 0.236, −11.665)
3	(−3.033, −0.351, −13.619)	(1.608, −0.359, 10.952)

As can be seen from the comparison of Tables 5 and 8, the flight directions of the satellites were almost parallel to the North-South direction owing to the SAR satellites operating in the near-polar orbit. Thus, the method of motion decomposition was insensitive to the North-South deformation retrieval, which led to a large estimation error. As mentioned by the authors of [8], precise unambiguous retrieval of the North-South component is not possible when only using the geometry configuration of current SAR satellites. The experimental results of the above simulation also draw the same conclusion. Therefore, the feasibility of the proposed algorithm in this paper for retrieving the 3-D deformation components was further verified by the analysis of the compared experiment.

According to Table 4, the estimation errors were not very small, especially for v_y. Nevertheless, the accuracy of the estimation in the azimuth (North-South) direction also greatly improved compared with the motion decomposition method [8] shown in Table 8. In addition, compared with the decomposition method, the proposed algorithm only needed two sets of SAR acquisitions with different oblique angles without changing the orbit of the SAR satellite, which is conducive to practical application. On the other hand, as can be seen from Table 1 and Figure 4, the higher diversity of the squint angles between the two sets of SAR acquisitions, the more precise the deformation estimation in North-South was. In the above simulation, the estimated results of Table 4 were obtained by using the SAR acquisitions with squint angles of 5 degrees and 21.3 degrees, respectively. In order to further improve the accuracy of the deformation estimation in the azimuth direction, it is necessary to increase the diversity of the squint angles between the two sets of SAR acquisitions. To illustrate this point, an additional experiment was performed. The parameters used in this experiment were similar to those in Table 2, except that the squint angle of Satellite 2 increased from 21.3 degrees to 45 degrees. Then, the same simulation scenario was adopted, and the estimation results of the 3-D deformation in the slant range, azimuth, and elevation directions are shown in Table 9. The experimental results show that the accuracy of the deformation estimation in the azimuth direction improved with the increase in the diversity of the squint angles, which verifies the correctness of the above conclusions.

Table 9. Estimation of 3-D deformation velocity of three scatterers in large diversity of squint angles between two sets of SAR acquisitions.

Scatterer	Estimation Value	Estimation Error
	3-D Deformation (v_r, v_y, v_s) (cm/year)	3-D Deformation (v_r, v_y, v_s) (cm/year)
1	$(-2.176, 1.612, -2.108)$	$(-0.124, -0.212, -0.092)$
2	$-1.409, -2.257, -3.248)$	$(-0.091, 0.157, -0.052)$
3	$(1.128, -2.327, 1.464)$	$(0.072, -0.173, -0.064)$

In addition, we set the variation of the SNR of SAR imaging in the range of $[-10\ \text{dB}, 15\ \text{dB}]$ to evaluate the effect of noise on the elevation and the 3-D deformation retrieval. For each SNR, 250 simulations were performed. The parameters of scatterers are shown in Table 2. Figure 7 presents the three scatterers' average estimation errors of elevation and 3-D deformation velocities as a function of the SAR imaging SNR, showing the performance improvement of the estimation when increasing the SNR of SAR imaging. For SNR = 5dB, the average error of elevation estimation was less than 0.4 m, and the error of deformation velocity estimations were no more than 0.5 cm/year in the slant range and elevation directions. The error of deformation estimation in the azimuth direction was larger than that in the other two directions due to the inadequate angular diversity of the squint angles used in the two combined acquisitions. Table 10 summarizes the elevation and 3-D deformation retrieval in the different SNRs. The experimental results show that the elevation and 3-D deformation velocities could be still estimated accurately and robustly by the proposed algorithm at a low SNR.

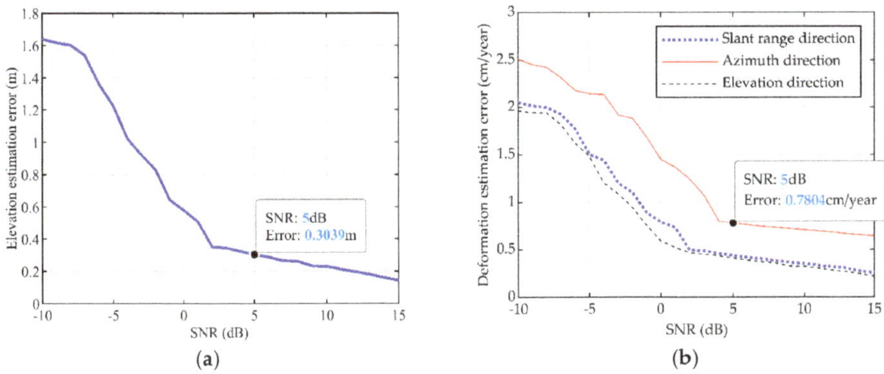

(a)　　　　　　　　　　　　　　　　**(b)**

Figure 7. Average estimation errors of elevation and 3-D deformation velocities. (**a**) Average estimation error of elevation. (**b**) Average estimation errors of 3-D deformation velocities in slant range, azimuth, and elevation directions.

Table 10. Elevation and 3-D deformation retrieval errors in different signal noise ratios (SNRs).

SNR (dB)	−10	−5	0	5	10	15
Elevation error (m)	1.641	1.224	0.580	0.304	0.231	0.144
Deformation error in range (cm/year)	2.055	1.508	0.796	0.439	0.353	0.256
Deformation error in azimuth (cm/year)	2.509	2.145	1.455	0.780	0.707	0.643
Deformation error in elevation (cm/year)	1.964	1.486	0.592	0.411	0.324	0.218

4.2. Experiment with Semi-Real Data

In this part, an experiment was performed to verify the effectiveness of the proposed method for the scene target by using the semi-real data. In this experiment, we used Giorgio Franceschetti's method [33] to generate the SAR raw data. The digital elevation model (DEM) data provided by Shuttle Radar Topography Mission (SRTM) were used as the terrain data, as shown in Figure 8a, and the deformation velocity maps in slant range, azimuth, and elevation directions were simulated in the corresponding scene, respectively, as shown in Figure 8b–d. The parameters of SAR systems are shown in Table 2.

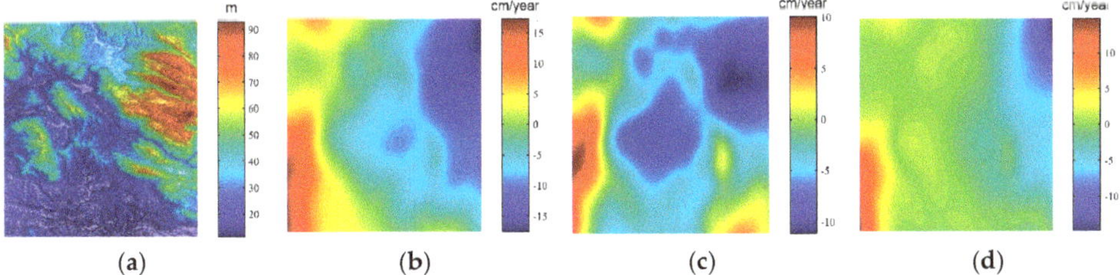

(a)　　　　　　　　**(b)**　　　　　　　　**(c)**　　　　　　　　**(d)**

Figure 8. The terrain and the simulation of the deformation maps: (**a**) digital elevation model (DEM) data from SRTM. (**b–d**) are the simulated deformation velocity maps in slant range, azimuth, and elevation direction, respectively.

Figure 9a illustrates one of the semi-real SAR images as an example. The elevation and deformation parameters are estimated by the proposed algorithm, and the results are as follows: Figure 9b is the estimation of elevation; Figure 9c–e are the estimations of deformation velocities in slant range, azimuth, and elevation directions, respectively. It can be seen that the estimated deformation in the three directions had the same trend as the real deformation map. The black line in Figure 9b shows the position of the analysis slice, and the estimation errors of 3-D deformation velocities for the scatterers located in this line are presented in Figure 10. Similar to the experimental results in the previous sub-section, the estimation errors in the slant range and elevation direction were also no more than 0.5 cm/year, and the accuracy of deformation retrieval in the azimuth direction was worse than that in the other two directions. Nevertheless, according to Figure 4b, the accuracy of the estimation in the azimuth direction could be improved by increasing the angular diversity of the squint angles used in the two combined acquisitions. Experimental results show the potential of the proposed algorithm for the reconstruction of the elevation and deformation parameters from the full SAR image.

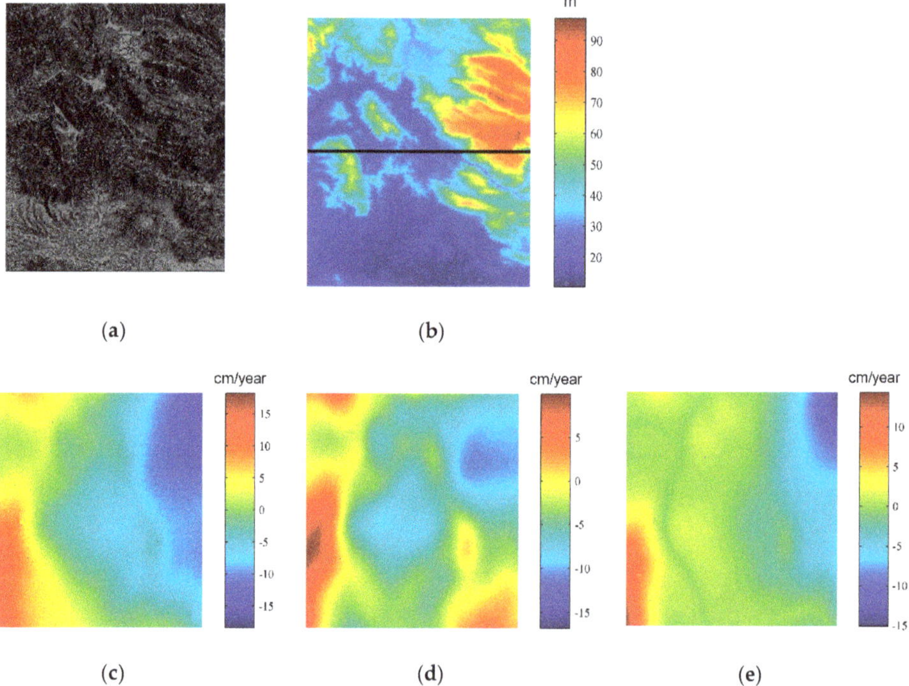

Figure 9. Results of the estimations: (**a**) SAR simulated image generated by Giorgio Franceschetti's method. (**b**) is the estimation result of elevation. (**c–e**) are the estimations of deformation velocity in the direction of slant range, azimuth, and elevation.

In the above experiment with semi-real data, the DEM data were used to generate the SAR raw data in a natural scene, and the layover phenomenon was ignored. However, the D-TomoSAR was mainly applied to monitor the scenario with layover phenomenon such as urban areas and forests. Therefore, a semi-real SAR raw data of the urban area was simulated to further verify the effectiveness of the proposed algorithm. The DEM data of Shanghai was adopted to simulate the urban scene in this experiment, as shown in Figure 11. The red box in Figure 11 is the region of interest (ROI), which contains some buildings. Thus, the layover phenomenon occurs when imaging for the ROI. Assuming that each slant range-azimuth resolution cell of the SAR image in the ROI contains two scatterers, we call them the dominant scatterer and the secondary scatterer. Accordingly, the elevation and deformation maps of the ROI are simulated, shown as in Figure 12, where Figure 12a,e are the elevation of dominant scatterer and secondary scatterer, respectively. Figure 12b–d show the deformation velocities of the dominant scatterer along the slant range, azimuth, and elevation directions respectively, while Figure 12f–h correspond to the secondary scatterer.

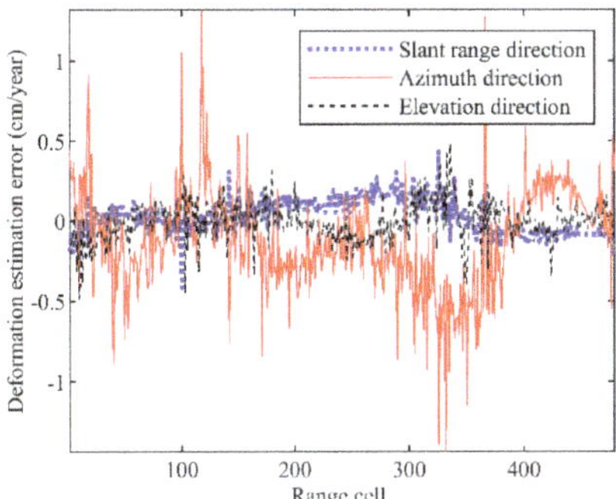

Figure 10. The estimation errors of 3-D deformation velocities for the scatterers located in the slice.

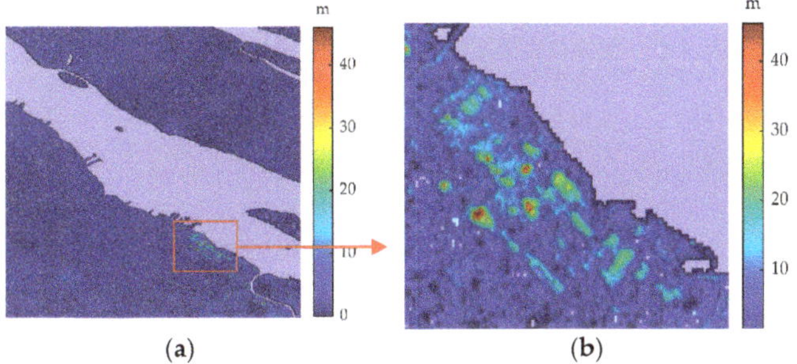

Figure 11. DEM data of Shanghai, where the red box area is the region of interest (ROI). (**a**) DEM data, (**b**) DEM of the ROI.

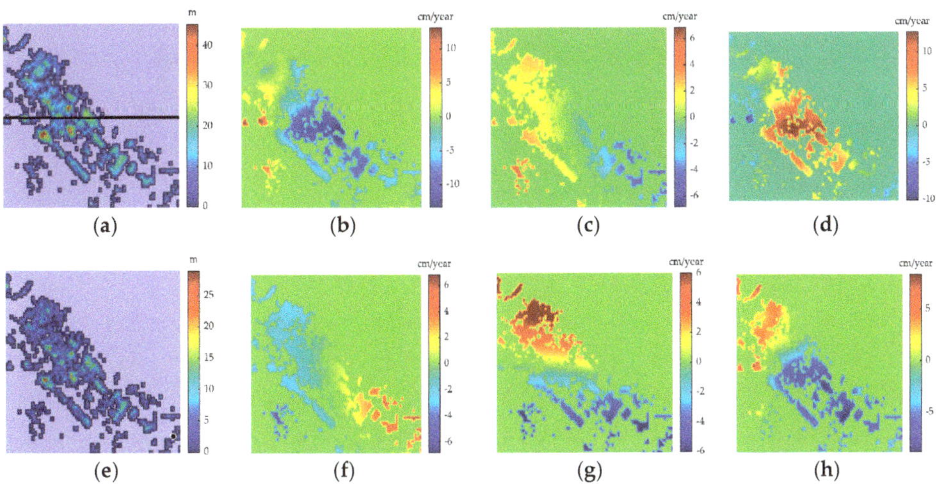

Figure 12. The simulation of elevation and deformation maps. (**a**–**d**) are the simulated elevation of the dominant scatterer and its simulated deformation velocities along the slant range, azimuth, and elevation directions, respectively. (**e**–**h**) correspond to the secondary scatterer.

The proposed algorithm was used to estimate the elevation and the deformation velocity of the ROI, and the estimation results are shown in Figure 13. It can be seen from the experimental results that the estimated elevation and the 3-D deformation velocities of the dominant scatterer and secondary scatterer had the same trend as the true values. Similarly, the estimation errors of elevation and deformation velocity of scatterers located at the slice in Figure 12a were calculated, as shown in Figure 14. The experimental results show that the estimation results were consistent with our expectation, which validates the ability of the proposed algorithm to retrieve the elevation and 3-D deformation parameters in the scenario with layover phenomenon.

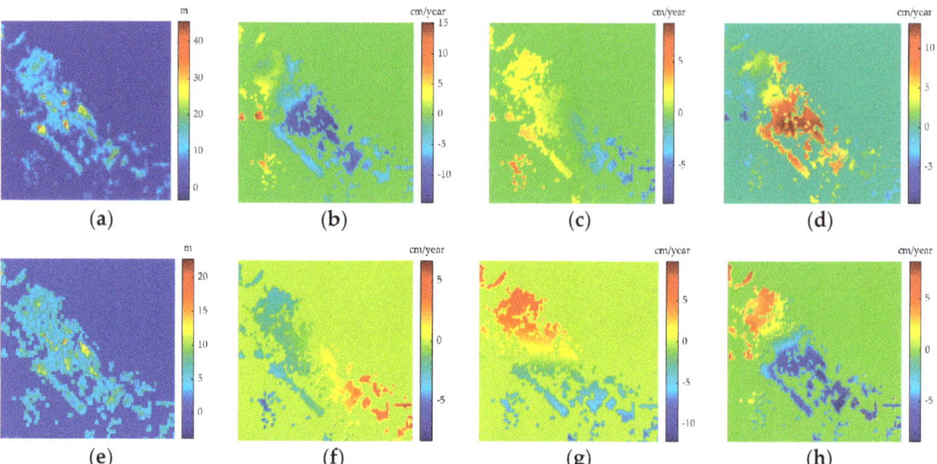

Figure 13. Estimations of elevation and deformation maps. (**a–d**) are the estimated elevation of the dominant scatterer and its estimated deformation velocities along the slant range, azimuth, and elevation directions, respectively. (**e–h**) correspond to the secondary scatterer.

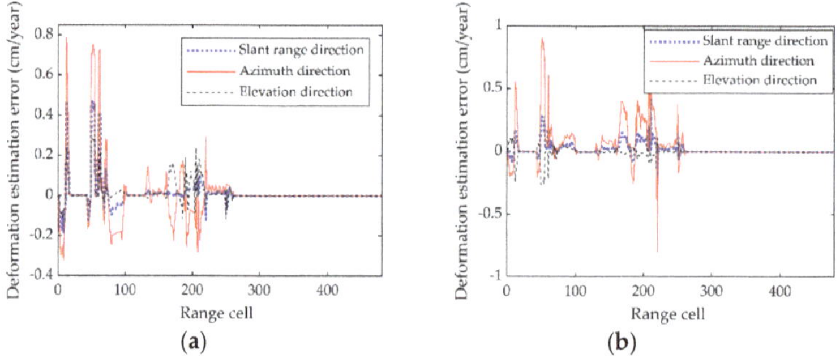

Figure 14. The estimation errors of 3-D deformation velocities for the scatterers located in the slice of Figure 12a. (**a**) dominant scatterer. (**b**) secondary scatterer.

5. Conclusions

In this paper, a method is proposed for retrieving the elevation and 3-D deformation velocities of ground from an improved D-TomoSAR system model. Firstly, the relationship between the phase term of the improved D-TomoSAR and the 3-D deformation displacements is established from the imaging geometry of the D-TomoSAR system. The improved D-TomoSAR signal model can be regarded as a 2-D PPS, thus, the 2-D PHAF with the RELAX algorithm is introduced to estimate the elevation and 3-D deformation velocities of scatterers. Subsequently, the theoretical accuracy of 3-D deformation retrieval of the improved D-TomoSAR system is analyzed with respect to the squint angle of SAR imaging. In addition, the performance to assess the 3-D deformation retrieval of the proposed algorithm and the traditional method is compared. Results show that increasing the squint angle of SAR imaging and changing the satellite heading angel have a comparable effect on improving the accuracy of deformation retrieval in North-South direction. Finally, simulation and semi-real data results demonstrate the effectiveness and accuracy of the proposed method.

Author Contributions: Z.W. conceived and designed the method; M.L. guided the students to complete the research; Z.W. and K.L. performed the experiments and analyzed the performance; and Z.W. wrote the paper.

Funding: This research was funded by the National Natural Science Foundation of China, grant number 61771164 and the Innovation of Science and Technology Program of Shanghai Aerospace Science and Technology Corporation 2017.

Acknowledgments: The authors would like to thank the anonymous reviewers for their constructive comments and suggestions which are helpful in improving the readability of this paper.

Conflicts of Interest: The authors declare no conflict of interest.

References

1. Lombardini, F. Differential tomography: A new framework for SAR interferometry. *IEEE Trans. Geosci. Remote Sens.* **2005**, *43*, 37–44. [CrossRef]
2. Zhu, X.X.; Wang, Y.; Montazeri, S.; Ge, N. A Review of Ten-Year Advances of Multi-Baseline SAR Interferometry Using TerraSAR-X Data. *Remote Sens.* **2018**, *10*, 1374. [CrossRef]
3. Drews, R.; Rack, W.; Wesche, C.; Helm, V. A spatially adjusted elevation model in Dronning Maud Land, Antarctica, based on differential SAR interferometry. *IEEE Trans. Geosci. Remote Sens.* **2009**, *47*, 2501–2509. [CrossRef]
4. Ferretti, A.; Prati, C.; Rocca, F. Permanent scatterers in SAR interferometry. *IEEE Trans. Geosci. Remote Sens.* **2001**, *39*, 8–20. [CrossRef]
5. Ferretti, A.; Fumagalli, A.; Novali, F.; Prati, C.; Rocca, F.; Rucci, A. A new algorithm for processing interferometric data-stacks: SqueeSAR. *IEEE Trans. Geosci. Remote Sens.* **2011**, *49*, 3460–3470. [CrossRef]
6. Berardino, P.; Fornaro, G.; Lanari, R.; Sansosti, E. A new algorithm for surface deformation monitoring based on small baseline differential SAR interferograms. *IEEE Trans. Geosci. Remote Sens.* **2002**, *40*, 2375–2383. [CrossRef]
7. Pepe, A.; Solaro, G.; Calo, F.; Dema, C. A minimum acceleration approach for the retrieval of multiplatform InSAR deformation time series. *IEEE J. Sel. Top. Appl. Earth Obs. Remote Sens.* **2016**, *9*, 3883–3898. [CrossRef]
8. Montazeri, S.; Zhu, X.X.; Eineder, M.; Bamler, R. Three-dimensional deformation monitoring of urban infrastructure by tomographic SAR using multitrack TerraSAR-X data stacks. *IEEE Trans. Geosci. Remote Sens.* **2016**, *54*, 6868–6878. [CrossRef]
9. Liang, C.; Fielding, E.J. Measuring azimuth deformation with L-band ALOS-2 ScanSAR interferometry. *IEEE Trans. Geosci. Remote Sens.* **2017**, *55*, 2725–2738. [CrossRef]
10. Milillo, P.; Minchew, B.; Simons, M.; Agram, P.; Riel, B. Geodetic imaging of time-dependent three-component surface deformation: Application to tidal-timescale ice flow of Rutford ice stream, West Antarctica. *IEEE Trans. Geosci. Remote Sens.* **2017**, *55*, 5515–5524. [CrossRef]
11. Razi, P.; Sumantyo, J.T.S.; Perissin, D.; Kuze, H.; Chua, M.Y.; Panggabean, G.F. 3D Land Mapping and Land Deformation Monitoring Using Persistent Scatterer Interferometry (PSI) ALOS PALSAR: Validated by Geodetic GPS and UAV. *IEEE Access* **2018**, *6*, 12395–12404. [CrossRef]
12. Ansari, H.; De Zan, F.; Parizzi, A.; Eineder, M.; Goel, K.; Adam, N. Measuring 3-D surface motion with future SAR systems based on reflector antennae. *IEEE Geosci. Remote Sens. Lett.* **2016**, *13*, 272–276. [CrossRef]
13. Rocca, F. 3D motion recovery with multi-angle and/or left right interferometry. In Proceedings of the Third International Workshop on ERS SAR, Frascati, Italy, 3–5 December 2003.
14. Kou, L.; Wang, X.; Xiang, M.; Zhu, M. Interferometric estimation of three-dimensional surface deformation using geosynchronous circular SAR. *IEEE Trans. Aerosp. Electron. Syst.* **2012**, *48*, 1619–1635. [CrossRef]
15. Liu, F.; Fan, X.; Zhang, T.; Liu, Q. GNSS-Based SAR Interferometry for 3-D Deformation Retrieval: Algorithms and Feasibility Study. *IEEE Trans. Geosci. Remote Sens.* **2018**, *56*, 5736–5748. [CrossRef]

16. Jung, H.S.; Lu, Z.; Shepherd, A.; Wright, T. Simulation of the SuperSAR multi-azimuth synthetic aperture radar imaging system for precise measurement of three-dimensional Earth surface displacement. *IEEE Trans. Geosci. Remote Sens.* **2015**, *53*, 6196–6206. [CrossRef]

17. Mittermayer, J.; Wollstadt, S.; Prats-Iraola, P.; López-Dekker, P.; Krieger, G.; Moreira, A. Bidirectional SAR Imaging Mode. *IEEE Trans. Geosci. Remote Sens.* **2013**, *51*, 601–614. [CrossRef]

18. Fornaro, G.; Lombardini, F.; Serafino, F. Three-dimensional multipass SAR focusing: Experiments with long-term spaceborne data. *IEEE Trans. Geosci. Remote Sens.* **2005**, *43*, 702–714. [CrossRef]

19. Reigber, A.; Moreira, A. First demonstration of airborne SAR tomography using multibaseline L-band data. *IEEE Trans. Geosci. Remote Sens.* **2000**, *38*, 2142–2152. [CrossRef]

20. Fornaro, G.; Serafino, F.; Soldovieri, F. Three-dimensional focusing with multipass SAR data. *IEEE Trans. Geosci. Remote Sens.* **2003**, *41*, 507–517. [CrossRef]

21. Zhu, X.X.; Bamler, R. Tomographic SAR inversion by $L1$-norm regularization-The compressive sensing approach. *IEEE Trans. Geosci. Remote Sens.* **2010**, *48*, 3839–3846. [CrossRef]

22. Budillon, A.; Evangelista, A.; Schirinzi, G. Three-dimensional SAR focusing from multipass signals using compressive sampling. *IEEE Trans. Geosci. Remote Sens.* **2011**, *49*, 488–499. [CrossRef]

23. Liu, M.; Wang, Z.; Wang, P. Extension of D-TomoSAR for multi-dimensional reconstruction based on polynomial phase signal. *IET Radar Sonar Navig.* **2018**, *12*, 449–457. [CrossRef]

24. Simeunović, M.; Djurović, I. Parameter estimation of multicomponent 2D polynomial-phase signals using the 2D PHAF-based approach. *IEEE Trans. Signal Process.* **2016**, *64*, 771–782. [CrossRef]

25. Hu, C.; Li, Y.; Dong, X.; Wang, R.; Cui, C. Optimal 3D deformation measuring in inclined geosynchronous orbit SAR differential interferometry. *Sci. China Inf. Sci.* **2017**, *60*, 060303. [CrossRef]

26. Sun, X.; Yu, A.; Dong, Z.; Liang, D. Three-dimensional SAR focusing via compressive sensing: The case study of angel stadium. *IEEE Geosci. Remote Sens. Lett.* **2012**, *9*, 759–763.

27. Zhu, X.X.; Bamler, R. Let's do the time warp: Multicomponent nonlinear motion estimation in differential SAR tomography. *IEEE Geosci. Remote Sens. Lett.* **2011**, *8*, 735–739. [CrossRef]

28. Lazecky, M.; Hlavacova, I.; Bakon, M.; Sousa, J.J.; Perissin, D.; Patricio, G. Bridge displacements monitoring using space-borne X-band SAR interferometry. *IEEE J. Sel. Top. Appl. Earth Obs. Remote Sens.* **2017**, *10*, 205–210. [CrossRef]

29. Wang, Y.; Zhu, X.X.; Bamler, R. An efficient tomographic inversion approach for urban mapping using meter resolution SAR image stacks. *IEEE Geosci. Remote Sens. Lett.* **2014**, *11*, 1250–1254. [CrossRef]

30. Ge, N.; Gonzalez, F.R.; Wang, Y.; Shi, Y.; Zhu, X.X. Spaceborne Staring Spotlight SAR Tomography—A First Demonstration with TerraSAR-X. *IEEE J. Sel. Top. Appl. Earth Obs. Remote Sens.* **2018**, *11*, 3743–3756. [CrossRef]

31. Krieger, G.; Moreira, A.; Fiedler, H.; Hajnsek, I.; Werner, M.; Younis, M.; Zink, M. TanDEM-X: A satellite formation for high-resolution SAR interferometry. *IEEE Trans. Geosci. Remote Sens.* **2007**, *45*, 3317–3341. [CrossRef]

32. Li, Z.; Liang, Y.; Xing, M.; Huai, Y.; Zeng, L.; Bao, Z. Focusing of highly squinted SAR data with frequency nonlinear chirp scaling. *IEEE Geosci. Remote Sens. Lett.* **2016**, *13*, 23–27. [CrossRef]

33. Franceschetti, G.; Iodice, A.; Perna, S.; Riccio, D. Efficient simulation of airborne SAR raw data of extended scenes. *IEEE Trans. Geosci. Remote Sens.* **2006**, *44*, 2851–2860. [CrossRef]

Article

Narrowband Interference Separation for Synthetic Aperture Radar via Sensing Matrix Optimization-Based Block Sparse Bayesian Learning

Guojing Li [1,*]**, Wei Ye** [2]**, Guochao Lao** [3]**, Shuya Kong** [4] **and Di Yan** [1]

[1]   Graduate School, Space Engineering University, Beijing 101416, China; yandimail@126.com
[2]   Space Engineering University, Beijing 101416, China; yeyuhan@sina.com
[3]   The 96901 Unit of PLA, Beijing 100094, China; laoguochao@mail.sdu.edu.cn
[4]   The 66135 Unit of PLA, Beijing 100144, China; skysalen@163.com
*   Correspondence: leeguojing1014@mail.dlut.edu.cn; Tel.: +86-010-6636-8478

Received: 28 March 2019; Accepted: 23 April 2019; Published: 25 April 2019

Abstract: High-resolution synthetic aperture radar (SAR) operating with a large bandwidth is subject to impacts from various kinds of narrowband interference (NBI) in complex electromagnetic environments. Recently, many radio frequency interference (RFI) suppression approaches for SAR based on sparse recovery have been proposed and demonstrated to outperform traditional ones in preserving the signal of interest (SOI) while suppressing the interference by exploiting their intrinsic structures. In particular, the joint recovery strategy of SOI and NBI with a cascaded dictionary, which eliminates the steps of NBI reconstruction and time-domain cancellation, can further reduce unnecessary system complexity. However, these sparsity-based approaches hardly work effectively for signals from an extended target or NBI with a certain bandwidth, since neither of them is sparse in a prescient domain. Moreover, sub-dictionaries corresponding to different components in the cascaded matrix are not strictly independent, which severely limits the performance of separated reconstruction. In this paper, we present an enhanced NBI separation algorithm for SAR via sensing matrix optimization-based block sparse Bayesian learning (SMO-BSBL) to solve these problems above. First, we extend the block sparse Bayesian learning framework to a complex-valued domain for the convenience of radar signal processing with lower computation complexity and modify it to deal with the separation problem of NBI in the contaminated echo. For the sake of improving the separated reconstruction performance, we propose a new block coherence measure by defining the external and internal block structure, which is used for optimizing the observation matrix. The optimized observation matrix is then employed to reconstruct SOI and NBI simultaneously under the modified BSBL framework, given a known and fixed cascaded dictionary. Numerical simulation experiments and comparison results demonstrate that the proposed SMO-BSBL is effective and superior to other advanced algorithms in NBI suppression for SAR.

Keywords: synthetic aperture radar; narrowband interference separation; block sparse Bayesian learning; sensing matrix optimization; block coherence measure

---

1. Introduction

High-resolution synthetic aperture radar (SAR) is an active remote sensing modality for real-time information acquisition. It plays a significant role in the field of civil exploration and military reconnaissance owing to its capability of all-weather, all-time, and high-resolution imaging. A SAR system usually operates at a wide range of microwave frequencies and it is inevitably subject to various kinds of electromagnetic interference. These kinds of interference with the characteristics of high power and narrowband may seriously degrade the quality of SAR images and cause trouble for the subsequent interpretation.

Multi-channel technology and signal processing are two typical methods of interference suppression for SAR. A multi-channel processing method [1,2] uses the space information and extracts the signal of interest (SOI) from the contaminated echo by zeroing the interference direction, and this method outperforms that of a single channel. However, this special multiplex architecture increases the complexity of the radar system and cannot be directly applied to existing devices.

From the perspective of pure signal processing, narrowband interference suppression can be mainly divided into parametric, non-parametric, and semi-parametric methods. The parametric methods such as high-order ambiguity function [3] and complex empirical mode decomposition [4] are based on interference modeling with multi-order terms. However, it is heavily dependent on model accuracy and has a large amount of calculation in the process of parameter searching. A non-parametric method such as notched filtering (NF) [5,6], least mean square (LMS) filtering [7], eigen-subspace filtering (ESF) [8], independent component analysis (ICA) [9], independent subspace analysis (ISA) [10,11], and robust principal component analysis (RPCA) [12] can suppress the interference from raw data without any prior knowledge or parametric model. Notched filtering and LMS filtering are actually equivalent to adding a band-stop filter where the interference is located, regardless of whether there is a signal component in this frequency range. The basic idea of ESF, ICA, ISA, and RPCA is the singular value decomposition (SVD) of the data matrix, and the signal or interference is reconstructed by inverse transform after extracting the dominant components. The main problem of the non-parametric method is the signal distortion, since the SOI is also suppressed when the interference is eliminated.

Sparse recovery, as a typical semi-parametric method for interference suppression, is state-of-the-art, especially in terms of reducing signal distortion. It can be considered as an optimization problem of reconstructing few coefficients with a given dictionary. The sparsity-based method is mainly used for suppressing RFI that appears in the form of spikes in a large frequency range. Considering the sparse property in the range-frequency domain and the low-rank property in the azimuth, in References [13,14], RFI was extracted and suppressed based on a sparse and low-rank model. In Reference [15], the matrix factorization technique was introduced into the sparse and low-rank model to avoid large residuals after SVD and further reduce the computational complexity at the same time. In our previous work [16], we proposed an RFI suppression method for SAR based on morphological component analysis (MCA), in which a stepwise reconstruction algorithm was adopted to the reconstruction. Given that the steps of interference reconstruction and cancellation may limit the suppression performance and increase the system complexity, the alternating direction multiplier method (ADMM) [17] was adopted to reconstruct the signal and the interference simultaneously in Reference [18]. The premise of this method is that both the SOI and interference are sparse in their respective domains.

The observed scene in most SAR images is not sparse and it is difficult to find a proper dictionary to represent the echo signal with few nonzero coefficients. Moreover, the narrowband interference (NBI) of a noise-modulated type with a certain bandwidth is not sparse either in the frequency domain. Classical recovery algorithms [19,20] such as basis pursuit (BP), matching pursuit (MP), and orthogonal matching pursuit (OMP) fail to recover the signal accurately. The block MP (BMP) and block OMP (BOMP) algorithm proposed in Reference [21] can improve the reconstruction probability with a slight requirement for sparsity by exploiting the block sparse structure. Still, with the increase in scene complexity and interference bandwidth, the reconstruction probability decreases, since the block sparse feature gradually weakens. The global minimum of the above algorithm is not really the sparsest solution, unless strict conditions are satisfied. Hence, sparse Bayesian learning (SBL) [22], which considers all unknown parameters as random variables and adds appropriate prior distributions according to the sparse structure, is no doubt a better choice. Derived from the SBL framework, block sparse Bayesian learning (BSBL) [23,24] is a robust recovery algorithm for both sparse and non-sparse signals from a low-dimensional space by exploiting the temporal correlation of intra-block data. In

Reference [25], the BSBL framework is first used and modified for RFI suppression where the target or observed scene is not strictly sparse but block sparse, and the S-BSBL and A-BSBL algorithms are, respectively, proposed to improve reconstruction performance and reduce the amount of computation. Judging from the results of interference suppression, the BSBL-based approach is indeed superior to other advanced ones and can be used more widely.

Nevertheless, there remain several problems to be solved. As is known, radar signals are complex-valued in most processing steps, so the BSBL algorithm cannot be directly applied. A widely accepted trade-off approach is to concatenate the real and imaginary parts of the signal into a new vector. There are two main shortcomings in this scheme. One is that the length of the new real-valued vector is twice as long as the original complex-valued vector and the corresponding sensing matrix will expand in square with the signal length increasing, which will result in a huge amount of computational burden. The other is that the reconstruction performance may be degraded due to the loss of structural information, since the real and imaginary part of the signal are processed separately. In addition, while the BSBL framework is robust to the interatomic coherence in the sensing matrix for the reconstruction of a clean signal, the block coherence of sub-dictionaries corresponding to different components in the contaminated echo has a great impact on the separated reconstruction performance, since the diagonal block of the covariance matrix cannot be effectively distinguished.

To solve these problems above, our goal is to reduce the amount of calculation with a modified BSBL algorithm, which can be applied to the complex-valued signal directly and further improve the performance of NBI separation by optimizing the cascaded sensing matrix.

The main contents of this paper are divided into three parts. In Section 2, the problem of separated reconstruction for SOI and NBI based on complex-valued block sparse Bayesian learning framework is formulated. In Section 3, the optimal sensing matrix is designed by minimizing the newly defined block coherence measure, and the SMO-BSBL algorithm for NBI separation, which is embedded in the entire procedure of SAR imaging, is presented. In Section 4, numerical experiments with simulated data are carried out, and results of the proposed algorithm in this paper are compared with existing BSBL-based algorithms.

2. Problem Formulation

2.1. Sparse Model and Joint Recovery

The raw SAR echo is usually considered as the convolution of the scattering points and the transmitted signal from radar. The most commonly used signal type is the linear frequency-modulated (LFM) signal and the ideal echo signal in the analogy domain can be expressed as [26]

$$s_r(\tau, t) = \sum_{p=1}^{P} \sigma_p w_r\left(t - 2R_p(\tau)/c\right) \exp\left(-j4\pi R_p(\tau)/\lambda\right) \exp\left[j\pi K_r\left(t - 2R_p(\tau)/c\right)^2\right] \tag{1}$$

where t is the fast time in range direction; τ is the slow time in azimuth direction; P is the number of scattering points in observed scene; σ_p is the backscatter coefficient of the p-th point; c is the speed of light; λ is the electromagnetic wavelength; R_p is the oblique distance between scattering point and SAR platform; K_r is the frequency modulation slope; $w_r(\cdot)$ denotes the rectangular window function with the length of r.

In real-world environments, SAR may be subjected to various forms of interference, including the natural radiation and the man-made interference. The former is also subject to RFI and commonly modeled in the form of multi-tone complex sine, which adds bright stripes to SAR images. This type of NBI in analog time domain can be expressed as

$$n_1(t) = \sum_{i=1}^{L} A_i(t) \exp\{j[2\pi f_i(t)t + \varphi_i(t)]\} \tag{2}$$

where $A(t)$, $f(t)$, and $\phi(t)$ are, respectively, the amplitude, carrier frequency, and phase varying over time; L is the number of interference tone. The latter, which is often generated artificially by modulating a narrowband noise into the frequency band of the signal, will add speckles similar to salt and pepper on the image. This type of NBI mainly includes the amplitude-modulated type and frequency-modulated type, and its general mathematical model in the analog time domain can be expressed as

$$n_2(t) = [U_0 + U_n(t)] \exp\left\{ j\left[2\pi f t + 2\pi K_{FM} \int_0^1 U_n(\tau)d\tau + \varphi(t) \right] \right\} \tag{3}$$

where U_0 is the constant amplitude; $U_n(t)$ is the band-limited noise whose amplitude may change over time; K_{FM} is the frequency-modulated slope. The waveform and spectrum diagram of two types of NBI added to the LFM signal are shown in Figure 1.

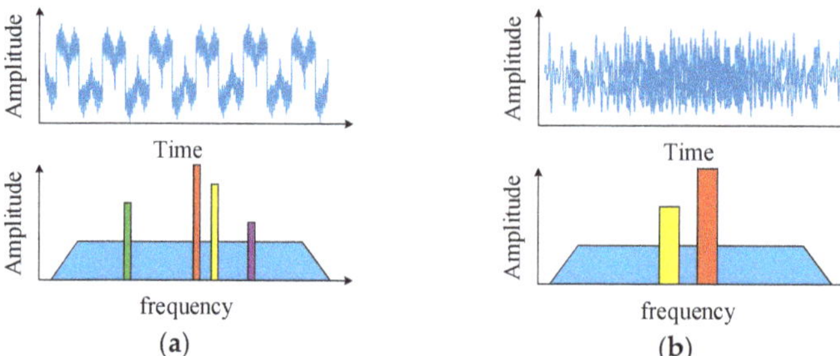

Figure 1. Waveform and spectrum diagram for two types of narrowband interference (NBI): (**a**) complex sine model; (**b**) noise-modulated model.

As shown in Figure 1a, the NBI based on the complex sine model is sparse in the frequency domain, since there are only a few dominant, scattered interference components with an extremely narrow bandwidth. In contrast, the narrowband noise-modulated NBI in Figure 1b is not strictly sparse in frequency because it densely occupies a segment of the spectrum. Moreover, the high-power characteristic of this kind of NBI is not as outstanding as that of RFI, which indicates that it is more difficult to extract and separate from the SOI. Therefore, we mainly focus on this narrowband noise-modulated NBI and seek effective approaches to suppress it.

For the convenience of theoretical analysis, the echo contaminated with NBI in each pulse is analyzed in the discrete domain, which is expressed in the form of N-dimensional complex-valued vectors, i.e.,

$$\widetilde{x} = \widetilde{s} + \widetilde{n} + \widetilde{w} \tag{4}$$

where $\widetilde{s} \in \mathbb{C}^N$ is the SOI component; $\widetilde{n} \in \mathbb{C}^N$ is the NBI component; $\widetilde{w} \in \mathbb{C}^N$ is the additional white noise assumed to satisfy complex Gaussian distribution $\widetilde{w} \sim \mathcal{CN}(0, \sigma^2 \mathbf{I}_w)$, where $\mathbf{I}$ is an identity matrix.

If the observed scene is divided into grids, the echo $\widetilde{x}$ can be considered as the accumulation of transmitted signals with a different delay of range. In addition, the SAR system is a collaborative platform and its signal form and parameters are known. Therefore, the basis dictionary $\widetilde{\mathbf{\Psi}}_s$ of the SOI can be constructed by the reference signal with delays of range, and the SOI can be expressed as $\widetilde{s} = \widetilde{\mathbf{\Psi}}_s \widetilde{\alpha}_s$, where $\widetilde{\alpha}_s$ is the coefficient vector with few nonzero elements in the sparse case. Similarly, the NBI component in each pulse can be represented with few aggregated atoms on a specific basis $\widetilde{\mathbf{\Psi}}_n$, since it is sparse or block sparse in the frequency domain, which can be expressed as $\widetilde{n} = \widetilde{\mathbf{\Psi}}_n \widetilde{\alpha}_n$, where $\widetilde{\alpha}_n$ is the coefficient vector used for representing the NBI.

Compressed sensing theory has demonstrated that an N-dimensional vector with a sparse structure can be accurately recovered from M-dimensional ($M < N$) compressed measurements via nonlinear optimization with a high probability. The SAR echo contaminated with NBI after compressed sampling can be expressed as the following cascaded matrix form:

$$\widetilde{y} = \widetilde{\Phi}\left(\widetilde{\Psi}_s \widetilde{\alpha}_s + \widetilde{\Psi}_n \widetilde{\alpha}_n\right) + \widetilde{w} = \begin{bmatrix} \widetilde{\Theta}_s & \widetilde{\Theta}_n \end{bmatrix} \begin{bmatrix} \widetilde{\alpha}_s \\ \widetilde{\alpha}_n \end{bmatrix} + \widetilde{w} \tag{5}$$

where $\widetilde{\Phi}$ is the compressed observation matrix; $\widetilde{\Theta}_s$ and $\widetilde{\Theta}_n$ are the sensing matrix of SOI and NBI, respectively.

The separated optimization problem in Equation (5) can be expressed as

$$\{\widetilde{\alpha}_s^*, \widetilde{\alpha}_n^*\} = \underset{\{\widetilde{\alpha}_s, \widetilde{\alpha}_n\}}{\operatorname{argmin}} = \|\widetilde{\alpha}_s\|_0 + \|\widetilde{\alpha}_n\|_0 \tag{6}$$

where $\|\cdot\|_p$ denotes the l_p-norm. Since minimizing the l_0-norm needs to list all possible combinations of non-zero elements in a sparse vector, which will take an enormous amount of time, a relaxed form of (6) can be expressed as

$$\{\widetilde{\alpha}_s^*, \widetilde{\alpha}_n^*\} = \underset{\{\widetilde{\alpha}_s, \widetilde{\alpha}_n\}}{\operatorname{argmin}} = \|\widetilde{\alpha}_s\|_1 + \lambda\|\widetilde{\alpha}_n\|_1 \quad s.t. \quad \|\widetilde{y} - \widetilde{\Psi}_s \widetilde{\alpha}_s - \widetilde{\Psi}_n \widetilde{\alpha}_n\|_2 < \varepsilon \tag{7}$$

where λ is a constant regularization parameter.

The ADMM algorithm [17] is widely used for this joint optimization problem as long as components are sparse in their respective domain. However, when the target is not sparse in the observed scene or the spectrum of NBI occupies a certain amount of bandwidth, the simultaneous reconstruction performance of components via ADMM degrades or even fails. The BSBL performs better for highly underdetermined problems compared to existing algorithms, which can obtain the sparsest solution by modeling the temporal correlation, even in non-sparse cases. In Reference [25], it was verified that BSBL is superior to ADMM in terms of joint reconstruction when the target is not sparse. It is worth noting that the BSBL-based NBI suppression is still implemented in real-valued signals formed by splicing the real and imaginary part of the complex signals. This approach not only destroys the phase structure, which is significant for SAR, but also increases the computational cost. Next, we modify the original BSBL framework to enable it to deal with the complex-valued signal directly.

2.2. Complex BSBL Framework

The initial BSBL framework is generally applicable to real-valued signal processing. The most common way to deal with the complex-valued radar signal is to process the real and imaginary part separately, and this bi-channel signal observation model can be expressed as

$$\begin{bmatrix} \operatorname{Re}(\widetilde{y}) \\ \operatorname{Im}(\widetilde{y}) \end{bmatrix} = \begin{bmatrix} \operatorname{Re}(\widetilde{\Theta}_s) & \operatorname{Re}(\widetilde{\Theta}_n) & -\operatorname{Im}(\widetilde{\Theta}_s) & -\operatorname{Im}(\widetilde{\Theta}_n) \\ \operatorname{Im}(\widetilde{\Theta}_s) & \operatorname{Im}(\widetilde{\Theta}_n) & \operatorname{Re}(\widetilde{\Theta}_s) & \operatorname{Re}(\widetilde{\Theta}_n) \end{bmatrix} \begin{bmatrix} \operatorname{Re}(\widetilde{\alpha}_s) \\ \operatorname{Re}(\widetilde{\alpha}_n) \\ \operatorname{Im}(\widetilde{\alpha}_s) \\ \operatorname{Im}(\widetilde{\alpha}_n) \end{bmatrix} + \begin{bmatrix} \operatorname{Re}(\widetilde{n}) \\ \operatorname{Im}(\widetilde{n}) \end{bmatrix} \tag{8}$$

where $\operatorname{Re}(\cdot)$ and $\operatorname{Im}(\cdot)$ denote the real part and imaginary part of the complex vector. Here, we attempt to modify the BSBL so that it can be directly adopted to complex signal processing with less computation instead of the above approach.

To solve the optimization problem in Equation (6) via a complex BSBL framework, the $2N$-dimensional cascaded coefficient vector to be reconstructed is divided into cascaded blocks of the same length, i.e.,

$$\widetilde{\alpha} = \begin{bmatrix} \underbrace{\widetilde{\alpha}_s^1, \ldots, \widetilde{\alpha}_s^{d_1}}_{d_{1s}}, \ldots, \underbrace{\widetilde{\alpha}_s^{d_{gs-1}+1}, \ldots, \widetilde{\alpha}_s^N}_{d_{gs}} & \underbrace{\widetilde{\alpha}_n^1, \ldots, \widetilde{\alpha}_n^{d_1}}_{d_{1n}}, \ldots, \underbrace{\widetilde{\alpha}_s^{d_{gn-1}+1}, \ldots, \widetilde{\alpha}_s^N}_{d_{gn}} \end{bmatrix}^T \tag{9}$$

where d_i is the length of the *i-th* block; g_s and g_n are the number of blocks for SOI and NBI. Similarly, the sensing matrix is divided into blocks corresponding to the coefficient vector, i.e.,

$$\widetilde{\Theta} = [\ \underbrace{\widetilde{\Theta}_{s_1}, \ldots, \widetilde{\Theta}_{sg_s}}_{g_s}\ \underbrace{\widetilde{\Theta}_{n_1}, \ldots, \widetilde{\Theta}_{ng_n}}_{g_n}\ ]. \tag{10}$$

Given that the covariance matrix is a semi-positive Hermitian matrix and the imaginary part of each diagonal element is zero, we assume that blocks are independent of each other and $\widetilde{\alpha}_i$ in each block satisfies a multivariate complex Gaussian distribution $\widetilde{\alpha}_i \sim C\mathcal{N}(0, \mathbf{C}_{0,i})$, where $\mathbf{C}_{0,i} = \gamma_i \widetilde{\mathbf{B}}_i \in \mathbb{C}^{d_i \times d_i}$ is the prior covariance matrix of $\widetilde{\alpha}_i$; $\widetilde{\mathbf{B}}$ is a Hermitian matrix used for characterizing the correlation structure of $\widetilde{\alpha}_i$; γ_i is a real non-negative correlation coefficient. Most γ_i will approach zero in the process of Bayesian learning owing to automatic relevance determination [22]. In other words, a sparse solution is obtained by changing $\widetilde{\alpha}_i$ into an irrelevant zero-valued vector or complex Gaussian noise with low variance. Thus, the prior covariance of $\widetilde{\alpha}$ can be expressed as

$$\mathbf{C}_0 = \gamma \otimes \widetilde{\mathbf{B}} \tag{11}$$

where $\gamma = diag(\gamma_1, \ldots, \gamma_{g_s+g_n})$; $\widetilde{\mathbf{B}} = diag(\widetilde{\mathbf{B}}_i, \ldots, \widetilde{\mathbf{B}}_{g_s+g_n})$; $\otimes$ denotes the Kronecker product. Under the parameters γ and $\widetilde{\mathbf{B}}$, the prior probability density function of $\widetilde{\alpha}$ can be expressed as [27] (p. 504)

$$p(\widetilde{\alpha}; \gamma, \widetilde{\mathbf{B}}) = \frac{1}{\pi^N |\mathbf{C}_0|} \exp(-\widetilde{\alpha}^H \mathbf{C}_0^{-1} \widetilde{\alpha}) \tag{12}$$

where $|\cdot|$ denotes the determinant value; $(\cdot)^H$ denotes the conjugate transposition. The Gaussian likelihood function of compressed observation $\widetilde{y}$ is

$$p(\widetilde{y}|\widetilde{\alpha}; \sigma^2) = \frac{1}{\pi^M \sigma^{2M}} \exp\left[-\frac{1}{\sigma^2}(\widetilde{y} - \widetilde{\Theta}\widetilde{\alpha})^H(\widetilde{y} - \widetilde{\Theta}\widetilde{\alpha})\right]. \tag{13}$$

According to the Bayesian criterion, under the parameters γ, $\widetilde{\mathbf{B}}$, and σ^2, the posterior probability density function of $\widetilde{\alpha}$ is

$$p(\widetilde{\alpha}|\widetilde{y}; \gamma, \widetilde{\mathbf{B}}, \sigma^2) = \frac{p(\widetilde{y}|\widetilde{\alpha}; \gamma, \widetilde{\mathbf{B}}, \sigma^2)p(\widetilde{\alpha}; \gamma, \widetilde{\mathbf{B}}, \sigma^2)}{p(\widetilde{y})} \tag{14}$$

where $p(\widetilde{y}) = \int p(\widetilde{y}|\widetilde{\alpha}; \gamma, \widetilde{\mathbf{B}}, \sigma^2)p(\widetilde{\alpha}; \gamma, \widetilde{\mathbf{B}}, \sigma^2)d\widetilde{\alpha}d\gamma d\widetilde{\mathbf{B}}d\sigma^2$.

Considering that it is difficult to give an analytical expression of the above integral formula, we decompose it into another form based on Bayesian rule and Gaussian identity [28], i.e.,

$$p(\widetilde{\alpha}|\widetilde{y}; \gamma, \widetilde{\mathbf{B}}, \sigma^2) = \frac{p(\widetilde{y}|\widetilde{\alpha}; \sigma^2)p(\widetilde{\alpha}; \gamma, \widetilde{\mathbf{B}})}{p(\widetilde{y}; \gamma, \widetilde{\mathbf{B}}, \sigma^2)} \tag{15}$$

where $p(\widetilde{y}; \gamma, \widetilde{\mathbf{B}}, \sigma^2) = \int p(\widetilde{y}|\widetilde{\alpha}; \sigma^2)p(\widetilde{\alpha}; \gamma, \widetilde{\mathbf{B}})d\widetilde{\alpha}$. The likelihood function of $\widetilde{y}$ is then

$$p(\widetilde{y}; \gamma, \widetilde{\mathbf{B}}, \sigma^2) = \frac{1}{\pi^M |\sigma^2 \mathbf{I}_{\widetilde{n}} + \widetilde{\Theta}\mathbf{C}_0\widetilde{\Theta}^H|} \exp\left[-\widetilde{y}^H(\sigma^2 \mathbf{I}_{\widetilde{n}} + \widetilde{\Theta}\mathbf{C}_0\widetilde{\Theta}^H)^{-1}\widetilde{y}\right]. \tag{16}$$

Thus, the posterior probability density function of $\widetilde{\alpha}$ can be expressed as

$$p(\widetilde{\alpha}|\widetilde{y}; \gamma, \widetilde{\mathbf{B}}, \sigma^2) = \frac{1}{\pi^N |\mathbf{C}_{\widetilde{\alpha}}|} \exp\left[-(\widetilde{\alpha} - \mu_{\widetilde{\alpha}})^H \mathbf{C}_{\widetilde{\alpha}}^{-1}(\widetilde{\alpha} - \mu_{\widetilde{\alpha}})\right] \tag{17}$$

where $\mathbf{C}_{\widetilde{\alpha}} = \left(\mathbf{C}_0^{-1} + \sigma^{-2}\widetilde{\Theta}^H\widetilde{\Theta}\right)^{-1}$; $\mu_{\widetilde{\alpha}} = \sigma^{-2}\mathbf{C}_{\widetilde{\alpha}}\widetilde{\Theta}^H\widetilde{y}$. Here, the maximum posterior estimation of the complex coefficient vector $\widetilde{\alpha}$ is

$$\widetilde{\alpha}^* = \mu_{\widetilde{\alpha}} = \sigma^{-2}\left(\mathbf{C}_0^{-1} + \sigma^{-2}\widetilde{\Theta}^H\widetilde{\Theta}\right)^{-1}\widetilde{\Theta}^H\widetilde{y}. \tag{18}$$

Expectation maximum (EM) is a typical optimization algorithm for BSBL, according to which the parameters γ, $\widetilde{\mathbf{B}}$, and σ^2 of each block can be updated as described in Reference [23,24]. In the EM algorithm, the goal is to maximize the likelihood function of $p(\widetilde{y}; \gamma, \widetilde{\mathbf{B}}, \sigma^2)$, which is equivalent to minimizing the following cost function, which can be expanded according to the matrix inverse operation, i.e.,

$$\begin{aligned}\mathcal{L}(\gamma, \widetilde{\mathbf{B}}, \sigma^2) &\triangleq -2\log \int p(\widetilde{y}|\widetilde{\alpha}; \sigma^2)p(\widetilde{\alpha}; \gamma, \widetilde{\mathbf{B}})d\widetilde{\alpha} \\ &= \log|\gamma \otimes \widetilde{\mathbf{B}}| + N\log\sigma^2 + \log|\mathbf{C}_{\widetilde{\alpha}}^{-1}| + \sigma^{-2}\|\widetilde{y} - \widetilde{\Theta}\mu_{\widetilde{\alpha}}\|_2^2 + \mu_{\widetilde{\alpha}}^H\mathbf{C}_0\mu_{\widetilde{\alpha}}\end{aligned} \tag{19}$$

What needs to be noticed here is that both the correlation matrix $\widetilde{\mathbf{B}}$ and the prior covariance matrix $\mathbf{C}_{\widetilde{\alpha}}$ are defined as a complex-valued matrix, but their diagonal elements are real. In order to ensure that all γ_i are real-valued as we initially defined, we set an absolute constraint on them. We then calculate these parameters by calculating the partial derivative of the cost function. The noise variance was deduced in Reference [23], and here we focus on the other two parameters, which can be deduced by employing the complex-valued matrix derivation rules. Based on Proposition 3.14 in Reference [29], we can update σ^2, γ_i, and $\widetilde{\mathbf{B}}_i$, which are located in a complex variable function as follows:

$$\sigma^2 = \frac{\|y - \widetilde{\Theta}\mu_{\widetilde{\alpha}}\|_2^2 + tr\left(\mathbf{C}_{\widetilde{\alpha},i}\widetilde{\Theta}^H\widetilde{\Theta}\right)}{N} \tag{20}$$

$$\gamma_i = \left|\frac{tr\left[\widetilde{\mathbf{B}}_i^{-1}\left(\mu_{\widetilde{\alpha},i}\mu_{\widetilde{\alpha},i}^H + \mathbf{C}_{\widetilde{\alpha},i}\right)\right]}{d_i}\right| \tag{21}$$

$$\widetilde{\mathbf{B}}_i = \frac{\left(\mu_{\widetilde{\alpha},i}\mu_{\widetilde{\alpha},i}^H + \mathbf{C}_{\widetilde{\alpha},i}\right)}{\gamma_i} \tag{22}$$

where $tr(\cdot)$ denotes the trace operation.

Given that $\mathbf{B}$ can be modeled as a first-order auto-regressive (AR) process and constrained in a Toeplitz form to avoid over-fitting [24], we assign different AR coefficients for the correlation matrix corresponding to each component. In other words, $\widetilde{\mathbf{B}}_{ii}(i = 1, \ldots, g_s)$ and $\widetilde{\mathbf{B}}_{ii}(i = g_s + 1, \ldots, g_s + g_n)$ are updated, respectively, in order to reconstruct SOI and NBI simultaneously from the contaminated echo, since the intra-block correlation of the two components are not similar. The Toeplitz form of the correlation matrix can be expressed uniformly as

$$\widetilde{\mathbf{B}}_i \sim \text{Toeplitz}\left(1, \widetilde{r}, \ldots, \widetilde{r}^{d-1}\right) \tag{23}$$

where $\widetilde{r}$ denotes the average AR coefficient. Generally, $\widetilde{r}$ is obtained by empirical formula, which can be defined as the mean value ratio of all minor and principal diagonal elements [24], i.e.,

$$\widetilde{r} = \frac{1}{g}\sum_{i=1}^{g}\frac{tr\left(\widetilde{\mathbf{B}}_{i,sub,1}\right)/(d_i - 1)}{tr\left(\widetilde{\mathbf{B}}_{i,main}\right)/d_i}, \quad |\widetilde{r}| \leq 0.9 \tag{24}$$

where $\widetilde{\mathbf{B}}_{i,main}$ and $\widetilde{\mathbf{B}}_{i,sub,1}$ denote the principle and the first minor diagonal elements.

3. NBI Separation Based on SMO-BSBL

As stated above, the final coefficients of SOI and NBI were reconstructed simultaneously with the cascaded dictionary via the complex BSBL framework. However, the effectiveness of actual separation was not satisfactory. In this section, we analyze the problem and put forward an effective approach to suppress these adverse effects.

There are two main reasons why the separation performance was not as good as expected. First, the cascaded dictionary was directly built by splicing the SOI and the NBI dictionary, the coherence of which is not taken into account. The coherence of these two sub-dictionaries has great impact on reconstruction, especially on separated reconstruction. It was demonstrated in References [30,31] that a well-designed observation matrix or optimized dictionary used for reducing the coherence can improve reconstruction performance. The sub-dictionaries were predefined and fixed according to the signal model, so the only approach is to carefully design the observation matrix. Second, the correlation characteristic of noise-modulated NBI was generally far weaker than that of the SOI. If the threshold for pruning out the blocks of each component were assigned with the same value, the global convergence rate of the algorithm would decrease. Therefore, the pruning threshold for the correlation coefficient should be adaptively adjusted.

3.1. Block Coherence Measure

Setting the AR coefficients in $\widetilde{\mathbf{B}}$ corresponding to SOI and NBI to different values is equivalent to dividing the cascaded sensing matrix with its corresponding coefficient vector into two parts. We call each part an external block. Blocks in each external part are internal blocks. The noise-modulated NBI model is not sparse but block sparse in the frequency domain. Therefore, more general conclusions can be obtained by analyzing the block coherence, since the traditional coherence is a special case when the block size is 1. The diagram of the structural relationship between the external block and the internal block is shown in Figure 2. The hierarchical block structure shown in Figure 2a can be extended to cases of more interference components rather than just limited to one type of NBI.

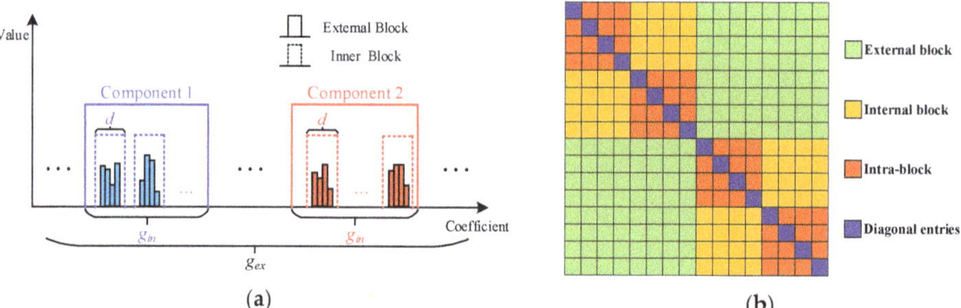

Figure 2. Diagram of the structure relationship between external block and internal block. (**a**) The hierarchical block structure with g_{ex} external blocks and g_{in} internal blocks with the size of d. The solid wireframe represents the external block, the dashed wireframe represents the internal block, and different components are in different colors. (**b**) The Gram form of the sensing matrix with two external blocks and two internal blocks of size four in each external block. Entries belonging to different external blocks are in green. Entries belonging to the same external block, but different internal blocks, are in yellow. Entries belonging to the same internal block are in red, and diagonal entries are in blue.

In Reference [32], the dictionary coherence, which is used for measuring the similarity of atoms is defined as

$$\mu = \max_{u,v \neq u} \left| \psi_u^H \psi_v \right| \tag{25}$$

where ψ_u is the u-th column of dictionary $\mathbf{\Psi}$; $|\cdot|$ denotes the absolute operation. The sensing matrix coherence is extended to a block structure in Reference [21]. The inter-block coherence is mainly used for quantifying the global property, which is defined as

$$\mu_B = \max_{i,j\neq i}\frac{1}{d}\rho\big(G_{i,j}\big) = \max_{i,j\neq i}\frac{1}{d}\rho\big(\mathbf{\Psi}_i^H\mathbf{\Psi}_j\big) \tag{26}$$

where $\mathbf{\Psi}_i$ is the i-th block of $\mathbf{\Psi}$ with d normalized columns; $G_{i,j}$ denotes the (i,j)-th Gram matrix with the dimension of $d \times d$ (yellow entries in Figure 2b); ρ denotes the spectral norm expressed as

$$\rho(G) = \sqrt{\lambda_{\max}\big(G^H G\big)} \tag{27}$$

where $\lambda_{\max}$ is the maximum eigen-value of a positive-semidefinite matrix. Correspondingly, the intra-block coherence used for quantifying the local property is defined as

$$v_B = \max_i \max_{u,v\neq u}\left|G_{i,i}^{u,v}\right| \tag{28}$$

where $G_{i,i}^{u,v}$ is the (u,v)-th block in the i-th Gram matrix with the dimension of $d \times d$ (the red entries in Figure 2b).

For the purpose of improving the average performance of reconstruction as well as separation, all pairs of external blocks and all pairs of internal blocks should be as orthogonal as possible. Therefore, we further generalize the block structure and redefine the block coherence measure.

Assume that there are g_{ex} external blocks and g_{in} internal blocks in each external block. Meanwhile, the number of elements in each internal block is d. The cascaded sensing matrix and the corresponding coefficient vector should then be divided into g_{ex} parts. Similar to the definitions above, we define the total external block coherence to measure their orthogonality, which can be expressed as

$$\mu_{B,ex}^{total} = \sum_{p=1}^{g_{ex}}\sum_{q\neq p}\big\|G_{p,q}\big\|_F^2 \tag{29}$$

where $G_{p,q}$ is the (p,q)-th external Gram matrix with the dimension of $dg_{in} \times dg_{in}$ (green entries in Figure 2b); $\|\cdot\|_F^2$ denotes the Frobenius norm, which is calculated by the sum of the square of all elements in the matrix. For each internal block, we define the total internal coherence measure as

$$\mu_{B,in}^{total} = \sum_{p=1}^{g_{ex}}\sum_{i=1}^{g_{in}}\sum_{j\neq i}\big\|G_p^{i,j}\big\|_F^2 = \sum_{p=1}^{g_{ex}}\big\|G_p\big\|_F^2 - \sum_{p=1}^{g_{ex}}\sum_{i=1}^{g_{in}}\big\|G_p^i\big\|_F^2 \tag{30}$$

where $G_p^{i,j}$ is the (i,j)-th block in the p-th external Gram matrix (the yellow entries in Figure 2b).

3.2. Sensing Matrix Optimization

The optimization of sensing matrix $\widetilde{\Theta}$ in this paper can be defined as the design of an optimal observation matrix $\widetilde{\Phi}$ that improves the performance of NBI separation with a given cascaded block dictionary $\widetilde{\Psi}$, where $\widetilde{\Theta} = \widetilde{\Phi}\widetilde{\Psi}$. The total block coherence measure is

$$\mu_{B,ex}^{total} + \mu_{B,in}^{total} = \sum_{p=1}^{g_{ex}}\sum_{q\neq p}\big\|G_{p,q}\big\|_F^2 + \sum_{p=1}^{g_{ex}}\big\|G_p\big\|_F^2 - \sum_{p=1}^{g_{ex}}\sum_{i=1}^{g_{in}}\big\|G_p^i\big\|_F^2 = \|G - \mathbf{I}\|_F^2 - \xi \tag{31}$$

where $\xi = \sum_{p=1}^{g_{ex}}\sum_{i=1}^{g_{in}}\big\|G_p^i - \mathbf{I}\big\|_F^2$ denotes the penalty for each internal block to measure the normalization error. If the penalty is also taken into account, the problem becomes one of finding an optimal observation matrix $\widetilde{\boldsymbol{\Phi}}^{*}$ to minimize $\|\mathbf{G} - \mathbf{I}\|_F^2$. Of course, we expect both the internal and external block coherence to be as small as possible so that the reconstruction and separation performance are synchronously optimal. However, it was demonstrated in Reference [33] that there is a lower bound in the process of minimizing the equivalent objective function. Inspired by Reference [30], we define a total block coherence measure by weighting $\mu_{B,in}^{total}$ and $\mu_{B,ex}^{total}$, and build an objective function with the weighted block coherence measure as well as the block normalization penalty, i.e.,

$$\widetilde{\boldsymbol{\Phi}}^{*} = \underset{\widetilde{\boldsymbol{\Phi}}}{\arg\min}\ (1-\eta)\mu_{B,ex}^{total}(\widetilde{\boldsymbol{\Phi}}) + \eta\mu_{B,in}^{total}(\widetilde{\boldsymbol{\Phi}}) + \frac{1}{2}\xi\left(\widetilde{\boldsymbol{\Phi}}\right) \tag{32}$$

where $\eta(0 < \eta < 1)$ is the parameter controlling the weight of the external and internal block coherence. To obtain the optimal solution of Equation (32), we first initialize the observation matrix by minimizing $\|\mathbf{G} - \mathbf{I}\|_F^2$, which can be implemented by the eigen-value decomposition of $\widetilde{\boldsymbol{\Psi}}\widetilde{\boldsymbol{\Psi}}^{H}$, i.e.,

$$\widetilde{\boldsymbol{\Psi}}\widetilde{\boldsymbol{\Psi}}^{H} = \widetilde{\mathbf{U}}\Lambda\widetilde{\mathbf{U}}^{H} \tag{33}$$

where Λ is a real diagonal matrix composed of eigen-values; the columns of $\widetilde{\mathbf{U}}$ are the eigen-vectors corresponding to the eigen-values. The initial observation matrix is

$$\widetilde{\boldsymbol{\Phi}}^{(0)} = \mathbf{I}_{M,0}\Lambda^{1/2}\widetilde{\mathbf{U}}^{H} \tag{34}$$

where $\mathbf{I}_{M,0}$ denotes the augmentation matrix of $\mathbf{I}_M$ with zero-valued column vectors.

We define the objective function in the form of Gram matrix as

$$f(\mathbf{G}) = (1-\eta)\|\mathbf{G} - g_{B,ex}(\mathbf{G})\|_F^2 + \eta\|\mathbf{G} - g_{B,in}(\mathbf{G})\|_F^2 + \frac{1}{2}\|\mathbf{G} - g_{\xi}(\mathbf{G})\|_F^2 \tag{35}$$

where

$$g_{\xi}\left(\mathbf{G}_{p,q}^{i,j}\right) = \begin{cases} \mathbf{I}, & p=q, i=j \\ \mathbf{G}_{p,q}^{i,j}, & else \end{cases}$$

$$g_{B,ex}\left(\mathbf{G}_{p,q}^{i,j}\right) = \begin{cases} \mathbf{0}, & p \neq q \\ \mathbf{G}_{p,q}^{i,j}, & else \end{cases} \tag{36}$$

$$g_{B,in}\left(\mathbf{G}_{p,q}^{i,j}\right) = \begin{cases} \mathbf{0}, & p=q, i \neq j \\ \mathbf{G}_{p,q}^{i,j}, & else \end{cases}$$

Then, according to Proposition 1 in Reference [30], the updated observation matrix at the n-th iteration can be obtained by

$$\widetilde{\boldsymbol{\Phi}}^{(n)} = {\Lambda'}_M^{1/2}\widetilde{\mathbf{V}}_M^{H}\Lambda^{1/2}\widetilde{\mathbf{U}}^{H} \tag{37}$$

where ${\Lambda'}_M$ and $\widetilde{\mathbf{V}}_M$ are the top M eigen-values and the corresponding eigen-vectors of $\widetilde{\mathbf{P}}\mathbf{H}\widetilde{\mathbf{P}}^{H}$; $\widetilde{\mathbf{P}} = \Lambda^{-1/2}\mathbf{U}^{H}\widetilde{\boldsymbol{\Psi}}$; $\mathbf{H} = \frac{2}{3}\left[(1-\eta)g_{B,ex}(\mathbf{G}) + \eta g_{B,in}(\mathbf{G}) + \frac{1}{2}g_{\xi}(\mathbf{G})\right]$.

3.3. SMO-BSBL Algorithm

Based on the above analysis and derivation, we provide the detailed sensing matrix optimization-based block sparse Bayesian learning (SMO-BSBL) algorithm for NBI separation and SOI reconstruction in Table 1. Given that the correlation coefficients of NBI are much smaller than that of SOI and that they are calculated separately, we employ the cell-averaging constant false-alarm rate (CA-CFAR) [34] to update the pruning threshold of correlation coefficients of SOI adaptively to avoid low convergence rate when the fixed threshold is initially set too low. The adaptive threshold can be calculated by

$$\gamma_T = T_h \cdot \frac{1}{N_c} \sum_{i=1}^{N_c} \gamma_i \tag{38}$$

where $T_h = N_c \cdot \left(P_{fa}^{-1/N_c} - 1\right)$ denotes the threshold product factor; N_c is the number of detection cells; P_{fa} is the false-alarm rate. From Equation (38), the pruning threshold is affected by the product factor determined by false-alarm rate as well as the number of detection cells. The product factor values under different false-alarm rates are shown in Table 2. If the factor were set too high, over-pruning would occur, which led to serious distortion of the reconstructed SOI. If it was set too low, the convergence rate would not be guaranteed. In this paper, we choose $P_{fa} = 10^{-2}$ as a trade-off value to determine the pruning threshold according to empirical results of tentative experiments without interference.

Table 1. NBI separation algorithm based on sensing matrix optimization-based block sparse Bayesian learning (SMO-BSBL).

Task: Find an Optimal Observation Matrix to Improve NBI Separation and SOI Reconstruction Based on Block Sparse Bayesian Learning.

Inputs:
1. Random observation matrix $\widetilde{\Phi}$;
2. Compressed measurement $\widetilde{y}$;
3. Cascaded dictionary $\widetilde{\Psi} = \left[\widetilde{\Psi}_s \ \widetilde{\Psi}_n\right]$;
4. Block size $d_i (i = 1,\ldots,g)$;
5. Number of external and internal blocks g_{ex}, g_{in};
6. Coherence weight η;

Outputs:
1. Optimal observation matrix $\widetilde{\Phi}^*$;
2. Reconstructed NBI-free signal $\widehat{s}^*$;

Initialization:
1. Initialize $\widetilde{\Phi}$ as an M×N Gaussian random matrix;
2. Initialize the maximum number of optimizing iterations as $N_{max} = 500$;
3. Initialize the public parameters to be estimated as $\sigma^2 = 10^{-3}$;
4. Initialize the parameters for separation as $\mathbf{B} = [\mathbf{B}_s \ \mathbf{B}_n] = diag[\mathbf{eye}(d_1),\ldots,\mathbf{eye}(d_g)]$, $\gamma = [\gamma_s \gamma_n] = 1$;
5. Initialize the threshold for pruning out γ as $\gamma_{T_s} = \gamma_{T_n} = 10^{-2}$;
6. Initialize the iteration stop condition as $\Delta\gamma_{stop} = 10^{-5}$;
7. Initialize the maximum number of reconstructive iterations as $K_{max} = 1000$;

A. Sensing matrix optimizing stage
1. Calculate the total block coherence by Equations (29)–(31);
2. Build objective function for optimizing by Equation (32);
3. Calculate a new initialized observation matrix $\widetilde{\Phi}^{(0)}$ by Equations (33)–(34);
   Repeat from $n = 0$ until $n = N_{max} - 1$
   (1) Calculate the Gram matrix by $G^{(n)} = \left(\widetilde{\Phi}^{(n)}\widetilde{\Psi}\right)^H \widetilde{\Phi}^{(n)}\widetilde{\Psi}$;
   (2) Build the equivalent objective function by Equations (35)–(36);
   (3) Update the optimal $\widetilde{\Phi}^{(n)}$ by Equation (37);
   (4) $n = n + 1$;
4. Set $\widetilde{\Phi}^* = \widetilde{\Phi}^{(n)}$;

B. Separation and reconstruction stage:
1. Reset the sensing matrix by $\widetilde{\Theta}^* = \widetilde{\Phi}^*\widetilde{\Psi}$;
   Repeat from $k = 1$ until K_{max} or $\left\|\gamma^k - \gamma^{k-1}\right\|_\infty < \Delta\gamma_{stop}$;
   (1) Update the prior covariance matrix $\mathbf{C}_0$ by Equation (11);
   (2) Update the covariance matrix by $\mathbf{C}_{\widetilde{\alpha}} = \left(\mathbf{C}_0^{-1} + \sigma^{-2}\widetilde{\Theta}^H\widetilde{\Theta}\right)^{-1}$;
   (3) Update the expectation by $\mu_{\widetilde{\alpha}} = \sigma^{-2}\mathbf{C}_{\widetilde{\alpha}}\widetilde{\Theta}^H\widetilde{y}$;
   (4) Update the parameters $\sigma^2, \gamma_s^{(k)}, \widetilde{\mathbf{B}}_s^{(k)}, \gamma_n^{(k)}, \widetilde{\mathbf{B}}_n^{(k)}$ by Equations (20)–(24);
   (5) Update the threshold of γ_{T_s} by Equation (38)
   (6) $k = k + 1$
2. Calculate coefficient by $\widetilde{\alpha}^* = [\widetilde{\alpha}_s^*; \widetilde{\alpha}_n^*] = \mu_{\widetilde{\alpha}}^*$;
3. Reconstruct the NBI-free signal by $\widehat{s}^* = \widetilde{\Psi}\widetilde{\alpha}_s^*$.

Table 2. Product factor values under different false-alarm rates.

	$P_{fa} = 10^{-1}$	$P_{fa} = 10^{-2}$	$P_{fa} = 10^{-3}$	$P_{fa} = 10^{-4}$
$N_c = 4$	3.11	8.65	18.49	36.00
$N_c = 8$	2.69	6.23	10.97	17.30
$N_c = 16$	2.48	5.34	8.64	12.45
$N_c = 32$	2.39	4.95	7.71	10.67

3.4. SAR Imaging Procedure with NBI Separation

The flowchart of SAR imaging with NBI separation based on SMO-BSBL is shown in Figure 3. It is obvious that the proposed algorithm can be embedded in the imaging process with excellent compatibility. As indicated by the omissible procedure in the dashed wireframe in Figure 3, the steps of NBI reconstruction and cancellation that increase the system complexity are not necessary in the presented procedure, and the reconstructed coefficients corresponding to the SOI can be directly used for clean image formation. Furthermore, the echo data in each pulse can be processed in parallel, and the formed two-dimensional matrix will then be used for range-azimuth imaging with range cell migration correction (RCMC). The two-dimensional imaging process with compressed measurements is explained in Reference [35], so we will not discuss it in detail in this paper.

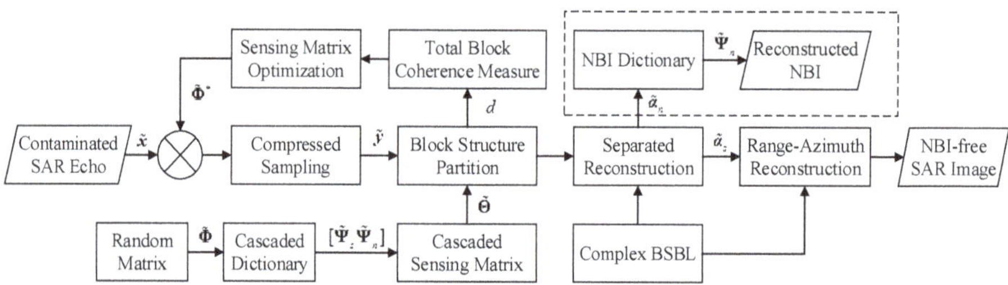

Figure 3. Synthetic aperture radar (SAR) imaging procedure with NBI separation via the SMO-BSBL algorithm (the procedure in the dashed wireframe can be omitted when it is unnecessary).

4. Experiments

4.1. Experiment Setup

4.1.1. Simulation Specification

In order to demonstrate the effectiveness and superiority of the proposed algorithm, we carried out multiple numerical experiments with simulated data. We started with a simple case where the proposed SMO-BSBL was adopted to one-dimensional range profile imaging of a multi-points target that is non-sparse in range cells. We then extended the case to the range-azimuth imaging of an aircraft target. At the same time, we analyzed the performance under different parameters and compared it with that of other advanced algorithms. Simulations were carried out in window7(64bit) system on the computer with 3.4GHz Intel Core i7-4770 CPU and 16GB memory.

4.1.2. Performance Indicators

To benchmark the performance of NBI suppression via different methods comprehensively, we employed multiple indicators in both the signal and image domains. In the signal domain, the following indicators can be used for evaluating the performance of SOI reconstruction and NBI separation.

(1) Normalized mean square error (NMSE).

The mean square error (MSE) is widely used to measure how much the reconstructed signal deviates from the original. Considering that the MSE may be affected by signal type and power, we normalize it as

$$NMSE = \frac{\|x - \hat{x}\|_2^2}{\|x\|_2^2} \tag{39}$$

where x and $\hat{x}$ are, respectively, the original and reconstructed signal; $\|\cdot\|_2$ denotes the l_2-norm. In this paper, when x represents the SOI, the NMSE, which reflects the degree of signal distortion, can also be considered as the ratio of the constructed residual error to the real signal.

(2) Interference suppression degree (ISD).

We also employ the ISD to check the NBI suppression effectiveness, which is defined as

$$ISD = 20 \log_{10} \frac{\|x - s\|_2}{\|\hat{s} - s\|_2} \tag{40}$$

where x is the contaminated signal; s and $\hat{s}$ are the original and reconstructed SOI, respectively. The ISD reflects the ratio of undesirable components in the SOI before and after interference suppression. It is a comprehensive indicator in which both the interference suppression performance and signal distortion are considered. A larger ISD indicates a better performance of NBI separation.

As mentioned at the beginning of this paper, NBI may cause serious damage to SAR images. Therefore, the performance can also be evaluated from the perspective of image quality. In Reference [36], we proposed several performance indicators for quality evaluation of SAR image. In this paper, we select the peak signal-to-noise ratio (PSNR), the equivalent number of looks (ENL), and the image entropy as the main indicators for the image quality evaluation.

(1) Peak signal-to-noise ratio (PSNR).

The PSNR is a common indicator for evaluating the reconstructed image quality, which is often defined by the MSE. Given that SAR images are more discrete than optical ones, here we redefine part of the physical meaning and apply it to the reconstructed SAR image evaluation. The PSNR for a SAR image can be defined as

$$PSNR = 10 \log_{10} \left(\frac{\sum_{p=1}^{P} \max_P |A_{i,j}|^2}{\frac{1}{N_a N_r - P} \left(\sum_{i=1}^{N_a} \sum_{j=1}^{N_r} |A_{i,j}|^2 - \sum_{p=1}^{P} \max_P |A_{i,j}|^2 \right)} \right) \tag{41}$$

where N_a and N_r are the number of azimuth and range cells of a SAR image; P is the number of scattering points; $A_{i,j}$ denotes the complex value of the point at the (i,j)-th position; $|\cdot|$ denotes the modulus value; $\max_P$ represents picking out P largest values. The PSNR reflects the extent to which the SAR image is affected by noise or interference, and a larger value of PSNR indicates better image quality. It is worth noting that the PSNR specifically redefined for SAR can evaluate the quality of a reconstructed SAR image without any prior information of the original one as long as the number of target points is known or probably known.

(2) Equivalent number of looks (ENL).

The ENL is often used for measuring the relative intensity of speckle noise for SAR. Considering that SAR data is complex-valued, it is necessary to convert it to a grayscale one in advance. The ENL of a SAR image is defined as

$$ENL = 10 \log_{10} \frac{\mu^2}{\sigma^2} \tag{42}$$

where μ and σ are the mean and standard deviation of SAR image grayscale. The ENL can reflect the contrast ratio of the image, and a larger ENL indicates that there is more noise or interference in the SAR image, which leads to substantial blurring.

(3) Image entropy.

The image entropy is a statistical form used for representing the aggregation characteristic of the grayscale distribution and for measuring the average amount of information in an image. It can be expressed as

$$E = -\sum_{i}^{N_G} p_i \log_2 p_i \tag{43}$$

where p_i is the probability of the i-th grayscale level; N_G is the total number of all grayscale levels in the image. For traditional images, a larger entropy indicates that the image contains more information and is of higher quality. However, the principle of SAR imaging is different from that of conventional optical imaging, and a non-uniform grayscale histogram distribution can highlight the texture or the contour in the observation scene. Therefore, we would rather obtain a SAR image with a smaller entropy after the noise-modulated NBI suppression.

4.2. Simulation and Analysis

4.2.1. Range Profile Imaging

Given that the BSBL framework is capable of reconstructing signals in non-sparse cases, it is obviously a better option for recovering the extended target or signal with a certain bandwidth. To verify the effectiveness and superiority of our proposed algorithm, we simulated range profile imaging for an extended target.

First, we modeled an extended target by generating 30 scattering points with random normalized backscattering coefficients from 0 to 1 and random locations within the range of 256 m. The signal transmitted from a 3-km-high radar was modeled as an LFM waveform, and the signal bandwidth and pulse width were 100 MHz and 1 µs, respectively. Since the number of valid range cells was 265 and the theoretical range resolution was 1.5 m according to the above parameters, we considered a target with more than 20 scattering points as an extended target. Here, the number of range cell was set to 512.

We then generated the NBI data by modulating a band-limited noise signal with Rayleigh distribution to the carrier frequency of SAR and aligning it with the central band of transmitted signal, and added the NBI to the raw echo. The interference-to-signal ratio (ISR) was set to 15 dB and the additive signal-to-noise ratio (SNR) was set to 30 dB. The bandwidth of NBI was successively set to 10 MHz and 20 MHz. The waveform in the time domain and the spectrum in the frequency domain of SOI and NBI as well as the range distribution of the extended target are shown in Figure 4.

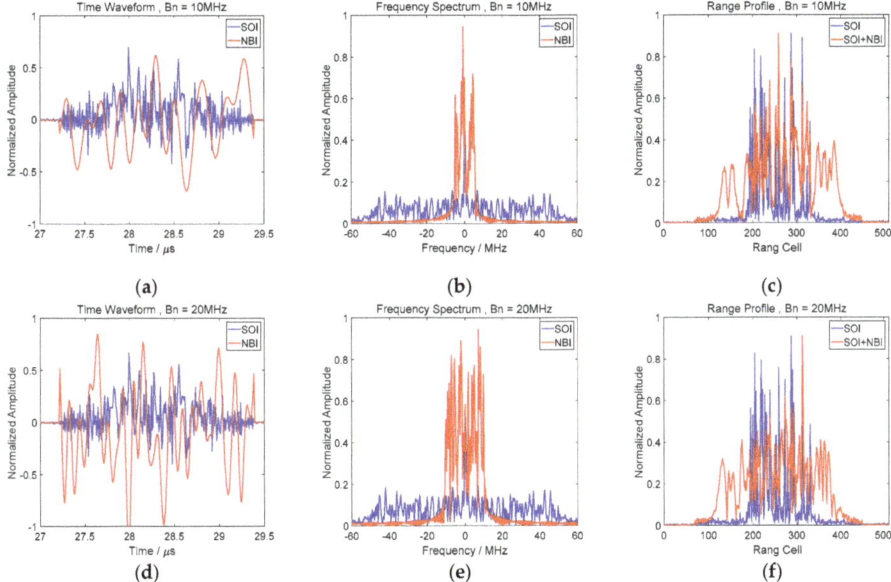

Figure 4. Characteristic of the signal of interest (SOI) and NBI in different domains. (**a**) Waveform in time domain, B_n = 10 MHz; (**b**) spectrum in the frequency domain, B_n = 10 MHz; (**c**) range distribution, B_n = 10 MHz; (**d**) waveform in time domain, B_n = 20 MHz; (**e**) spectrum in frequency domain, B_n = 20 MHz; (**f**) range distribution, B_n = 20 MHz.

It is obvious that the SOI and NBI is non-sparse in both the time and frequency domains. The distribution of the extended target in range cells is not sparse either. In other words, it is almost impossible to represent the signal and NBI with a small number of non-zero coefficients via traditional sparsity-based methods such as BP, MP, and OMP. Therefore, we attempted to compare the NBI separation performance of our proposed algorithm only with other advanced BSBL framework-based algorithms.

Next, we performed the separated reconstruction of SOI and NBI from the contaminated echo with the proposed SMO-BSBL algorithm in this paper. The initial dictionary used for representing SOI was composed of a reference signal with delays, and the Fourier basis was used as the initial dictionary for NBI. The number of external blocks was 2, and each external block contained 16 internal blocks. The internal block size was set to 16, and the maximum iteration was set to 500 to guarantee convergence. In order to find a better weight for sensing matrix optimization, we set η from 0.1 to 0.9 and carried out the reconstruction simulation 100 times. The Gram matrices optimized with different weights and signal reconstruction performance are shown in Figures 5 and 6.

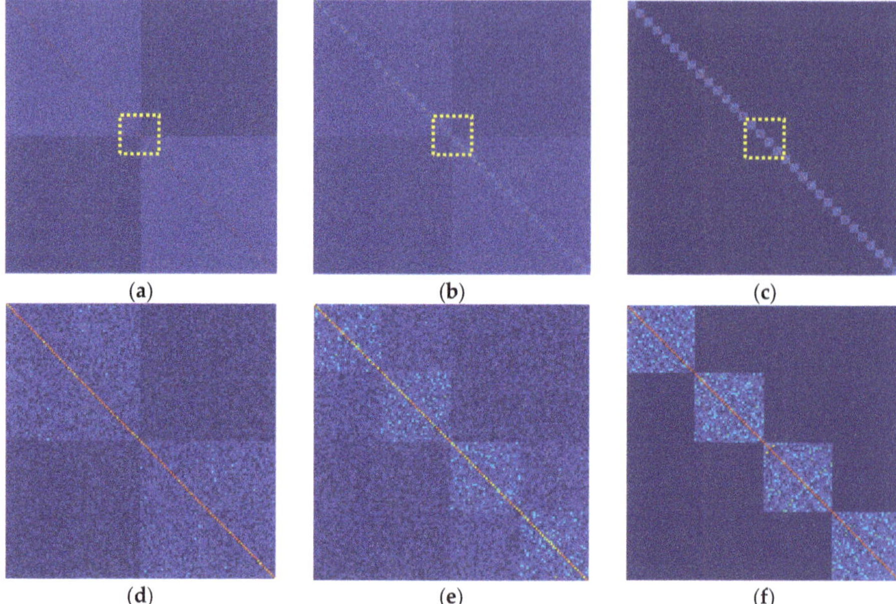

Figure 5. Gram matrices optimized with different weights. (**a**) $\eta = 0.1$; (**b**) $\eta = 0.5$; (**c**) $\eta = 0.9$; (**d**) enlarged view of $\eta = 0.1$ in the dashed wireframe; (**e**) enlarged view of $\eta = 0.5$ in the dashed wireframe; (**f**) enlarged view of $\eta = 0.5$ in the dashed wireframe.

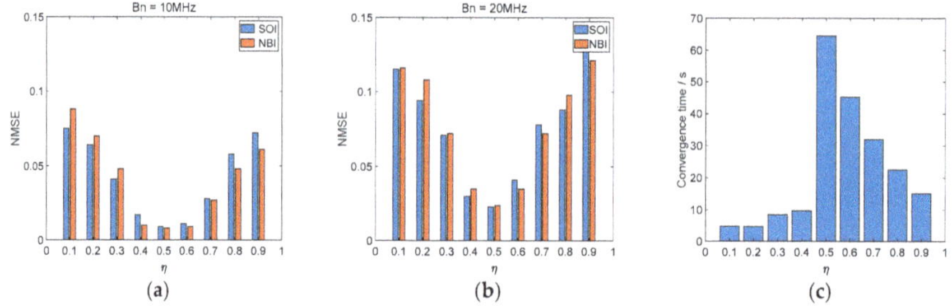

Figure 6. Histogram of reconstruction performance under different weights. (**a**) Reconstruction error with B_n = 10 MHz; (**b**) reconstruction error with B_n = 20 MHz; (**c**) SMO convergence time.

There is an irreconcilable contradiction between the external and internal block coherence. When the weight controlling parameter η is set to a small value, as shown in Figure 5a,d, we attempted to minimize the coherence measure between external blocks as far as possible, ignoring the internal block coherence. On the contrary, in Figure 5c,f, η with a larger value indicates that minimizing the internal block coherence measure is more dominant, regardless of which component the internal block belongs to. Therefore, a trade-off weight value is set in Figure 5b,e to take both the two block structures into account. From the statistical result of reconstruction simulation in Figure 6a,b, we obtained the minimum reconstruction error for both SOI and NBI when η is close to 0.5. In Figure 6c, however, when η reaches 0.5, the convergence time increases dramatically. Thus, we set η to 0.4 as a trade-off value for subsequent experiments.

In addition, we adopted the basic BSBL combined with a cascaded dictionary in our previous work [16], which makes no distinction when updating the covariance matrix of different components, and the S-BSBL [25], which takes this distinction into account but ignores the block coherence of the cascade dictionary, to make a comparison under the same parameters. The spectrum of reconstructed NBI and the range profile of the extended target via different algorithms are shown in Figures 7 and 8.

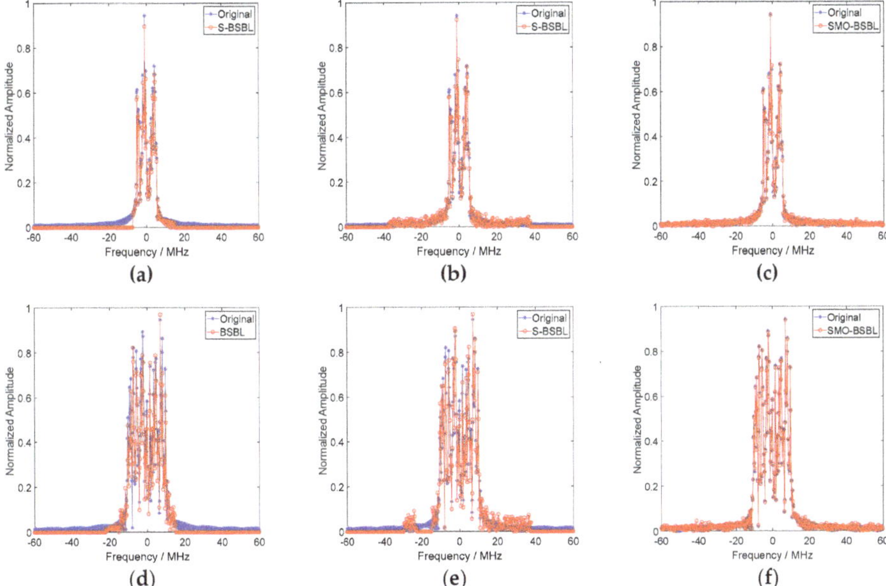

Figure 7. Spectrum of the reconstructed NBI. (**a**) BSBL, B_n = 10 MHz; (**b**) S-BSBL, B_n = 10 MHz; (**c**) SMO-BSBL, B_n = 10 MHz; (**d**) BSBL, B_n = 20 MHz; (**e**) S-BSBL, B_n = 20 MHz; (**f**) SMO-BSBL, B_n = 20 MHz.

In Figure 7a–c, the NBI with a bandwidth of 10 MHz is successively reconstructed via BSBL, S-BSBL, and SMO-BSBL. It can be seen from the spectra that the proposed SMO-BSBL algorithm performs better than the other two. In Figure 7d–e, with the NBI bandwidth increasing, performance degradation occurs for each algorithm, which is due to the expansion of overlap between NBI and SOI in the frequency domain. Nevertheless, the proposed algorithm is still superior to the other two algorithms.

As shown in Figure 8, the proposed algorithm outperforms the others with a smaller error from range profile imaging through pulse compression of the reconstructed SOI. Specifically, it is obvious that the distortion of the range profile after NBI suppression with BSBL is substantial. The comparison shows the significance of block coherence to separated reconstruction.

We then calculate the interference suppression degree (ISD) with 500 numerical simulations for each value of ISR ranged from 0 to 30 dB. The statistical results are shown in Figure 9.

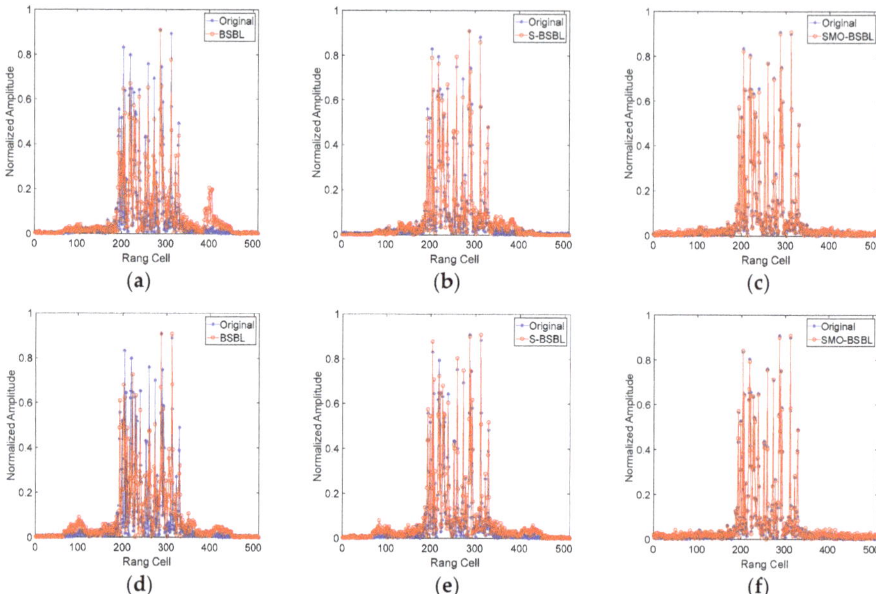

Figure 8. Range profile of the reconstructed SOI. (**a**) BSBL, B_n = 10 MHz; (**b**) S-BSBL, B_n = 10 MHz; (**c**) SMO-BSBL, B_n = 10 MHz; (**d**) BSBL, B_n = 20 MHz; (**e**) S-BSBL, B_n = 20 MHz; (**f**) SMO-BSBL, B_n = 20 MHz.

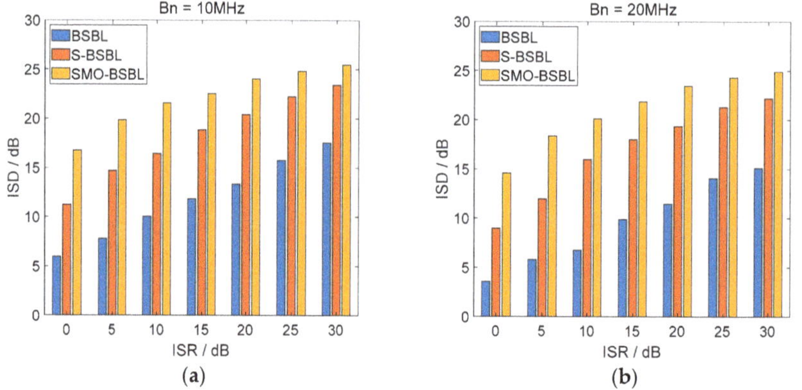

Figure 9. Interference suppression degree under different interference-to-signal ratio (ISR). (**a**) B_n = 10 MHz; (**b**) B_n = 20 MHz.

As shown in Figure 9, the ISD increases with ISR, and the average ISD of the proposed SMO-BSBL is nearly 5 dB higher than that of S-BSBL and 10 dB higher than that of BSBL. The SMO-BSBL and the S-BSBL are less affected by the bandwidth of NBI than is BSBL.

In addition, we analyzed the influence of compressed ratio (CR), which is defined as the ratio of the actual sampling rate to the Nyquist rate, and the influence of block parameters on the separated reconstruction performance of our proposed SMO-BSBL algorithm. We set 1 and 0.5, respectively, for CR to check signal reconstruction robustness with compressed measurement and calculated the average NMSE under different ISRs. To benchmark the algorithm complexity, we set the block size from 8 to 64 empirically according to signal length, and calculated the average iteration time under different block size. We also compared the time performance of the widely used Bi-channel BSBL, the complex BSBL in this paper, and the case of compressed sampling. The statistical results of 100 simulation experiments are shown in Figure 10.

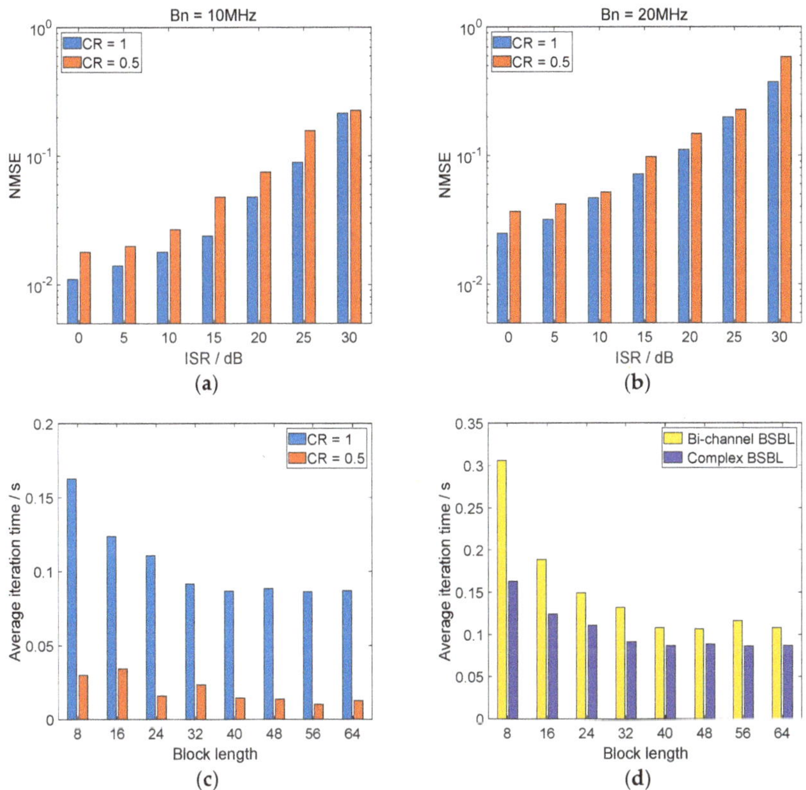

Figure 10. Reconstruction and time performance under different compressed ratio (CR) and block length. (**a**) Signal reconstruction error, $B_n = 10$ MHz; (**b**) signal reconstruction error, $B_n = 20$ MHz; (**c**) average iteration time under different CR; (**d**) average iteration time under different methods.

As shown in Figure 10a,b, when CR is set to 0.5, which indicates that only half of all the data were used to reconstruct the SOI and NBI, it is inevitable that the performance will degrade relative to full sampling. However, the degradation of performance is acceptable, especially when the bandwidth of NBI increases. Another benefit of compressed sampling is the reduction of the data dimension; as can be seen in Figure 10c, the average iteration time using compressed data is significantly lower than that using full data. We admit that it takes a certain amount of time in the process of sensing matrix optimization to improve reconstruction performance; nevertheless, our algorithm, as shown in Figure 10d, is more time-efficient than the widely used bi-channel approach in the process of signal reconstruction.

4.2.2. Range-Azimuth Imaging

To further verify the performance of NBI separation in range-azimuth imaging, we modeled an aircraft target with multiple scattering points and carried out simulation experiments with the proposed algorithm. For the convenience of analysis and verification, we assumed that the SAR platform operates in airborne strip-map mode. Referring to the parameters in our previous work, the main simulation parameters in this section were set using parameters listed in Table 3.

Table 3. Main simulation parameters of SAR imaging.

Parameter Class	Parameter Name	Parameter Value
Platform	Platform height	3000 m
	Pitch angle	45°
	Squint Angle	0°
Target	Number of points	1932
	Scene vertical range	−128 to 128 m
	Scene parallel range	−128 to 128 m
Signal	Carrier frequency	3 GHz
	Bandwidth	100 MHz
	Pulse width	1 μs
	Pulse repetition frequency	125 Hz
	Oversampling coefficient	1.2
	Size of Range-Azimuth Cells	512 × 512

The imaging results without NBI via BSBL at different CRs are shown in Figure 11. The CR is set to 1, 0.5, and 0.25 in turn, and the block size is 16. It can be seen that the BSBL-based imaging algorithm avoids the sidelobe effect by replacing the matched filtering with sparsity-based reconstructing and improves the quality of the SAR image. As the CR decreases, the imaging quality degrades slightly; however, when CR drops to 0.25, the quality is still high enough to distinguish the target.

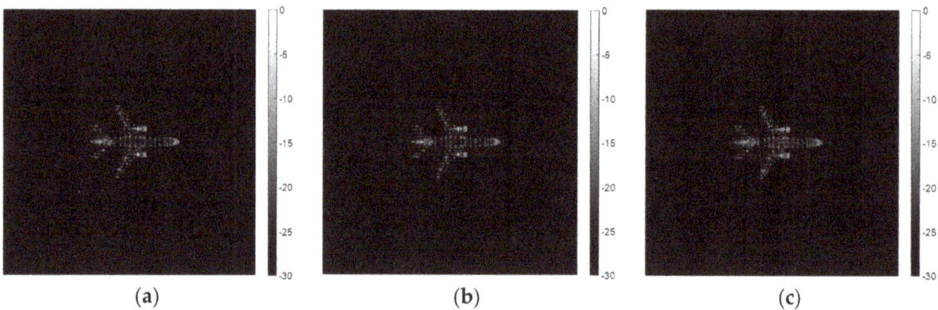

Figure 11. Range-azimuth imaging results via BSBL without NBI. (**a**) CR = 1; (**b**) CR = 0.5; (**c**) CR = 0.25.

On the basis of the above results, we carried out numerical experiments of NBI separation for the simulated aircraft target. Narrowband noise-modulated interference was added to the raw echo data. We assumed that the pulse width of the NBI was equal to the entire duration of the signal for each pulse. The ISR was set to 15 dB, the CR was set to 0.5, and the bandwidth of the NBI was set to 10 MHz and 20 MHz. Imaging results before and after NBI suppression via our proposed SMO-BSBL as well as other advanced algorithms are shown in Figure 12.

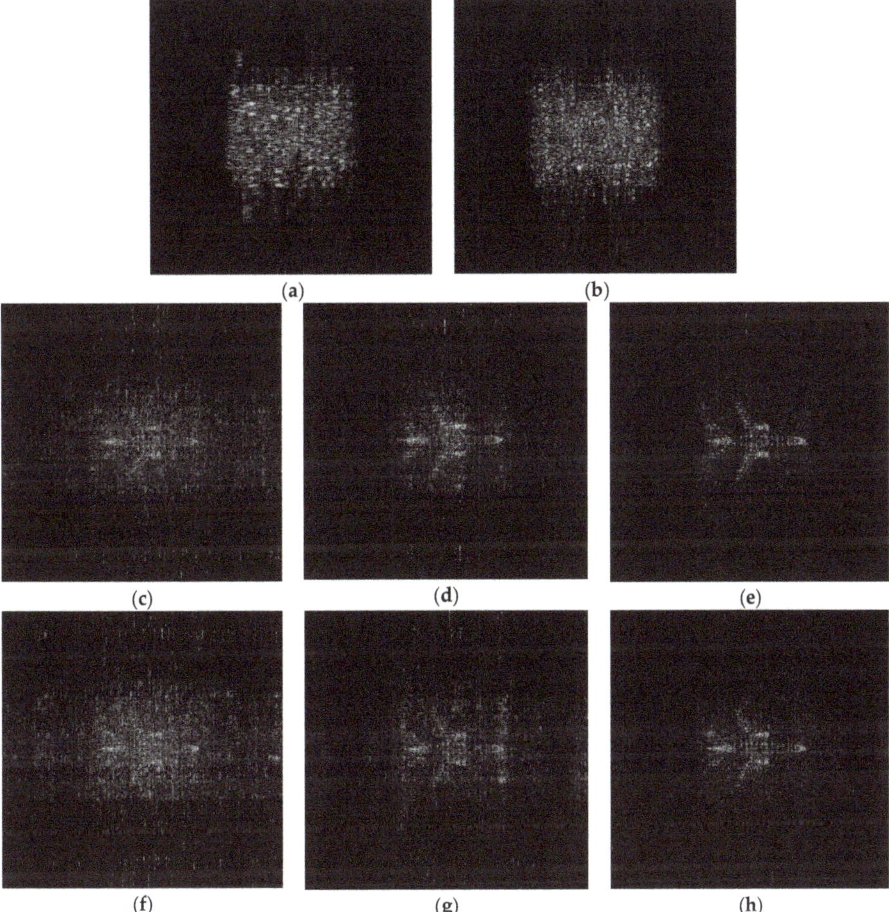

Figure 12. Range-azimuth imaging results of the simulated aircraft target. (**a**) Without NBI suppression (B_n = 10 MHz); (**b**) without NBI suppression (B_n = 20 MHz); (**c**) BSBL (B_n = 10 MHz); (**d**) S-BSBL (B_n = 10 MHz); (**e**) SMO-BSBL (B_n = 10 MHz); (**f**) BSBL (B_n = 20 MHz); (**g**) S-BSBL (B_n = 20 MHz); (**h**) SMO-BSBL (B_n = 20 MHz).

In Figure 12a,b, without any suppression approach, NBI completely obscures the target in the SAR image when its bandwidth is 10% of the signal bandwidth. In Figure 12c–e, BSBL, S-BSBL, and SMO-BSBL are adopted to separate the NBI. The BSBL algorithm, which only builds a cascaded dictionary to separate NBI and the SOI, fails to recover the target effectively, since it makes no distinction between the two components when updating the covariance matrix and correlation coefficients. The S-BSBL algorithm takes this distinction into account and improves the separation and reconstruction quality. However, the suppression result is still not satisfactory, since the coherence between different sub-dictionaries will seriously disrupt the reconstruction process. In contrast, the SMO-BSBL algorithm proposed in this paper shows a superior performance in this kind of NBI separation owing to the optimal sensing matrix, which is designed to minimize the total block coherence measure. In Figure 12f–h, with the bandwidth of the NBI increasing, the target is more substantially covered. It is difficult for BSBL and S-BSBL to extract and separate NBI from the contaminated echo. The SMO-BSBL is still capable of separating NBI despite some performance degradation.

To benchmark the NBI separation performance in the image domain, we calculated the PSNR, ENL, and image entropy of the reconstructed SAR images using Equations (41)–(43). The statistical results are shown in Table 4.

Table 4. Statistical results of simulated SAR image quality.

	Original	B_n	Contaminated	BSBL	S-BSBL	SMO-BSBL
PSNR (dB)	18.809	10 MHz	10.489	11.441	15.617	16.322
		20 MHz	10.479	11.319	12.770	14.915
ENL (dB)	1.537	10 MHz	3.840	2.950	2.366	2.140
		20 MHz	3.681	2.941	3.463	2.636
Entropy	3.902	10 MHz	6.006	5.404	4.435	4.292
		20 MHz	5.877	5.122	4.950	4.577

As shown in Table 4, the quality of the reconstructed image via the proposed SMO-BSBL is superior to the other two advanced algorithms according to the statistical results of indicators.

5. Conclusions

In this paper, we present an enhanced NBI separation algorithm for SAR data on the basis of a sparse Bayesian learning framework. The proposed sensing matrix optimization-based block sparse Bayesian learning, which is abbreviated as SMO-BSBL, is focused on reducing the block coherence between the sensing matrix of the SOI and NBI in order to improve the separated reconstruction performance. First, we review the NBI suppression problem based on the sparse recovery model, and we then extend the basic BSBL framework to a complex-valued domain for the radar signal to reduce computational complexity. For the sake of enhancing the separability, we propose a new block coherence measure that is calculated by the newly defined external and internal block structure. We obtained an optimal sensing matrix by minimizing the optimization objective function and adopted it to the modified BSBL framework for sparse reconstruction. Moreover, we described the entire procedure of NBI separation for SAR imaging where the proposed algorithm can be embedded with excellent compatibility. Finally, we carried out simulation experiments including range imaging and range-azimuth imaging of extended targets to verify the effectiveness and superiority of our proposed algorithm. The statistical results of different indicators demonstrate that the SMO-BSBL in this paper outperforms other advanced BSBL-based algorithms for NBI separation. It is necessary to note here that, while the reconstruction time reduces under a complex BSBL framework, the total computational complexity still increases, since the sensing matrix optimization process requires additional time as a cost. Therefore, how to accelerate the convergence rate of optimization process to improve the real-time performance is our future work. In addition, given that the practical SNR condition for SAR system is not ideal in real environment, and the estimation rules of noise variance are different with different SNR ranges in basic BSBL framework, analyses of the impact of SNR on the algorithm performance will also be the focus in the subsequent studies.

Author Contributions: Methodology and formal analysis: G.L. (Guojing Li); investigation: G.L. (Guochao Lao) and S.K.; simulation: G.L. (Guojing Li), S.K., and D.Y.; writing—review and editing: G.L. (Guojing Li); supervision: W.Y.

Funding: This research was funded by a Research Project of State Key Laboratory of Complex Electromagnetic Environment Effects on Electronics and Information System, grant number 2017Z0203B.

Conflicts of Interest: The authors declare no conflict of interest.

References

1. Lin, X.; Li, X.; Man, X.; Tian, W. Narrow-band interference suppression method in multichannel SAR based on beamforming technique and sparse recovery. *Electron. Lett.* **2018**, *54*, 1189–1191. [CrossRef]

2. Meller, M.; Niedźwiecki, M. Multichannel self-optimizing narrowband interference canceller. *Signal Process* **2014**, *98*, 396–409. [CrossRef]

3. Djukanovic, S.; Popovic, V. A Parametric Method for Multicomponent Interference Suppression in Noise Radars. *IEEE Trans. Aerosp. Electron. Syst.* **2012**, *48*, 2730–2738. [CrossRef]

4. Zhou, F.; Xing, M.; Bai, X.; Sun, G.; Bao, Z. Narrow-Band interference suppression for SAR based on complex empirical mode decomposition. *IEEE Geosci. Remote S.* **2009**, *6*, 423–427. [CrossRef]

5. Lamont-Smith, T.; Hill, R.D.; Hayward, S.D.; Yates, G.; Blake, A. Filtering approaches for interference suppression in low-frequency SAR. *IEEE Proc. Radar Sonar. Navig.* **2006**, *153*, 338–344. [CrossRef]

6. Wu, P.; Yang, L.; Zhang, Y. A modified notch filter for suppressing radio-frequency-interference in P-band SAR data. In Proceedings of the 2016 IEEE International Geoscience and Remote Sensing Symposium (IGARSS), Beijing, China, 10–15 July 2016.

7. Lord, R.T.; Inggs, M.R. Efficient RFI suppression in SAR using LMS adaptive filter integrated with range/Doppler algorithm. *Electron. Lett.* **1999**, *35*, 629–630. [CrossRef]

8. Zhou, F.; Wu, R.; Xing, M.; Bao, Z. Eigensubspace-Based filtering with application in Narrow-Band interference suppression for SAR. *IEEE Geosci. Remote S.* **2007**, *4*, 75–79. [CrossRef]

9. Zhou, F.; Tao, M.; Bai, X.; Liu, J. Narrow-Band interference suppression for SAR based on independent component analysis. *IEEE Trans. Geosci. Remote* **2013**, *51*, 4952–4960. [CrossRef]

10. Tao, M.; Zhou, F.; Liu, J.; Liu, Y.; Zhang, Z.; Bao, Z. Narrow-Band interference mitigation for SAR using independent subspace analysis. *IEEE Trans. Geosci. Remote* **2014**, *52*, 5289–5301.

11. Zhou, F.; Tao, M. Research on methods for narrow-band interference suppression in synthetic aperture radar data. *IEEE J.-STARS* **2015**, *8*, 3476–3485. [CrossRef]

12. Su, J.; Tao, H.; Tao, M.; Wang, L.; Xie, J. Narrow-Band interference suppression via RPCA-Based signal separation in time–frequency domain. *IEEE J.-STARS* **5016**, *10*, 5016–5025. [CrossRef]

13. Nguyen, L.H.; Tran, T.D. Efficient and robust RFI extraction via sparse recovery. *IEEE J.-STARS* **2016**, *9*, 2104–2117. [CrossRef]

14. Huang, Y.; Liao, G.; Li, J.; Xu, J. Narrowband RFI suppression for SAR system via fast implementation of joint sparsity and Low-Rank property. *IEEE Trans. Geosci. Remote* **2018**, *56*, 2748–2761. [CrossRef]

15. Huang, Y.; Liao, G.; Zhang, Z.; Xiang, Y.; Li, J.; Nehorai, A. Fast narrowband RFI suppression algorithms for SAR systems via Matrix-Factorization techniques. *IEEE Trans. Geosci. Remote* **2019**, *57*, 250–262. [CrossRef]

16. Li, G.; Liu, G.; Ye, W. RFI mitigation for SAR based on compressed sensing and morphological component analysis. In Proceedings of the 2017 IEEE International Conference on Signal Processing, Communications and Computing (ICSPCC), Xiamen, China, 22–25 October 2017.

17. Boyd, S.; Parikh, N.; Chu, E.; Peleato, B.; Eckstein, J. *Distributed Optimization and Satistical Learning via the Alternating Direction Method of Mmultipliers*; Now Publishers Inc.: Hanover, MA, USA, 2011; pp. 13–24.

18. Liu, H.; Li, D. RFI suppression based on sparse frequency estimation for SAR imaging. *IEEE Geosci. Remote Sens.* **2016**, *13*, 63–67. [CrossRef]

19. Donoho, D. Compressed sensing. *IEEE Trans. Inform. Theory* **2006**, *52*, 1289–1306. [CrossRef]

20. Tropp, J.A.; Gilbert, A.C. Signal recovery from random measurements via orthogonal matching pursuit. *IEEE Trans. Inform. Theory* **2007**, *53*, 4655–4666. [CrossRef]

21. Eldar, Y.C.; Kuppinger, P.; Bolcskei, H. Block-Sparse signals: Uncertainty relations and efficient recovery. *IEEE Trans. Signal Proces.* **2010**, *58*, 3042–3054. [CrossRef]

22. Tipping, M.E. Sparse Bayesian learning and the relevance vector machine. *J. Mach. Learn. Res.* **2001**, *1*, 211–244.

23. Zhang, Z.; Rao, B.D. Sparse signal recovery with temporally correlated source vectors using sparse bayesian learning. *IEEE J.-STSP* **2011**, *5*, 912–926. [CrossRef]

24. Zhang, Z.; Rao, B.D. Extension of SBL algorithms for the recovery of block sparse signals with Intra-Block correlation. *IEEE Trans. Signal Proces.* **2013**, *61*, 2009–2015. [CrossRef]

25. Lu, X.; Su, W.; Yang, J.; Gu, H.; Zhang, H.; Yu, W.; Yeo, T.S. Radio frequency interference suppression for SAR via block sparse Bayesian learning. *IEEE J.-STARS* **2018**, *11*, 4835–4847. [CrossRef]

26. Cumming, I.G.; Wong, F.H. *Digital Processing of Synthetic Aperture Radar Data: Algorithms and Implementation*; Publishing House of Electronics Industry: Beijing, China, 2012; pp. 75–104.

27. Kay, S.M. *Fundamentals of Statistical Signal Processing: Estimation Theory*; Prentice Hall PTR: Upper Saddle River, NJ, USA, 1993; p. 504.

28. Mahler, R. *Statistical Multisource-Multitarget Information Fusion*; Artech House: Norwood, MA, USA, 2007; pp. 699–703.
29. Hjørungnes, A. *Complex-Valued Matrix Derivatives: With Applications in Signal Processing and Communications*; Cambridge University Press: New York, NY, USA, 2011; pp. 43–59.
30. Zelnik-Manor, L.; Rosenblum, K.; Eldar, Y.C. Sensing matrix optimization for Block-Sparse decoding. *IEEE Trans. Signal Proces.* **2011**, *59*, 4300–4312. [CrossRef]
31. Zelnik-Manor, L.; Rosenblum, K.; Eldar, Y.C. Dictionary optimization for Block-Sparse representations. *IEEE Trans. Signal Proces.* **2012**, *60*, 2386–2395. [CrossRef]
32. Elad, M.; Bruckstein, A.M. A generalized uncertainty principle and sparse representation in pairs of bases. *IEEE Trans. Inform. Theory* **2002**, *48*, 2558–2567. [CrossRef]
33. Duarte-Carvajalino, J.M.; Sapiro, G. Learning to sense sparse signals: Simultaneous sensing matrix and sparsifying dictionary optimization. *IEEE Trans. Image Process.* **2009**, *18*, 1395–1408. [CrossRef] [PubMed]
34. Gandhi, P.; Kassam, S. Analysis of CFAR processors in nonhomogeneous background. *IEEE Trans. Aerosp. Electron. Syst.* **1988**, *24*, 427–445. [CrossRef]
35. Fang, J.; Xu, Z.; Zhang, B.; Hong, W.; Wu, Y. Fast Compressed Sensing SAR Imaging Based on Approximated Observation. *IEEE J.-STARS* **2014**, *7*, 352–363.
36. Lao, G.; Ye, W.; Li, G.; Zhang, W. A quality evaluation method of SAR image based on grayscale image and electromagnetic scattering characteristics. In Proceedings of the 8th International Conference on Digital Image Processing, Chengdu, China, 29 August 2016; Society of Photo-Optical Instrumentation Engineers (SPIE): Bellingham, WA, USA, 2016. [CrossRef]

Article

Bistatic-ISAR Linear Geometry Distortion Alleviation of Space Targets

Lin Shi [1], Baofeng Guo [1,*], Ning Han [2,*], Juntao Ma [1], Xiaoxiu Zhu [1] and Chaoxuan Shang [1]

[1]   Department of Electronic and Optical Engineering, Army Engineering University Shijiazhuang Campus, Shijiazhuang 050003, China; shilin85@foxmail.com (L.S.); tm2001@sina.com (J.M.); zhuxiaoxiu13@163.com (X.Z.); scx1207@sina.com (C.S.)
[2]   PLA 32181 Unit, Shijiazhuang 050003, China
*   Correspondence: guobao_feng870714@126.com (B.G.); haning1103@163.com (N.H.); Tel.: +86-031-187-994245 (B.G.); +86-031-181-563176 (N.H.)

Received: 28 March 2019; Accepted: 16 May 2019; Published: 20 May 2019

Abstract: The linear geometry distortion caused by time variant bistatic angles induces the sheared shape of the bistatic inverse synthetic aperture radar (bistatic-ISAR) image. A linear geometry distortion alleviation algorithm for space targets in bistatic-ISAR systems is presented by exploiting prior information. First, we analyze formation mathematics of linear geometry distortions in the Range Doppler (RD) domain. Second, we estimate coefficients of first-order polynomial of bistatic angles by least square error (LSE) method through exploiting the imaging geometry and orbital information of space targets. Third, we compensate the linear spatial-variant terms to restore the linear geometry distortions. Consequently, the restored bistatic-ISAR image with real shape is obtained. Simulated results of the ideal point scatterers dataset and electromagnetic numerical dataset verify the performance of the proposed algorithm.

Keywords: bistatic inverse synthetic aperture radar; linear geometry distortion; prior information; least square error

---

1. Introduction

The monostatic inverse synthetic aperture radar (ISAR) system provides electromagnetic images of targets in the Range Doppler (RD) domain [1–4], which is suitable for the target recognition [5]. However, in monostatic ISAR systems, the image cannot be obtained when targets only move along the line of sight (LOS) of radar within the coherent process interval (CPI). To overcome this limitation, the bistatic configuration is introduced for the ISAR system [6]. In bistatic radar systems, the transmitting station and receiving station are separated spatially, and the length of the baseline is comparable to the distance of targets. The bistatic configuration is capable of obtaining complementary information of the target and providing better anti-jamming ability [6]. Hence, the bistatic-ISAR system has been an effective solution for space targets surveillance [7–12]. The bistatic-ISAR research, with respect to application and algorithm, has been studied in recent years [6,13–19].

Synchronization issues between the transmitting station and receiving station are inherent for the practical bistatic-ISAR configuration. Both the back-projection (BP) algorithm [20] and the polar-format-algorithm (PFA) [21] are sensitive to synchronization accuracy. Thus, applicability of those two algorithms is limited in bistatic-ISAR systems [13]. The RD imaging algorithm, with low requirement for synchronization accuracy and concise physical meaning, is widely used for simulated and real data process in bistatic-ISAR systems [6,8,14]. The bistatic equivalent monostatic (BEM) radar, an approximation of bistatic-ISAR systems, is derived, subject to certain constraints [14]. It effectively simplifies the bistatic-ISAR signal processing. The bistatic angle is the angle between the LOS of the transmitting station and receiving station. The time variant bistatic angles are caused by the bistatic configuration. However, the linear geometry/quadratic-defocusing distortions are caused by time variant bistatic angles when using the RD imaging algorithm [6,13,22].

In [16], a distortions mitigation algorithm based on "linked scatterers" to alleviate the distortions are introduced. However, this algorithm requires the transmitting station to work in duplex mode. Recently, a distortions mitigation algorithm was proposed in [22] by combining CPI reduction with super-resolution algorithms. However, the reduction factor of CPI needs to be deliberately calculated to guarantee that the distortions are less than corresponding resolution. Moreover, the tradeoff between distortion alleviation and resolution is required, even if using super resolution algorithms. In other words, higher resolution can be achieved in the same case, if the alleviation can be completed without reduction of CPI. Furthermore, the problem induced by the linear geometry distortion is more serious than the quadratic-defocusing one [23] in generic bistatic-ISAR systems. The linear geometry distortion causes a sheared target shape, since the linear Doppler shift of scatters in different range-bins change with the range position. It is almost impossible to perform target recognition with the sheared shape image. To alleviate the linear geometry distortions, the interferometric technique is adopted in bistatic-ISAR systems [24]. However, a super-receiver with at least three antennas needs to be configured for the interferometric technique. Both structure and complexity of the system are increased fast with an increasing antenna number, as compared with the classical bistatic-ISAR system with one transmitting station and one receiving station. Conversely, linear geometry distortion alleviation algorithms by exploiting prior information are promising solutions. In [25] we estimate the image distortion angle for space targets by exploiting prior information. By calculating the image distortion angle that we defined in [26], the linear geometry distortion alleviation is conducted in each range-bin. However, the assumption in [25] is that the space target position and the corresponding bistatic angle are completely accurate. In practical systems, there are errors of the satellite position data obtained by the telemetry network. Those errors are accumulated with slow time when calculating the image distortion angle. Hence, little error of the target position affects the subsequent image distortion alleviation. Further consideration of the linear geometry alleviation method should be discussed for practical bistatic-ISAR systems.

In this paper, we focus on the alleviation of linear geometry distortions and propose a corresponding alleviation algorithm of space targets through exploiting prior information in the classical bistatic-ISAR system. First, we calculate the bistatic angles through exploiting prior information (the imaging geometry and orbital information of space targets). Second, we obtain the coefficients of first-order polynomial of bistatic angles using the least square error (LSE) method. Then, we construct compensation in terms of the linear spatial-variant terms and conduct restoration the process based on phase compensation along each range-bin. Finally, we obtain the well-restored image with real shape via the compression and rescaling of cross range.

This paper is organized as follows. The bistatic-ISAR theories are revisited and related distortions are analytically treated in Section 2. The linear geometry distortion alleviation algorithm is discussed in detail in Section 3. In Section 4, simulated results using both the ideal point scatterers dataset and the electromagnetic numerical dataset are presented, respectively. In Section 5, we draw the conclusions.

2. Signal Model with Distortions

The generic bistatic radar configuration is shown in Figure 1. The transmitting station and receiving station are separated. R_T and R_R denote the distances between the target and the transmitting station and receiving station, respectively. The length of baseline L is comparable with R_T and R_R. The bistatic angle formed by bistatic radar geometry is referred as β.

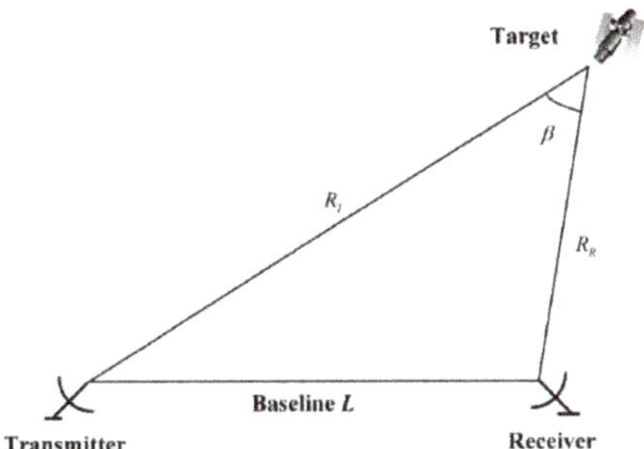

Figure 1. Generic bistatic radar configuration.

In the generic bistatic-ISAR system of the space target, assumptions of both the short CPI and far field are satisfied. The bistatic angle β_m at t_m and the distortion term $\cos\beta_m/2$ caused by the bistatic configuration can be represented as Equations (1) and (2) by first-order polynomial, respectively [6]:

$$\beta_m \approx \beta_0 + \Delta\beta t_m \tag{1}$$

where t_m is slow time, $\beta_0 = \beta(t_0)$ and $\Delta\beta = \left.\frac{d\beta(t_m)}{dt_m}\right|_{t_m=0}$.

$$\cos\left(\frac{\beta_m}{2}\right) \approx \cos\left(\frac{\beta_0}{2}\right) - \frac{\Delta\beta}{2}\sin\left(\frac{\beta_0}{2}\right)t_m = K_0 + K_1 t_m \tag{2}$$

where

$$K_0 = \cos\left(\frac{\beta_0}{2}\right) \quad K_1 = -\frac{\Delta\beta}{2}\sin\left(\frac{\beta_0}{2}\right) \tag{3}$$

The transmitted linear chirp modulation signal is as follows:

$$s_t(\hat{t}, t_m) = \text{rect}\left(\frac{\hat{t}}{T_p}\right)\exp\left[j2\pi\left(f_c t + \frac{1}{2}\mu\hat{t}^2\right)\right] \tag{4}$$

where pulse repetition period is T_{PRT}, the rectangle function is $\text{rect}(u) = \begin{cases} 1 & |u| \le \frac{1}{2} \\ 0 & |u| > \frac{1}{2} \end{cases}$. $\hat{t} = t - t_m$ denotes fast time. T_p denotes pulse width. f_c denotes carrier frequency. μ denotes chirp rate.

If constraint of the range-bin migration is satisfied, after successful signal pre-processing and translational motion compensation, the phase change between each period caused by both translational motion and the propagation of electromagnetic wave e.g., refractive effect, is compensated. The signal of the nth range-bin is written as Equation (5). More details are available in [6].

$$s_n(t_m) = \sum_{i=1}^{L_n} A_i \exp\left(\frac{-j4\pi f_c}{c}\left(x_i\omega_0 t_m + y_i\left(1 - \frac{\omega_0^2 t_m^2}{2}\right)\right)(K_0 + K_1 t_m)\right) \tag{5}$$

where (x_i, y_i) are the coordinates of the ith scatterer, L_n is the scatterers number of the nth range-bin, A_i is the complex amplitude and ω_0 is the rotation velocity (RV).

The positions of range in Equation (5) are the same $(y_1 = y_2 = \cdots = y_{L_n} = y_n)$. Neglecting the constant and third-order terms, $s_n(t_m)$ can be rewritten as follows:

$$s_n(t_m) = \sum_{i=1}^{L_n} A_i \exp\left(\frac{-j2\pi f_c}{c}\left(2y_n K_1 t_m + 2\omega_0 x_i K_0 t_m + \gamma_i t_m^2\right)\right) \tag{6}$$

where $\gamma_i = 2\omega_0 x_i K_1 - y_n K_0 \omega_0^2$.

Apply Fourier transform (FT) to $s_n(t_m)$ in Equation (6) along the slow-time direction.

$$
\begin{aligned}
S_n(f_d) &= \int_0^T s_n(t_m)\exp(-j2\pi f_d t_m)\,dt_m \\
&= \sum_{i=1}^{L_n} A_i psf_i\!\left(f_d - \left(f_i + \frac{2f_c K_1 y_n}{c}\right)\right)
\end{aligned}
\tag{7}
$$

T denotes the CPI. f_d denotes the Doppler frequency. $f_i = 2f_c\omega_0 x_i K_0/c$ is the Doppler frequency of ith scatterer. $psf_i(f_d) = T\sin c(f_d)\otimes D_i(f_d)$ is the point spread function of ith scatterer. Symbol $\otimes$ denotes the convolution operator. $D_i(f_d)$ is the FT of the quadratic distortion terms of the ith scatterer.

$$
D_i(f_d) = \int_0^T A_i\exp\!\left(\frac{-j2\pi f_c}{c}\gamma_i t_m^2\right)\exp(-j2\pi f_d t_m)dt_m
\tag{8}
$$

Because the quadratic distortion terms are relatively small in generic bistatic-ISAR systems [23], we focus on the linear distortion term. As expressed in Equation (7), scatterers cross-range positions are shifted by $2f_c K_1 y_n/c$ along the cross-range direction in the nth range-bin. The shift of Doppler depends on the range position y_n and leads to the sheared shape of bistatic-ISAR image.

3. Linear Geometry Distortion Alleviation Algorithm

3.1. Exploiting Prior Information

The configuration of bistatic-ISAR systems is shown in Figure 2. Tx and Rx represent the transmitting station and receiving station, respectively. The baseline length is referred to L. BEM radar is the instantaneous approximation of bistatic-ISAR system under certain constraints [6,14]. The target velocity is referred as V. O is the target's mass center. The distances between O and the transmitting station and receiving station are R_{TO} and R_{RO} respectively. β_0 denotes initial bistatic angle at t_0. ζ_0 denotes the initial view angle of the transmitting station at t_0. We establish the coordinate system xOy according to the right hand rule. O is the origin of xOy. Bisector of β_0 is the y axis. (x_P, y_P) are the coordinates of the scatterer P in xOy. d denotes the length of OP. The target's mass center O moves to O_m at t_m. And the new coordinate system $x'O_m y'$ is translational motion of the old coordinate system xOy. β_m denotes the bistatic angle of slow time t_m. ζ_m denotes the view angle of the transmitting station of slow time t_m. With the bisector of β_m as v axis and O_m as the origin, we establish the coordinate system $uO_m v$ according to the right hand rule. (x_{Pm}, y_{Pm}) are the coordinates of the scatterer P in the $uO_m v$. α_m is the angle between $O_m P_m$ and u axis. θ_m is the change of equivalent view angle of BEM radar. The distances between P_m and the transmitting station and receiving station are R_{TPm} and R_{RPm} respectively. The distances between O_m and the transmitting station and receiving station are R_{TOm} and R_{ROm} respectively.

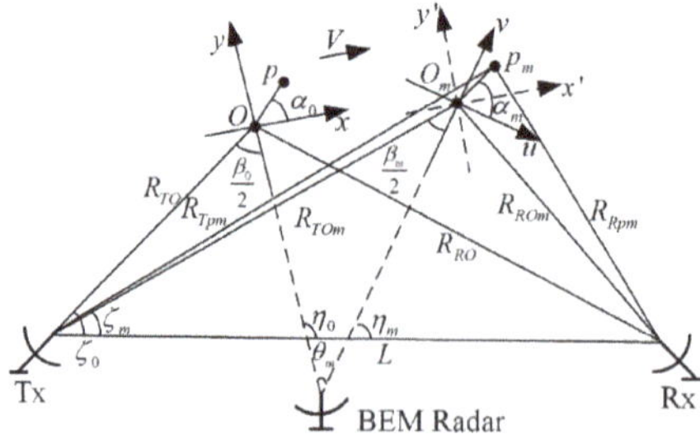

Figure 2. Imaging geometry of bistatic inverse synthetic aperture radar (bistatic-ISAR) system.

Two positions of the transmitting station and receiving station are known previously in the bistatic-ISAR system. Positions of the space targets can be calculated by combining the orbital motion model with precise ephemeris, as we mentioned in [10]. The precise ephemeris can be calculated from the telemetry data. Telemetry data can be achieved through fusing the radar and optical sensor results in telemetry network. R_{TOm}, R_{ROm} and L can be calculated according to the geometry shown in Figure 2. Thus, the time variant bistatic angle β_0 and β_m, the corresponding view angles ζ_0 and ζ_m can be obtained by the following equations:

$$\beta_0 = \arccos\left(\frac{R_{TO}^2 + R_{RO}^2 - L^2}{2R_{TO}R_{RO}}\right) \tag{9}$$

$$\beta_m = \arccos\left(\frac{R_{TOm}^2 + R_{ROm}^2 - L^2}{2R_{TOm}R_{ROm}}\right) \tag{10}$$

$$\zeta_0 = \arccos\left(\frac{R_{TO}^2 + L^2 - R_{RO}^2}{2R_{TO}L}\right) \tag{11}$$

$$\zeta_m = \arccos\left(\frac{R_{TOm}^2 + L^2 - R_{ROm}^2}{2R_{TOm}L}\right) \tag{12}$$

The corresponding η_0 and η_m can be obtained by

$$\eta_0 = \zeta_0 + \beta_0/2 \tag{13}$$

$$\eta_m = \zeta_m + \beta_m/2 \tag{14}$$

The change of equivalent view angle of BEM radar θ_m (the equivalent rotation angle $(\alpha_m - \alpha_0)$) is calculated by

$$\theta_m = \eta_0 - \eta_m \tag{15}$$

As mentioned, the bistatic angle can be approximated by Equation (1) $[\beta_m \approx \beta_0 + \Delta\beta t_m]$. Therefore, we can estimate the β_0 and $\Delta\beta$ as $\hat{\beta}_0$ and $\hat{\Delta\beta}$ using the LSE method based on β_m, respectively. Then, we estimate K_0 and K_1 through the following equations respectively:

$$\hat{K}_0 = \cos\left(\frac{\hat{\beta}_0}{2}\right) \tag{16}$$

$$\hat{K}_1 = -\frac{\hat{\Delta\beta}}{2}\sin\left(\frac{\hat{\beta}_0}{2}\right) \tag{17}$$

We obtain the estimated RV $\hat{\omega}_0$ by the CPI and θ_m. We should note $\hat{K}_0$, $\hat{K}_1$ and $\hat{\omega}_0$ can be obtained with high accuracy. $\xi_{R_{TPm}}$ and $\xi_{R_{RPm}}$ (the errors of the distance R_{TPm} and R_{RPm}) are relatively small compared with R_{TPm} and R_{RPm} and L in the generic bistatic-ISAR system observing space targets. ξ_{β_m} (the relative error of β_m) and ξ_{θ_m} (the relative error of θ_m) are calculated according to the geometry shown in Figure 2. In this scenario, those two ξ_{β_m} and ξ_{θ_m} are relatively small, e.g., $\xi_{\beta_m} = 8.13 \times 10^{-6}$ and $\xi_{\theta_m} = 8.61 \times 10^{-6}$, when $\xi_{R_{TPm}} = 2$ m, $\xi_{R_{RPm}} = 2$ m, $R_{TPm} = 305.29$ km, $R_{RPm} = 603.45$ km, and $L = 800$ km. The LSE method can find the optimum fitting coefficients of given data set. For more details, see our previous discussion in [27].

3.2. Compensation of Linear Spatial-Variant Phase Terms

As mentioned, the first order term $\exp\left(\frac{-j4\pi f_c}{c} y_n K_1 t_m\right)$ of the phase term depends on y_n of the scatterers. It is the linear spatial-variant phase term. The range position y_n is written with respect of range-bin index:

$$y_n = (n - n_c)\rho_y \tag{18}$$

where n represents the range-bin index of y_n and n_c represents the unknown index of range-bin of equivalent RC. Length of a sampling point along the direction of range ρ_y can be written as [6]:

$$\rho_y = \frac{c}{2f_s \cos(\beta_0/2)} = \frac{c}{2f_s K_0} \tag{19}$$

where the f_s is the sampling rate.

By substituting Equation (18) into Equation (6), the signal of the nth range-bin $s_n(t_m)$ becomes

$$s_n(t_m) = \sum_{i=1}^{L_n} A_i \exp\left(\frac{-j2\pi f_c}{c}\left(2n\rho_y K_1 t_m - 2n_c\rho_y K_1 t_m + 2\omega_0 x_i K_0 t_m + \gamma_i t_m^2\right)\right) \tag{20}$$

According to Equation (20), the linear geometry distortion is related to the $f_c n \rho_y K_1 t_m/c$. We can estimate the coefficients K_0, K_1 and the ω_0 in advanced by exploiting prior information. Therefore, we can construct the spatial-variant compensation term φ_c in Equation (21) to alleviate the distortion.

$$\varphi_c = \exp\left(\frac{j4\pi f_c}{c} n\rho_y K_1 t_m\right) \tag{21}$$

Multiplying Equation (20) by Equation (21), the restoration of $s_n(t_m)$ can be written as

$$\bar{s}_n(t_m) = \sum_{i=1}^{L_n} A_i \exp\left(\frac{-j2\pi f_c}{c}\left(-2n_c\rho_y K_1 t_m + 2\omega_0 x_i K_0 t_m + \gamma_i t_m^2\right)\right) \tag{22}$$

The FT of the restored $\bar{s}_n(t_m)$ are obtained as follows

$$\bar{S}_n(f_d) = \sum_{i=1}^{L_n} A_i ps f_i(f_d - f_i - f_{rc}) \tag{23}$$

where f_i represents the Doppler frequency of ith scatterer defined in Equation (7) and $f_{rc} = -2f_c K_1 n_c \rho_y/c$. f_{rc} only leads to corresponding amount of shift of the slow time compression data over all range-bins. In turn, the shape of the target does not change with the unknown index of the range-bin of equivalent RC n_c. It does not need to estimate n_c. Thus, the restored image with correct shape can be obtained by the spatial-variant compensation term φ_c in Equation (21).

3.3. Algorithm Summary

The overall algorithm of the linear geometry distortion alleviation is summarized.

Step 1: Perform pulse compression along range direction and translational motion compensation on the received bistatic-ISAR signal and obtain the one-dimensional range profile expressed in Equation (6).

Step 2: Calculate time variant bistatic angles β_m and change of the equivalent view angle of BEM radar θ_m (equivalent rotation angle), according to the geometry shown in Figure 2 and the positions of transmitting station, receiving station and space target (prior information).

Step 3: Estimate $\hat{\beta}_0$ and $\hat{\Delta\beta}$ based on β_m by the LSE method, respectively. Calculate RV $\hat{\omega}_0$ according to θ_m/CPI. Calculate $\hat{K}_0$ and $\hat{K}_1$ according to Equations (16) and (17), respectively.

Step 4: Construct phase terms for compensation according to Equation (21) and compensate the signal $s_n(t_m)$ of each range-bin with the corresponding compensation term.

Step 5: Apply FT to the restored data along slow-time direction and obtain the rescale the image with $\hat{K}_0$ and $\hat{\omega}_0$ to display the real shape of the target.

4. Results and Discussion

We conduct the simulations based on an ideal point scatterers dataset using a space target model and an electromagnetic scattering dataset using a typical satellite model.

4.1. Simulation Setting

The simulation scenario is selected as follows. The transmitting station and receiving station are located at Beijing and Shanghai respectively. The International Space Station orbit is chosen as the simulation orbit. The two-line elements (TLE) of the International Space Station are shown in Table 1.

Table 1. Two-line elements (TLE) of the International Space Station (12 September 2018).

1 25544U 98067A   18255.09915832 .00001088 00000-0 23933-4 0 9999
2 25544 51.6419 305.5808 0005084 148.3817 299.1230 15.53835622132031

The International Space Station orbit is determined by its TLE data (provided by the Space-Surveillance-Network of America). The epoch time in the initial orbital elements is on 12 September 2018 at 02:22:47. The visible time-window of the bistatic-ISAR system is from 14:28:15 to 14:37:09 on 12 September 2018. We chose the particular CPI with the bistatic angle linear change from the visible time window as the imaging segment.

The simulation scenario is illustrated in Figure 3.

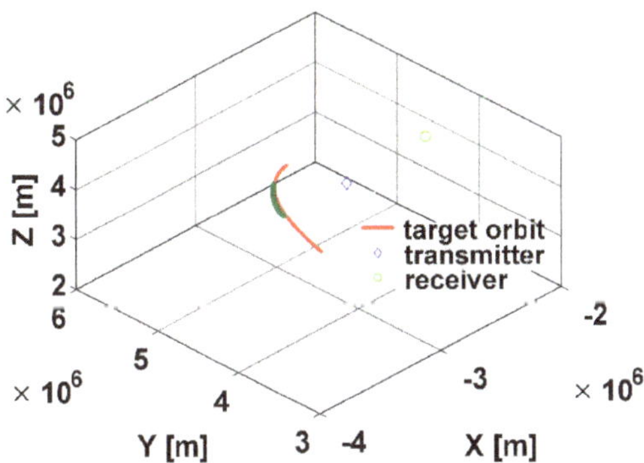

Figure 3. Simulation scenario.

The parameters of the bistatic-ISAR system are listed in Table 2.

Table 2. Parameters of the bistatic inverse synthetic aperture radar (bistatic-ISAR) system.

Parameter	Value	Parameter Name	Value
Carrier frequency	10 GHz	Sample frequency	800 MHz
Signal bandwidth	600 MHz	Pulse repetition frequency	100 Hz
Pulse width	20 us	Accumulated pulses	512
Integration angle	4.68°	Envelope alignment	Cross-correlation accumulation
Phase compensation	Phase gradient auto-focus	Image algorithm	Range Doppler
$\hat{K}_0$	0.7425	$\hat{K}_1$	−0.0073

4.2. Simulation Based on Two Datasets

The bistatic angles are calculated by the imaging geometry and the positions of the space target, transmitting station and receiving station. Figure 4 shows the change of bistatic angles with the pulse number within the CPI. Bistatic angles β_m within the CPI almost change linearly with slow time. That is because the imaging segment is chosen with the bistatic angle linear change. Figure 5 shows the change of θ_m with the pulse number within the CPI. The rotation angle also changes linearly with slow time. The variation of the rotation angle is 4.68°. The equivalent RV is 0.016 rad/s.

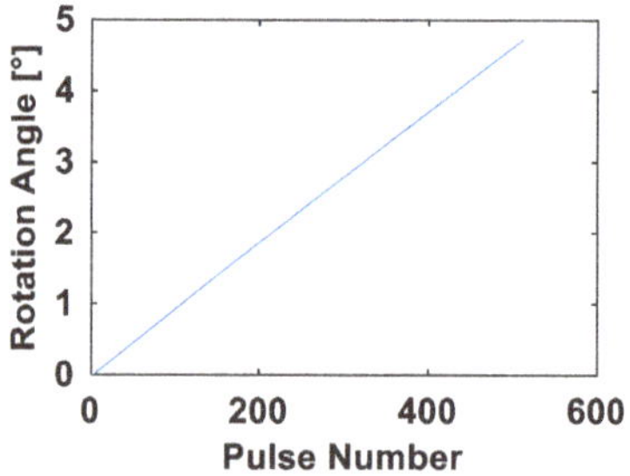

Figure 4. Bistatic angle β_m.

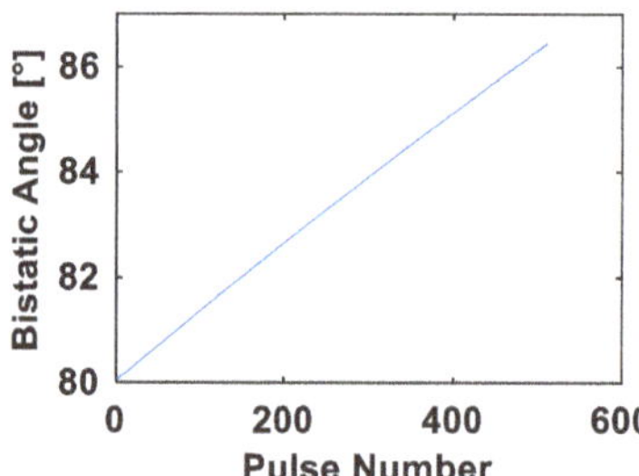

Figure 5. Equivalent rotation angle θ_m.

The ideal point scatterer dataset of the space target shown in Figure 6a is comprised of 307 point scatterers. For the ideal point scatterer, scattering coefficient of each scatterer is 1. The typical satellite shown in Figure 7a has x, y and z extends of 40.09 m, 30.37 m, and 20.74 m, respectively. The ISAR image is the electromagnetic reflection of targets in the RD domain. It is dependent on the instantaneous radar cross section (RCS) distribution on the imaging plane. For further assessment of the proposed algorithm, the bistatic electromagnetic RCS dataset of the typical satellite is obtained by the numerical physical optics technique. The echo data corresponding to the two datasets respectively are generated by the method we proposed in [8].

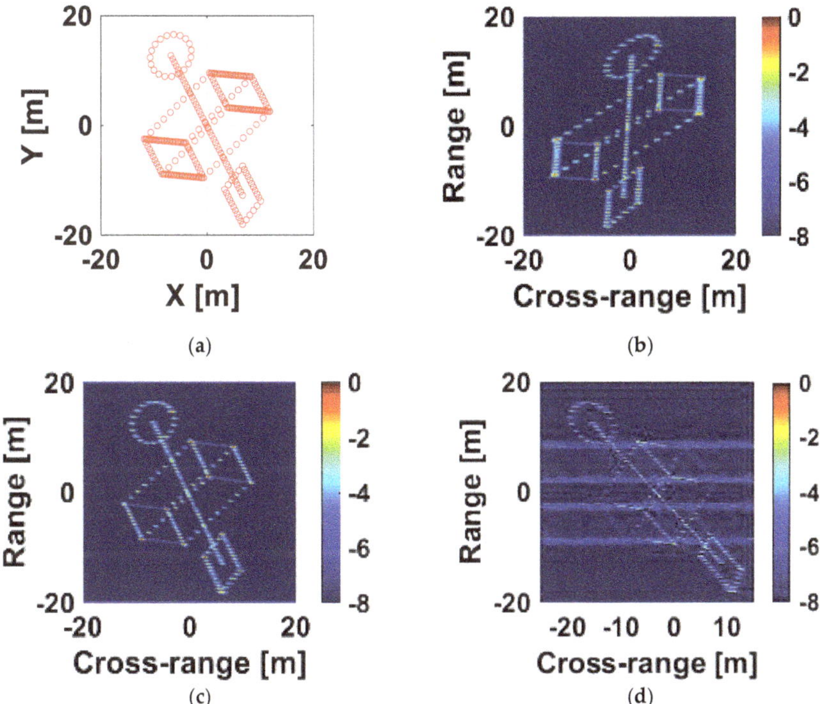

Figure 6. Results using the ideal point scatterers dataset: (**a**) The ideal point scatterers dataset (view of the line of sight (LOS) of bistatic equivalent monostatic (BEM) radar); (**b**) Before linear geometry distortion alleviation; (**c**) After linear geometry distortion alleviation with the proposed algorithm; (**d**) After linear geometry distortion alleviation with the algorithm in [25].

The envelope alignment and the auto-focusing are achieved by the maximum cross-correlation algorithm and phase gradient auto-focus (PGA) algorithm, respectively. Then, the images of Figures 6b and 7b are obtained by the RD imaging algorithm.

Due to the linear geometry distortion caused by time variant bistatic angles, the targets shape in Figures 6b and 7b are sheared. Using the proposed linear geometry distortion alleviation algorithm, the linear geometry distortions are alleviated in Figures 6c and 7c (compare Figure 6b with Figure 6c, compare Figure 7b with Figure 7c). The shape of the targets in the restored Figures 6c and 7c are consistent with the corresponding ones in Figures 6a and 7a (view of the LOS of BEM radar), respectively. It is beneficial to targets recognition.

The real shape of the target cannot be obtained by the algorithm in [25] and the image of several range-bins is defocused (Figures 6d and 7d). The reason is that the assumption of complete accurate prior information is not satisfied in practical system and the image distortion angle is sensitive to the errors of prior information. Meanwhile, the results of the electromagnetic numerical dataset of typical satellite in Figure 7 and the comparison between proposed algorithm and the algorithm in [25] further verified the robustness of the presented algorithm.

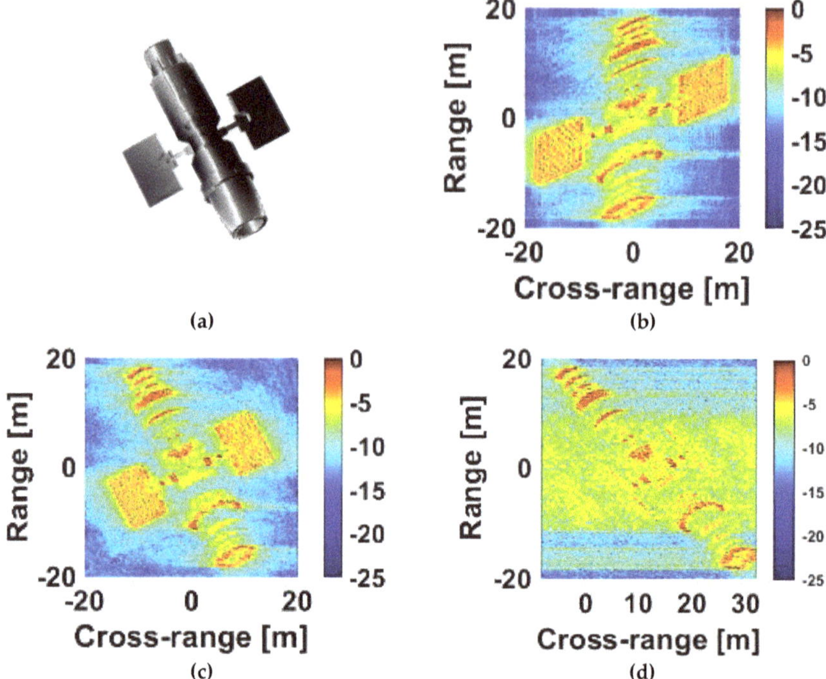

Figure 7. Results using the electromagnetic numerical radar cross section (RCS) dataset: (**a**) Computer aided design model of the satellite (the view of LOS of BEM radar); (**b**) Before linear geometry distortion alleviation; (**c**) After linear geometry distortion alleviation with the proposed algorithm; (**d**) After linear geometry distortion alleviation with the algorithm in [25].

5. Conclusions

We present a clear procedure of the linear geometry distortion alleviation algorithm based on prior information for better utilization of the bistatic-ISAR image. Simulated results of both ideal point scatterer dataset of space target model and electromagnetic numerical dataset of the typical satellite model verify the feasibility and robustness of proposed algorithm. The comparisons of the results before and after the alleviation indicate that our algorithm is capable of restoring the linear geometry distortion and providing the real shape of the target. The restored results are beneficial to further target classification and recognition.

Author Contributions: Conceptualization, L.S. and C.S.; methodology, L.S., B.G. and N.H.; software, J.M.; validation, L.S. and X.Z.; formal analysis, L.S. and X.Z.; investigation, B.G. and N.H.; resources, J.M.; data curation, L.S. and B.G.; writing—original draft preparation, L.S.; writing—review and editing, B.G. and N.H.; supervision, C.S.; funding acquisition, C.S.

Funding: The research was funded by the National Natural Science Foundation of China under Grant 61601496 and 61701544.

Acknowledgments: We appreciate the editors and peer reviewers for their valuable comments and suggestions.

Conflicts of Interest: The authors declare no conflicts of interest.

References

1. Walker, J.L. Range-doppler imaging of rotating objects. *IEEE Trans. Aerosp. Electron. Syst.* **1980**, *16*, 23–52. [CrossRef]

2. Xing, M.; Wu, R.; Li, Y.; Bao, Z. New ISAR imaging algorithm based on modified Wigner-Ville distribution. *IET Radar Sonar Navig.* **2008**, *3*, 70–80. [CrossRef]

3. Zhao, J.; Zhang, M.; Wang, X.; Cai, Z.; Nie, D. Three-dimensional super resolution ISAR imaging based on 2D unitary ESPRIT scattering centre extraction technique. *IET Radar Sonar Navig.* **2017**, *11*, 98–106. [CrossRef]

4. Lv, Y.; Wu, Y.; Wang, H.; Qiu, L.; Jiang, J.; Sun, Y. An Inverse Synthetic Aperture Ladar Imaging Algorithm of Maneuvering Target Based on Integral Cubic Phase Function-Fractional Fourier Transform. *Electronics* **2018**, *7*, 148. [CrossRef]

5. Giusti, E.; Martorella, M.; Capria, A. Polarimetrically-Persistent-Scatterer-Based Automatic Target Recognition. *IEEE Trans. Geosci. Remote Sens.* **2011**, *49*, 4588–4599. [CrossRef]

6. Martorella, M.; Palmer, J.; Homer, J.; Littleton, B.; Longstaff, D. On bistatic inverse synthetic aperture radar. *IEEE Trans. Aerosp. Electron. Syst.* **2007**, *43*, 1125–1134. [CrossRef]

7. Bai, X.; Zhou, F.; Xing, M.; Bao, Z. Scaling the 3-D Image of Spinning Space Debris via Bistatic Inverse Synthetic Aperture Radar. *IEEE Geosci. Remote Sens. Lett.* **2010**, *7*, 430–434. [CrossRef]

8. Guo, B.F.; Shang, C.X.; Wang, J.L. Bistatic ISAR echo simulation of space target based on two-body model. *Syst. Eng. Electron.* **2016**, *38*, 1771–1779. [CrossRef]

9. Ma, J.; Gao, M.; Hu, W.; Di, X.; Shi, L. Optimum Distribution of Multiple Location ISAR and Multi-angles Fusion Imaging for Space Target. *J. Electron. Inf. Technol.* **2017**, *39*, 2834–2843. [CrossRef]

10. Ma, J.T.; Gao, M.G.; Guo, B.F.; Dong, J.; Xiong, D.; Feng, Q. High resolution inverse synthetic aperture radar imaging of three-axis-stabilized space target by exploiting orbital and sparse priors. *Chin. Phys. B* **2017**, *26*, 108401. [CrossRef]

11. Tian, B.; Zou, J.; Xu, S.; Chen, Z. Squint model interferometric ISAR imaging based on respective reference range selection and squint iteration improvement. *IET Radar Sonar Navig.* **2015**, *9*, 1366–1375. [CrossRef]

12. Han, N.; Li, B.; Wang, L.; Tong, J.; Guo, B. Algorithm for autofocusing of bistatic ISAR of space target based on sparse decomposition. *Acta Aeronaut. Astronaut. Sin.* **2018**, *39*, 322037. [CrossRef]

13. Martorella, M. Analysis of the Robustness of Bistatic Inverse Synthetic Aperture Radar in the Presence of Phase Synchronisation Errors. *IEEE Trans. Aerosp. Electron. Syst.* **2011**, *47*, 2673–2689. [CrossRef]

14. Martorella, M.; Cataldo, D.; Brisken, S. Bistatically equivalent monostatic approximation for bistatic ISAR. In Proceedings of the IEEE Radar Conference, Ottawa, ON, Canada, 29 April–3 May 2013; pp. 1–5.

15. Shi, L.; Guo, B.F.; Ma, J.T.; Shang, C.X.; Zeng, H.Y. A Novel Channel Calibration Method for Bistatic ISAR Imaging System. *Appl. Sci.* **2018**, *8*, 2160. [CrossRef]

16. Sun, S.B.; Jiang, Y.C.; Yuan, Y.S.; Hu, B.; Yeo, T.S. Defocusing and distortion elimination for shipborne bistatic ISAR. *Remote Sens. Lett.* **2016**, *7*, 523–532. [CrossRef]

17. Xiong, D.; Zhang, X.; Wang, J.; Zhao, H.; Gao, M. Reception and calibration of bistatic SF ISAR imaging system with wideband receiver. *IET Radar Sonar Navig.* **2017**, *11*, 379–385. [CrossRef]

18. Zhang, S.S.; Sun, S.B.; Zhang, W.; Zong, Z.L.; Yeo, T.S. High-Resolution Bistatic ISAR Image Formation for High-Speed and Complex-Motion Targets. *IEEE J. Sel. Top. Appl. Earth Obs. Remote Sens.* **2015**, *8*, 3520–3531. [CrossRef]

19. Zhao, L.; Gao, M.; Martorella, M.; Stagliano, D. Bistatic three-dimensional interferometric ISAR image reconstruction. *IEEE Trans. Aerosp. Electron. Syst.* **2015**, *51*, 951–961. [CrossRef]

20. Chen, S.; Zhao, H.C.; Zhang, S.N.; Chen, Y. An improved back projection imaging algorithm for dechirped missile-borne SAR. *Acta Phys. Sin.* **2014**, *62*, 218405. [CrossRef]

21. Horvath, M.S.; Gorham, L.A.; Rigling, B.D. Scene Size Bounds for PFA Imaging with Postfiltering. *IEEE Trans. Aerosp. Electron. Syst.* **2013**, *49*, 1402–1406. [CrossRef]

22. Cataldo, D.; Martorella, M. Bistatic ISAR Distortion Mitigation via Superresolution. *IEEE Trans. Aerosp. Electron. Syst.* **2018**, *54*, 2143–2157. [CrossRef]

23. Jiang, Y.; Sun, S.; Yeo, T.S.; Yuan, Y. Bistatic ISAR distortion and defocusing analysis. *IEEE Trans. Aerosp. Electron. Syst.* **2016**, *52*, 1168–1182. [CrossRef]

24. Ma, C.Z.; Yeo, T.S.; Guo, Q.; Wei, P.J. Bistatic ISAR Imaging Incorporating Interferometric 3-D Imaging Technique. *IEEE Trans. Geosci. Remote Sens.* **2012**, *50*, 3859–3867. [CrossRef]

25. Guo, B.F.; Shang, C.X.; Wang, J.L.; Gao, M.G.; Fu, X.J. Correction of migration through resolution cell in bistatic inverse synthetic aperture radar in the presence of time variant bistatic angle. *Acta Phys. Sin.* **2014**, *63*, 238406.

26. Dong, J.; Gao, M.G.; Shang, C.X.; Fu, X.J. The Image Plane Analysis and Echo Model Amendment of Bistatic ISAR. *J. Electron. Inf. Technol.* **2010**, *32*, 1855–1862. [CrossRef]
27. Zhu, X.X.; Hu, W.H.; Ma, J.T.; Guo, B.F.; Xue, D.F. ISAR autofocusing imaging with sparse apertures and time variant bistatic angle. *Acta Aeronaut. Astronaut. Sin.* **2018**, *39*, 322059. [CrossRef]

 *electronics*

Article

Focusing of Ultrahigh Resolution Spaceborne Spotlight SAR on Curved Orbit

Yulei Qian * and Daiyin Zhu

Key Laboratory of Radar Imaging and Microwave Photonics, Ministry of Education, College of Electronic Information Engineering, Nanjing University of Aeronautics and Astronautics, 29 Jiangjun Avenue, Nanjing 210016, China; zhudy@nuaa.edu.cn
* Correspondence: qian_yulei@nuaa.edu.cn; Tel.: +86-025-8489-2410

Received: 11 April 2019; Accepted: 30 May 2019; Published: 3 June 2019

Abstract: Aiming to acquire ultrahigh resolution images, algorithms for spaceborne spotlight synthetic aperture radar (SAR) imaging typically confront challenges of curved orbit and azimuth spectral aliasing. In order to conquer these difficulties, a method is proposed in this paper to obtain ultrahigh resolution spaceborne SAR images on a curved orbit, which is composed of the modified RMA (Range Migration Algorithm) and the modified deramping-based approach. The modified RMA is developed to deal with the effect introduced by a curved orbit and the modified deramping-based approach is utilized to handle the problem of azimuth spectral aliasing. In the modified RMA, the polynomial expression of SAR two-dimensional spectrum on a curved orbit is derived with fourth-order azimuth phase history model and series reversion. Then, the singular value decomposition (SVD) is applied to decompose the expression of SAR two-dimensional spectrum numerically in order to acquire coordinates for Stolt interpolation in the scenario of curved orbit. In addition, the modified deramping-based approach is derived by introducing orbital state vectors in order to accommodate the situation of curved orbit in the proposed method. Experiments are implemented on point target simulation in order to verify the effectiveness of the presented method. In experiments, the range and azimuth resolution can achieve 0.15 m and 0.14 m, with focused scene size of 3 km by 3 km.

Keywords: ultrahigh resolution; synthetic aperture radar (SAR); spaceborne; curved orbit; series reversion; singular value decomposition (SVD); deramping-based approach

---

1. Introduction

Synthetic aperture radar (SAR) has become an indispensable means for remote sensing, which is competent to provide high resolution images and videos for monitoring various targets under any weather condition [1–7]. With the desire for higher resolution, the future task of spaceborne SAR challenges the development of SAR imaging algorithms [8]. However, traditional SAR imaging algorithms perform disappointedly with the requirement of ultrahigh resolution in a spaceborne scenario. Various influences occur when the resolution achieves decimeter level in low earth orbit (LEO) spaceborne SAR, which have responsibility for the unsatisfied focusing performance [9]. Curved orbit and azimuth spectral aliasing are two of these influences which remain to be solved.

Traditional SAR imaging algorithms are derived from an assumption that the radar platform has linear movement with constant velocity [10]. Nevertheless, this assumption is no longer accurate when spaceborne SAR is expected to achieve ultrahigh resolution at the decimeter level. Therefore, the difference between the theoretical assumption and real orbit degrades the imaging quality in the azimuth direction. Aiming to solve this problem, a sequence of algorithms have been developed in order to cope with spaceborne spotlight SAR on a curved orbit, which describe the relative satellite-earth motion using Taylor expansion with azimuth time [11–14]. In the literature [11], a slant range model based on fourth-order Taylor expansion is firstly proposed and is implemented to process spaceborne spotlight SAR data on a curved orbit. An advanced RMA (Range Migration Algorithm) is proposed in [12] for dealing with high resolution spaceborne spotlight SAR imaging. The RMA is also known as the ωK algorithm [15]. In [12], a slant range is derived according to the model in [11] so as to accommodate the curved orbit. Moreover, a method for achieving the analytical expression of two-dimensional (2D) point target spectrum for bistatic SAR is presented in [16] and is applied for bistatic SAR imaging in [17], which is obtained via associating the slant range model in [11] with series reversion [18]. Subsequently, a method, which is named as generalized ωK, is proposed by combining multivariable Taylor expansion with the derivation of 2D spectrum in [16]. The generalized ωK presented in [13] is utilized to deal with geosynchronous spotlight SAR on a curved orbit. Consequently, the generalized ωK is extended to adapt for SAR data with squint angle in [14]. Although, the algorithms presented in [13,14] perform satisfactorily in the case of geosynchronous spotlight SAR, the phase error induced by the first-order multivariable Taylor expansion cannot be neglected when resolution attains decimeter level. Apart from algorithms derived from the slant range model in [11], the singular value decomposition Stolt (SVDS) presented in [19] offers another approach for settling the problem caused by curved orbit. The SVDS associates the singular value decomposition (SVD) with ωK algorithm so as to deal with spaceborne spotlight SAR in the scenario of curved orbit.

In order to deal with ultrahigh resolution spaceborne spotlight SAR on a curved orbit, the modified RMA is proposed in this paper. The modified RMA is developed to handle the effect introduced by curved orbit in the scenario of spaceborne spotlight SAR. In the modified RMA, the 2D spectrum of echo data on a curved orbit is derived via the range model in [11] and series reversion. Then, the expression of 2D spectrum of reference point is utilized to perform reference function multiplication (RFM). After RFM operation, point at reference range has been fully focused, and a residual phase exists for targets at other ranges. Afterwards, SVD is utilized to numerically decompose the expression of 2D spectrum in order to acquire coordinates for Stolt interpolation. Subsequently, Stolt interpolation is implemented to perform the residual range cell migration correction, residual secondary range compression and residual azimuth compression. Then, focused image can be obtained via inverse 2D Fourier transform. Although the generalized ωK in [13] is capable of processing echo data on a curved orbit, the phase error caused by the first-order multivariable Taylor expansion seriously degrades image quality when resolution achieves 0.14 m. Meanwhile, the phase error induced by SVD is much smaller than the first-order multivariable Taylor expansion on numerical decomposition of 2D spectrum. As a result, the modified RMA performs better than the generalized ωK when resolution achieves 0.14 m in azimuth direction. The modified RMA is presented in this paper as a part of the proposed method.

In addition to curved orbit, azimuth spectral aliasing is another problem which stems from high azimuth resolution requirement under the circumstance of LEO. The deramping-based approach for solving azimuth spectral aliasing is firstly proposed in [20] and extended to squinted spotlight SAR imaging in [21]. However, deramping-based approaches in [20,21] are only discussed under the condition that the SAR platform moves uniformly in a linear track. Orbital state vectors, which are capable of estimating Doppler parameters in the scenario of curved orbit, are implemented to modify deramping-based approach in order to accommodate the curved orbit in this paper. As a consequence, a modified deramping-based approach is presented in this paper as a part of the proposed method.

The proposed method comprises of modified RMA and modified deramping-based approach. The traditional RMA and the traditional deramping-based approach is only capable of dealing with the scenario of uniform linear motion. The modified RMA extends the traditional RMA to the scenario of curved orbit and the modified deramping-based approach achieves the same purpose. In addition, the modified RMA is developed based on the generalized ωK and has much smaller phase error than the generalized ωK when azimuth resolution achieves 0.14 m. Consequently, the proposed method performs better than the generalized ωK when azimuth resolution achieves 0.14 m in the situation of LEO on a curved orbit.

The proposed method is the improvement on the generalized ωK in [13]. Although the generalized ωK performs well in the scenario of geosynchronous SAR on a curved orbit, it is unable to cope with ultrahigh resolution spaceborne spotlight SAR in the scenario of LEO on a curved orbit. The generalized ωK lacks means to solve the azimuth spectral aliasing problem when azimuth resolution is expected to achieve 0.14 m. Additionally, the phase error in the generalized ωK becomes unbearable under the ultrahigh resolution requirement. The modified deramping-based approach aims to cope with azimuth spectrum aliasing and the modified RMA has much less phase error when azimuth resolution is desired to achieve 0.14 m. Point target simulation is operated to confirm the validity of the proposed method under the circumstances of LEO. In simulation, the swath width of the illuminated scene achieves 3 km in range direction and 3 km in azimuth direction. Furthermore, resolution of focused targets can achieve 0.15 m and 0.14 m in range and azimuth direction, respectively. The proposed method is compared with the generalized ωK. Results in simulation show that the proposed method has smaller phase error and performs better than the generalized ωK. Vectors and matrices are in bold italic while variables are in italic in this paper. The rest of this paper is organized as follows. Section 2 describes the proposed method. Then traditional RMA is briefly presented. Then, the modified RMA and the modified deramping-based approach are derived and described in details. In addition, procedure of the proposed method is also illustrated in this section. Section 3 presents experimental results of simulation in order to validate the effectiveness of the proposed method. Discussion of experiments is drawn in Section 4. Finally, Section 5 gives the conclusion.

2. Methodology

2.1. Traditional RMA

In this part, a brief introduction on traditional RMA is demonstrated. The geometry of spaceborne spotlight SAR is shown in Figure 1. The black dashed track with an arrow in Figure 1 denotes approximate straight path while the red ellipse represents curved orbit of true satellite motion. In Figure 1, the blue dashed lines and circle illustrate that the beam illuminates the target all the time during collecting SAR data. T denotes the illuminated target and O denotes the position of zero Doppler on the curved orbit. $R_a(\eta)$ is the slant range between the radar and a target at azimuth time η and r represents the closest slant range.

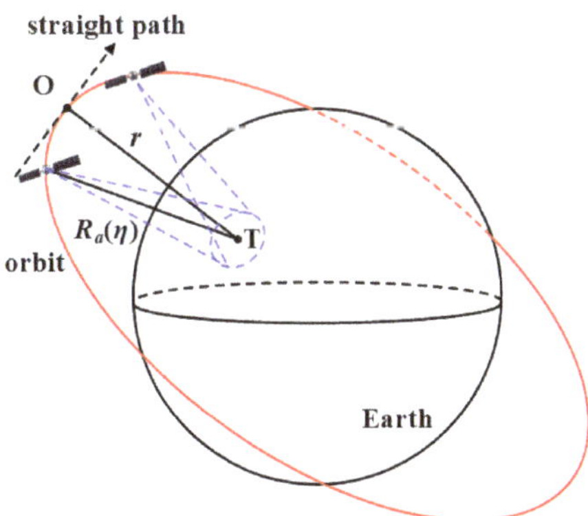

Figure 1. Geometry of spotlight spaceborne SAR.

The traditional RMA is acquired under the consideration of uniform linear motion. After modulation to baseband, the echo signal of point target T is able to be described in terms of range time τ and azimuth time η as follows:

$$s(\tau, \eta; T) = w_r\left[\tau - \frac{2R_a(\eta)}{c}\right] w_a(\eta) \exp\left[\frac{-j4\pi f_0 R_a(\eta)}{c}\right] \exp\left\{ j\pi K_r\left[\tau - \frac{2R_a(\eta)}{c}\right]^2 \right\} \tag{1}$$

The amplitude factors have been ignored. $w_r(\bullet)$ is the range envelope, $w_a(\bullet)$ is the azimuth envelope, c is the velocity of light, K_r is the range frequency modulation rate, f_0 is the carrier frequency and j is the imaginary unit.

In the scenario of uniform linear motion, the slant range is modelled by the hyperbolic equation with the velocity v and closest slant range r as follows:

$$R_a(\eta) = \sqrt{r^2 + v^2\eta^2} \tag{2}$$

By performing 2D Fourier transform (FT) on Equation (1) via the method of stationary phase with the slant range model in Equation (2), the 2D spectrum of echo signal can be represented as follows:

$$S_{2df}(f_\tau, f_\eta; T) = W_r(f_\tau) W_a(f_\eta) \exp\left[-j\frac{4\pi r}{c}\sqrt{(f_0 + f_\tau)^2 - \frac{c^2 f_\eta^2}{4v^2}} - j\pi\frac{f_\tau^2}{K_r}\right] \tag{3}$$

In Equation (3), f_τ and f_η are range and azimuth frequency, respectively. And $W_r(\bullet)$ is the envelope of the range frequency while $W_a(\bullet)$ is the envelope of the azimuth frequency.

The first step in traditional RMA is the RFM. The reference function in 2D frequency domain can be chosen as conj$[S_{2df}(f_\tau, f_\eta; T_{ref})]$, which is the conjugated 2D spectrum of the reference target. T_{ref} denotes the reference point target and conj$[\bullet]$ denotes conjugation operation.

The RFM operation can be expressed as follows:

$$\begin{aligned} S_{RFM}(f_\tau, f_\eta; T) &= S_{2df}(f_\tau, f_\eta; T) \mathrm{conj}\left[S_{2df}(f_\tau, f_\eta; T_{ref})\right] \\ &= W_r(f_\tau) W_a(f_\eta) \exp\left[-j\frac{4\pi(r - r_{ref})}{c}\sqrt{(f_0 + f_\tau)^2 - \frac{c^2 f_\eta^2}{4v^2}}\right] \end{aligned} \tag{4}$$

The RFM operation completely cancels the range migration of all targets at the reference range. Nevertheless, the RFM only partly corrects the range migration of targets at other ranges. Therefore, a subsequent Stolt interpolation is implemented so as to compensate the residual quadratic and higher order phase modulation. In the scenario of uniform linear motion, The Stolt interpolation is defined as follows:

$$\sqrt{(f_0 + f_\tau)^2 - \frac{c^2 f_\eta^2}{4v^2}} = f_0 + f_\tau' \tag{5}$$

After implementation of Stolt interpolation, the SAR data is transformed into the new domain (f_τ', f_η) as follows:

$$\begin{aligned} &S'_{RFM}(f_\tau', f_\eta; T) \\ &= W_r(f_\tau') W_a(f_\eta) \exp\left[-j\frac{4\pi(r - r_{ref})}{c}(f_0 + f_\tau')\right] \end{aligned} \tag{6}$$

By performing 2D Inverse FT to Equation (6), the echo data can be transformed into time domain. As a result, a finely focused SAR image is acquired as follows:

$$I'_{RFM}(\tau, \eta; T) = \sin c\left[\tau - \frac{2(r - r_{ref})}{c}\right]\sin c(\eta) \tag{7}$$

2.2. Modified RMA

When spaceborne spotlight SAR is desired to achieve ultrahigh resolution in decimeter level, the motion of platform of SAR is unable to be treated as uniform linear motion. Therefore, curved orbit is required to be considered in algorithms for SAR imaging. The modified RMA is proposed to deal with SAR data on curved orbit. In the modified RMA, the expression of 2D spectrum of echo data is derived by combining the range model in [11] with series reversion. Subsequently, the formula of 2D spectrum of reference point is used to implement RFM operation. Then, SVD is applied for numerically decomposing the expression of 2D spectrum so as to obtain coordinates for Stolt interpolation. After Stolt interpolation, focused image is able to be acquired via 2D inverse Fourier transform.

In comparison with the traditional RMA, the slant range model and the expression of the 2D spectrum in modified RMA are capable of accommodating the curved orbit while the traditional RMA is only able to deal with the scenario of uniform linear motion. Additionally, although the generalized ωK is able to cope with the scenario of curved orbit, the modified RMA has much smaller phase error than the generalized ωK when azimuth bandwidth is larger than 49 kHz. SVD, which is applied in modified RMA, is able to provide numerical decomposition with much smaller phase error than the first-order multivariable Taylor expansion in the generalized ωK. Therefore, the modified RMA performs better than the generalized ωK when the azimuth resolution achieves 0.14 m. The modified RMA is derived and presented in this part.

In the scenario of curved orbit, the slant range is inaccurate to be modelled as the hyperbolic equation in Equation (2). Therefore, it is necessary to utilize more accurate slant range model for curved orbit. The slant range model in the literature [11] is capable of modelling the slant range between radar and target in the scenario of curved orbit. According to the literature [11], the slant range between the radar and a target can be expressed in terms of azimuth time η as follows:

$$R_a(\eta) = \sqrt{e_0 + e_1\eta + e_2\eta^2 + e_3\eta^3 + e_4\eta^4} \tag{8}$$

The definitions of e_0, e_1, e_2, e_3 and e_4 are presented from Equations (9)–(13). And $\circ$ denotes inner product operation.

$$e_0 = R \circ R \tag{9}$$

$$e_1 = 2V \circ R \tag{10}$$

$$e_2 = A \circ R + V \circ V \tag{11}$$

$$e_3 = A \circ V + \frac{1}{3}(R \circ B) \tag{12}$$

$$e_4 = \frac{1}{3}(V \circ B) + \frac{1}{12}(R \circ C) + \frac{1}{4}(A \circ A) \tag{13}$$

The R, V, A, B and C present relative position, velocity, acceleration, rate of acceleration, and rate of rate of acceleration 3-dimentional vectors between the radar and a target, respectively.

With a fourth-order Taylor expansion in azimuth time η, the $R_a(\eta)$ can be expressed as follows:

$$R_a(\eta) = g_0 + g_1\eta + g_2\eta^2 + g_3\eta^3 + g_4\eta^4 \tag{14}$$

The coefficients in Equation (14) are obtained according to the formula in Equation (15). And the expressions of g_0, g_1, g_2, g_3 and g_4 are given in Equations (16)–(20).

$$g_q = \frac{1}{q!} \bullet \left. \frac{dR_a(\eta)}{d\eta^q} \right|_{\eta=0} , q = 0, 1, 2, 3, 4 \tag{15}$$

$$g_0 = e_0^{1/2} \tag{16}$$

$$g_1 = \frac{e_1}{2e_0^{1/2}} \tag{17}$$

$$g_2 = \frac{e_2}{2e_0^{1/2}} - \frac{e_1^2}{8e_0^{3/2}} \tag{18}$$

$$g_3 = \frac{e_1^3}{16e_0^{5/3}} + \frac{e_3}{2e_0^{1/2}} - \frac{e_1 e_2}{4e_0^{3/2}} \tag{19}$$

$$g_4 = \left(\frac{e_4}{2e_0^{1/2}} - \frac{5e_1^4}{128e_0^{7/2}} - \frac{e_2^2}{8e_0^{3/2}} - \frac{e_1 e_3}{4e_0^{3/2}} + \frac{3e_1^2 e_2}{16e_0^{5/2}} \right) \tag{20}$$

By implementing a FT along the range direction to the echo signal in Equation (1) with the method of stationary phase, the signal can be given as follows:

$$\begin{aligned} S(f_\tau, \eta; T) &= W_r(f_\tau)w_a(\eta)\exp\left[-j\frac{4\pi(f_0+f_\tau)R_a(\eta)}{c} - j\pi\frac{f_\tau^2}{K_r}\right] \\ &= W_r(f_\tau)w_a(\eta)\exp\left[-j\frac{4\pi(f_0+f_\tau)R_c(\eta)}{c} - j\pi\frac{f_\tau^2}{K_r}\right]\exp\left[-j\frac{4\pi(f_0+f_\tau)g_1\eta}{c}\right] \end{aligned} \tag{21}$$

The definition of $R_c(\eta)$ is presented in Equation (22) as follows:

$$R_c(\eta) = g_0 + g_2\eta^2 + g_3\eta^3 + g_4\eta^4 \tag{22}$$

The second exponential term in Equation (21) represents linear range cell migration (LRCM). In order to derive the 2D spectrum via series reversion, the exponential term of LRCM is temporarily removed. After removal of LRCM, the point target signal in range frequency and azimuth time domain is presented as follows:

$$S_c(f_\tau, \eta; T) = W_r(f_\tau)w_a(\eta)\exp\left[-j\frac{4\pi(f_0 + f_\tau)R_c(\eta)}{c} - j\pi\frac{f_\tau^2}{K_r}\right] \tag{23}$$

The k is defined as follows:

$$k = \frac{2(f_\tau + f_0)}{c} \tag{24}$$

Via the method of stationary phase, the azimuth frequency f_η is associated to the azimuth time η as follows:

$$-\frac{f_\eta}{k} = 2g_2\eta + 3g_3\eta^2 + 4g_4\eta^3 \tag{25}$$

By using series reversion, the azimuth time η can be expressed in terms of the azimuth frequency f_η as follows:

$$\eta(f_\eta) = \frac{1}{2g_2}\left(-\frac{f_\eta}{k}\right) - \frac{3g_3}{8g_2^3}\left(-\frac{f_\eta}{k}\right)^2 + \frac{18g_3^2 - 8g_2 g_4}{32g_2^5}\left(-\frac{f_\eta}{k}\right)^3 \tag{26}$$

Using Equation (26) with Equation (23), the 2D spectrum of $s_c(\tau, \eta; T)$ can be expressed as follows:

$$S_c(f_\tau, f_\eta; T) = W_r(f_\tau)W_a(f_\eta)\exp\left\{-j2\pi k R_c\left[\eta(f_\eta)\right]\right\}\exp\left[-j2\pi f_\eta \eta(f_\eta) - j\pi\frac{f_\tau^2}{K_r}\right] \tag{27}$$

According to the shift property of FT, the 2D spectrum of $s(\tau, \eta; T)$ can be obtained as follows:

$$\begin{aligned} S_{2df}(f_\tau, f_\eta; T) &= FT_a\left[Sc(f_\tau, \eta; T)\exp(-2j\pi k g_1\eta)\right] = S_c(f_\tau, f_\eta + kg_1; T) \\ &= W_r(f_\tau)W_a(f_\eta)\exp(-j2\pi\Theta)\exp\left[-j\pi\frac{f_\tau^2}{K_r}\right] \end{aligned} \tag{28}$$

$$\begin{aligned}
\Theta = \quad & -\left(\frac{9g_3^2}{64g_2^5k^3} - \frac{g_4}{16g_2^4k^3}\right)f_\eta^4 - \left(\frac{g_3}{8g_2^3k^2} + \frac{9g_1g_3^2}{16g_2^5k^2} - \frac{g_1g_4}{4g_2^4k^2}\right)f_\eta^3 \\
& -\left(\frac{1}{4g_2k} - \frac{3g_1^2g_4}{8g_2^4k} + \frac{27g_1^2g_3^2}{32g_2^5k} + \frac{3g_1g_3}{8g_2^3k}\right)f_\eta^2 - \left(\frac{g_1}{2g_2} + \frac{3g_1^2g_3}{8g_2^3} - \frac{g_1^3g_4}{4g_2^4} + \frac{9g_1^3g_3^2}{16g_2^5}\right)f_\eta \\
& -\left(\frac{g_1^2k}{4g_2} - g_0k + \frac{9g_1^4g_3^2k}{64g_2^5} + \frac{g_1^3g_3k}{8g_2^3} - \frac{g_1^4g_4k}{16g_2^4}\right)
\end{aligned}$$

$$(29)$$

The FT$_a$ denotes FT operation along azimuth direction.

After the expression of 2D spectrum of echo signal has been acquired, the RFM operation can be presented as follows:

$$\begin{aligned}
S_{RFM}(f_\tau, f_\eta; T) &= S_{2df}(f_\tau, f_\eta; T)\text{conj}\left[S_{2df}(f_\tau, f_\eta; T_{ref})\right] \\
&= W_r(f_\tau)W_a(f_\eta)\exp\left[j\theta_{RFM}(f_\tau, f_\eta; T)\right]
\end{aligned} \tag{30}$$

$$\theta_{RFM}(f_\tau, f_\eta; T) = -2\pi\left(\Theta - \Theta_{ref}\right) \tag{31}$$

The T$_{ref}$ denotes reference target, the r_{ref} represents reference closest slant range and the Θ_{ref} denotes Θ of reference target. Subsequently, for each azimuth frequency f_η, it is assumed that there exists a decomposition for θ_{RFM} as follows:

$$\theta_{RFM}\left(f_\tau, f_\eta; T\right) = -2\pi\beta(f_\tau, f_\eta)\cdot 2\gamma(f_\eta; T)/c \tag{32}$$

In Equation (32), β takes charge of Stolt interpolation and γ denotes the difference of the closest slant range between the target T and the reference target T$_{ref}$. Then, the Stolt interpolation is defined as follows:

$$\beta(:, f_\eta) = f_0 + f_\tau' \tag{33}$$

Here, $\beta(:, f_\eta)$ denotes any column in matrix β. After Stolt interpolation, the SAR data is transformed into the new (f_τ', f_η) domain as follows:

$$\begin{aligned}
& S'_{RFM}(f_\tau', f_\eta; T) \\
& = W_r(f_\tau')W_a(f_\eta)\exp\left[-j\frac{4\pi\gamma(f_\eta; T)}{c}(f_0 + f_\tau')\right]
\end{aligned} \tag{34}$$

The variation of γ with azimuth frequency is much smaller than the pixel cell in range direction when resolution achieves decimeter level. As a result, the influence caused by variation of γ on SAR image focusing can be ignored. And the following approximation can be given:

$$\gamma(f_\eta; T) \approx r - r_{ref} \tag{35}$$

As a result, by operating 2D IFFT to Equation (34), the data can be transformed into time domain, and then a well-focused image can be acquired as follows:

$$I'_{RFM}(\tau, \eta; T) = \text{sinc}\left[\tau - \frac{2(r - r_{ref})}{c}\right]\text{sinc}(\eta) \tag{36}$$

The critical point of modified RMA is to obtain the decomposition in Equation (32). Then, SVD is competent to obtain the numerical approximation of decomposition in Equation (32). The purpose of this decomposition is to acquire a matrix for Stolt interpolation.

Actually, β is unable to be acquired directly with SVD. The matrix Φ introduced in the following part is prepared for Stolt interpolation, which can be acquired directly with SVD. In the sense of calculating coordinate for Stolt interpolation, the matrix Φ is consistent with the matrix β. As a result, the matrix Φ can be obtained with the decomposition in Equation (32).

Range and azimuth sampling point number of echo data are defined as N_r and N_a, respectively. For each f_η, a matrix $\mathbf{\Psi}_i$ with size $N_r \times N_E$ can be constituted by the phase θ_{RFM} of different target points, and N_E is the number of point targets for SVD. The matrix $\mathbf{\Psi}_i$ can be acquired as follows:

$$\mathbf{\Psi}_i = [\theta_{RFM}(:, f_\eta; T_1), \theta_{RFM}(:, f_\eta; T_2), \ldots, \theta_{RFM}(:, f_\eta; T_{N_E})]$$
$$f_\eta = (i - N_a/2)\cdot PRF/N_a, i = 1, 2, \ldots, N_a \tag{37}$$

The $\theta_{RFM}(:, f_\eta; T_m)$ denotes the column in θ_{RFM} for each f_η of the m-th target. Here, m = 1, 2, ... , N_E. SVD($\bullet$) is defined as the operation of SVD, and SVD is performed to $\mathbf{\Psi}_i$ as follows:

$$SVD(\mathbf{\Psi}_i) = U_i \Sigma_i V_i^H \tag{38}$$

In Equation (38), superscript H denotes conjugate transpose operation, U_i is left singular vector matrix with size of $N_r \times N_r$, Σ_i is an $N_r \times N_E$ matrix with singular values on the diagonal and V_i is the right singular vector matrix with size of $N_E \times N_E$.

Figure 2 presents a demonstration of singular values. Figure 2a presents amplitudes of the first three singular values in decibels, which are normalized by the amplitude of the largest first singular value. Amplitudes of the first singular values in each azimuth frequency are shown in Figure 2b. Singular values in Figure 2 are acquired with the parameters in Tables 1 and 2. As shown in Figure 2a, it is apparent that first singular values are much larger than second, and third singular values in each azimuth frequency. Consequently, the decomposition of $\mathbf{\Psi}_i$ can be approximated as follows:

$$SVD(\mathbf{\Psi}_i) = U_i \Sigma_i V_i^H \approx u_{i,1} \sigma_{i,1} v_{i,1}^H \tag{39}$$

In (39), $u_{i,1}$ denotes the first column of matrix U_i, $\sigma_{i,1}$ denotes the first singular value corresponding to $\mathbf{\Psi}_i$ and $v_{i,1}$ denotes the first column of matrix V_i.

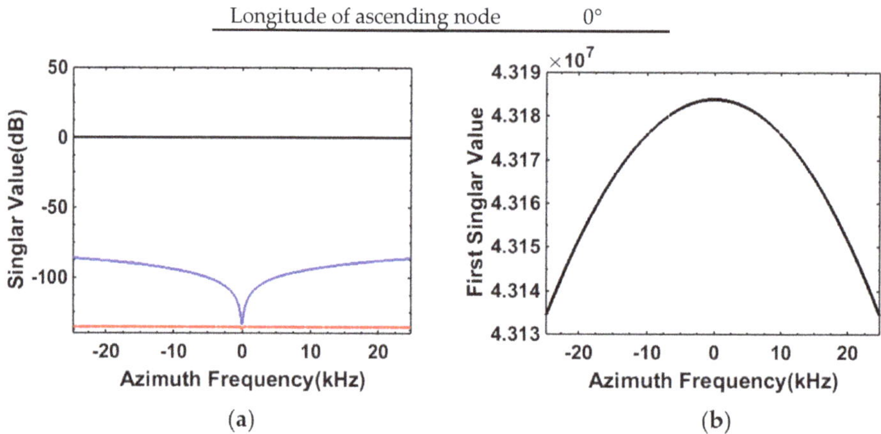

Figure 2. Demonstration of singular values. (a) First three singular values in decibels; (b) First singular values.

Table 1. SAR system parameters.

Parameter Name	Value
Radar centre frequency	9.65 GHz
Signal bandwidth	1 GHz
Range sampling rate	1.2 GHz
Pulse Duration	8 μs
Azimuth processed bandwidth	49.43 kHz
Pulse repetition frequency	4.5 kHz
Look angle	35°
Synthetic aperture time	9 s
Range/Azimuth scene size	3 km/3 km
Closest slant range	629.913 km

Table 2. Orbit parameters.

Parameter Name	Value
Semi-major axis	6870.140 km
Eccentricity	0.0011
Inclination	97.423°
Argument of perigee	90°
Longitude of ascending node	0°

Figure 2b illustrates that first singular values in different azimuth frequency are not constant and vary with azimuth frequency. In essence, performing SVD to $\mathbf{\Psi}_i$ in each f_η is intended to obtain numerical results of decomposition in Equation (32). By comparing Equation (32) with Equation (39) and considering information indicated in Figure 2b, it is reasonable that singular values should not be ignored in generating matrix for Stolt interpolation. Different from SVDS, modified RMA in this paper takes the singular value into consideration and acquires φ_i as follows:

$$\varphi_i = u_{i,1}\sigma_{i,1} \tag{40}$$

After SVD operation in each f_η, the obtained φ_i can be arrayed to form a matrix as follows:

$$\Phi = [\varphi_1, \varphi_2, \ldots, \varphi_i, \ldots, \varphi_{N_a}] \tag{41}$$

The procedure of acquiring matrix Φ is presented in Figure 3. And Φ is a matrix with size of $N_r \times N_a$.

The connection between the decomposition in Equation (32) and the decomposition in Equation (39) can be presented as follows:

$$\begin{aligned}
\theta_{RFM}\left(:, f_\eta; T_1\right) &= -2\pi\beta(:, f_\eta) \cdot 2\gamma(f_\eta; T_1)/c \\
&\approx u_{i,1}\bullet\sigma_{i,1}\bullet v_{i,1}^H\{1\} \\
&= -2\pi\frac{u_{i,1}\bullet\sigma_{i,1}}{\alpha} \cdot \frac{2\alpha\bullet v_{i,1}^H\{1\}}{c}
\end{aligned} \tag{42}$$

$$\beta(:, f_\eta) \approx \frac{u_{i,1}\bullet\sigma_{i,1}}{\alpha} = \frac{\varphi_i}{\alpha} \tag{43}$$

$$\gamma(f_\eta; T_1) \approx \alpha\bullet v_{i,1}^H\{1\} \tag{44}$$

In Equation (42), $\theta_{RFM}(:,f_\eta;T_1)$ denotes any column in θ_{RFM} of target T_1, $v_{i,1}^H\{1\}$ denotes the first element in $v_{i,1}^H$ and α is a constant for associating the decomposition in Equation (32) with decomposition in Equation (39). As α is a constant, the interpolation coordinates calculated from Φ is the same as the interpolation coordinates calculated from β. Therefore, the matrix Φ is able to calculate interpolation coordinates for Stolt interpolation. As a result, acquisition and derivation of reference function and Stolt interpolation coordinates in the Modified RMA have been presented in this part.

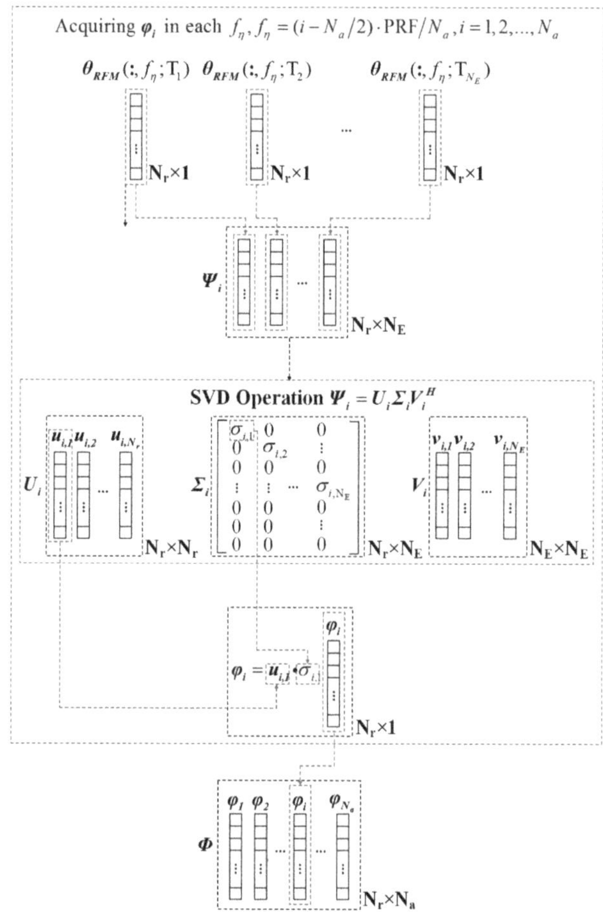

Figure 3. Acquisition of matrix Φ.

2.3. Modified Deramping-Based Approach

According to the Nyquist sampling theory, the pulse repetition frequency (PRF) should be larger than the azimuth bandwidth of echo data so as to avoid azimuth spectral aliasing problem. Nevertheless, in the scenario of high resolution spaceborne spotlight SAR, the PRF is usually smaller than the azimuth bandwidth. As a result, the phenomenon of azimuth spectral aliasing occurs in the situation of high resolution. Therefore, the traditional deramping-based approach is proposed to deal with the azimuth spectral aliasing problem under the consideration of uniform linear motion.

However, curved orbit is desired to be taken into consideration in spaceborne SAR when azimuth resolution achieves 0.14 m. Consequently, the traditional deramping-based approach is not suitable for the situation of curved orbit. Therefore, a modified deramping-based approach is proposed to solve azimuth spectral aliasing problem on curved orbit in this part. Orbital state vectors are utilized to estimate the Doppler parameter, the azimuth frequency modulation rate K_a, in the scenario of curved orbit. With the application of orbital state vectors, deramping-based approach is modified in order to accommodate the curved orbit in this paper.

The azimuth frequency modulation rate K_a can be obtained as follows when SAR platform is in uniform linear motion:

$$K_a == 2v^2/(\lambda r) \tag{45}$$

In Equation (45), v is the velocity of the SAR platform, λ is the wavelength of carrier wave and r is the closest slant range.

In the scenario of curved orbit, K_a cannot be acquired with Equation (45). So as to solve this problem, orbital state vectors are introduced to the azimuth frequency modulation rate K_a in the scenario of curved orbit. The K_a can be acquired with orbital state vectors as follows:

$$K_a = -\frac{2}{\lambda}\left[\frac{R \circ A + V \circ V}{g_0} - \frac{(V \circ R)^2}{g_0^3}\right] \tag{46}$$

The R, V and A present relative position, velocity and acceleration 3-dimentional orbital state vectors between the radar and a target, respectively.

After K_a has been acquired, azimuth deramping can be implemented to $s(\tau,\eta)$ as follows:

$$s'(\tau,\eta) = s(\tau,\eta) *_{az} \exp(j\pi K_a \eta^2) \tag{47}$$

The $*_{az}$ denotes convolution operation in azimuth direction. $s(\tau,\eta)$ is the echo data after demodulation to baseband.

The convolution in Equation (47) can be implemented in an another approach which contains a chirp multiplication of the azimuth signal $h_1(\eta)$, a subsequent FT operation and a residual phase $h_2(\eta')$ multiplication. In other words, the alternate way for azimuth convolution in Equation (47) can be expressed by Equation (48) as follows:

$$s'(\tau,\eta') = h_2(\eta') \cdot \text{FFTa}[s(\tau,\eta)h_1(\eta)] \tag{48}$$

where the FFTa($\bullet$) denotes azimuth fast Fourier transform operation. Two quadratic phase signals, $h_1(\eta)$ and $h_2(\eta')$, are defined in Equations (49) and (50) as follows:

$$h_1(\eta) = \exp(j\pi K_a \eta^2) \tag{49}$$

$$h_2(\eta') = \exp\left[j\pi K_a (\eta')^2\right] \tag{50}$$

In Equations (49) and (50), the η and η' are defined in Equations (51) and (52) as follows:

$$\eta = n/\text{PRF}, n = -N_a/2 + 1, \ldots, N_a/2 \tag{51}$$

$$\eta' = n \cdot \text{PRF}/(K_a P), n = -P/2 + 1, \ldots, P/2 \tag{52}$$

As $P > N_a$, a zero padding operation of $s(\tau,\eta)$ in azimuth direction is required. As the η' is determined by PRF, K_a and P, the P can be chosen according to the requirement of η'. In addition, the P can also be selected depending on the efficient implementation of FFT codes. Through application of modified deramping-based approach, azimuth spectral aliasing problem is solved.

2.4. Implementation of Proposed Method

The proposed method consists of modified RMA and modified deramping-based approach. The flowchart of the proposed method is illustrated in this part. As illustrated in Figure 4, the flowchart of the proposed method can be separated into two steps: azimuth deramping and precise focusing.

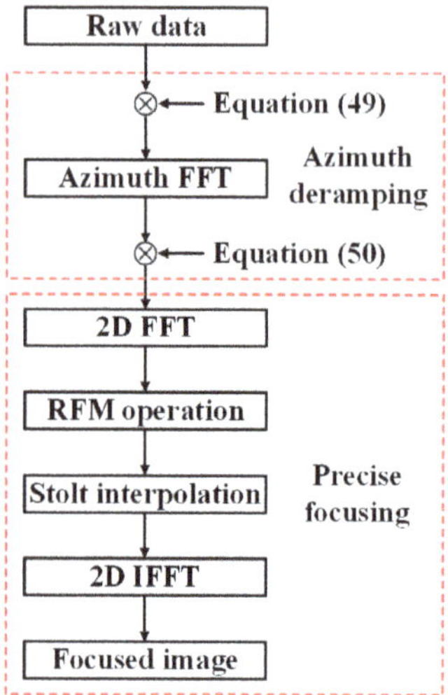

Figure 4. Flowchart of the proposed method.

The step of azimuth deramping aims to solve the azimuth spectral aliasing problem in the scenario of curved orbit. In this step, modified deramping-based approach is implemented to SAR raw data according to Equation (48). The modified deramping-based approach is implemented through a phase multiplication of the signal $h_1(\eta)$, a subsequent azimuth FT operation and another phase multiplication of the signal $h_2(\eta')$. With the implementation of modified deramping-based approach, the azimuth spectral aliasing problem has been solved.

After application of azimuth deramping, the step of precise focusing is utilized to obtain focused image. In this step, the raw data after azimuth deramping is firstly transformed into 2D frequency domain via 2D fast Fourier transform (FFT). Then, RFM operation is implemented to totally compensate the range migration of all targets at the reference range. The procedure of RFM operation is presented in Equation (30). After RFM operation, a residual phase exists for targets at other ranges. So as to cancel the residual phase, a subsequent Stolt interpolation is performed to cope with it. The Stolt interpolation is implemented in light of the matrix $\boldsymbol{\Phi}$. The procedure of obtaining matrix $\boldsymbol{\Phi}$ is demonstrated in Figure 3 and is described in Section 2.2. After implementation of Stolt interpolation, 2D IFFT is operated to transform the data into the time domain and the data is eventually focused in time domain.

3. Results

In this section, point targets simulation is conducted to assess the effectiveness of the proposed method. The proposed method is compared with the generalized ωK in [13]. The results of simulation represent that the proposed method has smaller phase error and performs better than the generalized ωK when azimuth resolution achieves 0.14 m.

The simulation is under the consideration of monostatic spotlight spaceborne SAR with transmitting pulse chirp signal. The SAR system parameters for simulation are listed in Table 1. In simulation, the echo data is an $80,000 \times 40,500$ matrix with 80,000 range sampling points and 40,500 azimuth sampling pulses. After zero padding along azimuth direction in modified deramping-based approach, the size of data for processing becomes $80,000 \times 44,500$. The resolution of point target is expected to achieve 0.15 m and 0.14 m in range and azimuth direction, respectively. The swath width of scene is set as 3 km and 3 km in range and azimuth direction, respectively.

As the platform of SAR is considered to be a satellite in this paper, orbit parameters are taken into account in the simulation. A total of six independent parameters are required to describe the motion of a satellite around the earth [22]. The mean anomaly is time variant, which defines the position of satellite along the orbit. The other five constant orbit parameters are listed in Table 2. Orbit parameters in Table 2 are utilized in simulation so as to generate the curved orbit of spaceborne spotlight SAR. Nine point targets are positioned for simulation, which are labelled from T_1 to T_9. The distribution of point targets is shown in Figure 5.

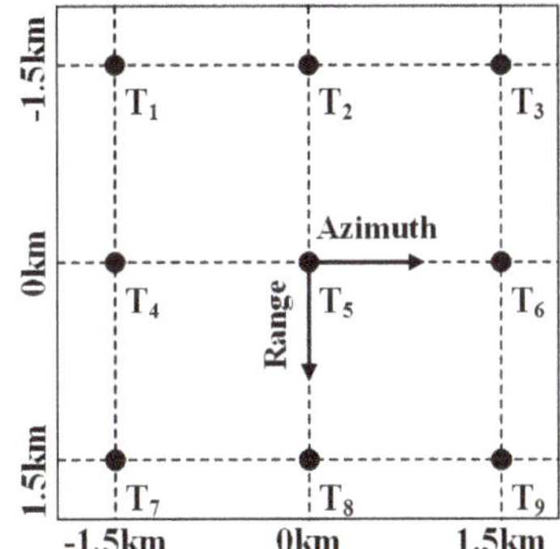

Figure 5. Distribution of point targets.

The effectiveness of modified deramping-based approach is shown in Figures 6 and 7. Figure 6a demonstrates 2D spectrum of single point target without application of modified deramping-based approach. And Figure 6b displays 2D spectrum of single point target with application of modified deramping-based approach. The row, which is at 0 Hz in the range frequency of each 2D spectrum, is chosen to illustrate the one dimensional (1D) azimuth profile of each 2D spectrum in Figure 6. Figure 7a shows the 1D azimuth profile of 2D spectrum without application of modified deramping-based approach. Figure 7b depicts the 1D azimuth profile of 2D spectrum with application of modified deramping-based approach. With the modified deramping-based approach, the PRF is enlarged from 4.5 kHz to 54.33 kHz. Meanwhile, the processed azimuth bandwidth is 49.43 kHz. Consequently, the enlarged PRF is larger than the azimuth spectrum bandwidth and the azimuth spectrum aliasing problem is solved with the application of modified deramping-based approach. As demonstrated in Figures 6b and 7b, the azimuth spectrum aliasing problem is solved with the implementation of the modified deramping-based approach.

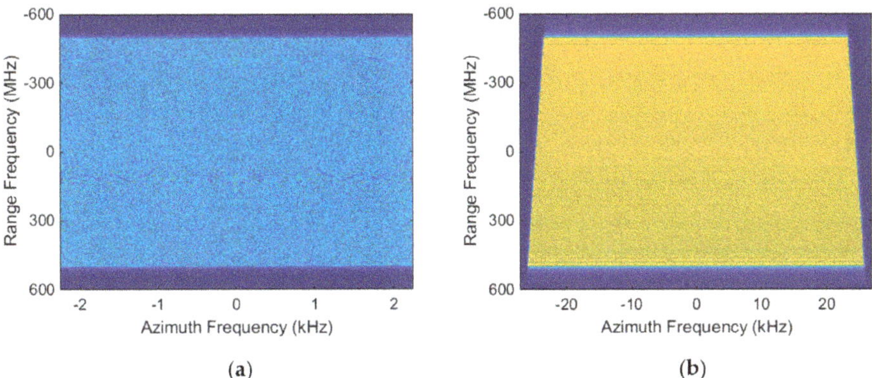

Figure 6. 2D Spectrum of single point target. (**a**) 2D spectrum without application of modified deramping-based approach; (**b**) 2D spectrum with application of modified deramping-based approach.

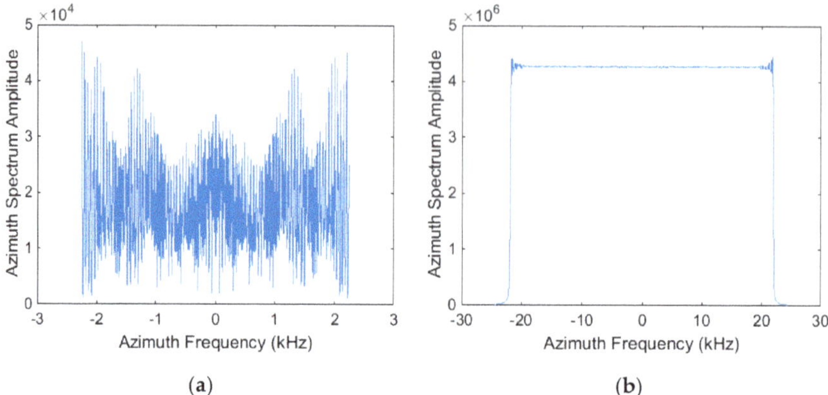

Figure 7. Demonstration of 1D azimuth profile of 2D spectrum. (**a**) 1D azimuth profile of 2D spectrum without application of modified deramping-based approach; (**b**) 1D azimuth profile of 2D spectrum with application of modified deramping-based approach.

Figure 8a,b show the 2D spectrum phase error in the generalized ωK and the proposed method, respectively. The phase error is mainly caused by the numerical decomposition which is applied to the phase of the data after RFM operation. The phase errors shown in Figure 8a,b are obtained using parameters in Tables 1 and 2. The first-order multivariable Taylor expansion, which is utilized for the numerical decomposition in the generalized ωK method, is responsible for the phase error of the generalized ωK. It is apparent that phase error in Figure 8a is larger than 0.8π rad at the edges of the spectrum with the parameters in Tables 1 and 2. In other words, Figure 8a indicates that phase error of the generalized ωK is too large for SAR focusing when resolution is desired to achieve 0.15 m and 0.14 m in range and azimuth direction, respectively. As a result, such phase error in Figure 8a seriously degrades the imaging quality of SAR raw data. The SVD, which is used for numerical decomposition in the proposed method, is responsible for the phase error of the proposed method. In Figure 8b, phase error of the proposed method is less than $5 \times 10^{-4}\,\pi$ rad. The phase error shown in Figure 8b indicates that the proposed method is more suitable for SAR focusing under the imaging requirements in this paper.

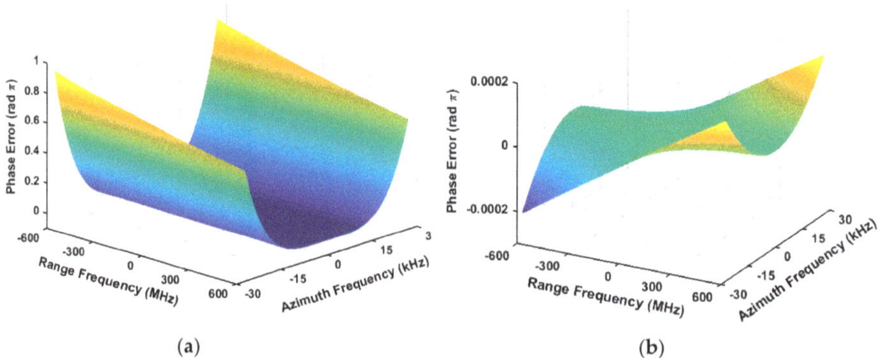

Figure 8. Phase error. (**a**) Phase error caused by first-order multivariable Taylor expansion in the generalized ωK; (**b**) Phase error caused by SVD in the proposed method.

The imaging results of point targets via the generalized ωK and the proposed method are presented in Figures 9 and 10. Nine target points are located according to the distribution in Figure 5. In Figure 9, imaging results of point targets acquired with the generalized ωK, which are labelled from T_1 to T_9, are depicted Figure 9a–i, respectively. In Figure 10, imaging results of point targets acquired with the proposed method, which are labelled from T_1 to T_9, are depicted Figure 10a–i, respectively.

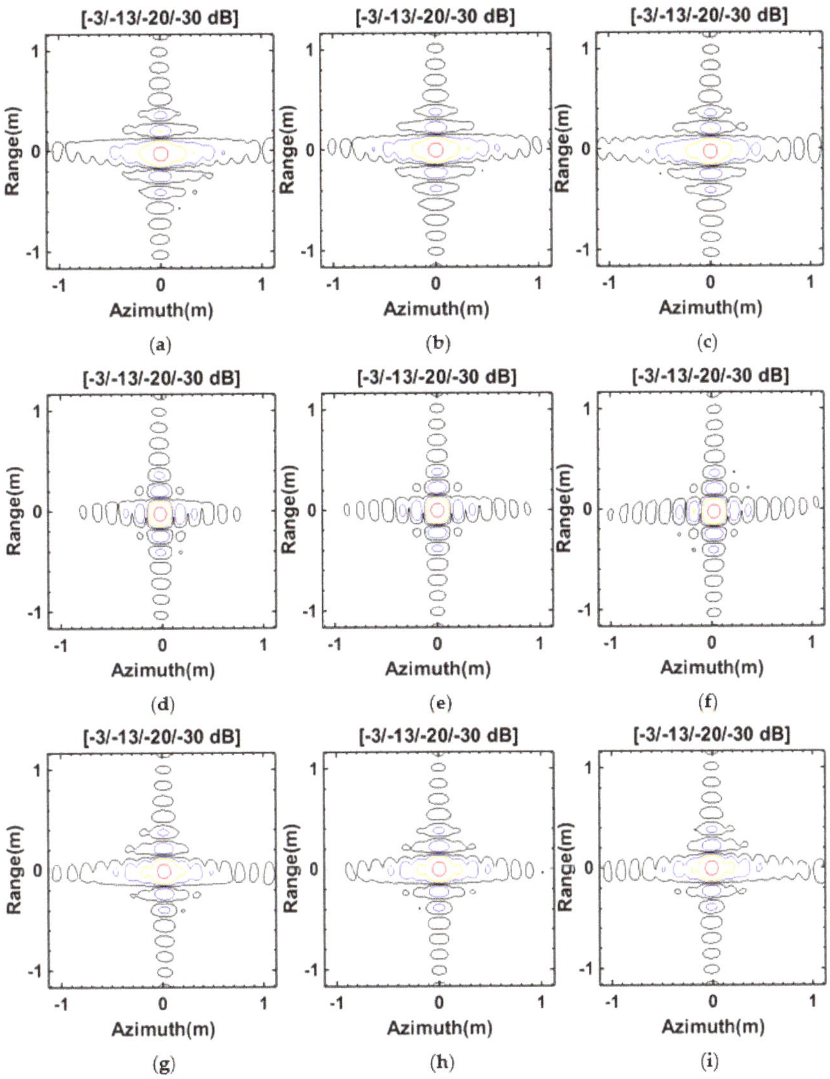

Figure 9. Imaging results with the generalized ωK (**a**) Imaging result of target T_1; (**b**) Imaging result of target T_2; (**c**) Imaging result of target T_3; (**d**) Imaging result of target T_4; (**e**) Imaging result of target T_5; (**f**) Imaging result of target T_6; (**g**) Imaging result of target T_7; (**h**) Imaging result of target T_8; (**i**) Imaging result of target T_9.

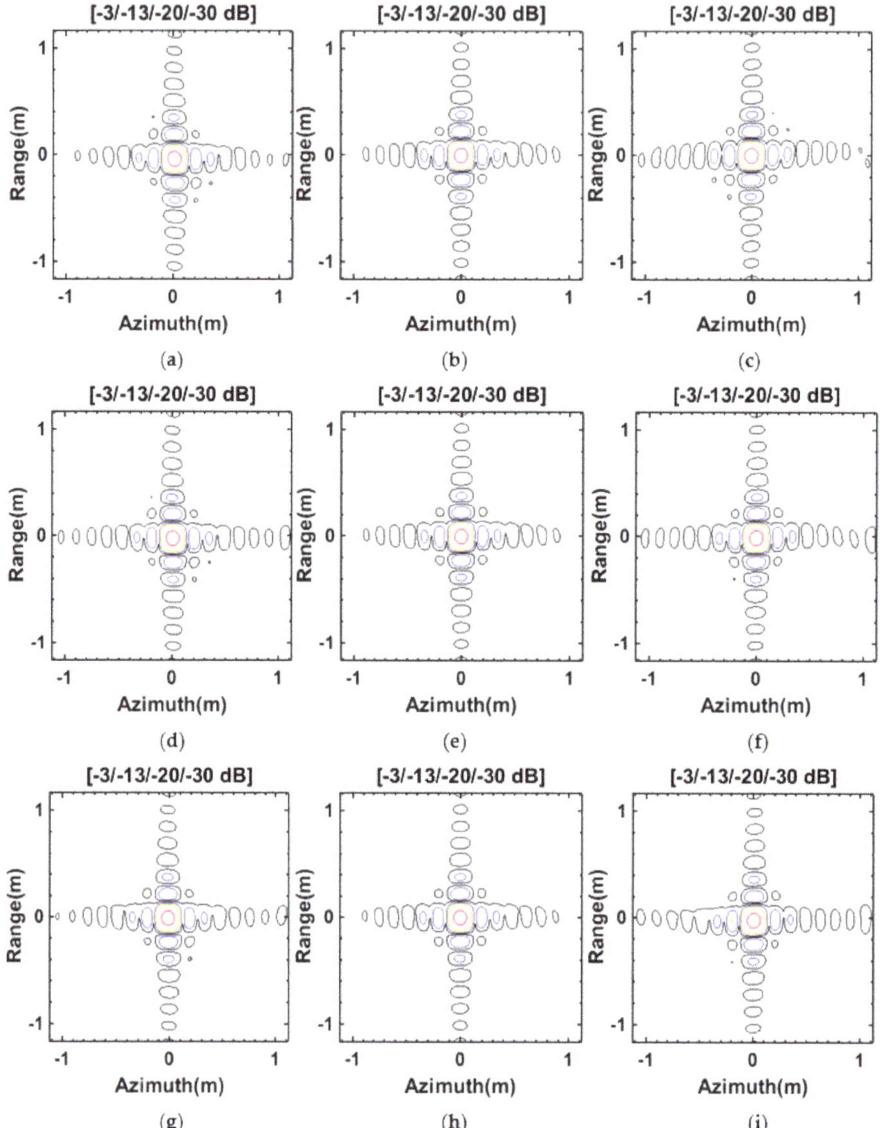

Figure 10. Imaging results with the proposed method. (**a**) Imaging result of target T_1; (**b**) Imaging result of target T_2; (**c**) Imaging result of target T_3; (**d**) Imaging result of target T_4; (**e**) Imaging result of target T_5; (**f**) Imaging result of target T_6; (**g**) Imaging result of target T_7; (**h**) Imaging result of target T_8; (**i**) Imaging result of target T_9.

As shown in Figure 9, it is that imaging results of T_1, T_2, T_3, T_7, T_8 and T_9 are not well-focused and suffer from azimuth defocusing. It can be concluded that the generalized ωK is unable to focus the echo data of spaceborne spotlight SAR when azimuth resolution attains 0.14 m. As illustrated in Figure 8a, such defocusing phenomenon in azimuth direction is caused by the unbearable phase error of the first-order multivariable Taylor expansion when resolution achieves 0.15 m and 0.14 m in range and azimuth direction. In contrast, the phase error of 2D spectrum in Figure 8b is much smaller than phase error in Figure 8a. Consequently, it is apparent that all the point targets presented in Figure 10 are focused much better than point targets in Figure 9. It can be concluded that the proposed method performs better than the generalized ωK when resolution achieves 0.15 m and 0.14 m in range and azimuth direction, with focused scene size of 3 km by 3 km.

4. Discussion

In order to further demonstrate the effectiveness of the proposed method, impulse response width (IRW), peak sidelobe ratio (PSLR) and integrated sidelobe Ratio (ISLR) are chosen as criteria for evaluating the quality of imaging results. The analysis for imaging results of point targets with the generalized ωK is listed in Table 3. And the analysis for imaging results of point targets via the proposed method is listed in Table 4.

Table 3. Analysis of results with the generalized ωK.

T	Range			Azimuth		
	IRW (m)	PSLR (dB)	ISLR (dB)	IRW (m)	PSLR (dB)	ISLR (dB)
1	0.1334	−13.10	−10.43	0.1457	−	−5.73
2	0.1331	−13.09	−10.31	0.1459	−	−6.15
3	0.1336	−13.11	−10.40	0.1474	−	−5.67
4	0.1332	−13.03	−10.50	0.1298	−13.30	−9.69
5	0.1334	−13.24	−10.31	0.1272	−13.24	−10.31
6	0.1335	−13.02	−10.49	0.1269	−13.33	−10.32
7	0.1338	−13.32	−10.46	0.1330	−	−7.91
8	0.1339	−13.12	−10.31	0.1344	−	−7.51
9	0.1335	−13.22	−10.47	0.1331	−	−8.05

The '−' denotes that azimuth defocusing makes measurement meaningless.

Table 4. Analysis of results with the proposed method.

T	Range			Azimuth		
	IRW (m)	PSLR (dB)	ISLR (dB)	IRW (m)	PSLR (dB)	ISLR (dB)
1	0.1336	−13.08	−10.43	0.1302	−12.80	−9.61
2	0.1335	−13.23	−10.31	0.1273	−13.27	−10.23
3	0.1334	−13.05	−10.40	0.1274	−13.08	−10.18
4	0.1338	−13.32	−10.50	0.1282	−13.05	−9.85
5	0.1334	−13.23	−10.31	0.1273	−13.22	−10.24
6	0.1335	−13.28	−10.49	0.1279	12.96	9.94
7	0.1334	−13.25	−10.46	0.1280	−13.07	−9.83
8	0.1336	−13.23	−10.31	0.1282	−13.14	−10.18
9	0.1335	−13.26	−10.47	0.1281	−13.02	−9.85

In SAR processing, the IRW refers to as the image resolution. Namely, the IRW in both range and azimuth direction should satisfy the requirement of resolution. Furthermore, in order to guarantee the quality of the focused point target, PSLR should be less than −13 dB and ISLR should be about −10 dB.

The analysis in Table 3 shows that IRW, PSLR and ISLR of processed results obtained by the method in [13] satisfy the requirements of criteria in range direction. However, analysis in Table 3 indicates that IRW, PSLR and ISLR of processed results fail to satisfy the requirements of criteria in azimuth direction. As shown in Table 3, T_1, T_2 and T_3 fail to achieve 0.14 m in azimuth resolution. Meanwhile, the analysis in Table 3 is consistent with the imaging results in Figure 9 on the unsatisfactory focused performance in azimuth direction. In Table 3, the '−' denotes that azimuth defocusing makes measurement of PSLR in azimuth direction meaningless. In comparison with the analysis in Table 3, Table 4 indicates that IRW, PSLR and ISLR of imaging results obtained by the proposed method almost satisfy the requirements of criteria in both range and azimuth direction. It is obvious that the proposed method has superior focusing performance in azimuth direction. Analysis in Table 4 and imaging results in Figure 10 verify that the proposed method performs better than the generalized ωK. In conclusion, the proposed method performs effectively on focusing raw data of high resolution spotlight spaceborne SAR when resolution achieves 0.15 m and 0.14 m in range and azimuth direction.

5. Conclusions

A method is proposed to deal with ultrahigh resolution spotlight spaceborne SAR imaging in this paper. The proposed method consists of modified RMA and modified deramping-based approach. The modified RMA method is developed for accommodating the scenario of curved orbit. The modified deramping-based approach is utilized to solve the azimuth spectral aliasing problem in curved orbit scenario. Point targets simulation and analysis operated on spaceborne spotlight SAR parameters validate the effectiveness of the proposed method. The focused results obtained by the proposed method generally obtain the expecting performance. Analysis demonstrates that resolution of focused results can achieve 0.15 m in range direction and 0.14 m in azimuth direction. Furthermore, the swath width of focused scene can achieve 3 km and 3 km in range and azimuth direction, respectively.

Author Contributions: Y.Q. and D.Z. conceived and designed the method; Y.Q. performed the simulations and experiments; and Y.Q. wrote the paper.

Funding: This research was funded by National Key R&D Program of China, grant number 2017YFB0502700 and by Postgraduate Research & Practice Innovation Program of Jiangsu Province, grant number KYCX17_0266.

Conflicts of Interest: The authors declare no conflict of interest.

References

1. Yang, M.; Zhu, D. Efficient Space-Variant Motion Compensation Approach for Ultra-High-Resolution SAR Based on Subswath Processing. *IEEE J. Sel. Top. Appl. Earth Obs. Remote Sens.* **2018**, *11*, 2090–2103. [CrossRef]

2. Qian, Y.; Zhu, D. High-resolution SAR imaging from azimuth periodically gapped raw data via generalised orthogonal matching pursuit. *Electron. Lett.* **2018**, *54*, 1235–1236. [CrossRef]

3. Li, N.; Wang, R.; Deng, Y.; Yu, W.; Zhang, Z.; Liu, Y. Autofocus Correction of Residual RCM for VHR SAR Sensors with Light-Small Aircraft. *IEEE Trans. Geosci. Remote Sens.* **2017**, *55*, 441–452. [CrossRef]

4. Zhang, Y.; Zhu, D. Height Retrieval in Postprocessing-Based VideoSAR Image Sequence Using Shadow Information. *IEEE Sens. J.* **2018**, *18*, 8108–8116. [CrossRef]

5. Penner, J.F.; Long, D.G. Ground-Based 3D Radar Imaging of Trees Using a 2D Synthetic Aperture. *Electronics* **2017**, *6*, 11. [CrossRef]

6. Lao, G.; Yin, C.; Ye, W.; Sun, Y.; Li, G.; Han, L. An SAR-ISAR Hybrid Imaging Method for Ship Targets Based on FDE-AJTF Decomposition. *Electronics* **2018**, *7*, 46. [CrossRef]

7. Qian, Y.; Zhu, D. High Resolution Imaging from Azimuth Missing SAR Raw Data via Segmented Recovery. *Electronics* **2019**, *8*, 336. [CrossRef]

8. Villano, M.; Krieger, G.; Moreira, A. Staggered SAR: High-Resolution Wide-Swath Imaging by Continuous PRI Variation. *IEEE Trans. Geosci. Remote Sens.* **2014**, *52*, 4462–4479. [CrossRef]

9. Prats-Iraola, P.; Scheiber, R.; Rodriguez-Cassola, M.; Mittermayer, J.; Wollstadt, S.; De Zan, F.; Brautigam, B.; Schwerdt, M.; Reigber, A.; Moreira, A. On the Processing of Very High Resolution Spaceborne SAR Data. *IEEE Trans. Geosci. Remote Sens.* **2014**, *52*, 6003–6016. [CrossRef]

10. Cumming, I.G.; Wong, F.H. *Digital Processing of Synthetic Aperture Radar Data: Algorithms and Implementation*, 1st ed.; Artech House Press: Norwood, MA, USA, 2005; pp. 113–168.

11. Eldhuset, K. A new fourth-order processing algorithm for spaceborne SAR. *IEEE Trans. Aerosp. Electron. Syst.* **1998**, *34*, 824–835. [CrossRef]

12. Zeng, T.; Yang, W.; Ding, Z.; Liu, L. Advanced range migration algorithm for ultra-high resolution spaceborne synthetic aperture radar. *IET Radar Sonar Navig.* **2013**, *7*, 764–772. [CrossRef]

13. Hu, B.; Jiang, Y.; Zhang, S.; Zhang, Y.; Yeo, T. Generalized Omega-K Algorithm for Geosynchronous SAR Image Formation. *IEEE Geosci. Remote Sens. Lett.* **2015**, *12*, 2286–2290. [CrossRef]

14. Wu, J.; An, H.; Zhang, Q.; Sun, Z.; Li, Z.; Du, K.; Huang, Y.; Yang, J. Two-dimensional frequency decoupling method for curved trajectory synthetic aperture radar imaging. *IET Radar Sonar Navig.* **2018**, *12*, 766–773. [CrossRef]

15. Mao, X.; He, X.; Li, D. Knowledge-Aided 2-D Autofocus for Spotlight SAR Range Migration Algorithm Imagery. *IEEE Trans. Geosci. Remote Sens.* **2018**, *56*, 5458–5470. [CrossRef]

16. Neo, Y.L.; Wong, F.; Cumming, I.G. A Two-Dimensional Spectrum for Bistatic SAR Processing Using Series Reversion. *IEEE Geosci. Remote Sens. Lett.* **2007**, *4*, 93–96. [CrossRef]

17. Neo, Y.L.; Wong, F.; Cumming, I.G. Processing of Azimuth-Invariant Bistatic SAR Data Using the Range Doppler Algorithm. *IEEE Trans. Geosci. Remote Sens.* **2008**, *46*, 14–21. [CrossRef]

18. Dwight, H.B. *Table of Integrals and Other Mathematical Data*, 3rd ed.; Macmillan Press: New York, NY, USA, 1957; pp. 10–11.

19. D'Aria, D.; Guarnieri, A.M. High-Resolution Spaceborne SAR Focusing by SVD-Stolt. *IEEE Geosci. Remote Sens. Lett.* **2007**, *4*, 639–643. [CrossRef]

20. Lanari, R.; Tesauro, M.; Sansosti, E.; Fornaro, G. Spotlight SAR data focusing based on a two-step processing approach. *IEEE Trans. Geosci. Remote Sens.* **2001**, *39*, 1993–2004. [CrossRef]

21. An, D.; Huang, X.; Jin, T.; Zhou, Z. Extended Two-Step Focusing Approach for Squinted Spotlight SAR Imaging. *IEEE Trans. Geosci. Remote Sens.* **2012**, *50*, 2889–2900. [CrossRef]

22. Montenbruck, O.; Gill, E. *Satellite Orbits: Models, Methods and Applications*, 1st ed.; Springer Press: Berlin, Germany, 2000; pp. 15–46.

Article

Application of S-Transform in ISAR Imaging

Bo Zang, Mingzhe Zhu *, Xianda Zhou and Lu Zhong

School of Electronic Engineering, Xidian University, Xi'an 710126, China; bzang@mail.xidian.edu.cn (B.Z.);
zhouxianda999@gmail.com (X.Z.); zhonglu@protonmail.com (L.Z.)
* Correspondence: zhumz@mail.xidian.edu.cn

Received: 25 April 2019; Accepted: 11 June 2019; Published: 14 June 2019

Abstract: In inverse synthetic aperture radar (ISAR) imaging, time-frequency analysis is the basic method for processing echo signals, which are reflected by the results of time-frequency analysis as each component changes over time. In the time-frequency map, a target's rigid body components will appear as a series of single-frequency signals in the low-frequency region, and the micro-Doppler components generated by the target's moving parts will be distributed in the high-frequency region with obvious frequency modulation. Among various time-frequency analysis methods, S-transform is especially suitable for analyzing these radar echo signals with micro-Doppler (m-D) components because of its multiresolution characteristics. In this paper, S-transform and the corresponding synchrosqueezing method are used to analyze the ISAR echo signal and perform imaging. Synchrosqueezing is a post-processing method for the time-frequency analysis result, which could retain most merits of S-transform while significantly improving the readability of the S-transformation result. The results of various simulations and actual data will show that S-transform is highly matched with the echo signal for ISAR imaging: the better frequency-domain resolution at low frequencies can concentrate the energy of the rigid body components in the low-frequency region, and better time resolution at high frequencies can better describe the transformation of the m-D component over time. The combination with synchrosqueezing also significantly improves the effect of time-frequency analysis and final imaging, and alleviates the shortcomings of the original S-transform. These results will be able to play a role in subsequent work like feature extraction and parameter estimation.

Keywords: time-frequency analysis; S-transformation; ISAR; micro-Doppler; synchrosqueezing

---

1. Introduction

In inverse synthetic aperture radar (ISAR) imaging, to obtain the final image, two steps of Fourier transform are necessary for the original echo signal. Through the Fourier transform, the periodic characteristics of the time domain signal are found and reflected in the frequency domain. With the periodic characteristics in different dimensions combined with the appropriate resolution, the shape information of the target can be obtained [1].

In these radar echo signals, the rigid body part of the target and the moving part will produce two types of component. The latter will be referred to as the micro-Doppler (m-D) component with frequency modulation, which may generate from rotation of propellers or rotor wings on plane, surface vibration caused by engine, and swinging arms when human walk may cause the m-D effect [2,3].

The m-D component usually interferes with the final imaging result, but on the other hand, it can also be used for feature extraction and parameter estimation of the target, which is useful information in subsequent work [4]. However, in the face of such components, the simple Fourier transform cannot analyze the transformation of their frequency domain characteristics with time. For this reason, time-frequency analysis, which was born to characterize the transformation of signals in the frequency domain over time, has been applied to this field and has become one of the most basic analysis methods for m-D components [1,5,6].

Among many time-frequency analysis methods, S-transform (ST) is especially suitable for analyzing radar echo signals for imaging [7]. The multiresolution characteristics of S-transform are consistent with ISAR's signal type. Its high frequency domain resolution in the low-frequency region can achieve accurate imaging of the target's rigid body component, while the time domain resolution advantage of the high frequency region can effectively provide target micro-Doppler information.

S-transform can be further combined with the synchrosqueezing transform (SST) method to further improve the concentration of the time-frequency analysis results [8]. The algorithm performs synchronous squeezing of the time-frequency map in the frequency domain, which improves the concentration while inverse transformation, in a strict mathematical sense, can be ensured. Therefore, SST can not only do analysis from the time domain to the time-frequency domain, but can also hold the synthesis from the time-frequency domain to the time domain. Recently, by adopting the second-order operator, second-order synchrosqueezing (SST2) has been able to analyze a strong frequency-modulated signal [9–11].

In this paper, S-transform and second-order synchrosqueezing methods are applied to ISAR imaging. Both the simulation and the actual signal will have the benefit of the multiresolution characteristics of the S-transform when analyzing radar echo signals containing micro-Doppler components. Synchrosqueezing, on the basis of retaining these characteristics, greatly improves the readability of time-frequency analysis results, which brings more favorable conditions for subsequent work. The arrangement of this paper is as follows. A brief introduction of ISAR imaging and m-D effects are given in Section 2. The principles of time-frequency and S-transform is given in Section 3. Synchrosqueezing and second-order synchrosqueezing is introduced in Section 4. The results of numerical simulations and real data tests are presented in Section 5.

2. Principle of ISAR Imaging

A model of the ISAR echo signal is described in Figure 1.

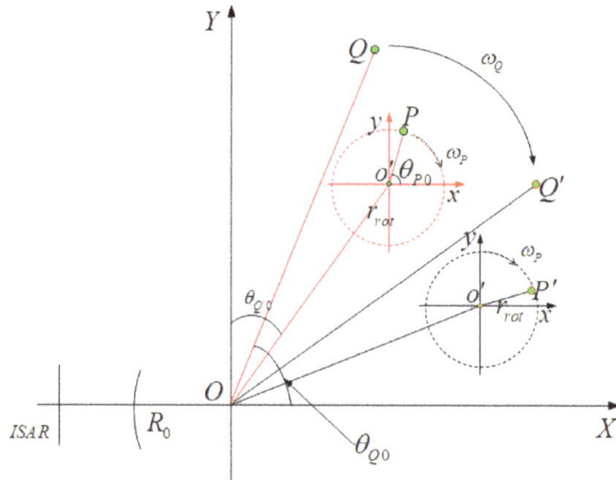

Figure 1. ISAR geometric model.

The micro-Doppler effect is illustrated by the rigid body portion scattering point Q and the rotational scattering point P of the target in Figure 1. The echo signal after range compression is defined as

$$S(f_r, t_m) = AT_pc \cdot \sin[T_p(f_r + 2\frac{\gamma}{c}R_\Delta(t_m))]\exp(-j2\pi\frac{2}{\lambda}R_\Delta(t_m)), \qquad (1)$$

where A is the amplitude of the echo signal, T_p is the pulse width, γ is the modulation rate, c is the velocity of light, λ is the wavelength of carrier frequency, and $R_\Delta(t_m) = R - R_0$ is the distance difference from scattering point Q to the radar displaced phase center and reference point O. f_r and t_m correspond to the frequency domain and slow time of the pulse compression result, respectively.

For the rigid scattering point Q,

$$
\begin{aligned}
R_{\Delta Q}(t_m) &= \sqrt{R_0^2 + r_Q^2 - 2R_0 r_Q \cos(\theta(t_m) + \theta_{Q0} + \tfrac{\pi}{2})} \\
&\approx r_Q \sin(\theta_Q(t_m) + \theta_{Q0}),
\end{aligned}
\tag{2}
$$

where r_Q is the distance from the rigid rotation center to point Q, θ_{Q0} is the angle between the X axis and OQ in coordinate XOY, and $\theta_Q(t_M) = \omega t_m$. Based on the assumption of the uniform circular motion for point Q, it is centered at O with angular velocity ω_Q in the coherent accumulative time.

Since the accumulation on cross-range is very small here, it has $\cos(\theta_Q(t_m)) \approx 1$, $\sin(\theta_Q(t_m)) \approx 0$. The Doppler frequency of Q is given by

$$
f_d Q = \frac{2}{\lambda} \frac{dR_{\Delta Q}(t_m)}{dt_m} \approx \frac{2}{\lambda} \omega_Q x_{Q0}.
\tag{3}
$$

For high-speed rotating scattering point P on target, while it has the same rotation center to point O as Q, another rotation with center point o' of moving parts makes it different. As in Figure 1, it has

$$
R_{\Delta P} \approx r_{o'} \sin(\theta_{o'}(t_m) + \theta_{o'0}) + r_{rot} \sin(\theta_p(t_m) + \theta_{P0}),
\tag{4}
$$

where r_o' is the distance from o' to O, $\theta_{o'0}$ is the angle between the X axis and Oo' at zero time, and they have $\theta_{o'}(t_m) = \omega t_m$. Similarly, r_{rot} is the radius of the rotation of P, θ_{P0} is the angle between the X axis and $o'P$ at zero time, and they are related as $\theta_P(t_m) = \omega_P t_m$. In comparison to Q, one more rotation of P introduces a term of sinusoidal modulation to it's Doppler frequency:

$$
\begin{aligned}
f_d P &= \frac{2}{\lambda} \frac{dR_{\Delta P}(t_m)}{dt_m} \\
&= \frac{2}{\lambda} \frac{d[r_{o'} \sin(\theta_{o'}(t_m) + \theta_{o'0}) + r_{rot} \sin(\theta_p(t_m) + \theta_{P0}))]}{dt_m} \\
&\approx \frac{2}{\lambda} [x_{o'} \omega_Q + r_{rot} \cos(\theta_p(t_m) + \theta_{P0}) \omega_P],
\end{aligned}
\tag{5}
$$

which is called m-D information.

In the above model of the m-D effect, in coherent accumulative time, rigid parts appear as superpositions of sinusoidal waves in ISAR's cross-range. Then, moving parts add modulation. When there are Nr scattering points in rigid parts, and Nm in moving parts, the echo signal could be written as

$$
\begin{cases}
s = s_1 + s_2 \\[2mm]
s_1 = \displaystyle\sum_{p=1}^{Nr} A_{1p} \exp(j2\pi f_{1p} t) \\[2mm]
s_2 = \displaystyle\sum_{q=1}^{Nm} A_{2q} \exp(j2\pi (f_{2q} t + \lambda_q \sin(\omega_q t))),
\end{cases}
\tag{6}
$$

where s_1 represents the components from rigid parts, and A_{1p} and f_{1p} are the amplitude and frequency of the cross-range Doppler signal for the pth scattering point, respectively. s_2, A_{2q} and f_{2q} have the same meaning for moving parts, and $\lambda_q \sin(\omega_q t)$ is the m-D information.

Figure 2 shows the process of ISAR imaging.

Originally, in the dotted line parts of Figure 2, signals from different range cells only need to do the Fourier transform then arrange results in order, the imaging result would then be obtained. In the frequency domain of those signals, the energy of the sinusoidal frequency modulated (FM) component brought by moving parts will be distributed in a range where the spectrum is larger than the rigid body component. In the final imaging results, they seemed to be unwanted interference information relative to the rigid body part. The Fourier transform only shows the distribution of these interferences in the frequency domain but does not indicate the information of these m-D components over time. In Figure 1, Fourier transform is replaced by time-frequency analysis, and the m-D information is reflected in a more specific form on the time-frequency plane.

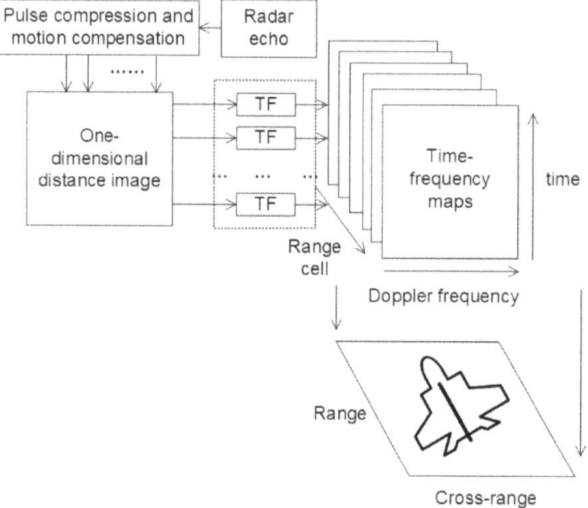

Figure 2. The process of ISAR time-frequency imaging.

3. Time-Frequency Analysis and S-Transform

Generally, time-frequency analysis methods can be divided into two categories: one-order time-frequency transform and second-order time-frequency distribution [12]. Although the latter works much better than the former for the single-component signals, it would bring cross-terms that are inconvenient to handle when processing multicomponent signals. As mentioned above, the signals used for imaging in this paper are multicomponent (6), so the time-frequency distribution method is not desirable. Thus, here we will start with the most basic short-time Fourier transform (STFT) in the one-order time-frequency transform.

The multicomponent (MC) FM signal is considered as

$$x(t) = \sum_{m=1}^{M} A_m \cdot \exp(j(\phi_m(t))),\qquad(7)$$

here $\phi_m(t)$ is the phase function of the *m*th signal, and A_m is the amplitude.

As the signal is inserted into the time-frequency plane by short time Fourier transform (STFT), we have

$$STFT(t,f) = \int_{-\infty}^{+\infty} x(\tau)w(\tau - t)e^{-2i\pi f\tau}d\tau,\qquad(8)$$

where $w(t)$ is usually a Gaussian window function with fixed parameters. This would bring a fixed time-frequency resolution to the analysis result. Even if it is adjusted, it is difficult to get the overall optimality.

Similarly, the S-transform is defined as

$$ST_x(t,f) = \int_{-\infty}^{+\infty} x(\tau)h(\tau - t, f)e^{-2i\pi f\tau}d\tau, \tag{9}$$

and it holds a variable Gaussian window function.

$$h(t,f) = \frac{|f|}{\sqrt{2\pi}}e^{\frac{-f^2 t^2}{2}} \tag{10}$$

For the component located in the low frequency region, the window function is longer and has better frequency resolution. The high-frequency area is the opposite, with better time resolution.

This feature matches the radar echo signal form (6) used for imaging in this paper: The better frequency resolution at low frequencies concentrates the energy of the rigid body component, while the better time resolution at high frequencies depicts the variation of the m-D component over time.

All of the above would make it possible for ST to get better results than STFT overall. In application, to keep the frequency resolution of the high frequency region, a lower bound should be set for the window length.

$$h(t,f) = \begin{cases} \dfrac{|f|}{\sqrt{2\pi}}e^{\frac{-f^2 t^2}{2}} & ,f \leq f_{max} \\ \dfrac{|f_{max}|}{\sqrt{2\pi}}e^{\frac{-f_{max}^2 t^2}{2}} & ,f > f_{max} \end{cases} \tag{11}$$

4. Synchrosqueezing Method

The results obtained from ST analysis are still somewhat insufficient: the energy would spread in the frequency direction of the high-frequency part, which affects the readability of the time-frequency map. Here, the synchrosqueezing method will be introduced for this problem. It is a post-processing method for time-frequency analysis results. The energy is compressed from the frequency direction, the final map is significantly improved, and the beneficial characteristics of the ST can be preserved.

If we consider a harmonic signal $x(t) = Ae^{i2\pi f_1 t}$, its Fourier Transform is shown as

$$X(f) = 2\pi A\delta(f - f_1), \tag{12}$$

where δ is the Dirac-delta function. Hence, the S-transform of $x(t)$ can be expressed as

$$ST_x(t,f) = Ae^{\frac{-2\pi^2(f_1-f)^2}{f^2}}e^{2\pi i(f_1-f)t}. \tag{13}$$

This illustrates that $e^{\frac{-2\pi^2(f_1-f)^2}{f^2}}e^{2\pi i(f_1-f)t}$ are distributed into the ambiguity area in (9). Ideally, the frequency of $x(t)$ is concentrated around f_1. In practice, the energy of $ST_x(t,f)$ spreads out in a range of frequency. To eliminate the effect of modulated items, in the ST spectrum, the signal can calculate the instantaneous frequency (IF) whenever $ST_x(t,f) \neq 0$ for any (t,f) by

$$\hat{\omega}_x(t,f) = f + Im(\frac{\partial_t ST_x(t,f)}{ST_x(t,f)}). \tag{14}$$

The form of the IF is suitable for S-transform [13]. Here, $\partial_t ST_x(t,f)$ means $ST_x(t,f)$ is partial to t. By using (9), we can obtain $\hat{\omega} = f_1$. The multicomponent signals can be defined as a superposition of AM-FM components

$$x(t) = \sum_{K=1}^{N} A_K(t)e^{2\pi i\phi_k(t)}, \tag{15}$$

where the A_K and ϕ_k are, respectively, time-varying amplitude and phase functions satisfying $A_K(t) > 0, \phi_k'(t) > 0$ and $\phi_{k+1}'(t) > \phi_k'(t)$ for any t. The goal is to recover the instantaneous frequencies ϕ_k' and the instantaneous amplitudes $A_K(t)$. If there is some distance between the different components, i.e.,

$$\phi_{k+1}'(t) - \phi_k'(t) > 2\Delta, \tag{16}$$

where $\Delta \in (0,1)$, it is called the well-separated multicomponent signal. $\hat{\omega}(t, f)$ can also perform effectively for IF estimation [14].

4.1. Synchrosqueezing S-Transform

Before introducing synchrosqueezing transform, it is necessary to revisit the reassignment technique (RM) [15]. The aim of RM is to sharpen the time frequency (TF) representation. There are two meaningful quantities that are called reassignment operators, $\hat{\omega}_x(t, f)$ and $\hat{t}_x(t, f)$. The former is the same as Equation (14) and the latter is defined as

$$\hat{t}_x(t, f) = t - Re\left(\frac{\partial_f ST_x(t, f)}{2i\pi ST_x(t, f)}\right). \tag{17}$$

Here, $\partial_f ST_x(t, f)$ means $ST_x(t, f)$ is partial to f. While RM as a useful post-processing technique that moves the coefficients according to the map $(t, f) \rightarrow (\hat{t}_x, \hat{\omega}_x)$ in the TF plane, no mode reconstruction method is available.

Synchrosqueezing transform was originally introduced for analyzing auditory signals in [16] and developed further in [8]. It can be viewed as a special reassignment method. The coefficients are reassigned according to the map $(t, f) \rightarrow (t, \hat{\omega}_x(t, f))$, making it remain invertible. The synchrosqueezing S-transform (SSST) is defined as follows:

$$\begin{aligned} SSST(t, \omega) \\ = \int_{\{\omega_l | ST_x(t, \omega_l)| > \gamma\}} ST_x(t, \omega_l)\delta(\omega - \hat{\omega}_x(t, \omega_l))d\omega_l, \end{aligned} \tag{18}$$

where γ is an adjustable threshold.

By squeezing the frequency components within a certain range of instantaneous frequency, an energy-concentrated TF representation can be obtained. Then, the reconstruction of the original signal is computed by

$$x(t) = \frac{1}{C_h} \int_{-\infty}^{+\infty} SSST(t, \omega)\frac{d\omega}{\omega}, \tag{19}$$

where $C_h = Re(C_\varphi C_\psi), C_\varphi = e^{-i2\pi ft}f^2$, and $C_\psi = \frac{1}{2}\int_0^{+\infty} \overline{\hat{\psi}(\xi)}\frac{1}{\xi}d\xi$. $\overline{\hat{\psi}(\xi)}$ is the complex conjugate of the Fourier transform of the mother wavelet $\psi(t) = \frac{1}{\sqrt{w\pi}}e^{-\frac{t^2}{2}}e^{i2\pi t}$.

More details of the proof can be found in [13].

4.2. Second-Order Synchrosqueezing S-Transform

Although SST gives a good TF representation and mode reconstruction for multicomponent signals, it is restricted to analysis signals made of weakly modulated modes [17]. For a mode $x(t) = A_K(t)e^{2\pi i\phi_k}$, only when $\phi_k''(t)$ is approximate to zero or very small compared to $\phi_k'(t)$, the instantaneous frequency estimation $\hat{\omega}(t, f)$ is close to $\phi_k'(t)$. When considering the strong frequency-modulated signal, for instance a quadratic chirp, the ϕ_k'' is no longer negligible.

In order to deal with highly modulated signals like m-D components, second-order synchrosqueezing S-transform (SST2) was introduced by a more accurate IF estimate, based on the second-order operator. The operator corresponds to the second-order derivatives of phase $Re\{\hat{q}_x(t,f)\} = \phi_k''(t)$. The second-order local complex modulation operator $\hat{q}_x(t,f)$ is defined as

$$
\begin{aligned}
\hat{q}_x(t,f) &= \frac{\partial_t \hat{\omega}_x(t,f)}{\partial_t \hat{t}_x(t,f)} \\
&= \frac{ST_x \partial_{tt}^2 ST_x - (\partial_t ST_x)^2}{2i\pi ST_x^2 - ST_x \partial_{tf}^2 ST_x + \partial_t ST_x \partial_f ST_x},
\end{aligned}
\tag{20}
$$

where $\partial_{tt}^2 ST_x$ represents ST_x to t for second-order partial derivatives, and $\partial_{tf}^2 ST_x$ means that ST_x is partial to t and then partial to f. Owing to the operator, a more precise IF estimate $\hat{\omega}_x^{(2)}(t,f)$ can be obtained as

$$
\hat{\omega}_x^{(2)}(t,f) = \begin{cases} \hat{\omega}_x + f + \hat{q}_x(2t - \tilde{\tau}), & \partial_t \hat{t}_x \neq 0 \\ \hat{\omega}_x + f, & \partial_t \hat{t}_x = 0. \end{cases}
\tag{21}
$$

Then, the second-order synchrosqueezing S-transform (SSST2) is modified by replacing $\hat{\omega}_x$ by $\hat{\omega}_x^{(2)}$ in (15).

$$
\begin{aligned}
&SSST2(t,\omega) \\
&= \int_{\{\omega_l, |ST_x(t,\omega_l)| > \gamma\}} ST_x(t,\omega_l)\delta(\omega - \hat{\omega}_x^{(2)}(t,\omega_l))d\omega_l
\end{aligned}
\tag{22}
$$

5. Numerical Experimental Results

For the sake of clarity, the abbreviations used below will be detailed here. They are S-transform (ST), short-time Fourier transform (STFT), synchrosqueezing transform (SST), synchrosqueezing S-transform (SSST), and second-order synchrosqueezing S-transform (SSST2).

In this section, the simulation experiment with adjustable parameters will be given first. Results are compared using different time-frequency analysis methods. Subsequently, the actual ISAR signal is also tested.

First, ISAR data is simulated on the helicopter miniaturization model in Figure 3. The simulation parameters are shown in Table 1. Its resolution will be more than enough for the target. Three different propeller speeds are set, which plays a decisive role in the performance of the m-D component.

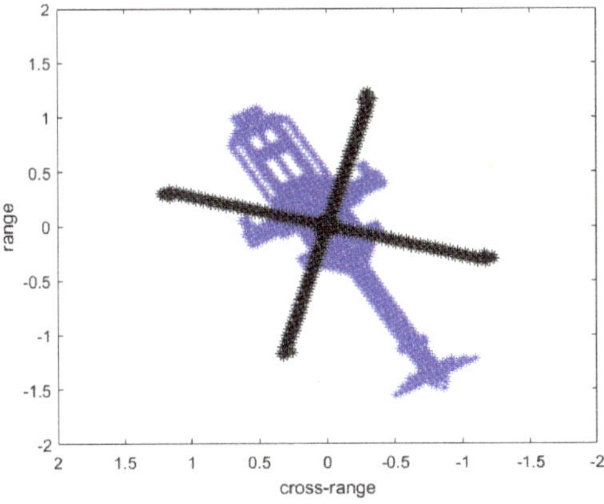

Figure 3. ISAR geometric model.

Table 1. Simulation parameters.

Parameter/Variable	Value	Parameter/Variable	Value
Center wavelength	10.6×10^{-6} m	Target size	2.5 m $\times$ 2.5 m
Transmitting bandwidth	20×10^9 Hz	Velocity	100 m/s
Pulse repetition frequency	7×10^3 Hz	Main rotor speed	20/10/5 r/s
Coherent accumulation time	0.1 s	Route angle relative to radar ray	30°
Transmitting width	1 μs	Imaging distance	10 Km
Range sampling points	620	Cross-range sampling points	700

In Figure 4a,c, the imaging results using STFT and ST are respectively shown. Although both of them roughly represent the outline of the target rigid body part due to the setting of the experimental parameters, it is obvious that the ST subsection is more accurate and reflects some details. The reason can be seen from the time-frequency analysis results of the single distance unit (300th) using two methods: for the rigid body component, ST can obtain a more gradual result for the low-frequency part, while it would jitter in the STFT (Figure 4b,d).

After applying the synchrosqueezing method, the imaging effects of both methods have been significantly improved. The SST imaging result (Figure 4e) is more concentrated, and the contour of the target becomes clearer. In the time-frequency map, the energy is concentrated in the low-frequency part, but the jitter of the rigid body component still exists (Figure 4f).

For relatively clear ST imaging results, the SSST changes are more reflected in the further depiction of the details (Figure 4g). Since the application of the ST changes the energy distribution of the signal in the frequency domain, the difference in energy between the rigid body imaging result and the m-D information is not as sharp as the SST result. But this also makes the m-D component not too weak (Figure 4h). In addition, in the results of SSST, the readability of the m-D component is also significantly improved, although this should mainly be attributed to ST. But the effect of the original ST in the high-frequency part is not ideal (Figure 4b).

When the rotational speed is increased to 10 r/s and 20 r/s, the same analysis results as those of Figure 4 are also correspondingly placed in Figures 5 and 6.

The difference between these results is the same as when the rotating speed is 5 r/s, but the m-D component brings more interference than before, which makes the imaging result blurred (Figures 5a,c,e,g and 6a,c,e,g).

The characteristics of the results of each time-frequency analysis are not significantly different from the previous ones; the jitter range of the rigid body components of STFT and SST seems to be larger (Figures 5b,f and 6b,f). In the results of ST and SSST, the energy is more distributed at high frequencies (Figure 5c,d,g,h). These changes become more pronounced after increasing the rotational speed to 20 r/s (Figure 6c,d,g,h).

After the second-order synchrosqueezing is further applied (Figure 7), it mainly affects the time resolution of the m-D separation in each time-frequency map (Figure 7b,d,f). This is easier to observe in the zoomed red rectangle region of Figures 6h and 7f,g,h. The imaging result has no obvious gap (Figure 7a,c,e) with the SSST result.

With simple energy accumulation criteria, the rigid body component can be roughly separated from the imaging results. The advantage of ST-based results (Figure 8b,d,f) relative to STFT (Figure 8a,c,e) is obvious. The former almost restored the details of the rigid part of the original model (Figure 3). It is worth mentioning that, although the contour of the rigid body part of the SSST2 imaging result is blurred at the rotational speeds of 10 r/s and 20/s (Figure 7c,e), the details of the extraction result can still be compared, with SST having a better performance (Figure 8d,f).

For the computing speed, SSST is not very different from the more complex SSST2, and the latter is even slightly faster (SSST: 1227s; SSST2: 934s). Moreover, due to the use of more matrix operations, SST is much faster than the other two methods (SST: 23s) that use more loop statements, as they hold similar levels of complexity. The above results show that there is still a lot of optimization space for the SSST and SSST2 programs used in this paper, and the calculation of operator (20) for SSST2 does not significantly change the computational complexity.

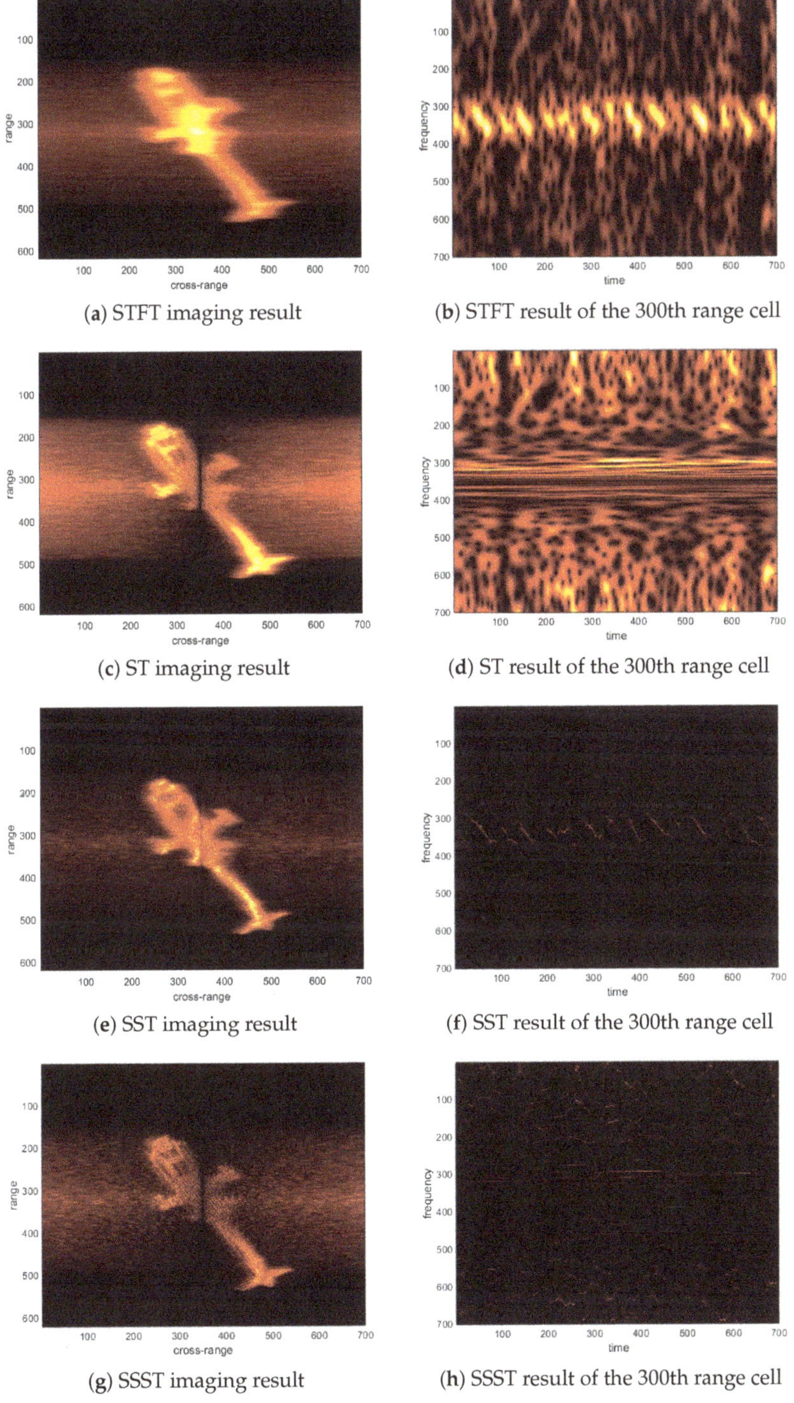

Figure 4. Results of target with a main rotor speed of 5 r/s. ST—S-transform; SST—synchrosqueezing transform; SSST—synchrosqueezing S-transform; STFT—short time Fourier transform.

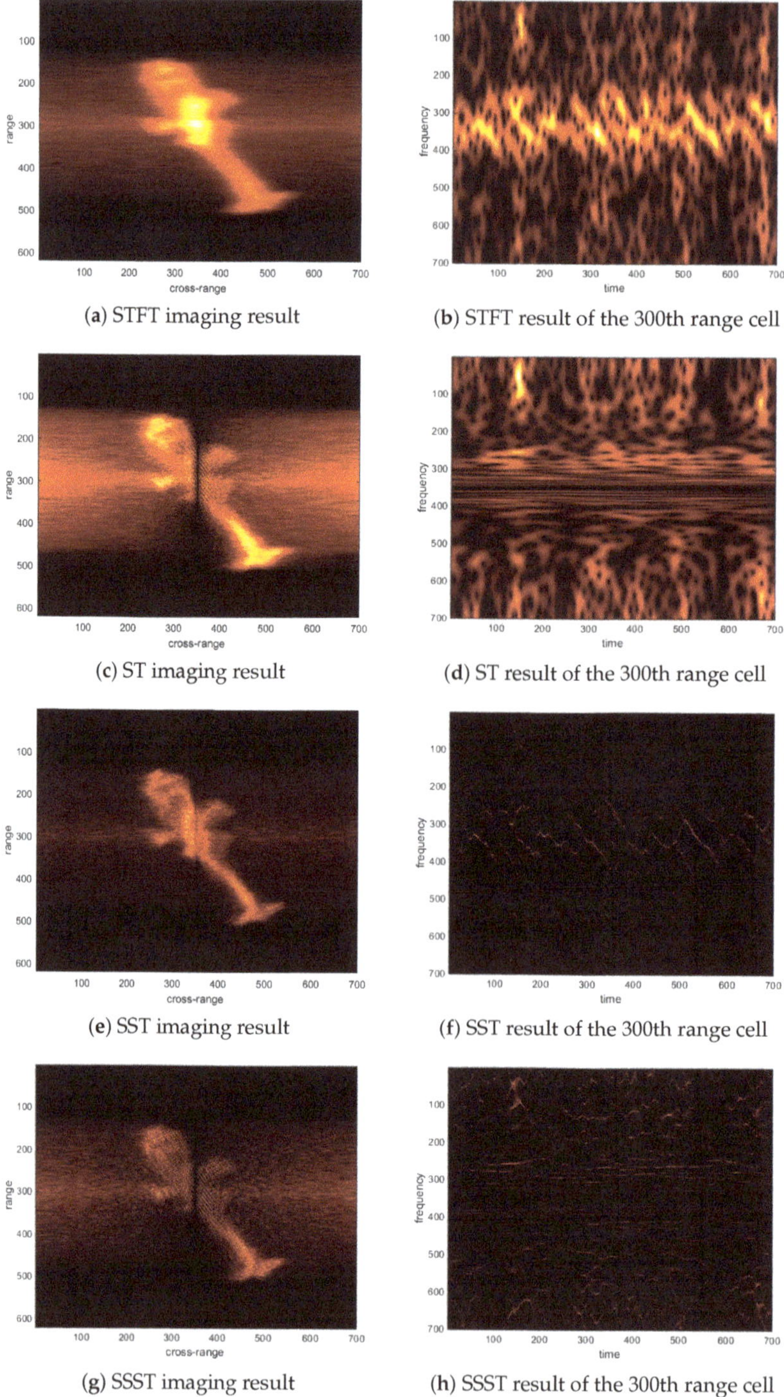

(**a**) STFT imaging result

(**b**) STFT result of the 300th range cell

(**c**) ST imaging result

(**d**) ST result of the 300th range cell

(**e**) SST imaging result

(**f**) SST result of the 300th range cell

(**g**) SSST imaging result

(**h**) SSST result of the 300th range cell

Figure 5. Results of target with a main rotor speed of 10 r/s.

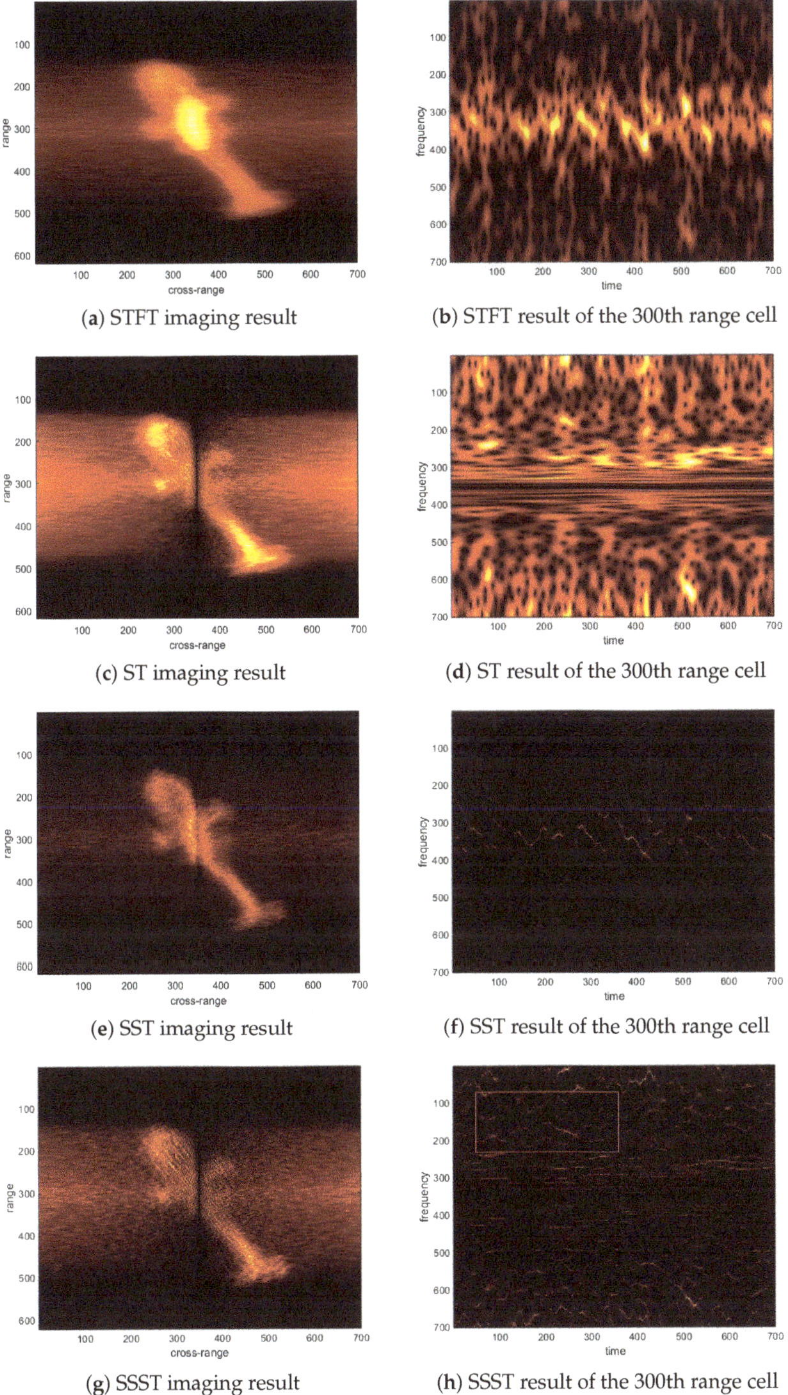

(**a**) STFT imaging result

(**b**) STFT result of the 300th range cell

(**c**) ST imaging result

(**d**) ST result of the 300th range cell

(**e**) SST imaging result

(**f**) SST result of the 300th range cell

(**g**) SSST imaging result

(**h**) SSST result of the 300th range cell

Figure 6. Results of target with a main rotor speed of 20 r/s.

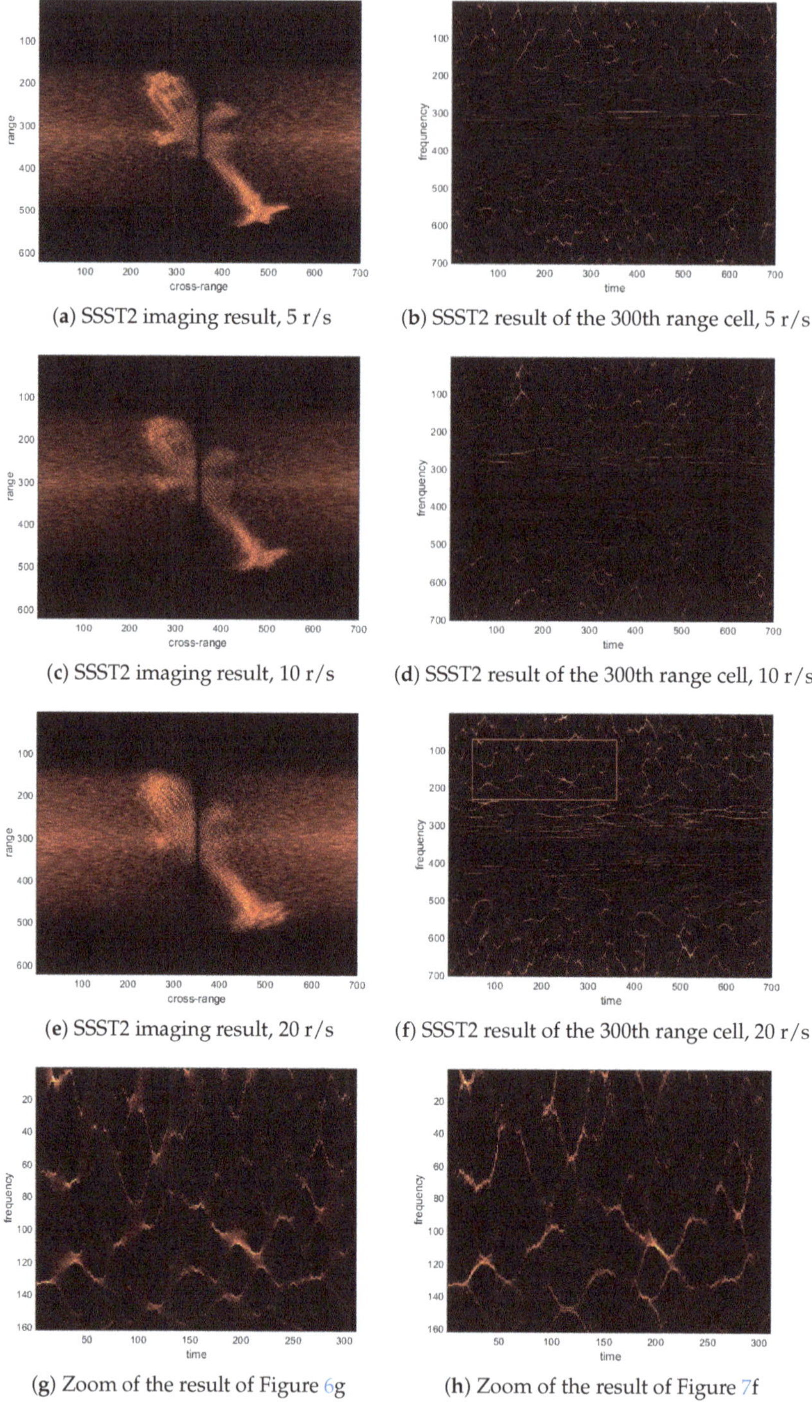

(a) SSST2 imaging result, 5 r/s

(b) SSST2 result of the 300th range cell, 5 r/s

(c) SSST2 imaging result, 10 r/s

(d) SSST2 result of the 300th range cell, 10 r/s

(e) SSST2 imaging result, 20 r/s

(f) SSST2 result of the 300th range cell, 20 r/s

(g) Zoom of the result of Figure 6g

(h) Zoom of the result of Figure 7f

Figure 7. Second-order synchrosqueezing S-transform (SSST2) results of target with different main rotor speed.

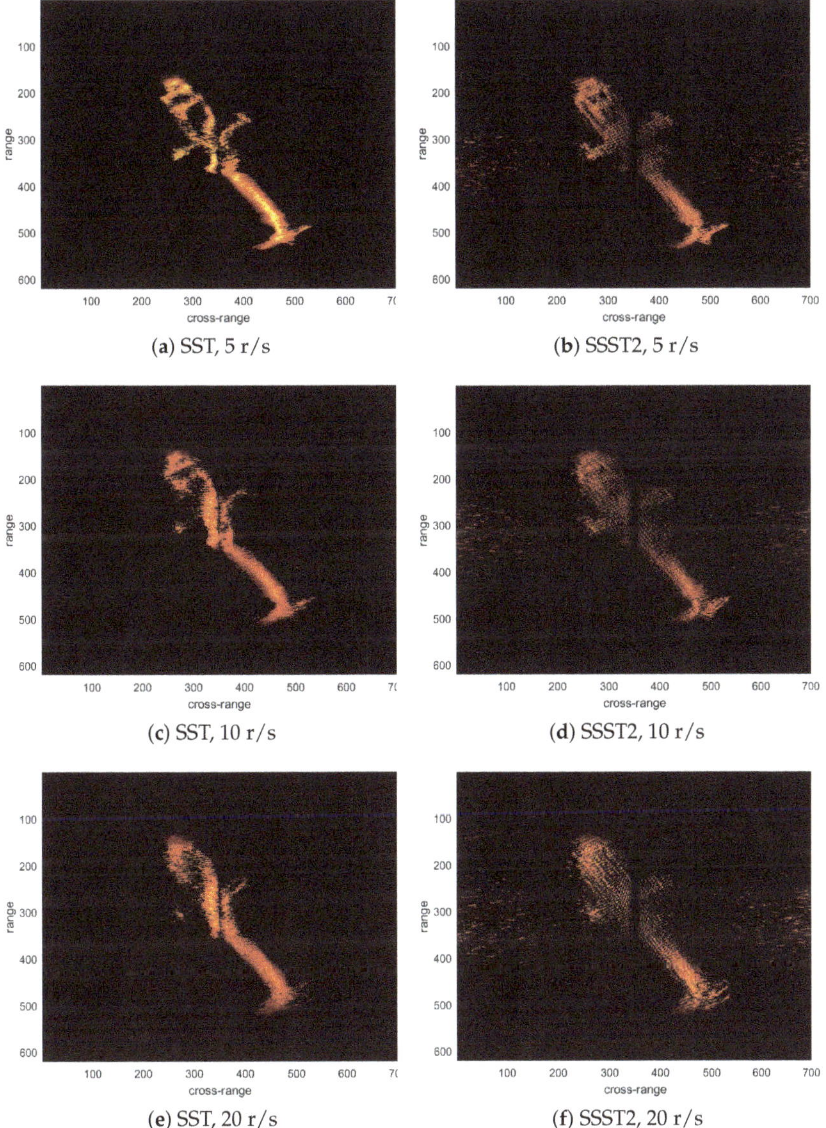

Figure 8. Imaging separation results of simulated signals.

Actual An-26 aircraft data is also subjected to similar experiments. Similar to Figures 4–6, Figure 9 shows the imaging effects of different methods (Figure 9a,c,e,g) and the time-frequency analysis results of a certain distance unit (Figure 9b,d,f,h). The difference between the results is still consistent with the previous simulation data. In addition, the imaging effects of SSST and SSST2 are still similar (Figures 9g and 10a), and their differences are reflected in the time resolution of high-frequency components (Figures 9h and 10b). For real data, STFT-based imaging is even worse; the more concentrated energy allows the latter to achieve a better separation result (Figure 11a,b).

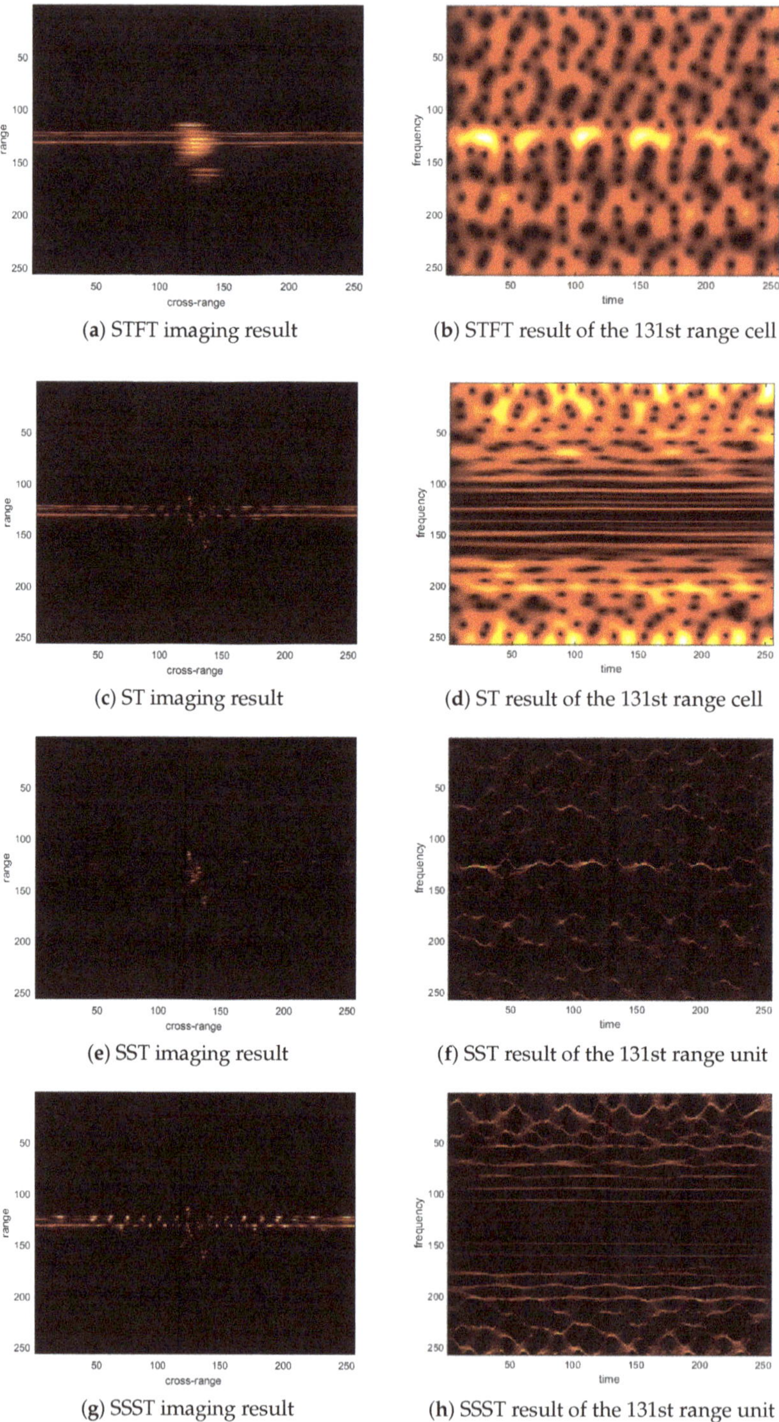

(**a**) STFT imaging result

(**b**) STFT result of the 131st range cell

(**c**) ST imaging result

(**d**) ST result of the 131st range cell

(**e**) SST imaging result

(**f**) SST result of the 131st range unit

(**g**) SSST imaging result

(**h**) SSST result of the 131st range unit

Figure 9. Time-frequency and imaging analysis results of real An-26 data.

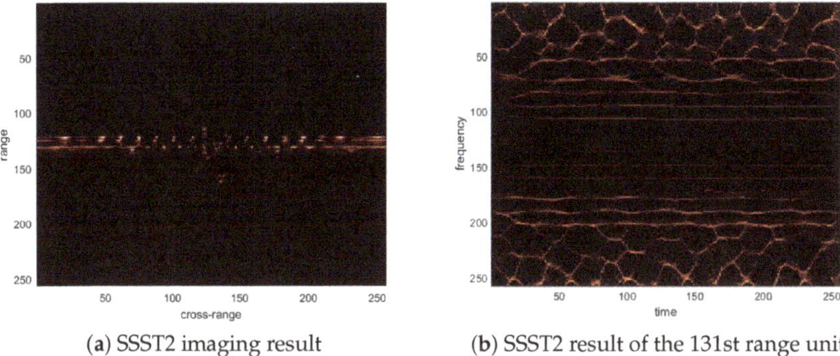

(**a**) SSST2 imaging result   (**b**) SSST2 result of the 131st range unit

Figure 10. Time-frequency and imaging analysis results of real An-26 data.

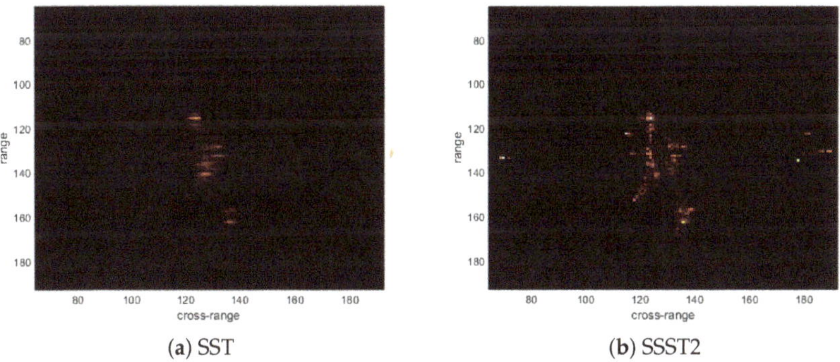

(**a**) SST   (**b**) SSST2

Figure 11. Imaging separation results of real An-26 data.

6. Conclusions

In this paper, S-transform is applied to ISAR imaging. Since the multiresolution characteristics of S-transform and the characteristics of radar echo signals match each other, better analysis and imaging results are achieved compared with STFT; this contrast is obvious in both simulation and actual data results. In the simulated helicopter model, although the micro-Doppler component brings more interference to the imaging results as the rotational speed of the rotor increases, after the application of the synchrosqueezing method, the quality of the time-frequency analysis results is significantly improved. Especially in the ST-based synchronous compression results, the change in the micro-Doppler component over time in the frequency direction is clearly depicted. All of the above will play an important role in subsequent work such as feature extraction, parameter estimation, etc.

Author Contributions: The contributions of the authors are as follows. Data curation, B.Z.; Formal analysis, M.Z.; Funding acquisition, X.Z.; Investigation, L.Z.

Funding: This research was funded by National Natural Science Foundation of China grant number 61701374.

Conflicts of Interest: The authors declare no conflict of interest.

References

1. Chen, V.C. Joint time-frequency analysis for radar signal and imaging. In Proceedings of the IEEE International Geoscience and Remote Sensing Symposium, Barcelona, Spain, 23–28 July 2007; pp. 5166–5169.
2. Chen, V.C.; Li, F.; Ho, S.-S.; Wechsler, H. Analysis of micro-Doppler signatures. *IEEE Proc. Radar Sonar Navig.* **2003**, *150*, 271. [CrossRef]

3. Chen, V.C.; Li, F.; Ho, S.-S.; Wechsler, H. Micro-Doppler Effect in Radar: Phenomenon, Model, and Simulation Study. *IEEE Trans. Aerosp. Electron. Syst.* **2006**, *42*, 2–21. [CrossRef]

4. Luo, Y.; Zhang, Q.; Qiu, C.; Liang, X.; Li, K. Micro-Doppler Effect Analysis and Feature Extraction in ISAR Imaging With Stepped-Frequency Chirp Signals. *IEEE Trans. Geosci. Remote Sens.* **2010**, *48*, 2087–2098.

5. Stankovi'c, L.J.; Stankovi'c, S.; Thayaparan, T.; Dakovi'c, M.; Orovi'c, I. Separation and Reconstruction of the Rigid Body and Micro-Doppler Signal in ISAR Part I—Theory. *IET Radar, Sonar Navig.* **2015**, *9*, 1147–1154. [CrossRef]

6. Stankovi'c, L.J.; Stankovi'c, S.; Thayaparan, T.; Dakovi'c, M.; Orovi'c, I. Separation and Reconstruction of the Rigid Body and Micro-Doppler Signal in ISAR Part II—Statistical Analaysis. *IET Radar Sonar Navig.* **2015**, *9*, 1155–1161. [CrossRef]

7. Stockwell, R.G.; Mansinha, L.; Lowe, R.P. Localization of the complex spectrum: The S transform. *IEEE Trans. Signal Process.* **1996**, *44*, 998–1001. [CrossRef]

8. Daubechies, I.; Lu, J.; Wu, H.T. Synchrosqueezed wavelet transforms: An empirical mode decomposition-like tool. *Appl. Comput. Harmon. Anal.* **2011**, *30*, 243–261. [CrossRef]

9. Brajovic, M.; Popovic-Bugarin, V.; Djurovic, I.; Djukanovic, S. Post-processing of Time-Frequency Representations in Instantaneous Frequency Estimation Based on Ant Colony Optimization. *Signal Process.* **2017**, *138*, 195–210. [CrossRef]

10. Oberlin, T.; Meignen, S.; Perrier, V. Second-Order Synchrosqueezing Transform or Invertible Reassignment? Towards Ideal Time-Frequency Representations. *IEEE Trans. Signal Process.* **2015**, *63*, 1335–1344. [CrossRef]

11. Pham, D.H.; Meignen, S. High-Order Synchrosqueezing Transform for Multicomponent Signals Analysis—With an Application to Gravitational-Wave Signal. *IEEE Trans. Signal Process.* **2017**, *65*, 3168–3178. [CrossRef]

12. Sejdic, E.; Djurovic, I.; Jiang, J. Time-frequency feature representation using energy concentration: An overview of recent advances. *Digit. Signal Process.* **2009**, *19*, 153–183. [CrossRef]

13. Huang, Z.; Zhang, J.; Zhao, T.; Sun, Y. Synchrosqueezing S-Transform and Its Application in Seismic Spectral Decomposition. *IEEE Trans. Geosci. Remote Sens.* **2016**, *54*, 817–825. [CrossRef]

14. Thakur, G.; Wu, H.T. Synchrosqueezing-based Recovery of Instantaneous Frequency from Nonuniform Samples. *Siam J. Math. Anal.* **2010**, *43*, 2078–2095. [CrossRef]

15. Auger, F.; Flandrin, P. Improving the readability of time-frequency and time-scale representations by the reassignment method. *IEEE Trans. Signal Process.* **1995**, *43*, 1068–1089. [CrossRef]

16. Daubechies, I.; Maes, S. A nonlinear squeezing of the continuous wavelet transform based on auditory nerve models. In *Wavelets in Medicine and Biology*; CRC Press: Boca Raton, FL, USA, 1996; pp. 527–546

17. Auger, F.; Flandrin, P.; Lin, Y.T.; McLaughlin, S.; Meignen, S.; Oberlin, T.; Wu, H.-T. Time-Frequency Reassignment and Synchrosqueezing: An Overview. *IEEE Signal Process. Mag.* **2013**, *30*, 32–41. [CrossRef]

Article

Wideband Noise Interference Suppression for Sparsity-Based SAR Imaging Based on Dechirping and Double Subspace Extraction

Guojing Li [1,*], Qinglin Lu [1], Guochao Lao [2] and Wei Ye [3]

[1]   Graduate School, Space Engineering University, Beijing 101416, China; ql_lu54570@163.com
[2]   The 96901 Unit of PLA, Beijing 100094, China; laoguochao@mail.sdu.edu.cn
[3]   Space Engineering University, Beijing 101416, China; yeyuhan@sina.com
*   Correspondence: leeguojing1014@mail.dlut.edu.cn

Received: 23 August 2019; Accepted: 9 September 2019; Published: 11 September 2019

Abstract: Sparsity-based synthetic aperture radar (SAR) imaging has attracted much attention since it has potential advantages in improving the image quality and reducing the sampling rate. However, it is vulnerable to deliberate blanket disturbance, especially wideband noise interference (WBNI), which severely damages the imaging quality. This paper mainly focuses on WBNI suppression for SAR imaging from a new perspective—sparse recovery. We first analyze the impact of WBNI on signal reconstruction by deducing the interference energy projected on the real support set of the signal under different observation parameters. Based on the derived results, we propose a novel WBNI suppression algorithm based on dechirping and double subspace extraction (DDSE), where the signal of interest (SOI) is reconstructed by exploiting the known geometric prior and waveform prior, respectively. The experimental results illustrate that the DDSE-based WBNI suppression algorithm for sparsity-based SAR imaging is effective and outperforms the other algorithms.

Keywords: synthetic aperture radar; sparse recovery; wideband noise interference; dechirping; subspace extraction; denoising detection; orthogonal matching pursuit

---

1. Introduction

Synthetic aperture radar (SAR) is an active remote sensing modality for real-time information acquisition. It plays a significant role in the fields of civil exploration and military reconnaissance, owing to its capabilities of all-weather, all-time, and high-resolution imaging. Traditional SAR imaging technology is based on matched filtering in the Nyquist sampling framework, which is by far the most common but performs with some limitations. On the one hand, the increasing system bandwidth proportional to the radar resolution poses a great challenge in signal acquisition and data storage. On the other hand, the side lobe effect, which is caused by window functions in the process of pulse compression, affects the visual quality of SAR images. Sparsity-based SAR imaging [1,2], as a new radar imaging mechanism, has potential advantages in improving the image quality by introducing sparse signal processing into the SAR system. Since the concept of compressed sensing [3] has been proposed, radar imaging with incomplete data has become realizable by exploiting prior information. This kind of imaging method is also known as compressed sensing radar imaging [4,5].

Most SAR systems operate in the microwave band, and they are inevitably subject to various types of electromagnetic interference, including natural radiation and man-made interference, the latter of which is used for deliberately protecting important targets or scenes by damaging the image quality. Man-made SAR interference is divided into various categories based on different criteria. For example, the narrowband interference (NBI) and the wideband interference (WBI) are discriminated in terms of the range of frequency band occupied, while incoherent and coherent interference can be measured by structural similarity with the signal of interest (SOI). Generally speaking, coherent interference, usually generated by digital radio frequency memory (DRFM) [6], theoretically performs with higher efficiency, since the processing gain can be obtained after pulse compression. However, this coherence is difficult to guarantee strictly in practical applications due to the estimation error of motion and other signal parameters. Moreover, under the conditions of a large scene, echoes of scattering points within the observation area overlap in the time domain, increasing the coherent interference power required [7]. By contrast, incoherent interference is easier to implement by a universal jammer with a simple structure, which directly sends disturbance waveforms to the SAR system instead of intercepting, modulating, and repeating.

Wideband noise interference (WBNI) is one of the most typical incoherent interference types, which blankets a specific area in an SAR image by enhancing the background noise level [8]. Intuitively, this kind of interference is inefficient, since it has a wide spectrum characteristic compared to narrowband types, provided that the total power is constant. However, once the power is no longer limited, which is actually possible because of current high-power microwave technology, it becomes extremely difficult to deal with. For the sparsity-based SAR imaging system, this type of interference is particularly destructive to signal reconstruction. There are three main reasons for this statement. First, WBNI is characterized as the receiving noise [9] and is widely considered as the optimal choice used for raising the false-alarm threshold and disrupting the potential attributes of SOI, such as the sparsity, according to the information theory. Second, given that the scheme based on matched filtering and a high analog-to-digital converter (ADC) rate is replaced by that based on nonlinear optimization (and low ADC rate if compressed observation is considered [10]) in the sparsity-based system, the original coherent accumulation for improving the signal-to-noise ratio (SNR) in the process of pulse compression may no longer exist, or it may be transformed into other agnostic forms. Besides, when the observation dimension is less than the Nyquist requirement, the noise folding effect [11,12] occurs, making the signal recovery more sensitive to the change of SNR.

Compared to narrowband interference, there is less literature on wideband interference suppression for SAR, especially for sparsity-based imaging. Judging from the existing research results, they can be grouped into parametric, non-parametric, and semi-parametric methods. Parametric methods such as high-order ambiguity function (HAF) [13,14], fractional Fourier transform (FrFT) [15], empirical mode decomposition (EMD) [16], and time-frequency analysis (TFA) [17] are based on polynomial signal modeling, the performances of which are heavily dependent on the order of the model. Non-parametric methods such as the time–frequency filtering (TFF) [18,19] and the iteration-adaptive approach (IAA) [20] separate the signal and the interference by utilizing their respective concentrations in the time–frequency domain, where the TFF method simplifies the WBI to a series of instantaneous NBIs by short time Fourier transform (STFT). Sparse recovery, as a kind of semi-parametric method, is state-of-the-art, especially in terms of reducing signal distortion. It can be considered as an optimization problem of reconstructing a few coefficients with a given dictionary. This kind of method is mainly used for suppressing structured WBI that can be sparsely represented on a specified domain. For example, WBI based on sinusoidal-modulated or chirp-modulated models is sparse on an inverse STFT basis, and it can be separated from SOI by time-varying filtering [21] or alternative optimization algorithms [22,23].

Unfortunately, the methods mentioned above are not competent to suppress the blanket WBI modulated by noise, i.e., WBNI, since it is difficult to find a suitable domain to effectively separate the WBNI from the SOI. For the suppression of noise-like interference from the sparse perspective, there are two alternative approaches. From the aspect of sensing recovery, basis pursuit denoising (BPDN) [24] takes the disturbance component into account in the reconstruction model and weakens the noise by decomposing the observed data into signal and residual components. It is a common method, provided that the signal has been contaminated but the SNR is not quite low. The other way is to filter the interference directly in the process of compressed observation, in which a feedback loop composed of interference detection and adaptive selective sampling is introduced into the entire echo acquisition and processing procedures. The advantage of this adaptive compressed sampling (ACS) method [25] is that the interference entrance is cut off from the source. Nevertheless, the signal distortion and the high system complexity are the main limitations, since the prior is not fully utilized.

In this paper, we focus on the suppression of incoherent wideband noise interference for SAR imaging from the perspective of sparse signal processing. Given that WBNI is hardly sparse on any known domain, existing methods based on interference reconstruction and elimination are no longer applicable. Fortunately, there is an incoherent relation between the WBNI and the SOI, making it possible to extract useful components by exploiting the prior information of observation geometry and transmitted signal, respectively. The geometric prior can be used to obtain a more compact subspace from the Fourier basis that minimizes the projected energy of interference. The waveform prior can be used to perform denoising detection and then extract possible atoms corresponding to SOI, making the reconstruction more accurate and efficient. Based on the above considerations, we propose a novel WBNI suppression approach based on dechirping and double subspace extraction (DDSE) algorithms that can be applied to sparsity-based SAR imaging.

The main contents of this paper are divided into four parts. In Section 2, we provide a brief review of sparsity-based SAR imaging. In Section 3, we analyze the impact of WBNI on sparse recovery by theoretical derivation. In Section 4, we propose the DDSE algorithm for WBNI suppression and present the detailed procedure. In Section 5, we carry out numerical experiments to investigate the performance of the proposed algorithm.

2. A Brief Review of Sparsity-Based SAR Imaging

The raw echo of SAR is usually considered as the convolution of the scattering points and the transmitted signal. The linear frequency-modulated (LFM) pulse is the most commonly used signal type, since it has a larger time-bandwidth product to ensure resolution. The ideal receiving signal in the analogy domain can be expressed as [26]:

$$s_r(t, \tau) = \sum_{l=1}^{L} \sigma_l w_r\left(t - \frac{2R_l(\tau)}{c}\right) w_a(\tau - \tau_c) \exp\left[j2\pi f_c\left(t - \frac{2R_l(\tau)}{c}\right) + j\pi K_r\left(t - \frac{2R_l(\tau)}{c}\right)^2\right] \tag{1}$$

where t is the fast time in the range direction; τ is the slow time in the azimuth direction; τ_c is the zero-Doppler time; L is the number of scattering points in observed scene; σ_l is the backscatter coefficient of the l-th point; f_c is the carrier frequency; c is the speed of light; R_l is the oblique distance between scattering point and SAR platform; K_r is the frequency modulation slope; and $w_r(\cdot)$ and $w_a(\cdot)$ denote the rectangular window function in range and azimuth, respectively.

In sparsity-based SAR imaging models, the observed scene is assumed to be uniformly divided into grids and composed of discrete scattering points, as shown in Figure 1. If the number of points with large scattering coefficients is much smaller than that of grids, the scene can be considered sparse in the space domain. Thus, the ergodic scattering matrix can be expressed as:

$$\mathbf{A} = \begin{bmatrix} \sigma(1,1) & \cdots & \sigma(X,1) \\ \vdots & \ddots & \vdots \\ \sigma(1,Y) & \cdots & \sigma(X,Y) \end{bmatrix} \tag{2}$$

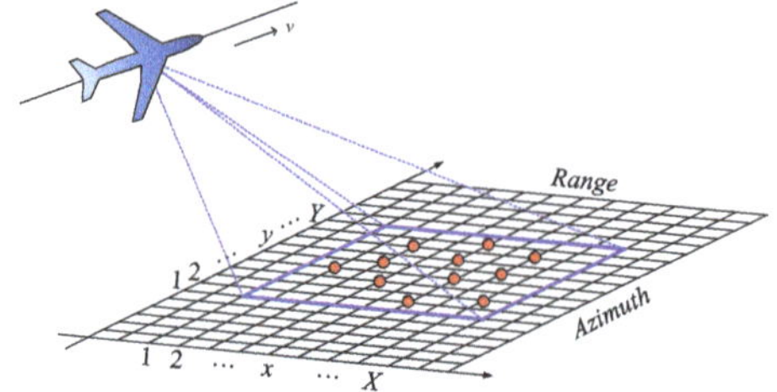

Figure 1. Diagram of a gridded synthetic aperture radar (SAR) observation model. The blue box represents the observed area. The red points represent discrete scattering points with large coefficients. X and Y are the number of grids in range and azimuth directions, respectively.

In order to intuitively analyze the mechanism of sparse imaging, the observed data matrix is usually vectorized into a one-dimensional vector with a length of XY in the following model:

$$s = \mathbf{\Psi}\alpha + \varepsilon \tag{3}$$

where $\alpha = \text{vec}(\mathbf{A}^T)$ is the coefficient column vector obtained by cascading rows of $\mathbf{A}$; ε is the additive noise; and $\mathbf{\Psi}$ is the $PQ \times XY$ mapping matrix expressed as:

$$\mathbf{\Psi} = \begin{bmatrix} \Psi(1,1,1) & \cdots & \Psi(1,1,XY) \\ \vdots & \ddots & \vdots \\ \Psi(P,Q,1) & \cdots & \Psi(P,Q,XY) \end{bmatrix} \tag{4}$$

where P and Q are the number of samples in range and azimuth, respectively, and each row of $\mathbf{\Psi}$ can be considered as the discrete form of Equation (1). To simplify the description, we now consider a simple but representative case where the azimuth dimension is assumed to be one, i.e., $Q = 1$. Then, the corresponding coefficient vector is reduced to $\alpha = [\sigma(1), \ldots, \sigma(X)]^T$. Each column in $\mathbf{\Psi}$ is the transmitted signal with a specific delay determined by the oblique range of the scattering point from the SAR platform, reflecting the weight information of the backscattering coefficient of a specific target to each echo sample in the scene. Sometimes, we prefer to call the mapping matrix an echo dictionary or a basis, in which each column is called an atom.

The process of sparsity-based radar imaging is essentially a kind of parameter estimation, based on which regularization introduces prior information to improve the estimation performance. Therefore, imaging is realized by solving a constrained optimization problem, i.e.:

$$\hat{\alpha} = \min_{\alpha}\|s - \mathbf{\Psi}\alpha\|_2 + \lambda\|\alpha\|_p \tag{5}$$

where λ is the regularization parameter; and $\|\cdot\|_p$ denotes the Euclid norm ($0 \leq p \leq 1$). When $p = 0$, Equation (5) is specialized to the compressed sensing radar imaging problem, i.e.:

$$\hat{\alpha} = \min_{\alpha}\|\alpha\|_0 \quad s.t. \quad \|y - \mathbf{\Phi}\mathbf{\Psi}\alpha\|_2 < \delta \tag{6}$$

where $\delta > 0$; $\mathbf{\Phi}$ is an underdetermined observation matrix; and y is the compressed measurement vector. In compressed sensing, the problem of solving the ill-conditioned l_0-norm is usually relaxed to convex optimization or greedy pursuit as long as the sensing matrix $\mathbf{\Theta} = \mathbf{\Phi}\mathbf{\Psi}$ satisfies the restricted isometry property (RIP) [3,27].

3. Impact of WBNI on Sparse Recovery

3.1. Sparse Models for Interference

The bandwidth of WBNI is generally not less than that of SAR for the purpose of blanketing the entire spectrum of signal. The unified mathematical model in the analog time domain can be expressed as:

$$n(t) = [U_0 + K_{AM}U_n(t)] \exp\left[j2\pi f_c t + j2\pi K_{FM} \int_0^t U_n(\eta)d\eta + \varphi(t) \right] \tag{7}$$

where U_0 is a constant; $U_n(t)$ is the time-varying and band-limited noise; K_{AM} and K_{FM} are the amplitude-modulated and the frequency-modulated coefficients, respectively; and $\varphi(t)$ is the random phase uniformly distributed in $[0, 2\pi]$. Notably, when $K_{FM} = 0$ and $U_0 = 0$, $n(t)$ represents the radio frequency interference; when $K_{FM} = 0$, $n(t)$ represents the amplitude-modulated noise interference; and when $K_{AM} = 0$, $n(t)$ represents the frequency-modulated noise interference.

From sparse point of view, WBNI cannot be sparsely represented in any known signal dictionary. We recall Equation (3) and express the echo signal x in the presence of interference as:

$$x = \mathbf{\Psi}\alpha + \mathbf{F}v + \varepsilon \tag{8}$$

where $\mathbf{F}$ and v are the interference dictionary and the coefficient vector, respectively. The Fourier basis, as shown in Figure 2a, is commonly used for sparsely representing the narrowband interference. It can be expressed as the following $N \times N$ normalized orthogonal basis:

$$\Omega = \frac{1}{\sqrt{N}}\begin{bmatrix} 1 & 1 & \cdots & 1 \\ 1 & W_N^{1\cdot 1} & \cdots & W_N^{1\cdot(N-1)} \\ \vdots & \vdots & \ddots & \vdots \\ 1 & W_N^{(N-1)\cdot 1} & \cdots & W_N^{(N-1)\cdot(N-1)} \end{bmatrix} \tag{9}$$

where $W_N = \exp(-j2\pi/N)$. For radio frequency interference (RFI) or narrowband noise interference, their coefficients are sparsely or block-sparsely distributed in terms of both location and energy, as shown in Figure 2b,c. For WBNI, however, its coefficients present a compact distribution throughout the entire dictionary, as shown in Figure 2d.

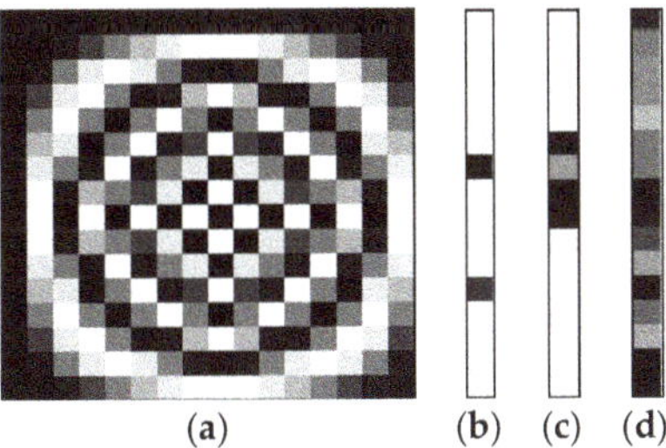

Figure 2. Diagram of Fourier basis and coefficient distribution of different interference types. (**a**) Fourier basis (real part); (**b**) radio frequency interference; (**c**) narrowband noise interference; and (**d**) wideband noise interference.

We assume that the signal of interest (SOI) is on a K-dimensional subspace $\mathbf{\Psi}_{\mathcal{A}}$ composed of columns corresponding to nonzero coefficients in $\mathbf{\Psi}$, and the WBNI is approximately modeled as a zero-mean Gaussian random vector, i.e., $v \sim \mathcal{N}(0, \sigma_v^2 \mathbf{I}_N)$. These nonzero coefficients containing location and value information are also called a support set, denoted by $\alpha_{\mathcal{A}}$ and indexed by $\mathcal{A}$.

Therefore, the process of SAR image reconstruction can be transformed into a support estimation problem. If $\mathbf{\Psi}$ satisfies the K-RIP property with a constant $0 < \delta_K < 1$ for any K-sparse vector α, i.e.:

$$(1 - \delta_K)\|\alpha\|_2^2 \leq \|\mathbf{\Psi}\alpha\|_2^2 \leq (1 + \delta_K)\|\alpha\|_2^2 \tag{10}$$

and $\mathbf{\Psi}_{\mathcal{A}}$ is full-rank, then the support set can be estimated by:

$$\hat{\alpha}_{\mathcal{A}} = \alpha_{\mathcal{A}} + \mathbf{\Psi}_{\mathcal{A}}^{\dagger}\Omega v \tag{11}$$

where $(\cdot)^{\dagger}$ denotes the pseudo-inverse operation with the property of $\mathbf{\Psi}_{\mathcal{A}}^{\dagger}\mathbf{\Psi}_{\mathcal{A}} = \mathbf{I}_K$; and $\mathbf{\Psi}_{\mathcal{A}}^{\dagger}\Omega v$ is the interference component projected on the real support.

3.2. Impact of WBNI on Signal Recovery

From Equation (11), the estimated support vector contains a real component and the projected WBNI component, the latter of which depends on the structure of the observation matrix and the support index. We next discuss the following cases for the observation matrix by introducing the coherent measure, defined as [28]:

$$\mu_{\mathbf{\Psi}} = \max_{i,j \neq i} \frac{|\psi_i^H \psi_j|}{\|\psi_i\|_2\|\psi_j\|_2} \tag{12}$$

where ψ_i is the i-th column of $\mathbf{\Psi}$; and $\|\cdot\|_2$ denotes the l_2-norm.

3.2.1. Case 1: $\mu_{\mathbf{\Psi}} = 0$

When x is sparse on an orthogonal basis, such as the Fourier basis shown in Figure 2a, the estimated support set in Equation (11) can be rewritten as:

$$\hat{\alpha}_{\mathcal{A}} = \alpha_{\mathcal{A}} + \hat{v}_{\mathcal{A}} \tag{13}$$

where $\hat{v}_{\mathcal{A}} = \Omega_{\mathcal{A}}^{\dagger}\Omega v$. Then, we have the following proposition.

Proposition 1. *If the observation matrix $\mathbf{\Psi}$ is an orthogonal Fourier basis Ω, i.e., $\mu_{\mathbf{\Psi}} = 0$, the interference energy projected on the signal support holds after estimation.*

Proof. Based on the property of Gaussian distribution and Fourier basis, we have:

$$\mathbb{E}(\Omega v) = 0, \quad \mathbb{E}\left[(\Omega v)(\Omega v)^H\right] \approx \sigma_v^2 \mathbf{I}_N \tag{14}$$

and the interference energy on the support set of K-sparse signal $v_{\mathcal{A}}$ is:

$$\mathbb{E}\left(\|v_{\mathcal{A}}\|_2^2\right) = \text{tr}\left[\mathbb{E}\left(v_{\mathcal{A}}v_{\mathcal{A}}^H\right)\right] = K\sigma_v^2 \tag{15}$$

where $\text{tr}(\cdot)$ denotes the trace. Drawing support from the equivalent relation [29], we have:

$$\text{tr}\left(\mathbf{C}\mathbf{C}^H\right) = \|\mathbf{C}\|_F^2 \tag{16}$$

where $\|\cdot\|_F$ denotes the Frobenius norm. The projected energy of interference component in Equation (13) after estimation can be calculated by:

$$\begin{aligned}
\mathbb{E}\left(\|\hat{v}_{\mathcal{A}}\|_2^2\right) &= \mathbb{E}\left(\|\Omega_{\mathcal{A}}^{\dagger}\Omega v\|_2^2\right) = \text{tr}\left\{\Omega_{\mathcal{A}}^{\dagger}\mathbb{E}\left[\Omega v(\Omega v)^H\right]\left(\Omega_{\mathcal{A}}^{\dagger}\right)^H\right\} \\
&= \sigma_v^2 \text{tr}\left\{\Omega_{\mathcal{A}}^{\dagger}\left(\Omega_{\mathcal{A}}^{\dagger}\right)^H\right\} = \sigma_v^2\|\Omega_{\mathcal{A}}^{\dagger}\|_F^2 = K\sigma_v^2
\end{aligned} \tag{17}$$

□

Therefore, the projected energy of interference remains the same before and after estimation. On this basis, we can also obtain the following corollary.

Corollary 1. *When the observation matrix* $\mathbf{\Psi}$ *is orthogonal, i.e., the coherence measure is zero, the interference energy projected on the signal support will be minimum.*

The proof of this corollary is given in the following Case 2.

3.2.2. Case 2: $\mu_{\mathbf{\Psi}} \neq 0$

Considering that the WBNI presents a Gaussian distribution on any basis, when $\mathbf{\Psi}$ is not an orthogonal basis, i.e., $\mu_{\mathbf{\Psi}} \neq 0$, Equation (11) can be rewritten as:

$$\hat{\alpha}_{\mathcal{A}} = \alpha_{\mathcal{A}} + \mathbf{\Psi}_{\mathcal{A}}^{\dagger} \mathbf{\Psi} u \tag{18}$$

where $u \sim \mathcal{N}(0, \sigma_u^2 \mathbf{I}_N)$ is a zero-mean Gaussian random vector. Then, the interference energy on the support set of K-sparse signal $u_{\mathcal{A}}$ is:

$$\mathbb{E}\left(\|u_{\mathcal{A}}\|_2^2\right) = \mathrm{tr}\left[\mathbb{E}\left(u_{\mathcal{A}} u_{\mathcal{A}}^H\right)\right] = K\sigma_u^2 \tag{19}$$

The projected energy of interference in Equation (18) after estimation can be calculated by:

$$\begin{aligned}
\mathbb{E}\left(\|\hat{u}_{\mathcal{A}}\|_2^2\right) &= \mathbb{E}\left(\|\mathbf{\Psi}_{\mathcal{A}}^{\dagger} \mathbf{\Psi} u\|_2^2\right) = \mathrm{tr}\left\{\mathbf{\Psi}_{\mathcal{A}}^{\dagger} \mathbb{E}\left[\mathbf{\Psi} u (\mathbf{\Psi} u)^H\right]\left(\mathbf{\Psi}_{\mathcal{A}}^{\dagger}\right)^H\right\} \\
&= \sigma_u^2 \mathrm{tr}\left\{\mathbf{\Psi}_{\mathcal{A}}^{\dagger} \mathbf{\Psi}\left(\mathbf{\Psi}_{\mathcal{A}}^{\dagger} \mathbf{\Psi}\right)^H\right\} = \sigma_u^2 \|\mathbf{\Psi}_{\mathcal{A}}^{\dagger} \mathbf{\Psi}\|_F^2
\end{aligned} \tag{20}$$

We define $\mathbf{\Psi}_{\mathcal{A}}^{C}$ as the complement to $\mathbf{\Psi}_{\mathcal{A}}$. Equation (20) can be further calculated by dividing the observation matrix into two parts, i.e.:

$$\mathbb{E}\left(\|\hat{u}_{\mathcal{A}}\|_2^2\right) = \sigma_u^2 \|\mathbf{\Psi}_{\mathcal{A}}^{\dagger} \mathbf{\Psi}\|_F^2 = \sigma_u^2 \|\mathbf{\Psi}_{\mathcal{A}}^{\dagger} \langle \mathbf{\Psi}_{\mathcal{A}} | \mathbf{\Psi}_{\mathcal{A}}^{C} \rangle \|_F^2 = \sigma_u^2\left(K + \|\mathbf{\Psi}_{\mathcal{A}}^{\dagger} \mathbf{\Psi}_{\mathcal{A}}^{C}\|_F^2\right) \tag{21}$$

where $\langle \cdot | \cdot \rangle$ denotes the operation that divides a matrix into a submatrix and its complementary.

Since the coherence measure $\mu_{\mathbf{\Psi}} \neq 0$, the elements on non-diagonal lines of $\mathbf{\Psi}_{\mathcal{A}}^{\dagger} \mathbf{\Psi}_{\mathcal{A}}^{C}$ are not all zero. Then, we have the following inequality:

$$\mathbb{E}\left(\|\hat{u}_{\mathcal{A}}\|_2^2\right) > \sigma_u^2 \|\mathbf{I}_K\|_F^2 = K\sigma_u^2 \tag{22}$$

Therefore, when $\mathbf{\Psi}$ is orthogonal, the interference energy projected on the signal support is minimal.

To further investigate the projected energy gain (PEG) of interference after signal recovery under different parameters, including the sparsity level, the measurement dimension, and the sampling mode, we performed numerical simulations using the LFM reference signal with delays and the Fourier matrix as their respective basis. The signal length N was set to 512, and the Monte-Carlo time was set to 1000.

Figure 3a shows the simulation results of PEG under different values of sparsity level K, where the measurement dimension M is equal to N. It can be seen that the PEG presented nonlinear growth when the coherence measure was nonzero, while it remained zero when an orthogonal basis was adopted. Figure 3b illustrates variations of PEG with the measurement dimension M and the sampling mode, where the sparsity level K was set to 100. In this simulation, we set up two sampling modes; one was to directly observe the echo data using a randomly generated Gaussian matrix, and the other was to extract the echo data with a random sampling matrix. The latter has the smallest PEG with the change of the measurement dimension when the coherence measure is zero.

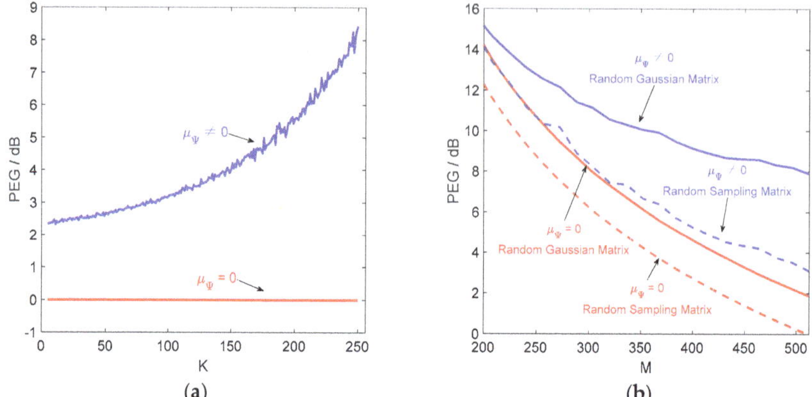

Figure 3. The projected energy gain (PEG) of wideband noise interference (WBNI) on a signal support set after sparse recovery. (**a**) Under different sparsity levels (M = N); (**b**) under different observation parameters (K = 100).

According to the above analysis, we can infer that, when the SAR echo signal is contaminated with WBNI, the coherence measure of the basis matrix for sparse recovery should be minimized as much as possible, and the random sampling matrix is a better choice for the compressed observation (if reducing the data rate is necessary). In the next section, we develop an effective WBNI suppression algorithm based on these results.

4. WBNI Suppression Based on Dechirping and Double Subspace Extraction

4.1. Dechirping Observation

From the derived results in Section 3.2, the projected energy of WBNI on signal support depends on the sparsity level and the observation parameters. It reaches a minimum when the basis satisfies the orthogonality. As is known, however, most SAR systems usually observe targets by transmitting wideband linear frequency-modulated (LFM) waveforms that are non-sparse on an orthogonal basis. Therefore, we first need to find a linear transformation that meets both sparsity and orthogonality requirements.

Dechirping, also called stretch, is a specific approach for processing LFM signals [30]. It utilizes a time-fixed reference waveform with the same frequency-modulated slope as the transmitted signal and performs the mixing with the raw echo. For sparsity-based SAR, the basis is generally composed of reference sequences with specific delays determined by the distance from scattering points to the radar platform. Assuming that the range in an azimuth is divided into N grids, then the observation model in Equation (4) can be specified to the following expression after demodulation, i.e.:

$$\Psi = \begin{bmatrix} s_{r,1}^T(n) & s_{r,2}^T(n) & \cdots & s_{r,N}^T(n) \end{bmatrix}, \quad n = 1, \ldots, N \tag{23}$$

where $s_{r,l}(n) = w_r\left(nT_s - \frac{2R_l}{c}\right)\exp\left[-j\frac{4\pi R_l}{\lambda} + j\pi K_r\left(nT_s - \frac{2R_l}{c}\right)^2\right]$, $l = 1, \ldots, N$ is the echo atom of the l-th grid; λ is the radar wavelength; and T_s is the sampling interval.

The $N \times N$ dechirping observation matrix Φ_D is composed of reference signals, each column of which can be expressed as:

$$s_{ref}(n) = w_{ref}\left(nT_s - \frac{2R_{ref}}{c}\right)\exp\left[-j\frac{4\pi R_{ref}}{\lambda} + j\pi K_{ref}\left(nT_s - \frac{2R_{ref}}{c}\right)^2\right], \quad n = 1, \ldots, N \tag{24}$$

where K_{ref} is the frequency modulation slope; R_{ref} is the reference distance referring to the nearest oblique distance from the scene center to the SAR platform; and $w_{ref}(\cdot)$ denotes the reference rectangular window function, the length of which is not less than that of $w_r(\cdot)$.

Then, the dechirped sensing matrix is:

$$\boldsymbol{\Psi}_D = \boldsymbol{\Phi}_D^* \circ \boldsymbol{\Psi} = \begin{bmatrix} \boldsymbol{s}_{de,1}^T(n) & \boldsymbol{s}_{de,2}^T(n) & \cdots & \boldsymbol{s}_{de,L}^T(n) \end{bmatrix}, \quad n = 1, \ldots, N \tag{25}$$

where $\circ$ denotes the Hadamard product. The l-th dechirped atom is:

$$s_{de,l}(n) = w_r\left(nT_s - \frac{2R_l}{c}\right)\exp\left(-\frac{j4\pi K_r \Delta R_l}{c} nT_s\right)C(l), l = 1, \ldots, N \tag{26}$$

where $C(l) = \exp\left[-j4\pi\left(\frac{\Delta R_l}{\lambda} - K_r \frac{\Delta R_l^2 + 2\Delta R_l R_{ref}}{c^2}\right)\right]$ is a constant term determined by $\Delta R_l = R_l - R_{ref}$.

4.2. Double Subspace Extraction

Given that the dechirped measurement can be considered as a series of single frequency signals, we employ the orthogonal Fourier basis as the initial dictionary for signal reconstruction. To further reduce the proportion of WBNI projection as much as possible, we propose the double subspace extraction algorithm, the diagram of which is shown in Figure 4.

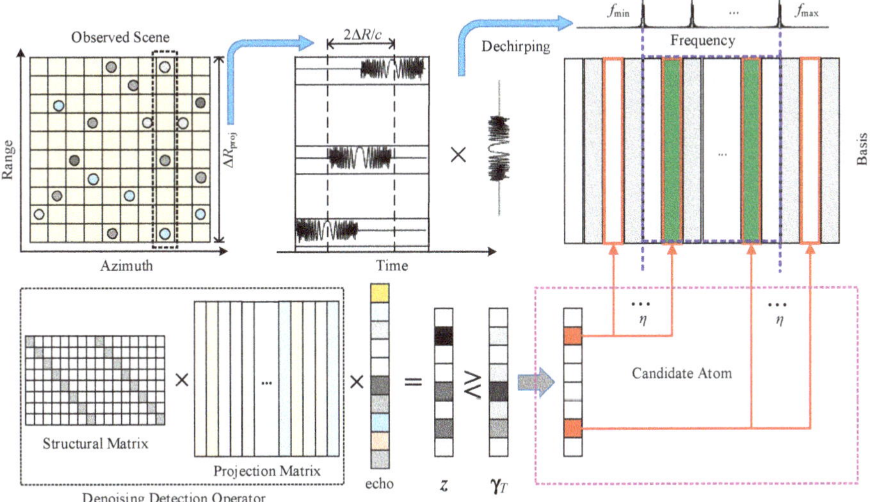

Figure 4. The diagram of double subspace extraction.

For small observed scenes, the dechirped echo of SAR occupies a smaller part of the frequency spectrum than the raw. Meanwhile, WBNI is still distributed throughout the entire spectrum after dechirping because of its incoherence. Since most SAR systems are cooperative, the geometric prior known in advance can be exploited to eliminate redundant information. That is, the relationship between target position and frequency spectrum makes it possible to extract the subspace spanned by effective columns in the basis. The others that contribute little to signal reconstruction and aggravate the interference projection are pruned out.

According to the observation geometry, the spectrum of the dechirped signal is within the range of:

$$\Delta B_{de} = \frac{2(R_{max} - R_{min})}{cT_r}B \tag{27}$$

where B is the signal bandwidth; T_r is the time width; and $R_{\max}$ and $R_{\min}$ are, respectively, the maximum and the minimum oblique distances depending on the range of the observed scene. Assuming that the geometric and the signal parameters satisfy the condition of $\Delta B_{de} < 0.5B$, we first extract the subspace matrix Ω_I composed of columns indexed by the set I (blue dashed box in Figure 4) according to Equation (27).

Moreover, there still exist redundant atoms in the preliminarily extracted subspace, since the observed scene is sparse. We adopt constant false-alarm ratio (CFAR) detection [31] to extract the candidate index set composed of possible atoms (red solid box in Figure 4). This process can be considered as the subspace extraction by exploiting the waveform prior.

Assuming that the raw echo contains only the SOI and the WBNI, then the contaminated signal can be modeled as:

$$x = s + n = \Psi\alpha + n \tag{28}$$

where α is a K-sparse vector; and n is a non-sparse vector, which increases the decision threshold and reduces the detection performance.

The waveform prior can be utilized to project the raw echo to K sparse coefficients. More advantageously, this projection has little effect on changing the distribution of WBNI, making it possible to further suppress most of the interference component by designing an observation matrix with a specific structure.

We model the following detection problem based on a binary hypothesis test:

$$\begin{aligned} \mathcal{H}_0: & \quad z = \Gamma n \\ \mathcal{H}_1: & \quad z = \Gamma(s+n) \end{aligned} \tag{29}$$

where Γ denotes the denoising detection operator expressed as:

$$\Gamma = \Phi_0 \Psi_{\text{proj}} \tag{30}$$

where $\Psi_{\text{proj}} = \left(\Psi^H \Psi\right)^{-1}\Psi^H$ is the projection matrix. To suppress the projection of the interference component as much as possible, the structural observation matrix Φ_0 is designed in the following form:

$$\Phi_0 = \begin{bmatrix} 1 & & & 1 & \\ & 1 & \cdots & & 1 \\ & & \ddots & & & \ddots \\ & & 1 & & & 1 \end{bmatrix}^{M_0 \times N} \tag{31}$$

where $\eta = N/M_0$ is an integer. A greater value of M_0 is better for reducing the interference projection, since there are more zero-valued elements in each row of Φ_0. Moreover, the cell-averaging constant false-alarm rate (CA-CFAR) structure [32] is utilized to determine the decision threshold, i.e.:

$$\gamma_T = \left(P_{fa}^{-1/N_c} - 1\right)\sum_{j=1}^{N_c}\left|z_j\right|^2 \tag{32}$$

where N_c is the number of detection cells; and P_{fa} is the false-alarm rate.

There are some points that need to be specified in the above detection model. First, when the structural observation matrix is underdetermined, i.e., $M_0 < N$, one detection result corresponds to η candidate atoms (purple dashes in Figure 4). Second, we also include a certain number of atoms adjacent to the precise candidate in the index set to reduce signal distortion as much as possible, since offset may exist or detection is missed under low signal-to-interference ratio conditions. We denote the extracted subspace indexed by candidate set $\mathcal{J}$ as $\Omega_{\mathcal{J}}$. Therefore, the final subspace matrix can be expressed as the intersection of these two extraction results, i.e.:

$$\Omega_S = \Omega_I \cap \Omega_{\mathcal{J}} \tag{33}$$

4.3. Algorithm and Procedure Details

Based on the above derivation, we designed the following compressed dechirping matrix (CDM) to perform the echo observation:

$$\Phi_{Dc} = \Xi\Phi_D^* \tag{34}$$

where Ξ is the random observation matrix; and $(\cdot)^T$ denotes the transposition. Then, the measurement vector can be expressed as:

$$y = \Phi_{Dc}x = s_{Dc} + n_{Dc} \tag{35}$$

where s_{Dc} is the dechirped SOI; and n_{Dc} denotes the WBNI. Hence, the target reconstruction has been transformed into a sparse recovery problem in Equation (6) that can be solved by the classical orthogonal matching pursuit (OMP) algorithm [27].

It is worth noting that three phase terms exist and cannot be ignored in the dechirped echo, since they introduce doppler and make the subsequent azimuth processing more complex. Hence, the reconstructed signal should be further compensated with the de-oblique factor $C^*(n)$, $n = 1, \ldots, N$, i.e., the complex conjugate form of $C(n)$ in Equation (26).

Based on the above analysis and derivation, we present the detailed steps of DDSE in Algorithm 1 and the flowchart of the WBNI suppression procedure in Figure 5.

Algorithm 1 Dechirping and double subspace extraction (DDSE)

Inputs: raw echo x with WBNI, random sampling matrix Ξ, Fourier basis Ω;
Outputs: WBNI-free signal x^*;

1. Construct the compressed dechirping matrix Φ_{Dc} by Equations (25) and (34);
2. Construct the sensing matrix by $\Theta = \Phi_{Dc}\Omega$;
3. Obtain the measurements y by Equation (35);
4. Extract the subspace Ω_I based on Equation (27);
5. Construct the denoising detection operator Γ by Equations (30)–(31);
6. Extract the subspace $\Omega_{\mathcal{J}}$ by denoising CFAR detection based on Equation (32);
7. Determine the final subspace matrix ΩS by Equation (33);
8. Repeat from $t = 1$ until N_{iter};

    (1)   Initialize the residual $r_0 = y$, index set $\Lambda_0 = \emptyset$, and augment matrix $A_0 = \emptyset$;
    (2)   Find the index Λ_t that maximizes the inner product $\Lambda_t^* = \max\langle r_{t-1}, \theta_n \rangle$, $n = 1, \ldots, N$, where is the n-th column of Θ, and r_t is the current residual vector;
    (3)   If $\Lambda_t^* \in S$, continue; else, set $\theta_n = 0$ and return to (2);
    (4)   Update the augment matrix by $A_t = A_{t-1} \cup \theta_n$;
    (5)   Estimate the least square solution by $\alpha_t^* = \operatorname{argmin} \|y\text{-}A_t\alpha_t\|_2$;
    (6)   Update the residual by $r_t = y - A_t\alpha_t^*$;
    (7) $t = t + 1$;

9. Obtain the compensation operator C with the de-oblique factor in Equation (26);
10. Reconstruct the WBNI-free signal by $x^* = C\Omega\alpha_t^*$.

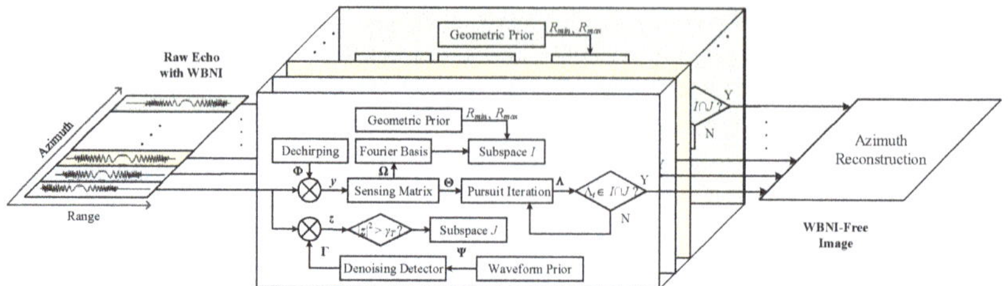

Figure 5. The flowchart of the WBNI suppression procedure based on the double subspace extraction (DDSE) algorithm.

5. Experiments

5.1. Experiment Specifications

To verify the performance of WBNI suppression for sparsity-based SAR imaging based on the proposed algorithm in this paper, we carried out multiple experiments with simulated data. First, we adopted the DDSE algorithm to range profile reconstruction of a multi-point target, where the interference suppression performance was mainly investigated from the perspective of signal reconstruction. Then, we extended the case to range-azimuth imaging of an aircraft target, where the interference suppression effect was mainly evaluated by visual quality and statistical characteristics of reconstructed SAR images. Moreover, we also compared the proposed algorithm to other advanced ones to analyze its superiority. Simulations were implemented with Matlab R2018b on a computer running Windows 7 with 3.4GHz Intel Core i7-4770 CPU and 16 GB RAM.

5.2. Simulation and Analysis

5.2.1. Range Profile Reconstruction

In this part, we first set up a simulation environment for range profile imaging of point targets, where geometric and waveform parameters for SAR observations are listed in Table 1 [33,34]. Given that the distribution of the dechirped signal in the frequency domain depends on time width and distance differences according to Equation (27), the theoretical bandwidth ratio after dechirping observations is about 0.3 of the original. This means that a large amount of redundant information can be reduced in the frequency domain after coherent processing with the signal prior.

Table 1. Main parameters for range profile imaging simulations.

Parameter Class	Parameter Name	Parameter Value
Geometric Parameters	Platform height	3 km
	Pitch angle	45°
	Scene Range	128 m
Signal Parameters	Carrier frequency	3 GHz
	Bandwidth	100 MHz
	LFM pulse width	2 µs
	Oversampling rate	1.2

*LFM= linear frequency-modulated.

For the simulation, a five-point target with normalized amplitudes and fixed locations within the scene range was modeled. The WBNI data were generated by modulating a band-limited noise to the carrier frequency and aligning them with the center of the signal spectrum, the bandwidth of which was set as equal to that of SAR. The interference-to-signal ratio (ISR) was set to 15 dB, and the additive signal-to-noise ratio (SNR) was set to 30 dB. We added the generated WBNI to the raw echo of SAR, and the signal characteristics in time, frequency, and range domain are respectively shown in Figure 6.

It was apparent that the WBNI covered the entire pulse and spectrum of the SOI, making it impossible to obtain accurate range information by sparse reconstruction. Hence, it was necessary to introduce some suppression approaches into the process of signal acquisition and reconstruction to reduce the impact of interference as much as possible.

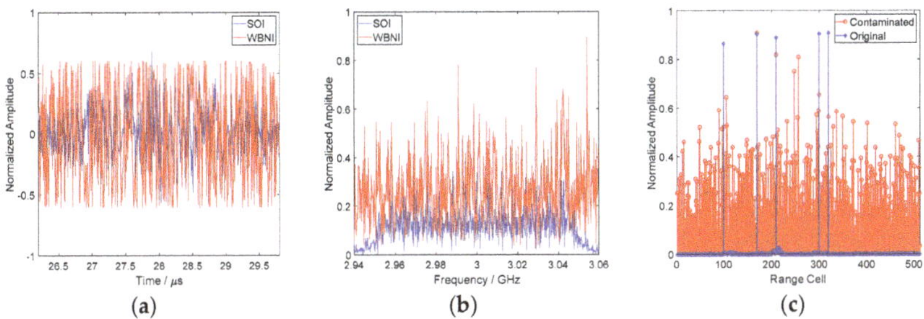

Figure 6. Signal characteristics. (**a**) Waveform; (**b**) frequency spectrum; and (**c**) range reconstruction.

We adopted the proposed DDSE-based WBNI suppression algorithm to the process of range profile reconstruction. As mentioned above, to avoid signal distortion resulting from missing detections, we also added the atoms adjacent to the detected position to the subspace matrix. Figure 7 shows the detection results under different η, where the probability of false-alarm was set to 10^{-3}. As can be seen, the WBNI introduced false alert into the process of detection and added more undesired atoms in the subspace matrix. When $\eta > 1$, the subspace dimension increases, since one compressed range cell corresponds to η atoms.

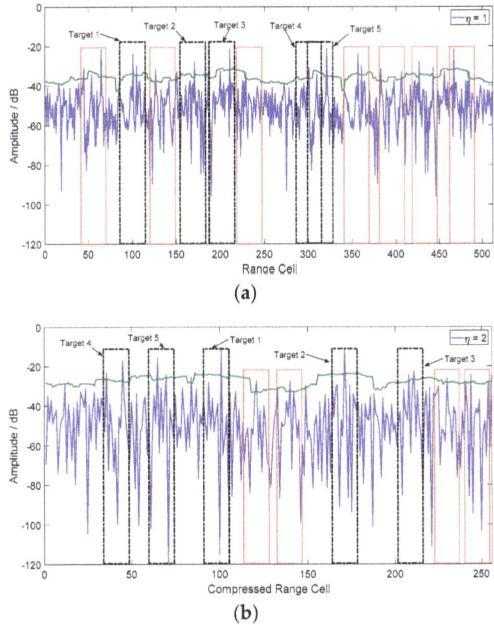

Figure 7. Subspace extraction results by denoising constant false-alarm ratio (CFAR) detection. (**a**) $\eta = 1$; (**b**) $\eta = 2$. The green line represents the CFAR threshold. The black dashed box represents the extracted index set where the real target is located. The red dashed box represents the extracted false-alert index set.

To verify the superiority of the DDSE, we compared it to other algorithms, including basic pursuit denoising (BPDN) [24], adaptive compressed sampling (ACS) [25], and block sparse Bayesian learning (BSBL) [33], the results of which are shown in Figure 8.

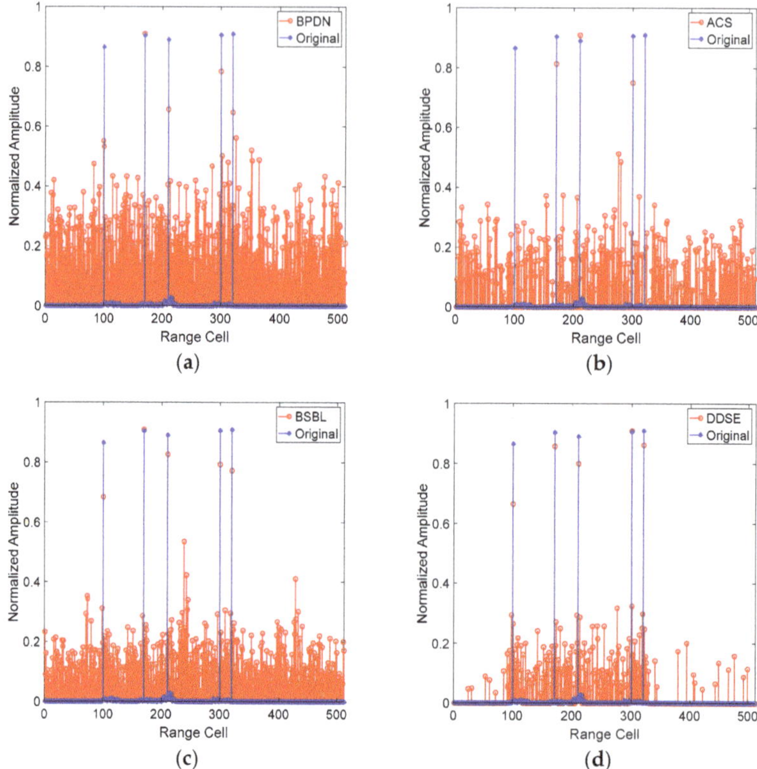

Figure 8. Range profile reconstruction based on different WBNI suppression algorithms. (**a**) basic pursuit denoising (BPDN); (**b**) adaptive compressed sampling (ACS); (**c**) block sparse Bayesian learning (BSBL); and (**d**) DDSE.

As shown in Figure 8, the proposed DDSE algorithm outperformed the others in terms of both signal distortion and WBNI suppression effects. The BPDN algorithm was hardly effective for WBNI suppression under low SNR conditions except for a little contribution to noise reduction. The ACS algorithm eliminated large amounts of interference but also useful information, leading to serious distortion since the signal prior was not fully exploited. The BSBL algorithm also generated undesired components in the process of WBNI suppression, though it was superior in narrowband interference (NBI) separation by utilizing structural information and time correlation [33,34].

To further benchmark the interference suppression performance of signal reconstruction, we employed the interference suppression degree (ISD) and the signal distortion degree (SDD) as main indicators [19]. ISD is usually employed to measure the ability to eliminate interference, which is defined as the energy ratio of the contaminated signal to the reconstructed one after interference suppression, i.e.:

$$ISD = 10\log_{10}\frac{\|\boldsymbol{x}_c\|_2^2}{\|\boldsymbol{x}_s\|_2^2} \tag{36}$$

where x_c is the contaminated signal; and x_s is the reconstructed signal after interference suppression. Since interference suppression inevitably leads to signal distortion, SDD is also utilized as an assisted but significant indicator, which is defined as the degree of energy loss of the reconstructed signal to the undisturbed one, i.e.:

$$SDD = 10 \log_{10} \frac{\|x_s - x_0\|_2^2}{\|x_0\|_2^2} \tag{37}$$

where x_0 is the original signal without interference.

We also investigated and analyzed the impact of different parameters on WBNI suppression performance by performing 100 numerical simulations. Figure 9 shows the statistical average of ISD and SDD under different ISRs from 0 dB to 30 dB. Tables 2 and 3 respectively show the changes of ISD and SDD based on the proposed DDSE algorithm with the sparsity level and the compression ratios (CR), where the ISR was set to 15 dB.

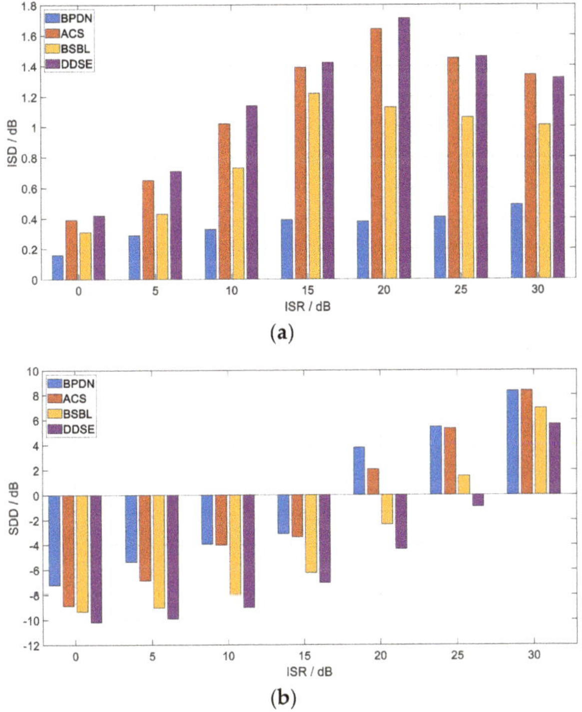

Figure 9. Signal reconstruction performance under different interference-to-signal ratio (ISRs). (**a**) interference suppression degree (ISD); (**b**) signal distortion degree (SDD).

Table 2. ISDs under different sparsity levels and compression ratios (CR) (dB).

	$K = 5$	$K = 10$	$K = 20$	$K = 40$
CR = 1/1	1.422	1.354	1.247	1.102
CR = 1/2	1.417	1.352	1.226	1.097
CR = 1/4	1.410	1.346	1.240	1.080
CR = 1/8	1.373	1.279	1.192	1.025

Table 3. SDDs under different sparsity levels and compression ratios (dB).

	$K = 5$	$K = 10$	$K = 20$	$K = 40$
CR = 1/1	−7.035	−6.685	−6.104	−5.217
CR = 1/2	−6.490	−5.782	−4.878	−3.632
CR = 1/4	−5.852	−5.046	−4.220	−3.096
CR = 1/8	−1.507	−0.879	−0.121	−0.274

From the results in Figure 9, the proposed DDSE algorithm outperformed the others within the given parameter range in terms of both ISD and SDD. For the former indicator, the BPDN algorithm changed little, and the others presented a trend of increasing first and then decreasing. The ACS and the DDSE algorithms started to decrease when ISR reached 20 dB, while the BSBL algorithm started to decrease at 15 dB, which shows that DDSE and ACS are better able to suppress WBNI. For the latter indicator, all algorithms presented an upward trend with the ISR, where the DDSE and BSBL algorithms had better performances, indicating higher stability of signal recovery.

In Table 2, the ISD mainly depended on the sparsity and changed little with the compression ratio, since the subspace for signal recovery was constructed according to target distribution and CFAR detection results in the DDSE algorithm, which meant that more atoms were eliminated from the subspace when the target was sparser. In Table 3, the SDD was more affected by the compression ratio, since the low-dimensional observation introduced measurement noise into the process of signal recovery, leading to more serious signal distortion.

From the perspective of time efficiency, it is not difficult to see that the convergence iteration number in our proposed DDSE algorithm, which is the main factor of the running time, is determined by the dimensions of the extracted subspace. However, the detection results cannot be analyzed by a specific formula, since the WBNI is unpredictable. Thus, we calculated the average running time of each algorithm with the sparsity parameters listed above by 30 repeated simulations, the results of which are shown in Table 4. It can be seen that the running time of our proposed DDSE algorithm was at a minimum under low sparsity conditions, but it increased with the sparsity just like ACS, while the BPDN and the BSBL changed little.

Table 4. Average running time of different algorithms (second).

	BPDN	ACS	BSBL	DDSE
$K = 5$	16.825	5.325	6.827	3.364
$K = 10$	17.081	6.811	6.795	4.162
$K = 20$	17.630	8.739	6.949	6.593
$K = 40$	18.533	12.447	6.952	−199.259

5.2.2. Range-Azimuth Imaging

To further investigate the WBNI suppression effects with the proposed DDSE algorithm, we extended the simulation to the case of range-azimuth reconstruction, where an aircraft target with multiple scattering points was modeled and utilized for SAR imaging. The main parameters in this part are listed in Table 5 [33,34], and the intuitive results of range-azimuth imaging based on different WBNI suppression algorithms are shown in Figure 10.

Table 5. Main parameters for range-azimuth imaging simulations.

Parameter Class	Parameter Name	Parameter Value
Geometric Parameters	Platform height	3 km
	Scene range	128 m × 128 m
	Pitch angle	45°
	Squint angle	0°
	Parallel velocity	150 m/s
Signal Parameters	Carrier frequency	3 GHz
	Bandwidth	100 MHz
	LFM pulse width	2 µs
	Oversampling coefficient	1.2
	Pulse repetition frequency	125 Hz

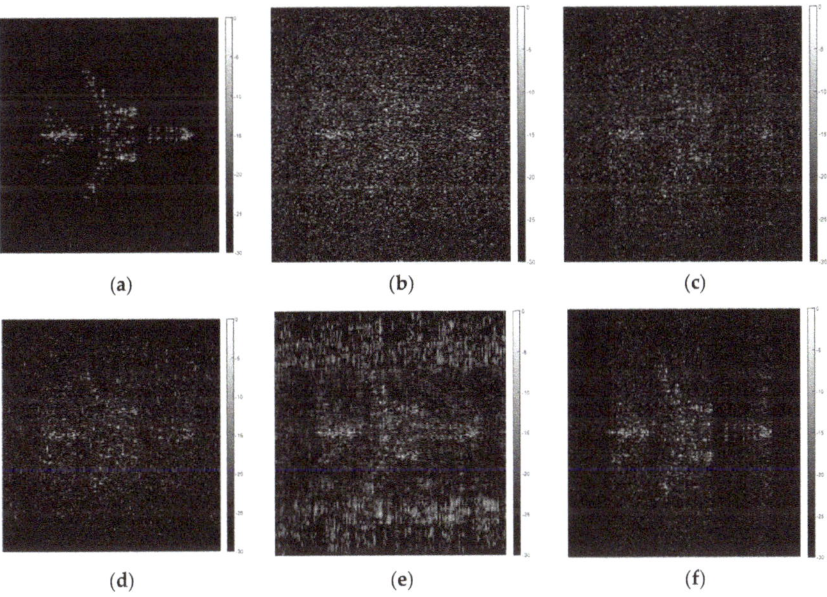

Figure 10. Range-azimuth imaging based on different WBNI suppression algorithms. (**a**) Original;
(**b**) contaminated; (**c**) BPDN; (**d**) ACS; (**e**) BSBL; and (**f**) DDSE.

In Figure 10a,b, the reconstructed aircraft target is almost covered by WBNI and can hardly
be distinguished if no measures are taken. In Figure 10c,d, the BPDN algorithm has little effect on
interference suppression, and the ACS algorithm leads to serious signal distortion. The BSBL algorithm
in Figure 10e, which is effective for narrowband interference (NBI) separation, increases the adverse
effect for image reconstruction. In contrast, the proposed DDSE algorithm in Figure 10f performs better
than the others in terms of visual quality.

We employed the peak signal-to-noise ratio (PSNR) and image entropy to perform a
quantitative evaluation of WBNI suppression performance for range-azimuth imaging, and then
we carried out multiple range-azimuth imaging simulations to compare these two indicators under
different parameters.

PSNR is a common indicator for evaluating image quality, which is often defined by the mean
square error. It reflects the extent to which the SAR image is affected by noise or interference, and a
larger value of PSNR indicates better image quality. Given the discrete property of sparse SAR images,
we redefined it as:

$$PSNR= 10\log_{10}\frac{\frac{1}{L}\sum_{l=1}^{L}\max_L\left|A_{i,j}\right|^2}{\frac{1}{N_aN_r-L}\left(\sum_{i=1}^{N_a}\sum_{j=1}^{N_r}\left|A_{i,j}\right|^2-\sum_{l=1}^{L}\max_L\left|A_{i,j}\right|^2\right)} \tag{38}$$

where N_a and N_r are, respectively, the number of cells in azimuth and range of an SAR image; L is the number of scattering points; $A_{i,j}$ denotes the complex value of the (i,j)-th point; $|\cdot|$ denotes the modulus value; and $\max_L$ represents picking out L largest values.

The image entropy is a statistical form used for representing the aggregation characteristics of the grayscale distribution and for measuring the average amount of information in an image. Since the principle of SAR imaging is different from that of conventional optical imaging, and a non-uniform grayscale histogram distribution can highlight the texture or the contour of the observation scene, we would rather obtain an SAR image with a lower entropy after suppression. This indicator can be calculated by:

$$IE= -\sum_{i=1}^{N_G}p_i\log_2 p_i \tag{39}$$

where p_i is the probability of the i-th grayscale level; and N_G is the total number of all grayscale levels in the image.

Figure 11 shows the statistical average of the peak-signal-to-noise ratio (PSNR) and image entropy under different ISRs from 0 dB to 30 dB by 100 numerical simulations. Figures 12 and 13 show the range-azimuth imaging results based on the DDSE algorithm under different sparsity levels and compression ratios, where the number of points in the simulated aircraft model were set to 174, 348, and 696, and the CRs were set to 1/2, 1/4, and 1/8, respectively.

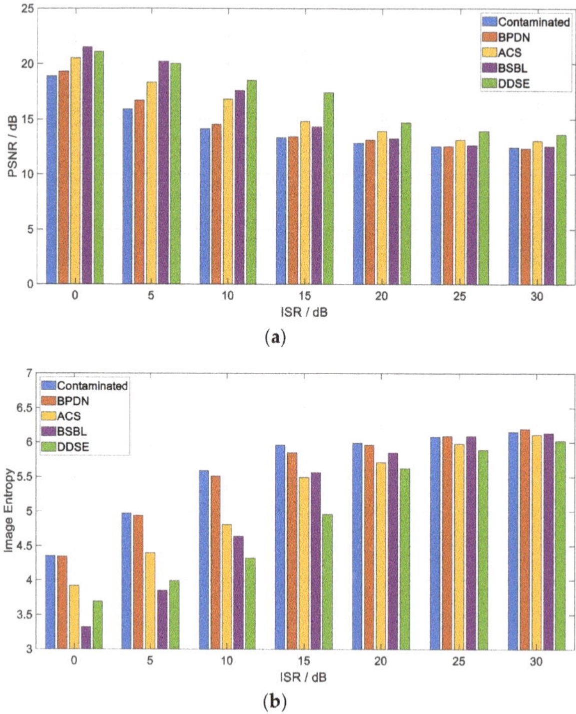

Figure 11. Range-azimuth imaging performances under different ISRs. (**a**) peak signal-to-noise ratio (PSNR); (**b**) image entropy.

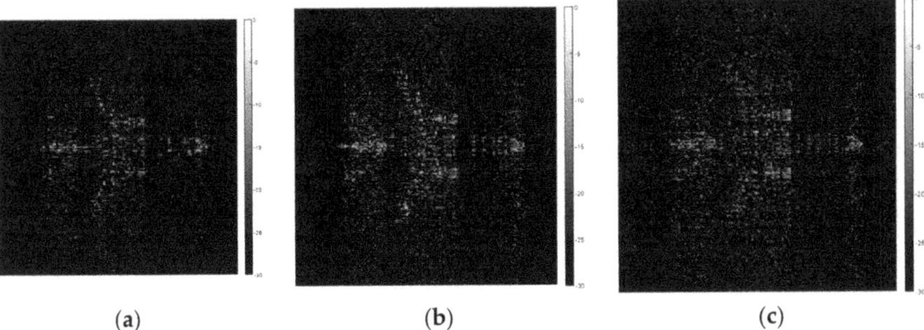

(a) (b) (c)

Figure 12. Effects of WBNI suppression for range-azimuth imaging based on the DDSE algorithm under different sparsity levels. (**a**) $K = 174$; (**b**) $K = 348$; and (**c**) $K = 696$.

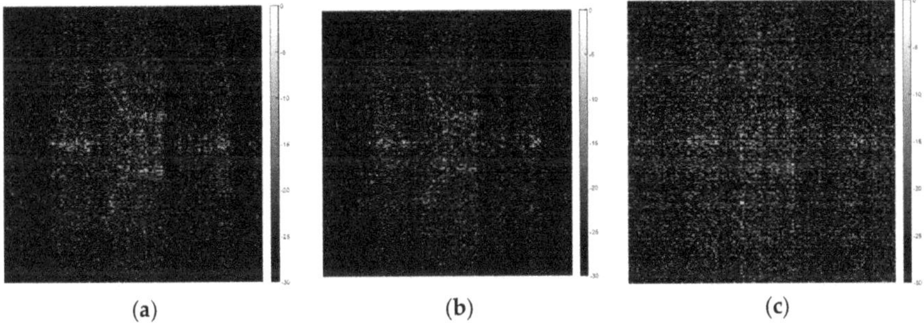

(a) (b) (c)

Figure 13. Effects of WBNI suppression for range-azimuth imaging based on the DDSE algorithm under different compression ratios. (**a**) CR = 1/2; (**b**) CR = 1/4; and (**c**) CR = 1/8.

As can be seen from the results in Figure 11, the PSNR decreases with an increasing interference-to-signal ratio (ISR), while the image entropy increases. The BSBL algorithm performs better when the ISR is lower than 10 dB, but it also presents a rapid deterioration in performance. The proposed DDSE algorithm in this paper is superior to the other ones from the aspects of both WBNI suppression and stability. However, this superiority gradually weakens as the interference power increases further.

As shown in Figures 12 and 13, WBNI suppression for range-azimuth imaging is influenced by both the sparsity level and the compression ratio. Under the same conditions, the proposed DDSE algorithm performs better for the observed scene with a lower sparsity level, since the dimensions of subspace corresponding to the SOI are smaller, leading to more interference components being suppressed in the process of signal reconstruction. When the compression ratio is higher than 1/4, the image quality with WBNI suppression changes little with the reduction of measurement number, but when the compression ratio is reduced to 1/8, it begins to decline seriously.

6. Conclusions

In this paper, we proposed a novel WBNI suppression approach for sparsity-based SAR imaging based on dechirping and double subspace extraction (DDSE) algorithms, the starting point of which was based on the derived conditions for minimizing energy projection of the interference. The dechirping observations were utilized to transform the raw echo to a series of single frequency signals, making it possible to be sparsely represented on an orthogonal basis, which was proven to be the optimal one. The subspace for accurate signal reconstruction was extracted from two separate and parallel steps, where both the geometric prior and the waveform prior were exploited, and then determined as the intersection of these two obtained index sets. The experimental results show that the proposed

DDSE algorithm outperforms the others both in suppressing WBNI and in reducing signal distortion. It is necessary to note here that our proposed algorithm is effective under the assumptions of a small observed scene. Therefore, how to extend this algorithm to large scene conditions is our main area of focus in future research.

Author Contributions: Conceptualization and supervision: W.Y.; methodology and investigation: G.L. (Guojing Li) and G.L. (Guochao Lao); experiment: G.L. (Guojing Li) and Q.L.; writing: G.L. (Guojing Li).

Funding: This research was funded by the Research Project of State Key Laboratory of Complex Electromagnetic Environment Effects on Electronics and Information System, grant number 2017Z0203B.

Conflicts of Interest: The authors declare no conflict of interest.

References

1. Çetin, M.; Stojanovic, I.; Onhon, O.; Varshney, K.; Samadi, S.; Karl, W.C.; Willsky, A.S. Sparsity-driven synthetic aperture radar imaging: Reconstruction, autofocusing, moving targets, and compressed sensing. *IEEE Signal Process. Mag.* **2014**, *31*, 27–40. [CrossRef]
2. Zhang, B.; Hong, W.; Wu, Y. Sparse microwave imaging: Principles and applications. *Sci. China Inf. Sci.* **2012**, *55*, 1722–1754. [CrossRef]
3. Donoho, D.L. Compressed sensing. *IEEE Trans. Inf. Theory* **2006**, *52*, 1289–1306. [CrossRef]
4. Patel, V.M.; Easley, G.R.; Healy, D.M.; Chellappa, R. Compressed synthetic aperture radar. *IEEE J. Sel. Top. Signal Process.* **2010**, *4*, 244–254. [CrossRef]
5. Fang, J.; Xu, Z.; Zhang, B.; Hong, W.; Wu, Y. Fast Compressed sensing SAR imaging based on approximated observation. *IEEE J. Sel. Top. Appl. Earth Obs. Remote Sens.* **2014**, *7*, 352–363. [CrossRef]
6. Soumekh, M. SAR-ECCM using phase-perturbed LFM chirp signals and DRFM repeat jammer penalization. *IEEE Trans. Aerosp. Electron. Syst.* **2006**, *42*, 191–205. [CrossRef]
7. Yang, L.; Gao, S.; Hu, R.; Wei, H. Performance analysis of coherent jamming and non-coherent jamming against SAR. *Syst. Eng. Electron.* **2018**, *40*, 2444–2449.
8. Li, B.; Hong, W. Study of noise jamming to SAR. *Acta Electron. Sin.* **2004**, *32*, 2035–2037.
9. Orlando, D. A novel noise jamming detection algorithm for radar applications. *IEEE Signal Process. Lett.* **2017**, *24*, 206–210. [CrossRef]
10. Mishali, M.; Eldar, Y.C.; Elron, A. Xampling: Signal acquisition and processing in union of subspaces. *IEEE Trans. Signal Process.* **2011**, *59*, 4719–4734. [CrossRef]
11. Arias-Castro, E.; Eldar, Y.C. Noise folding in compressed sensing. *IEEE Signal Process. Lett.* **2012**, *18*, 478–481. [CrossRef]
12. Davenport, M.A.; Laska, J.N.; Treichler, J.R.; Baraniuk, R.G. The pros and cons of compressive sensing for wideband signal acquisition: Noise folding versus dynamic range. *IEEE Trans. Signal Process.* **2012**, *60* 4628–4642. [CrossRef]
13. Li, D.; Zhan, M.; Fang, Z.; Xiong, H.; Jiang, Q. Parameterized wideband interference suppression for SAR imaging based on HAF. *Syst. Eng. Electron.* **2017**, *39*, 514–521.
14. Su, J.; Tao, M.; Xie, J.; Wang, L. Interference Suppression for SAR Based on Ambiguity Function Iteration Decomposition. In Proceedings of the 2018 IEEE International Geoscience and Remote Sensing Symposium (IGARSS), Valencia, Spain, 22–27 July 2018. [CrossRef]
15. Chen, R.; Wang, Y. Universal FRFT-based algorithm for parameter estimation of chirp signals. *J. Syst. Eng. Electron.* **2012**, *23*, 495–501. [CrossRef]
16. Elgamel, S.A.; Soraghan, J.J. Using EMD-FrFT filtering to mitigate very high power interference in chirp tracking radars. *IEEE Signal Process. Lett.* **2011**, *18*, 263–266. [CrossRef]
17. Su, J.; Tao, H.; Song, D.; Rao, X.; Xie, J. Interference suppression algorithm for SAR based on WD and sliding window masking technique in time-frequency domain. *Acta Electron. Sin.* **2015**, *43*, 2345–2351.
18. Zhang, S.; Xing, M.; Guo, R.; Zhang, L.; Bao, Z. Interference suppression algorithm for SAR based on time–Frequency transform. *IEEE Trans. Geosci. Remote Sens.* **2011**, *49*, 3765–3779. [CrossRef]
19. Tao, M.; Zhou, F.; Zhang, Z. Wideband interference mitigation in high-resolution airborne synthetic aperture radar data. *IEEE Trans. Geosci. Remote Sens.* **2016**, *54*, 74–87. [CrossRef]

20. Yang, Z.; Du, W.; Liu, Z.; Liao, G. WBI suppression for SAR using iterative adaptive method. *IEEE J. Sel. Top. Appl. Earth Obs. Remote Sens.* **2016**, *9*, 1008–1014. [CrossRef]
21. Lu, X.; Yang, J.; Ma, C.; Gu, H.; Su, W. Wide-band interference mitigation algorithm for SAR based on time-varying filtering and sparse recovery. *Electron. Lett.* **2018**, *54*, 165–167. [CrossRef]
22. Liu, H.; Li, D.; Zhou, Y.; Truong, T. Joint wideband interference suppression and SAR signal recovery based on sparse representations. *IEEE Geosci. Remote Sens. Lett.* **2017**, *14*, 1542–1546. [CrossRef]
23. Liu, H.; Li, D.; Zhou, Y.; Truong, T. Simultaneous Radio Frequency and Wideband Interference Suppression in SAR Signals via Sparsity Exploitation in Time-Frequency Domain. *IEEE Trans. Geosci. Remote Sens.* **2018**, *56*, 5780–5793. [CrossRef]
24. Chen, S.S.; Donoho, D.L.; Saunders, M.A. Atomic decomposition by basis pursuit. *SIAM Rev.* **2001**, *43*, 129–159. [CrossRef]
25. Kang, R.; Tian, P.; Yu, Y. An adaptive compressed sensing method based on selective measure. *Acta Phys. Sin.* **2014**, *63*, 143–150.
26. Cumming, I.G.; Dettwiler, M. *Digital Processing of Synthetic Aperture Radar Data: Algorithms and Implementation*; Artech House: Norwood, MA, USA, 2004.
27. Tropp, J.A.; Gilbert, A.C. Signal recovery from random measurements via orthogonal matching pursuit. *IEEE Trans. Inf. Theory* **2007**, *53*, 4655–4666. [CrossRef]
28. Elad, M.; Bruckstein, A.M. A generalized uncertainty principle and sparse representation in pairs of bases. *IEEE Trans. Inf. Theory* **2002**, *48*, 2558–2567. [CrossRef]
29. Zhang, X. *Matrix Analysis and Applications*, 2nd ed.; Tsinghua University Press: Beijing, China, 2004.
30. Caputi, W.J. Stretch: A time-transformation technique. *IEEE Trans. Aerosp. Electron. Syst.* **1971**, *AES-7*, 269–278. [CrossRef]
31. Ma, J.; Liu, C.; Gan, L. CFAR target detection algorithm based on compressive sensing. *J. Electron. Inf. Technol.* **2017**, *39*, 2899–2904.
32. Gandhi, P.; Kassam, S. Analysis of CFAR processors in nonhomogeneous background. *IEEE Trans. Aerosp. Electron. Syst.* **1988**, *24*, 427–445. [CrossRef]
33. Lu, X.; Su, W.; Yang, J.; Gu, H.; Zhang, H.; Yu, W.; Yeo, T.S. Radio frequency interference suppression for SAR via block sparse Bayesian learning. *IEEE J. Sel. Top. Appl. Earth Obs. Remote Sens.* **2018**, *11*, 4835–4847. [CrossRef]
34. Li, G.; Ye, W.; Lao, G.; Kong, S.; Yan, D. Narrowband interference separation for synthetic aperture radar via sensing matrix optimization-based block sparse Bayesian learning. *Electronics* **2019**, *8*, 458. [CrossRef]

 *electronics*

Article

Designing Constant Modulus Sequences with Good Correlation and Doppler Properties for Simultaneous Polarimetric Radar

Fulai Wang *, Chen Pang, Yongzhen Li and Xuesong Wang

The State Key Laboratory of Complex Electromagnetic Environment Effects on Electronics and Information System, College of Electronic Science, National University of Defense Technology, Changsha 410073, China; pangchen_2017@163.com (C.P.); liyongzhen@nudt.edu.cn (Y.L.); wxs_2017@163.com (X.W.)
* Correspondence: wflmadman@outlook.com; Tel.: +86-1561-6264-857

Received: 12 July 2018; Accepted: 17 August 2018; Published: 20 August 2018

Abstract: Simultaneous polarimetric radar transmits a pair of orthogonal waveforms both of which must have good auto- and cross-correlation properties. Besides, high Doppler tolerance is also required in measuring the highly maneuvering targets. A new method for the design of sequences with good correlation and Doppler properties is proposed. We formulate a fourth-order polynomial, but unconstrained, minimization problem. An iterative algorithm based on the gradient method on the phases is applied to solve it. Numerical results demonstrate the superiority of the proposed algorithm compared to the previous state-of-the-art method.

Keywords: simultaneous polarimetric radar; constant modulus sequences; correlation properties; doppler tolerance

---

1. Introduction

In recent years, the simultaneous polarimetric scheme has been widely used to obtain accurate polarization features, which can be described by a second-order polarization scattering matrix (PSM), of targets [1–4]. A pair of waveforms transmitted in this mode is required to be stringently orthogonal, which is usually evaluated by the Isolation of the waveforms, to reduce the interference caused by simultaneous transmission and reception [4]. Meanwhile, the autocorrelation properties of the sequences, which usually represent the peak side-lobe level (PSL) and the integrated side lobe level (ISL), are also of great importance [5]. Generally, the peak of the sidelobes corresponds to falsely detected objects, while high PSL result in masking of the weak targets with small radar cross-section (RCS) next to high signature targets in nearby range cells. Besides, for typical use of the polarimetric radar in meteorology, the ISL is an important metric. In [6–8], it has been pointed out that the targets for weather radar are extended volume scatterers and range sidelobes are a major source of error for quantitative applications. Therefore, designing waveforms with good correlation (in the rest of the article, correlation is used to denote both auto- and cross-correlation) properties is of great significance for simultaneous polarimetric radar.

Sequences design with good correlation properties is a traditional problem in radar and communication systems [9–12]. Focusing on the designing problem, numerous algorithms have been proposed to design the sequences. Among these techniques, a class of typical methods to design the sequences is to use the intelligent algorithms directly. Deng et al. designed unimodular sequences by using Simulated Annealing algorithm and analysed the performance of the sequences in applications of orthogonal netted radar [13]. In addition, Liu et al. used genetic algorithm to obtain sequences with good correlation properties [14]. However, the correlation properties, including PSL and Isolation of the results designed by the above-mentioned algorithms will become worse than the theoretical limiting performance with the sequences length increasing [15,16]. Besides, many researchers have also proposed computationally efficient cyclic optimization algorithms for the design of unimodular sequences. Following a similar line of derivation, Stoica et al. proposed a series of four algorithms containing CAP, CAN, WeCAN and CAD [17], which are based on the minimization of ISL with high computational efficiency. Meanwhile, these procedures can optimize the specified part of the correlation function. It should be pointed out that in some cases, the interest lies in making partial sidelobes small rather than making all sidelobes small [5,15]. However, the practical convergence rate of these four algorithms becomes slow with the sequences length increasing. Based on a customized Limited-Memory Broyden Fletcher Goldfarb and Shanno (LBFGS) algorithm, the problem of minimizing the concerned sidelobes of the correlation function is addressed by Wang et al. in [18]. By defining a new correlation matrix and using the Fast Fourier Transforms (FFT) algorithm, they moderated the computational complexity and improved the convergence rate of the method compared to the WeCAN algorithm. In a recent work [5], Palomar et al. developed the general majorization-minimization (MM) method to tackle the optimization problem arising from the sequences design. By means of the FFT algorithm, they solve the problem of designing sets of very long sequences. However, like LBFGS and MM methods, the Doppler tolerance of the sequences is not considered. In other words, they lack the ability to optimize the Doppler tolerance of the sequence sets.

In [19], Kretschmer et al. have proved that the correlation performance of the phase-coded sequences is sensitive to the target velocity. Even if the target velocity is low, the PSL and Isolation of the waveforms will seriously deteriorate compared with the same metrics of the static target's echoes. Focusing on this problem, Pezeshki et al. investigated the Doppler tolerant waveforms designing problem for the polarimetric radar [20–22]. The constituent waveforms are Golay complementary codes, which have been used in many active sensing and communication systems, for instance, radar pulse compression [23], orthogonal frequency-division multiplexing (OFDM) [24] and channel estimation [25], because of the perfect autocorrelation properties along the zero Doppler shift. Based on the Prouhet-Thue-Morse sequence, they constructed Doppler resilient sequences and the range sidelobes almost vanished at modest Doppler shifts. However, the cross-correlation properties of the sequences cannot be optimized by their methods. Meanwhile, the correlation properties of the sequences keep good only in a small Doppler shifts interval around the zero Doppler shift. Their method lacks the ability to design sequences with good correlation properties in specific range and Doppler bins of the ambiguity function. In [26], Cui et al. considered the local ambiguity function shaping and proposed an accelerate iterative sequential optimization (AISO) algorithm to minimized the average value of the weighted integrated sidelobe level (WISL) over specific Doppler bins and range bins of interest. However, the orthogonality of the sequences that is of great importance for simultaneous polarimetric radar is not considered in AISO algorithm.

In this article, based on the gradient method shown in [18,27], we propose a new cyclic algorithm that can design sequences not only with good correlation properties but also with high Doppler tolerance. The obtained sequences can be used for the space surveillance radar to improve the measurement accuracy of the PSM of the highly maneuvering targets, including satellites, spacecraft, etc. The proposed method can be summarized as follows:

- Suppressing a specific part of the correlation function
- Good correlation properties under motion states of interest
- Better sidelobes both in terms of PSL and ISL

The remainder of this paper is organized as follows: the problem for designing sequences with good correlation and Doppler properties is formulated in Section 2; in Section 3, the Iterative Algorithm–Gradient (IAG) algorithm is proposed to solve the design problem; simulations are presented to validate the method in Section 4; and Section 5 concludes the paper.

Mathematical Notations: In this paper, it is assumed that a lower-case letter (e.g., *a*) denotes a scalar; a boldface lower-case letter (e.g., **a**) denotes a vector; and a boldface upper-case letter (e.g., **A**) indicates a matrix. Additionally, the symbols $\mathbf{A}^{\mathrm{T}}$ and $\mathbf{A}^{\mathrm{H}}$ represent transpose and conjugate transpose of the matrix $\mathbf{A}$, respectively. $|a|$ and $(a)^{*}$ indicate the absolute value and the conjugate of the scalar a. Besides, $\|\cdot\|_{F}$ denotes the Frobuinous norm of a matrix and $\odot$ represents the Hadamard product of two matrices with the same dimension. $\mathrm{Re}\,(a)$ represents the real part of the scalar a and $\mathrm{Tr}\,(\mathbf{A})$ denotes the trace of the matrix $\mathbf{A}$.

2. Problem Formulation

A pair of constant modulus sequences used for simultaneous polarimetric radar can be written as

$$\begin{cases} s_{\mathrm{H}}\,(t) = \dfrac{1}{\sqrt{N\tau}} \displaystyle\sum_{n=1}^{N} \alpha\,[t - (n-1)\,\tau]\,x_{\mathrm{H}}\,(n) \\[4mm] s_{\mathrm{V}}\,(t) = \dfrac{1}{\sqrt{N\tau}} \displaystyle\sum_{n=1}^{N} \alpha\,[t - (n-1)\,\tau]\,x_{\mathrm{V}}\,(n) \end{cases} \tag{1}$$

where

$$x_p\,(n) = e^{j\phi_p(n)},\ p = \mathrm{H,V}\ \text{and}\ n = 1,2,\cdots,N \tag{2}$$

are the sequences to be designed (it is assumed that the phases $\{\phi_p\,(n)\}$ can be arbitrary values from $[-\pi,\pi]$), H and V represent the horizontal and vertical polarized channels, τ is the time duration of one subpulse and $\alpha\,(t)$ is the shaping function, e.g., a rectangular pulse with amplitude 1. When the target moves, the Doppler shift of the received signals in different times is

$$\phi_{\mathrm{d},l}\,(n) = 2\pi f_{\mathrm{d},l}\,(n)\,n\tau = 4\pi \frac{v_l + a_l n\tau}{\lambda} n\tau \tag{3}$$

$$n = 1,2,\cdots,N\ \text{and}\ l = 1,2,\cdots,L,$$

where λ is the carrier wavelength, v_l and a_l are the radial velocity and acceleration, respectively, the subscript l represents the lth motion state of interest, and L is the amount of the concerning motion states. The (aperiodic) correlation of $\{x_p\,(n)\}_{n=1}^{N}$ and $\{x_q\,(n)\}_{n=1}^{N}$ (in the rest of the paper, p,q are both used to denote H and V) is defined as

$$r_{pq,l}\,(k) = \begin{cases} \displaystyle\sum_{n=k+1}^{N} x_p\,(n)\left[x_q\,(n-k)\,e^{j\phi_{\mathrm{d},l}(n-k)}\right]^{*},\ 0 \le k \le N-1 \\[4mm] \displaystyle\sum_{n=1}^{N+k} x_p\,(n)\left[x_q\,(n-k)\,e^{j\phi_{\mathrm{d},l}(n-k)}\right]^{*},\ -N+1 \le k < 0 \end{cases} \tag{4}$$

When $p = q$, (4) becomes the autocorrelation of $\{x_p\,(n)\}_{n=1}^{N}$. Denote the transmitted sequences by an N-by-2 matrix

$$\mathbf{X} = [\mathbf{x}_{\mathrm{H}}\ \ \mathbf{x}_{\mathrm{V}}], \tag{5}$$

and the Doppler echoes sequences by an N-by-$2L$ matrix

$$\mathbf{X}_{\mathrm{d}} = [\mathbf{x}_{\mathrm{H}} \odot \mathbf{D}_1\ \ \mathbf{x}_{\mathrm{V}} \odot \mathbf{D}_1\ \ \cdots\ \ \mathbf{x}_{\mathrm{H}} \odot \mathbf{D}_L\ \ \mathbf{x}_{\mathrm{V}} \odot \mathbf{D}_L] \tag{6}$$

where

$$\mathbf{x}_p = [x_p\,(1),\cdots,x_p\,(N)]^{\mathrm{T}},\ p = \mathrm{H,V} \tag{7}$$

and

$$\mathbf{D}_l = \left[e^{j\phi_{\mathrm{d},l}(1)},\cdots,e^{j\phi_{\mathrm{d},l}(N)}\right]^{\mathrm{T}},\ l = 1,2,\cdots,L. \tag{8}$$

Then, the correlation entries for the kth lag in (4) are given by the elements of the following correlation matrix [18]

$$\mathbf{R}_k = \mathbf{X}_{\mathrm{d}}^{\mathrm{H}} \mathbf{J}_k \mathbf{X} \tag{9}$$

where

$$\mathbf{J}_k = \begin{bmatrix} \mathbf{0} & \mathbf{I}_{N-k} \\ \mathbf{0} & \mathbf{0} \end{bmatrix}_{N \times N} = \mathbf{J}_{-k}^{\mathrm{H}} \tag{10}$$

and $\mathbf{I}$ is the identity matrix. It should be noted that the elements of $\mathbf{R}_k$ are made up of the autocorrelations and cross-correlations of the sequences. A compact optimization model for deigning constant modulus sequences with low correlation sidelobes levels can be formulated as

$$\min_{\mathbf{x}_H, \mathbf{x}_V} \sum_{k=-G}^{G} w_k \left\| \mathbf{X}_{\mathrm{d}}^{\mathrm{H}} \mathbf{J}_k \mathbf{X} - N\delta(k)\mathbf{Y} \right\|_F^2 \tag{11}$$

$$\text{subject to } \left| x_p(n) \right| = 1, \; n = 1, 2, \cdots, N.$$

where

$$\mathbf{Y} = \begin{bmatrix} \mathbf{I}_2 & \mathbf{I}_2 & \cdots & \mathbf{I}_2 \end{bmatrix}_{2L \times 2}^{\mathrm{T}}, \tag{12}$$

and $\delta(k)$ is the impulse function, i.e., $\delta(k) = 1$ for $k = 0$ and otherwise $\delta(k) = 0$. In (11), $\{w_k\}_{k=-G}^{G}$ are positive real weights chosen by the user, and $G < N$ is also a preset positive integer, which represents the lag interval of interest. The model (11) involves minimizing a fourth-order polynomial with nonlinear equality constraints, which is numerically difficult to handle. However, using the phases $\{\phi_p(n)\}_{n=1}^{N}$ as new variables, the constant modulus constraints can be dropped and (11) can be formulated as an unconstrained fourth-order polynomial minimization problem

$$\min_{\phi} f(\phi) = \sum_{k=-G}^{G} w_k \left\| \mathbf{X}_{\mathrm{d}}^{\mathrm{H}}(\phi) \mathbf{J}_k \mathbf{X}(\phi) - N\delta(k)\mathbf{Y} \right\|_F^2 \tag{13}$$

where

$$\phi = \begin{bmatrix} \phi_{\mathrm{H}}(1) & \cdots & \phi_{\mathrm{H}}(N) & \phi_{\mathrm{V}}(1) & \cdots & \phi_{\mathrm{V}}(N) \end{bmatrix}^{\mathrm{T}}. \tag{14}$$

Compared with (11), obtaining the global optimal solution of (13) is still a NP-hard problem. However, considering the local optimal solutions, it is easier to handle the optimization problem. In this paper, an iterative algorithm based on the gradient method, which can be applied directly, is proposed to solve the problem of (13). In the gradient method, the consuming time of each iteration is determined by the calculation of $f(\phi)$ and $\nabla f(\phi)$. The computational costs of calculating $f(\phi)$ and $\nabla f(\phi)$ directly are $O(NGL)$ and $O(N^2GL)$, respectively, according to (13) [18]. In the following, an efficient method is proposed to compute them.

3. Solving the Model by IAG Algorithm

Introducing the notation, the matrix $\mathbf{M}$ is defined as

$$\mathbf{M} = \begin{bmatrix} \mathbf{m}_{-N+1} \cdots \underbrace{\mathbf{m}_{-G} \cdots \mathbf{m}_{-1}}_{\mathbf{M}_{-G}} \; \mathbf{m}_0 \; \underbrace{\mathbf{m}_1 \cdots \mathbf{m}_G}_{\mathbf{M}_G} \cdots \mathbf{m}_{N-1} \end{bmatrix} \tag{15}$$

where $\mathbf{m}_k = \sqrt{w_k}\,\mathrm{vec}\left(\mathbf{X}_{\mathrm{d}}^{\mathrm{H}}(\phi) \mathbf{J}_k \mathbf{X}(\phi) \right)$, and $\mathrm{vec}(\cdot)$ vectorizes a matrix by stacking its columns on top of one another. Then the function $f(\phi)$ can be rewritten as

$$f(\phi) = \left\| \mathbf{m}_0 - \mathbf{h} \right\|_2^2 + \left\| \mathbf{M}_{-G} \right\|_F^2 + \left\| \mathbf{M}_G \right\|_F^2 \tag{16}$$

where $\mathbf{h} = \sqrt{w_0}\mathrm{vec}\,(N\mathbf{Y})$ and $\|\cdot\|_2$ denotes the 2-norm of a vector. According to (16), the gradient $\nabla f\,(\phi)$ can be computed as follows:

$$\frac{\partial f\,(\phi)}{\partial \phi_p\,(n)} = 2\,\mathrm{Re}\left[(\mathbf{m}_0 - \mathbf{h})^{\mathrm{H}}\frac{\partial \mathbf{m}_0}{\partial \phi_p\,(n)} + \mathrm{Tr}\left(\mathbf{M}_{-G}^{\mathrm{H}}\frac{\partial \mathbf{M}_{-G}}{\partial \phi_p\,(n)} + \mathbf{M}_G^{\mathrm{H}}\frac{\partial \mathbf{M}_G}{\partial \phi_p\,(n)}\right)\right] \tag{17}$$

Furthermore, it should be noted that each row vector of $\mathbf{M}$ can be computed by the convolution product, $\mathbf{x}_p \otimes \left(\left(\mathbf{x}_q \odot \mathbf{D}_l\right)^{\mathrm{r}}\right)^*$, where the sequence $\mathbf{x}_q^{\mathrm{r}}$ can be obtained by reversing the order of the entries of $\mathbf{x}_q$. In other words, the entries in $\mathbf{m}_0$, $\mathbf{M}_{-G}$ and $\mathbf{M}_G$ can be computed by truncating $\mathbf{x}_p \otimes \left(\left(\mathbf{x}_q \odot \mathbf{D}_l\right)^{\mathrm{r}}\right)^*$ from the indices $-G$ to G. Besides, the convolution operation in the time domain corresponds to the product operation in the frequency domain. Thus, the FFT algorithm can be applied to obtain $\mathbf{m}_0$, $\mathbf{M}_{-G}$ and $\mathbf{M}_G$ by $12L\,(2N-1)\log_2\,(2N-1)$ complex multiplication operations at most. Computing $\|\mathbf{m}_0 - \mathbf{h}\|_2^2 + \|\mathbf{M}_G\|_F^2 + \|\mathbf{M}_{-G}\|_F^2$ takes $4L\,(2G+1)$ complex multiplication operations. So the computation cost of obtaining $f\,(\phi)$ is $O\,(LG + LN\log_2 N)$. As for $\nabla f\,(\phi)$, there are $2N$ entries. According to (4), for $k \geq 0$ (the same as $k < 0$), it should be noticed that

$$\begin{aligned}
r_{pq,l}\,(k) = &\sum_{m=k+1,m\neq n}^{N} x_p\,(m)\left[x_q\,(m-k)\,e^{j\phi_{\mathrm{d},l}(m-k)}\right]^* \\
&+ x_p\,(n)\left[x_q\,(n-k)\,e^{j\phi_{\mathrm{d},l}(n-k)}\right]^*
\end{aligned} \tag{18}$$

$$\begin{aligned}
\frac{\partial r_{pq,l}\,(k)}{\partial \phi_p\,(n)} &= \left[x_q\,(n-k)\,e^{j\phi_{\mathrm{d},l}(n-k)}\right]^*\frac{\partial x_p\,(n)}{\partial \phi_p\,(n)} \\
&= \left[x_q\,(n-k)\,e^{j\phi_{\mathrm{d},l}(n-k)}\right]^* j x_p\,(n)
\end{aligned} \tag{19}$$

and

$$\begin{aligned}
r_{pp,l}\,(k) = &\sum_{\substack{m=k+1 \\ m\neq n,m\neq n+k}}^{N} x_p\,(m)\left[x_p\,(m-k)\,e^{j\phi_{\mathrm{d},l}(m-k)}\right]^* \\
&+ x_p\,(n)\left[x_p\,(n-k)\,e^{j\phi_{\mathrm{d},l}(n-k)}\right]^* + x_p\,(n+k)\left[x_p\,(n)\,e^{j\phi_{\mathrm{d},l}(n)}\right]^*
\end{aligned} \tag{20}$$

$$\begin{aligned}
\frac{\partial r_{pp,l}\,(k)}{\partial \phi_p\,(n)} = &\left[x_p\,(n-k)\,e^{j\phi_{\mathrm{d},l}(n-k)}\right]^* j x_p\,(n) \\
&- x_p\,(n+k)\,e^{-j\phi_{\mathrm{d},l}(n)} j x_p^*\,(n)
\end{aligned} \tag{21}$$

The entries in $\mathbf{m}_0^{\mathrm{H}}\frac{\partial \mathbf{m}_0}{\partial \phi_p(n)}$, $\mathbf{M}_{-G}^{\mathrm{H}}\frac{\partial \mathbf{M}_{-G}}{\partial \phi_p(n)}$ and $\mathbf{M}_G^{\mathrm{H}}\frac{\partial \mathbf{M}_G}{\partial \phi_p(n)}$ can be obtained according to (19) and (21), and correspondingly, it takes $2L\,(2G+1) + 1$ complex multiplication operations. Thus, the computational cost of $\nabla f\,(\phi)$ is $O\,(NGL)$. The computational complexities to compute $f\,(\phi)$ and $\nabla f\,(\phi)$ using the original expression (13) and new expression (16) are shown in Table 1. Furthermore, using the gradient $\nabla f\,(\phi)$, the following IAG algorithm shown in Table 2 can be performed to obtain the sequences with good correlation properties and high Doppler tolerance. One thing should be pointed out is that for the classical gradient method, the line search rule is based on the Wolfe conditions which can ensure the stability of the iterations. In this paper, to improve the efficiency of the algorithm, a low computation complexity line search rule, named Armijo-rule, is used [28].

Table 1. The computational complexity using original expression (13) and new expression (16).

	Original (13)	New (16)
$f\,(\phi)$	$O\,(NGL)$	$O\,(LG + LN\log_2 N)$
$\nabla f\,(\phi)$	$O\,(N^2 GL)$	$O\,(NGL)$

Table 2. Steps for the IAG algorithm.

Step 0:	Set **X** to initial sequences, and fix the motion states of interest, which means the v_l and a_l should be determined.
Step 1:	Compute the gradient $\nabla f(\phi)$ according to (17), (19) and (21).
Step 2:	Renew the phases of the sequences using $\text{vec}\left(\mathbf{X}^{i+1}\right) = \text{vec}\left(\mathbf{X}^{i}\right) \cdot e^{-j\mu^i \nabla f(\phi)}$, where the step length μ^i can be obtained according to the line search algorithm and the index i represents the ith iteration [28].
Step 3:	Repeat **Step 1** and **Step 2** until a termination criterion is satisfied, e.g., $\left| f(\phi)^{i+1} - f(\phi)^{i} \right| \leq \varepsilon$, where ε is a predefined threshold.

4. Numerical Results

Here, we provide numerical examples to illustrate the performance of the proposed IAG algorithm. The state-of-the-art WeCAN algorithm proposed in [17] is considered for comparison. The metrics Peak Side-lobe Level (*PSL*), Integrated Side-lobe Level (*ISL*) and the Isolation (*I*) are used to evaluate the performance of the sequences which are defined as follows [4]

$$PSL = 20\log_{10} \frac{\max\left(\left|r_{pp,l}(k)\right|\right)}{\left|r_{pp,l}(0)\right|}, k \in [-G, -1] \cup [1, G], \tag{22}$$

$$ISL = \frac{2\sum_{k=1}^{G}\left(\left|r_{HH,l}(k)\right|^2 + \left|r_{VV,l}(k)\right|^2\right) + \sum_{k=-G}^{G}\left(\left|r_{HV,l}(k)\right|^2 + \left|r_{VH,l}(k)\right|^2\right)}{2\left|r_{HH,l}(0)\right|^2} \tag{23}$$

$$I = 20\log_{10} \frac{\max\left(\left|r_{pq,l}(k)\right|\right)}{\left|r_{pp,l}(0)\right|}, p \neq q \text{ and } k \in [-G, G]. \tag{24}$$

The running time is obtained using the Matlab 2016b version, running on a standard PC (with a 2.6 GHz Core i7 CPU and 4-GB RAM).

In the first numerical example, we consider the measurement of the stationary target, meaning the Doppler shift in (3) is zero. Simulation parameters are shown in Table 3. Assume the sequence length is $N = 256$ and the lag interval of interest is $G = 39$. The weight coefficients are $\{w_k\}_{k=-G}^{G} = 1$ for the IAG method. For the WeCAN algorithm, the parameter w_0 should be large enough to guarantee the stability of the algorithm. Thus, the weights are set as 30 while $k = 0$ and 1 while $k \neq 0$ for the WeCAN method. The predefined threshold of the predefined threshold is $\varepsilon = 10^{-14}$ for the two methods. To avoid unreasonable solution, the minimum iteration number is set to 1000 in the simulations.

Figure 1a,b show the convergence curves of the two algorithms. The convergence rate of the *PSL* and *I* of the IAG method is much faster than that of the WeCAN method. The *PSL* and *I* of the sequences designed by the IAG method achieve about -325 dB after 700 iterations, however, the corresponding number of iterations is over 10^5 for the WeCAN method. Further, Figure 2a,b illustrate the correlation level of the sequences designed by the two methods. The *PSL* and *I* of the IAG sequences are about -325 dB better than that of the WeCAN sequences with -265 dB under this situation. Meanwhile, the *ISL* is 8.92×10^{-32} for IAG and 8.92×10^{-32} for WeCAN, which indicates the energy of the sidelobes is lower for IAG sequences. In other words, the IAG sequences are suitable for observing the volume targets that extend over range areas. The consuming time and iteration number are shown in the Table 4. It can be observed that the average execution time of per iteration of the IAG method is shorter than that of the WeCAN method. As mentioned before, by the new expression (16) and the using of FFT, the consuming time for computing the objective function and the corresponding gradient becomes less, leading the average execution time of the IAG method shorter.

Meanwhile, the IAG method takes much less iterations than the WeCAN method. Thus, the total execution time is less than one in a hundred in comparison with that of the WeCAN method.

Table 3. Simulation parameters for the first example.

N	G	w_k	ε
256	39	$\{w_k\}_{k=-G}^{G} = 1$	10^{-14}

Table 4. Comparison between IAG and WeCAN.

	Iteration Number	Total Execution Time (s)	Execution Time per Iteration (s)
IAG	1000	196.7	0.194
WeCAN	112, 432	25, 881.5	0.231

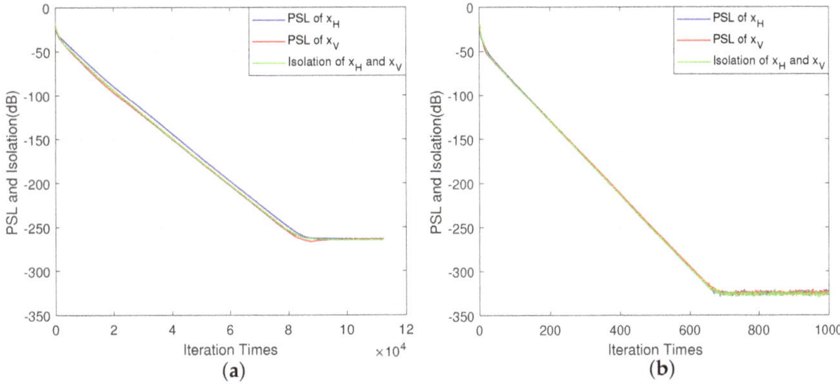

Figure 1. The convergence curves of the WeCAN approach and the IAG approach. (**a**) WeCAN; (**b**) IAG.

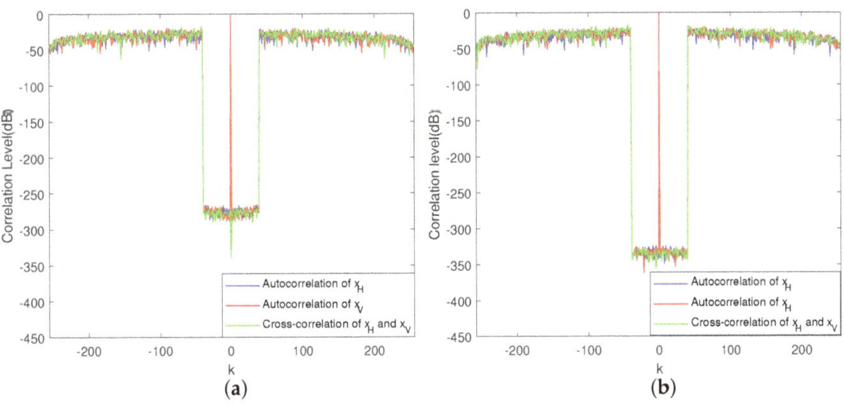

Figure 2. The correlation levels of the WeCAN approach and the IAG approach. (**a**) WeCAN; (**b**) IAG.

In the second numerical example, designing sequences for measuring the polarization features of the highly maneuvering target is simulated. In [29], Li has pointed out that typical highly maneuvering targets contain satellites, spacecrafts, etc. The velocity of these targets is usually about

Mach 10 or even higher. Meanwhile, the acceleration is about $10g$ (g is the acceleration of gravity). Thus, in the simulations, the velocity and the acceleration of the target are assumed to be within $v \in (2000, 3000)$ m/s and $a \in (50, 200)$ m/s^2, respectively. We uniformly discretize the velocity and acceleration intervals into L bins with the grid size $\Delta v = 50$ m/s and $\Delta a = 10$ m/s^2. Then the motion states of interest in (3) can be expressed as the column of the following matrix $\mathbf{S}$.

$$
\begin{aligned}
\mathbf{S} &= \begin{bmatrix} v_1 & v_2 & \cdots & v_L \\ a_1 & a_2 & \cdots & a_L \end{bmatrix}_{2 \times L} \\
&= \begin{bmatrix} 2000 & 2050 & \cdots & 3000 & \cdots\cdots & 2000 & 2050 & \cdots & 3000 \\ 50 & 50 & \cdots & 50 & \cdots\cdots & 200 & 200 & \cdots & 200 \end{bmatrix}_{2 \times L}
\end{aligned}
\tag{25}
$$

where the amount of concerning motion states L can be computed as

$$
L = \left(\frac{3000 - 2000}{50} + 1 \right) \cdot \left(\frac{200 - 50}{10} + 1 \right) = 336.
\tag{26}
$$

Other simulation parameters are shown in Table 5. The first four parameters are same as those in Table 3. Besides, taking typical simultaneous polarimetric radars, including the MERIC radar [30] and the CSU-CHILL radar [31], as references, the carrier frequency is set to $f_0 = 10$ GHz. The carrier wavelength in (3) can be computed by $\lambda = c/f_0$, where c is the speed of light. The time duration of the subpulse is $\tau = 5 \times 10^{-9}$ s. Then the IAG algorithm shown in Table 2 can be performed to obtain the sequences with good correlation properties under the motion states of interest. Figures 3 and 4 show the correlation properties of the designed sequences designed by the IAG and WeCAN algorithms through the metrics PSL and I, respectively. It clearly demonstrates that the correlation properties of the sequences designed by the IAG method are better than that of the sequences designed by the WeCAN method under the given motion states. Correspondingly, the PSL and I of the IAG sequences are under -53 dB, however, the same metrics of the WeCAN sequences are about -25 dB in this situation. The ISL as a function of velocity and acceleration is shown in Figure 5. It can be observed that under the same conditions, the ISL of the IAG sequences is about one in a thousand in comparison with that of the WeCAN sequences. The reason is that the influence of the velocity and the acceleration of the target on the correlation properties of the sequences, which is not considered in the WeCAN method, is taken into account in the objective function (11) of the IAG method. Besides, it can be observed that the metrics, including PSL, ISL and I, change slightly with the acceleration increasing. Since the time duration of the subpulse τ is set to 5×10^{-9} s in the simulation, the time length of the sequence is equal to $N\tau$, which is quite small in this situation, leading the slight change of the velocity. Thus, the properties of the sequences change slightly with the increasing of the acceleration.

Table 5. Simulation parameters for the second example.

N	G	w_k	ε	f_0	τ
256	39	$\{w_k\}_{k=-G}^{G} = 1$	10^{-14}	10 GHz	5×10^{-9} s

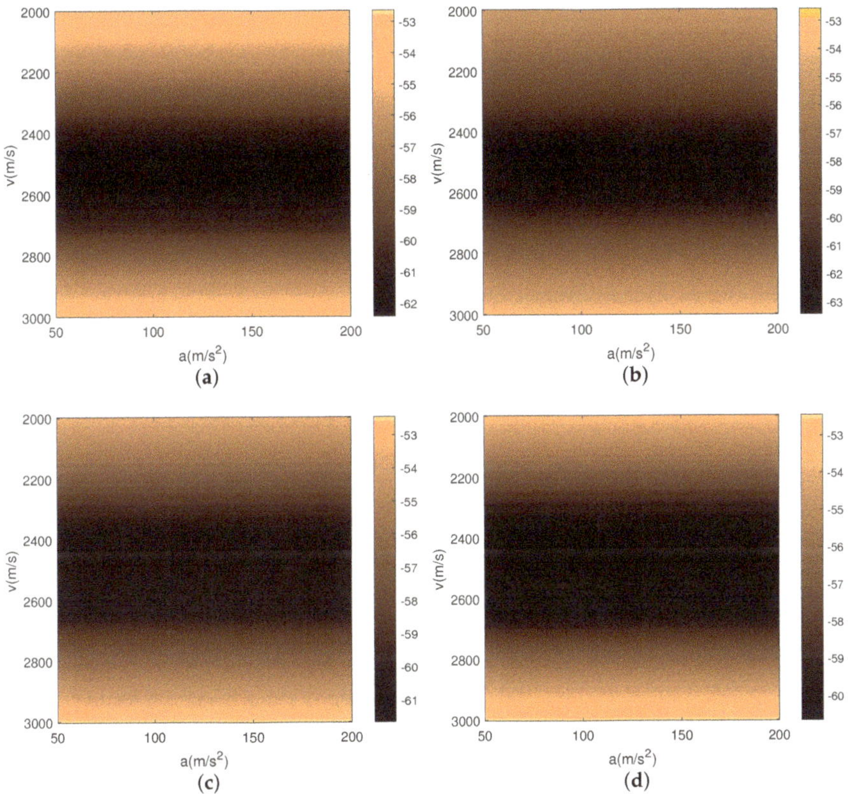

Figure 3. The *PSL* and *I* of the sequences designed by the IAG algorithm (dB). (**a**) The *PSL* of $r_{\mathrm{HH},l}$ under different motion states; (**b**) The *PSL* of $r_{\mathrm{VV},l}$ under different motion states; (**c**) The *I* of $r_{\mathrm{HV},l}$ under different motion states; (**d**) The *I* of $r_{\mathrm{VH},l}$ under different motion states.

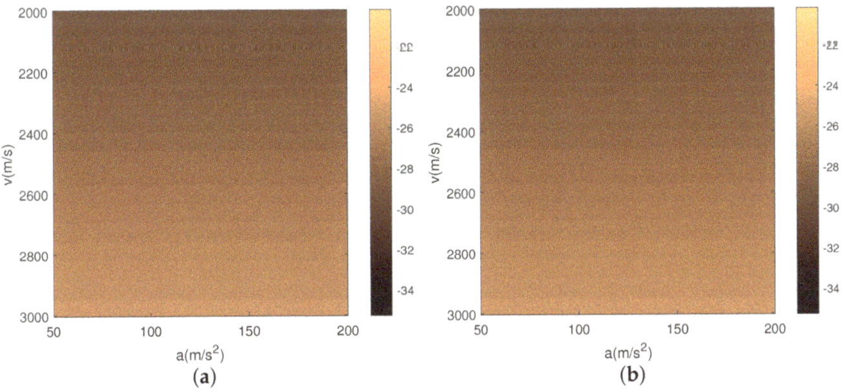

Figure 4. *Cont.*

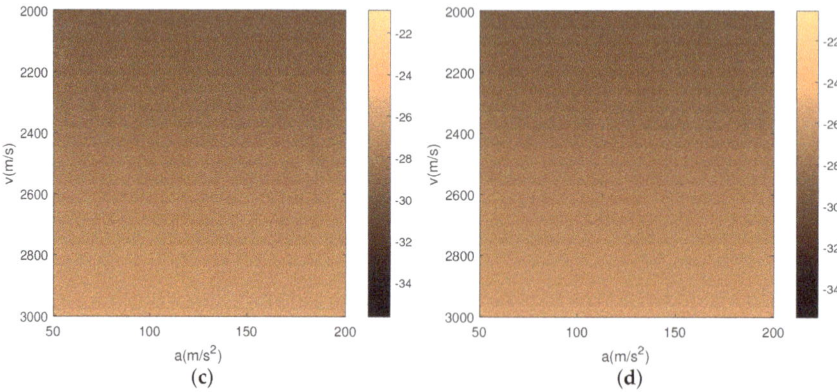

Figure 4. The *PSL* and *I* of the sequences designed by the WeCAN algorithm (dB). (**a**) The *PSL* of $r_{\mathrm{HH},l}$ under different motion states; (**b**) The *PSL* of $r_{\mathrm{VV},l}$ under different motion states; (**c**) The *I* of $r_{\mathrm{HV},l}$ under different motion states; (**d**) The *I* of $r_{\mathrm{VH},l}$ under different motion states.

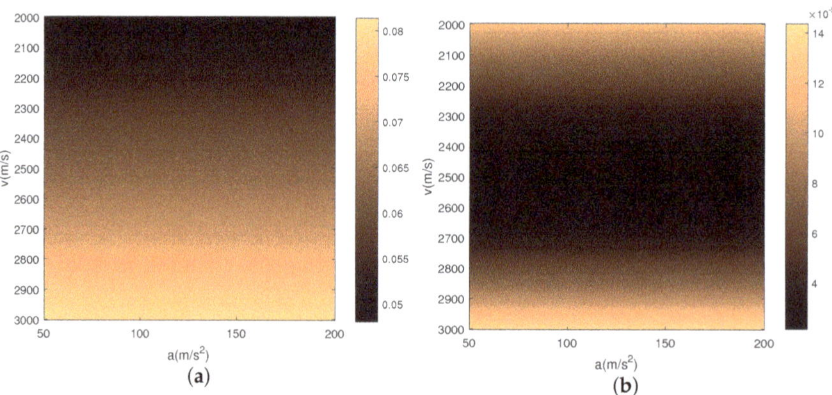

Figure 5. The *ISL* of the sequences designed by the WeCAN algorithm and the IAG algorithm under different motion states. (**a**) WeCAN; (**b**) IAG.

5. Conclusions

In this paper, we propose an iterative algorithm based on the gradient method to design the constant modulus sequences with good correlation and Doppler properties for simultaneous polarimetric radar. By transforming the objective function to a new expression, the computation complexities of the gradient and the objective function are reduced by using the FFT algorithm. Compared with the state-of-the-art WeCAN approach, the proposed IAG method has a better performance in terms of correlation sidelobes levels, Doppler tolerance and execution time.

Author Contributions: F.W. proposed the original idea of this paper; X.W. and C.P. conceived and designed the experiments; F.W. performed the experiments; F.W. and Y.L. analyzed the data; F.W. wrote the paper.

Funding: This research was funded by the National Natural Science Foundation of China, grant numbers 61490690, 61490694 and 61701512.

Conflicts of Interest: The authors declare no conflict of interest.

References

1. Wang, F.; Li, C.; Pang, C.; Li, Y.Z.; Wang, X.S. A Method for Estimating the Polarimetric Scattering Matrix of Moving Target for Simultaneous Fully Polarimetric Radar. *Sensors* **2018**, *18*, 1418. [CrossRef] [PubMed]
2. Yanovsky, F.J.; Russchenberg, H.W.J.; Unal, C.M.H. Retrieval of information about turbulence in rain by using Doppler-polarimetric radar. *IEEE Trans. Microw. Theory Tech.* **2005**, *53*, 444–450. [CrossRef]
3. Wang, F.; Pang, C.; Li, Y.Z.; Wang, X.S. Algorithms for Designing Unimodular Sequences with High Doppler Tolerance for Simultaneous Fully Polarimetric Radar. *Sensors* **2018**, *18*, 905. [CrossRef] [PubMed]
4. Giuli, D.; Fossi, M.; Facheris, L. Radar target scattering matrix measurement through orthogonal signals. *IEE Proc. F Radar Signal Process.* **1993**, *140*, 233–242. [CrossRef]
5. Song, J.; Babu, P.; Palomar, D.P. Sequence set design with good correlation properties via majorization-minimization. *IEEE Trans. Signal Process.* **2016**, *64*, 2866–2879. [CrossRef]
6. Cilliers, J.E.; Smit, J.C. Pulse Compression Sidelobe Reduction by Minimization of Lp-Norms. *IEEE Trans. Aerosp. Electron. Syst.* **2007**, *43*, 1238–1247. [CrossRef]
7. Mishra, K.V. Frequency Diversity Wideband Digital Receiver and Signal Processor for Solid-State Dual-Polarimetric Weather Radars. Ph.D. Thesis, Colorado State University, Fort Collins, CO, USA, 2012.
8. Bharadwaj, N.; Mishra, K.V.; Chandrasekar, V. Waveform considerations for dual-polarization Doppler weather radar with solid-state transmitters. In Proceedings of the IEEE International Geoscience and Remote Sensing Symposium, Cape Town, South Africa, 12–17 July 2009; pp. 267–270.
9. Boehmer, A.M. Binary pulse compression codes. *IEEE Trans. Inf. Theory* **1967**, *13*, 156–167. [CrossRef]
10. Baden, J.M. Efficient optimization of the merit factor of long binary sequences. *IEEE Trans. Inf. Theory* **2011**, *57*, 8084–8094. [CrossRef]
11. Xiong, T.; Hall, J. Construction of even length binary sequences with asymptotic merit factor 6. *IEEE Trans. Inf. Theory* **2008**, *54*, 931–935. [CrossRef]
12. Chu, D. Polyphase codes with good periodic correlation properties (corresp.). *IEEE Trans. Inf. Theory* **1972**, *18*, 531–532. [CrossRef]
13. Deng, H. Polyphase code design for orthogonal netted radar systems. *IEEE Trans. Signal Process.* **2004**, *52*, 3126–3135. [CrossRef]
14. Liu, B.; He, Z.S. Orthogonal Discrete Frequency-Coding Waveform Set Design with Minimized Autocorrelation Sidelobes. *IEEE Trans. Aerosp. Electron. Syst.* **2009**, *45*, 1650–1657. [CrossRef]
15. Mohammad, A.K.; Augusto, A.; Antonio, D.; Mohammad, M.N.; Mahmoud, M.H. A Coordinate-Descent Framework to Design Low PSL/ISL Sequences. *IEEE Trans. Signal Process.* **2017**, *65*, 5942–5956.
16. Nasrabadi, M.; Bastani, M. *A Survey on the Design of Binary Pulse Compression Codes with Low Autocorrelation*; INTECH: Rijeka, UK, 2010.
17. He, H.; Stoica, P.; Li, J. Designing unimodular sequences sets with good correlations-Including an application to MIMO radar. *IEEE Trans. Signal Process.* **2009**, *57*, 4391–4405. [CrossRef]
18. Wang, Y.C.; Dong, L.; Xue, X.; Yi, K.C. On the Design of Constant Modulus Sequences with Low Correlation Sidelobes Levels. *IEEE Signal Process. Lett.* **2012**, *16*, 462–465. [CrossRef]
19. Kretschmer, F.F.; Lewis, B.L. Doppler properties of poly-phase coded pulse compression waveforms. *IEEE Trans. Aerosp. Electron. Syst.* **1983**, *19*, 521–531. [CrossRef]
20. Pezeshki, A.; Calderbank, A.R.; Moran, W.; Howard, S.D. Doppler Resilient Golay Complementary Waveforms. *IEEE Trans. Inf. Theory* **2008**, *54*, 4254–4266. [CrossRef]
21. Pezeshki, A.; Calderbank, A.R.; Howard, S.D.; Moran, W. Doppler Resilient Golay Complementary Pairs for Radar. In Proceedings of the IEEE/SP, Workshop on Statistical Signal Processing, Madison, WI, USA, 26–29 August 2007; pp. 483–487.
22. Pezeshki, A.; Calderbank, A.R.; Moran, W.; Howard, S.D. Doppler Resilient Waveforms with Perfect Autocorrelation. *arXiv* **2007**, arXiv:cs/0703057. Available online: https://arxiv.org/abs/cs/0703057 (accessed on 12 July 2018).
23. Levanon, N. Noncoherent radar pulse compression based on complementary sequences. *IEEE Trans. Aerosp. Electron. Syst.* **2009**, *45*, 742–747. [CrossRef]
24. Schmidt, K. Complementary sets, generalized Reed-Muller codes, and power control for OFDM. *IEEE Trans. Inf. Theory* **2007**, *53*, 808–814. [CrossRef]

25. Mishra, K.V.; Eldar, Y.C. Sub-Nyquist channel estimation over IEEE 802.11ad link. In Proceedings of the International Conference on Sampling Theory and Applications, Tallinn, Estonia, 3–7 July 2017; pp. 355–359.
26. Cui, G.; Fu, Y.; Yu, X.; Li, J. Local Ambiguity Function Shaping via Unimodular Sequence Design. *IEEE Signal Process. Lett.* **2017**, *24*, 977–981. [CrossRef]
27. Arriaga, I.A.; Orozco, A.; Flores, J. Design of Unimodular Sequences with Good Autocorrelation and Good Complementary Autocorrelation Properties. *IEEE Signal Process. Lett.* **2017**, *24*, 1153–1157. [CrossRef]
28. Yuan, Y.X.; Sun, W.Y. *Optimization Theory and Method*; Science Press: Beiing, China, 2006; pp. 94–100.
29. Li, W.C. Radar Signal Parameter Estimation and Image Processing of High-speed Maneuvering Target. Ph.D. Thesis, National University of Defense Technology, Hunan, China, 2009.
30. Titinschnaider, C.; Attia, S. Calibration of the MERIC full-polarimetric radar: theory and implementation. *Aerosp. Sci. Technol.* **2003**, *7*, 633–640. [CrossRef]
31. Hubbert, J.; Bringi, V.N.; Carey, L.D.; Bolen, S. CSU-CHILL Polarimetric Radar Measurements from a Severe Hail Storm in Easter. *J. Appl. Meteorol.* **1996**, *37*, 749–775. [CrossRef]

Article

Unmanned Aerial Vehicle Recognition Based on Clustering by Fast Search and Find of Density Peaks (CFSFDP) with Polarimetric Decomposition

Hao Wu, Bo Pang *, Dahai Dai, Jiani Wu and Xuesong Wang

State Key Laboratory of Complex Electromagnetic Environment Effects on Electronics and Information System, National University of Defense Technology, Changsha 410073, China; jsczjtwh@126.com (H.W.); daidahai@nudt.edu.cn (D.D.); wujiani06@nudt.edu.cn (J.W.); wxs1019@vip.sina.com (X.W.)
* Correspondence: pangbo84826@126.com; Tel.: +86-137-3907-4648

Received: 8 October 2018; Accepted: 23 November 2018; Published: 1 December 2018

Abstract: Unmanned aerial vehicles (UAV) have become vital targets in civilian and military fields. However, the polarization characteristics are rarely studied. This paper studies the polarization property of UAVs via the fusion of three polarimetric decomposition methods. A novel algorithm is presented to classify and recognize UAVs automatically which includes a clustering method proposed in *"Science"*, one of the top journals in academia. Firstly, the selection of the imaging algorithm ensures the quality of the radar images. Secondly, local geometrical structures of UAVs can be extracted based on Pauli, Krogager, and Cameron polarimetric decomposition. Finally, the proposed algorithm with clustering by fast search and find of density peaks (CFSFDP) has been demonstrated to be better than the original methods under the various noise conditions with the fusion of three polarimetric decomposition methods.

Keywords: unmanned aerial vehicle; clustering methods; man-made targets; synthetic aperture radar (SAR); inverse synthetic aperture radar (ISAR); polarimetric decomposition

1. Introduction

Radar is an electronic system that utilizes electromagnetic waves to obtain the attribute information of the object which consists of velocity and range [1,2]. The potential targets that are often sensed with radar are vessels, aircraft, spacecraft, civilian vehicles, military tanks, terrain, and so on [3]. The automatic target recognition (ATR) [4–6] system takes advantage of the information reflected from the targets, by analyzing electromagnetic characteristic and extracting features to classify and recognize the objects. Radar automatic target recognition continues to advance with the development of radar hardware and information processing technology. Applications of broadband and multi-polarization technologies not only enhance the comprehension of target attribute information, but also bring a challenge of information processing [7].

The research on man-made targets has attained significant achievements in SAR/ISAR imaging and processing [8,9]. Dungan et al. focused on civilian vehicle radar data and investigated the signature of cars which had been adopted to the reconstruction of elevation [10]. In Reference [11], the study reveals the characteristic of civilian cars in SAR imaging with noise and classifies ten kinds of vehicles under various signal-to-noise ratio (SNR) conditions. Fuller et al. established a high-frequency model to solve the problems of parameter estimation and scatter classification in the spatial domain with simulated data [12]. In addition, the literature [13] established a forward approach of parametric scattering center model in the ATR system which had achieved great agreement between the simulation and experimental data. In the application of wide-angle SAR/ISAR, Jianxiong Zhou et al. reconstructed 3D tank and slice-like targets with a single elevation [14]. Dal-Jae Yun presented a 3D scattering center extraction algorithm with a fast Fourier transform-based scheme which has been applied to tank models [15]. Though great progress has been made for the analysis of those targets with simple structures, more work should be made for analyzing objects with complex structures or even with various materials like UAVs.

The polarization characteristic can be utilized to extract the structure characteristic of the man-made target. A Huynen Parametric decomposition technique has been applied to estimate the height profile of civilian vehicles and demonstrate the target structural mechanism [16]. Reference [17] developed a method of feature extraction and parametric estimation for scattering centers which can be utilized to simplify the description of the electromagnetic property of the target. Reference [18] proposed a novel algorithm to retrieve the geometrical structure of man-made targets in images with Cameron decomposition and had shown the potential to estimate the coordinates and the types of scattering centers. An original polarimetric coherent target decomposition method in SAR images which consists of data simulated by XPATCH software with different noise and the robustness is proved under strong noise [19]. Despite polarimetric decomposition methods bring convenience to the analysis of the target, the limitation of the single decomposition method needs to be addressed.

The study on UAVs, a typical kind of man-made targets, has become a frontier issue in recent years. As far as UAVs have been concerned, a motion compensation algorithm was proposed by Xing et al. to minimize the 3D motion error [20]. In Reference [21], micro-Doppler signatures of UAVs with rotating rotors were analyzed which could be utilized for the recognition of unmanned gyroplanes. X-band tracking radar was applied to measure the radar cross-section of the UAV in flight and the dynamic effects were considered [22]. In Reference [23], different micro-UAVs in flight were measured in various realistic environments and micro-Doppler characteristics were discussed. Pieraccini from the University of Florence reconstructed the 2D and 3D images of a small quadcopter with the blades not rotating [24]. However, there are few studies on the polarization characteristics of the UAV target and its component structure. This paper focuses on the polarization features of simulated and measured electromagnetic data of UAVs and a novel algorithm has been proposed to classify UAVs.

In this paper, on the basis of CFSFDP and polarimetric decomposition methods, we propose a novel clustering algorithm to recognize UAVs. The algorithm can not only obtain the physical structure, but also classify UAVs precisely. The advantages of the proposed method are shown as follows:

Classifying the UAVs automatically without prior information;

- Better performance under conditions of different SNRs in contrast with other algorithms;
- Extraction of the local geometrical structure of UAVs based on polarimetric decomposition.

The framework of contents in this paper is shown in Figure 1. Section 1 introduces the background of UAV recognition and the structure of this paper. Sections 2 and 3 present the imaging algorithms and coherent polarimetric decomposition methods; Section 4 introduces the methodology of clustering by fast search and find of density peaks; Section 5 provides the procedure of the proposed novel algorithm which is designed based on contents from Sections 2–4; Experimental results and conclusions are shown in Sections 6 and 7 respectively.

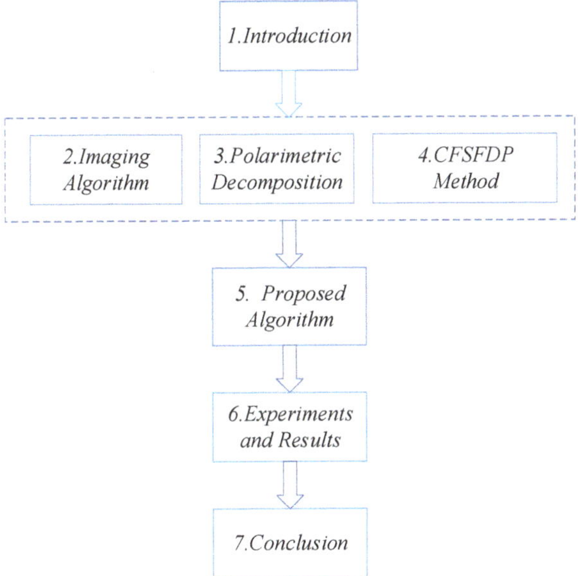

Figure 1. The framework of the contents.

2. Imaging Algorithm

Imaging is an indispensable operation before extracting features and recognizing targets. Two-dimensional Fourier transform and Convolution Back-projection (CBP) algorithm have been employed in this paper. These two methods are widely used in the research of ISAR imaging and have their advantages respectively. The process of 2D Fourier transform is straightforward while the consumptions of computing resources are considerably huge. The methodology of the convolution back-projection algorithm has been reviewed in Reference [25].

2.1. Polarimetric Matrix and the Signal Model of Radar Imaging

Polarimetric inverse synthetic aperture radar (PolISAR) obtains more information in the imaging scene than conventional radar. Conventional radar has only one or two polarimetric channels, whereas PolISAR has four channels. For PolISAR [26], in the horizontal and vertical polarization bases (H, V), the acquired fully polarimetric data could form a scattering matrix with representation as

$$S = \begin{bmatrix} S_{HH} & S_{HV} \\ S_{VH} & S_{VV} \end{bmatrix} \tag{1}$$

where S_{HV} is the backscattered coefficient from the vertical transmitting and horizontal receiving polarization, S_{HH} is the backscattered coefficient transmitted and received both from the horizontal polarization. The other backscattered coefficients are similarly defined.

According to the definition of radar cross section σ_{pq} and the reciprocity theorem [27]

$$S_{HV} = S_{VH} \tag{2}$$

$$\sigma_{pq} = 4\pi|S_{pq}|^2 \qquad p,q = H,V \tag{3}$$

2.2. 2D Fourier Transform Algorithm

In Cartesian coordinates, $g(x, y)$ denotes the ground reflectivity function and the 2D Fourier transform [28] is defined as follows:

$$G(X, Y) = \int_{-\infty}^{+\infty} \int_{-\infty}^{+\infty} g(x, y) e^{-j(Xx + Yy)} dx dy \tag{4}$$

And

$$g(x, y) = \frac{1}{4\pi^2} \int_{-\infty}^{+\infty} \int_{-\infty}^{+\infty} G(X, Y) e^{-j(Xx + Yy)} dX dY \tag{5}$$

Let (w, θ) represent the polar coordinates in the (u, v) plane. $G(w, \theta)$ denotes the value of $G(X, Y)$ along a line at an angle θ with y-axis.

The reconstructed image $\hat{g}(x, y)$ can be expressed as

$$\hat{g}(x, y) = \int_{w_1}^{w_2} \int_{-(\theta_m/2)}^{\theta_m/2} G(w, \theta) e^{-jw(x \cos \theta + y \sin \theta)} d\theta dw \tag{6}$$

$$\hat{g}(r, \theta) = \int_{w_1}^{w_2} \int_{-(\theta_m/2)}^{\theta_m/2} G(w, \theta) e^{-jwr} d\theta dw \tag{7}$$

3. Polarimetric Decomposition Methods

The physical structural characteristics of the target can be obtained effectively by polarimetric decomposition. The polarimetric decomposition methods of the target are mainly divided into coherent decomposition and incoherent decomposition. The incoherent decomposition discusses the characteristics of the change of the target under time-varying conditions, and the model is quite complex. This paper focuses on the UAV target with coherent decomposition methods. Coherent decomposition methods consist of Pauli decomposition, Krogager decomposition, and Cameron decomposition.

3.1. Pauli Decomposition

The Pauli decomposition decomposes the scattering matrix S into three simple scattering mechanisms that represent odd-bounce scattering, even-bounce scattering, and asymmetric component scattering [29–31]. The even-bounce scattering can be further divided into dihedral targets with azimuth angles of $0°$ and $45°$.

The scattering matrix of the target can be rewritten as

$$S = \frac{\alpha}{\sqrt{2}} \begin{bmatrix} 1 & 0 \\ 0 & 1 \end{bmatrix} + \frac{\beta}{\sqrt{2}} \begin{bmatrix} 0 & 1 \\ 1 & 0 \end{bmatrix} + \frac{\chi}{\sqrt{2}} \begin{bmatrix} 1 & 0 \\ 0 & -1 \end{bmatrix} + \frac{\delta}{\sqrt{2}} \begin{bmatrix} 0 & -j \\ j & 0 \end{bmatrix} \tag{8}$$

As mentioned in the previous section, the polarimetric scattering matrix consists of four elements: S_{HH}, S_{HV}, S_{VH} and S_{VV}. When the reciprocity condition is satisfied, $S_{HV} = S_{VH}$ can be obtained. The four complex numbers are α, β, χ and δ, they are thus given by

$$\alpha = \frac{S_{HH} + S_{VV}}{\sqrt{2}} \tag{9}$$

$$\beta = \frac{S_{HV} + S_{VH}}{\sqrt{2}} \tag{10}$$

$$\chi = \frac{S_{HH} - S_{VV}}{\sqrt{2}} \tag{11}$$

$$\delta = j\frac{S_{HV} - S_{VH}}{\sqrt{2}} \tag{12}$$

When the limit of monostatic and reciprocity is satisfied, (8) can be simplified as

$$S = \frac{\alpha}{\sqrt{2}}\begin{bmatrix} 1 & 0 \\ 0 & 1 \end{bmatrix} + \frac{\beta}{\sqrt{2}}\begin{bmatrix} 0 & 1 \\ 1 & 0 \end{bmatrix} + \frac{\chi}{\sqrt{2}}\begin{bmatrix} 1 & 0 \\ 0 & -1 \end{bmatrix} \tag{13}$$

3.2. Krogager Decomposition

The Krogager decomposition [29,30] characterizes the scattering electromagnetic properties of the complex object with three basic scattering mechanisms, namely, the sphere, the dihedral with azimuth angle φ, and the helix.

$$S = e^{j\psi}\left\{ k_H S_{he} + e^{j\psi_s} k_S S_{sp} + k_D S_{di} \right\} \tag{14}$$

where

$$S_{sp} = \begin{bmatrix} 1 & 0 \\ 0 & 1 \end{bmatrix}, S_{he} = \begin{bmatrix} 1 & \mp j \\ \mp j & -1 \end{bmatrix}, S_{di} = \begin{bmatrix} \cos 2\varphi & \sin 2\varphi \\ \sin 2\varphi & -\cos 2\varphi \end{bmatrix}.$$

S_{sp}, S_{he}, and S_{di} represent the scattering matrices corresponding to a sphere, helix, and dihedral. k_S, k_D and k_H, are denoted as the contribution of three scattering mechanisms of the specific target. φ is the azimuth of the dihedral, ψ represents the absolute phase, ψ_s shows the relative phase difference between the scattering component corresponding to the minimum unit sphere and other scattering mechanisms.

After the operation under the circular polarization [29], the scattering matrix S is represented as

$$S = e^{j\psi}\begin{bmatrix} e^{j\psi_s} k_S + k_H e^{\mp j2\varphi} + k_D \cos 2\varphi & jk_H e^{\mp j2\varphi} \pm k_D \sin 2\varphi \\ jk_H e^{\mp j2\varphi} \pm k_D \sin 2\varphi & e^{j\psi_s} k_S - k_H e^{\mp j2\varphi} - k_D \cos 2\varphi \end{bmatrix} \tag{15}$$

The Krogager decomposition decomposes the target into three basic scattering mechanisms that simplify the interpretation and analysis of the target physical properties.

3.3. Cameron Decomposition

Cameron decomposition of radar targets has two characteristics: symmetry and reciprocity [31,32]. It is different from Pauli decomposition. Pauli decomposition decomposes the target scatter into several simple scattering mechanisms. Cameron decomposition first decomposes the scattering matrix into reciprocal and nonreciprocal parts. The reciprocal part is then divided into symmetrical and asymmetrical parts based on symmetry. The expression of Cameron decomposition is as follows:

$$S = a\left\{ \hat{S}_{nonrec} \sin \theta_{rec} + \cos \theta_{rec} \left\{ \hat{S}_{sym}^{max} \cos \tau_{sym} + \hat{S}_{sym}^{min} \sin \tau_{sym} \right\} \right\} \tag{16}$$

where

$$a = \|\vec{S}\|_2^2 = Span(S) \tag{17}$$

The angle θ_{rec} represents the proportion of reciprocal scatter and τ_{sym} shows the proportion of symmetric part of the scatter. The vector $\hat{S}_{nonrec}$ is the representation of the normalized non-reciprocal scatter, $\hat{S}_{sym}^{max}$ corresponds to the portion of the normalized symmetric scatter, $\hat{S}_{sym}^{min}$ is denoted as the vector form of the normalized asymmetric scatter.

Cameron decomposition first uses the Pauli decomposition to weight the sum of the target's scattering matrix, then transforms the scattering matrix into the vector form [33]:

$$\vec{S} = \frac{\alpha}{\sqrt{2}}\begin{bmatrix} 1 \\ 0 \\ 0 \\ 1 \end{bmatrix} + \frac{\beta}{\sqrt{2}}\begin{bmatrix} 1 \\ 0 \\ 0 \\ -1 \end{bmatrix} + \frac{\chi}{\sqrt{2}}\begin{bmatrix} 0 \\ -1 \\ 1 \\ 0 \end{bmatrix} + \frac{\delta}{\sqrt{2}}\begin{bmatrix} 0 \\ 1 \\ 1 \\ 0 \end{bmatrix} \tag{18}$$

The reciprocity of the target scatter is given by θ_r and it decreases with the increasing value.

$$\theta_r = \arccos\left(\frac{\|P_r\hat{S}\|}{\|\hat{S}\|}\right) \tag{19}$$

where $P_{rec} = I - P_C,\ P_C = \hat{S}_C \cdot \hat{S}_C{}^T,\ \hat{S} = \frac{\vec{S}}{\|\vec{S}\|}$.

The process of Cameron Decomposition is shown in Figure 2. The specific type can be determined after the reciprocal test, symmetric test, and the distance classification. The different types include dihedral, trihedral, cylinder, dipole, narrow diplane, quarter wave device, right helix, left helix, asymmetric scatter and non-reciprocal scatter. More details of Cameron decomposition can be obtained from the literature [31–34].

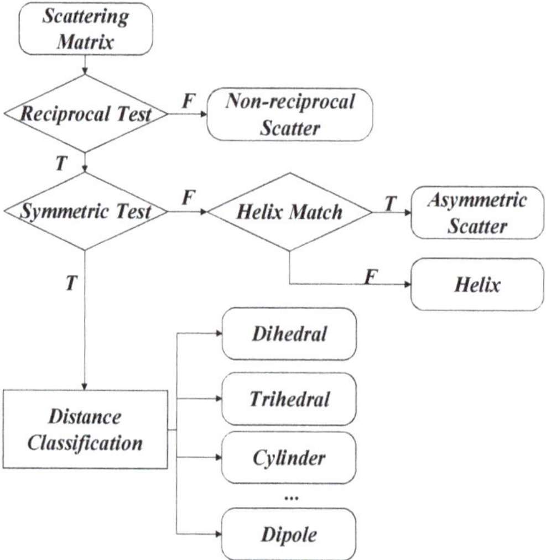

Figure 2. The classification flow chart of Cameron Decomposition [31].

4. Clustering by Fast Search and Find of Density Peaks (CFSFDP)

Clustering by fast search and find of density peaks (CFSDP) algorithm [35] assumes that cluster centers are surrounded by neighbors with lower local density and that they are at a relatively large distance from the point with higher local density. Local density and the distance from any points with higher local density are required to compute.

The definition of a local density of point i is defined by the following formula

$$\rho_i = \sum_{j \in I_s, j \neq i} \exp\left(-(d_{ij}/d_c)^2\right) \tag{20}$$

where d_c represents the cutoff distance, and the distance from point i to point j is d_{ij}. The set of the points is defined as I_s where all the points are included.

The minimum distance between the point i and the other points with higher density is defined as δ_i:

$$\delta_i = \begin{cases} \min(d_{ij}), \rho_j > \rho_i \\ \max(d_{ij}), otherwise \end{cases} \tag{21}$$

when the point is with the highest density, δ_i is defined as $\max(d_{ij})$. δ_i tends to be larger than the typical nearest neighbor distance when the points are local maxima in density. The steps for the CFSFDP algorithm are shown in Table 1.

Table 1. The steps for the CFSFDP algorithm [35–37].

Algorithm Clustering by Fast Search and Find of Density Peaks (CFSFDP)

---

1. Input: $d_{ij}, i < j, i, j \in I_S$
2. Initialization: $d_c, n_i = 0, i \in I_S$
3. The computation of $\{\rho_i\}_{i=1}^N$ and $\{q_i\}_{i=1}^N$ (subscripts of $\{\rho_i\}_{i=1}^N$ in descending order).
4. For $i = 2, 3, \cdots, N$
$\{\delta_{q_i} = d_{\max};$
For $j = 1, 2, \cdots, i - 1$
$\{$IF $(dist(X_{q_i}, X_{q_j}) < \delta_{q_i})$
$\{\delta_{q_i} = dist(X_{q_i}, X_{q_j}); n_{q_i} = q_j;\}$
$\}$
$\}$
5. $\delta_{q_i} = \max_{j \geq 2}\{\delta_j\};$

6. Computation of the cluster centers $\{m_j\}_{j=1}^{n_c}$ where n_c represents the number of clustering centers

7. $c_i = \begin{cases} k, if \ X_i \ is \ the \ cluster \ center \ k \\ -1, otherwise \end{cases}$
8. For $j = 1, 2, \cdots, N$
$\{$
IF $\left(c_{q_j} = -1\right)\{c_{q_j} = c_{n_{q_j}}\}$
$\}$
9. Initialization: $h_l = 0, l \in I_S$
10. The computation of thresholds of mean local density for cluster centers $\{\rho_l^b\}_{l=1}^{n_c}$
11. Label cluster halos
For $l = 1, 2, \cdots, N$
$\{$
IF $\left(\rho_l < \rho_{c_l}^b\right)\{h_l = 1\}$
$\}$
end

---

5. The Flowchart of the Proposed Algorithm

The proposed algorithm shown in Figure 3 is organized as follows:

Step 1: Full polarimetric data of the UAVs are measured in an anechoic chamber or is obtained by EM simulations. Different from the single polarimetric data, more detail information includes the structure that enables us to classify and recognize the targets.

Step 2: Radar imaging algorithm affects the subsequent processing procedure, 2D Fourier transform and Convolution Back-projection Algorithm is applied to form the imaging. The different images produced by two algorithms are compared for the selection of the better one.

Step 3: In order to simplify the processing procedure, strong scattering points (point cloud) in the image are extracted based on the amplitude. It not only reduces the computational load but also ascends the following operation speed. Additionally, these points are conducive to the estimation of size information.

Step 4: Length, width, and the oriental angle of the UAV targets are obtained according to the point cloud by step 3. Size information is utilized for the coarse classification of the UAV targets (mainly for excluding objects of unusual size).

Step 5: Pauli, Krogager, and Cameron decomposition methods are applied to study the property of the targets. Details of the different parts of the unmanned aerial vehicle are analyzed and discussed, the decomposition results show the structural feature at various oriental angles.

Step 6: The fusion of multiple polarimetric decomposition methods and azimuth angles are adopted to realize the fine classification. We utilize CFSFDP to classify UAV targets automatically without training and testing.

Step 7: Finally, the results of the novel clustering algorithm based on polarimetric decomposition are discussed and studied.

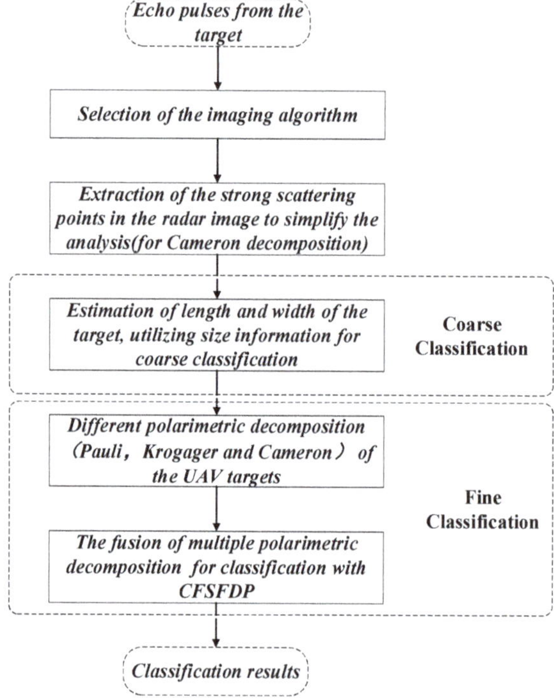

Figure 3. The flowchart of the classification algorithm with CFSFDP for UAVs.

6. Experiments and Results

This paper focuses on two kinds of UAVs: "Frontier" UAV and "MQ-1" UAV. The wide-band frequency ranges from 8 to 12 GHz and the frequency interval is 20 MHz. A linear frequency modulation signal is utilized in radar measurement with horn antennas. The "Frontier" UAV mainly consists of a composite material which contains plastic and metal. The "MQ-1" UAV which is measured by electromagnetic software is composed of metal.

In this paper, the head of the UAV is toward the +X axis and the back of the fuselage is toward the +Z axis. Figure 4a shows the picture of the "Frontier" UAV taken in the microwave anechoic chamber, (b) reveals the computer-aided design (CAD) model of the "MQ-1" UAV. The experimental data of the "Frontier" UAV and electromagnetic simulation data of the "MQ-1" UAV are thoroughly studied and investigated. What should be emphasized is that the electromagnetic model of the "MQ-1" UAV is 1:4 the scale of a real UAV. The length of the simulated "MQ-1" UAV is 4.2 m and the width of that is 2.1 m. The length and width of the "Frontier" UAV are 2.7 m and 2.4 m, respectively.

The imaging results of the "Frontier" UAV with the FFT and CBP algorithms are shown in Figure 5. Figure 5a–c shows images of the "Frontier" UAV under HH, HV, and VV polarization respectively, which utilizes two-dimension Fourier transform. Similarly, the full-polarization images of the "Frontier" UAV with the CBP algorithm are indicated in Figure 5d–f.

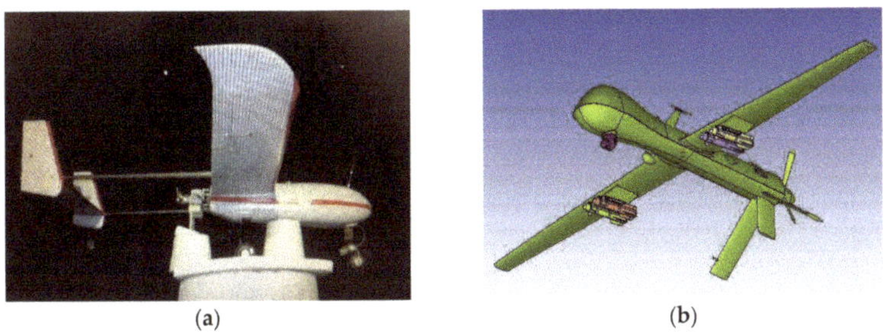

Figure 4. The two kinds of UAVs studied in this paper. (**a**) "Frontier" UAV; (**b**) "MQ-1" UAV.

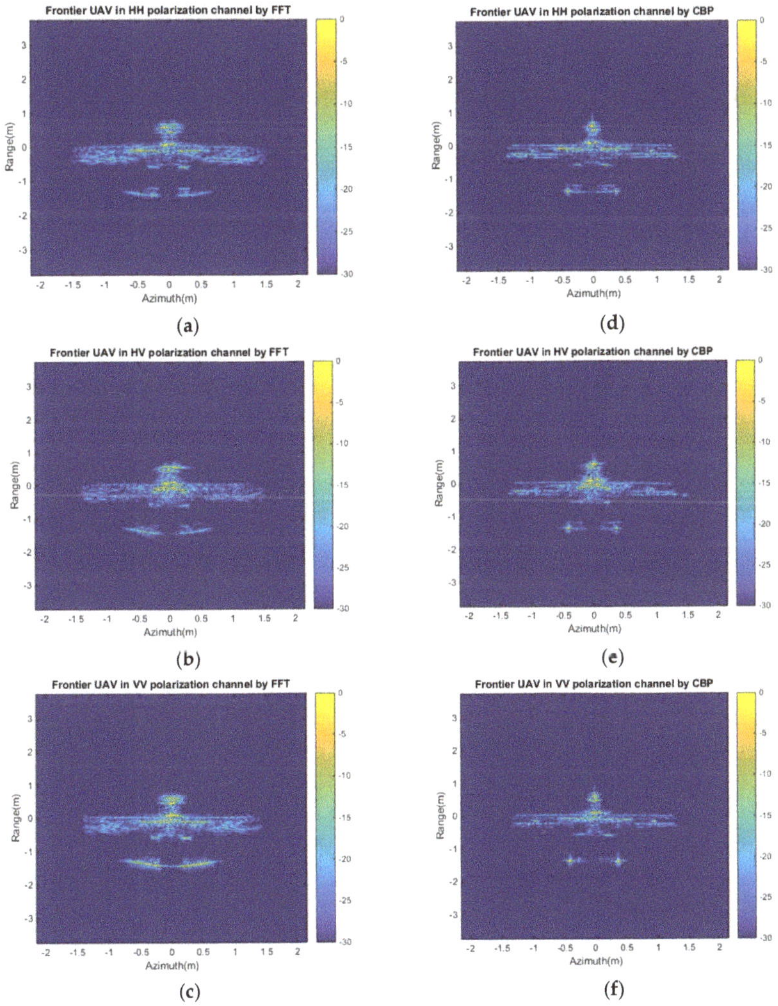

Figure 5. The full-polarization imaging results of the "Frontier" UAV with the FFT and CBP algorithm (**a**) "Frontier" in HH polarization by FFT; (**b**) "Frontier" in HV polarization by FFT; (**c**) "Frontier" in VV polarization by FFT; (**d**) "Frontier" in HH polarization by CBP; (**e**) "Frontier" in HV polarization by CBP; (**f**) "Frontier" in VV polarization by CBP.

Figure 6 shows the full-polarization images of the "MQ-1" UAV with the two algorithms. From Figures 5 and 6, the difference between images of "Frontier" and "MQ-1" with different algorithms are obvious. The 2D Fourier transform can roughly describe the outline of the UAV under HH and VV imaging conditions, however, the shape of "MQ-1" UAV components cannot be identified accurately with the HV channel such as the wing and empennage. The image results of the cross-polarization channel are worse than the other two channels, and the overall structure of "MQ-1" UAV can hardly be identified. Images utilizing the convolution back-projection algorithm show better effect than images with 2D Fourier transform. With the information given by HH and VV channel, not only can the basic outline of UAV be extracted, but the size and structural information can also be obtained. Although the intensity of HH polarization is weaker than that of the other two channels, the missile and tail parts are still able to be recovered.

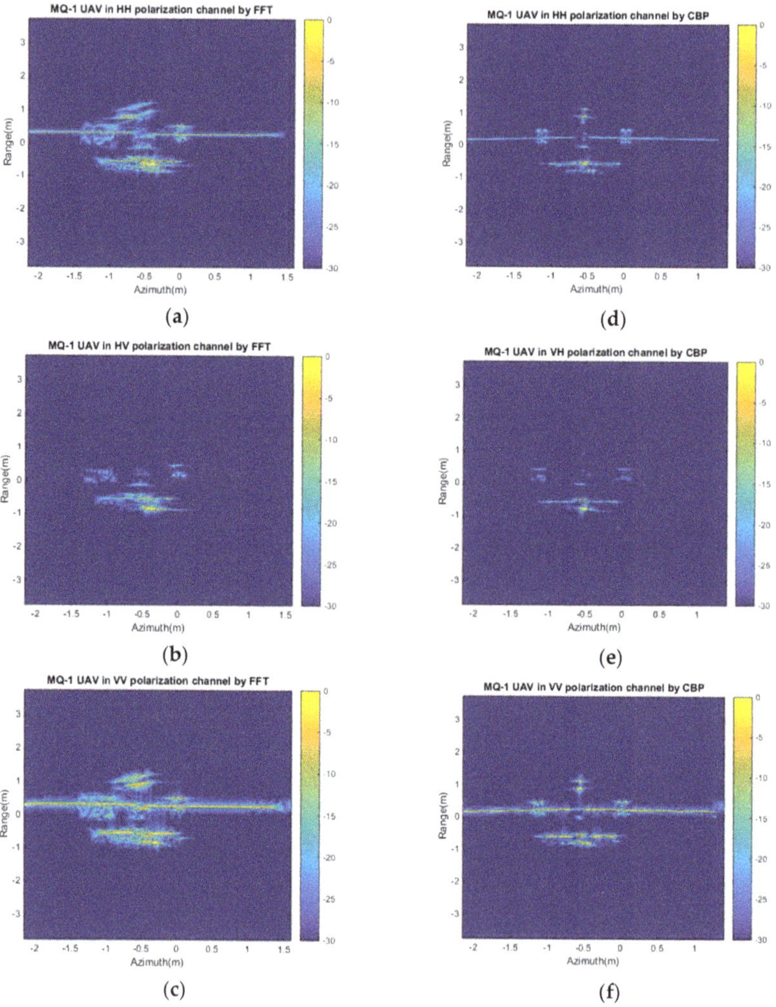

Figure 6. The full-polarization imaging results of "MQ-1" UAV with FFT and CBP algorithm. (**a**)"MQ-1" in HH polarization by FFT; (**b**) "MQ-1" in HV polarization by FFT; (**c**) "MQ-1" in VV polarization by FFT; (**d**) "MQ-1" in HH polarization by CBP; (**e**) "MQ-1" in HV polarization by CBP; (**f**) "MQ-1" in VV polarization by CBP.

In order to show the differences between the 2D Fourier transform and CBP algorithm meticulously, empennage and missile parts of two UAVs are selected. The empennage is an arrangement of stabilizing surfaces at the tail of the UAV. As shown in Figure 7a,b, empennage generates a relative large distortion by the 2D FFT method, while the image clearly shows the details of the component with the CBP algorithm. Figure 7c,d reveal the missile part of "MQ-1" UAV in HH polarization channel with FFT and CBP algorithms where the details of missiles of "MQ-1" can be obtained in the image generated by CBP algorithm. Nevertheless, the terrible image formation of missiles with FFT algorithm is difficult to be recognized. Based on the results, the CBP algorithm is chosen to carry out the follow-up study.

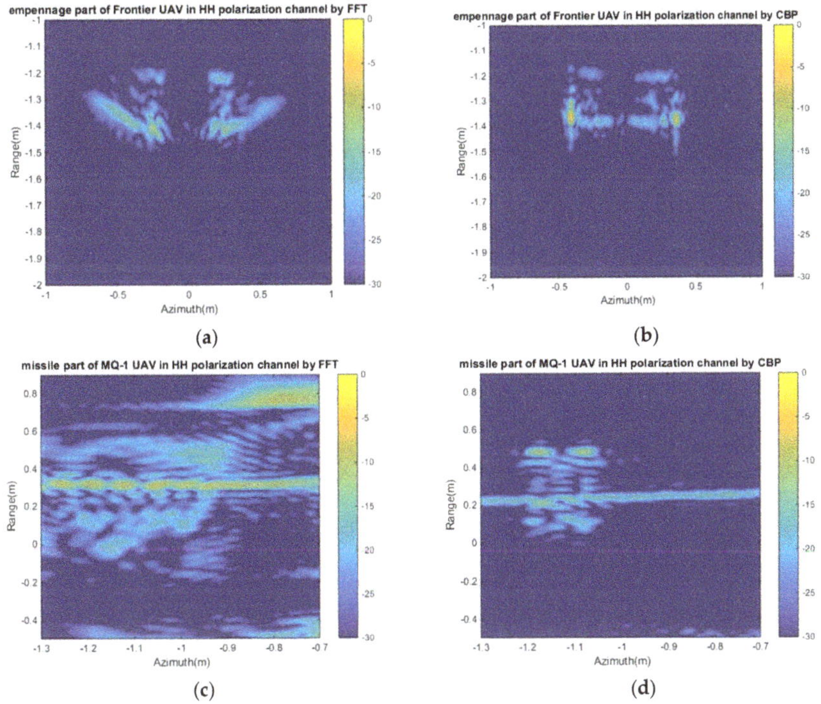

Figure 7. The empennage and missile parts of "MQ-1" UAV with FFT and CBP algorithm. (**a**) Empennage part of "MQ-1" in HH polarization by FFT; (**b**) Empennage part of "MQ-1" in HH polarization by CBP; (**c**) Missile parts of "MQ-1" in HH polarization by FFT; (**d**) Missile parts of "MQ-1" in HH polarization by CBP.

Firstly, the extraction of strong scattering points of the "Frontier" UAV is carried out which are chosen based on the magnitude of the imaging results and the application of the sliding window (20 by 20). Two hundred strong scattering points are filtered out from the image of the "Frontier" UAV that are utilized to estimate the length and width information. Additionally, these scattering points will be employed to inverse structural information with polarimetric decomposition methods. Principal Component Analysis (PCA) is an effective method for calculating target size in a two-dimensional Cartesian coordinate space and estimating the length and width of "Frontier" UAV with the point clouds. Figure 8a shows the extraction of scattering centers in the image of "Frontier" UAV which is represented by blue-black diamonds. The scattering centers cover the entire fuselage of the UAV which provides the basis for the length and width estimation of the UAV. The directions of two straight lines in Figure 8b represent the orientations of length and width according to the point clouds in Figure 8a.

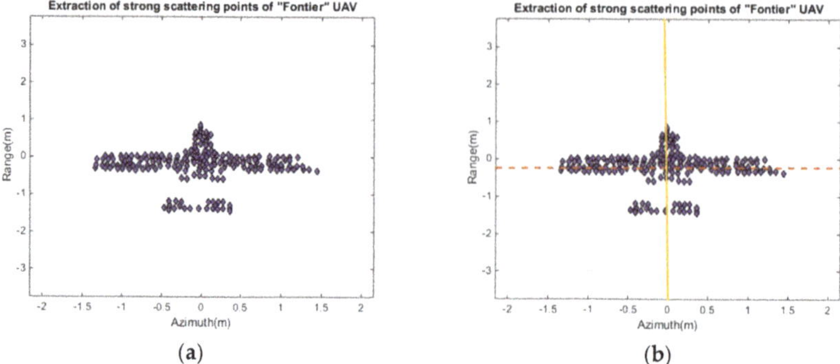

(a) (b)

Figure 8. The extraction of scattering centers and the size estimation of "Frontier" UAV. (**a**) Extraction of scattering centers in the image of "Frontier" UAV; (**b**) Orientations of length and width of "Frontier" UAV.

The estimated length and width of "Frontier" and "MQ-1" UAVs with an azimuth ranging from 0–9° are shown in Figure 9. The estimated width of "Frontier" UAV remains steady at about 2.3 m with the azimuth rotation. The estimated length of "Frontier" UAV fluctuates between 2.5 and 3 m. The estimated length of "MQ-1" UAV is more than 4 m which is greater than that of the "Frontier" UAV.

The estimated length and width of "Frontier" and "MQ-1" UAVs are applied for the coarse classification of UAVs which can be further adapted for more types of UAVs.

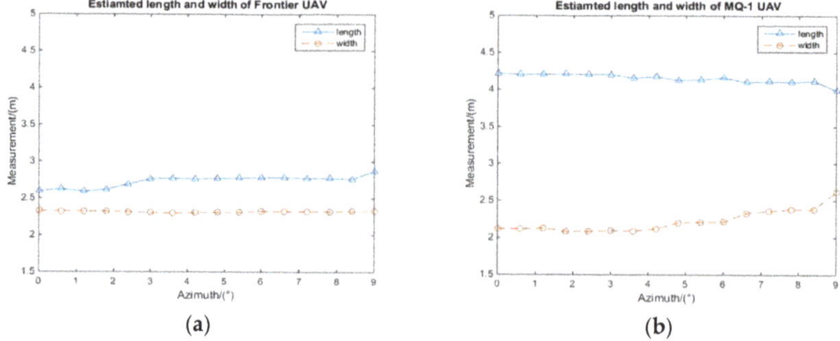

(a) (b)

Figure 9. The estimated length and width of the "Frontier" and "MQ-1" UAVs. (**a**) Estimated length and width of the "Frontier" UAV; (**b**) Estimated length and width of the "MQ-1" UAV.

Pauli decomposition can mainly decompose the UAV target into three scattering mechanisms: odd scattering mechanism and two kinds of even scattering mechanism with an azimuth angle of 0 degrees and 45 degrees. Pauli decomposition of "MQ-1" UAV with an azimuth of 0° is shown in Figure 10. The odd scattering mechanism of "MQ-1" is the main mechanism when the azimuth is 0°. The outline and components of the "MQ-1" UAV can be clearly reflected in decomposition. The power of the odd scattering mechanism accounts for more than 65% of that of the whole image. The empennage of "MQ-1" UAV is shown in an even scattering mechanism with an azimuth of 45° while other components are not displayed in Figure 10c. Even the scattering mechanism with an azimuth of 0° also shows the body of the UAV, however, the intensity and ability that displaying details are weaker than that of odd scattering mechanism. In order to study the polarization characteristics in depth mathematically, the proportions of the scattering mechanisms are analyzed with different azimuth angles.

Figure 11 reveals the proportions of the Pauli decomposition scattering mechanisms of "MQ-1" and "Frontier" UAVs. When the azimuth angle ranges from 0 to 25 degrees, odd scattering is the major scattering mechanism of the two UAVs. With the increase of the azimuth angle of the "MQ-1" UAV, the even scattering mechanism with an azimuth of 45° increases gradually. The even scattering mechanism with an azimuth angle of 0° is greater than the even scattering mechanism with an azimuth of 45° in the image of the "MQ-1" UAV. With regard to the "Frontier" UAV, three proportions of scattering mechanisms have a stable fluctuation from −5 to 25 degrees and odd scattering is the main scattering mechanism which is greater than the other two kinds of mechanisms. The decomposition results of the two UAV targets in the Pauli decomposition show the difference of the polarization characteristics between "MQ-1" and "Frontier" UAVs.

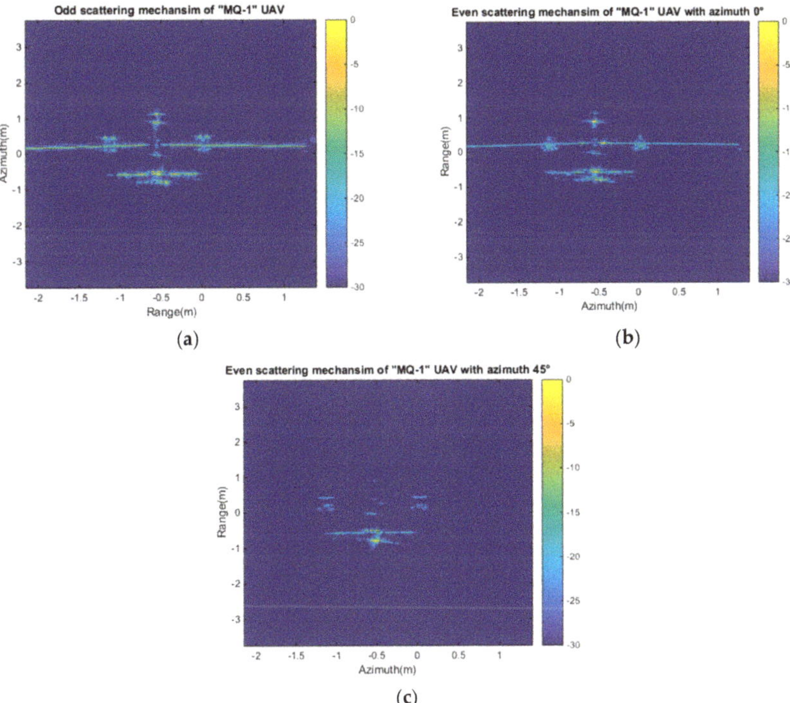

Figure 10. The Pauli decomposition of "MQ-1" UAV with an azimuth of 0°. (**a**) Odd scattering mechanism of "MQ-1" UAV; (**b**) Even Scattering mechanism with an azimuth of 0° of "MQ-1" UAV, (**c**) Even Scattering mechanism with an azimuth of 45° of "MQ-1" UAV.

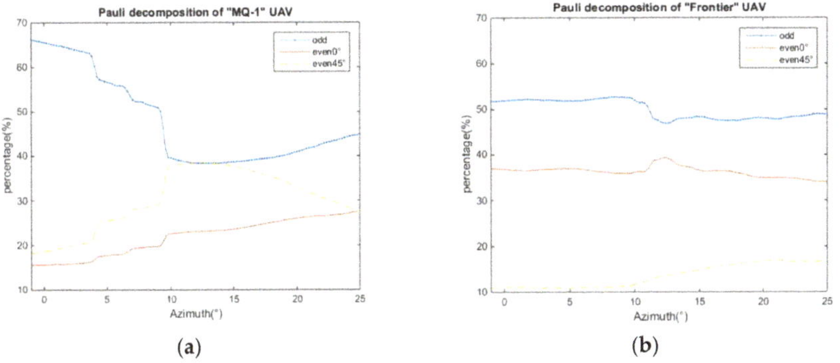

Figure 11. The proportions of scattering Pauli decomposition mechanisms of "MQ-1" and "Frontier" UAVs. (**a**) "MQ-1" UAV; (**b**) "Frontier" UAV.

As shown in Figure 12, the sphere, diplane, and helix scattering mechanisms are utilized to analyze the "MQ-1" and "Frontier" UAVs. The elements of Krogager decomposition of "Frontier" UAV are relatively stable, while the results of the Krogager decomposition of "MQ-1" UAV vary greatly. For "MQ-1" UAV, the mechanisms of the sphere and diplane scattering are greater than that of helix scattering. The scattering mechanism of helix ascends with the increase of the azimuth angle.

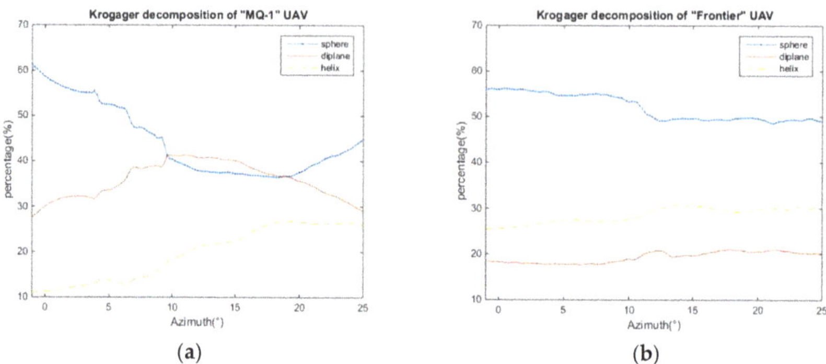

Figure 12. The proportions of the scattering mechanisms of Krogager decomposition of "MQ-1" and "Frontier" UAVs. (**a**) "MQ-1" UAV; (**b**) "Frontier" UAV.

The elements of "MQ-1" and "Frontier" UAVs with Cameron decomposition are shown in Figure 13. The wings of "MQ-1" UAV with a 0-degree azimuth angle are presented mainly by the scattering mechanism of Trihedral. The strong scattering points in "Frontier" UAV are mainly centered at the front of the body. Quarter wave device scattering emerges in the front wing of "Frontier" UAV which may be caused by the composite material.

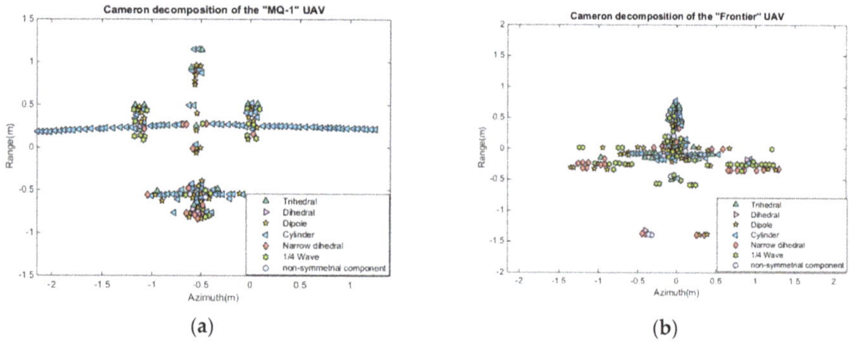

Figure 13. The Cameron decomposition of the "MQ-1" and "Frontier" UAVs. (**a**) "MQ-1" UAV; (**b**) "Frontier" UAV.

"MQ-1" UAV and "Frontier" UAV with different imaging azimuth angles are chosen to test the algorithm performance. Definitions of T1, T2, and T3 are shown in Table 2. Each target contains 151 samples which include elements by multiple polarimetric decomposition methods (such as odd scattering, trihedral, dihedral, etc.). The structure of a UAV can be inverted and three targets will be classified according to those scattering mechanisms.

Table 2. The three targets for the experiment.

Abbreviation	Target
T1	"MQ-1" UAV with azimuth angle from -5 to $25°$
T2	"Frontier" UAV with azimuth angle from 0 to $30°$
T3	"Frontier" UAV with azimuth angle from 75 to $105°$

Figure 14a shows the decision graph of targets which can decide the number of clustering centers automatically according to the thresholds of ρ and δ without training and testing. In the experiment, thresholds of ρ and δ are chosen as 12 and 22 respectively. Three clustering centers are selected correctly which are marked in yellow, green, and blue. Figure 14b reveals the clustering results of three targets which are processed by the multidimensional scaling for the presence of visualization and all the samples are classified accurately.

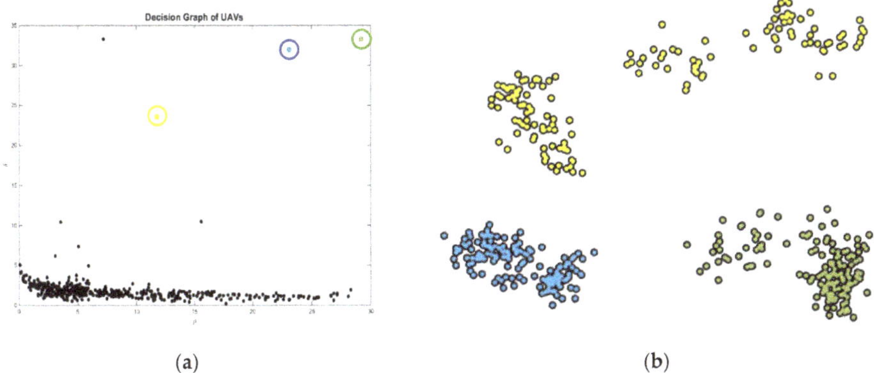

(a) (b)

Figure 14. The classification of three targets by CFSFDP with polarimetric decomposition. (**a**) Decision graph of targets; (**b**) Clustering results for three targets.

To further verify the effectiveness of the proposed algorithm, samples with different SNRs have been utilized. As the SNR increases, all four algorithms reveal the improvement of the classification results. DBSCAN (density-based spatial clustering of applications with noise), K-means and K-medoids are typical clustering algorithms in machine learning. K-means and K-medoids methods classify elements based on the distance to the cluster centers, while DBSCAN method is on the basis of local density. As shown in Table 3, the performance of the proposed algorithm is superior to the other three algorithms. When the SNR is 5 dB, the classification result of DBSCAN is 33.33%, whereas the value of our algorithm is 77.26%. All the classification results become stable as the SNR is greater than 20 dB which has shown the effectiveness of the polarimetric scattering mechanisms with different classifiers. The proposed algorithm not only has a satisfying performance with high SNRs, but also behaves robustly under severe conditions. Different UAVs or even the same UAV with different azimuth angles can be recognized precisely and automatically with no prior information.

Table 3. The classification accuracy of original algorithms and proposed algorithm with different SNRs.

	5 dB	10 dB	20 dB	30 dB	40 dB
DBSCAN [38]	33.33%	66.67%	83.66%	83.66%	83.66%
K-means [39,40]	53.20%	82.78%	83.00%	83.00%	83.00%
K-medoids [41]	76.16%	82.78%	83.00%	83.00%	83.00%
Proposed algorithm	**77.26%**	**83.22%**	**100.00%**	**100.00%**	**100.00%**

7. Conclusions

UAVs have become a significantly important issue in both civilian and military fields, nevertheless, the electromagnetic and polarization properties are barely studied in recent years. In this paper, the polarization characteristics of unmanned aerial vehicles are analyzed according to electromagnetic data of the "MQ-1" and "Frontier" UAVs. A novel clustering algorithm has been proposed to classify the UAVs automatically and recognize the same UAV with different azimuth angles via polarization characteristics.

More kinds of UAVs could be utilized for classification and recognition to further prove the effectiveness of the proposed algorithm. The fusion of infrared and optical information could also be considered as the compensation for the polarization analysis, which may enable the improvements of classification and recognition.

Author Contributions: H.W. and D.D. designed the algorithm; B.P. and J.W. performed the algorithm; B.P. proposed important modified scheme of the paper; H.W. wrote the paper; X.W. revised the paper.

Funding: This research was funded by the National Natural Science Foundation of China (No. 61501473, 61490690) and Excellent Youth Foundation of Hu'nan Scientific Committee (No. 2017JJ1006).

Conflicts of Interest: The authors declare no conflict of interest.

References

1. Cloude, S.R.; Pottier, E. A review of target decomposition theorems in radar polarimetry. *IEEE Trans. Geosci. Remote Sens.* **1996**, *34*, 498–518. [CrossRef]
2. Lee, J.S.; Jurkevich, L.; Dewaele, P.; Oosterlinck, A. Speckle filtering of synthetic aperture radar images: A review. *Remote Sens. Rev.* **1994**, *8*, 255–267. [CrossRef]
3. Skolnik, M.I. Introduction to radar. In *Radar Handbook*, 3rd ed.; Mc Graw Hill: New York, NY, USA, 2008; pp. 2–6.
4. Zhao, Q.; Principe, J.C. Support vector machines for SAR automatic target recognition. *IEEE Trans. Aerosp. Electron. Syst.* **2001**, *37*, 643–654. [CrossRef]
5. Jacobs, S.P.; Sullivan, J.A.O. Automatic target recognition using sequences of high resolution radar range-profiles. *IEEE Trans. Aerosp. Electron. Syst.* **2000**, *36*, 364–381. [CrossRef]
6. Sun, Y.; Liu, Z.; Todorovic, S.; Li, J. Adaptive boosting for SAR automatic target recognition. *IEEE Trans. Aerosp. Electron. Syst.* **2007**, *43*, 112–125. [CrossRef]
7. Narayanan, R.M.; Xu, X. Principles and applications of coherent random noise radar technology. *Proc. SPIE Int. Soc. Opt. Eng.* **2003**, *5113*, 503–514.
8. Novak, L.M. Performance of a high-resolution polarimetric SAR automatic target recognition system. *Linc. Lab. J.* **1993**, *6*, 11–24.
9. Ainsworth, T.L.; Schuler, D.L.; Lee, J.S. Polarimetric SAR characterization of man-made structures in urban areas using normalized circular-pol correlation coefficients. *Remote Sens. Environ.* **2008**, *112*, 2876–2885. [CrossRef]
10. Dungan, K.E.; Potter, L.C. 3D imaging of vehicles using wide aperture radar. *IEEE Trans. Aerosp. Electron. Syst.* **2011**, *47*, 187–200. [CrossRef]
11. Saville, M.A.; Saini, D.K.; Smith, J. Commercial vehicle classification from spectrum parted linked image test-attributed synthetic aperture radar imagery. *IET Radar Sonar Navig.* **2016**, *10*, 569–576. [CrossRef]
12. Fuller, D.F.; Saville, M.A. A high-frequency multipeak model for wide-angle SAR imagery. *IEEE Trans. Geosci. Remote Sens.* **2013**, *51*, 4279–4291. [CrossRef]
13. He, Y.; He, S.Y.; Zhang, Y.H.; Wen, G.J.; Yu, D.F.; Zhu, G.Q. A forward approach to establish parametric scattering center models for known complex radar targets applied to SAR ATR. *IEEE Trans. Antennas Propag.* **2014**, *62*, 6192–6205. [CrossRef]
14. Zhou, J.; Shi, Z.; Fu, Q. Three-dimensional scattering center extraction based on wide aperture data at a single elevation. *IEEE Trans. Geosci. Remote Sens.* **2014**, *53*, 1638–1655. [CrossRef]

15. Yun, D.J.; Lee, J.I.; Bae, K.U.; Yoo, J.H.; Kwon, K.I.; Myung, N.H. Improvement in Computation Time of 3D Scattering Center Extraction Using the Shooting and Bouncing Ray Technique. *IEEE Trans. Antennas Propag.* **2017**, *65*, 4191–4199. [CrossRef]

16. Li, Y.; Jin, Y.Q. Imaging and structural feature decomposition of a complex target using multi-aspect polarimetric scattering. *Sci. China Inf. Sci.* **2016**, *59*, 082308. [CrossRef]

17. Potter, L.C.; Moses, R.L. Attributed scattering centers for SAR ATR. *IEEE Trans. Image Process.* **1997**, *6*, 79–91. [CrossRef] [PubMed]

18. Wu, J.; Chen, Y.; Dai, D.; Chen, S.; Wang, X. Clustering-based geometrical structure retrieval of man-made target in SAR images. *IEEE Geosci. Remote Sens. Lett.* **2017**, *14*, 279–283. [CrossRef]

19. Duan, J.; Zhang, L.; Xing, M.; Wu, Y.; Wu, M. Polarimetric target decomposition based on attributed scattering center model for synthetic aperture radar targets. *IEEE Geosci. Remote Sens. Lett.* **2014**, *11*, 2095–2099. [CrossRef]

20. Xing, M.; Jiang, X.; Wu, R.; Zhou, F.; Bao, Z. Motion compensation for UAV SAR based on raw radar data. *IEEE Trans. Geosci. Remote Sens.* **2009**, *47*, 2870–2883. [CrossRef]

21. Jian, M.; Lu, Z.; Chen, V.C. Experimental study on radar micro-Doppler signatures of unmanned aerial vehicles. In Proceedings of the IEEE Radar Conference, Seattle, WA, USA, 8–12 May 2017; pp. 0854–0857.

22. Guay, R.; Drolet, G.; Bray, J.R. Measurement and modelling of the dynamic radar cross-section of an unmanned aerial vehicle. *IET Radar Sonar Navig.* **2017**, *11*, 1155–1160. [CrossRef]

23. Harman, S. Analysis of the radar return of micro-UAVs in flight. In Proceedings of the IEEE Radar Conference, Seattle, WA, USA, 8–12 May 2017; pp. 1159–1164.

24. Pieraccini, M.; Rojhani, N.; Miccinesi, L. 2D and 3D-ISAR Images of a Small Quadcopter. In Proceedings of the IEEE European Radar Conference, London, UK, 3–7 October 2017; pp. 307–310.

25. Gorham, L.A.; Moore, L.J. SAR image formation toolbox for MATLAB. In *Proceedings of Volume 7699, Algorithms for Synthetic Aperture Radar Imagery XVII, SPIE Defense, Security, and Sensing, Orlando, FL, USA, 5–9 April, 2010*; SPIE: Bellingham, WA, USA, 2010; Volume 7699, p. 769906. [CrossRef]

26. Chen, S.W. Polarimetric coherence pattern: A visualization and characterization tool for PolSAR data investigation. *IEEE Trans. Geosci. Remote Sens.* **2018**, *56*, 286–297. [CrossRef]

27. Dai, D.H. *Study on Polarimetric Radar Imaging and Target Feature Extraction*; National University of Defense Technology: Changsha, China, 2008; Volume 1, pp. 50–56.

28. Munson, D.C.; O'brien, J.D.; Jenkins, W.K. A tomographic formulation of spotlight-mode synthetic aperture radar. *Proc. IEEE* **1983**, *71*, 917–925. [CrossRef]

29. Krogager, E. New decomposition of the radar target scattering matrix. *Electron. Lett.* **1990**, *26*, 1525–1527. [CrossRef]

30. Krogager, E.; Boerner, W.M.; Madsen, S.N. Feature-motivated Sinclair matrix (sphere/diplane/helix) decomposition and its application to target sorting for land feature classification. *Int. Soc. Opt. Photonics* **1997**, *3120*, 144–155.

31. Cameron, W.L.; Leung, L.K. Feature motivated polarization scattering matrix decomposition. In Proceedings of the IEEE Radar Conference, Arlington, VA, USA, 7–10 May 1990; pp. 549–557.

32. Cameron, W.L.; Youssef, N.N.; Leung, L.K. Simulated polarimetric signatures of primitive geometrical shapes. *IEEE Trans. Geosci. Remote Sens.* **1996**, *34*, 793–803. [CrossRef]

33. Lee, J.S.; Pottier, E. *Polarimetric Radar Imaging: From Basics to Applications*; CRC Press: Boca Raton, FL, USA, 2009.

34. PolSARpro. Available online: https://earth.esa.int/web/polsarpro/polarimetry-tutorial (accessed on 30 September 2018).

35. Rodriguez, A.; Laio, A. Clustering by fast search and find of density peaks. *Science* **2014**, *344*, 1492–1496. [CrossRef] [PubMed]

36. CFSFDP in Science. Available online: http://science.sciencemag.org/content/344/6191/1492 (accessed on 30 September 2018).

37. An Introduction to CFSFDP. Available online: https://blog.csdn.net/itplus/article/details/38926837 (accessed on 30 September 2018).

38. Ester, M.; Kriegel, H.P.; Sander, J.; Xu, X. A density-based algorithm for discovering clusters in large spatial databases with noise. In Proceedings of the 2nd International Conference on Knowledge Discovery and Data Mining, Portland, OR, USA, 2–4 August 1996; pp. 226–231.

39. Wu, X.; Kumar, V.; Quinlan, J.R.; Ghosh, J.; Yang, Q.; Motoda, H.; Zhou, Z.H. Top 10 algorithms in data mining. *Knowl. Inf. Syst.* **2008**, *14*, 1–37. [CrossRef]
40. Hartigan, J.A.; Wong, M.A. Algorithm AS 136: A k-means clustering algorithm. *J. R. Stat. Soc. Ser. C (Appl. Stat.)* **1979**, *28*, 100–108. [CrossRef]
41. Kaufman, L.; Rousseeuw, P.J. *Finding Groups in Data: An Introduction to Cluster Analysis*; John Wiley and Sons: Hoboken, NJ, USA, 2009; Volume 344.

 *electronics*

Article

Atomic Norm-Based DOA Estimation with Dual-Polarized Radar

Min Han * and Wenbin Dou

State Key Laboratory of Millimeter Waves, Southeast University, Nanjing 210096, China; njdouwb@163.com
* Correspondence: hanmin@seu.edu.cn; Tel.: +86-189-9408-3357

Received: 18 August 2019; Accepted: 17 September 2019; Published: 19 September 2019

Abstract: In the dual-polarized radar system, the horizontally and vertically polarized signals can be exploited to improve the direction of arrival (DOA) estimation performance. In this paper, the DOA estimation problem is considered in the dual-polarized radar. By exploiting the target sparsity in the spatial domain, the sparse-based method is proposed after formulating the DOA estimation problem as a sparse reconstruction problem. In the traditionally sparse methods using the compressed sensing (CS) theory, the spatial domain is discretized into grids to establish a dictionary matrix and solve the sparse reconstruction problem, but the off-grid error is introduced in the discretized grids. Therefore, we formulate a novel definition of atomic norm for the dual-polarized signals and give an atomic norm-based method to denoise the received signals. Then, an efficient semidefinite program (SDP) is derived, and the DOA is estimated by searching the peak values of the denoised signals. Simulation results show that the proposed method can significantly improve the DOA estimation performance in the dual-polarized radar. Additionally, compared with the state-of-art methods, the proposed method has better estimation performance with relatively low computational complexity.

Keywords: dual-polarized radar; DOA estimation; atomic norm; off-grid sparse problem

1. Introduction

The direction of arrival (DOA) estimation problem has been studied in applications including wireless communication, radar, and sonar and array signal processing [1–3]. Usually, the discrete Fourier transform (DFT) method is used and formulates the DOA estimation problem as a spatial sampling reconstruction problem [4–7]. However, the resolution of traditional methods is limited by the *Rayleigh criterion* [8,9]. To achieve a better estimation performance than the Rayleigh criterion, super-resolution methods have been proposed. The most important super-resolution methods are subspace-based methods, including the multiple signal classification (MUSIC) method [10] and the estimating signal parameters via rotational invariance techniques (ESPRIT) method [11]. Since then, extension algorithms based on the MUSIC and ESPRIT methods have been proposed in more recent papers, such as the Root-MUSIC method [12], space-time MUSIC method [13], G-MUSIC method [14], higher-order ESPRIT and virtual ESPRIT [15], etc. In the subspace-based methods, the signal and noise subspaces are estimated from the covariance of received signals, so multiple measurements are needed to obtain the corresponding covariance matrix. However, for the fast moving target, multiple measurements cannot be obtained, so the subspace-based methods cannot be used in this scenario.

To improve the DOA estimation performance, the target sparsity in the spatial domain can also be exploited [16]. Therefore, the sparse-based methods, especially the compressed sensing (CS)-based methods [17–20], were proposed to transform the DOA estimation problem into a sparse reconstruction problem. In multiple-input and multiple-output (MIMO) radar systems, a CS-based DOA estimation method was proposed in [18,21]. Additionally, in [22], a compressed sparse array scheme was proposed.

The spatial domain is discretized into grids to obtain a dictionary matrix, and the DOA estimation problem is formulated as a sparse reconstruction problem. However, the targets can be not exactly at the discretized angles, so the discretized angles introduce the *off-grid*. Some off-grid methods have been proposed to solve the off-grid problem [23,24]. For example, reference [25] considered the structured dictionary mismatch and gave a sparse reconstruction method with off-grid. A sparse Bayesian inference was given in [26] with the off-grid consideration. Moreover, an iterative reweighted method estimated the off-grid and sparse signals jointly in [27]. Many papers have studied the off-grid problem. For example, in [28], the mismatch problem—including the sampling jitter in A/D conversion and model errors—was investigated and the perturbed orthogonal matching pursuit method was proposed. In [29], a new parameter-refined orthogonal matching pursuit (OMP) method was proposed to jointly estimate the off-grid positions and reflectivities of true scatterers. A new OMP-based sparse reconstruction method with parameter perturbation, named as PPOMP, was proposed in [30], in the delay-Doppler radar with the off-grid error. Additionally, in [31], a new image-focusing algorithm was given for sparsity-driven radar to image the rotating targets, and also to consider the off-grid scatterers. Moreover, in [23], the atomic norm-theory is formulated for the sparse reconstruction in the compressed sensing problem, which inspires us to address the DOA estimation problem using the atomic norm theory. However, different from the method in [23], we formulate a new type of atomic norm for the dual-polarized radar, and derive the expressions of DOA estimation. As shown in [23], the atomic norm theory is more suitable for the off-grid, underdetermined, and structured linear inverse problems based on the convex method. Therefore, we study the off-grid problem in the dual-polarized radar system based on the atomic norm theory.

In this paper, we consider the DOA estimation problem in the dual-polarized radar system. To exploit the target sparsity in the spatial domain, we formulate the DOA estimation problem as a sparse reconstruction problem. Then, we formulate a novel definition of atomic norm, which can be used to describe the sparsity of the dual-polarized signals. Based on the proposed atomic norm, we establish a denoising method for the received signal, but the denoising method is convex and cannot be solved efficiently. Therefore, a semidefinite program (SDP) optimization method is proposed to transform the nonconvex problem into a convex problem using the Schur complement theory, and can then be solved efficiently. Finally, the DOA of the target is estimated by searching the peak values of the polynomial function.

The remainder of this paper is organized as follows. The system model in the dual-polarized radar is formulated in Section 2. A novel definition of atomic norm and a DOA estimation method are given in Section 3. Simulation results are given in Section 4. Finally, Section 5 concludes the paper.

Notations: $(\cdot)^{\mathrm{H}}$ denotes the Hermitian transpose. $\|\cdot\|_1$, $\|\cdot\|_F$, and $\|\cdot\|_2$ denote the ℓ_1 norm, the Frobenius norm, and the ℓ_2 norm, respectively. $\|\cdot\|^*$ denotes the dual norm. $\boldsymbol{I}_N$ denotes an $N \times N$ identity matrix. $\mathrm{Tr}\{\cdot\}$ denotes the trace of a matrix. $\mathcal{R}\{a\}$ denotes the real part of complex value a.

2. Dual-Polarized Radar System

In the array radar system, the dual-polarized signals are transmitted and N antennas receive the echoed signals. As shown in Figure 1, we denote the horizontally polarized signal as $s_{\mathrm{H}}(t)$ and the vertically polarized signal as $s_{\mathrm{V}}(t)$. Therefore, with K far-field targets, the received waveform for the horizontally polarized signal in the n-th antenna can be expressed as

$$r_{\mathrm{H},n}(t) = \sum_{k=0}^{K-1} e^{j2\pi(n-1)\frac{d}{\lambda}\sin\theta_k} \alpha_{\mathrm{H},k} s_{\mathrm{H}}(t) + w_{\mathrm{H},n}(t), \tag{1}$$

where d denotes the distance between adjacent antennas, λ is the wavelength, and $w_{\mathrm{H},n}(t)$ denotes the additive white Gaussian noise (AWGN) in the horizontally polarized antenna. θ_k is the DOA of the

k-th target, and $\alpha_{H,n}$ is the corresponding scattering coefficient for the horizontally polarized signal. Similarly, the received signal for the vertically polarized signal can be expressed as

$$r_{V,n}(t) = \sum_{k=0}^{K-1} e^{j2\pi(n-1)\frac{d}{\lambda}\sin\theta_k}\alpha_{V,k}s_V(t) + w_{V,n}(t). \tag{2}$$

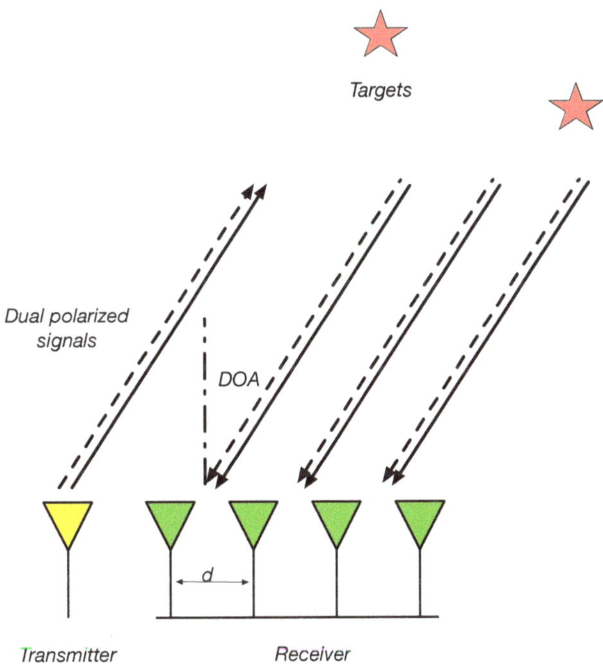

Figure 1. The dual-polarized radar system.

In this paper, we will estimate the DOA θ_k from the received signals $r_{H,n}(t)$ and $r_{V,n}(t)$. In the scenario with a fast moving target, the stationary assumption for the received signals in the antenna array no longer holds, so the traditional super resolution methods based on the subspace decomposition, such as the MUSIC and ESPRIT methods, cannot be used in scenarios where multiple measurements are needed to obtain the covariance matrix for the stationary signals. Therefore, we consider the DOA estimation problem in the scenario with only one measurement for the polarized signals. For the horizontally polarized signal, we sample the received signals in the antennas and formulate the following vector form:

$$r_H \triangleq \left[r_{H,0}(T_s), r_{H,1}(T_s), \ldots, r_{H,N-1}(T_s)\right]^{\mathrm{T}} = As_H + w_H, \tag{3}$$

where T_s is the sampling frequency. A denotes the steering matrix $A \triangleq \left[a(\theta_0), a(\theta_1), \ldots, a(\theta_{K-1})\right]$, and $a(\theta)$ is the steering vector for the receiving antennas $a(\theta) \triangleq \left[1, e^{j2\pi\frac{d}{\lambda}\sin\theta}, \ldots, e^{j2\pi\frac{(N-1)d}{\lambda}\sin\theta}\right]^{\mathrm{T}}$. $s_H \in \mathbb{C}^{K\times 1}$, the k-th entry of s is $\alpha_{H,k}s_H(T_s)$ and $w_H \triangleq \left[w_{H,0}, w_{H,1}, \ldots, w_{H,N-1}\right]^{\mathrm{T}}$. Similarly, the vertically polarized signal in the receiving antennas can be expressed as

$$r_V \triangleq \left[r_{V,0}(T_s), r_{V,1}(T_s), \ldots, r_{V,N-1}(T_s)\right]^{\mathrm{T}} = As_V + w_V. \tag{4}$$

Finally, collect all the polarized signals and we can obtain

$$R \triangleq \left[r_{\mathrm{H}}, r_{\mathrm{V}}\right] = AS + W, \tag{5}$$

where $S \triangleq \left[s_{\mathrm{H}}, s_{\mathrm{V}}\right]$ and $W \triangleq \left[w_{\mathrm{H}}, w_{\mathrm{V}}\right]$. In this paper, we will estimate the DOA of target θ_k from the received signal R using the dual-polarized radar system.

3. Novel Atomic Norm-Based Method for DOA Estimation

In this section, we will propose a novel atomic norm for the DOA estimation problem in the scenario with dual-polarized array radar. Based on the proposed atomic norm, an efficient method will be given to obtain the DOA using the convex optimization theory.

For the dual-polarized array radar, we formulate the following atomic norm:

$$\|D\|_{\mathcal{A}} \triangleq \inf\left\{d \geq 0 : D \in d \operatorname{conv}\{\mathcal{A}\}\right\} \tag{6}$$

$$= \inf\left\{\|d\|_1 : D = \sum_k d_k a(\theta_k) t_k^{\mathrm{H}}, d = [d_0, d_1, \ldots d_{K-1}]^{\mathrm{T}}, \|t_k\|_2 = 1, t_k \in \mathbb{C}^{2\times 1}\right\}.$$

Then, with the received signal R, we have the following optimization problem to estimate the echoed signal:

$$\min_X \frac{1}{2}\|R - X\|_F^2 + \mu\|X\|_{\mathcal{A}}. \tag{7}$$

To solve the optimization problem Equation (7), we have the following proposition.

Proposition 1. *With the definition of the atomic norm in Equation (6), the optimization problem in Equation (7) can be rewritten as a dual-optimization problem:*

$$\min_P \|R - P\|_F^2 \tag{8}$$

$$\text{s.t.}\|P\|_{\tilde{\mathcal{A}}} \leq \mu,$$

where $\|P\|_{\tilde{\mathcal{A}}}$ is the dual norm of $\|P\|_{\mathcal{A}}$.

Proof. The optimization problem in Equation (7) can be rewritten as

$$\min_X \frac{1}{2}\|R - Z\|_F^2 + \mu\|X\|_{\mathcal{A}} \tag{9}$$

$$\text{s.t. } Z = X.$$

Therefore, we have the Lagrange function of X and Z with the Lagrange parameter P as

$$L(X, Z, P) \triangleq \frac{1}{2}\|R - Z\|_F^2 + \mu\|X\|_{\mathcal{A}} + \langle Z - X, P\rangle, \tag{10}$$

where $\langle A, B \rangle \triangleq \mathcal{R}\{\text{Tr}(B^H A)\}$ denotes the inner product between matrices. Therefore, the dual-optimization problem of Equation (9) can be obtained as

$$\max_{P} \min_{X,Z} L(X, Z, P) = \max_{P} \min_{X,Z} \frac{1}{2}\|R - Z\|_F^2 + \mu\|X\|_{\mathcal{A}} + \langle Z - X, P \rangle \tag{11}$$

$$= \max_{P} \left\{ \min_{Z} \underbrace{\frac{1}{2}\|R - Z\|_F^2 + \langle Z, P \rangle}_{g_1(P,Z)} + \min_{X} \underbrace{\mu\|X\|_{\mathcal{A}} - \langle X, P \rangle}_{g_2(P,X)} \right\}.$$

Since we have

$$\frac{\partial g_1(P, Z)}{\partial Z} = -(R - Z) + \frac{1}{2}\frac{\partial \text{Tr}(P^H Z) + \text{Tr}^*(P^H Z)}{\partial Z}$$

$$= Z - R + P, \tag{12}$$

the Z minimizers $g_1(P, Z)$ can be obtained by $\frac{\partial g_1(P,Z)}{\partial Z} = 0$, so we have $Z = R - P$. Therefore, we can obtain

$$\min_{Z} g_1(P, Z) = \frac{1}{2}\|P\|_F^2 + \langle R - P, P \rangle = \frac{1}{2}\left(\|R\|_F^2 - \|R - P\|_F^2\right). \tag{13}$$

With the definition of proposed atomic norm in Equation (6), we can define the dual norm of atomic norm as

$$\|D\|_{\bar{\mathcal{A}}} \triangleq \sup_{\|X\|_{\mathcal{A}} \leq 1} \langle D, X \rangle. \tag{14}$$

Therefore, we can obtain

$$\min_{X} g_2(P, X) = -\mu \left(\max_{X} \langle X, \frac{1}{\mu}P \rangle - \|X\|_{\mathcal{A}} \right) = I\left(\|P\|_{\bar{\mathcal{A}}} \leq \mu\right), \tag{15}$$

where $I(\cdot)$ is an indicate function.

Finally, the dual optimization problem can be expressed as

$$\min_{P} \|R - P\|_F^2 \tag{16}$$

$$\text{s.t. } \|P\|_{\bar{\mathcal{A}}} \leq \mu,$$

and the proposition is proofed. $\square$

In Proposition 1, the dual-optimization problem is obtained from the atomic norm minimization (ANM) problem Equation (7), but the dual problem cannot be solved efficiently. We will show that the dual problem can be rewritten as an SDP problem. With the definition of dual norm, the constraint in Proposition 1 can be given as

$$\|P\|_{\tilde{\mathcal{A}}} = \sup_{\|X\|_{\mathcal{A}} \leq 1} \langle P, X \rangle \tag{17}$$

$$= \sup_{\theta_k \in [0,2\pi), \|d\|_1 \leq 1, \|t_k\|_2 = 1} \left\langle P, \sum_k d_k a(\theta_k) t_k^{\mathrm{H}} \right\rangle$$

$$= \sup_{\theta \in [0,2\pi), \|t\|_2 = 1} \left\langle P, a(\theta) t^{\mathrm{H}} \right\rangle$$

$$= \sup_{\theta \in [0,2\pi), \|t\|_2 = 1} \mathcal{R} \left\{ \mathrm{Tr} \left(t a(\theta)^{\mathrm{H}} P \right) \right\}$$

$$= \sup_{\theta \in [0,2\pi), \|t\|_2 = 1} \mathcal{R} \left\{ b(\theta)^{\mathrm{H}} P t \right\}$$

$$= \sup_{\theta \in [0,2\pi)} \mathcal{R} \left\{ a(\theta)^{\mathrm{H}} P \frac{P^{\mathrm{H}} a(\theta)}{\|P^{\mathrm{H}} a(\theta)\|_2} \right\}$$

$$= \sup_{\theta \in [0,2\pi)} \|P^{\mathrm{H}} a(\theta)\|_2.$$

Hence, if we have $\sup_{\theta \in [0,2\pi)} \|P^{\mathrm{H}} a(\theta)\|_2 \leq \mu$, the constraint $\|P\|_{\tilde{\mathcal{A}}} \leq \mu$ can be satisfied. We have the Schur complement theory as follows:

Lemma 1. *For a matrix* $G = \begin{bmatrix} A & B \\ C & D \end{bmatrix}$, *we have* $G \succeq 0$ *if and only if we have*

$$A \succeq 0, \tag{18}$$

$$A - BD^{-1}C \succeq 0. \tag{19}$$

Therefore, we can formulate a semidefinite positive matrix

$$\begin{bmatrix} W & P \\ P^{\mathrm{H}} & \mu^2 I \end{bmatrix} \succeq 0, \tag{20}$$

so we have $W \succeq 0$ and $W - \mu^{-2} P P^{\mathrm{H}} \succeq 0$. For any vector l, we have

$$l^{\mathrm{H}} W l - \mu^{-2} l^{\mathrm{H}} P P^{\mathrm{H}} l \geq 0. \tag{21}$$

When we choose $l = a(\theta)$, we can obtain

$$\|P^{\mathrm{H}} a(\theta)\|_2^2 \leq \underbrace{a(\theta)^{\mathrm{H}} W a(\theta)}_{g_3(W,\theta)} \mu^2. \tag{22}$$

By letting $g_3(W,\theta) \leq 1$, we can finally satisfy the constraint $\sup_{\theta \in [0,2\pi)} \|P^{\mathrm{H}} a(\theta)\|_2 \leq \mu$. We can formulate the matrix W as a Hermitian matrix, satisfying the following condition:

$$\sum_q W_{q,q+k} = \begin{cases} 0, & k \neq 0 \\ 1, & k = 0 \end{cases}, \tag{23}$$

where $W_{q,q+k}$ is the entry of W at the q-th row and $q + k$-th column. Therefore, we can simplify $g_3(W, \theta)$ as

$$g_3(W, \theta) = a^H(\theta)Wa(\theta) \tag{24}$$
$$= \sum_{n_1}\sum_{n_2} a_{n_1}^H(\theta)a_{n_2}(\theta)W_{n_1,n_2}$$
$$\leq 1.$$

Therefore, the dual optimization problem in Equation (7) can be rewritten as an SDP problem:

$$\min_{P,W}\|R - P\|_F^2 \tag{25}$$

$$\text{s.t.} \quad \begin{bmatrix} W & P \\ P^H & \mu^2 I \end{bmatrix} \succeq 0$$

$$\sum_q W_{q,q+k}^{n,n} = \begin{cases} 0, & k \neq 0 \\ 1, & k = 0 \end{cases}$$

$$W \text{ is Hermitian.}$$

By solving this SDP problem, the denoised signal P can be obtained. Then, the DOA is estimated by the peak search of $\|P^H a(\theta)\|_2$.

4. Simulation Results

In this section, the simulation results for the DOA estimation in dual-polarized radar are given, and the corresponding simulation parameters are given in Table 1. All the simulation results are obtained with 10^4 Monte Carlo trials. The simulation results are carried out in a personal computer with 16 GB RAM and an Intel i7 CPU. Additionally, we compare the following state-of-art methods for the DOA estimation:

- With denoising method—this method is realized by calculating the correlation between the received signals and the steering vector to obtain the spatial spectrum, where the spatial spectrum is $\|R^H a(\theta)\|$ ($\theta \in [-60°, 60°]$).
- Simultaneous orthogonal matching pursuits (SOMP) method [32]—this method is proposed for the sparse reconstruction in the scenario with multiple measurements with lower computational complexity. In the SOMP method, the column of dictionary matrix indicating the corresponding DOA is selected iteratively.
- Sparse Bayesian learning (SBL) method [26]—SBL method is the sparse Bayesian learning method and achieves good reconstruction performance in the scenario with correct distribution assumptions for the received signals, noise, and target scattering coefficients. However, the computational complexity of SBL is much higher.

Table 1. Simulation Parameters.

Parameter	Value
The signal-to-noise ratio (SNR) of received signals	20 dB
The number of antennas N	20
The number of targets K	4
The distance between adjacent antennas d	0.5 wavelength
The detection DOA range	$[-60°, 60°]$
The type of antennas	dual-polarized antennas

First, we show the DOA estimation for 4 targets, with the DOA being $-29.474°$, $0.685\,32°$, $10.836°$, and $30.654°$, respectively. The spatial spectrum for the DOA estimation is shown in Figure 2, where all the methods can estimate the DOA from the received signals. The spatial spectrum of the proposed method is different from other methods, since we calculate the polynomial values $\|\boldsymbol{P}^{H}\boldsymbol{a}(\theta)\|_2$. As shown in Equation (25), if θ is the DOA of the target, we have $\|\boldsymbol{P}^{H}\boldsymbol{a}(\theta)\|_2 = 1$, so we can choose the corresponding DOA with $\|\boldsymbol{P}^{H}\boldsymbol{a}(\theta)\|_2$ being 1 from the polynomial values. The DOA with different methods can be obtained from Figure 2, and the estimated values are given in Table 2. The DOA estimation performance is measured by the root-mean-squared error (RMSE)

$$\text{RMSE} = \sqrt{\sum_{m=0}^{M-1} \frac{\|\boldsymbol{\theta}_m - \hat{\boldsymbol{\theta}}_m\|_2^2}{MK}}, \tag{26}$$

where $\boldsymbol{\theta}_m$ denotes the target DOA in the m-th Monte Carlo trial, and $\hat{\boldsymbol{\theta}}_m$ is the estimated DOA. As shown in Table 2, the RMSEs of without denoising method, SOMP method, and SBL method are 0.21075, 0.15609, and 0.14631 in degree, respectively. The RMSE of the proposed method is 0.048264 in degree, which is much lower than existing methods. Therefore, the proposed method achieves better performance of DOA estimation.

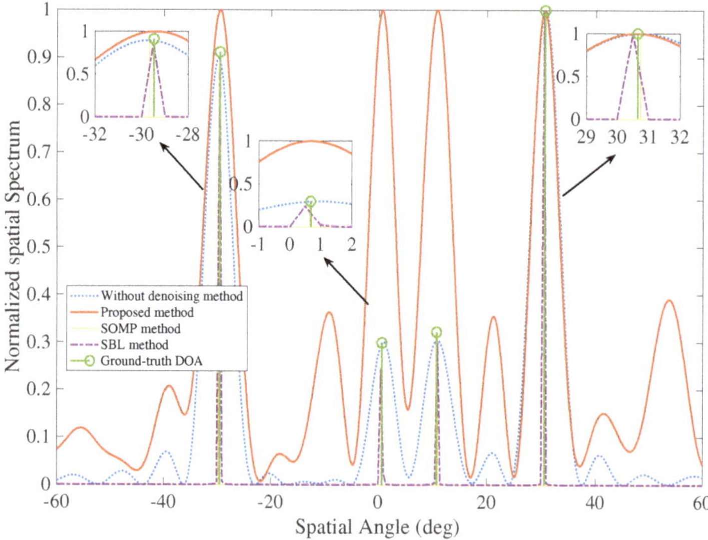

Figure 2. The spatial spectrum for direction of arrival (DOA) estimation.

Table 2. DOA Estimation Performance.

Methods	Target 1	Target 2	Target 3	Target 4	RMSE (deg)
Ground-truth DOA	$-29.474°$	$0.685\,32°$	$10.836°$	$30.654°$	–
Without denoising method	$-29.723°$	$0.992°$	$10.953°$	$30.743°$	0.21075
Simultaneous orthogonal matching pursuits (SOMP) method	$-29.663°$	$0.7233°$	$11.062°$	$30.75°$	0.15609
Sparse Bayesian learning (SBL) method	$-29.5°$	$0.5°$	$11°$	$30.5°$	0.14631
Proposed method	$-29.405°$	$0.6917°$	$10.876°$	$30.6°$	0.048264

Additionally, the computational complexity is shown in Table 3, where we show the computational time for one Monte Carlo trial. The without denoising method has the lowest computational complexity, with the computational time being 0.0475 s; and that of the SOMP method is 0.6570. The SBL method has the highest computational complexity with the computational time being 6.1734 s. The computational time of the proposed method is 2.0176, so the computational complexity is not very high and acceptable. The memory of MATLAB is 1.29 GB when the proposed method is running, 1.29 GB for the SOMP method, and 1.29 GB for the SBL method. Therefore, the memory used in the proposed method is more than the SOMP method, but almost the same as the SBL method.

Table 3. Computational Time.

Methods	Time (s)
Without denoising method	0.0475
SOMP method	0.6570
SBL method	6.1734
Proposed method	2.0176

Then, the DOA estimation performance with different SNR in the dual-polarized radar is shown in Figure 3, where the SNR of received signals is from 0 dB to 40 dB. As shown in this figure, the proposed method achieves the best DOA estimation performance, especially at SNR $\geq$ 15 dB. When SNR $<$ 15 dB, the proposed method has almost the same performance as the SBL method, and is better than both the SOMP method and the without denoising method.

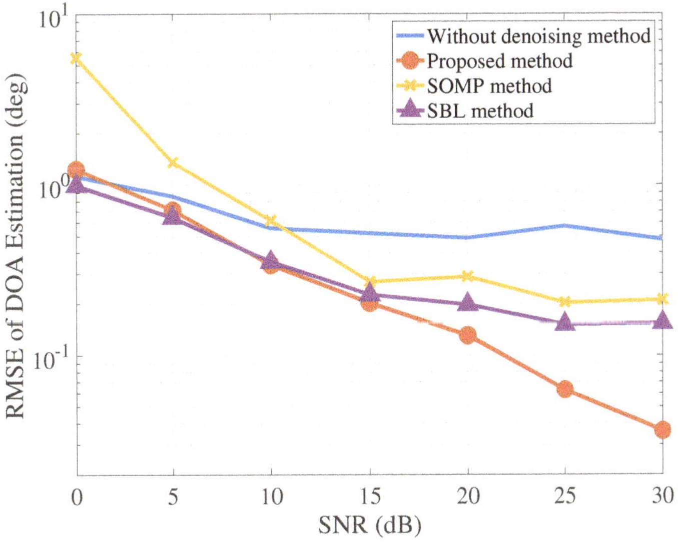

Figure 3. The DOA estimation with different SNRs.

Moreover, the DOA estimation performance with different numbers of antenna is shown in Figure 4. As shown in this figure, the DOA estimation performance is improved with increasing the number of antennas. When the antenna number is larger than 10, the proposed method achieves the best estimation performance. However, when the number of antennas is less than 10, the DOA estimation performance of the proposed method is the same as that of the SBL method.

Finally, we show the resolution performance in Figure 5, where the minimum separation is the minimum DOA difference between adjacent targets. As shown in this figure, when the separation is larger than $4°$, all the methods can have better performance of DOA estimation. The performance cannot be further improved when the separation is larger than $4°$. The proposed method has the best estimation performance with different DOA separations, and achieves the lower RMSE floor.

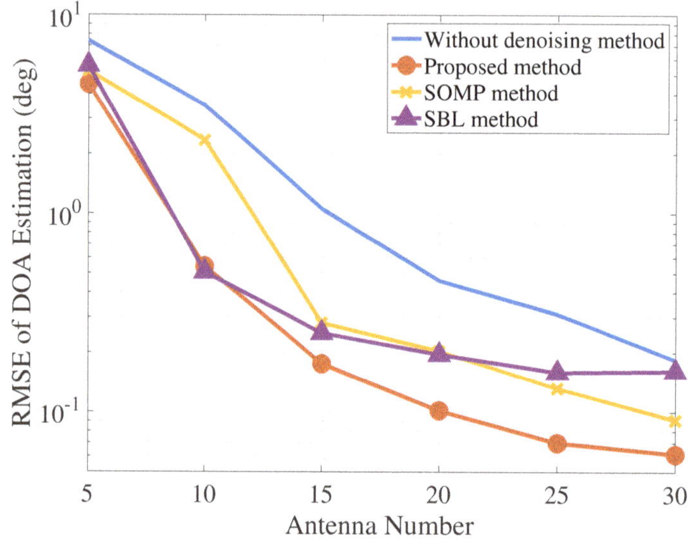

Figure 4. The DOA estimation with different numbers of antennas.

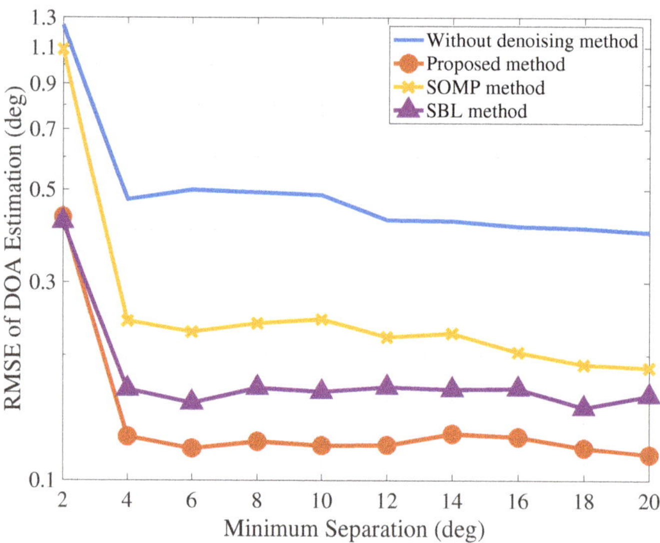

Figure 5. The DOA estimation with different DOA separations.

5. Conclusions

In the dual-polarized radar system, the DOA estimation problem has been addressed, and the target sparsity has been exploited to improve the DOA estimation performance. Additionally, the novel atomic norm has been defined in the scenario with dual-polarized signals, so the denoising method has been formulated based on the proposed atomic norm. Then, the convex SDP problem has been derived to solve the DOA estimation problem efficiently. Simulation results show that the better performance of DOA estimation for the dual-polarized signals is obtained with relatively lower computational complexity, compared with the state-of-art methods.

Author Contributions: Conceptualization, M.H. and W.D.; methodology, M.H.; software, M.H.; validation, W.D.; formal analysis, W.D.; investigation, M.H.; resources, W.D.; data curation, M.H.; writing—original draft preparation, M.H.; writing—review and editing, W.D.; visualization, M.H.; supervision, W.D.; project administration, W.D.; funding acquisition, W.D.

Funding: This work was supported by the National Natural Science Foundation of China (Grant No. 61171025)

Conflicts of Interest: The authors declare no conflict of interest.

References

1. Zheng, L.; Lops, M.; Wang, X. Adaptive Interference Removal for Uncoordinated Radar/Communication Coexistence. *IEEE J. Sel. Top. Signal Process.* **2018**, *12*, 45–60. [CrossRef]
2. Chen, P.; Cao, Z.; Chen, Z.; Liu, L.; Feng, M. Compressed Sensing-Based DOA Estimation with Unknown Mutual Coupling Effect. *Electronics* **2018**, *7*, 424. [CrossRef]
3. Cao, Z.; Geng, H.; Chen, Z.; Chen, P. Sparse-Based Millimeter Wave Channel Estimation with Mutual Coupling Effect. *Electronics* **2019**, *8*, 358. [CrossRef]
4. Burintramart, S.; Sarkar, T.; Zhang, Y.; Salazar-Palma, M. Nonconventional least squares optimization for DOA estimation. *IEEE Trans. Antennas Propag.* **2007**, *55*, 707–714. [CrossRef]
5. Kim, S.; Oh, D.; Lee, J. Joint DFT-ESPRIT estimation for TOA and DOA in vehicle FMCW radars. *IEEE Antennas Wirel. Propag. Lett.* **2015**, *14*, 1710–1713. [CrossRef]
6. Liu, L.; Liu, H. Joint estimation of DOA and TDOA of multiple reflections in mobile communications. *IEEE Access* **2016**, *4*, 3815–3823. [CrossRef]
7. Chen, P.; Qi, C.; Wu, L.; Wang, X. Waveform Design for Kalman Filter-Based Target Scattering Coefficient Estimation in Adaptive Radar System. *IEEE Trans. Veh. Technol.* **2018**, *67*, 11805–11817. [CrossRef]
8. Rueckner, W.; Papaliolios, C. How to beat the Rayleigh resolution limit: A lecture demonstration. *Am. J. Phys.* **2002**, *70*, 587. [CrossRef]
9. Chen, P.; Cao, Z.; Chen, Z.; Yu, C. Sparse off-grid DOA estimation method with unknown mutual coupling effect. *Digit. Signal Process.* **2019**, *90*, 1–9. [CrossRef]
10. Schmidt, R.O. Multiple emitter location and signal parameter estimation. *IEEE Trans. Antennas Propag.* **1986**, *34*, 276–280. [CrossRef]
11. Roy, R.; Kailath, T. ESPRIT-estimation of signal parameters via rotational invariance techniques. *IEEE Trans. Acoust. Speech Signal Process.* **1989**, *37*, 984–995. [CrossRef]
12. Zoltowski, M.; Kautz, G.; Silverstein, S. Beamspace Root-MUSIC. *IEEE Trans. Signal Process.* **1993**, *41*, 344–364. [CrossRef]
13. Claudio, E.D.D.; Parisi, R.; Jacovitti, G. Space time MUSIC: Consistent signal subspace estimation for wideband sensor arrays. *IEEE Trans. Signal Process.* **2018**, *66*, 2685–2699. [CrossRef]
14. Vallet, P.; Mestre, X.; Loubaton, P. Performance analysis of an improved MUSIC DoA estimator. *IEEE Trans. Signal Process.* **2015**, *63*, 6407–6422. [CrossRef]
15. Yuen, N.; Friedlander, B. Asymptotic performance analysis of ESPRIT, higher order ESPRIT, and virtual ESPRIT algorithms. *IEEE Trans. Signal Process.* **1996**, *44*, 2537–2550. [CrossRef]
16. Chen, P.; Chen, Z.; Zhang, X.; Liu, L. SBL-Based Direction Finding Method with Imperfect Array. *Electronics* **2018**, *7*, 426. [CrossRef]
17. Li, W.T.; Lei, Y.J.; Shi, X.W. DOA estimation of time-modulated linear array based on sparse signal recovery. *IEEE Antennas Wirel. Propag. Lett.* **2017**, *16*, 2336–2340. [CrossRef]
18. Yu, Y.; Petropulu, A.P.; Poor, H.V. Measurement matrix design for compressive sensing-based MIMO radar. *IEEE Trans. Signal Process.* **2011**, *59*, 5338–5352. [CrossRef]
19. Xiong, W.; Greco, M.; Gini, F.; Zhang, G.; Peng, Z. SFMM design in colocated CS-MIMO radar for jamming and interference joint suppression. *IET Radar Sonar Navig.* **2018**, *12*, 702–710. [CrossRef]

20. Yang, Z.; Xie, L. Exact joint sparse frequency recovery via optimization methods. *IEEE Trans. Signal Process.* **2016**, *64*, 5145–5157. [CrossRef]
21. Chen, P.; Zheng, L.; Wang, X.; Li, H.; Wu, L. Moving target detection using colocated MIMO radar on multiple distributed moving platforms. *IEEE Trans. Signal Process.* **2017**, *65*, 4670–4683. [CrossRef]
22. Guo, M.; Zhang, Y.D.; Chen, T. DOA estimation using compressed sparse array. *IEEE Trans. Signal Process.* **2018**, *66*, 4133–4146. [CrossRef]
23. Tang, G.; Bhaskar, B.N.; Shah, P.; Recht, B. Compressed Sensing Off the Grid. *IEEE Trans. Inf. Theory* **2013**, *59*, 7465–7490. [CrossRef]
24. Chen, P.; Cao, Z.; Chen, Z.; Wang, X. Off-Grid DOA Estimation Using Sparse Bayesian Learning in MIMO Radar With Unknown Mutual Coupling. *IEEE Trans. Signal Process.* **2019**, *67*, 208–220. [CrossRef]
25. Tan, Z.; Yang, P.; Nehorai, A. Joint Sparse Recovery Method for Compressed Sensing with Structured Dictionary Mismatches. *IEEE Trans. Signal Process.* **2014**, *62*, 4997–5008. [CrossRef]
26. Yang, Z.; Xie, L.; Zhang, C. Off-grid direction of arrival estimation using sparse bayesian inference. *IEEE Trans. Signal Process.* **2012**, *61*, 38–43. [CrossRef]
27. Fang, J.; Li, J.; Shen, Y.; Li, H.; Li, S. Super-resolution compressed sensing: An iterative reweighted algorithm for joint parameter learning and sparse signal recovery. *IEEE Signal Process. Lett.* **2014**, *21*, 761–765. [CrossRef]
28. Teke, O.; Gurbuz, A.C.; Arikan, O. Perturbed Orthogonal Matching Pursuit. *IEEE Trans. Signal Process.* **2013**, *61*, 6220–6231. [CrossRef]
29. Nguyen, N.H.; Dogancay, K.; Tran, H.; Berry, P.E. Parameter-Refined OMP for Compressive Radar Imaging of Rotating Targets. *IEEE Trans. Aerosp. Electron. Syst.* **2019**. [CrossRef]
30. Teke, O.; Gurbuz, A.C.; Arikan, O. Sparse delay-Doppler image reconstruction under off-grid problem. In Proceedings of the 2014 IEEE 8th Sensor Array and Multichannel Signal Processing Workshop (SAM), A Coruna, Spain, 22–25 June 2014; pp. 409–412.
31. Nguyen, N.H.; Doğançay, K.; Tran, H.T.; Berry, P. An Image Focusing Method for Sparsity-Driven Radar Imaging of Rotating Targets. *Sensors* **2018**, *18*, 1840. [CrossRef]
32. Determe, J.; Louveaux, J.; Jacques, L.; Horlin, F. On the Noise Robustness of Simultaneous Orthogonal Matching Pursuit. *IEEE Trans. Signal Process.* **2017**, *65*, 864–875. [CrossRef]

 *electronics*

Article

Position Estimation of Automatic-Guided Vehicle Based on MIMO Antenna Array

Zhimin Chen [1,*], Xinyi He [2,*], Zhenxin Cao [3], Yi Jin [4] and Jingchao Li [1]

[1] School of Electronic and Information Engineering, Shanghai Dianji University, Shanghai 201306, China; lijc@sdju.edu.cn
[2] Science and Technology on Electromagnetic Scattering Laboratory, Shanghai 200438, China
[3] State Key Laboratory of Millimeter Waves, School of Information Science and Engineering, Southeast University, Nanjing 210096, China; caozx@seu.edu.cn
[4] Xi'an Branch of China Academy of Space Technology, Xi'an 710000, China; john.0216@163.com
* Correspondence: chenzm@sdju.edu.cn (Z.C.); hexinyiseu@126.com (X.H.)

Received: 5 August 2018; Accepted: 7 September 2018; Published: 11 September 2018

Abstract: The existing positioning methods for the automatic guided vehicle (AGV) in the port can not achieve high location precision, Therefore, a novel multiple input multiple output (MIMO) antenna radar positioning scheme is proposed in this paper. The positioning problem for AGV is considered, and the joint estimation problem for direction of departure (DoD) and direction of arrival (DoA) is addressed in the multiple-input multiple-output (MIMO) radar system. With the radar detect the transponder and estimate the DoA/DoD, the relative location between the transponder and the AGV can be obtained. The corresponding Cramér–Rao lower bounds (CRLBs) for the target parameters are also derived theoretically. Finally, we compare the positioning accuracy of the traditional global position system (GPS) with the proposed MIMO radar system. Simulation results show that the proposed method can achieve better performance than the traditional GPS.

Keywords: antenna array; automatic guided vehicle; DoA/DoD estimation; MIMO radar

1. Introduction

The automatic guided vehicle (AGV) is widely used in modern intelligent ports or other industrial applications, which can safely moving materials to the rightful destination under the control of local area network [1,2]. The navigation of AGV mainly using vision, magnets, or lasers. Before the early 1980s, the embedded electromagnetic induction method was always the main guiding technology of AGV. With the development of electronic technology, new guiding technologies are constantly being researched and promoted. At present, the navigation of AGV including two types, the fixed path method and the free path method. The fixed path method is represented by magnetic navigation technology. Since the route is fixed, the path change and expansion is inconvenient. The free path method mainly includes laser position, visual position, millimeter wave radar method, inertial navigation system and global position system (GPS).

The laser position is based on the triangle localization method. By using some reflection sign, the laser scan the surrounding area and obtain the location information of the sign. The laser position has high precision, but is susceptible to weather and needs to install a large number of reflective signs. The machine vision navigation provides guidance through visual image processing that suitable for various scenes with high positioning accuracy. However, the cost of the systems is too high and the performance is greatly affected by image sensors. The inertial navigation system (INS) is a relative positioning method. A gyroscope is installed on the vehicle to accurately obtain the direction and speed of the trolley. When the coordinates of the starting position is known, the data of the trolley can be calculated. The system is simple and flexible with low cost and good real-time performance.

The disadvantage is that the error accumulates due to various reasons, and long-term operation can lead to loss of precision completely. The GPS is a wireless navigation system that is positioned by navigation satellites. It can provide global, All-weather, continuous and real-time navigation and positioning. However, ordinary users can only use the standard positioning service (SPS) provided by GPS. The SPS accuracy is ± 10 m on the horizontal plane, and the accuracy of AGV positioning requirement is 0.03 m$\sim$0.1 m. The millimeter wave radar (MMWR) is place a millimeter wave radar on an AGV, the radar rotates to find a beacon installed at a known position, then uses the relative position information of the beacon to determine and continuously update the position of the AGV. The navigation accuracy of the method can reach ± 0.1 m.

In this paper, the position of the AGV is addressed. Different from the above methods, a novel positioning method based on multiple input multiple output (MIMO) antenna array is proposed, where the direction of arrival (DoA) and direction of departure (DoD) is estimated to determine the position of the AGV. Additionally, we theoretically derive the corresponding Cramér-Rao lower bounds (CRLBs) for the target parameters. Then we compare the estimation performance of the traditional global position system (GPS) with the proposed MIMO radar system.

The remainder of the manuscript is organized as follows. The description of the proposed system is given in Section 2, and the system model is given in Section 3, the compressed sensing-based DoD and DoA estimation method is proposed in Section 4, The CRLB is given in Section 5. The simulation results are given in Section 6. Section 7 concludes the paper.

Notations: $\mathcal{E}\{\cdot\}$ denotes the expectation operation, $\boldsymbol{I}_P$ denotes a $P \times P$ identity matrix, $(\cdot)^\mathsf{T}$ denotes the matrix transposeand, $(\cdot)^\mathrm{H}$ denotes the Hermitian transpose, $\mathcal{CN}(\boldsymbol{\mu}, \boldsymbol{R})$ denotes the complex Gaussian distribution with the mean being $\boldsymbol{\mu}$ and the variance matrix being $\boldsymbol{R}$, $\| \cdot \|_2$, $\| \cdot \|_0$, $\| \cdot \|$, $\otimes$, vec$\{\cdot\}$ denote the ℓ_2 norm, the ℓ_0 norm, Frobenius norms, the Kronecker product, the vectorization of a matrix, respectively.

2. System Description

Figure 1 shows the AGV with an antenna frame for track guidance. With the aid of the transponder (L1) the deviation from the predetermined track is determined. With this information, an external computer is able to determine the new direction required to return to the predetermined track as soon as possible (the external computer is not part of the system). Rotary encoders enable changing the direction of travel whenever necessary. Thus, it is possible to switch tracks at predetermined points. Again, the AGV corrects its position independently upon reaching the next transponder.

When the antenna crosses a transponder, the transponder is supplied by a 128 KHz energy field and transmits its code number back to the reading antenna at half frequency. The relative transponder position rectangular to the direction of travel is measured. From this relative position it is not possible to derive a world coordinate system without further effort due to the fact, that the transponder field is rotationally symmetrical to the longitudinal axis of the transponder. The internal interpreter decodes the transponder code. Each exceeding of the coordinate axes in direction of travel generates a positioning impulse with adjustable duration. Due to the measuring principle different signal strengths of transponders and altitude variations of the antenna have hardly any influence on the output signal.

In 3-dimensional (3-D) space, in order to obtain the position of the AGV, three parameters including the x coordinate, y coordinate and the z coordinate should be determined. Since the height h between AGV and floor is a constant, We only need to estimate the x coordinate and y coordinate. As the distance between the AGV and transponder is very close (about 30 cm), the positioning accuracy can not be assured by measuring the distance or angles in the polar coordinates. Therefore, in this paper, multi-antenna array is adopted to estimate the DoD/DoA for precise positioning.

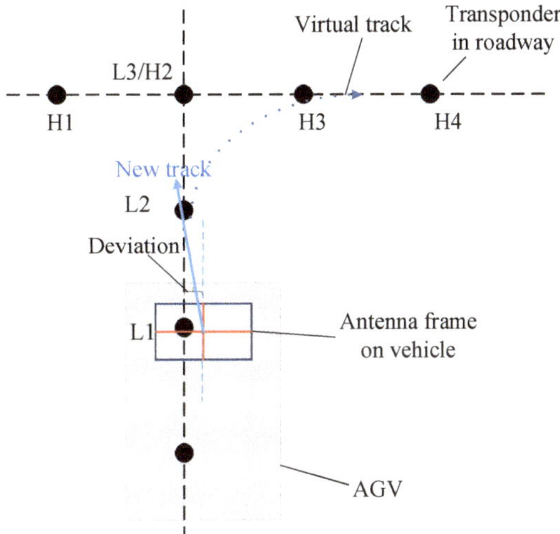

Figure 1. AGV with an antenna frame for track guiding.

3. System Model

Consider a collocated MIMO radar with M element ULA as the transmitter and N element ULA as the receiver [3–7], as Figure 2 shows, the transmit element lined in the x-axis and the receive element lined in the y-axis. The transmitter transmits M orthogonal waveforms, each through one antenna which are separately extracted through matched filtering at the receiver. The transmitted waveform for the m-th transmitting antenna is denoted as $s_m(t,p)$ ($m = 0,1,\ldots,M-1$) in the time domain, where p denotes the pulse index ($p = 0,1,\ldots,P-1$) and the number of pulses is P. Therefore, for the transmitted waveforms, we have

$$\int_{t \in T_p} s_m(t,p) s_{m'}^{H}(t,p)\, dt = \begin{cases} 0, & m \neq m' \\ 1, & m = m' \end{cases}, \tag{1}$$

where T_p denotes the pulse duration.

As the distance between the AGV and transponder is very close, the transponders can be considered as near-field targets. Assuming that there are K near-field targets, the DoD and the DoA for the k-th target ($k = 0,1,\ldots,K-1$) of the n-th antenna are denoted as α_k^n and β_k^n, respectively. In each target, we assume that the scattering coefficient is a type of Swelling II radar cross section (RCS) [8] and follows the independent and identically distribution (i.i.d.) between pulses. Therefore, during the p-th pulse, the scattering coefficient of the k-th target can be denoted as $\mu_k(p)$.

The received signals in the n-th antenna ($n = 0,1,\ldots,N-1$) can be expressed as

$$r_n(t,p) = \sum_{k=0}^{K-1} \sum_{m=0}^{M-1} \mu_k(p) s_m(t,p) e^{-j2\pi\frac{d}{\lambda}\left(n \sin \beta_k^n + m \sin \alpha_k^n\right)} + w_n(t,p), \tag{2}$$

where d denotes the fundamental antenna spacing, λ denotes the wavelength, $w_n(t,p)$ denotes the additive white Gaussian noise (AWGN) in the n-th receiving antenna during the p-th pulse, and $w_n(t,p) \sim \mathcal{CN}(0,\sigma_n^2)$.

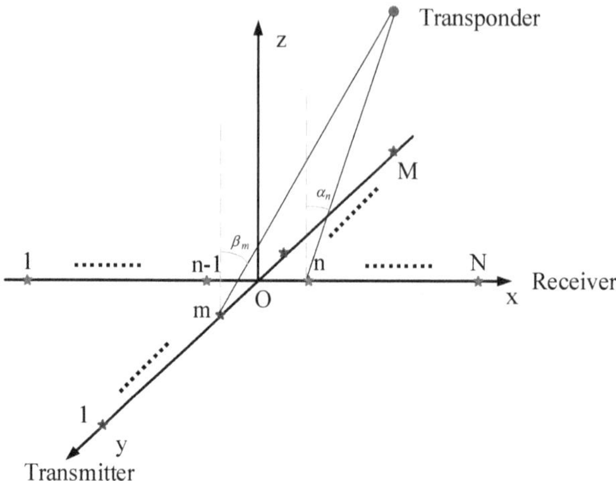

Figure 2. System model of the MIMO antenna for AGV positioning.

After the matched filter $h_m(t, p) \triangleq s_m^*(t_0 - t, p)$ for the m-th transmitted waveform and sampling at time t_0, we can obtain the pulse compression result

$$
\begin{aligned}
r_{n,m}(p) &\triangleq \int_{t \in T_p} r_n(t, p) s_m^*(t, p)\, dt \\
&= \int_t \sum_{k=0}^{K-1} e^{-j2\pi \frac{nd}{\lambda} \sin \alpha_k^n} \sum_{m'=0}^{M-1} \mu_k(p) s_{m'}(t, p) e^{-j2\pi \frac{m'd}{\lambda} \sin \beta_k^n} s_m^*(t, p)\, dt + \int w_n(p, t) s_m^*(t, p)\, dt \quad (3) \\
&= \sum_{k=0}^{K-1} \mu_k(p) e^{-j2\pi \frac{d}{\lambda}\left(n \sin \beta_k^n + m \sin \alpha_k^n\right)} + w_{n,m}(p),
\end{aligned}
$$

where we define

$$
w_{n,m}(p) \triangleq \int_{t \in T_p} w_n(t, p) s^*(t, p)\, dt, \tag{4}
$$

and $w_{n,m}(p) \sim \mathcal{CN}(0, \sigma_n^2)$. By collecting $r_{n,m}(p)$ into a vector

$$
\boldsymbol{r}_n(p) \triangleq [r_{n,0}(p), r_{n,1}(p), \ldots, r_{n,M-1}(p)]^T, \tag{5}
$$

the vector form of received signal can be obtained as

$$
\boldsymbol{r}_n(p) = \sum_{k=0}^{K-1} e^{-j2\pi \frac{nd}{\lambda} \sin \beta_k^n} \mu_k(p) \boldsymbol{a}(\alpha_k^n) + \boldsymbol{w}_n(p), \tag{6}
$$

where the noise vector is defined as

$$
\boldsymbol{w}_n(p) \triangleq [w_{n,0}(p), r_{n,1}(p), \ldots, w_{n,M-1}(p)]^T, \tag{7}
$$

and the steering vector in the transmitter is defined as

$$
\boldsymbol{a}(\alpha_k^n) \triangleq \left[1, e^{-j2\pi \frac{d}{\lambda} \sin \alpha_k^n}, \ldots, e^{-j2\pi \frac{(M-1)d}{\lambda} \sin \alpha_k^n}\right]^T. \tag{8}
$$

Collect all the received signals into a matrix, and we can obtain

$$\begin{aligned}
\boldsymbol{R}(p) &\triangleq \left[\boldsymbol{r}_0(p), \boldsymbol{r}_1(p), \ldots, \boldsymbol{r}_{N-1}(p)\right] \\
&= \sum_{k=0}^{K-1} \mu_k(p)\boldsymbol{a}(\alpha_k^n)\boldsymbol{b}^T(\beta_k^n) + \boldsymbol{W}(p),
\end{aligned} \tag{9}$$

where the steering vector in the receiver is defined as

$$\boldsymbol{b}(\beta_k^n) \triangleq \left[1, e^{-j2\pi\frac{d}{\lambda}\sin\beta_k^n}, \ldots, e^{-j2\pi\frac{(N-1)d}{\lambda}\sin\beta_k^n}\right]^T, \tag{10}$$

and the noise matrix is defined as

$$\boldsymbol{W}(p) \triangleq \left[\boldsymbol{w}_0(p), \boldsymbol{w}_1(p), \ldots, \boldsymbol{w}_{N-1}(p)\right]. \tag{11}$$

Vectorizing the matrix of received signals into a vector $\boldsymbol{r}(p) \triangleq \mathrm{vec}\{\boldsymbol{R}(p)\}$, the received signals can be expressed as the following form

$$\begin{aligned}
\boldsymbol{r}(p) &= \sum_{k=0}^{K-1} \mu_k(p)\,\mathrm{vec}\left\{\boldsymbol{a}(\alpha_k^n)\boldsymbol{b}^T(\beta_k^n)\right\} + \boldsymbol{w}(p) \\
&= \sum_{k=0}^{K-1} \mu_k(p)\boldsymbol{b}(\beta_k^n) \otimes \boldsymbol{a}(\alpha_k^n) + \boldsymbol{w}(p) \\
&= \boldsymbol{G}\boldsymbol{\mu}(p) + \boldsymbol{w}(p),
\end{aligned} \tag{12}$$

where $\boldsymbol{\mu}(p) \triangleq \left[\mu_0(p), \mu_1(p), \cdots, \mu_{K-1}(p)\right]^T$, $\boldsymbol{G} \triangleq \left[\boldsymbol{g}_0, \boldsymbol{g}_1, \ldots, \boldsymbol{g}_{K-1}\right]$, and $\boldsymbol{g}_k \triangleq \boldsymbol{b}(\beta_k^n) \otimes \boldsymbol{a}(\alpha_k^n)$. $\boldsymbol{w}(p) \triangleq \mathrm{vec}\{\boldsymbol{W}(p)\}$, and $\boldsymbol{w}(p) \sim \mathcal{CN}\left(0, \sigma_n^2 \boldsymbol{I}_{MN}\right)$.

Collect the received signals from all pulses, and the matrix form of all received signals can be obtained as

$$\begin{aligned}
\boldsymbol{R} &\triangleq \left[r(0), r(1), \cdots, r(P-1)\right] \\
&= \boldsymbol{G}\boldsymbol{\Gamma} + \boldsymbol{W},
\end{aligned} \tag{13}$$

where

$$\boldsymbol{W} \triangleq \left[\omega_0(p), \omega_1(p), \cdots, \omega_{K-1}(p)\right], \tag{14}$$

$$\boldsymbol{\Gamma} \triangleq \left[\mu(0), \mu(1), \cdots, \mu(K-1)\right]. \tag{15}$$

The vector form of all received signals can be written as

$$\begin{aligned}
\boldsymbol{r} &\triangleq \left[r(0)^T, r(1)^T, \cdots, r(P-1)^T\right]^T \\
&= (\boldsymbol{I}_P \otimes \boldsymbol{G})\boldsymbol{\mu} + \boldsymbol{\omega},
\end{aligned} \tag{16}$$

where

$$\boldsymbol{\omega} \triangleq \left[\omega(0)^T, \omega(1)^T, \cdots, \omega(P-1)^T\right]^T, \tag{17}$$

$$\boldsymbol{\mu} \triangleq \mathrm{vec}\{\boldsymbol{\Gamma}\}. \tag{18}$$

Therefore, the problem of DoD/DoA estimation is formulated in Equation (16), where both α_k^n and β_k^n will be estimated from the received signal $\boldsymbol{r}$ without the knowledge of target scattering coefficient $\mu_k(p)$, in the next section, we will develop an compressed sensing based algorithm to estimate the DoA/DoD.

4. Compressed Sensing Based DoA/DoD Estimation Algorithm

The DoA estimation algorithms with array antennas are widely applied in many fields, which are known as spectral estimation, angle of arrival (AoA) estimation, and bearing estimation. Much of the state-of-the-art in DoA estimation has its roots in time series analysis, spectrum analysis, periodograms, eigenstructure methods, parametric methods, linear prediction methods, beamforming, array processing, and adaptive array methods [9–12]. These estimation algorithms which are having different capabilities and limitations. In array signal processing, most high-resolution algorithms need to accurately estimate the signal subspace or noise subspace, so subspace estimation plays an important role. The conventional subspace estimation methods are obtained by eigenvalue decomposition of the covariance matrix of the received data or singular value decomposition of the received data. In the case of a large number of array elements, the two methods have high computational complexity. It is difficult to meet the requirements of real-time processing. Besides, in the signal environment of small sample support, due to the influence of noise, the sampling covariance matrix is difficult to reflect the real signal characteristics, resulting in the performance of the subspace estimated based on the eigenvalue decomposition method is significantly limited. Compressive Sensing (CS) theory is a new theory that has been proposed in the field of signal processing in recent years [13–17]. Using the CS theory to reduce the dimension of the high-dimensional original feature set can reduce the amount of underlying feature calculations and, thus, improve the algorithm speed. Besides, in the field of target tracking, the tracking algorithm based on covariance matrix can fuse multiple underlying features while maintaining low-dimensional characteristics, which reduces the computational complexity of the target matching process and maintains the balance between algorithm efficiency and robustness.

The transponder can be considered as a sparse target to the MIMO antenna array, therefore, a sparse-based method is proposed to estimate the DoAs/DoDs [13,18–22]. Discretize the angle of detection area into B grids, and the possible DoDs/DoAs are from the following two discretized sets

$$\mathbb{S}_\alpha \triangleq \{\alpha_b | b = 0, 1, \ldots, B - 1\}, \tag{19}$$

$$\mathbb{S}_\beta \triangleq \{\beta_b | b = 0, 1, \ldots, B - 1\}, \tag{20}$$

where $\alpha_b \leq \alpha_{b+1}$, $\beta_b \leq \beta_{b+1}$, and we can define D as the dictionary matrix as following

$$D \triangleq \left[d_{0,0}, d_{1,0}, \ldots, d_{B-1,0}, d_{0,1}, \ldots, d_{B-1,B-1} \right], \tag{21}$$

where $d_{b_1,b_2} \triangleq b(\beta_k^n) \otimes a(\alpha_k^n)$, the b-th column can be denoted as d_b [11,21–24]. Then, the estimation of DoA/DoD can be addressed by solve the following sparse reconstruction problem

$$\begin{aligned} \min_A \quad & \|A\|_{2,0} \\ \text{s.t.} \quad & \|r - DA\|_F^2 \leq \varepsilon, \end{aligned} \tag{22}$$

where ε is the factor to control the estimation accuracy, the non-zero entries of the sparse vector A denote the targets scattering coefficients, and the positions of the non-zero entries denote the DoAs/DoDs.

The orthogonal matching pursuit (OMP) method can be adopted to reconstruct the sparse vector A [20,25–30]. Details of the OMP algorithm is shown in Algorithm 1, we can estimate the target scattering coefficients and the DoDs/DoAs from the non-zero entries of set $\mathcal{R}$.

Algorithm 1 The DoA/DoD estimation method via OMP.

1: *Input:* The dictionary matrix D, the received signal r and the iteration number T.

2: *Initialization:* $t = 0$, $\mathcal{R} = \varnothing$, $r' = r$.

3: **while** $t \leq T - 1$ **do**

4:      $\mathcal{F} = \arg\max_b |d_b^H r'|$.

5:      $\mathcal{R} = \mathcal{R} \cup \mathcal{F}$.

6:      $\hat{A}_{\mathcal{R}} = \arg\min_{A_{\mathcal{R}}} \|r - D_{\mathcal{R}} A_{\mathcal{R}}\|_2^2$.

7:      $r' = r - D_{\mathcal{R}} \hat{A}_{\mathcal{R}}$.

8: **end while**

9: *Output:* The estimated sparse vector $A_{\mathcal{R}}$, and the non-zero entries index set $\mathcal{R}$.

---

5. The Theoretical Carmér-Rao Lower Bound (CRLB)

In this section, we will derive the Carmér-Rao Lower Bound (CRLB) theoretically [31–33]. The parameter vector can be denoted as $v \triangleq \left[\alpha^T, \beta^T, \mu^T\right]^T$, from the distribution of the received signals r we can obtain the joint CRLB of v. According to (12),the received signal r follows the Gaussian distribute, and the probability density function (PDF) of r can be expressed as [34]

$$f(r|v) = \frac{1}{\pi^L \det(\sigma_w^2 I)} e^{-(r-G\mu)^H \sigma_w^{-2} I (r-G\mu)}, \tag{23}$$

where $L \triangleq MN$, and we can calculate the Fisher information matrix (FIM) as

$$F \triangleq -\mathcal{E}\left\{\frac{\partial^2 \ln f(r|v)}{\partial^2 v}\right\} = \begin{bmatrix} F_{\beta\beta} & F_{\beta\alpha} & F_{\beta\mu} \\ F_{\alpha\beta} & F_{\alpha\alpha} & F_{\alpha\mu} \\ F_{\mu\beta} & F_{\mu\alpha} & F_{\mu\mu} \end{bmatrix}, \tag{24}$$

where

$$F_{\beta\beta} \triangleq -\mathcal{E}\left\{\frac{\partial^2 \ln f(r|v)}{\partial^2 \beta}\right\}, \quad F_{\beta\alpha} \triangleq -\mathcal{E}\left\{\frac{\partial^2 \ln f(r|v)}{\partial\beta\partial\alpha}\right\},$$

$$F_{\beta\mu} \triangleq -\mathcal{E}\left\{\frac{\partial^2 \ln f(r|v)}{\partial\beta\partial\mu}\right\}, \quad F_{\alpha\beta} \triangleq -\mathcal{E}\left\{\frac{\partial^2 \ln f(r|v)}{\partial\alpha\partial\beta}\right\},$$

$$F_{\alpha\alpha} \triangleq -\mathcal{E}\left\{\frac{\partial^2 \ln f(r|v)}{\partial^2 \alpha}\right\}, \quad F_{\alpha\mu} \triangleq -\mathcal{E}\left\{\frac{\partial^2 \ln f(r|v)}{\partial\alpha\partial\mu}\right\},$$

$$F_{\mu\beta} \triangleq -\mathcal{E}\left\{\frac{\partial^2 \ln f(r|v)}{\partial\mu\partial\beta}\right\}, \quad F_{\mu\alpha} \triangleq -\mathcal{E}\left\{\frac{\partial^2 \ln f(r|v)}{\partial\mu\partial\alpha}\right\},$$

$$F_{\mu\mu} \triangleq -\mathcal{E}\left\{\frac{\partial^2 \ln f(r|v)}{\partial^2 \mu}\right\}.$$

The detailed calculation of the entries in F is given in the Appendix. With the FIM, we can define $f' \triangleq \text{diag}\left\{F^{-1}\right\}$, and f_k' denotes the k-th entry of f'. Therefore, we can obtain the CRLBs of α and β

$$\text{CRLB}_\alpha = \sum_{i=K}^{2K-1} f_i', \tag{25}$$

$$\text{CRLB}_\beta = \sum_{i=0}^{K-1} f_i'. \tag{26}$$

6. Simulation Results

The simulation results are given in this section, and the parameters are set as follows: the echo signals signal-to-noise ratio (SNR) is 20 dB, the pulses number is $P = 100$, the space between antennas is $d = \frac{1}{2}\lambda$, the number of the vertical and horizontal receiving antennas is $M = 24$ and $N = 40$, respectively, the detection angle range is $[-90°, 90°]$. Since the size of antenna array is $2260 \times 1160 \times 185$ mm, the distance between two transponders is 25 feet (7.62 m), and the sensing area of the antenna array is about $x - y$ axes 500×1560 mm, the AGV can only detect one or two transponders at one time, for this consideration, the targets number is 2. Additionally, the root mean square error (RMSE) can be defined as following

$$\text{RMSE} \triangleq \sqrt{\frac{\sum_{q=0}^{Q-1} \left\| [\boldsymbol{\beta}^T, \boldsymbol{\alpha}^T]^T - [\hat{\boldsymbol{\beta}}_q^T, \hat{\boldsymbol{\alpha}}_q^T]^T \right\|_2^2}{Q}}. \tag{27}$$

where $\hat{\boldsymbol{\beta}}_q^T, \hat{\boldsymbol{\alpha}}_q^T$ denote the estimated DoD and DoA in the q-th simulation, respectively. $Q = 10^3$ denotes the total number of the simulations.

The two targets randomly distribute in the area $\beta_k \in [0, \frac{\pi}{2}]$ and $\alpha_k \in [-\frac{\pi}{2}, 0]$, set the numbers of discretized DoD/DoA are both 100, therefore, only two entries are non-zeros, the DoD/DoA estimation problem can be considered as a sparse reconstruction problem, and the OMP method can be used. The simulation results of the OMP estimation method is shown in Figure 3. From Figure 3 we can see that the proposed method can exactly estimate the DoDs/DoAs.

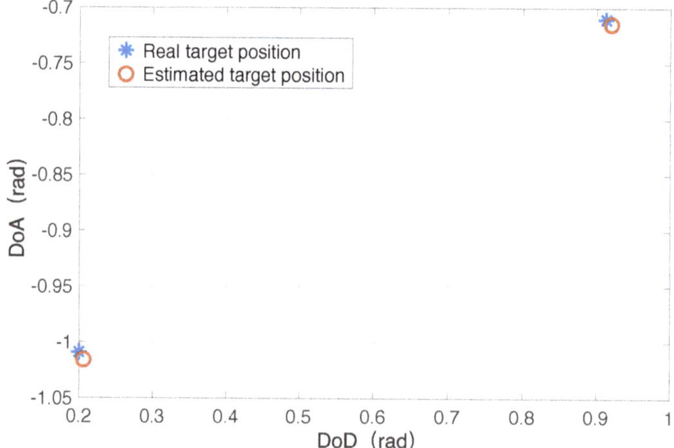

Figure 3. The estimated and the real targets position.

The DoD/DoA estimation performance of two targets is given in Figure 4, and we compare the estimated results with the CRLB derived in this paper. As shown in Figure 4, the RMSEs between simulation and CRLB shows a significant discrepancy in the lower SNR region, and as the SNR increases, the discrepancy can be converged to 0.1. The discrepancy is mainly caused by the grid mismatch problem with a discrete potential angle space. We can improve it by increasing the discrete grid number. However, it will increase the complexity of the system.

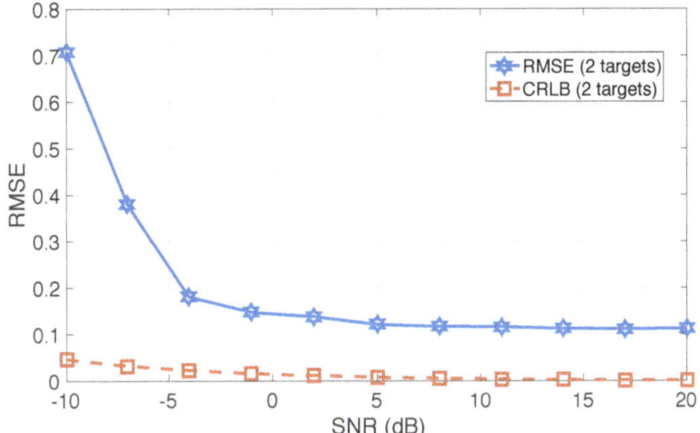

Figure 4. The 3D positioning error map of GPS.

We also give the comparison of the proposed angular estimation positioning system and the traditional GPS. GPS is a wireless navigation system that is positioned by navigation satellites. It consists of space, ground monitoring and user receivers. It can achieve omnipotence (marine, terrestrial, aerospace, aerospace), global, all-weather, continuous and real-time navigation and positioning capabilities provide precise position and speed information. It is one of the navigation systems suitable for port AGV applications.

The AGV positioning geometry is shown in Figure 5. One of the two satellite detectors is used as the coordinate origin $(0,0)$. The distance between the two detectors is L, and the coordinates of the other detector are $(L,0)$. Assuming that the coordinates of the AGV are (x, y), the angles of the AGV signals received by the two satellite detectors can be estimated as α and β by the spatial spectrum algorithm. Then, the coordinates of the AGV can be obtained by the following geometric relations

$$x = \frac{L \tan \alpha}{\tan \alpha - \tan \beta}, \tag{28}$$

$$y = \frac{L}{\tan \alpha - \tan \beta}. \tag{29}$$

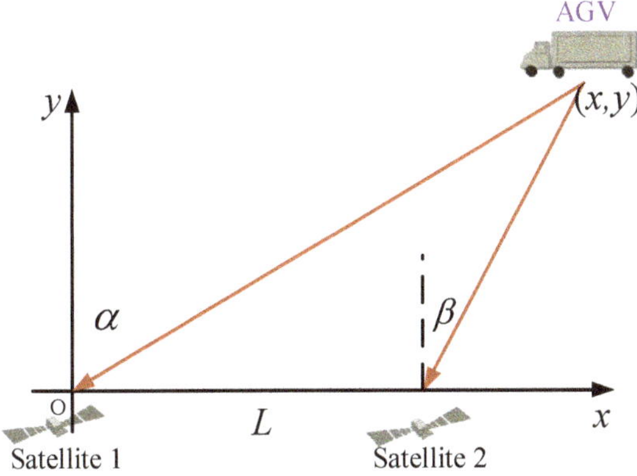

Figure 5. Geometry relationship of AGV positioning via GPS.

From Equations (28) and (29), we can obtain the AGV $x - y$ coordinate, and the angle can be derived directly. Figure 6 shows the RMSE of both the proposed MIMO antenna positioning method and GPS method, where the SNR of the received GPS signal is the equivalent value after despreaded and the actual SNR of GPS signal is 30 dB lower than that. Additionally, the estimation performance of the proposed method and GPS is shown in Figure 7, we can see that the proposed CS-based positioning method outperforms the traditional GPS method.

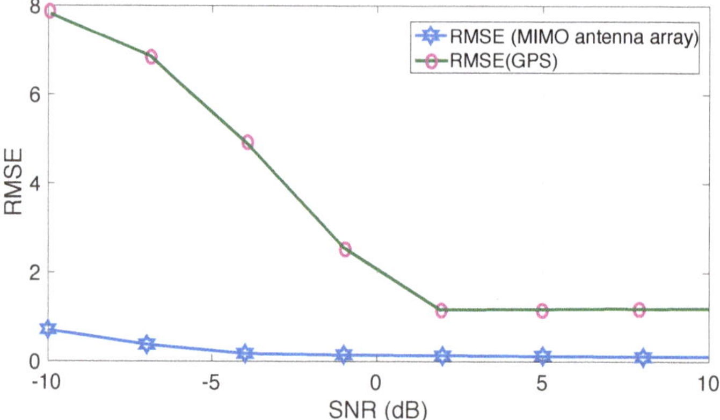

Figure 6. RMSE as a function of SNR with different positioning method.

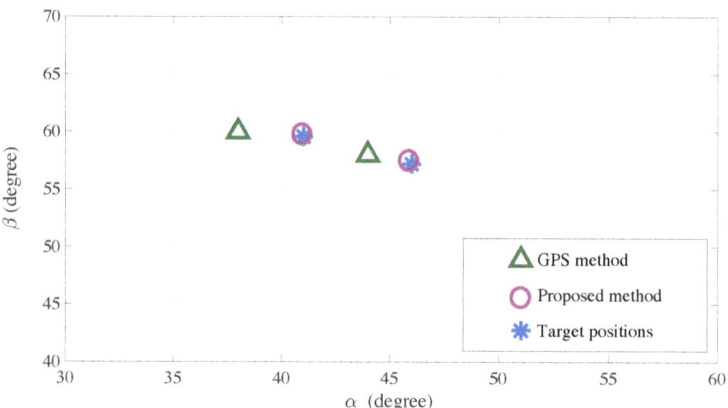

Figure 7. The estimation performance of the proposed method and GPS.

7. Conclusions

This paper utilizes the antenna array to acquire the arrival angle information of the transponder signal reaching each antenna, thereby obtaining the position information of the transponder and performing the AGV positioning by using the relative position of the transponder and the AGV. According to the method provided by the present invention, high-precision positioning of the AGV can be realized in an outdoor environment where there are many obstacles such as ports and docks. Compared with the GPS, the positioning method can ignore the influence of obstacles on signal propagation, the system structure can be simplified and the cost can be reduced.

Author Contributions: Z.C. conceived and designed the work that led to the submission, X.H. and Z.X.C. acquired data and played important roles in interpreting the results, Y.J. and J.L. revised the manuscript and approved the final version.

Funding: This research was funded by [the National Natural Science Foundation of China] grant number [61601281, 61501298, 61471117, 61603239,61801377]. [The Open Program of State Key Laboratory of Millimeter Waves] grant number [Southeast University, Z201804]. [The National Key Research and Development Program of China] grant number [2016YFB0501301].

Acknowledgments: This work was supported in part by the National Natural Science Foundation of China (Grant No. 61601281, 61501298, 61471117, 61603239), the Open Program of State Key Laboratory of Millimeter Waves (Southeast University, Z201804), and the National Key Research and Development Program of China (2016YFB0501301).

Conflicts of Interest: The authors declare no conflict of interest.

Appendix A. The Expressions of FIM Entries

Firstly, the following results can be obtained

$$
\begin{aligned}
l_\beta \;&\triangleq\; -\frac{\partial \ln f(r|v)}{\partial \beta} \\
&= \sigma_w^{-2}\frac{\partial (r-G\mu)^H(r-G\mu)}{\partial \beta} \\
&= -2\sigma_w^{-2}(r-G\mu)^H\frac{\partial G\mu}{\partial \beta} \\
&= -2\sigma_w^{-2}(r-G\mu)^H\frac{\partial \sum_{k=0}^{K-1} g_k\mu_k}{\partial \beta} \\
&= -2\sigma_w^{-2}(r-G\mu)^H\left[\frac{\partial g_0\mu_0}{\partial \beta_0},\ldots,\frac{\partial g_{K-1}\mu_{K-1}}{\partial \beta_{K-1}}\right],
\end{aligned}
\tag{A1}
$$

$$
\begin{aligned}
l_\alpha \;&\triangleq\; -\frac{\partial \ln f(r|v)}{\partial \alpha} \\
&= -2\sigma_w^{-2}(r-G\mu)^H\left[\frac{\partial g_0\mu_0}{\partial \alpha_0},\ldots,\frac{\partial g_{K-1}\mu_{K-1}}{\partial \alpha_{K-1}}\right],
\end{aligned}
\tag{A2}
$$

$$
\begin{aligned}
l_\mu \;&\triangleq\; -\frac{\partial \ln f(r|v)}{\partial \mu} \\
&= -2\sigma_w^{-2}(r-G\mu)^H\frac{\partial G\mu}{\partial \mu} \\
&= -2\sigma_w^{-2}(r-G\mu)^H G.
\end{aligned}
\tag{A3}
$$

The i-th row and j-th column entry of the FIM can be written as

$$
\begin{aligned}
\left[F_{\beta\beta}\right]_{i,j} \;&= -2\sigma_w^{-2}\mathcal{E}\left\{\frac{\partial (r-G\mu)^H\left(\mu_i a(\alpha_i)\otimes\frac{\partial b(\beta_i)}{\partial \beta_i}\right)}{\partial \beta_j}\right\} \\
&= -2\sigma_w^{-2}\mu_i\mathcal{E}\left\{-\left(\frac{\partial G\mu}{\partial \beta_j}\right)^H\left(a(\alpha_i)\otimes\frac{\partial b(\beta_i)}{\partial \beta_i}\right)\right. \\
&\qquad\left.+(r-G\mu)^H\frac{\partial\left(a(\alpha_i)\otimes\frac{\partial b(\beta_i)}{\partial \beta_i}\right)}{\partial \beta_j}\right\} \\
&= 2\sigma_w^{-2}\mu_i\mu_j^H\left(a(\alpha_j)\otimes\frac{\partial b(\beta_j)}{\partial \beta_j}\right)^H\left(a(\alpha_i)\otimes\frac{\partial b(\beta_i)}{\partial \beta_i}\right) \\
&= 2\sigma_w^{-2}\mu_i\mu_j^H a^H(\alpha_j)a(\alpha_i)\frac{\partial b^H(\beta_j)}{\partial \beta_j}\frac{\partial b(\beta_i)}{\partial \beta_i},
\end{aligned}
\tag{A4}
$$

$$
\begin{aligned}
\left[F_{\beta\alpha}\right]_{i,j} \;&= -2\sigma_w^{-2}\mathcal{E}\left\{\frac{\partial (r-G\mu)^H\left(\mu_i a(\alpha_i)\otimes\frac{\partial b(\beta_i)}{\partial \beta_i}\right)}{\partial \alpha_j}\right\} \\
&= 2\sigma_w^{-2}\mu_i\mu_j^H\frac{\partial a^H(\alpha_j)}{\partial \alpha_j}a(\alpha_i)b^H(\beta_j)\frac{\partial b(\beta_i)}{\partial \beta_i},
\end{aligned}
\tag{A5}
$$

$$
\begin{aligned}
\left[F_{\beta\mu}\right]_{i,j} \;&= -2\sigma_w^{-2}\mathcal{E}\left\{\frac{\partial (r-G\mu)^H\left(\mu_i a(\alpha_i)\otimes\frac{\partial b(\beta_i)}{\partial \beta_i}\right)}{\partial \mu_j}\right\} \\
&= 2\sigma_w^{-2}\mu_i a^H(\alpha_j)a(\alpha_i)b^H(\beta_j)\frac{\partial b(\beta_i)}{\partial \beta_i},
\end{aligned}
\tag{A6}
$$

$$\begin{aligned}
\left[\boldsymbol{F}_{\alpha\alpha}\right]_{i,j} &= -2\sigma_w^{-2}\mathcal{E}\left\{\frac{\partial(\boldsymbol{r}-\boldsymbol{G\mu})^H\left(\mu_i\frac{\partial\boldsymbol{a}(\alpha_i)}{\partial\alpha_i}\otimes\boldsymbol{b}(\beta_i)\right)}{\partial\alpha_j}\right\} \\
&= 2\sigma_w^{-2}\mu_i\frac{\partial(\boldsymbol{g}_j\mu_j)^H}{\partial\alpha_j}\left(\frac{\partial\boldsymbol{a}(\alpha_i)}{\partial\alpha_i}\otimes\boldsymbol{b}(\beta_i)\right) \\
&= 2\sigma_w^{-2}\mu_i\mu_j^H\frac{\partial\boldsymbol{a}^H(\alpha_j)}{\partial\alpha_j}\frac{\partial\boldsymbol{a}(\alpha_i)}{\partial\alpha_i}\boldsymbol{b}^H(\beta_j)\boldsymbol{b}(\beta_i),
\end{aligned} \tag{A7}$$

$$\begin{aligned}
\left[\boldsymbol{F}_{\alpha\beta}\right]_{i,j} &= -2\sigma_w^{-2}\mathcal{E}\left\{\frac{\partial(\boldsymbol{r}-\boldsymbol{G\mu})^H\left(\mu_i\frac{\partial\boldsymbol{a}(\alpha_i)}{\partial\alpha_i}\otimes\boldsymbol{b}(\beta_i)\right)}{\partial\beta_j}\right\} \\
&= 2\sigma_w^{-2}\mu_i\frac{\partial(\boldsymbol{g}_j\mu_j)^H}{\partial\beta_j}\left(\frac{\partial\boldsymbol{a}(\alpha_i)}{\partial\alpha_i}\otimes\boldsymbol{b}(\beta_i)\right) \\
&= 2\sigma_w^{-2}\mu_i\mu_j^H\boldsymbol{a}^H(\alpha_j)\frac{\partial\boldsymbol{a}(\alpha_i)}{\partial\alpha_i}\frac{\partial\boldsymbol{b}^H(\beta_j)}{\partial\beta_j}\boldsymbol{b}(\beta_i),
\end{aligned} \tag{A8}$$

$$\begin{aligned}
\left[\boldsymbol{F}_{\alpha\mu}\right]_{i,j} &= -2\sigma_w^{-2}\mathcal{E}\left\{\frac{\partial(\boldsymbol{r}-\boldsymbol{G\mu})^H\left(\mu_i\frac{\partial\boldsymbol{a}(\alpha_i)}{\partial\alpha_i}\otimes\boldsymbol{b}(\beta_i)\right)}{\partial\mu_j}\right\} \\
&= 2\sigma_w^{-2}\mu_i\frac{\partial(\boldsymbol{g}_j\mu_j)^H}{\partial\mu_j}\left(\frac{\partial\boldsymbol{a}(\alpha_i)}{\partial\alpha_i}\otimes\boldsymbol{b}(\beta_i)\right) \\
&= 2\sigma_w^{-2}\mu_i\boldsymbol{a}^H(\alpha_j)\frac{\partial\boldsymbol{a}(\alpha_i)}{\partial\alpha_i}\boldsymbol{b}^H(\beta_j)\boldsymbol{b}(\beta_i),
\end{aligned} \tag{A9}$$

$$\begin{aligned}
\left[\boldsymbol{F}_{\mu\mu}\right]_{i,j} &= -2\sigma_w^{-2}\mathcal{E}\left\{\frac{\partial(\boldsymbol{r}-\boldsymbol{G\mu})^H\boldsymbol{g}_i}{\partial\mu_j}\right\} \\
&= 2\sigma_w^{-2}\boldsymbol{g}_j^H\boldsymbol{g}_i,
\end{aligned} \tag{A10}$$

$$\begin{aligned}
\left[\boldsymbol{F}_{\mu\beta}\right]_{i,j} &= -2\sigma_w^{-2}\mathcal{E}\left\{\frac{\partial(\boldsymbol{r}-\boldsymbol{G\mu})^H\boldsymbol{g}_i}{\partial\beta_j}\right\} \\
&= 2\sigma_w^{-2}\mu_j^H\frac{\partial\boldsymbol{g}_j^H}{\partial\beta_j}\boldsymbol{g}_i,
\end{aligned} \tag{A11}$$

$$\begin{aligned}
\left[\boldsymbol{F}_{\mu\alpha}\right]_{i,j} &= -2\sigma_w^{-2}\mathcal{E}\left\{\frac{\partial(\boldsymbol{r}-\boldsymbol{G\mu})^H\boldsymbol{g}_i}{\partial\alpha_j}\right\} \\
&= 2\sigma_w^{-2}\mu_j^H\frac{\partial\boldsymbol{g}_j^H}{\partial\alpha_j}\boldsymbol{g}_i.
\end{aligned} \tag{A12}$$

References

1. Le-Anh, T.; De Koster, M.B.M. A review of design and control of automated guided vehicle systems. *Eur. J. Oper. Res.* **2006**, *171*, 1–23. [CrossRef]
2. Lei Wang, J.; Shu, T.; Kumagai, M. A 3D scanning laser rangefinder and its application to an autonomous guided vehicle. In Proceedings of the IEEE Vehicular Technology Conference, Boston, MA, USA, 24–28 September 2000; pp. 331–335.
3. Li, J.; Stoica, P. MIMO radar with colocated antennas. *IEEE Signal Process. Mag.* **2007**, *24*, 106–114. [CrossRef]
4. Davis, M.; Showman, G.; Lanterman, A. Coherent MIMO radar: The phased array and orthogonal waveforms. *IEEE Aerosp. Electron. Syst. Mag.* **2014**, *29*, 76–91. [CrossRef]
5. Chen, Z.; Cao, Z.; He, X.; Jin, Y.; Li, J.; Chen, P. DoA and DoD Estimation and Hybrid Beamforming for Radar-Aided mmWave MIMO Vehicular Communication Systems. *Electronics* **2018**, *3*, 40. [CrossRef]
6. Chen, P.; Wu, L.; Qi, C. Waveform optimization for target scattering coefficients estimation under detection and peak-to-average power ratio constraints in cognitive radar. *Circ. Syst. Signal Process.* **2016**, *35*, 163–184. [CrossRef]
7. Chen, P.; Wu, L. Coding matrix optimization in cognitive radar system with EBPSK-based MCPC signal. *J. Electromagn. Waves Appl.* **2015**, *29*, 1497–1507. [CrossRef]
8. Skolnik, M. *Radar Handbook*, 3rd ed.; McGraw-Hill: New York, NY, USA, 2008.
9. Lemma, A.N.; van der Veen, A.; Deprettere, E.F. Analysis of joint angle-frequency estimation using ESPRIT. *IEEE Trans. Signal Process.* **2003**, *51*, 1264–1283. [CrossRef]

10. Chen, Z.; Chen, P.; Li, J.; Miao, P. Non-orthogonal multi-carrier MIMO communication system using M-ary efficient modulation. *Digit. Signal Process.* **2018**, *76*, 14–21. [CrossRef]

11. Zamani, H.; Zayyani, H.; Marvasti, F. An iterative dictionary learning-based algorithm for DOA estimation. *IEEE Commun. Lett.* **2016**, *20*, 1784–1787. [CrossRef]

12. Chen, Z.; Wu, L.; Chen, P. Efficient modulation and demodulation methods for multi-carrier communication. *IET Commun.* **2016**, *10*, 567–576. [CrossRef]

13. Petropulu, A.P.; Poor, H.V. Measurement matrix design for compressive sensing–based MIMO radar. *IEEE Trans. Signal Process.* **2011**, *59*, 5338–5352.

14. Donoho, D.L. Compressed sensing. *IEEE Trans. Inf. Theory* **2006**, *52*, 1289–1306. [CrossRef]

15. Candes, E.J.; Wakin, M.B. An Introduction To Compressive Sampling. *IEEE Signal Process. Mag.* **2008**, *25*, 21–30. [CrossRef]

16. Donoho, D.L.; Tsaig, Y.; Drori, I.; Starck, J. Sparse Solution of Underdetermined Systems of Linear Equations by Stagewise Orthogonal Matching Pursuit. *IEEE Trans. Inf. Theory* **2012**, *58*, 1094–1121. [CrossRef]

17. Massa, A.; Rocca, P.; Oliveri, G. Compressive Sensing in Electromagnetics–A Review. *IEEE Antennas Propag. Mag.* **2015**, *57*, 224–238. [CrossRef]

18. Baraniuk, R. Compressive sensing. *IEEE Signal Process. Mag.* **2007**, *24*, 118–121. [CrossRef]

19. Yao, Y.; Petropulu, A.P.; Poor, H.V. MIMO radar using compressive sampling. *IEEE J. Sel. Areas Signal Process.* **2010**, *4*, 146–163.

20. Zhao, T.; Peng, Y.; Nehorai, A. Joint sparse recovery method for compressed sensing with structured dictionary mismatches. *IEEE Trans. Signal Process.* **2014**, *62*, 4997–5008.

21. Fishler, E.; Haimovich, A.; Blum, R.S.; Cimini, L.J.; Chizhik, D.; Valenzuela, R.A. Spatial diversity in radars-models and detection performance. *IEEE Trans. Signal Process.* **2006**, *54*, 823–838. [CrossRef]

22. Fishler, E.; Haimovich, A.; Blum, R.; Chizhik, D.; Cimini, L.; Valenzuela, R. MIMO radar: An idea whose time has come. In Proceedings of the IEEE Radar Conference, Philadelphia, PA, USA, 26–29 April 2004; pp. 71–78.

23. Yang, Z.; Xie, L. Exact joint sparse frequency recovery via optimization methods. *IEEE Trans. Signal Process.* **2016**, *64*, 5145–5157. [CrossRef]

24. Chen, P.; Qi, C.; Wu, L.; Wang, X. Estimation of Extended Targets Based on Compressed Sensing in Cognitive Radar System. *IEEE Trans. Veh. Technol.* **2017**, *66*, 941–951. [CrossRef]

25. Khwaja, A.S.; Cetin, M. Compressed Sensing ISAR Reconstruction Considering Highly Maneuvering Motion. *Electronics* **2017**, *6*, 21. [CrossRef]

26. Tropp, J.A.; Gilbert, A.C. Signal recovery from random measurements via orthogonal matching pursuit. *IEEE Trans. Inf. Theory* **2007**, *53*, 4655–4666. [CrossRef]

27. Chen, P.; Zheng, L.; Wang, X.; Li, H.; Wu, L. Moving target detection using colocated MIMO radar on multiple distributed moving platforms. *IEEE Trans. Signal Process.* **2017**, *65*, 4670–4683. [CrossRef]

28. Zheng, L.; Wang, X. Super-Resolution Delay-Doppler Estimation for OFDM Passive Radar. *IEEE Trans. Signal Process.* **2017**, *65*, 2197–2210. [CrossRef]

29. Davenport, M.; Wakin, M. Analysis of orthogonal matching pursuit using the restricted isometry property. *IEEE Trans. Inf. Theory* **2010**, *56*, 4395–4401. [CrossRef]

30. Pinchera, D.; Migliore, M.D.; Lucido, M.; Schettino, F.; Panariello, G. A Compressive-Sensing Inspired Alternate Projection Algorithm for Sparse Array Synthesis. *Electronics* **2018**, *6*, 3. [CrossRef]

31. Chen, P.; Qi, C.; Wu, L. Antenna placement optimisation for compressed sensing-based distributed MIMO radar. *IET Radar Sonar Navig.* **2017**, *11*, 285–293. [CrossRef]

32. Chen, Z.; Chen, P. Compressed sensing-based DOA and DOD estimation in bistatic co-prime MIMO arrays. In Proceedings of the 2017 IEEE Conference on Antenna Measurements & Applications (CAMA), Ibaraki, Japan, 4–6 December 2017; pp. 297–300.

33. Gogineni, S.; Nehorai, A. Target estimation using sparse modeling for distributed MIMO radar. *IEEE Trans. Signal Process.* **2011**, *59*, 5315–5325. [CrossRef]

34. Beckmann, P. Statistical distribution of the amplitude and phase of a multiply scattered field. *J. Res. Natl. Bur. Stand.* **1962**, *660*, 231–240. [CrossRef]

 *electronics*

Article

Sparse DOD/DOA Estimation in a Bistatic MIMO Radar With Mutual Coupling Effect

Peng Chen [1,*], Zhenxin Cao [1], Zhimin Chen [2] and Chunhua Yu [3]

[1] State Key Laboratory of Millimeter Waves, Southeast University, Nanjing 210096, China; caozx@seu.edu.cn
[2] School of Electronic and Information, Shanghai Dianji University, Shanghai 201306, China; chenzm@sdju.edu.cn
[3] School of Electronic Science and Engineering, Nanjing University, Nanjing 210096, China; yu506@hotmail.com
* Correspondence: chenpengseu@seu.edu.cn; Tel.: +86-158-9595-2189

Received: 1 November 2018; Accepted: 19 November 2018; Published: 21 November 2018

Abstract: The unknown mutual coupling effect between antennas significantly degrades the target localization performance in the bistatic multiple-input multiple-output (MIMO) radar. In this paper, the joint estimation problem for the direction of departure (DOD) and direction of arrival (DOA) is addressed. By exploiting the target sparsity in the spatial domain and formulating a dictionary matrix with discretizing the DOD/DOA into grids, compressed sensing (CS)-based system model is given. However, in the practical MIMO radar systems, the target cannot be precisely on the grids, and the unknown mutual coupling effect degrades the estimation performance. Therefore, a novel CS-based DOD/DOA estimation model with both the off-grid and mutual coupling effect is proposed, and a novel sparse reconstruction method is proposed to estimate DOD/DOA with updating both the off-grid and mutual coupling parameters iteratively. Moreover, to describe the estimation performance, the corresponding Cramér–Rao lower bounds (CRLBs) with all the unknown parameters are theoretically derived. Simulation results show that the proposed method can improve the DOD/DOA estimation in the scenario with unknown mutual coupling effect, and outperform state-of-the-art methods.

Keywords: bistatic MIMO radar; DOD/DOA estimation; mutual coupling; off-grid sparse problem

1. Introduction

In multiple-input multiple-output (MIMO) radar systems [1,2], the independent waveforms are adopted in different transmitting antennas, so compared with the traditional array radars, the better performance of target estimation and detection can be achieved by using more spatial and waveform diversities [3–5]. Usually, the MIMO radar systems can be categorized into the following two types with different antenna distances: (1) Colocated MIMO radar system: The antennas in receiver and transmitter are close to each other, so the waveform diversity can be exploited to improve the radar performance [1,6,7]; (2) Distributed MIMO radar system: The antennas in receiver and transmitter are widely separated, so the radar performance can be improved by exploiting the diversity of target's radar cross-section (RCS) [2,8].

Moreover, with different positions of transmitter and receiver, the colocated MIMO radar can also be categorized into monostatic and bistatic MIMO radar systems. The transmitter and receiver are close in the monostatic MIMO radar system [9], so more reliable beam-pattern design and target detection can be achieved. However, in the bistatic MIMO radar [9,10], the transmitter and receiver are widely separated, so the better performance of target localization can be achieved with the different view angles from transmitter and receiver. Therefore, in this paper, we consider the problem of the direction of departure (DOD) and the direction of arrival (DOA) estimation, and the bistatic MIMO radar system is adopted.

The DOD/DOA estimation problem in the MIMO radar system has been widely studied. For example, in the scenario with a non-uniform array, a novel method is proposed to construct a virtual MIMO array and estimate DOD/DOA in [11]. In [12,13], the DOD/DOA estimation method for the scenario with unknown correlated noise has been proposed, where the estimation method is based on the canonical correlation decomposition and the shift-invariance properties of Kronecker product. Additionally, some studies [14–17] have proposed algorithms based on the multiple signal classification (MUSIC) and the estimation of signal parameters via rotational invariance techniques (ESPRIT) to estimate DOD/DOA in MIMO radar systems. However, these studies have not considered the mutual coupling between antennas in transmitter and receiver. In [18], the mutual coupling has been studied in the problem of direction finding. Additionally, the DOD/DOA estimation method with unknown mutual coupling is proposed in [19]. Different from these present papers, we propose a novel method to estimate the DOD/DOA in the bistatic MIMO radar system, where the sparsity of targets has been exploited to improve the estimation performance.

In this paper, we consider the problem of estimating the DOD/DOA in the bistatic MIMO radar system with mutual coupling between antennas. A novel iterative method based on compressed sensing (CS) is proposed to estimate the parameters including DOD/DOA, mutual coupling matrices, and target scattering coefficients, by exploiting the sparsity of targets in the spatial domain. Additionally, to further improve the estimation performance, an off-grid problem is formulated, and the parameters are polished iteratively by solving the off-grid problem. Furthermore, the corresponding Cramér–Rao lower bounds (CRLBs) for the estimated target parameters are derived theoretically. To summarize, we make the contributions as follows:

- **Sparse DOD/DOA estimation model with mutual coupling effect:** In the bistatic MIMO radar system, the DOD/DOA estimation model is proposed based on the sparse reconstruction model, and the unknown mutual coupling effect between antennas is also considered.
- **Sparse DOD/DOA estimation method with off-grid effect:** In the sparse reconstruction methods, the detection area is discretized into grids to formulated the dictionary matrix, so the off-grid effect limits the reconstruction performance. Therefore, combining both off-grid effect and mutual coupling effect, the sparse DOD/DOA estimation method is proposed.
- **Theoretical CRLB expression for DOD/DOA estimation with mutual coupling effect:** The corresponding CRLB with the unknown mutual coupling effect is theoretically derived to describe the estimation performance.

The remainder of this paper is organized as follows. The system model of bistatic MIMO radar is given in Section 2. The estimation method for DOD/DOA and mutual coupling matrices is proposed in Section 3. Section 4 derives the Cramér–Rao lower bound (CRLB). Then, Section 5 gives the computational complexity. Simulation results are given in Section 6. Finally, Section 7 concludes the paper.

Notations: I_N denotes an $N \times N$ identity matrix. $\mathcal{E}\{\cdot\}$ denotes the expectation operation. $\mathcal{CN}(a, B)$ denotes the complex Gaussian distribution with the mean being a and the variance matrix being B. $\|\cdot\|_1$, $\|\cdot\|_2$, $\otimes$, $\text{Tr}\{\cdot\}$, $\text{vec}\{\cdot\}$, $(\cdot)^*$, $(\cdot)^T$ and $(\cdot)^H$ denote the ℓ_1 norm, the ℓ_2 norm, the Kronecker product, the trace of a matrix, the vectorization of a matrix, the conjugate, the matrix transpose and the Hermitian transpose, respectively. For a matrix A, $[A]_n$ denotes the n-th column of A, and for a vector a, $[a]_n$ denotes the n-th entry of a.

2. The System Model of Bistatic MIMO Radar

2.1. Bistatic MIMO Radar System without Mutual Coupling

In this paper, the bistatic MIMO radar system [12,20,21] is considered and the radar system is shown in Figure 1, where M transmitting antennas and N receiving antennas are adopted. In each transmitting antenna, the orthogonal signal is transmitted. The transmitted waveform for the m-th transmitting antenna is denoted as $s_m(t, p)$ ($m = 0, 1, \ldots, M - 1$) in the time domain, where p denotes the pulse index ($p = 0, 1, \ldots, P - 1$) and the number of pulses is P. Therefore, for the transmitted waveforms, we have

$$\int_{t \in T_p} s_m(t, p) s_{m'}^H(t, p) \, dt = \begin{cases} 0, & m \neq m' \\ 1, & m = m' \end{cases}, \tag{1}$$

where T_p denotes the pulse duration.

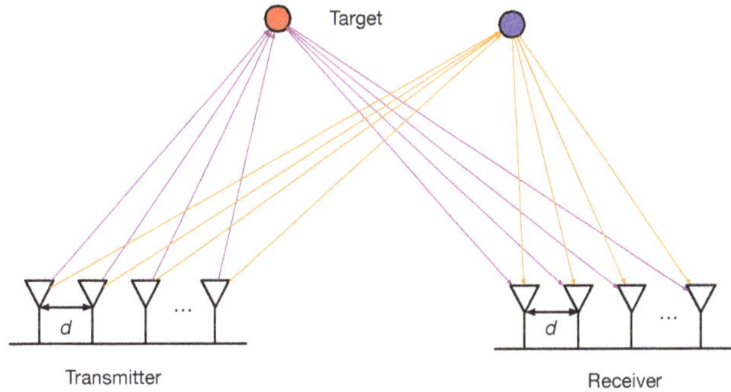

Figure 1. The system model of the bistatic MIMO radar.

Assuming that there are K far-field targets, the direction of departure (DOD) and the direction of arrival (DOA) for the k-th target ($k = 0, 1, \ldots, K - 1$) are denoted as ϕ_k and ψ_k, respectively. In each target, we assume that the the scattering coefficient is a type of Swerling II RCS [22] and follows the independent and identically distribution (i.i.d.) between pulses. Therefore, during the p-th pulse, the scattering coefficient of the k-th target can be denoted as $\alpha_k(p)$.

Without considering the mutual coupling between antennas, the received signals in the n-th antenna ($n = 0, 1, \ldots, N - 1$) can be expressed as

$$r_n(t, p) = \sum_{k=0}^{K-1} \sum_{m=0}^{M-1} \alpha_k(p) s_m(t, p) e^{-j2\pi \frac{d}{\lambda}(n \sin \psi_k + m \sin \phi_k)} + w_n(t, p), \tag{2}$$

where d denotes the fundamental antenna spacing, λ denotes the wavelength, and $w_n(l, p)$ denotes the additive white Gaussian noise (AWGN) in the n-th receiving antenna during the p-th pulse, and $w_n(t, p) \sim \mathcal{CN}(0, \sigma_n^2)$.

After the matched filter $h_m(t, p) \triangleq s_m^*(t_0 - t, p)$ for the m-th transmitted waveform and sampling at time t_0, we can obtain the result of pulse compression

$$\begin{aligned} r_{n,m}(p) &\triangleq \int_{t \in T_p} r_n(t, p) s_m^*(t, p) \, dt \\ &= \sum_{k=0}^{K-1} \alpha_k(p) e^{-j2\pi \frac{d}{\lambda}(n \sin \psi_k + m \sin \phi_k)} + w_{n,m}(p), \end{aligned} \tag{3}$$

where we define $w_{n,m}(p) \triangleq \int_{t \in T_p} w_n(t,p) s^*(t,p) \, dt$, and $w_{n,m}(p) \sim \mathcal{CN}(0, \sigma_n^2)$. By collecting $r_{n,m}(p)$ into a vector $\boldsymbol{r}_n(p) \triangleq [r_{n,0}(p), r_{n,1}(p), \ldots, r_{n,M-1}(p)]^T$, the vector form of received signal can be obtained as

$$\boldsymbol{r}_n(p) = \sum_{k=0}^{K-1} e^{-j2\pi \frac{nd}{\lambda} \sin \psi_k} \alpha_k(p) \boldsymbol{a}(\phi_k) + \boldsymbol{w}_n(p), \tag{4}$$

where the noise vector is defined as $\boldsymbol{w}_n(p) \triangleq [w_{n,0}(p), r_{n,1}(p), \ldots, w_{n,M-1}(p)]^T$, and the steering vector in the transmitter is defined as

$$\boldsymbol{a}(\phi_k) \triangleq \left[1, e^{-j2\pi \frac{d}{\lambda} \sin \phi_k}, \ldots, e^{-j2\pi \frac{(M-1)d}{\lambda} \sin \phi_k}\right]^T. \tag{5}$$

Collect all the received signals into a matrix, and we can obtain

$$\boldsymbol{R}(p) \triangleq \left[\boldsymbol{r}_0(p), \boldsymbol{r}_1(p), \ldots, \boldsymbol{r}_{N-1}(p)\right] = \sum_{k=0}^{K-1} \alpha_k(p) \boldsymbol{a}(\phi_k) \boldsymbol{b}^T(\psi_k) + \boldsymbol{W}(p), \tag{6}$$

where the steering vector in the receiver is defined as

$$\boldsymbol{b}(\psi_k) \triangleq \left[1, e^{-j2\pi \frac{d}{\lambda} \sin \psi_k}, \ldots, e^{-j2\pi \frac{(N-1)d}{\lambda} \sin \phi_k}\right]^T, \tag{7}$$

and the noise matrix is defined as $\boldsymbol{W}(p) \triangleq \left[\boldsymbol{w}_0(p), \boldsymbol{w}_1(p), \ldots, \boldsymbol{w}_{N-1}(p)\right]$.

Vectorizing the matrix of received signals into a vector $\boldsymbol{r}(p) \triangleq \text{vec}\{\boldsymbol{R}(p)\}$, the received signals can be expressed as the following vector form

$$\boldsymbol{r}(p) = \sum_{k=0}^{K-1} \alpha_k(p) \, \text{vec}\left\{\boldsymbol{a}(\phi_k) \boldsymbol{b}^T(\psi_k)\right\} + \boldsymbol{w}(p) = \sum_{k=0}^{K-1} \alpha_k(p) \boldsymbol{b}(\psi_k) \otimes \boldsymbol{a}(\phi_k) + \boldsymbol{w}(p) \tag{8}$$

where $\boldsymbol{w}(p) \triangleq \text{vec}\{\boldsymbol{W}(p)\}$, and $\boldsymbol{w}(p) \sim \mathcal{CN}(0, \sigma_n^2 \boldsymbol{I}_{MN})$. Therefore, without the mutual coupling effect between antennas, the problem of DOD/DOA estimation is formulated in (8), where both ϕ_k and ψ_k will be estimated from the received signal $\boldsymbol{r}(p)$ without the knowledge of target scattering coefficient $\alpha_k(p)$.

2.2. Bistatic MIMO Radar System With Mutual Coupling

However, in the practical radar system, when the mutual coupling between the antennas in both transmitter and receiver is considered [23], the system model developed in (8) cannot be used. Therefore, this subsection will discuss the system model with mutual coupling. Usually, the mutual coupling matrices in the transmitter and receiver are respectively defined as [18]

$$\boldsymbol{C}_T \triangleq (Z_{TA} + Z_{TL})(\boldsymbol{Z}_T + Z_{TL}\boldsymbol{I})^{-1}, \tag{9}$$

$$\boldsymbol{C}_R \triangleq (Z_{RA} + Z_{RL})(\boldsymbol{Z}_R + Z_{RL}\boldsymbol{I})^{-1}, \tag{10}$$

where Z_{TA} and Z_{TL} denote the antenna impedance and terminating load in transmitter, and Z_{RA} and Z_{RL} denote the antenna impedance and terminating load in receiver. $\boldsymbol{Z}_T$ and $\boldsymbol{Z}_R$ denote the mutual impedance matrix in transmitter and receiver, respectively.

The m_1-th row and m_2-th column of mutual impedance matrix $\boldsymbol{Z}_T$ can be expressed as [19,24,25]

$$Z_{T,m_1,m_2} = \begin{cases} 30(0.5772 + \ln(2\gamma L) - g_C(2\gamma L) + jg_S(2\gamma L)), & m_1 = m_2 \\ 30(g_R(m_1, m_2) + jg_X(m_1, m_2)), & m_1 \neq m_2 \end{cases} \tag{11}$$

where $\gamma \triangleq 2\pi/\lambda$, and L denotes the length of dipole antennas. $g_R(m_1, m_2)$ and $g_X(m_1, m_2)$ are defined respectively as

$$
\begin{aligned}
g_R(m_1, m_2) \triangleq \sin(\gamma L)\big[g_S(v_0) - g_S(\mu_0) + 2g_S(\mu_1) \\
- 2g_S(v_1) \big] + \cos(\gamma L)\big[g_C(\mu_0) + g_C(v_0) - 2g_C(\mu_1) \\
- 2g_C(v_1) + 2g_C(\gamma d(m_1, m_2)) \big] - \big[2g_C(\mu_1) + 2g_C(v_1) \\
- 4g_C(\gamma d(m_1, m_2)) \big],
\end{aligned}
\tag{12}
$$

$$
\begin{aligned}
g_X(m_1, m_2) \triangleq \sin(\gamma L)\big[g_C(v_0) - g_C(\mu_0) + 2g_C(\mu_1) \\
- 2g_C(v_1) \big] + \cos(\gamma L)\big[-g_S(\mu_0) - g_S(v_0) + 2g_S(\mu_1) \\
+ 2g_S(v_1) - 2g_S(\gamma d(m_1, m_2)) \big] + \big[2g_S(\mu_1) + 2g_S(v_1) \\
- 4g_S(\gamma d(m_1, m_2)) \big],
\end{aligned}
\tag{13}
$$

where $d(m_1, m_2)$ denotes the distance between the m_1-th antenna and the m_2-th antenna. μ_0, v_0, μ_1 and v_1 are defined respectively as

$$
\mu_0 = \gamma \left(\sqrt{d^2(m_1, m_2) + L^2} - L \right),
\tag{14}
$$

$$
v_0 = \gamma \left(\sqrt{d^2(m_1, m_2) + L^2} + L \right),
\tag{15}
$$

$$
\mu_1 = \gamma \left(\sqrt{d^2(m_1, m_2) + 0.25L^2} - 0.5L \right),
\tag{16}
$$

$$
v_1 = \gamma \left(\sqrt{d^2(m_1, m_2) + 0.25L^2} + 0.5L \right).
\tag{17}
$$

$g_C(x)$ and $g_S(x)$ are defined respectively as

$$
g_C(x) \triangleq \int_{-\infty}^{x} \frac{\cos(t)}{t}\, dt, \quad g_S(x) \triangleq \int_{0}^{x} \frac{\sin(t)}{t}\, dt.
\tag{18}
$$

Similarly, the mutual impedance matrix Z_R can be also obtained from the expression of Z_T.

However, the expresses for Z_T and Z_R in (11) are too complex to analysis. Since Z_T and Z_R depend on the length of dipole antennas and the distances between antennas, the mutual coupling matrices C_T and C_R can be approximated, respectively, by two symmetric Toeplitz matrices

$$
C_T \approx T(c_T), \quad C_R \approx T(c_R),
\tag{19}
$$

where $T(c_T) \in \mathbb{C}^{M \times M}$ is defined as

$$
T(c_M) \triangleq
\begin{bmatrix}
c_{T,0} & c_{T,1} & c_{T,2} & \cdots & c_{T,M-1} \\
c_{T,1} & c_{T,0} & c_{T,1} & \cdots & c_{T,M-2} \\
c_{T,2} & c_{T,1} & c_{T,0} & \cdots & c_{T,M-3} \\
\vdots & \vdots & \vdots & \ddots & \vdots \\
c_{T,M-1} & c_{T,M-2} & c_{T,M-3} & \cdots & c_{T,0}
\end{bmatrix},
\tag{20}
$$

and $T(c_R) \in \mathbb{C}^{N \times N}$ is defined as

$$T(c_R) \triangleq \begin{bmatrix} c_{R,0} & c_{R,1} & c_{R,2} & \cdots & c_{R,N-1} \\ c_{R,1} & c_{R,0} & c_{R,1} & \cdots & c_{R,N-2} \\ c_{R,2} & c_{R,1} & c_{R,0} & \cdots & c_{R,N-3} \\ \vdots & \vdots & \vdots & \ddots & \vdots \\ c_{R,N-1} & c_{R,N-2} & c_{R,N-3} & \cdots & c_{R,0} \end{bmatrix}. \tag{21}$$

Additionally, for the mutual coupling matrices, we also have

$$1 = |c_{T,0}| \geq |c_{T,1}| \geq \ldots \geq |c_{T,M-1}|, \tag{22}$$
$$1 = |c_{R,0}| \geq |c_{R,1}| \geq \ldots \geq |c_{R,N-1}|. \tag{23}$$

Therefore, in the scenario with mutual coupling between antennas, the received signal in (8) can be rewritten as

$$r(p) = \sum_{k=0}^{K-1} \alpha_k(p) \left[C_R b(\psi_k) \right] \otimes \left[C_T a(\phi_k) \right] + w(p) \tag{24}$$
$$= CA\alpha(p) + w(p),$$

where

$$C \triangleq C_R \otimes C_T, \tag{25}$$
$$\alpha(p) \triangleq \left[\alpha_0(p), \alpha_1(p), \ldots, \alpha_{K-1}(p) \right]^T, \tag{26}$$
$$A \triangleq \left[b(\psi_0) \otimes a(\phi_0), \ldots, b(\psi_{K-1}) \otimes a(\phi_{K-1}) \right]. \tag{27}$$

The orthogonal signals are affected by the mutual coupling effect, but we describe the corresponding effect by a matrix, and the non-orthogonality is transferred into the steering vectors by the mutual coupling matrix.

Finally, collect the received signals from all pulses, and the matrix form of all received signals can be obtained as

$$R \triangleq \left[r(0), r(1), \ldots, r(P-1) \right] = CA\Gamma + W, \tag{28}$$

where $W \triangleq \left[w(0), w(1), \ldots, w(P-1) \right], \Gamma \triangleq \left[\alpha(0), \alpha(1), \ldots, \alpha(P-1) \right]$. Then, the vector form of all received signals can be expressed as

$$r \triangleq \left[r^T(0), r^T(1), \ldots, r^T(P-1) \right]^T = (I_P \otimes C)(I_P \otimes A)\alpha + w, \tag{29}$$

where $w \triangleq \left[w^T(0), w^T(1), \ldots, w^T(P-1) \right]^T, \alpha \triangleq \text{vec}\{\Gamma\}$.

Therefore, considering the mutual coupling between antennas in both receiver and transmitter, we will develop an algorithm to estimate the DOD/DOA in A from the received signal r in (29) without the knowledge of mutual coupling matrix C and the scattering coefficient α.

3. DOA/DOD and Mutual Coupling Matrix Estimation

With the received signal R, we propose a novel sparse-based method to estimate the DOD/DOA in the scenario with unknown mutual coupling matrix. The possible DOD and DOA are, respectively, from the following two discretized sets

$$\mathbb{S}_\phi \triangleq \left\{ \phi_{D,z_1} | z_1 = 0, 1, \ldots, Z_1 - 1 \right\}, \tag{30}$$

$$\mathbb{S}_\psi \triangleq \left\{ \psi_{D,z_2} | z_2 = 0, 1, \ldots, Z_2 - 1 \right\}, \tag{31}$$

where $\phi_{D,z_1} \leq \phi_{D,z_1+1}$ and $\psi_{D,z_2} \leq \psi_{D,z_2+1}$.

Therefore, assume that the DOD and DOA of a target are respectively the z_1-th entry of $\mathbb{S}_\phi$, i.e., ϕ_{D,z_1}, and the z_2-th entry of $\mathbb{S}_\psi$, i.e., ψ_{D,z_2}, so the steering vector for this target can be expressed as

$$d_{z_1,z_2} = b(\psi_{D,z_2}) \otimes a(\phi_{D,z_1}). \tag{32}$$

Then, collecting the steering vectors for all the possible targets, a dictionary matrix can be formulated as

$$D \triangleq \left[d_{0,0}, d_{0,1}, \ldots, d_{0,Z_2-1}, d_{1,0}, \ldots, d_{Z_1-1,Z_2-1} \right]. \tag{33}$$

Consequently, we can formulate the following compressed sensing (CS)-based problem [26,27] for the DOD/DOA estimation

$$\min_X \|X\|_{2,0} \tag{34}$$

$$\text{s.t. } \|R - CDX\|_F^2 \leq \epsilon,$$

where the norm $\|X\|_{2,0}$ denotes the number of rows in X with the nonzero entries, and the parameter ϵ is adopted to control the accuracy of sparse reconstruction. As shown in Figure 2, $X \in \mathbb{C}^{Q \times P}$ denotes a sparse matrix and the nonzero entries are the scattering coefficients from Γ. The indexes of nonzero rows in X indicate the DOD/DOA of targets.

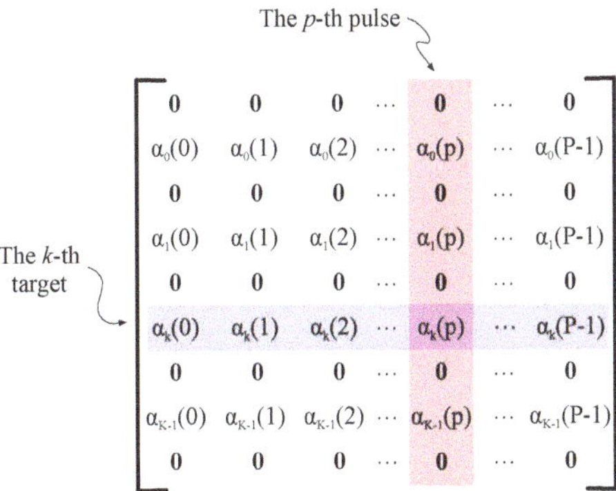

Figure 2. The structure of sparse matrix X.

In (34), both the sparse matrix X and the mutual coupling matrix C are unknown, so this paper proposes a novel method to estimate DOD/DOA with the unknown mutual coupling matrix. Additionally, the *off-grid* problem in DOD/DOA estimation is also considered, where the *off-grid* problem means that the actual values of DOD/DOA can be not exactly contained by the discretized DOD and DOA sets, i.e., $\phi_k \notin \mathbb{S}_\phi$ and $\psi_k \notin \mathbb{S}_\psi$, but $\phi_{D,0} \leq \phi_k \leq \phi_{D,Z_1-1}$ and $\psi_{D,0} \leq \psi_k \leq \psi_{D,Z_2-1}$, for $k = 0, 1, \ldots, K - 1$.

Unlike the traditional multiple measurement vectors (MMV) problem [28–30] in the CS theory, the mutual coupling matrix in the DOA/DOD estimation problem (34) is unknown, so the traditional MMV methods cannot be used directly. Therefore, a novel method is proposed to estimate DOD/DOA with the following objective function

$$\{\hat{\boldsymbol{\phi}}, \hat{\boldsymbol{\psi}}, \hat{\boldsymbol{X}}, \hat{\boldsymbol{C}}\} = \min_{\boldsymbol{\phi}, \boldsymbol{\psi}, \boldsymbol{X}, \boldsymbol{C}} f(\boldsymbol{\phi}, \boldsymbol{\psi}, \boldsymbol{X}, \boldsymbol{C}), \tag{35}$$

where X^q denotes the q-th row of X, and

$$f(\boldsymbol{\phi}, \boldsymbol{\psi}, \boldsymbol{X}, \boldsymbol{C}) \triangleq \mu \|\boldsymbol{X}\|_{2,1} + \|\boldsymbol{R} - \boldsymbol{CDX}\|_F^2, \tag{36}$$

$$\|\boldsymbol{X}\|_F \triangleq \sqrt{\sum_{p=0}^{P-1} \sum_{q=0}^{Q-1} X_{q,p}^2}, \tag{37}$$

$$\|\boldsymbol{X}\|_{2,1} \triangleq \sum_{q=0}^{Q-1} \sqrt{\sum_{p=0}^{P-1} X_{q,p}^2} = \sum_{q=0}^{Q-1} \|X^q\|_2. \tag{38}$$

In (35), the ℓ_1 norm is adopted as a relaxation form of ℓ_0 norm [31].

A novel iterative method is proposed to solve the problem (35), and the flow chart of the proposed method is shown in Figure 3. First, ignoring the effect of mutual coupling, the CS-based method is adopted to estimate the sparse matrix $\hat{\boldsymbol{X}}$ with assuming $\boldsymbol{C} = \boldsymbol{I}$. Second, based on the estimated $\hat{\boldsymbol{X}}$, the mutual coupling matrix $\boldsymbol{C}$ can be estimated as $\hat{\boldsymbol{C}}$ with the gradient descent method. Then, with the roughly estimated results, another gradient descent method is proposed to further polish the estimated results and solve the off-grid problem. Finally, Estimate DOD/DOA and mutual coupling matrix iteratively, and the estimated results are obtained when the estimation method is that of convergence. Details about the proposed method are given in the following subsections.

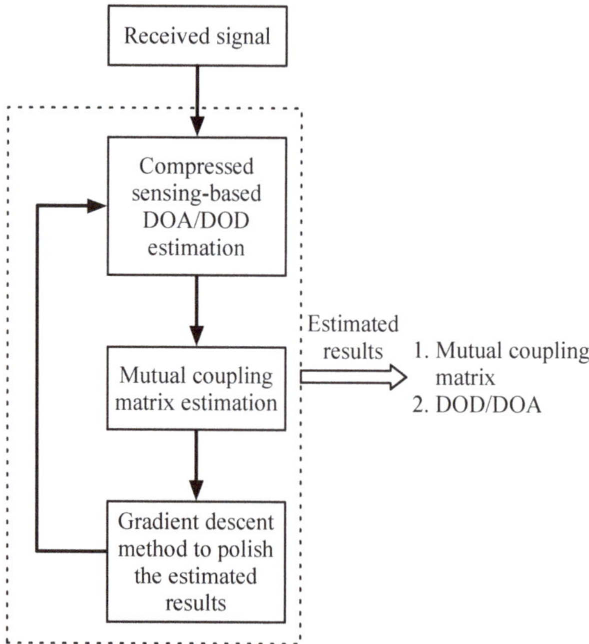

Figure 3. The flow chart of proposed method for DOD/DOA estimation with the unknown mutual coupling matrix.

3.1. CS-Based DOD/DOA Estimation

The CS-based method is adopted to estimate DOD/DOA. Since the multiple pulses are adopted in the bistatic MIMO radar system, the simultaneous orthogonal matching pursuit (SOMP) method [32] can be adopted with ignoring the mutual coupling between antennas. In Algorithm 1, the details of the SOMP method for DOD/DOA estimation is given. At Step 4 of Algorithm 1, the ℓ_1 norm is used to find the discretized DOD/DOA with the maximum correlation coefficient for all pulses.

Algorithm 1 Simultaneous orthogonal matching pursuit for DOD/DOA estimation

---

1: *Input:* received signal R, maximum number of targets K, dictionary matrix D.

2: *Initialization:* iteration $i = 0$, support $\mathbb{S}_x = \varnothing$, residual matrix $Z_i = R$, assuming $C = I$.

3: **for** $i = 0$ to $K - 1$ **do**

4:     $\{\hat{z}_1, \hat{z}_2\} = \arg\max_{z_1, z_2} \cdot \|Z_i^H d_{z_1, z_2}\|_1$.

5:     $j = z_1 Z_1 + z_2$.

6:     $\mathbb{S}_x \leftarrow \mathbb{S}_x \cup \{j\}$.

7:     $R_{i+1} = D_{\mathbb{S}_x} D_{\mathbb{S}_x}^\dagger R$, where $D_{\mathbb{S}_x}$ is formed by the columns from D with the column indexes from

     the support $\mathbb{S}_x$.

8:     $Z_{i+1} = R - R_{i+1}$.

9:     $i = i + 1$.

10: **end for**

11: *Output:* the estimated DOD/DOA from the support $\mathbb{S}_x$ and the estimated sparse matrix $\hat{X}$.

3.2. Gradient Decent-Based Mutual Coupling Matrix Estimation

With the estimated DOD/DOA and the sparse matrix $\hat{X}$, considering the symmetry characteristic of mutual coupling matrix, the mutual coupling vectors c_T and c_R can be estimated by the following objective function

$$\{\hat{c}_T, \hat{c}_R\} = \arg\min_{c_T, c_R} g(c_T, c_R), \tag{39}$$

where the objective function is defined as

$$g(c_T, c_R) \triangleq \left\| R - [T(c_R) \otimes T(c_T)] D\hat{X} \right\|_F^2. \tag{40}$$

Therefore, a gradient decent method is proposed in this paper to estimate the mutual coupling vectors c_T and c_R, and the details are given in Algorithm 2.

Here, the subgradients of objective function $g(c_T, c_R)$ can be obtained as

$$\nabla_{c_T^*} g(c_T, c_R) = \frac{\partial g(c_T, c_R)}{\partial c_T^*} = \left[\frac{\partial g(c_T, c_R)}{\partial c_{T,0}^*}, \ldots, \frac{\partial g(c_T, c_R)}{\partial c_{T,m}^*}, \ldots \right], \tag{41}$$

$$\nabla_{c_R^*} g(c_T, c_R) = \frac{\partial g(c_T, c_R)}{\partial c_R^*} = \left[\frac{\partial g(c_T, c_R)}{\partial c_{R,0}^*}, \ldots, \frac{\partial g(c_T, c_R)}{\partial c_{R,n}^*}, \ldots \right], \tag{42}$$

where the subgradients of $\dfrac{\partial g(c_T, c_R)}{\partial c_{T,m}^*}$ and $\dfrac{\partial g(c_T, c_R)}{\partial c_{R,n}^*}$ are given respectively as

$$\frac{\partial g(c_T, c_R)}{\partial c_{T,m}^*} = 2\mathcal{R}\left\{ \mathrm{vec}^T \left\{ R - CD\hat{X} \right\} \mathrm{vec}^* \left\{ - \left[T(c_R) \otimes \frac{\partial T(c_T)}{\partial c_{T,m}^*} \right] D\hat{X} \right\} \right\}, \tag{43}$$

$$\frac{\partial g(c_T, c_R)}{\partial c_{R,n}^*} = 2\mathcal{R}\left\{ \mathrm{vec}^T \left\{ R - CD\hat{X} \right\} \mathrm{vec}^* \left\{ - \left[\frac{\partial T(c_R)}{\partial c_{R,n}^*} \otimes T(c_T) \right] D\hat{X} \right\} \right\}. \tag{44}$$

Algorithm 2 Mutual coupling matrix estimation

1: *Input:* received signal R, estimated sparse matrix $\hat{X}$, dictionary matrix D, step size δ, stop threshold ϵ_S.

2: *Initialization:* $\quad \hat{c}_T = \left[1, 0_{1\times(M-1)} \right]^T$,
$$\hat{c}_R = \left[1, 0_{1\times(N-1)} \right]^T,$$
$$\hat{C} = T(\hat{c}_R) \otimes T(\hat{c}_T),$$
$$e = \| R - \hat{C}D\hat{X} \|_F^2.$$

3: **while** $e \leq \epsilon_S$ **do**

4: $\quad$ Obtain $\nabla_{c_R^*} g(c_T, c_R)$ and $\nabla_{c_R^*} g(c_T, c_R)$ from (41) and (42), respectively.

5: $\quad \hat{c}_T \leftarrow \hat{c}_T - \delta \nabla_{c_T^*} g(c_T, c_R)$.

6: $\quad \hat{c}_R \leftarrow \hat{c}_R - \delta \nabla_{c_R^*} g(c_T, c_R)$.

7: $\quad \hat{C} = T(\hat{c}_R) \otimes T(\hat{c}_T)$.

8: $\quad e' = \| R - \hat{C}D\hat{X} \|_F^2$.

9: $\quad$ **if** $e' > e$ **then**

10: $\quad\quad \delta \leftarrow \frac{\delta}{2}$.

11: $\quad$ **end if**

12: $\quad e = e'$.

13: **end while**

14: *Output:* the estimated mutual coupling matrix $\hat{C}$.

3.3. Polish the Estimated DOD/DOA and Mutual Coupling Matrix

The DOD/DOA are discretized and the dictionary matrix is formulated in Algorithm 1, so the estimated DOD/DOA must be in set $\mathbb{S}_\phi$ and $\mathbb{S}_\psi$. However, in the practical scenarios, the DOD/DOA of targets are continuous and can be not exact in the sets with discretized angles. Therefore, with the roughly estimated DOD/DOA and mutual coupling matrix from Algorithms 1 and 2, this subsection proposes a gradient descent method to further polish the estimated results and solve the off-grid problem. The details to polish the estimation results is given in Algorithm 3. The mutual coupling effect is compensated in Algorithm 3, where we estimate the mutual coupling coefficients. Then, the estimated coefficients can be used to improve the performance of DOD/DOA estimation.

Algorithm 3 Polish the estimated DOD/DOA and mutual coupling matrix

---

1: *Input:* received signal R, estimated sparse matrix $\hat{X}$, estimated mutual coupling matrxi $\hat{C}$, dictionary matrix D, step size δ, stop threshold ϵ_S.

2: *Initialization:* obtain $\hat{c}_T$ and $\hat{c}_R$ from $\hat{C}$; obtain $\hat{\phi}$ and $\hat{\psi}$ from $\hat{X}$; $\hat{x} = \mathrm{vec}\{\hat{X}\}$; $e = \|R - \hat{C}D\hat{X}\|_F^2$.

3: **while** $e \le \epsilon_S$ **do**

4:     Obtain $\nabla_\phi f(\phi,\psi,X,C)$, $\nabla_\psi f(\phi,\psi,X,C)$, $\nabla_{x^*} f(\phi,\psi,X,C)$, $\nabla_{c_T^*} f(\phi,\psi,X,C)$, and $\nabla_{c_R^*} f(\phi,\psi,X,C)$.

5:     $\hat{c}_T \leftarrow \hat{c}_T - \delta \nabla_{c_T^*} f(\phi,\psi,X,C)$.

6:     $\hat{c}_R \leftarrow \hat{c}_R - \delta \nabla_{c_R^*} f(\phi,\psi,X,C)$.

7:     $\hat{C} = T(\hat{c}_R) \otimes T(\hat{c}_T)$.

8:     $\hat{\phi} \leftarrow \hat{\phi} - \delta \nabla_\phi f(\phi,\psi,X,C)$.

9:     $\hat{\psi} \leftarrow \hat{\phi} - \delta \nabla_\psi f(\phi,\psi,X,C)$.

10:     $\hat{x} \leftarrow \hat{x} - \delta \nabla_{x^*} f(\phi,\psi,X,C)$.

11:     $e' = \|R - \hat{C}D\hat{X}\|_F^2$.

12:     **if** $e' > e$ **then**

13:         $\delta \leftarrow \frac{\delta}{2}$.

14:     **end if**

15:     $e = e'$.

16: **end while**

17: *Output:* the polished mutual coupling matrix $\hat{C}$, the polished DOD/DOA $\hat{\phi}$ and $\hat{\psi}$, and the polished sparse matrix $\hat{X}$.

The gradient descent method based on the subgradients of objective function $f(\phi,\psi,XC)$, which are given in Proposition 1.

Theorem 1. *The subgradients of $f(\phi,\psi,X,C)$ are*

$$\nabla_\phi f(\phi,\psi,X,C) = 2\mathcal{R}\left\{ [(I_P \otimes CD)x - r]^H (I_P \otimes C)\frac{\partial(I_P \otimes D)x}{\partial \phi} \right\}, \tag{45}$$

$$\nabla_\psi f(\phi,\psi,X,C) = 2\mathcal{R}\left\{ [(I_P \otimes CD)x - r]^H (I_P \otimes C)\frac{\partial(I_P \otimes D)x}{\partial \psi} \right\}, \tag{46}$$

$$\nabla_{x^*} f(\phi,\psi,X,C) = \frac{\mu}{2}x^T \sum_{q=0}^{K-1}\left(\sum_{p=0}^{P-1} x_{q,p}^2\right)^{-\frac{1}{2}} + [(I_P \otimes CD)x - r]^T (I_P \otimes CD)^*, \tag{47}$$

$$\nabla_{c_T^*} f(\phi,\psi,X,C) = [(I_P \otimes CD)x - r]^T \frac{\partial(I_P \otimes CD)^* x^*}{\partial c_T^*}, \tag{48}$$

$$\nabla_{c_R^*} f(\phi,\psi,X,C) = [(I_P \otimes CD)x - r]^T \frac{\partial(I_P \otimes CD)^* x^*}{\partial c_R^*}, \tag{49}$$

where

1. The k-th column of $\dfrac{\partial(I_P \otimes D)x}{\partial \phi}$ is

$$\left[\frac{\partial(I_P \otimes D)x}{\partial \phi}\right]_k = \left(I_P \otimes \left[0_{MN \times k}, b(\psi_k) \otimes \frac{\partial a(\phi_k)}{\partial \phi_k}, 0_{MN \times (K-1-k)}\right]\right) x;$$ (50)

2. The k-th column of $\dfrac{\partial(I_P \otimes D)x}{\partial \psi}$ is

$$\left[\frac{\partial(I_P \otimes D)x}{\partial \psi}\right]_k = \left(I_P \otimes \left[0_{MN \times k}, \frac{\partial b(\psi_k)}{\partial \psi_k} \otimes a(\phi_k), 0_{MN \times (K-1-k)}\right]\right) x;$$ (51)

3. The m-th column of $\dfrac{\partial(I_P \otimes CD)^* x^*}{\partial c_T^*}$ is

$$\left[\frac{\partial(I_P \otimes CD)^* x^*}{\partial c_T^*}\right]_m = \left(I_P \otimes C_R^* \otimes \frac{\partial C_T^*}{\partial c_{T,m}^*}\right)(I_P \otimes D^*) x^*;$$ (52)

4. The n-th column of $\dfrac{\partial(I_P \otimes CD)^* x^*}{\partial c_R^*}$ is

$$\left[\frac{\partial(I_P \otimes CD)^* x^*}{\partial c_R^*}\right]_n = \left(I_P \otimes \frac{\partial C_R^*}{\partial c_{R,n}^*} \otimes C_T^*\right)(I_P \otimes D^*) x^*;$$ (53)

5. The m-th entry of $\dfrac{\partial a(\phi_k)}{\partial \phi_k}$ is

$$\left[\frac{\partial a(\phi_k)}{\partial \phi_k}\right]_m = -j2\pi \frac{md}{\lambda} \cos \phi_k e^{-j2\pi \frac{md}{\lambda} \sin \phi_k};$$ (54)

6. The n-th entry of $\dfrac{\partial b(\psi_k)}{\partial \psi_k}$ is

$$\left[\frac{\partial b(\psi_k)}{\partial \psi_k}\right]_n = -j2\pi \frac{nd}{\lambda} \cos \psi_k e^{-j2\pi \frac{nd}{\lambda} \sin \psi_k}.$$ (55)

Here, we will proof this proposition.

Proof. The derivations for vectors or matrices are given in Appendix A. By defining $x \triangleq \text{vec}\{X\}$, we can obtain

$$\begin{aligned}
\nabla_\phi f(\phi, \psi, X, C) &= \frac{\partial \|r - (I_P \otimes CD)x\|_2^2}{\partial \phi} \\
&= -\frac{\partial r^H (I_P \otimes CD)x}{\partial \phi} - \frac{\partial x^H (I_P \otimes D^H C^H) r}{\partial \phi} + \frac{\partial x^H (I_P \otimes D^H C^H)(I_P \otimes CD)x}{\partial \phi} \\
&= -r^H \frac{\partial (I_P \otimes CD)x}{\partial \phi} - r^T \frac{\partial \left[(I_P \otimes CD)x\right]^*}{\partial \phi} + \left[(I_P \otimes CD)x\right]^T \frac{\partial \left(\left[(I_P \otimes CD)x\right]^*\right)}{\partial \phi} \\
&\quad + \left[(I_P \otimes CD)x\right]^H \frac{\partial \left[(I_P \otimes CD)x\right]}{\partial \phi} \\
&= 2\mathcal{R}\left\{\left[(I_P \otimes CD)x - r\right]^H (I_P \otimes C) \frac{\partial (I_P \otimes D)x}{\partial \phi}\right\}.
\end{aligned}$$ (56)

The k-th column of $\frac{\partial(I_P \otimes D)x}{\partial \phi}$ can be obtained as

$$\left[\frac{\partial(I_P \otimes D)x}{\partial \phi}\right]_k = \frac{\partial(I_P \otimes D)}{\partial \phi_k}x = \left(I_P \otimes \frac{\partial D}{\partial \phi_k}\right)x$$
$$= \left(I_P \otimes \left[0_{MN \times k}, b(\psi_k) \otimes \frac{\partial a(\phi_k)}{\partial \phi_k}, 0_{MN \times (K-1-k)}\right]\right)x, \tag{57}$$

where the m-th entry of $\frac{\partial a(\phi_k)}{\partial \phi_k}$ is

$$\left[\frac{\partial a(\phi_k)}{\partial \phi_k}\right]_m = -j2\pi\frac{md}{\lambda}\cos\phi_k e^{-j2\pi\frac{md}{\lambda}\sin\phi_k}. \tag{58}$$

Using the same method, $\nabla_\psi f(\phi, \psi, X, C)$ can be also obtained. Additionally, we also have

$$\nabla_{x^*} f(\phi, \psi, X, C) = \frac{\partial \mu \|X\|_{2,1} + \|r - (I_P \otimes CD)x\|_2^2}{\partial x^*}, \tag{59}$$

and

$$\frac{\partial \|r - (I_P \otimes CD)x^*\|_2^2}{\partial x^*}$$
$$= -\frac{\partial r^H(I_P \otimes CD)x}{\partial x^*} - \frac{\partial x^H(I_P \otimes D^H C^H)r}{\partial x^*} + \frac{\partial x^H(I_P \otimes D^H C^H)(I_P \otimes CD)x}{\partial x^*} \tag{60}$$
$$= -\left[(I_P \otimes D^H C^H)r\right]^T \frac{\partial x^*}{\partial x^*} + [(I_P \otimes CD)x]^T \frac{\partial([(I_P \otimes CD)x]^*)}{\partial x^*}$$
$$= [(I_P \otimes CD)x - r]^T [(I_P \otimes CD)]^*,$$

where the k-th entry of $\frac{\partial \|X\|_{2,1}}{\partial x^*}$ is

$$\left[\frac{\partial \|X\|_{2,1}}{\partial x^*}\right]_k = \frac{\partial \|X\|_{2,1}}{\partial x_k^*} = \frac{\partial \sum_{q=0}^{K-1} \sqrt{\sum_{p=0}^{P-1} x_{q,p}^2}}{\partial x_k^*} = \sum_{q=0}^{K-1} \frac{x_k}{2\sqrt{\sum_{p=0}^{P-1} x_{q,p}^2}}, \tag{61}$$

so we have

$$\frac{\partial \|X\|_{2,1}}{\partial x^*} = \frac{1}{2}x^T \sum_{q=0}^{K-1} \left(\sum_{p=0}^{P-1} x_{q,p}^2\right)^{-\frac{1}{2}}. \tag{62}$$

Therefore, we can obtain

$$\nabla_{x^*} f(\phi, \psi, X, C) = \frac{\mu}{2}x^T \sum_{q=0}^{K-1} \left(\sum_{p=0}^{P-1} x_{q,p}^2\right)^{-\frac{1}{2}} + [(I_P \otimes CD)x - r]^T (I_P \otimes CD)^*.$$

we can also obtain $\nabla_{c_T^*} f(\boldsymbol{\phi}, \boldsymbol{\psi}, X, C)$ as

$$
\begin{aligned}
\nabla_{c_T^*} f(\boldsymbol{\phi}, \boldsymbol{\psi}, X, C) &= \frac{\partial \mu \|X\|_{2,1} + \|r - (I_P \otimes CD)x\|_2^2}{\partial c_T^*} \\
&= -\frac{\partial r^H (I_P \otimes CD)x}{\partial c_T^*} - \frac{\partial x^H (I_P \otimes D^H C^H)r}{\partial c_T^*} + \frac{\partial x^H (I_P \otimes D^H C^H)(I_P \otimes CD)x}{\partial c_T^*} \\
&= -r^H \frac{\partial (I_P \otimes CD)x}{\partial c_T^*} - r^T \frac{\partial \left[(I_P \otimes CD)x\right]^*}{\partial c_T^*} + \left[(I_P \otimes CD)x\right]^T \frac{\partial \left(\left[(I_P \otimes CD)x\right]^*\right)}{\partial c_T^*} \\
&\quad + \left[(I_P \otimes CD)x\right]^H \frac{\partial \left[(I_P \otimes CD)x\right]}{\partial c_T^*} \\
&= \left[(I_P \otimes CD)x - r\right]^T \frac{\partial (I_P \otimes CD)^* x^*}{\partial c_T^*},
\end{aligned}
\tag{63}
$$

and the m-th column of $\dfrac{\partial (I_P \otimes CD)^* x^*}{\partial c_T^*}$ is

$$
\left[\frac{\partial (I_P \otimes CD)^* x^*}{\partial c_T^*}\right]_m = \left(I_P \otimes C_R^* \otimes \frac{\partial C_T^*}{\partial c_{T,m}^*}\right)(I_P \otimes D^*)\, x^*.
\tag{64}
$$

Using the same method, $\nabla_{c_R^*} f(\boldsymbol{\phi}, \boldsymbol{\psi}, X, C)$ can be also obtained. $\square$

4. Cramér–Rao Lower Bound

The CRLB is adopted to show the lower bound on the variance of the estimated parameters including DOD/DOA ($\boldsymbol{\phi}$ and $\boldsymbol{\psi}$), scattering coefficients ($\boldsymbol{\alpha}$), and mutual coupling coefficients c. CRLB can be obtained from the Fisher information matrix (FIM)

$$
e_{\mathrm{CRLB}} = \|d\|_2^2,
\tag{65}
$$

where d is a vector with the diagonal entries of $I^{-1}(\boldsymbol{\theta})$. $I(\boldsymbol{\theta})$ can be calculated as

$$
I(\boldsymbol{\theta}) \triangleq \mathcal{E}\left\{\left(\frac{\partial \ln p}{\partial \boldsymbol{\theta}}\right)^H \left(\frac{\partial \ln p}{\partial \boldsymbol{\theta}}\right)\right\},
\tag{66}
$$

where

$$
\frac{\partial \ln p}{\partial \boldsymbol{\theta}} = \left[\frac{\partial \ln p}{\partial \boldsymbol{\phi}}, \frac{\partial \ln p}{\partial \boldsymbol{\psi}}, \frac{\partial \ln p}{\partial \boldsymbol{\alpha}}, \frac{\partial \ln p}{\partial \boldsymbol{\alpha}^*}, \frac{\partial \ln p}{\partial c}, \frac{\partial \ln p}{\partial c^*}\right],
$$

$$
p \triangleq f(r|\boldsymbol{\phi}, \boldsymbol{\psi}, \boldsymbol{\alpha}, c) = \frac{1}{\pi^{MNP} \det(\sigma_n^2 I)} e^{-\sigma_n^{-2}(r - (I \otimes CA)\boldsymbol{\alpha})^H (r - (I \otimes CA)\boldsymbol{\alpha})},
$$

$$
c \triangleq \left[c_T^T, c_R^T\right]^T.
$$

The subgradients of $\ln p$ are calculated as follows:

1. $\dfrac{\partial \ln p}{\partial \boldsymbol{\phi}}$ is obtained as

$$
\begin{aligned}
\frac{\partial \ln p}{\partial \boldsymbol{\phi}} &= -\sigma_n^{-2} \frac{\partial (r - (I \otimes CA)\boldsymbol{\alpha})^H (r - (I \otimes CA)\boldsymbol{\alpha})}{\partial \boldsymbol{\phi}} \\
&= \sigma_n^{-2}\left[\left[r - (I \otimes CA)\boldsymbol{\alpha}\right]^T \frac{\partial (I \otimes CA)^* \boldsymbol{\alpha}^*}{\partial \boldsymbol{\phi}} + \left[r - (I \otimes CA)\boldsymbol{\alpha}\right]^H \frac{\partial (I \otimes CA)\boldsymbol{\alpha}}{\partial \boldsymbol{\phi}}\right] \\
&= 2\sigma_n^{-2}\mathcal{R}\left\{\left[r - (I \otimes CA)\boldsymbol{\alpha}\right]^H \frac{\partial (I \otimes CA)\boldsymbol{\alpha}}{\partial \boldsymbol{\phi}}\right\},
\end{aligned}
\tag{67}
$$

where the k-th column of $\dfrac{\partial(I \otimes CA)\alpha}{\partial\phi}$ is

$$\left[\frac{\partial(I \otimes CA)\alpha}{\partial\phi}\right]_k = \left(I \otimes C\frac{\partial A}{\partial\phi_k}\right)\alpha,$$ (68)

and

$$\frac{\partial A}{\partial\phi_k} = \left[\mathbf{0}_{MN\times k}, b(\psi_k) \otimes \frac{\partial a(\phi_k)}{\partial\phi_k}, \mathbf{0}_{MN\times(K-1-k)}\right].$$ (69)

With the same method, we can obtain $\dfrac{\partial \ln p}{\partial\psi}$.

2. $\dfrac{\partial \ln p}{\partial\alpha}$ is obtained as

$$\begin{aligned}\frac{\partial \ln p}{\partial\alpha} &= -\sigma_n^{-2}\frac{\partial(r - (I \otimes CA)\alpha)^H(r - (I \otimes CA)\alpha)}{\partial\alpha} \\ &= \sigma_n^{-2}\left[r - (I \otimes CA)\alpha\right]^H(I \otimes CA),\end{aligned}$$ (70)

and we have $\dfrac{\partial \ln p}{\partial\alpha^*} = \left(\dfrac{\partial \ln p}{\partial\alpha}\right)^*$.

3. $\dfrac{\partial \ln p}{\partial c}$ is obtained as

$$\begin{aligned}\frac{\partial \ln p}{\partial c} &= -\sigma_n^{-2}\frac{\partial(r - (I \otimes CA)\alpha)^H(r - (I \otimes CA)\alpha)}{\partial c} \\ &= \sigma_n^{-2}\left[\left[r - (I \otimes CA)\alpha\right]^T\frac{\partial([(I \otimes CA)\alpha]^*)}{\partial c} + \left[r - (I \otimes CA)\alpha\right]^H\frac{\partial(I \otimes CA)\alpha}{\partial c}\right] \\ &= \sigma_n^{-2}\left[r - (I \otimes CA)\alpha\right]^H\frac{\partial(I \otimes CA)\alpha}{\partial c},\end{aligned}$$ (71)

where the n-th column of $\dfrac{\partial(I \otimes CA)\alpha}{\partial c}$ is

$$\left[\frac{\partial(I \otimes CA)\alpha}{\partial c}\right]_n = \left(I_P \otimes \frac{\partial C}{\partial c_n}A\right)\alpha.$$ (72)

4. $\dfrac{\partial \ln p}{\partial c^*}$ is obtained as

$$\begin{aligned}\frac{\partial \ln p}{\partial c^*} &= \sigma_n^{-2}\left[\left[r - (I \otimes CA)\alpha\right]^T\frac{\partial([(I \otimes CA)\alpha]^*)}{\partial c^*} + \left[r - (I \otimes CA)\alpha\right]^H\frac{\partial(I \otimes CA)\alpha}{\partial c^*}\right] \\ &= \sigma_n^{-2}\left[r - (I \otimes CA)\alpha\right]^T\frac{\partial(I \otimes C^*A^*)\alpha^*}{\partial c^*},\end{aligned}$$ (73)

where the n-th column of $\dfrac{\partial(I \otimes C^*A^*)\alpha^*}{\partial c^*}$ is

$$\left[\frac{\partial(I \otimes C^*A^*)\alpha^*}{\partial c^*}\right]_n = \left(I \otimes \frac{\partial C^*}{\partial c_n^*}A^*\right)\alpha^*.$$ (74)

Finally, the FIM is obtained as

$$
\mathcal{E}\left\{\left(\frac{\partial \ln p}{\partial \boldsymbol{\theta}}\right)^{H}\left(\frac{\partial \ln p}{\partial \boldsymbol{\theta}}\right)\right\}
$$

$$
= \sigma_n^{-2}
\begin{bmatrix}
2\mathcal{R}\left\{\Omega_\phi^H \Omega_\phi\right\}, & 2\mathcal{R}\left\{\Omega_\phi^H \Omega_\psi\right\}, & \Omega_\phi^H G, & \Omega_\phi^T G^*, & \Omega_\phi^H \Omega_c, & \Omega_\phi^T \Omega_c^* \\
2\mathcal{R}\left\{\Omega_\psi^H \Omega_\phi\right\}, & 2\mathcal{R}\left\{\Omega_\psi^H \Omega_\psi\right\}, & \Omega_\psi^H G, & \Omega_\psi^T G^*, & \Omega_\psi^H \Omega_c, & \Omega_\psi^T \Omega_c^* \\
G^H \Omega_\phi, & G^H \Omega_\psi, & G^H G, & 0, & G^H \Omega_c, & 0 \\
G^T \Omega_\phi^*, & G^T \Omega_\psi^*, & 0, & G^T G^*, & 0, & G^T \Omega_c^* \\
\Omega_c^H \Omega_\phi, & \Omega_c^H \Omega_\psi, & \Omega_c^H G, & 0, & \Omega_c^H \Omega_c, & 0 \\
\Omega_c^T \Omega_\phi^*, & \Omega_c^T \Omega_\psi^*, & 0, & \Omega_c^T G^*, & 0, & \Omega_c^T \Omega_c^*
\end{bmatrix},
\tag{75}
$$

where $G \triangleq I_P \otimes CA$, $\Omega_\phi \triangleq \frac{\partial G\alpha}{\partial \phi}$, $\Omega_\psi \triangleq \frac{\partial G\alpha}{\partial \psi}$, $\Omega_c \triangleq \frac{\partial G\alpha}{\partial c}$. Then, with FIM, the corresponding CRLB can be obtained.

5. Computational Complexity

In Algorithm 1, to estimate the DOD/DOA using the SOMP method, the computational complexity is $\mathcal{O}(PMN + Q^3 + Q^2 MN + Q(MN)^2)$. In Algorithm 2 to estimate the mutual coupling matrix, the computational complexity is $\mathcal{O}(MNPQ + (MN)^2 Q)$. Additionally, in Algorithm 3, the estimation results are polished, and the computational complexity is $\mathcal{O}(MNPQ + (MN)^2 Q)$. Therefore, the computational complexity of the proposed method to estimate DOD/DOA and the mutual coupling matrix can be roughly expressed as $\mathcal{O}(Q^3 + MNQ^2 + (MN)^2 Q + MNPQ)$. Usually, we have $MN \leq Q$, so the roughly computational complexity can be simplified as $\mathcal{O}(Q^3 + PQ^2)$.

6. Simulation Results

In this section, the simulation results are given to show the performance of the proposed algorithm. The simulation parameters are given in Table 1. First, the reconstruction performance for the received signal is shown in Figure 4. The reconstruction error is defined as

$$
e_r = \frac{\|r - \hat{r}\|_2^2}{\|r\|_2^2},
\tag{76}
$$

where r is the received signal defined in (29), and $\hat{r}$ is the reconstruction signal with the estimated parameters including DOD/DOA, scattering coefficients and mutual coupling matrices. As shown in Figure 4, the proposed method polishes the estimated DOD/DOA and mutual coupling matrices iteratively, where Algorithm 2 is adopted to estimate mutual coupling matrices and Algorithm 3 is used to polished the estimated DOD/DOA and mutual coupling matrices. The relative reconstruction error is decreasing with increasing the number of iterations. Additionally, as shown in this figure, Algorithm 3 is more significant in improving the estimation performance than Algorithm 2. Therefore, it is efficient to polish the estimated results in the off-grid problem after the rough on-grid estimation.

Table 1. Simulation parameters.

Parameter	Value
Carrier frequency f_c	1 GHz
Speed of waveform c	3×10^8 m/s
Wavelength λ	0.3 m
Pulse number P	100
Antenna space d	0.15 m
Antenna number in transmitter M	20
Antenna number in receiver N	20
Dictionary resolution $|\phi_{D,z_1} - \phi_{D,z_1+1}|$	0.035
Detection angle range	$30° \sim 60°$
Iteration number	8×10^3
Target number K	2

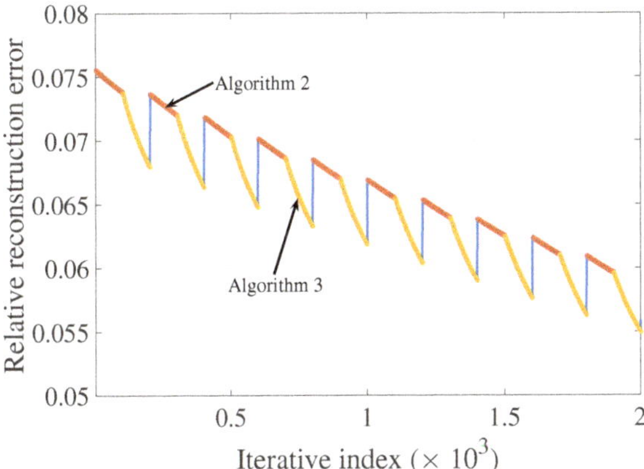

Figure 4. The reconstruction performance with the proposed Algorithms 2 and 3.

Figure 5 shows the estimated DOD/DOA using different methods, where ○ denotes the DOD/DOA of the target, × denotes the estimated DOD/DOA with the proposed method, and △ denotes the estimated DOD/DOA with the on-grid SOMP method [32]. As shown in this figure, when only the on-grid SOMP method is used to estimate the target DOD/DOA, the estimation error is larger than that using the proposed method. In the proposed method, we adopt the proposed off-grid method to further improve the on-grid result, so the proposed method can outperform the traditional on-grid method.

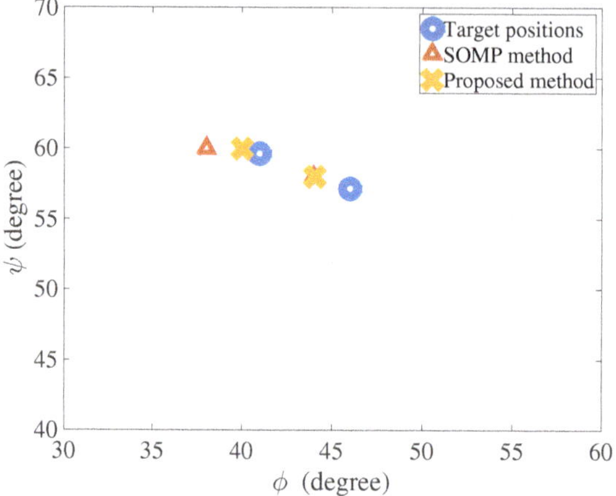

Figure 5. The estimated DOD/DOA using different methods.

Figure 6 shows the reconstruction performance with the proposed method where the signal-to-noise ratios (SNRs) are 5 dB, 10 dB and 20 dB. With different SNRs, the same waveforms are adopted, so the correlation between waveforms are the same. As shown in this figure, with increasing the SNR of the received signal, better reconstruction performance can be achieved. Additional, when SNR = 20 dB, the reconstruction performance is almost the same as the one without noise. After about 8×10^3 iterations, the reconstruction performance is convergence, so we can adopt 8×10^3 as the maximum number of iterations in the following simulations. In Figure 7, we also compare the estimated results with the CRLB derived in this paper. As shown in this figure, the proposed estimation method can approach the CRLB, so the estimation method is efficient.

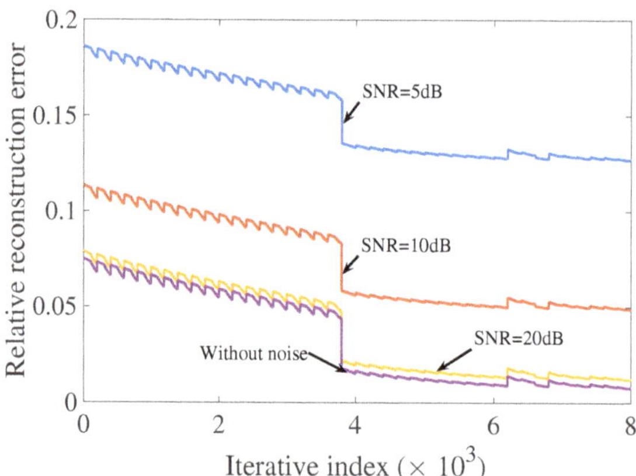

Figure 6. The iterative results for different SNRs.

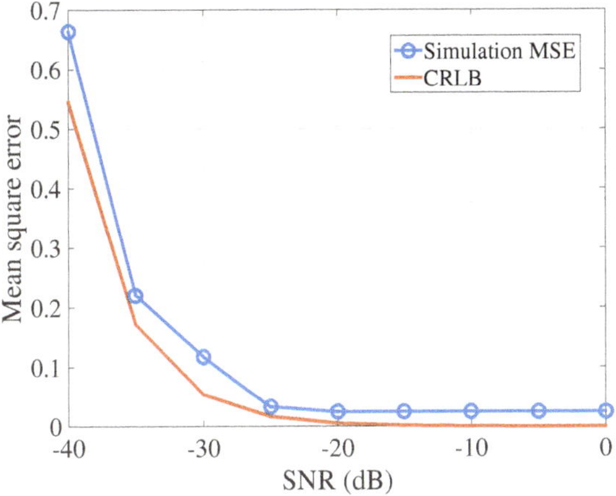

Figure 7. The CRLB and the simulation MSE of the proposed method.

In Figure 8, we show the effect of mutual coupling on the estimation performance. As shown in this figure, better estimation performance can be achieved by improving the SNR of the received signal. The curves "with perfect information" are the simulation results with the perfect information of mutual coupling effect. The best estimation performance can be achieved by the methods with perfect information. Moreover, the mutual coupling has great effect on the estimation performance, so better reconstruction performance can be achieved by estimating the mutual coupling matrices during the DOD/DOA estimation. With different target numbers, Figure 9 shows the reconstruction estimation performance. The targets are uniformly distributed in the angle range from 30° to 60°. When the target number is increasing, the reconstruction performance will be worse with the high correlation between the echoed waveforms from different targets. However, with better SNR, more targets can be estimated with the same reconstruction performance.

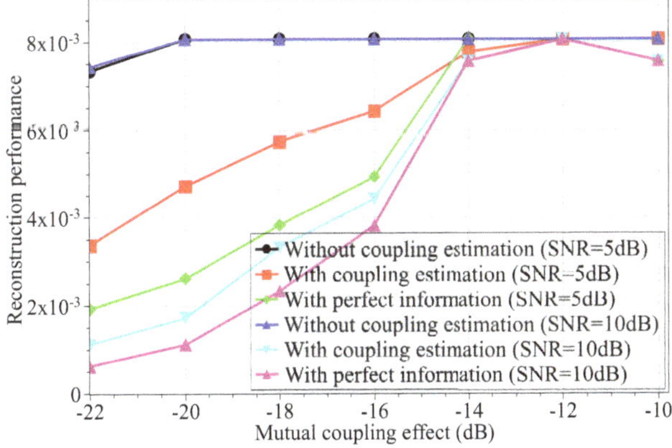

Figure 8. The reconstruction performance with and without coupling estimation.

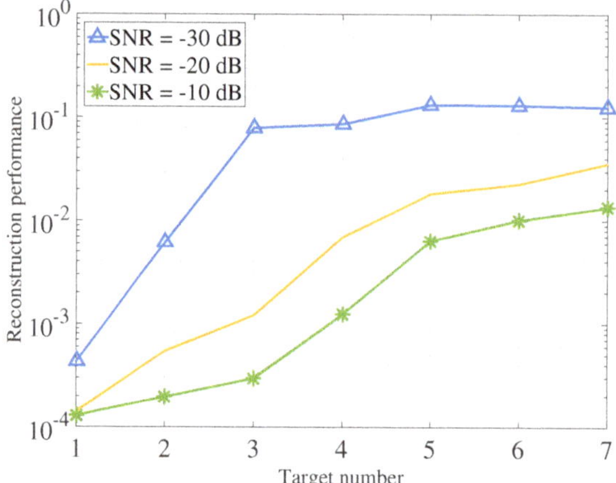

Figure 9. The reconstruction performance with different numbers of targets.

7. Conclusions

In the bistatic MIMO radar, the DOD/DOA estimation problem with mutual coupling effect between antennas has been addressed. After formulating the system model, the iterative method based on CS has been proposed to exploit the sparsity of targets in the detection area, where the estimation for DOD/DOA and mutual coupling has been polished by solving the off-grid problem. Then, the corresponding CRLBs for the parameters including DOD/DOA, mutual coupling matrices, and scattering coefficients, have been derived. Simulation results show that the proposed estimation method can approach the CRLB and achieve the better estimation performance than the traditional methods. Further work will focus on the estimation of moving targets in the MIMO radar system with mutual coupling.

Author Contributions: Conceptualization, P.C. and Z.C. (Zhenxin Cao); methodology, P.C.; software, Z.C. (Zhimin Chen); validation, C.Y.; formal analysis, Z.C. (Zhimin Chen); investigation, Z.C. (Zhimin Chen); resources, P.C.; data curation, Z.C. (Zhenxin Cao); writing—original draft preparation, P.C.; writing—review and editing, P.C.; visualization, Z.C. (Zhimin Chen); supervision, Z.C. (Zhenxin Cao); project administration, P.C.; funding acquisition, P.C.

Funding: This work was supported in part by the National Natural Science Foundation of China (Grant No. 61801112, 61471117, 61601281), the Natural Science Foundation of Jiangsu Province (Grant No. BK20180357), the Open Program of State Key Laboratory of Millimeter Waves (Southeast University, Grant No. Z201804).

Conflicts of Interest: The authors declare no conflict of interest.

Appendix A. The Derivations of Complex Vector and Matrix

Lemma A1. *With both the complex vectors ($u \in \mathbb{C}^{P \times 1}$, $v \in \mathbb{C}^{P \times 1}$) and the complex matrix $A \in \mathbb{C}^{M \times P}$ being the function of a complex vector $x \in \mathbb{C}^{N \times 1}$, the following derivations can be obtained [33]*

$$\frac{\partial u^H v}{\partial x} = v^T \frac{\partial (u^*)}{\partial x} + u^H \frac{\partial v}{\partial x}, \tag{A1}$$

$$\frac{\partial A u}{\partial x} = \left[\frac{\partial A}{\partial x_0} u + A \frac{\partial u}{\partial x_0}, \ldots, \frac{\partial A}{\partial x_n} u + A \frac{\partial u}{\partial x_n}, \ldots \right]. \tag{A2}$$

Proof.

$$
\begin{aligned}
\frac{\partial u^H v}{\partial x} &= \left[\frac{\partial u^H v}{\partial x_0}, \frac{\partial u^H v}{\partial x_1}, \ldots, \frac{\partial u^H v}{\partial x_{N-1}} \right] \\
&= \left[\frac{\partial \sum_{m=0}^{M-1} u_m^* v_m}{\partial x_0}, \ldots, \frac{\partial \sum_{m=0}^{M-1} u_m^* v_m}{\partial x_n}, \ldots \right] \\
&= \left[\ldots, \sum_{m=0}^{M-1} \frac{\partial u_m^*}{\partial x_n} v_m + u_m^* \frac{\partial v_m}{\partial x_n}, \ldots \right] \\
&= \left[\ldots, \left(\frac{\partial u^*}{\partial x_n} \right)^T v + u^H \frac{\partial v}{\partial x_n}, \ldots \right] \\
&= v^T \left[\frac{\partial u^*}{\partial x_0}, \ldots, \frac{\partial u^*}{\partial x_n}, \ldots \right] + u^H \left[\frac{\partial v}{\partial x_0}, \ldots, \frac{\partial v}{\partial x_n}, \ldots \right] \\
&= v^T d \frac{\partial (u^*)}{\partial x} + u^H \frac{\partial v}{\partial x}.
\end{aligned}
\tag{A3}
$$

With A and u being the function of x, we can obtain the entry in m-th row and n-th column of $\dfrac{\partial A u}{\partial x}$ as

$$
\begin{aligned}
\frac{\partial [Au]_m}{\partial x_n} &= \frac{\partial \sum_{p=0}^{P-1} A_{m,p} u_p}{\partial x_n} \\
&= \sum_{p=0}^{P-1} \frac{\partial A_{m,p}}{\partial x_n} u_p + A_{m,p} \frac{\partial u_p}{\partial x_n} \\
&= u^T \frac{\partial [A^T]_m}{\partial x_n} + [A^T]_m^T \frac{\partial u}{\partial x_n} \\
&= \left[\frac{\partial A}{\partial x_n} u + A \frac{\partial u}{\partial x_n} \right]_m,
\end{aligned}
\tag{A4}
$$

so the n-th column of $\dfrac{\partial A u}{\partial x}$ is

$$
\left[\frac{\partial Au}{\partial x} \right]_n = \frac{\partial A}{\partial x_n} u + A \frac{\partial u}{\partial x_n},
\tag{A5}
$$

and

$$
\frac{\partial Au}{\partial x} = \left[\frac{\partial A}{\partial x_0} u + A \frac{\partial u}{\partial x_0}, \ldots, \frac{\partial A}{\partial x_n} u + A \frac{\partial u}{\partial x_n}, \ldots \right].
\tag{A6}
$$

$\square$

References

1. Li, J.; Stoica, P. MIMO radar with colocated antennas. *IEEE Signal Process. Mag.* **2007**, *24*, 106–114. [CrossRef]
2. Haimovich, A.M.; Blum, R.S.; Cimini, L.J. MIMO radar with widely separated antennas. *IEEE Signal Process. Mag.* **2007**, *25*, 116–129. [CrossRef]
3. Fishler, E.; Haimovich, A.; Blum, R.; Chizhik, D.; Cimini, L.; Valenzuela, R. MIMO radar: An idea whose time has come. In Proceedings of the IEEE Radar Conference, Philadelphia, PA, USA, 26–29 April 2004; pp. 71–78.
4. Chen, P.; Qi, C.; Wu, L.; Wang, X. Estimation of Extended Targets Based on Compressed Sensing in Cognitive Radar System. *IEEE Trans. Veh. Technol.* **2017**, *66*, 941–951. [CrossRef]
5. Fishler, E.; Haimovich, A.; Blum, R.S.; Cimini, L.J.; Chizhik, D.; Valenzuela, R.A. Spatial diversity in radars-models and detection performance. *IEEE Trans. Signal Process.* **2006**, *54*, 823–838. [CrossRef]
6. Davis, M.; Showman, G.; Lanterman, A. Coherent MIMO radar: The phased array and orthogonal waveforms. *IEEE Aerosp. Electron. Syst. Mag.* **2014**, *29*, 76–91. [CrossRef]

7. Chen, P.; Wu, L.; Qi, C. Waveform Optimization for Target Scattering Coefficients Estimation Under Detection and Peak-to-Average Power Ratio Constraints in Cognitive Radar. *Circ. Syst. Signal Process.* **2016**, *35*, 163–184. [CrossRef]

8. Chen, P.; Qi, C.; Wu, L. Antenna placement optimisation for compressed sensing-based distributed MIMO radar. *IET Radar Sonar Navig.* **2017**, *11*, 285–293. [CrossRef]

9. Willis, N.J.; Griffiths, H.D. *Advances in Bistatic Radar*; Institution of Engineering and Technology: Stevenage, UK, 2007.

10. Zhang, J.; Wang, H.; Zhu, X. Adaptive waveform design for separated transmit/receive ULA-MIMO radar. *IEEE Trans. Signal Process.* **2010**, *58*, 4936–4942. [CrossRef]

11. Yao, B.; Wang, W.; Yin, Q. DOD and DOA estimation in bistatic non-uniform multiple-input multiple-output radar systems. *IEEE Commun. Lett.* **2012**, *16*, 1796–1799. [CrossRef]

12. Jiang, H.; Zhang, J.K.; Wong, K.M. Joint DOD and DOA estimation for bistatic MIMO radar in unknown correlated noise. *IEEE Trans. Veh. Technol.* **2015**, *64*, 5113–5125. [CrossRef]

13. Chen, P.; Qi, C.; Wu, L.; Wang, X. Waveform Design for Kalman Filter-Based Target Scattering Coefficient Estimation in Adaptive Radar System. *IEEE Trans. Veh. Technol.* **2018**. [CrossRef]

14. Zhang, X.; Xu, L.; Xu, L.; Xu, D. Direction of departure (DOD) and direction of arrival (DOA) estimation in MIMO radar with reduced-dimension MUSIC. *IEEE Commun. Lett.* **2010**, *14*, 1161–1163. [CrossRef]

15. Bencheikh, M.L.; Wang, Y. Joint DOD-DOA estimation using combined ESPRIT-MUSIC approach in MIMO radar. *Electron. Lett.* **2010**, *46*, 1796–1799. [CrossRef]

16. Zheng, G.; Tang, J.; Yang, X. ESPRIT and unitary ESPRIT algorithms for coexistence of circular and noncircular signals in bistatic MIMO radar. *IEEE Access* **2016**, *4*, 7232–7240. [CrossRef]

17. Oh, D.; Li, Y.C.; Khodjaev, J.; Chong, J.W.; Lee, J.H. Joint estimation of direction of departure and direction of arrival for multiple-input multiple-output radar based on improved joint ESPRIT method. *IET Radar Sonar Navig.* **2015**, *9*, 308–317. [CrossRef]

18. Zhang, C.; Huang, H.; Liao, B. Direction finding in MIMO radar with unknown mutual coupling. *IEEE Access* **2017**, *5*, 4439–4447. [CrossRef]

19. Zheng, Z.; Zhang, J.; Zhang, J. Joint DOD and DOA estimation of bistatic MIMO radar in the presence of unknown mutual coupling. *Signal Process.* **2012**, *92*, 3039–3048. [CrossRef]

20. Chen, P.; Zheng, L.; Wang, X.; Li, H.; Wu, L. Moving target detection using colocated MIMO radar on multiple distributed moving platforms. *IEEE Trans. Signal Process.* **2017**, *65*, 4670–4683. [CrossRef]

21. Liu, X.L.; Liao, G.S. Multi-target localisation in bistatic MIMO radar. *Electron. Lett.* **2010**, *46*, 945–946. [CrossRef]

22. Skolnik, M. *Radar Handbook*, 3rd ed.; McGraw-Hill: New York, NY, USA, 2008.

23. Maio, A.D.; Landi, L.; Farina, A. Adaptive radar detection in the presence of mutual coupling and near-field effects. *IET Radar Sonar Navig.* **2008**, *2*, 17–24. [CrossRef]

24. Lin, M.; Yang, L. Blind calibration and DOA estimation with uniform circular arrays in the presence of mutual coupling. *IEEE Antennas Wirel. Propag. Lett.* **2006**, *5*, 315–318. [CrossRef]

25. Liu, C.L.; Vaidyanathan, P.P. Super nested arrays: Linear sparse arrays with reduced mutual coupling—Part I: Fundamentals. *IEEE Trans. Signal Process.* **2016**, *64*, 3997–4012. [CrossRef]

26. Candes, E.J.; Wakin, M.B. An introduction to compressive sampling. *IEEE Signal Process. Mag.* **2008**, *25*, 21–30. [CrossRef]

27. Donoho, D.L. Compressed sensing. *IEEE Trans. Inf. Theory* **2006**, *52*, 1289–1306. [CrossRef]

28. Li, Y.; Chi, Y. Off-the-grid line spectrum denoising and estimation with multiple measurement vectors. *IEEE Trans. Signal Process.* **2016**, *64*, 1257–1269. [CrossRef]

29. Choi, J.W.; Shim, B. Statistical recovery of simultaneously sparse time-varying signals from multiple measurement vectors. *IEEE Trans. Signal Process.* **2015**, *22*, 6136–6148. [CrossRef]

30. Jin, Y.; Rao, B.D. Support recovery of sparse signals in the presence of multiple measurement vectors. *IEEE Trans. Inf. Theory* **2013**, *59*, 3139–3157. [CrossRef]

31. Zheng, L.; Maleki, A.; Weng, H.; Wang, X.; Long, T. Does ℓ_p-minimization outperform ℓ_1-minimization? *IEEE Trans. Inf. Theory* **2017**, *63*, 6896–6935. [CrossRef]

32. Determe, J.F.; Louveaux, J.; Jacques, L.; Horlin, F. On the noise robustness of simultaneous orthogonal matching pursuit. *IEEE Trans. Signal Process.* **2017**, *65*, 864–875. [CrossRef]

33. Petersen, K.B.; Pedersen, M.S. *The Matrix Cookbook*; Technical University of Denmark: Lyngby, Denmark, 2008; Volume 7, p. 15.

 *electronics*

Article

Direct Position Determination of Coherent Pulse Trains Based on Doppler and Doppler Rate

Guizhou Wu, Min Zhang *, Fucheng Guo and Xuebing Xiao

Key Laboratory of Complex Electromagnetic Environment Effects on Electronics and Information System, National University of Defense Technology, Changsha 410073, China; wuguizhouacademic@163.com (G.W.); gfcly@21cn.com (F.G.); xxbdbd90@hotmail.com (X.X.)
* Correspondence: zhangmin1984@126.com; Tel.: +86-0731-8700-3503

Received: 18 September 2018; Accepted: 18 October 2018; Published: 22 October 2018

Abstract: Direct Position Determination (DPD) of coherent pulse trains using a single moving sensor is considered in this paper. Note that when a large observation window and relative maneuvering course between emitter and receiver both exist, the localization accuracy of Doppler frequency shift only based DPD will decline because of the noticeable Doppler frequency shift variations. To circumvent this problem, a Doppler frequency shift and Doppler rate based DPD approach using a single moving sensor is proposed in this paper. First, the signal model of the intercepted coherent pulse trains is established where the Doppler rate is taken into consideration. Then, the Maximum Likelihood based DPD cost function is given, and the Cramer–Rao lower bound (CRLB) on localization is derived whereafter. At last, the Monto Carlo simulations demonstrate that in one exemplary scenario the Doppler frequency shift variations are noticeable with a large observation window and the proposed method has superior performance to the DPD, which is only based on the Doppler frequency shift.

Keywords: direct position determination; Doppler; Doppler rate; maximum likelihood estimator; coherent pulse trains; single moving sensor; Cramer–Rao lower bound

---

1. Introduction

Locating a stationary emitter from passive observations received by moving sensors is a problem that attracts much interest for both civil and defense-oriented applications in the signal-processing and underwater-acoustics literature, etc. Considering the relative motion between the emitter and the receiver, the Doppler effect is often used to estimate the position. The conventional methods usually estimate the intermediate parameters such as Frequency Difference of Arrival (FDOA) with respect to a reference receiver (also known as differential Doppler) [1,2], Doppler frequency shift [3,4] or Doppler Rate [5–7], etc in the first step independently, and determine the position in the second step [8–10]. However, these two-step methods are not guaranteed to yield optimal location results since they ignore the intrinsic constraint that all measurements should correspond to the same location [11]. Avoiding the step of estimating the intermediate parameters, a novel localization conception known as Direct Position Determination (DPD) was presented [12–18]. Emitter position is extracted directly through processing the raw signals in DPD. It has been proved that the localization accuracy of DPD is superior to the two-step method, especially at low signal-to-noise ratio (SNR) [19,20].

In the last decade, multiple Doppler effect based DPD algorithms have been presented. In [21], Doppler frequency shifts based DPD was presented for narrow-band radio emitters which provides better accuracy than the two-step differential Doppler based method at low SNR no matter if the intercepted signals are known or unknown. By using a Minimum Variance Distortion Response (MVDR) estimator, the high resolution DPD of narrow-band signals based on FDOA is studied in [22].

It can achieve higher resolution than the Maximum Likelihood (ML) type DPD since the MVDR is more sensitive to model errors in the steering vectors. In addition, some works also focus on the DPD of wide-band signals. Ref. [23] advocates a DPD approach of wide-band emitters based on time delay and FDOA, and the closed-form expression for Cramer–Rao lower bound (CRLB) is also presented. Different with [23], the received wide-band signals are partitioned into multiple non-overlapping short-time signal segments in [24]. By coherent summation and non-coherent summation of the multiple short-time signal segments, novel DPDs were derived in it. The results show that both coherent summation-based and non-coherent summation-based DPD exhibit improved localization accuracy when the number of short-time signal segments increases. Considering that the previous DPD methods only exploit a single pulse, which is treated as an independent interaction even if the sensors received multiple pulses, Ref. [25] proposed a multi-pulse coherent accumulation algorithm of DPD using the Time Difference of Arrival (TDOA) and FDOA for coherent short-pulse radar, which brings superior performance in accuracy and resolution.

Nevertheless, a multiple pulses accumulation will enlarge the observation window. Therefore, only using the Doppler frequency shifts cannot sufficiently reflect the Doppler effect since the Doppler frequency shift variation during each observation window is also noticeable. It is straightforward to infer that ignoring the Doppler frequency shift variations will result in a bias on the emitter location estimation in Doppler effect based DPD. Both the above inferences will be demonstrated in the simulations of this paper. To solve these problems, the Doppler rate, which is caused by relative acceleration, should be taken into consideration. Therefore, the utilization of Doppler rate in DPD is reasonable in two aspects. Firstly, it can enlarge observation window without the influence of the noticeable Doppler frequency shift variations. Secondly, it supplies additional information with respect to the emitter position which may result in a distinct increase in localization accuracy.

On the other hand, additional note should be set forth that these previous publications mainly focused on Doppler effect based DPD using multiple space separated sensors which has to synchronize and transfer data between different sensors. In this case, it is difficult to be implied in real applications especially for DPD which processes large amounts of raw signal data instead of the intermediate parameters in the two-step methods. To this end, DPD with a single moving receiver may be more practical.

Motivated by the above facts, this paper focuses on the Doppler and Doppler rate based DPD using a single moving sensor for coherent pulse trains. As studied in [26], coherent pulse trains are portions of a continuous wave and the phases from pulse to pulse are in phase with the original wave. Without the measurement errors, all pulses may be generated from a single pulse by extrapolation. It is well known that coherent technologies are widely used in modern radar systems since their excellent performance in estimating range, radial velocity, and acceleration of a target. The relevant research can be found in [27–30]. With the aim to localize such an emitter, the proposed approach is derived and analyzed.

The main contributions of the work in this paper can be concluded as three aspects. Firstly, we take a first look at the problem of the noticeable Doppler frequency shift variation with a large observation window in Doppler effect based DPD and illustrate that it will result in a degradation of the localization accuracy in a Doppler frequency shift only based DPD. Secondly, we proposed a Doppler and Doppler rate based DPD approach to solve this problem which was not previously available. Finally, the theoretical lower bound (CRLB) for localization is also derived as a reference to examine the performance of the proposed method.

The rest of this paper is organised as follows. Section 2 formulates the model of coherent pulse trains intercepted by a single moving sensor. In Section 3, we derive the ML type DPD cost function. Section 4 derives the CRLB, and Section 5 provides the numerical simulations. The discussion and conclusions are made in Sections 6 and 7, respectively.

2. Problem Formulation

The formulation is in a 2D plane for ease of illustration. It is straightforward to extend to three dimensions. Consider a stationary emitter and a sensor with a single receiving antenna is moving relative to the emitter. The position of the emitter is denoted by the vector of coordinates $p = [x \quad y]^T$. The sensor intercepts coherent pulse trains of the emitter at L interception intervals (also known as observation windows) along its trajectory. The starting velocity and acceleration of the sensor during the l-th interception interval are denoted as $\bar{v}_l$ and $\bar{a}_l$, respectively. The vector between the emitter and the sensor is $r_l(t)$, and, therefore, in a interception interval ($t \in [0, t_\triangle]$), the distance between the sensor and the emitter $\|r_l(t)\|$ can be expanded by Taylor series as a second-order binomial

$$\|r_l(t)\| \approx \|r_l\| + v_l t + 0.5 a_l t^2, \tag{1}$$

where $\|\alpha\|$ denotes the Euclidean norm of α, $r_l = p_l^0 - p$ where p_l^0 stands for the initial position of the sensor during the l-th interception interval, v_l is the radial velocity of the sensor at the start time of the l-th interception interval and a_l is supposed to be the constant radial acceleration during the l-th interception interval. Here, we do not further consider the higher order terms of $\|r_l(t)\|$ since generally they are quite small. Then, according to the principle of kinematics

$$v_l = \frac{\bar{v}_l^T r_l}{\|r_l\|}, a_l = \frac{\bar{v}_l^T \bar{v}_l - v_l^2}{\|r_l\|} + \frac{\bar{a}_l^T r_l}{\|r_l\|}, \tag{2}$$

where $(\alpha)^T$ stands for the transpose of α. In the second function of equation (2), we take the scene where the sensor is maneuvering into consideration. If the receiver motion is supposed to be Recti Linear Constant Speed (RLCS), which is conventional in radar research, we just need to set $\bar{a}_l = [0 \quad 0]^T$, resulting in $a_l = (\bar{v}_l^T \bar{v}_l - v_l^2)/\|r_l\|$. It should be noted that the radial acceleration which will result in the Doppler rate can be produced even though the receiver is on a nonmaneuvering course. A possible localization scenario is presented in Figure 1.

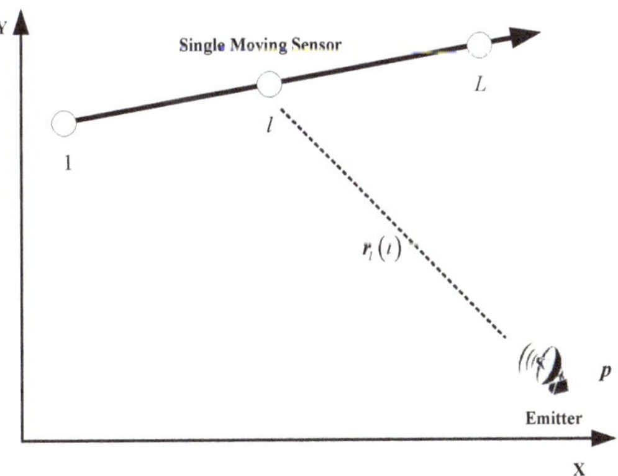

Figure 1. The localization scenario.

Hitherto, the relative motion relationship between the emitter and the sensor is established. Now, we formulate the model of the intercepted signals. At first, we define all the pulses during a single interception interval as a pulse train. Then, the emitted L coherent pulse trains can be formulated as

$$x_l(t) = \sum_{k=1}^{K} s(t - \bar{T}_{l,k}) exp\{j[2\pi f_c t + \phi_0]\} P(t - \bar{T}_{l,k}), \tag{3}$$

where we suppose that each interception interval has K pulses, $s(t - \bar{T}_{l,k})$ is the unknown envelope of the baseband transmitted signal of the coherent pulses as a function of the time t, delayed by $\bar{T}_{l,k}$, which is the transmitted time of the k-th pulse from the emitter to the sensor during the l-th interval. f_c is the nominal frequency of the transmitted signal, which is assumed to be known. ϕ_0 denotes the initial phase of the pulse trains, and $P(t)$ stands for a rectangular function which equals 1 at the interval $[0, t_p]$, where t_p is the pulse width.

To simplify the notations, consider each interception interval separately and replace t with $t_l = t - T_l \in [0, t_\triangle]$, where T_l is the start time of the l-th interval, $t_\triangle$ is the length of the interception interval which is supposed to be the same for every interval. Meanwhile, hypothesize that no multipath exists in the system so that the intercepted noise-free pulse train during the l-th interception interval appears as

$$z_l(t_l) = b_l x_l(t_l - \tau_l(t_l)) = \sum_{k=1}^{K} b_l s(t_l - T_{l,k}) exp\{j[2\pi f_c(t_l - \tau_l - \frac{v_l}{c}t_l - \frac{a_l t_l^2}{2c}) + \phi_0]\} P(t_l - T_{l,k}), \quad (4)$$

where b_l is an unknown complex scalar representing the channel attenuation during the l-th interval. Note that we do not assume any specific model for the relation between b_l and the location of the emitter, just suppose that b_l is fixed during the l-th interval. In addition, $\tau_l(t_l) = \|r_l(t_l)\|/c$ is the transmission delay where c is the signal propagation velocity, and $\tau_l = \|r_l\|/c$. Note that $T_{l,k} = \bar{T}_{l,k} + \tau_l - T_l$ is the known time of arrival of the k pulse during the l-th interval relative to T_l where we further ignore the phase change of $s(t)$ caused by the Doppler effect which is very little due to the extremely short duration of pulses. To show the model of the signals in (4) visibly, a schematic diagram is presented in Figure 2 where T_{ALL} denotes the total observation time in it.

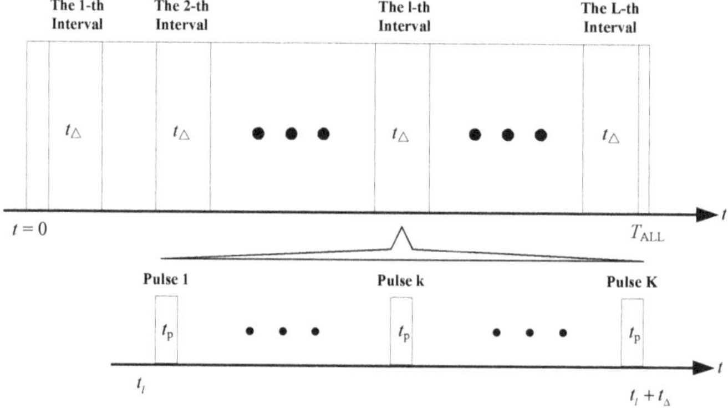

Figure 2. Schematic diagram of the signals.

Define

$$f_{d,l} \triangleq \frac{v_l}{c} f_c, a_{d,l} \triangleq \frac{a_l}{c} f_c, \quad (5)$$

where $f_{d,l}$ is the Doppler frequency shift at the start time of the l-th interception interval, and $a_{d,l}$ denotes the Doppler rate which is approximately stationary during the l-th interception interval. For the sake of simplicity, suppose that the receiver samples N snapshots during each interception interval. Then, the received pulse trains of discrete signal model after being down converted to intermediate frequency can be formulated as

$$z_l(n) = \sum_{k=1}^{K} b_l s(n - N_{l,k}) exp[j(2\pi f_l^l nT_s - \pi a_{d,l} n^2 T_s^2 + \theta_l)] P(n - N_{l,k}) + w_l(n), \quad (6)$$

where $n = 1, \ldots, N$, T_s is the sampling period, and $N_{l,k} = T_{l,k}/T_s$. The intermediate frequency influenced by Doppler frequency shift $f_l^I = f_c - f_{LF} - f_{d,l}$ where f_{LF} is the local frequency of the receiver, and $w_l(n)$ denotes the additive white Complex Gaussian noise.

Note that, in the previous research with respect to Doppler effect based DPD, $a_{d,l}$ is usually omitted since its small value compared with $f_{d,l}$. However, with a large observation window $t_\triangle$, the Doppler frequency shift variation may be noticeable even if the receiver is on a nonmaneuvering course. In this paper, we take $a_{d,l}$ into consideration which can compensate this problem well.

To simplify the notations, we assume that every pulse has M snapshots. Only consider the k-th pulse during the l-th interception interval, i.e., let $m = n - N_{l,k}$; then, (6) can be formulated as

$$z_{l,k}(m) = b_l' s(m) exp[j(2\pi f_l^I N_{l,k} T_s - \pi a_{d,l} N_{l,k}^2 T_s^2)] exp[j2\pi(f_l^I - a_{d,l} N_{l,k} T_s)mT_s] + w_l(m), \qquad (7)$$

where the parameter $b_l' = b_l exp[j(\theta_l)]$, $m = 1, \ldots, M$ which spread over the pulse. Note that, for simplicity, we omit $exp(-j\pi a_{d,l} m^2 T_s^2)$ in (7) since $\pi a_{d,l} m^2 T_s^2$ is much smaller than $2\pi(a_{d,l} N_{l,k} T_s)mT_s$. Next, we demonstrate it by an example. For a typical scenario, $a_{d,l} = 50$ Hz/s, assume that $K = 100$ coherent pulses with a constant pulse repetition interval (PRI) $T_{PRI} = 1$ ms are intercepted. Suppose the pulse width (PW) of the intercepted coherent pulse train $t_p = 1$ µs and the sampling period $T_s = 0.0025$ µs, resulting in $M = 400$. Then, we evaluate the terms $2\pi(a_{d,l} N_{l,k} T_s)MT_s$ and $\pi a_{d,l} M^2 T_s^2$, and find that $(2\pi(a_{d,l} N_{l,k} T_s)MT_s = 3.14 \times 10^{-5}) \gg (\pi a_{d,l} M^2 T_s^2 = 1.5708 \times 10^{-10})$. This result ends the proof.

To simplify the Formular (7), define

$$z_{l,k} \triangleq \begin{bmatrix} z_{l,k}(1) & \cdots & z_{l,k}(M) \end{bmatrix}^T,$$

$$s \triangleq \begin{bmatrix} s(1) & \cdots & s(M) \end{bmatrix}^T,$$

$$A_{l,k} \triangleq diag\{\gamma_{l,k}(1), \ldots, \gamma_{l,k}(M)\},$$

$$w_{l,k} \triangleq \begin{bmatrix} w_{l,k}(1) & \cdots & w_{l,k}(M) \end{bmatrix}^T, \qquad (8)$$

with

$$\gamma_{l,k}(m) \triangleq \kappa_{l,k} exp[j2\pi(f_l^I - a_{d,l} N_{l,k} T_s)mT_s],$$

$$\kappa_{l,k} \triangleq exp[j(2\pi f_l^I N_{l,k} T_s - \pi a_{d,l} N_{l,k}^2 T_s^2)],$$

where $diag\{\alpha_1, \ldots, \alpha_N\}$ denotes a diagonal matrix with $\alpha_1, \ldots, \alpha_N$ on the main diagonal; then, (7) can be calculated as

$$z_{l,k} = b_l' A_{l,k} s + w_{l,k}. \qquad (9)$$

It should be noted that the information with respect to the emitter position is embedded in $A_{l,k}$ which includes the Doppler frequency shift $f_{d,l}$ and the Doppler rate $a_{d,l}$, whereas the $A_{l,k}$ of Doppler frequency shift only based DPD in [21] omits the Doppler rate $a_{d,l}$.

Collecting the observations from all the interception intervals

$$z_k \triangleq \begin{bmatrix} z_{1,k}^T & \cdots & z_{L,k}^T \end{bmatrix}^T,$$

$$b \triangleq \begin{bmatrix} b_1' & \cdots & b_L' \end{bmatrix}^T,$$

$$A_k \triangleq \begin{bmatrix} b_1' A_{1,k} & \cdots & b_L' A_{L,k} \end{bmatrix}^T,$$

$$w_k \triangleq \begin{bmatrix} w_{1,k} & \cdots & w_{L,k} \end{bmatrix}^T. \qquad (10)$$

Finally, the received coherent pulse trains are given by

$$z_k = A_k s + w_k. \tag{11}$$

To conclude, the problem at hand now is to use the measurements z_k given in (11) to determine the position of the emitter p. To solve the problem of localization, the following assumptions are made:

1. The noise is temporally and spatially white and uncorrelated with the signals.
2. The coherent pulse trains have the same initial phase ϕ_0.
3. The envelope of the baseband $s(m)$ from pulse to pulse is the same.

All of these assumptions are justified and frequently made in the related studies on the coherent pulse trains (see details in [27–30]).

3. Direct Position Determination Approach

Consider the observation vectors in (11). The information of the emitter's position is embedded in A_k. We assume the position which best explains all the received data as the estimated position of the emitter. This is the main concept of DPD. We focus on the ML estimator because of its excellent asymptotic properties including consistency and efficiency. Without loss of generality, we may assume that $\|b\|^2 = 1$, since the missing factor can be included in the unknown vector s. Therefore,

$$A_k^H A_k = I_M, \tag{12}$$

where I_M denotes the $M \times M$ identity matrix. The ML estimator is given by minimizing the following cost function:

$$Q_{ML} = \sum_{k=1}^{K} \|z_k - A_k s\|^2, \tag{13}$$

where s is a nuisance vector. Note that, after utilizing (12), the nuisance vector that minimizes (13) is given by

$$\hat{s} = (A_k^H A_k)^{-1} A_k^H z = A_k^H z, \tag{14}$$

where $(\alpha)^H$ stands for the conjugate transpose of α. Substituting (14) into (13) and using (12) again gives

$$Q_{ML} = \sum_{k=1}^{K} \|z_k\|^2 - \|A_k^H z_k\|^2. \tag{15}$$

Instead of minimizing (15), we can maximize

$$Q_{ML2} = \sum_{k=1}^{K} \|A_k^H z_k\|^2. \tag{16}$$

According to the structure of A_k and z_k in (10), Q_{ML2} can be also given by

$$Q_{ML2} = \sum_{k=1}^{K} \sum_{m=1}^{M} |\gamma_{k,m}^H z_{k,m}|^2, \tag{17}$$

where

$$\gamma_{k,m} \triangleq A_{k,m}(p)b,$$
$$A_{k,m}(p) \triangleq diag\{\gamma_{1,k}(m), \gamma_{2,k}(m), \ldots, \gamma_{L,k}(m)\},$$
$$z_{k,m} \triangleq \begin{bmatrix} z_{1,k}(m) & \cdots & z_{L,k}(m) \end{bmatrix}^T. \tag{18}$$

Then, (17) can be written as

$$Q_{ML2} = \sum_{k=1}^{K} \sum_{m=1}^{M} b^H A_{k,m}^H(p) R_{k,m} A_{k,m}(p) b, \tag{19}$$

where $R_{k,m} = z_{k,m} z_{k,m}^H$. Since $\|b\|^2 = 1$, (19) will result in a criterion that depends only on p

$$Q_{ML2} = \lambda_{max}\Big(\sum_{k=1}^{K} \sum_{m=1}^{M} A_{k,m}^H(p) R_{k,m} A_{k,m}(p)\Big), \tag{20}$$

where $\lambda_{max}(M)$ denotes the maximum eigenvalue of matrix M. Note that $\sum_{k=1}^{K} \sum_{m=1}^{M} A_{k,m}^H R_{k,m} A_{k,m}$ is a $L \times L$ matrix. It is easy to calculate its maximum eigenvalue. The estimated emitter's position $\hat{p}$ will be at a maximum (20). In this paper, exhaustive searching is used to estimate the position of the emitter by calculating (20) with respect to every point in the solution space.

4. The CRLB

It is well known that the CRLB provides a low bound on the estimation accuracy. We now focus on the CRLB for estimation of all unknown (but deterministic) parameters of the problem although we are only interested in the CRLB for the estimation of p. To this end, we begin by defining the vector ζ of all real-valued parameters, composed of the real-part and the imaginary-part of s and b, and the "actual" parameters p, namely

$$\zeta = \begin{bmatrix} Re\{s^T\} & Im\{s^T\} & Re\{b^T\} & Im\{b^T\} & p^T \end{bmatrix}^T,$$

which is a $(2M + 2L + 2) \times 1$ vector.

Recap: s is a deterministic vector, whereas $w = [w_1 \dots w_K]^T$ is independent white Complex Gaussian vector, we can observe that the concatenated vector $z = [z_1 \dots z_K]^T$ is also a Complex Gaussian vector, and $z \backsim N(\mu, \Lambda)$ with mean and covariance

$$\mu \triangleq \begin{bmatrix} A_1 \\ \vdots \\ A_k \end{bmatrix} s, \Lambda \triangleq \sigma^2 I_{LKM}, \tag{21}$$

where I_{LKM} is the $LKM \times LKM$ identity matrix.

The Fisher Matrix (FIM) J_ζ for estimation of the vector ζ from the white Complex Gaussian noise w, where only the mean μ depends on ζ, is given by

$$J_\zeta = \frac{2}{\sigma^2} Re\{(\frac{\partial \mu}{\partial \zeta})^H (\frac{\partial \mu}{\partial \zeta}))\}. \tag{22}$$

Taking the derivative of μ with respect to ζ, we get the $LKM \times (2L + 2M + 2)$ (Jacobian) matrix

$$H(\zeta) \triangleq \frac{\partial \mu}{\partial \zeta} = \begin{bmatrix} A_1 & jA_1 & B_1 & jB_1 & G_1(p) \\ \vdots & \vdots & \vdots & \vdots & \vdots \\ A_K & jA_K & B_K & jB_K & G_K(p) \end{bmatrix}, \tag{23}$$

where $B_k = diag\{A_{1,k}, \dots, A_{L,k}\} \cdot (I_L \otimes s)$ is a $LM \times L$ matrix, $\alpha_1 \otimes \alpha_2$ denotes the Kronecker product of α_1 and α_2, and $G_k(p)$ is the derivative of μ with respect to p.

Define

$$C \triangleq \sum_{k=1}^{K} A_k^H B_k, D \triangleq \sum_{k=1}^{K} A_k^H G_k, E \triangleq \sum_{k=1}^{K} B_k^H G_k. \tag{24}$$

Substituting (23) into (22), consequently (exploiting (12) and (24)), we have

$$J_\zeta = \frac{2}{\sigma^2} \begin{bmatrix} Kb I_M & \mathbf{0}_M & Re\{C\} & -Im\{C\} & Re\{D\} \\ \mathbf{0}_M & Kb I_M & Im\{C\} & Re\{C\} & Im\{D\} \\ Re\{C^H\} & -Im\{C^H\} & Kp_s I_L & \mathbf{0}_L & Re\{E\} \\ Im\{C^H\} & Re\{C^H\} & \mathbf{0}_L & Kp_s I_L & Im\{E\} \\ Re\{D^H\} & -Im\{D^H\} & Re\{E^H\} & -Im\{E^H\} & \sum_{k=1}^{K} G_k^H G_k \end{bmatrix}, \tag{25}$$

where $b = \sum_{l=1}^{L} |b_l'|^2$, $p_s = s^H s$, and

$$G_k = \begin{bmatrix} \frac{\partial A_k s}{\partial x} & \frac{\partial A_k s}{\partial y} \end{bmatrix}.$$

In one embodiment, J_ζ can be calculated by (25). The CRLB for the estimation of p is obtained by the lower-right 2×2 block of the inverse of J_ζ. This concludes the derivation.

5. Numerical Simulations

To show the effectiveness of the proposed method, we conduct the Monte Carlo simulations in this section. Note that the Doppler frequency shift variation can be produced even though there is no acceleration of the receiver. Its value depends on the length of the observation window and the related motion between the emitter and the receiver (see the model derived in (2) and (5)). With a large observation window, we illustrate the areas in which Doppler frequency shift variation is noticeable with different related motion in Example1 below. Meanwhile, considering the information of the emitter position is embedded in the Doppler frequency shift in DPD, Example 1 also reveals how the Doppler frequency shift variation affects the localization accuracy of the Doppler Frequency shift only based DPD, which is henceforth denoted by D-DPD. Note that [25] has proposed a multiple pulse coherent accumulation based DPD using TDOA and FDOA with multiple moving sensors. As analyzed in it, coherent accumulation is mainly used to exploit the phase terms $f_{d,l} t_l$ during each interception interval, which is also involved in $\kappa_{l,k}$ in this paper. However, the TDOA between different sensors is nonexistent in the case of single moving sensor as considered in this paper. Thus, we carry out the D-DPD by following the approach in [25] but omitting the TDOA and enlarging the observation window in it.

Subsequently, in Example 2, we examine the performance of the proposed DPD approach, which is henceforth denoted by DDR-DPD, and compare it with D-DPD, and CRLB.

Note that the localization root mean square error (RMSE) and bias are adopted as the criteria of localization accuracy, which are defined by

$$RMSE = \sqrt{\frac{1}{N_{Mon}} \sum_{i=1}^{N_{Mon}} \|\hat{p}(i) - p\|^2},$$

$$Bias = \left\| \frac{1}{N_{Mon}} \sum_{i=1}^{N_{Mon}} \hat{p}(i) - p \right\|, \tag{26}$$

where N_{Mon} is the number of Monte Carlo trials and $\hat{p}(i)$ is the estimated position at the i-th trial. To obtain statistical results, N_{Mon} is set to be 250.

5.1. Example 1

Assume that a stationary emitter locates at coordinate $p = [0, 0]$ km. The Doppler frequency shift will be produced if the signals are intercepted by a moving sensor. Define the Doppler frequency shift variation as

$$f_v = \frac{|a_d|}{|f_d|} t_\triangle \times 100\%, \tag{27}$$

where a_d is the Doppler rate, $t_\triangle$ is the length of the observation window, and f_d is the initial Doppler frequency shift of the observation window.

To evaluate the areas where f_v is noticeable, we assume that the receiver is moving in a space where $x \in [-5, 5]$ km and $y \in [-5, 5]$ km, and calculate f_v for each position of the receiver. Three motion types of the receiver are given in this simulation. Type 1 and type 2 are both RLCS but with different velocities which are $[150, 0]$ m/s and $[300, 0]$ m/s, respectively. Type 3 is the uniform acceleration motion with the velocity and the acceleration being $[150, 0]$ m/s and $[5, 0]$ m/s^2, respectively. Set $t_\triangle = 100$ ms (e.g., 200 coherent pulses of constant pulse repetition interval $T_{PRI} = 0.5$ ms are intercepted), the filled contour plots of the Doppler frequency shift variations are given in Figures 3–5.

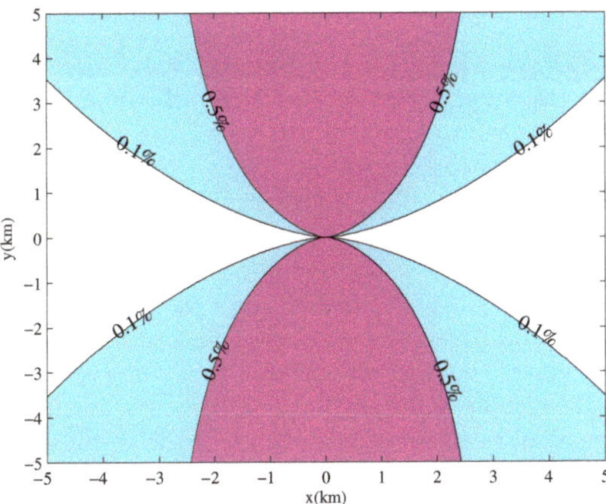

Figure 3. The filled contour plot of the Doppler frequency shift variation with motion type 1.

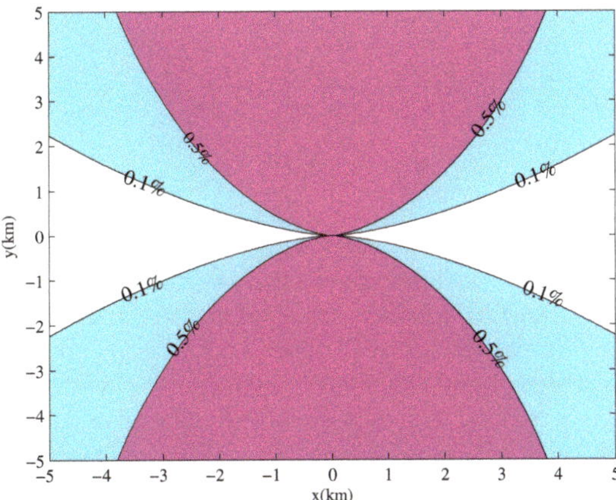

Figure 4. The filled contour plot of the Doppler frequency shift variance with motion type 2.

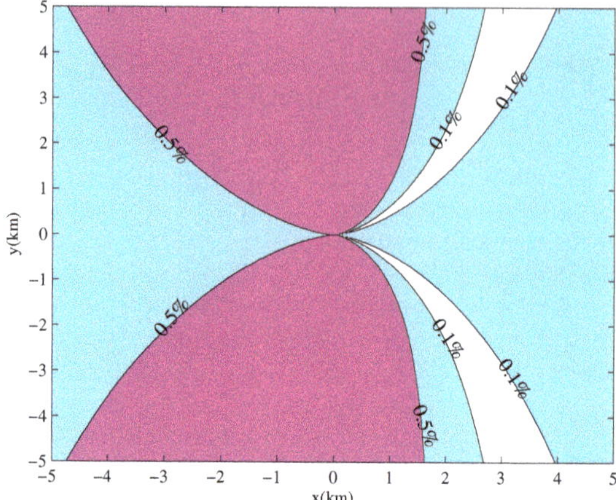

Figure 5. The filled contour plot of the Doppler frequency shift variance with motion type 3.

We observe that the areas where the Doppler frequency shift variation is noticeable ($f_v > 0.5\%$) is big enough so that it can not to be ignored especially when the receiver has a fast speed or an acceleration.

Next, we focus on how the Doppler frequency shift variation affects the localization accuracy of D-DPD. The layout of the system used during this simulation consists of a stationary emitter and a moving receiver which equips only a single receiving antenna. Concretely, the emitter locates at coordinate $p = [0,0]$ km as described before, and transmits coherent pulse trains. The baseband transmitted signals of each pulse are unknown sinusoidal waves. The constant PRI is $T_{PRI} = 1$ ms, the pulse width (PW) is $t_p = 1$ us, the number of the intercepted pulses at each train is K, and the initial phase ϕ of the coherent pulse trains is selected at random from a uniform distribution over $[0, 2\pi]$ for each Monte Carlo trial. The nominal frequency $f_c = 1$ GHz, and the signal received by the antenna is down converted to intermediate frequency $f_I = f_c - f_{LF} = 100$ MHz. For simplicity, the sensor moves straightly from $[-1, 5]$ km to $[0.5, 5]$ km. There are $L = 5$ interception intervals which start when the receiver is at coordinate $[-1, 5]$ km, $[-0.7, 5]$ km, $[-0.4, 5]$ km, $[-0.1, 5]$ km, $[0.2, 5]$ km, respectively. Three combinations of the initial velocity $\bar{v}_1$ and the constant acceleration $\bar{a}$ are displayed in Table 1.

Table 1. The set of $\bar{v}_1$ and $\bar{a}$.

Set Id	$\bar{v}_1$(m/s)	$\bar{a}$(m/s^2)
1	$[100, 0]^T$	$[0, 0]^T$
2	$[300, 0]^T$	$[0, 0]^T$
3	$[287.5, 0]^T$	$[5, 0]^T$

For Set Id $= 1, 2$, the receiver motion is supposed to be RLCS, and for Set Id $= 3$, the receiver has a constant acceleration during all the interception intervals. Moreover, the channel attenuation b_l is selected at random using normal distribution (mean = 1, std = 0.1), and the noise is complex white Gaussian whose amplitude is determined by the given SNR. Note that the location determination is based on all the LK intercepted pulses with the sample frequency $f_s = 400$ MHz.

At first, set the number of pulses K in each interception interval to be 200, resulting in $t_\triangle = 0.2$ s. For the above scene with Set Id $= 2$, the noise-free spectrums of D-DPD and DDR-DPD are calculated by (20). In order to display the peak of the spectrum clearly, the filled 2D contour plots of the spectrums are given in Figure 6. As expected, the D-DPD has a positioning bias which is greater than 10 m, but the estimated position of the emitter in DDR-DPD coincides with its true position.

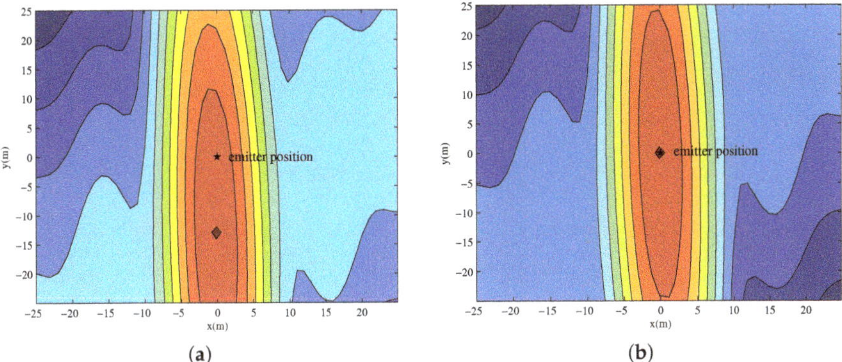

(a)   (b)

Figure 6. The filled 2D contour plot of the noise-free spectrum. (**a**) the filled contour plot of D-DPD's spectrum; (**b**) the filled contour plot of DDR-DPD's spectrum.

Afterwards, the bias and the RMSE are both calculated to analyze how the length of the observation window affects the localization accuracy. We fix the $SNR = 5$ dB, and vary the number of pulses K in each interception interval from 20 to 160 to change the length of the observation window, which is given by $t_{\triangle} = KT_{PRI}$. Then, we illustrate the biases in Figure 7 and the localization RMSE in Figure 8. The results are both calculated with three motion types given in Table 1. The bias and the localization RMSE of DDR-DPD are also plotted to be compared with D-DPD.

From Figures 7 and 8, we observe that the biases of D-DPD increase with the increase of the pulse number K of each interception interval, whereas the biases of DDR-DPD remain at small values. For Set Id = 2, 3, the performance of DDR-DPD improves as the observation window enlarges since more effective information is involved. On the other hand, the performance of D-DPD declines as K increases. For Set Id = 1, the two approaches yield similar localization performance when $K < 100$. This is because the Doppler frequency shift variation is still small. In this case, the performance of D-DPD and DDR-DPD are both improved as the observation window enlarges since more pulses with respect to the emitter position are intercepted. However, the accuracy of D-DPD still declines when $K > 140$, in contrast the performance of DDR-DPD improves. Note that there is a threshold of the observation window length T_{th}. When $t_{\triangle} > T_{th}$, the localization accuracy of D-DPD will be significantly worse than DDR-DPD. For Set Id = 1, 2, 3, the $T_{th} = 100$ ms, 40 ms, 20 ms, respectively.

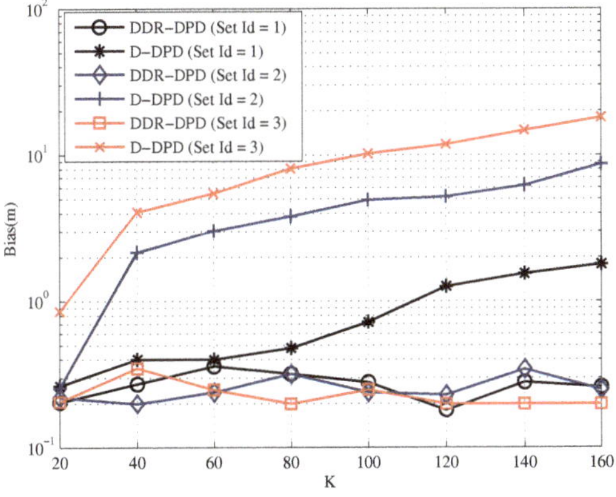

Figure 7. Bias vs. pulse number in each coherent pulse train.

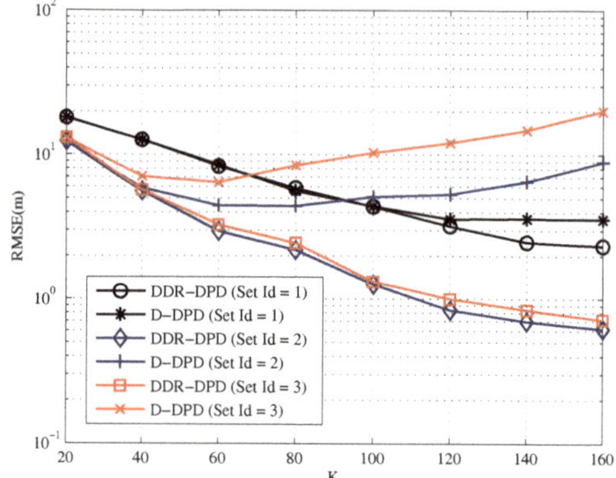

Figure 8. RMSE vs. pulse number in each coherent pulse train.

All the above results demonstrate that the performance of D-DPD and DDR-DPD will both improve as the observation window enlarges if the Doppler frequency shift variation is little. However, the Doppler frequency shift variation can not be ignored when large observation window and relative maneuvering course between emitter and receiver both exist. Large localization error will be produced if we omit the noticeable Doppler frequency shift variation.

5.2. Example 2

In this subsection, we take Set Id $= 2$ for example and examine the performance of the proposed approach. First, set the pulse number K of each coherent pulse train to be 100, and vary the SNR from -20 dB to 5 dB. The other parameters are the same with Example1. Calculate the RMSE of DDR-DPD and D-DPD, and the results are shown in Figure 9a. In addition, the CRLB of localization using DDR-DPD derived in Section 4 is also plotted which is denoted as CRLB in Figure 9a.

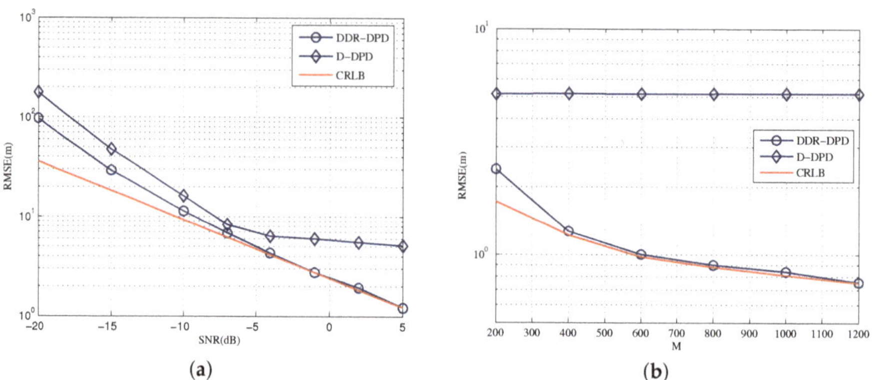

Figure 9. The performance of the proposed method with Set Id = 2. (**a**) the RMSE vs. SNR; (**b**) the RMSE vs. M.

We observe that the performance of DDR-DPD and D-DPD both improve as the SNR varied from -20 dB to 5 dB. However, the RMSE of DDR-DPD converges to the CRLB when $SNR > -5$ dB, and the RMSE of D-DPD approximately tends to be 5 m, which is exactly the bias of D-DPD as illustrated in Figure 7. The underlying reason could be that the Doppler frequency variation is noticeable with the large observation window ($t = 100$ ms), which is omitted by D-DPD. As illustrated in Example 1, it will result in a bias that can not be eliminated by improving the SNR. On the other hand, the performance of DDR-DPD remains superior for all SNR ($[-20, 5]$ dB) in this simulation. This is mainly caused by two reasons. At first, the bias which results from the noticeable Doppler frequency shift can be compensated in DDR-DPD but is involved in the RMSE of D-DPD. Secondly, the information with respect to the emitter position that is embedded in the Doppler rate is involved in DDR-DPD but is omitted in D-DPD.

Next, how the number of snapshots for each pulse affects the localization RMSE of the emitter is examined. We set the $SNR = 5$ dB, and the number of snapshots M for each pulse varies from 200 to 1200. Figure 9b shows the localization RMSE vs. M. The localization CRLB of DDR-DPD is also plotted. We observe that the localization accuracy of DDR-DPD improves as M increases. However, the performance of D-DPD is almost independent of the variation in M. Moreover, the localization accuracy of DDR-DPD can reach the CRLB with moderate snapshots.

Note that, in this example, $t_\triangle = 100$ ms $> T_{th} = 40$ ms. As expected in Example 1, the performance of DDR-DPD is superior to the performance of D-DPD. In addition, we observe that the superiority of the DDR-DPD will be more significant with the increase of the SNR and the number of snapshot M. Since RLCS model fits many kinds of real motions, it is justified to say that the proposed approach will have better applicability to the practical cases.

6. Discussion

We will discuss whether DDR-DPD has a good performance in practice from three aspects i.e., the necessity, the computation complexity and the possible implementation issues of the proposed method.

Firstly, as elaborated in the above sections, the proposed method shows superior performance in terms of localization accuracy compared with D-DPD when there is a noticeable Doppler frequency variation. In addition, the value of the Doppler frequency shift variation substantially depends on two factors i.e., the length of the observation window and the Doppler rate. To improve localization accuracy in DPD, multiple pulses accumulation can be used, which will unavoidably result in a large observation window (see Figure 8). Meanwhile, we have known that the Doppler rate can be produced even if the receiver is in RLCS from (2). In addition, the carrier frequency of the signals, its value depends on the position, the speed and the acceleration of the receiver and the emitter. Since the emitter is uncontrollable in a passive localization system, we can only decide the speed or the acceleration of the moving receiver. In a typical military or civilian scenario, the receiver can be equipped in an airplane or a satellite, etc. Generally speaking, the speed of the civil aircraft is between 150 m/s and 250 m/s, and the supersonic aircraft 340 m/s. The areas of the noticeable Doppler frequency shift variation with a speed similar to an aircraft have been illustrated in Figures 3 and 4. It is justified to say that, with an unavoidably large observation window, the Doppler frequency shift variation is noticeable even the receiver is equipped in an aircraft, let alone a satellite that has much faster speed or even a conspicuous acceleration.

Secondly, the computation complexity of the proposed method is considered. The number of multiplications N_{mul} required by the algorithm is imposed as an indicator of computation complexity. We compare DDR-DPD and D-DPD where exhaustive search are both used to estimate the emitter position. For the sake of simplicity, the small values e.g., the multiplications required by Eigenvalue Decomposition of a $L \times L$ matrix are ignored since it is independent from the number of the pulses and the snapshots. Then, $N_{mul} = N_g MK[2L^3 + (6D + 30)L]$ for DDR-DPD and $N_{mul} = N_g MK[2L^3 + (2D + 15)L]$ for D-DPD, where N_g stands for the number of the grid used in the exhaustive search and D denotes the dimension of the considered scene. Generally speaking, $D = 2$ or 3, which is a small value, and the value of L depends on the received signals which tends to be much larger than D. In this sense, we fortunately observe that DDR-DPD does not introduce much more calculation than D-DPD as the computational complexities of them mainly both depend on the high order term $2N_g MKL^3$. However, the computational complexity may still be too high especially when there are multiple emitters because of the inherent flaw of ML but not the method itself. To relax the restrictions of the proposed method on practice, Alternating Projection [15] can be utilized. Since the derivation is straightforward and not the main contribution of this paper, we will not present in-depth analysis of this problem.

Thirdly, we summarize two possible issues in terms of hardware based implementation of the proposed method i.e., the computation burden and the memory space. At first, the computation may be complex as discussed in the above paragraph. In addition, the DPD type approaches process the raw signals instead of the intermediate parameters as processed in the two-step method, which also increases the computational complexity. However, high-performance hardware equipment combined with a fast algorithm as also introduced in the above paragraph can deal with this problem very well. On the other hand, we also expect high demand for the memory space of the hardware as the proposed method needs to accumulate multiple interception intervals to estimate the emitter position. Nevertheless, large capacity FPGA are available easily now that can solve the problem straightway. In all of these three aspects, the proposed method may show a good performance in practice.

Moreover, inspired by [24], the large interception window with multiple pulses can be partitioned into multiple short time segments. In this case, we may further hypothesize that each short time segment has a single pulse and each pulse has its individual Doppler frequency shift. This is also an interesting approach to substantially compensate the noticeable Doppler frequency shift variation which may exhibit similar performance compared with the proposed approach. However, this approach introduces K unknown parameters in addition which will result in a complicated signal model and great computational complexity. We do not further provide a detailed study of this approach because it is beyond the scope of this paper. It may be an interesting topic for future investigations.

It also should be noted that DDR-DPD also maintains a superior performance for other general signals with a large observation window, although only the DDR-DPD of coherent pulse trains is considered in this paper. This is because the proposed model is more adapted to real situations when both a large observation window and relative acceleration between the emitter and the receiver exist.

7. Conclusions

Now, we conclude this paper. Maximum-likelihood direct location estimation for coherent pulse trains, using observations of a single moving sensor with only one passive antenna, is investigated in this paper. The proposed approach uses the Doppler rate as an additional parameter compared with the previous Doppler effect based DPD. Through the simulation and analysis, the following results are reported:

1. A large observation window can improve the accuracy of DPD by coherent or non-coherent accumulation, but also results in the noticeable Doppler frequency shift variation that cannot be omitted in most practical cases.
2. When the Doppler frequency shift variation is noticeable, the localization accuracy of the proposed method is superior to the ones only utilizing the Doppler frequency shift.
3. The proposed ML estimator can converge to the CRLB at a moderate SNR with moderate snapshots.

Author Contributions: G.W., M.Z. and F.G.; Data curation, G.W.; Formal analysis, G.W.; Funding Acquisition, M.Z.; Investigation, G.W. and M.Z.; Methodology, G.W. and X.X.; Project Administration, G.W. and X.X.; Resources, G.W. and X.X.; Visualization, G.W. and F.G.; Writing—Original Draft, G.W.; Writing—Review and Editing, G.W.

Funding: This research was funded by the National Defence Science and Technology Project Fund of China Grant No. 3101140, the Shanghai Aerospace Science and Technology Innovation Fund of China Grant No. SAST2015028, and the Equipment Prophecy Fund of China Grant No. 9140A21040115KG01001.

Conflicts of Interest: The authors declare no conflict of interest.

References

1. Ulman, R.; Geraniotis, E. Wideband TDOA/FDOA processing using summation of short-time CAF's. *IEEE Trans. Signal Process.* **1999**, *47*, 3193–3200, [CrossRef]
2. Ho, K.C.; Chan, Y.T. Geolocation of a known altitude object from TDOA and FDOA measurements. *IEEE Trans. Aerosp. Electron. Syst.* **1997**, *33*, 770–783. [CrossRef]
3. Bello, P. Joint estimation of delay, Doppler, and Doppler rate. *IRE Trans. Inf. Theory* **1960**, *6*, 330–341. [CrossRef]
4. Mazzenga, F.; Corazza, G.E. Blind least-squares estimation of carrier phase, Doppler shift, and Doppler rate for m-PSK burst transmission. *IEEE Commun. Lett.* **1998**, *2*, 73–75. [CrossRef]
5. Abatzoglou, T. "Fast maximum likelihood joint estimation of frequency and frequency rate". In Proceedings of the IEEE International Conference on Acoustics, Speech, and Signal Processing, Tokyo, Japan, 7–11 April 1986; Volume 11, pp. 1409–1412. [CrossRef]
6. Zhang, S.; Xing, M. A Novel Doppler Chirp Rate and Baseline Estimation Approach in the Time Domain Based on Weighted Local Maximum-Likelihood for an MC-HRWS SAR System. *IEEE Geosci. Remote Sens. Lett.* **2017**, *14*, 299–303. [CrossRef]
7. Zhang, W.Q. Fast Doppler rate estimation based on fourth-order moment spectrum. *Electron. Lett.* **2015**, *51*, 1926–1928. [CrossRef]
8. Chan, Y.T.; Jardine, F.L. Target localization and tracking from Doppler-shift measurements. *IEEE J. Ocean. Eng.* **1990**, *15*, 251–257. [CrossRef]
9. Chan, Y.T.; Towers, J.J. Passive localization from Doppler-shifted frequency measurements. *IEEE Trans. Signal Process.* **1992**, *40*, 2594–2598. [CrossRef]
10. Hu, D.; Huang, Z.; Chen, X.; Lu, J. A Moving Source Localization Method Using TDOA, FDOA and Doppler Rate Measurements. *IEICE Trans. Commun.* **2016**, *99*, 758–766. [CrossRef]
11. Amar, A.; Weiss, A.J. Advances in direct position determination. In Proceedings of the 2004 Sensor Array and Multichannel Signal Processing Workshop Proceedings, Barcelona, Spain, 18–21 July 2004; pp. 584–588. [CrossRef]
12. Weiss, A.J. Direct position determination of narrowband radio frequency transmitters. *IEEE Signal Process. Lett.* **2004**, *11*, 513–516. [CrossRef]
13. Alit Mendelson Reuven, A.J.W. Direct position determination of cyclostationary signals. *Signal Process.* **2009**, *89*, 2448–2464. [CrossRef]
14. Bar-Shalom, O.; Weiss, A.J. Efficient direct position determination of orthogonal frequency division multiplexing signals. *IET Radar Sonar Navig.* **2009**, *3*, 101–111. [CrossRef]
15. Oispuu, M.; Nickel, U. Direct detection and position determination of multiple sources with intermittent emission. *Signal Process.* **2010**, *90*, 3056–3064. [CrossRef]
16. Bialer, O.; Raphaeli, D.; Weiss, A.J. Maximum-Likelihood Direct Position Estimation in Dense Multipath. *IEEE Trans. Veh. Technol.* **2013**, *62*, 2069–2079. [CrossRef]
17. Tzafri, L.; Weiss, A.J. High-Resolution Direct Position Determination Using MVDR. *IEEE Trans. Wirel. Commun.* **2016**, *15*, 6449–6461. [CrossRef]
18. Tirer, T.; Weiss, A.J. High Resolution Direct Position Determination of Radio Frequency Sources. *IEEE Signal Process. Lett.* **2016**, *23*, 192–196. [CrossRef]
19. Weiss, A.J.; Amar, A. Direct position determination of multiple radio signals. *EURASIP J. Adv. Signal Process.* **2005**, *2005*, 37–49. [CrossRef]
20. Demissie, B.; Oispuu, M.; Ruthotto, E. Localization of multiple sources with a moving array using subspace data fusion. In Proceedings of the 2008 11th International Conference on Information Fusion, Cologne, Germany, 30 June–3 July 2008; pp. 1–7.
21. Amar, A.; Weiss, A.J. Localization of Narrowband Radio Emitters Based on Doppler Frequency Shifts. *IEEE Trans. Signal Process.* **2008**, *56*, 5500–5508. [CrossRef]
22. Tirer, T.; Weiss, A.J. High resolution localization of narrowband radio emitters based on doppler frequency shifts. *Signal Process.* **2017**, *141*, 288–298. [CrossRef]
23. Weiss, A.J. Direct Geolocation of Wideband Emitters Based on Delay and Doppler. *IEEE Trans. Signal Process.* **2011**, *59*, 2513–2521. [CrossRef]

24. Li, J.; Yang, L.; Guo, F.; Jiang, W. Coherent summation of multiple short-time signals for direct positioning of a wideband source based on delay and Doppler. *Digital Signal Process.* **2016**, *48*, 58–70. [CrossRef]
25. Zhou, L.; Zhu, W.; Luo, J.; Kong, H. Direct positioning maximum likelihood estimator using TDOA and FDOA for coherent short-pulse radar. *IET Radar Sonar Navig.* **2017**, *11*, 1505–1511. [CrossRef]
26. Becker, K. New algorithm for frequency estimation from short coherent pulses of a sinusoidal signal. *IEE Proc. F Radar Signal Process.* **1990**, *137*, 283–288. [CrossRef]
27. Rummler, W.D. Clutter Suppression by Complex Weighting of Coherent Pulse Trains. *IEEE Trans. Aerosp. Electron. Syst.* **1966**, *AES-2*, 689–699. [CrossRef]
28. Rihaczek, A.W.; Golden, R.M. Clutter Performance of Coherent Pulse Trains for Targets with Range Acceleration. *IEEE Trans. Aerosp. Electron. Syst.* **1971**, *AES-7*, 1093–1099. [CrossRef]
29. Mitchell, R.L. Resolution in Doppler and Acceleration with Coherent Pulse Trains. *IEEE Trans. Aerosp. Electron. Syst.* **1971**, *AES-7*, 630–636. [CrossRef]
30. Levanon, N. CW alternatives to the coherent pulse train-signals and processors. *IEEE Trans. Aerosp. Electron. Syst.* **1993**, *29*, 250–254. [CrossRef]

Article

Remote Sensing Image Fusion Based on Sparse Representation and Guided Filtering

Xiaole Ma [1,2], Shaohai Hu [1,2,*], Shuaiqi Liu [3,*], Jing Fang [1,2] and Shuwen Xu [4]

[1] Institute of Information Science, Beijing Jiaotong University, Beijing 100044, China; maxiaole@bjtu.edu.cn (X.M.); fangjing@sdnu.edu.cn (J.F.)

[2] Beijing Key Laboratory of Advanced Information Science and Network Technology, Beijing Jiaotong University, Beijing 100044, China

[3] College of Electronic and Information Engineering, Hebei University, Baoding 071002, China

[4] Research Institute of TV and Electro-Acoustics, Beijing 100015, China; xu_sw@263.com

* Correspondence: shhu@bjtu.edu.cn (S.H.); shdkj-1918@163.com (S.L.); Tel.: +86-010-5168-8646 (S.H.)

Received: 18 January 2019; Accepted: 5 March 2019; Published: 8 March 2019

Abstract: In this paper, a remote sensing image fusion method is presented since sparse representation (SR) has been widely used in image processing, especially for image fusion. Firstly, we used source images to learn the adaptive dictionary, and sparse coefficients were obtained by sparsely coding the source images with the adaptive dictionary. Then, with the help of improved hyperbolic tangent function (tanh) and $l_0 - \max$, we fused these sparse coefficients together. The initial fused image can be obtained by the image fusion method based on SR. To take full advantage of the spatial information of the source images, the fused image based on the spatial domain (SF) was obtained at the same time. Lastly, the final fused image could be reconstructed by guided filtering of the fused image based on SR and SF. Experimental results show that the proposed method outperforms some state-of-the-art methods on visual and quantitative evaluations.

Keywords: image fusion; sparse representation; hyperbolic tangent function; guided filter

---

1. Introduction

By making full use of the complementary information of the remote sensing images and other source images of the same scene, image fusion can be defined as the processing method for integrating this information together to obtain a fused image, which is more suitable for the human visual system [1]. Through image fusion, we can obtain one composite image, which contains more special features, and can provide more useful information. As a powerful tool for image processing, image fusion covers broad range of areas [2,3], such as computer vision, remote sensing, and so on [4].

Diversiform remote sensing image fusion methods have been proposed in recent years, which can be divided into three categories: Pixel-level fusion, feature-level fusion, and decision-level fusion [5]. Feature-level fusion mainly deals with the features of the source images, while decision-level fusion makes the decision after judging the information of the source images. Compared with the aforementioned levels, pixel-level fusion can serve more useful original information, although it has some shortcomings such as being time consuming. Despite complex computation, most researchers conduct image fusion based on pixel-fusion [6,7], such as the image fusion method based on the spatial domain, and the image fusion method based on the transform domain.

Recently, mainstream methods of image fusion have been based on the multi-scale transforms [8,9], such as image fusion based on object region detection and non-subsampled contourlet transform [10] and image fusion based on the complex shearlet transform with guided filtering [11]. For the image fusion method based on multi-scale transforms, the source images are represented by the fixed orthogonal basis functions, and the fused image can be obtained by fusing the coefficients of different sub-bands together in the transform domain. Although the multi-scale geometric transform can represent most features of the image, which are always complex and diverse, there are some features that cannot be represented sparsely. Thus, it cannot represent all the useful features accurately by limited fixed transforms.

The rapidly developing sparse representation methods can not only more sparsely represent the source images, but also effectively extract the potential information hidden in the source images and produce more accurate fused images, compared with the multi-scale transforms [12–14]. Based on these findings, scholars apply sparse representation to image fusion. Mitianoudis [13] and Yang [14] laid the foundation for image fusion based on SR. Yu [15] applied sparse representation with K-singular value decomposition (K-SVD) to medical image fusion, Yang [16] applied sparse representation and multi-scale decomposition to remote sensing image fusion, and Yin [17] applied a novel sparse-representation-based method to multi-focus image fusion.

In the sparse model, the generation of the dictionary and sparse coding is crucial for the image fusion [18]. Although the fixed over-complete dictionary can realize good fusion results, it usually takes a lot of time to obtain the sparse coefficients, resulting in inefficiency. In this paper, adaptive dictionary learning [19,20] is adopted for its simplicity and convenience. Motivated by the multi-strategy fusion rule based sigmoid function in reference [21] and the characteristics of the hyperbolic tangent function, the multifarious rule based on tanh and $l_0 - \max$ is proposed to fuse the sparse coefficients. Finally, by sparse reconstruction, the fused image based on SR is obtained, which is more suitable for the human visual system and subsequent image processing. However, there is more detailed information in the remote sensing images than other kinds of images. When performing image fusion by the method based on SR, it may lose some discontinuous edge features [22], which leads to the loss of some useful information of fused images. In addition, image fusion based on SR also ignores the spatial information, which can reflect the image structure more directly and accurately. As a result, we can simultaneously fuse the source remote sensing images by the method based on SR and SF, and obtain two different fused images, namely the fused image based on SR and the fused image based on SF. In this paper the two fused images above are processed by a guided filter to obtain the final image since a guided filter has good performance with edge preserving [23]. The main contributions of this paper can be summarized as follows.

(1) The learning of the dictionary is vital for sparse representation, and the adaptive dictionary of each source image can be generated in every step of dictionary learning. The final dictionary can be obtained by gathering together the sub-dictionaries. As a result, this work enriches the dictionary and can make the coefficients more sparsely.

(2) As is well known, the information in each source image is complementary and redundant. When fusing images to obtain the fused image, we need to consider the relationship between different source images. For the redundant information of the source images, the weighted rule would be better; on the other hand, the choose-max rule would result in a fused image with less block effect. Based on the above considerations and the characteristics of hyperbolic tangent function, the fusion rule based on tanh and $l_0 - \max$ is proposed in this paper.

(3) The image fusion methods based on SR can obtain the fused image by sparsely coding the source images and fusing the sparse coefficients. However, it ignores the correlation of the image information in the spatial domain and loses some important detailed information of the source images. In this paper, we adopt the image fusion method based on SF and filter the fused image based on SR and SF by the guided filter. By making full use of the information in the spatial and the sparse representation domain, the fused image can reserve more information of the source images.

The rest of this paper is organized as follows. The theory of the sparse representation is introduced briefly in Section 2. Adaptive dictionary learning is presented in Section 2.1, and the proposed fusion rule is given in Section 2.2. The flow chart of the remote sensing image fusion method based on SR and guided filtering is drawn in Section 3. In Section 4, some experiments and result analysis are done. Finally, conclusions are made in Section 5.

2. Sparse Representation

SR has been widely used in image processing, as one of the most powerful tools to represent signals especially image signals, such as image de-noising [24], image coding [25], object tracking [26], and image super resolution [27], etc.

In the SR model, the image is sparse and can be represented, or approximately represented, by one linear combination of a few atoms from the dictionary [14,28,29]. Suppose that the source image is $\mathbf{I}$, and the over-complete dictionary is $\mathbf{D} \in \mathbf{R}^{M \times k}$, the sparse representation model can be formulated as follows [16,22].

$$\hat{\alpha} = \arg\min_{\alpha} \|\alpha\|_0 \, s.t. \, \|\mathbf{I} - \mathbf{D}\alpha\|_2^2 \leq \varepsilon \tag{1}$$

where α denotes the sparse coefficients of the image and $\|\bullet\|_0$ denotes the $l_0 - norm$, respectively, which indicate the number of non-zero elements in the corresponding vector. Usually, $\|\alpha\|_0 \leq L << M$, and L is the maximal sparsity. ε indicates the limiting error.

For the image fusion method based on SR, there are two important steps: dictionary learning and sparse coding. Dictionary learning will be discussed in detail in Section 2.1. When performing sparse coding by orthogonal matching pursuit (OMP) [30] in this paper, Equation (1) can be replaced by Equation (2).

$$\hat{\alpha} = \arg\min_{\alpha} \|\mathbf{I} - \mathbf{D}\alpha\|_2^2 + \mu\|\alpha\|_0 \tag{2}$$

where, μ is the penalty factor.

2.1. Adaptive Dictionary Learning

When fuse the source images by the methods based on SR, dictionary learning is one of important processes. To make full use of the image information, we generate a dictionary based on the source images themselves. And the generation of the adaptive dictionary can be changed into the iteration of the dictionary atoms. By the iteration process, it can realize dictionary learning with the over-complete dictionary based on the source images.

Since dictionary learning is more efficient for small image blocks, if the dictionary updating step is processed by the original source images directly, the sparsity would be seriously influenced. Thus optimal sparse coefficients cannot be obtained [29]. In order to solve this problem, we divided the source images into image blocks, which can replace the dictionary atoms for better dictionary learning. The improved dictionary generation method can not only obtain the optimal sparse representation but also accelerates the efficiency and accuracy of the SR algorithm. However, since we perform dictionary learning on the image block rather than the whole image, the reshaped vector on every atom is not very large and it reduces the computation cost.

K-singular value decomposition (K-SVD) [31] is one of the most used image fusion methods based on SR. Here, we adopt the K-SVD model on the sub-dictionary of the image block by the following iteration process:

$$\hat{\mathbf{D}}_{ij}^{M} = \arg\min_{\hat{\mathbf{D}}_{ij}^{M}, \alpha_{ij}^{M}} \sum_{i,j} \left\| \mathbf{P}_{ij}^{M} - \hat{\mathbf{D}}_{ij}^{M}\alpha_{ij}^{M} \right\|_2^2 + \mu_{ij}^{M}\left\| \alpha_{ij}^{M} \right\|_0 \tag{3}$$

where ij denotes the position (i, j) in the image $\mathbf{M}$ and $\mathbf{P}_{ij}^{M}$ denotes the image block with the center pixel at the corresponding position (i, j).

Then, we can obtain the adaptive dictionary of the source image $\mathbf{M}$ shown in Equation (4).

$$\hat{\mathbf{D}}^{M} = \left\{ \hat{\mathbf{D}}_{ij}^{M} \right\} \tag{4}$$

At last, we can gather all the dictionaries of different source images by Equation (5), where n denotes the total number of the source images.

$$\mathbf{D} = [\mathbf{D}_1, \mathbf{D}_2, \dots, \mathbf{D}_n] \tag{5}$$

2.2. Fusion Rule Based on tanh and $l_0 - max$

As we all know, the fusion rules are vital for the final fusion results and for the sparse coefficients. In most cases, we always take the $l_1 - max$ rule to obtain the fused block vectors [7], where l_1 means the sum of absolute values of the vector elements. However, when there are noises or some unwanted pixels in the flat area of the source images, the unwanted portion will be included and lead to incorrect fusion [17]. The information in the source images is redundant and complementary for the image fusion shown in Figure 1. Figure 1a,b are one set of medical images, which contain complementary information, while Figure 1c,d are one set of multi-focus images, which contain redundant information. When the relationship of the image information is redundant, the weighted fusion rule is chosen, and the max fusion rule should be chosen for the complementary sparse coefficients [21]. The fused information would be lost and incomplete if the complementary information is multiplicative by the weighted factor. Based on these considerations, we proposed one new sparse coefficient fusion rule based on tanh and $l_0 - max$. We can obtain the fused coefficients by calculating $l_0 - norm$ and the weighting factor based on tanh.

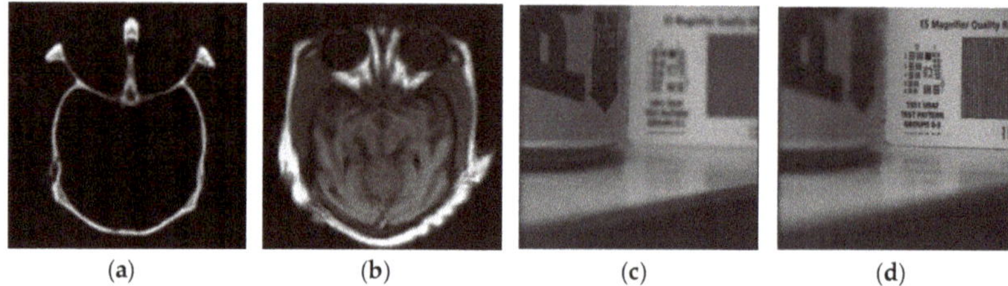

(a)   (b)   (c)   (d)

Figure 1. Images with different information: (**a**) CT image; (**b**) MRI image; (**c**) left-focus image; (**d**) right-focus image.

The hyperbolic tangent function is one of the hyperbolic functions, and derives from hyperbolic sine function and hyperbolic cosine function [32]. It can be calculated as follows:

$$\tanh(x) = \frac{\sinh(x)}{\cosh(x)} = \frac{e^x - e^{-x}}{e^x + e^{-x}} \tag{6}$$

where the hyperbolic sine function and hyperbolic cosine function can be defined as Equations (7) and (8), respectively.

$$\sinh(x) = \frac{e^x - e^{-x}}{2} \tag{7}$$

$$\cosh(x) = \frac{e^x + e^{-x}}{2} \tag{8}$$

Figure 2 shows the different hyperbolic functions. From Figure 2a,b we can see that tanh is symmetrical around the origin point. As x increases, the difference between the value of the hyperbolic sine function and the hyperbolic cosine function narrows, and the value of $\tanh(x)$ changes from -1 to 1. When there is redundant information in different source images and the weighted fusion rule is chosen, it would be better if different degrees of redundancy corresponded to different weights. Based on the aforementioned factors, we improve tanh shown in Figure 2c to obtain the weighted factor for fusing the sparse coefficients, and the corresponding equation is listed as Equation (9).

$$w_{ij} = \frac{1}{2} * [\tanh(a * (s_{ij} - 1)) + 1] \tag{9}$$

where s_{ij} denotes the sparse coefficient at the position (i, j) and w_{ij} denotes the corresponding weighted factor when adopting the fusion rule based on tanh. a denotes the sensitivity between the sparse coefficient and the weighted factor. According to the experiments on different image groups and values of the parameter a, we found that 3 is the best.

Compared with Figure 2b, the curve has a steeper slope in Figure 2c when s_{ij} is closer to 1, which means that the weighted factor is very sensitive to the sparse coefficients. When s_{ij} is near 0 or too large, the weighted factor w_{ij} is near 0 or 1, which means that the source images have complementary information, where the fusion rule based on $l_0 - \max$ is adopted.

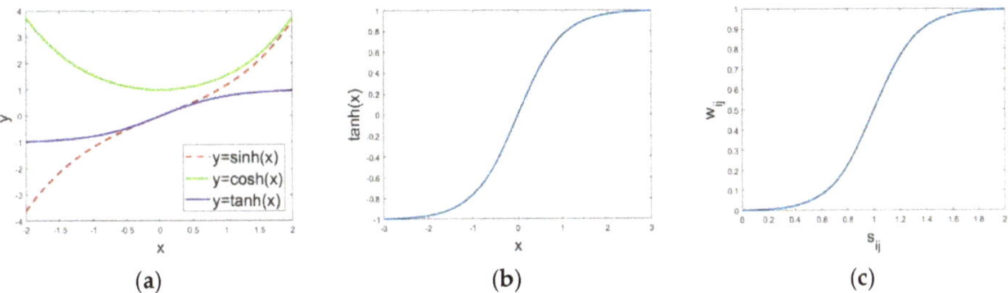

(a)   (b)   (c)

Figure 2. Different hyperbolic functions: (**a**) The hyperbolic functions; (**b**) tanh; (**c**) improved tanh.

Finally, we can obtain the fused sparse coefficients $\alpha_{F_{ij}}$ at the position (i, j) by Equation (10).

$$
\alpha_{F_{ij}} = \begin{cases} w_{ij} * \alpha_{A_{ij}} + (1 - w_{ij}) * \alpha_{B_{ij}} & if \ \alpha_{A_{ij}} \& \alpha_{B_{ij}} \neq 0 \\ \max(\alpha_{A_{ij}}, \alpha_{B_{ij}}) & else \end{cases} \tag{10}
$$

where $\alpha_{A_{ij}}$ and $\alpha_{B_{ij}}$ denotes the sparse coefficients in the source image A and B. $\alpha_{A_{ij}} \& \alpha_{B_{ij}} \neq 0$ means that both $\alpha_{A_{ij}}$ and $\alpha_{B_{ij}}$ are not zero. And w_{ij} can be calculated by Equation (9), where $s_{ij} = \alpha_{A_{ij}}$.

3. The Proposed Image Fusion Method

An interesting remote sensing fusion method based on sparse representation and guided filtering is presented in this paper, and the framework can be seen in Figure 3. It mainly includes three image processing elements: image fusion based on SR, image fusion based on SF, and guided filtering. The adaptive dictionary was learned by the source images themselves, and the fused sparse coefficients was obtained by the dictionary and proposed fusion rule. Then, the fused image based on SR was reconstructed by the obtained adaptive dictionary and fused sparse coefficients. At the same time, we fused the source images obtained by the image fusion method based on SF such as the gradient fusion. As shown in Figure 3, the guided filter was finally adopted to guide the fused images based on SR and SF. Since there was more detailed information in the fused image based on SF, in the last part of the proposed method, we made the fused image based on SF as the guidance image, and the other fused image served as the input image.

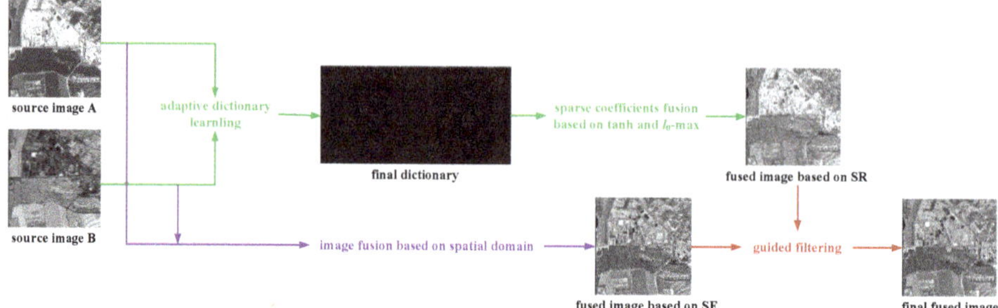

Figure 3. The framework of the proposed method.

4. The Experiments and Result Analysis

To testify the superiority of the proposed method, a series of experiments on the remote sensing and other source images were conducted in this section. We compared our method with some classical image fusion methods, including the multi-scale weighted gradient-based fusion (MWGF) [33], the image fusion with guided filtering (GuF) [34], image fusion based on Laplace transformation (LP) [35], multiresolution DCT decomposition for image fusion (DCT) [36], the image fusion algorithm in the nonsubsampled contourlet transform domain (NSCT) [37], image fusion with the joint sparsity model (SR) [1], and image fusion based on multi-scale and sparse representation(MST-SR) [8]. With adaptive dictionary learning, the size of every image block was 8×8. Experiments conducted on dictionary learning of different source images showed that when the number of iterations was 3, it guaranteed the convergence and stability of the coefficients. In addition, the experiments in this paper were carried out by Matlab code on an Intel Core i5-2450M (Acer, Beijing, China) 2.50 GHz with 6 GB RAM.

4.1. Objective Valuation Indexes

To evaluate the experimental results more objectively, we adopted some objective valuation indexes [37] to evaluate the fused images by different image fusion methods, which included entropy (EN), spatial frequency (SF), $Q^{AB/F}$, and structural similarity (SSIM).

When we want to balance the wealth of information in one image, EN is a wonderful choice. The larger the value of EN in the fused image is, the more information does the image contain, which means better image fusion result. And EN can be summarized as Equation (11).

$$EN = -\sum_{i=0}^{L-1} p_i \times \log_2 p_i \tag{11}$$

where L denotes the total number of pixels included in the image and p_i denotes the probability distribution of pixels for each gray level.

SF can detect the total active of the fused image in the spatial domain and it denotes the expression ability of one image for minor detail contrast. The equation of SF is shown as follows:

$$SF(i,j) = \sqrt{(RF)^2 + (CF)^2} \tag{12}$$

where RF stands for the horizontal frequency while CF stands for the vertical frequency. And they can be calculated by Equations (13) and (14).

$$RF = \sqrt{\frac{1}{M \times N} \sum_{x=1}^{M} \sum_{y=2}^{N} [F(x,y) - F(x,y-1)]^2} \tag{13}$$

$$CF = \sqrt{\frac{1}{M \times N} \sum_{x=2}^{M} \sum_{y=1}^{N} [F(x,y) - F(x-1,y)]^2} \tag{14}$$

where F denotes the fused image with the size of $M \times N$.

While $Q^{AB/F}$ can balance how much the edge information of the source images A and B does the fused image contain by Sobel operator. It can be defined as Equation (15).

$$Q^{AB/F} = \frac{\sum_{\forall n,m} \left(Q_{n,m}^{AF} w_{n,m}^{A} + Q_{n,m}^{BF} w_{n,m}^{B} \right)}{\sum_{\forall n,m} \left(w_{n,m}^{A} + w_{n,m}^{B} \right)} \tag{15}$$

where $w_{n,m}^{A} = [g_A(n,m)]^L$, $w_{n,m}^{B} = [g_B(n,m)]^L$. Normally, L is one constant and the value is 1. Taking the source image A as an example, edge information retention value $Q_{n,m}^{AF}$ and edge strength information $g_A(n,m)$ can be calculated by Equations (16) and (17).

$$Q_{n,m}^{AF} = \Gamma_g \Gamma_\alpha \left[1 + e^{K_g(G_{n,m}^{AF} - \sigma_g)} \right]^{-1} \left[1 + e^{K_\alpha(A_{n,m}^{AF} - \sigma_\alpha)} \right]^{-1} \tag{16}$$

$$g_A(n,m) = \sqrt{s_A^x(n,m)^2 + s_A^y(n,m)^2} \tag{17}$$

where $\Gamma_g, K_g, \sigma_g, \Gamma_\alpha, K_\alpha, \sigma_\alpha$ are constant and they affect the sigmoid function together. $(G_{n,m}^{AF}, A_{n,m}^{AF}) = [(\frac{g_{n,m}^F}{g_{n,m}^A})^M, 1 - \frac{|\alpha_A(n,m) - \alpha_F(n,m)|}{\pi/2}]$ and $M = \begin{cases} -1 & if \ g_A(n,m) \le g_F(n,m) \\ 1 & otherwise \end{cases}$. $\alpha_A(n,m) = \tan^{-1}[\frac{s_A^y(n,m)}{s_A^x(n,m)}]$ and $s_A^x(n,m)$, $s_A^y(n,m)$ denote the convolution results of Sobel model with the center pixel at the position (n,m) in the horizontal and vertical directions with the source image A.

SSIM is the structural similarity between the source images and the fused image. And the equation of SSIM is as follows:

$$SSIM(A,B,F) = \frac{1}{2}(SSIM(A,F) + SSIM(B,F)) \tag{18}$$

where $SSIM(A,F)$ denotes SSIM of the source image A and fused image F, and so is $SSIM(B,F)$. More detail of their calculation is shown in Equations (19) and (20).

$$SSIM(A,F) = \frac{(2\mu_A\mu_F + C_1) \cdot (2\sigma_{AF} + C_2)}{(\mu_A^2 + \mu_F^2 + C_1)(\sigma_A^2 + \sigma_F^2 + C_2)} \tag{19}$$

$$SSIM(B,F) = \frac{(2\mu_B\mu_F + C_1) \cdot (2\sigma_{BF} + C_2)}{(\mu_B^2 + \mu_F^2 + C_1)(\sigma_B^2 + \sigma_F^2 + C_2)} \tag{20}$$

where μ_A, μ_B, μ_F denote the average of pixels of the image A, B and F, respectively. $\sigma_A^2, \sigma_B^2, \sigma_F^2$ denote the variance and σ_{AF}, σ_{BF} denote the joint variance. For the convenience of calculation, we make $C_1 = C_2 = 0$.

The larger all the indexes above are, the better the fused image is. What's more, when obtaining the adaptive dictionary by the proposed method, there is slight deviation of the final results. We adopt the mean of the evaluation values in three times.

4.2. Large Scale Image Fusion of Optical and Radar Images

Figure 4 shows one SAR image of the harbor around Oslo with a size of 1131×942 and the registered optical image on a large scale for the whole scenery [38]. Due to the use of the high-resolution digital elevation model (DEM), the optical image fits onto the signatures of the buildings very well. Figure 4c,d are partially enlarged details of Figure 4a,b at the position of the red rectangle in Figure 4a. Figures 5 and 6 are the corresponding fused images obtained by the methods above, and partially enlarged views.

Since the optical image in Figure 4b is colorful, we processed the image fusion in the RGB dimension separately. Although the visual effect of Figure 6a is better, there was a greater color contrast in Figure 5a, which introduced some incorrect information in the left corner. In Figure 6, the partially enlarged detail images of Figure 5d by DCT and Figure 5f by SR are very blurred which seriously affects the fused images. Compared with Figure 6g, the left corner in Figure 6h contains more information of the remote sensing image in Figure 5c, which indicates that the fused image by our method is better.

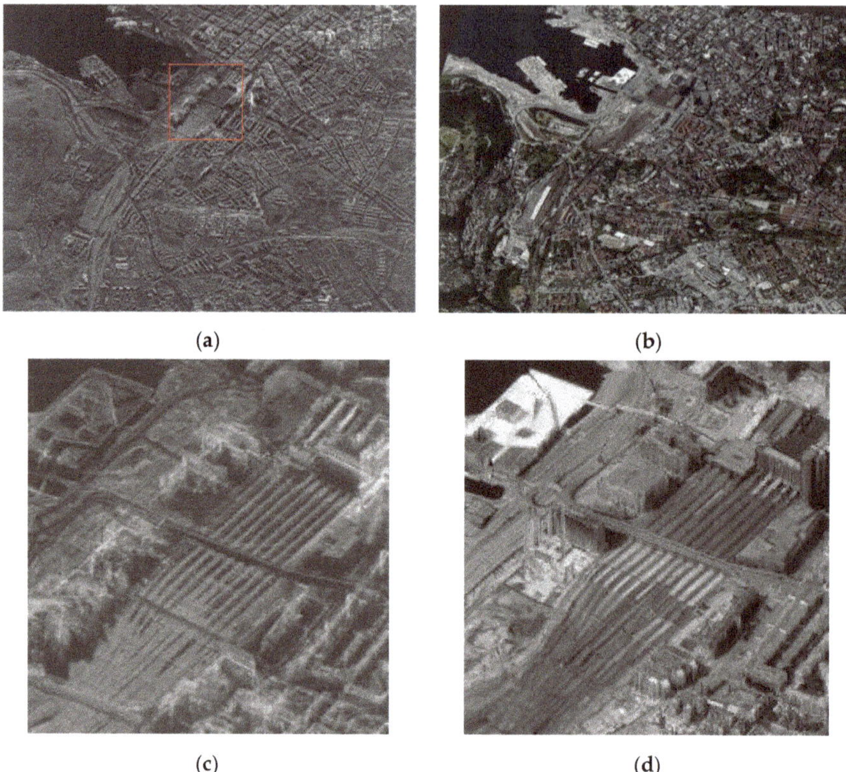

Figure 4. The large scale images: (**a**) TerraSAR-X staring spotlight image of Oslo; (**b**) optical image of Oslo; (**c**) part of (**a**); (**d**) part of (**b**).

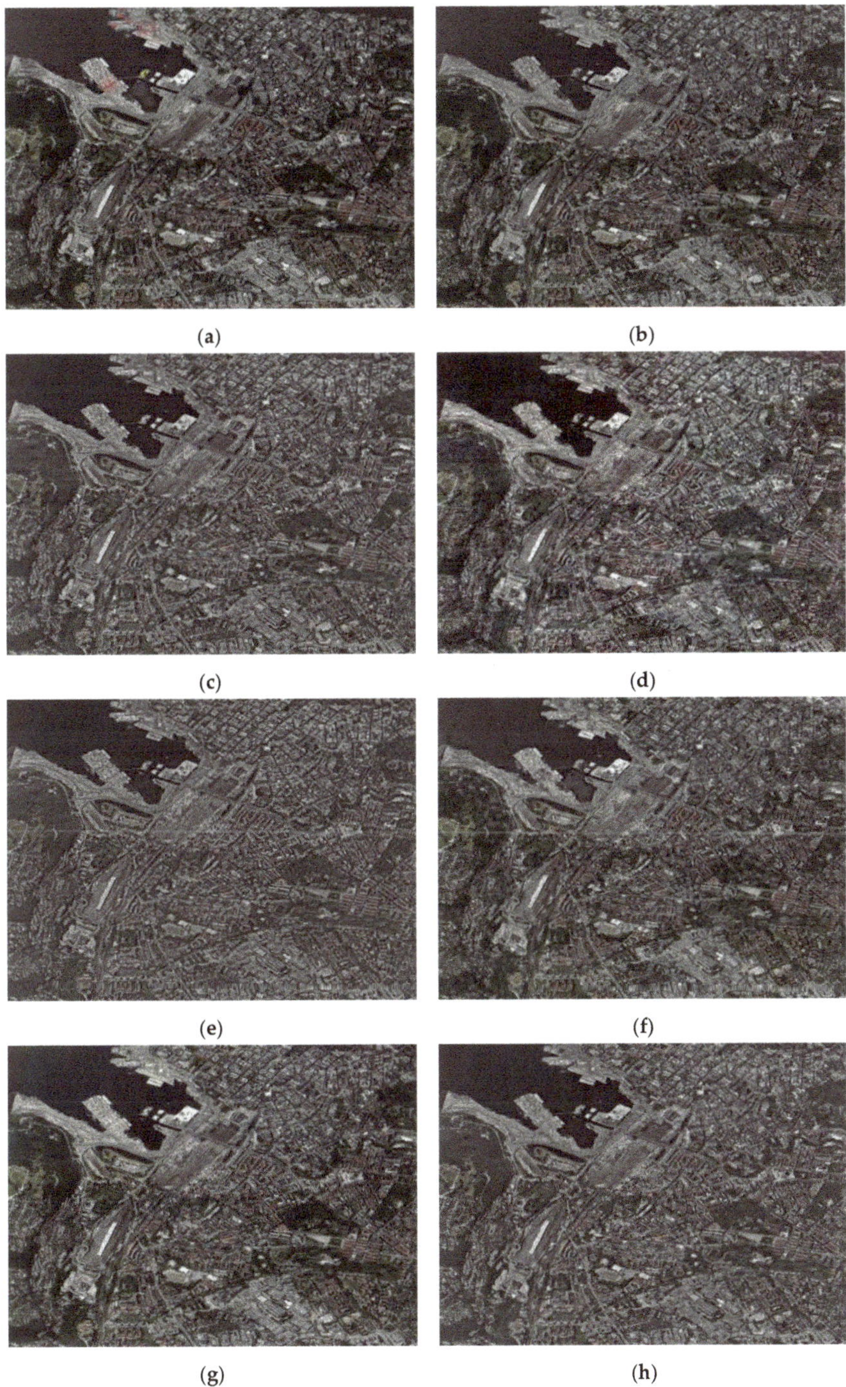

Figure 5. The fused images of Figure 4: (**a**) MWGF; (**b**) GuF; (**c**) LP; (**d**) DCT; (**e**) NSCT; (**f**) SR; (**g**) MST-SR; (**h**) the proposed.

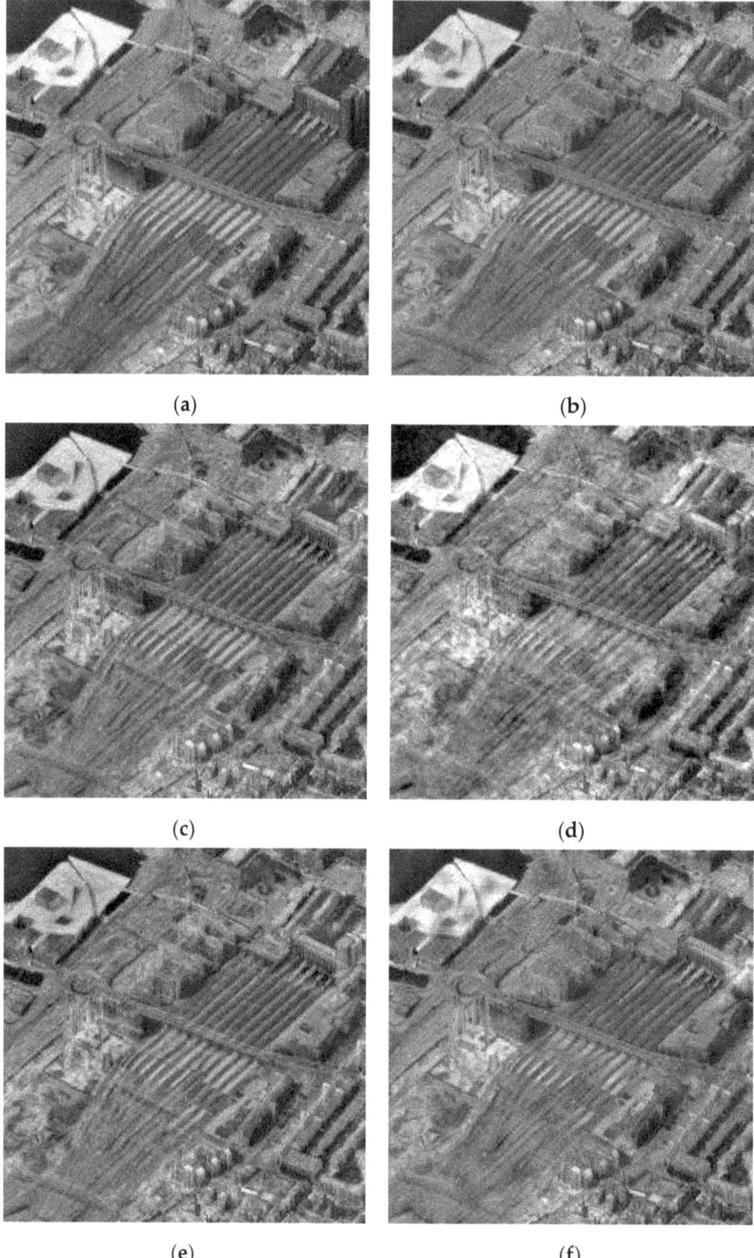

(a)

(b)

(c)

(d)

(e)

(f)

Figure 6. *Cont.*

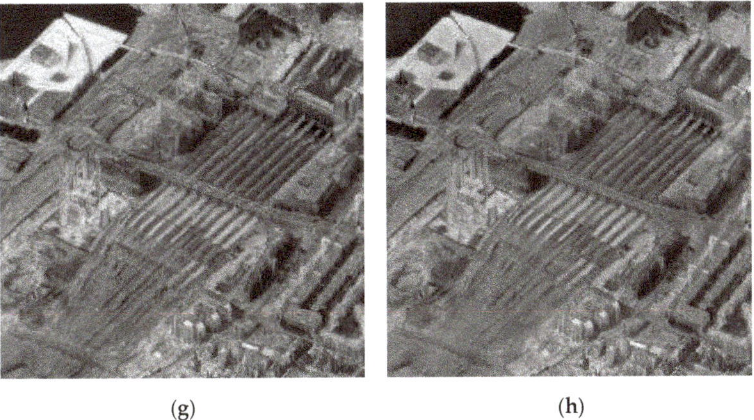

(g)   (h)

Figure 6. Part of Figure 5: (**a**) MWGF; (**b**) GuF; (**c**) LP; (**d**) DCT; (**e**) NSCT; (**f**) SR; (**g**) MST-SR; (**h**) the proposed.

Table 1 shows the corresponding index values of the fused images in Figure 5 and the best values are in bold. From Table 1, we can see the image fusion methods based on the spatial domain such as MWGF and GuF have big ability to preserve the spatial frequency, and MWGF has a better value of $Q^{AB/F}$. However, the visual result of MWGF is the worst. $Q^{AB/F}$ of the proposed method ranks third among the compared methods, which is worse than the methods based on the spatial domain. This explains why we adopt the image fusion method based on the spatial domain and guide it with the fused image-based SR in this paper. The values of EN, SF, and SSIM of the fused image obtained by the proposed method are better, which indicates that the proposed method has a better ability to fuse the remote sensing image.

Table 1. The evaluation index values of fused images in Figure 5.

Methods / Indexes	MWGF	GuF	LP	DCT	NSCT	SR	MST-SR	The Proposed
EN	7.3411	7.2547	7.2606	7.2158	7.1843	7.2711	7.3742	**7.5879**
SF	31.1834	30.1662	32.4682	30.1223	31.5884	30.2674	32.4878	**33.9360**
$Q^{AB/F}$	**0.6313**	0.6111	0.5753	0.3754	0.5453	0.5705	0.5767	0.5794
SSIM	0.5798	0.5981	0.5990	0.5410	0.5910	0.5829	0.5966	**0.6246**

4.3. Image Fusion of Remote Sensing Images

To testify the effectiveness and universality of the proposed method, the classical image pairs shared by Durga Prasad Bavirisetti (https://sites.google.com/view/durgaprasadbavirisetti/datasets) are used to test the performance of the fused algorithms. The dataset contains rich remote sensing images and we conduct our experiments on different kinds of image pairs, which contain the forest with greater high-frequency information, rivers with low-frequency information, and so on. To save space, we only show the four groups and the results analysis. The four groups include rich information with different types and are representative in the dataset, shown in Figure 7. Figures 8–11 are the fused images obtained by the diverse compared methods of the different source images.

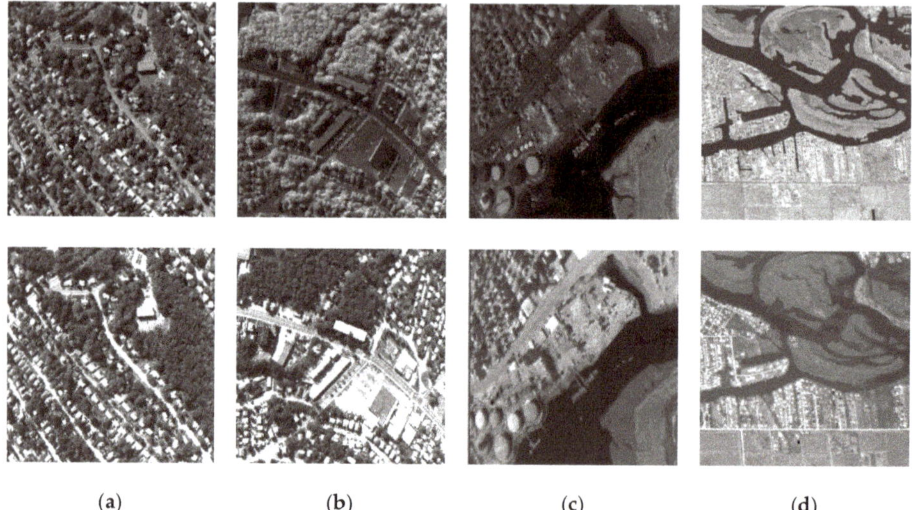

(a)   (b)   (c)   (d)

Figure 7. The source remote sensing images: (**a**) Group 1; (**b**) Group 2; (**c**) Group 3; (**d**) Group 4.

Figure 7a,b are forests and rural areas with fewer buildings, of which the top view is sharper and has richer detailed information. From Figure 8, we can see that the trees in Figure 8a–e is more darker than Figure 8f–g and has less information in the second line of Group 1 in Figure 7, which indicates that the image fusion based on SR is more powerful than the methods based on the spatial domain and transform domain. And there are some artificial textures in the roof of Figure 8f. Above all, the fused image of Figure 8h obtained by the proposed has better visual effect.

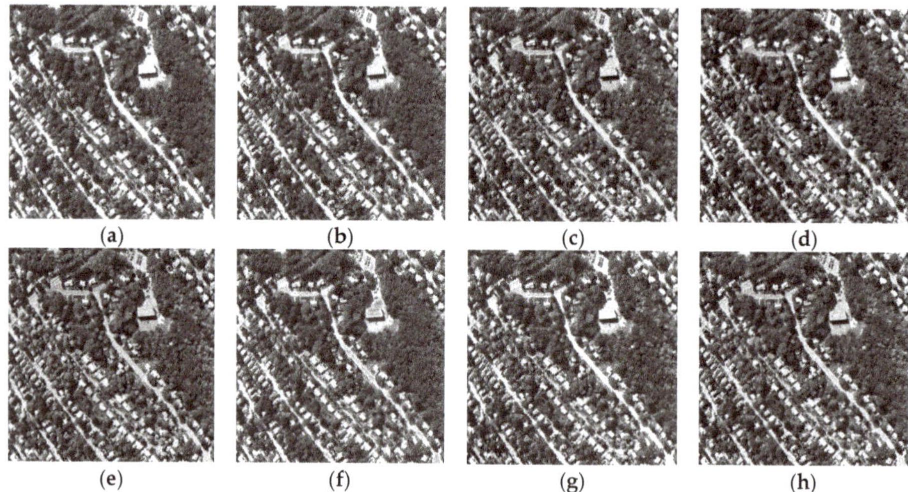

(a)   (b)   (c)   (d)

(e)   (f)   (g)   (h)

Figure 8. The fused images of Group 1: (**a**) MWGF; (**b**) GuF; (**c**) LP; (**d**) DCT; (**e**) NSCT; (**f**) SR; (**g**) MST-SR; (**h**) the proposed.

Compared with Group 2, there are some suburbs next to the forests in Group 1. And the contrast in Figure 9c,e,h looks better. From the roofs in the fused images shown in Figure 9, the flat area and edges in Figure 9h obtained by the proposed method look more comfortable and are more suitable for we human visual system, which indicates that the proposed method has powerful ability to fuse remote sensing images.

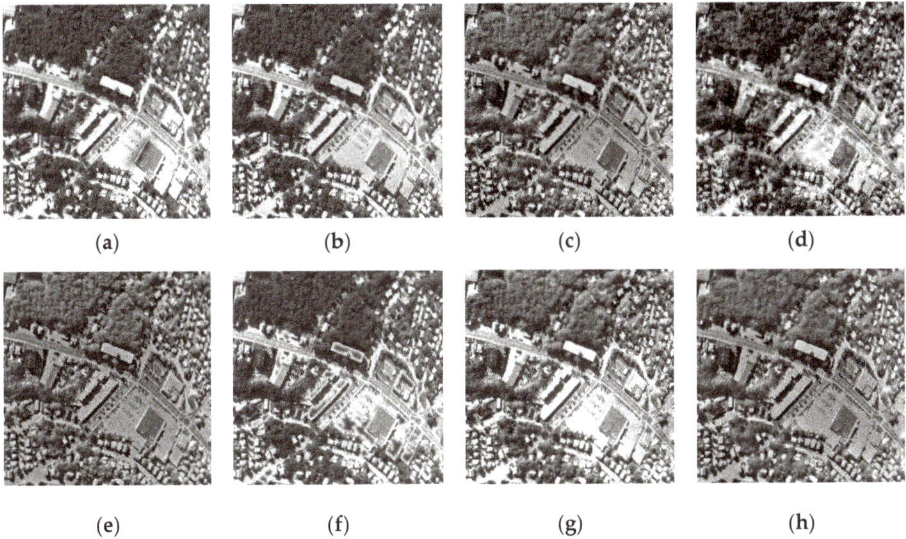

Figure 9. The fused images of Group 2: (**a**) MWGF; (**b**) GuF; (**c**) LP; (**d**) DCT; (**e**) NSCT; (**f**) SR; (**g**) MST-SR; (**h**) the proposed.

There are some river and coastal area in Group 3. And by comparing the fused images in Figure 10, the center in Figure 10a looks very bad and some areas in Figure 10g are too bright, which have the strong exposure. From these figures, we can see that there is less artificial texture in Figure 10h, which means the fused image obtained by the proposed method have better visual result.

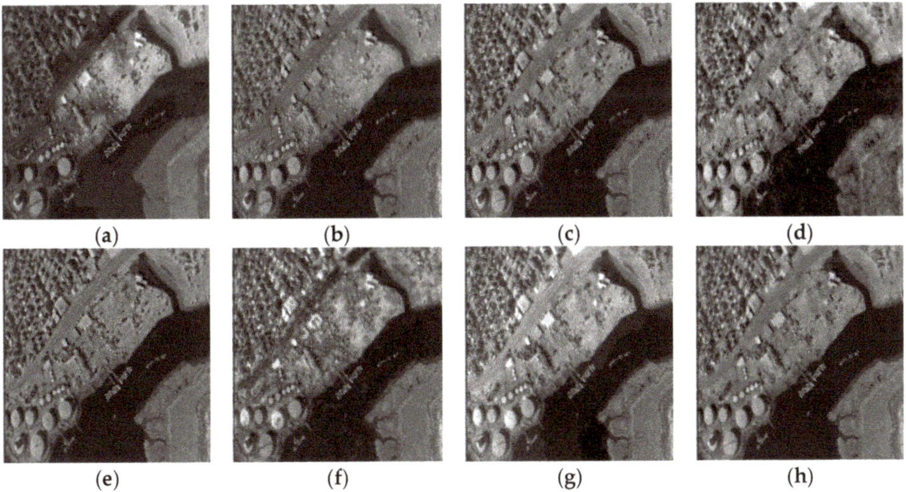

Figure 10. The fused images of Group 3: (**a**) MWGF; (**b**) GuF; (**c**) LP; (**d**) DCT; (**e**) NSCT; (**f**) SR; (**g**) MST-SR; (**h**) the proposed.

Group 4 is one set of classic multi-sensor image pair, which can be found in most of papers about remote sensing image fusion. By comparing the bottoms of the fused images in Figure 10, we can find that there are some unwanted spots and artificial texture in Figure 10d, and the small round black area is very blurred or even lost in Figure 10a–c,f. Since the rivers display as black areas like wide line or curve in the fused images, it has worst visual effect in Figure 10f, of which the detailed information has been lost. As a result, the fused image in Figure 10h looks more comfortable for our eyes and the proposed method has better ability to fuse remote sensing images.

Similarly, we use the aforementioned objection evaluation indexes to value the fused images in Figures 8–11 and the objective values are shown in Tables 2–5. As shown in Tables 2 and 3, the algorithm proposed in this paper has obtained the best results for Group 1 and Group 2 in Figure 7. This fully demonstrates that the proposed method has a better ability to perform remote sensing image fusion. Compared with Group 1 and Group 2, there is more low frequency information and less detail and edges in Group 3 and Group 4. However, the proposed method is more suitable for the images with great detail. As a result, the SSIM of the fused image by NSCT is better than others in Table 4, but other values of the proposed method are satisfactory. All these values demonstrate that the proposed method performs better in terms of remote sensing image fusion.

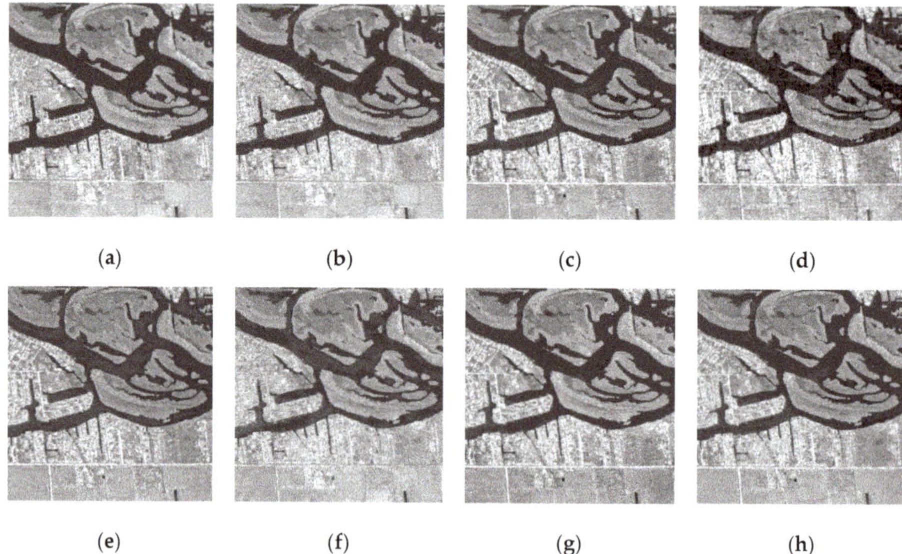

Figure 11. The fused images of Group 4: (**a**) MWGF; (**b**) GuF; (**c**) LP; (**d**) DCT; (**e**) NSCT; (**f**) SR; (**g**) MST-SR; (**h**) the proposed.

Table 2. The evaluation index values of fused images in Figure 8.

Methods / Indexes	MWGF	GuF	LP	DCT	NSCT	SR	MST-SR	The Proposed
EN	7.1741	7.4931	7.6687	7.6006	7.6935	7.4351	7.5925	**7.7076**
SF	54.1839	53.6728	55.0006	55.2435	54.6567	53.9868	54.8697	**56.3409**
$Q^{AB/F}$	0.7030	0.7104	0.7110	0.6105	0.7143	0.7089	0.7073	**0.7174**
SSIM	0.7933	0.7976	0.8077	0.7759	0.8170	0.7916	0.8060	**0.8235**

Table 3. The evaluation index values of fused images in Figure 9.

Methods Indexes	MWGF	GuF	LP	DCT	NSCT	SR	MST-SR	The Proposed
EN	6.9778	7.6668	7.8936	7.5537	7.8154	7.3634	7.6580	**7.8946**
SF	53.6844	53.3379	53.8149	53.2507	53.3216	53.1757	54.1247	**54.1590**
$Q^{AB/F}$	0.6668	0.6687	0.6310	0.4783	0.6190	0.6379	0.6346	**0.6710**
SSIM	0.6283	0.6314	0.6340	0.5433	0.6172	0.5969	0.6429	**0.6438**

Table 4. The evaluation index values of fused images in Figure 10.

Methods Indexes	MWGF	GuF	LP	DCT	NSCT	SR	MST-SR	The Proposed
EN	7.2073	7.0657	6.9885	6.9246	6.9359	7.2902	7.3556	**7.5855**
SF	15.4453	16.4780	18.7239	19.5068	18.3164	18.3152	18.9748	**19.8665**
$Q^{AB/F}$	0.5491	0.5640	0.5574	0.3619	0.5411	0.4741	0.5658	**0.5774**
SSIM	0.6641	0.6954	0.6898	0.4789	**0.6958**	0.6168	0.6797	0.6826

Table 5. The evaluation index values of fused images in Figure 11.

Methods Indexes	MWGF	GuF	LP	DCT	NSCT	SR	MST-SR	The Proposed
EN	6.0932	7.1620	7.3451	7.3081	7.3317	7.1194	7.0635	**7.6861**
SF	27.8713	27.0044	29.2465	28.9847	28.2961	25.7550	29.2601	**29.6650**
$Q^{AB/F}$	**0.6473**	0.6318	0.5857	0.4143	0.5653	0.5586	0.5886	0.5968
SSIM	0.6528	0.6647	0.6736	0.5548	0.6871	0.6590	0.6733	**0.6930**

5. Conclusions

Due to the good performance of sparse representation and the rich information in the spatial domain, this paper presents one new remote sensing image fusion method based on sparse representation and guided filtering. It also makes full use of the redundant and complementary information of different source images. Experimental results show that our method is more suitable for the human visual system and has better objective evaluation index values. However, the proposed image fusion method is very powerful for the details such as image edges. Although remote sensing images have rich detailed information, it would be inefficient if there is much more low frequency information than high frequency information. How to overcome this shortcoming will be investigated in future work.

Author Contributions: X.M. wrote the original draft and she performed the experiments with J.F., S.L. wrote the review & editing. S.H. provided Funding acquisition, and he also provided resources with S.X.

Funding: This research was funded by the Natural Science Foundation of China under Grant No. 61572063 and No. 61401308; Natural Science Foundation of Hebei Province under Grant No. F2016201142 and No. F2016201187; Science Research Project of Hebei Province under under Grant No. QN2016085. Opening Foundation of Machine vision Engineering Research Center of Hebei Province under Grant 2018HBMV02, Science Research Project of Hebei Province under Grant QN2016085, Natural Science Foundation of Hebei University under Grant 2014-303, Post-graduate's Innovation Fund Project of Hebei University under Grant hbu2018ss01.

Acknowledgments: Some of the images adopted in experiments are downloaded from the website of https://sites.google.com/view/durgaprasadbavirisetti/datasets. This work was supported by the High-Performance Computing Center of Hebei University. We also thank Qu XiaoBo, Zhou Zhiqiang, Li Shutao, and Liu Yu for their shared codes of image fusion. We also thank the Editor and Reviewers for the efforts made in processing this submission and we are particularly grateful to the reviewers for their constructive comments and suggestions which help us improve the quality of this paper.

Conflicts of Interest: The authors declare no conflict of interest.

References

1. Zhu, Z.; Yin, H.; Chai, Y.; Li, Y.; Qi, G. A novel multi-modality image fusion based on image decomposition and sparse representation. *Inf. Sci.* **2018**, *432*, 516–529. [CrossRef]
2. Anandhi, D.; Valli, S. An algorithm for multi-sensors image fusion using maximum a posteriori and nonsubsampled contourlet transform. *Comput. Electr. Eng.* **2017**, *65*, 139–152. [CrossRef]
3. Gao, Z.; Zhang, C. Texture clear multi-modal image fusion with joint sparsity model. *Optik* **2017**, *130*, 255–265. [CrossRef]
4. Hu, S.; Yang, D.; Liu, S.; Ma, X. Block-matching based mutimodal medical image fusion via PCNN with SML. In Proceedings of the 2016 IEEE 13th International Conference on Signal Processing (ICSP), Chengdu, China, 6–10 November 2016; pp. 13–18.
5. Li, H.; Chai, Y.; Ling, R.; Yin, H. Multifocus image fusion scheme using feature contrast of orientation information measure in lifting stationary wavelet domain. *J. Inf. Sci. Eng.* **2013**, *29*, 227–247.
6. Ghassemian, H. A review of remote sensing image fusion methods. *Inf. Fusion* **2016**, *32*, 75–89. [CrossRef]
7. Zhang, J.; Feng, X.; Song, B.; Li, M.; Lu, Y. Multi-focus image fusion using quality assessment of spatial domain and genetic algorithm. In Proceedings of the Conference on Human System Interactions, Krakow, Poland, 25–27 May 2008; pp. 71–75. [CrossRef]
8. Liu, Y.; Liu, S.; Wang, Z. A general framework for image fusion based on multi-scale transform and sparse representation. *Inf. Fusion* **2015**, *24*, 147–164. [CrossRef]
9. Li, H.; Qiu, H.; Yu, Z.; Li, B. Multifocus image fusion via fixed window technique of multiscale images and non-local means filtering. *Signal Process.* **2017**, *138*, 71–85. [CrossRef]
10. Meng, F.; Song, M.; Guo, B.; Shi, R.; Shan, D. Image fusion based on object detection and non-subsampled contourlet transform. *Comput. Electr. Eng.* **2017**, *62*, 375–383. [CrossRef]
11. Liu, S.; Shi, M.; Zhu, Z.; Zhao, J. Image fusion based on complex-shearlet domain with guided filtering. *Multidimens. Syst. Signal Process.* **2017**, *28*, 207–224. [CrossRef]
12. Donoho, D.L. Compressed Sensing. *IEEE Trans. Inf. Theory* **2006**, *52*, 1289–1306. [CrossRef]
13. Mitianoudis, N.; Stathaki, T. Pixel-based and region-based image fusion schemes using ICA bases. *Inf. Fusion* **2007**, *8*, 131–142. [CrossRef]
14. Yang, B.; Li, S. Multifocus image fusion and restoration with sparse representation. *IEEE Trans. Instrum. Meas.* **2010**, *59*, 884–892. [CrossRef]
15. Yu, N.; Qiu, T.; Liu, W. Medical image fusion based on sparse representation with K-SVD. In Proceedings of the World Congress on Medical Physics and Biomedical Engineering, Beijing, China, 26–31 May 2012; pp. 550–553. [CrossRef]
16. Yang, Y.; Wu, L.; Huang, S.; Wan, W.; Que, Y. Remote sensing image fusion based on adaptively weighted joint detail injection. *IEEE Access* **2018**, *6*, 6849–6864. [CrossRef]
17. Yin, H.; Li, Y.; Chai, Y.; Liu, Z.; Zhu, Z. A novel sparse-representation-based multi-foucs image fusion approach. *Neurocomputing* **2016**, *216*, 216–229. [CrossRef]
18. Zong, J.; Qiu, T. Medical image fusion based on sparse representation of classified image patches. *Biomed. Signal Process. Control* **2017**, *34*, 195–205. [CrossRef]
19. Zhu, Z.; Chai, Y.; Yin, H.; Li, Y.; Liu, Z. A novel dictionary learning approach for multi-modality medical image fusion. *Neurocomputing* **2016**, *214*, 471–482. [CrossRef]
20. Elad, M.; Yavneh, I. A plurality of sparse representation is better than the sparsest one alone. *IEEE Trans. Inf. Theory* **2009**, *55*, 4701–4714. [CrossRef]
21. Luo, X.; Zhang, Z.; Zhang, C.; Wu, X. Multi-focus image fusion using HOSVD and edge intensity. *J. Visual Commun. Image Represent.* **2017**, *45*, 46–61. [CrossRef]
22. Ma, X.; Hu, S.; Liu, S. SAR image de-noising based on invariant K-SVD and guided filter. *Remote Sens.* **2017**, *9*, 1311. [CrossRef]
23. He, K.; Sun, J.; Tang, X. Guided image filtering. *IEEE Trans. Pattern Anal. Mach. Intell.* **2013**, *35*, 1397–1409. [CrossRef]
24. Liu, S.; Liu, M.; Li, P.; Zhao, J.; Zhu, Z.; Wang, X. SAR image de-noising via sparse representation in shearlet domain based on continuous cycle spinning. *IEEE Trans. Geosci. Remote Sens.* **2017**, *55*, 2985–2992. [CrossRef]

25. Zhang, X.; Sun, J.; Ma, S.; Lin, Z.; Zhang, J.; Wang, S.; Gao, W. Globally variance-constrained sparse representation and its application in image set coding. *IEEE Trans. Image Process.* **2018**, *27*, 3753–3765. [CrossRef] [PubMed]
26. Sitani, D.; Subramanyam, A.V.; Majumdar, A. Online single and multiple analysis dictionary learning-based approach for visual object tracking. *J. Electron. Imag* **2019**, *28*, 013004. [CrossRef]
27. Peng, L.; Yang, J. Single-image super resolution via hashing classification and sparse representation. In Proceedings of the 2017 3rd IEEE International Conference on Computer and Communications (ICCC), Chengdu, China, 13–16 December 2017; pp. 1923–1927. [CrossRef]
28. Vanika, S.; Prerna, K.; Angshul, M. Class-wise deep dictionary learning. In Proceedings of the 2017 International Joint Conference on Neural Networks (IJCNN), Anchorage, AK, USA, 14–19 May 2017; pp. 1125–1132. [CrossRef]
29. Zhang, Z.; Xu, Y.; Yang, J.; Li, X.; Zhang, D. A survey of sparse representation: Algorithms and applications. *IEEE Access* **2015**, *3*, 490–530. [CrossRef]
30. Pati, Y.C.; Rezaiifar, R.; Krishnaprasad, P.S. Orthogonal matching pursuit: Recursive function approximation with applications to wavelet decomposition. In Proceedings of the 27th Annual Asilomar Conference on Signals Systems and Computers, Pacific Grove, CA, USA, 1–3 November 1993; pp. 40–44. [CrossRef]
31. Elad, M.; Aharon, M. Image denoising via sparse and redundant representations over learned dictionaries. *IEEE Trans. Image Process.* **2007**, *15*, 3736–3745. [CrossRef]
32. Wang, G. The inverse hyperbolic tangent function and Jacobian sine function. *J. Math. Anal. Appl.* **2017**, *448*, 498–505. [CrossRef]
33. Zhou, Z.; Li, S.; Wang, B. Multi-scale weighted gradient-based fusion for multi-focus images. *Inf. Fusion* **2014**, *20*, 60–72. [CrossRef]
34. Li, S.; Kang, X.; Hu, J. Image fusion with guided filtering. *IEEE Trans. Image Process.* **2013**, *22*, 2864–2875. [CrossRef]
35. Zhao, P.; Liu, G.; Hu, C.; Huang, H.; He, B. Medical image fusion algorithm based on the Laplace-PCA. In Proceedings of the 2013 Chinese Intelligent Automation Conference, CIAC 2013, Yangzhou, Jiangsu, China, 23–25 August 2013; Volume 256, pp. 787–794. [CrossRef]
36. Shreyamsha Kumar, B.K.; Swamy, M.N.S.; Ahmad, M.O. Multiresolution DCT decomposition for multifocus image fusion. In Proceedings of the 2013 26th IEEE Canadian Conference on Electrical and Computer Engineering, CCECE 2013, Regina, SK, Canada, 5–8 May 2013; pp. 1–4. [CrossRef]
37. Qu, X.; Yan, J.; Xiao, H.; Zhu, Z. Image fusion algorithm based on spatial frequency-motivated pulse coupled neural networks in nonsubsampled contourlet transform domain. *Acta Autom. Sin.* **2008**, *34*, 1508–1514. [CrossRef]
38. Available online: http://www.dlr.de/hr/en/desktopdefault.aspx/tabid-2434/3770_read-32515 (accessed on 6 March 2019).

Article

Weak Signal Extraction from Lunar Penetrating Radar Channel 1 Data Based on Local Correlation

Zhuo Jia [1,2], **Sixin Liu** [1,2,*], **Ling Zhang** [1,2], **Bin Hu** [1,2] and **Jianmin Zhang** [1,2]

1   College of Geo-exploration Science and Technology, Jilin University, Changchun 130026, China; jiazhuo16@mails.jlu.edu.cn (Z.J.); lingzhang16@mails.jlu.edu.cn (L.Z.); binhu16@mails.jlu.edu.cn (B.H.); zjm16@mails.jlu.edu.cn (J.Z.)
2   Ministry of Land and Resources Key Laboratory of Applied Geophysics, Jilin University, Changchun 130026, China
*   Correspondence: liusixin@jlu.edu.cn; Tel.: +86-136-0432-0072

Received: 30 April 2019; Accepted: 22 May 2019; Published: 23 May 2019

Abstract: Knowledge of the subsurface structure not only provides useful information on lunar geology, but it also can quantify the potential lunar resources for human beings. The dual-frequency lunar penetrating radar (LPR) aboard the Yutu rover offers a Special opportunity to understand the subsurface structure to a depth of several hundreds of meters using a low-frequency channel (channel 1), as well as layer near-surface stratigraphic structure of the regolith using high-frequency observations (channel 2). The channel 1 data of the LPR has a very low signal-to-noise ratio. However, the extraction of weak signals from the data represents a problem worth exploring. In this article, we propose a weak signal extraction method in view of local correlation to analyze the LPR CH-1 data, to facilitate a study of the lunar regolith structure. First, we build a pre-processing workflow to increase the signal-to-noise ratio (SNR). Second, we apply the K-L transform to separate the horizontal signal and then use the seislet transform (ST) to reserve the continuous signal. Then, the local correlation map is calculated using the two denoising results and a time–space dependent weighting operator is constructed to suppress the noise residuals. The weak signal after noise suppression may provide a new reference for subsequent data interpretation. Finally, in combination with the regional geology and previous research, we provide some speculative interpretations of the LPR CH-1 data.

Keywords: lunar penetrating radar; local correlation; SNR; K-L transform; seislet transform

---

1. Introduction

Chang'E-3 (CE-3) landed at 340.4875 °E, 44.1189 °N on the Moon on 14 December 2013, in a new region in the largest basin that had not been explored before, that is, the Mare Imbrium [1]. The dual-frequency lunar penetrating radar (LPR) aboard the Yutu Rover provides a special opportunity to understand the subsurface structure to a depth of several hundreds of meters from the low-frequency channel (CH-1, 60 MHz). This also includes mapping the layer near-surface stratigraphic structure of the regolith from the high-frequency channel (CH-2A and CH-2B, 500 MHz) [2].

The LPR data processing process and its preliminary results were first proposed by the National Astronomical Observatories (NAOC) [3]. Preliminary analysis of the LPR observations, especially observations from CH-1, showed more than nine subsurface layers from the surface to a depth of ~360 m [1]. The 114-m-long profile measurement of the onboard lunar penetrating radar measured the thickness of the lunar weathering layer at approximately 5 m and detected three basement basalt units with depths of 195, 215, and 345 m. The radar measurements show that other methods underestimate the thickness of the global lunar regolith layer, and it shows the large volume after the last volcanic eruption [4]. Fa et al., Lai et al., and Zhang et al. speculated on the near surface structure by processing the raw CH-2B data [5–7]. Meanwhile, Dong et al. and Zhang et al. calculated the parameters of the regolith [8,9].

Owing to the complex environment of moon acquisition, the LPR data, especially the CH-1 data, contains various types of noise. Useful weak signals that are distorted by this noise are difficult to identify, thereby limiting the subsequent data interpretation [10,11]. Moreover, the observable signals (3700 and 5800 ns) were proven to be false signals caused by the instrumentation [11]. To take full advantage of the CH-1 data, we focused on highlighting the weak signals to study the lunar structure. Previous researches indicate that the terrain of the LPR coverage is relatively flat [1], so the target signal is a horizontal signal with a certain continuity.

Regarding the LPR CH-1 data, there are still a lot of noise residuals in the denoised section owing to the high noise level. Liu et al. proposed a stacking method using local correlation to solve the problem of noise residuals in the seismic stack profile. The basic principle of the local correlation denoising method is to preprocess the common-midpoint (CMP) gathers by the conventional denoising method, and calculate the local correlation coefficients of the CMP gathers before and after denoising. They assume that the local correlation coefficient of the effective signal is much larger than the noise, and based on this difference, construct a time-space variable weight operator to suppress the noise represented by the small coefficient before stack. Local correlation [12] is a typical local attribute used to measure the local similarity of two signals. It has been utilized in several seismic signal processing fields, such as image contrast [13,14], time-frequency analysis [15], and random noise attenuation [16,17].

In this paper, we proposed a weak signal extraction method based on local correlation to deal with the LPR CH-1 data, and then we studied the structure of the lunar regolith. To extract these weak signals, we chose the K-L transform and seislet transform (ST) to process the LPR data. The K-L transform [18] can decompose 2D signals into sub-signals corresponding to a series of eigenvalues, which represent the strength of the horizontal coherence. It is often used to extract coherent signals or to eliminate random noise from the 2D seismic data [19]. The ST [20] is a sparse transform, which can utilize the local dip information to map signals into subsets with different frequencies and dips. After the transformation process, the continuous signal can be highlighted based on this property. Therefore, it is applicable to random noise suppression [21], deblending [22], and data reconstruction [23,24]. First, we built a pre-processing workflow to improve the signal-to-noise ratio (SNR). Second, we applied the K-L transform to reserve the horizontal signal, and the ST to reserve continuous signal, respectively. Then, the local correlation map was calculated using the two denoising results and a time–space dependent weighting operator was constructed to suppress the noise residuals. The weak signal after noise suppression can provide a new reference for subsequent data interpretation. Finally, combining with the regional geology and previous research, particularly on LPR data, we provide some speculative interpretations of the LPR CH-1 data.

2. Methods

2.1. Data Preprocessing

The lunar penetrating radar (LPR) track extends 114.8 m (Figure 1) near to Sinus Iridum. Two bottom sides of the top board of the moon rover mount two CH1 antennas respectively. The bottom board of the lunar rover mount the CH2 antenna. Each of the CH-1 antennas is mounted in a tubular radome. The monopole antenna has a radome for support and protection. The radome has a length of 1150 mm and a diameter of 12 mm. The spacing between the transmitting antenna and the receiving antenna is about 800 mm. In order to generate pulse waves to propagate along with the antenna without reflection, the Wu-King impedance loading method is used to load the antenna through a continuous resistance load from the feed point to the end of the antenna. In practice, it is usually difficult to generate a continuous distribution of resistance load. Therefore, a piecewise loading method which called concentrated resistive loading is usually used (Figure 2b).

The CH-2 transceiver antenna is mounted at the bottom of the lunar rover, which is approximately 30 cm from the ground. We can see the structure of the antenna from Figure 2c. The CH-2 antenna has three antenna elements. The antenna elements are arranged side by side in a metal back cavity which is divided into three separate cavities. One of the components is used to transmit EM waves, while the other two are used to receive EM waves [2].

In this section, the data preprocessing effects of the LPR CH-1 data are reported. As the focus is the subsurface structure, the CH-1 data was selected. Based on the acquisition parameters, the actual situation, and the data quality, an LPR data preprocessing flow was designed (Figure 3). The CH-1 image (an output of Figure 3) was accessible after data preprocessing.

Owing to various types of electromagnetic waves in lunar space, the complex terrain on the moon, and the harsh environment, LPR data has a low signal-to-noise ratio (SNR). In the shallow part (Figure 4, red box) of the pre-processed CH-1 data, there were several harsh horizontal noises. Irrespective of whether we used the averaging tracks or the sliding filter, etc., the noise could not be effectively eliminated and, therefore, it affected the extraction of useful information from the data. In the deep part, two obvious events were found at 3700 ns and 5800 ns (blue arrows). Regrettably, Li et al. proved that these were distortions from the instrumentation [11]. In summary, the CH-1 data of the LPR had a very low signal-to-noise ratio, raising the problem of how to extract the weak signals from the deep part.

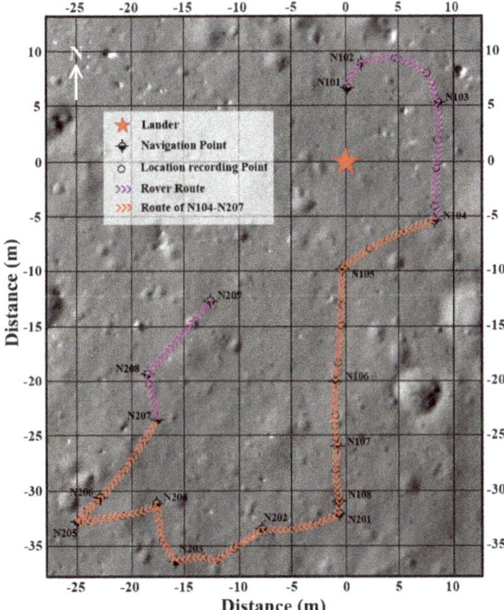

Figure 1. Yutu's path on the Moon. The background image was taken by the descent camera on the Chang'E-3 (CE-3) lander. The red star shows the landing site. The inset line shows the path (purple line and red line).

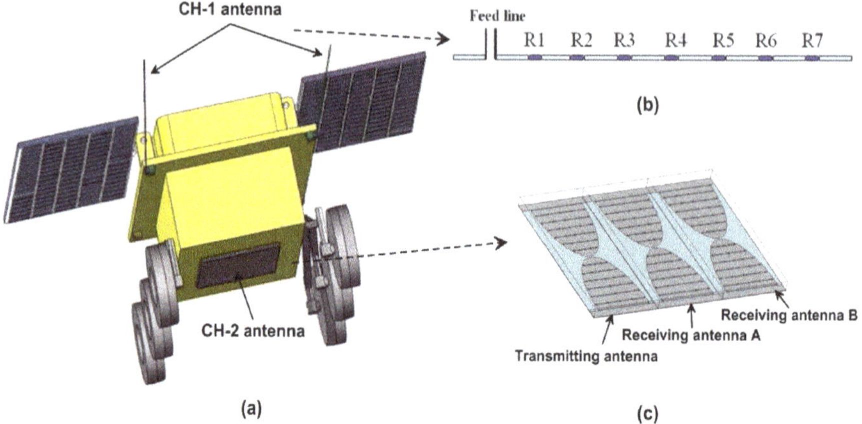

Figure 2. The antennas on Yutu Rover. (**a**) Yutu rover; (**b**) CH-1; (**c**) CH-2.

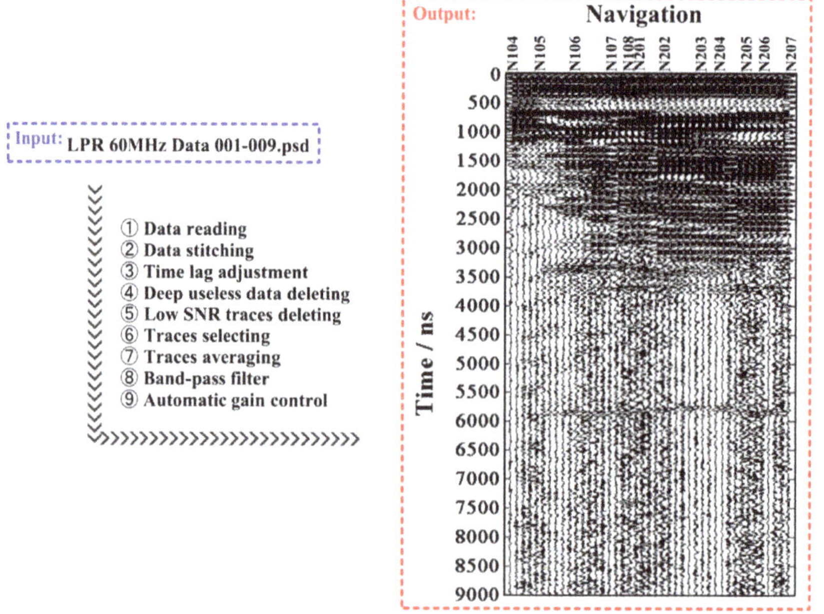

Figure 3. The flowchart of the CH-1 (lunar penetrating radar) LPR data preprocessing (left). The output is the LPR CH-1 data after preprocessing (right). N104–N207 denote the positions where the LPR was rebooted.

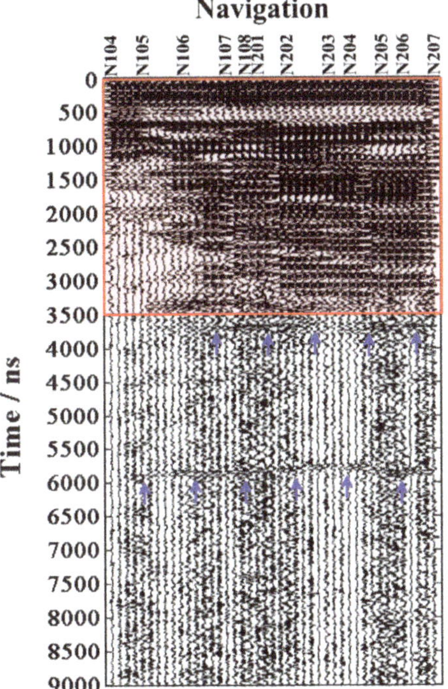

Figure 4. LPR CH-1 data with interpretation.

2.2. Weak Signal Extraction Method Based on Local Correlation

The acquired field data could be considered as a collection of signals and noise. Noise attenuation involves suppressing the noise as much as possible while preserving the signal. However, the denoising effect is limited by the incompleteness of the denoising assumption and the close amplitude of the weak signal and noise. Therefore, for most denoising methods, the choice of denoising parameters is a trade-off between signal preservation and noise attenuation. To preserve the weak signal and simplify parameter selection, we introduced local correlation to the LPR data processing. First, we reviewed the definition of local correlation [12]. The global correlation coefficients γ between two signals $\mathbf{a}_i$ and $\mathbf{b}_i$ can be defined as

$$\gamma = \frac{\sum\limits_{i=1}^{N} \mathbf{a}_i \mathbf{b}_i}{\sqrt{\sum\limits_{i=1}^{N} \mathbf{a}_i^2 \mathbf{b}_i^2}}, \tag{1}$$

where N is the number of signal elements. To measure the similarity between the two signals in a local way, the local correlation coefficients can be defined using a sliding window:

$$\gamma_w(t) = \frac{\sum\limits_{i=t-w/2}^{t+w/2} \mathbf{a}_i \mathbf{b}_i}{\sqrt{\sum\limits_{i=t-w/2}^{t+w/2} \mathbf{a}_i^2 \sum\limits_{i=t-w/2}^{t+w/2} \mathbf{b}_i^2}}, \tag{2}$$

where w is the size of the sliding window.

Fomel redefines the local correlation in a smoother way [12]. Based on linear algebra theory, Equation (1) can be rewritten as

$$\gamma^2 = \gamma_1 \gamma_2 \tag{3}$$

where γ_1 and γ_2 are obtained by solving the optimization problem in the least squares sense:

$$\gamma_1 = \arg\min_{\gamma_1} \|\mathbf{a} - \gamma_1 \mathbf{b}\|_2^2 = (\mathbf{a}^T \mathbf{a})^{-1} (\mathbf{a}^T \mathbf{b}), \tag{4}$$

$$\gamma_2 = \arg\min_{\gamma_2} \|\mathbf{a} - \gamma_2 \mathbf{b}\|_2^2 = (\mathbf{b}^T \mathbf{b})^{-1} (\mathbf{b}^T \mathbf{a}), \tag{5}$$

where $\mathbf{a},\mathbf{b}$ is the vector form of $\mathbf{a}_i,\mathbf{b}_i$. Meanwhile, $\mathbf{A}$ and $\mathbf{B}$ are written as two diagonal matrices where the main diagonal elements are $\mathbf{a}$ and $\mathbf{b}$, respectively. Then, this is followed by containing a shaping regularization [20]. The optimization problem in the least squares sense can then be modified as follows:

$$\mathbf{c}_1 = \left[\lambda_1^2 \mathbf{I} + \mathbf{S}_m (\mathbf{A}^T \mathbf{A} - \lambda_1^2 \mathbf{I})\right]^{-1} \mathbf{S}_m \mathbf{A}^T \mathbf{b}, \tag{6}$$

$$\mathbf{c}_2 = \left[\lambda_2^2 \mathbf{I} + \mathbf{S}_m (\mathbf{B}^T \mathbf{B} - \lambda_2^2 \mathbf{I})\right]^{-1} \mathbf{S}_m \mathbf{B}^T \mathbf{a}, \tag{7}$$

where $\mathbf{c}_1$, $\mathbf{c}_2$ are the vector form of γ_1 and γ_2 $\mathbf{S}_m$ is a function for smoothness promotion; λ_1 and λ_2 are the two stable parameters used in the process of inversion to accelerate the convergence speed. We can select λ_1 and λ_2 as follows:

$$\lambda_1 = \|\mathbf{A}^T \mathbf{A}\|_2, \tag{8}$$

$$\lambda_2 = \|\mathbf{B}^T \mathbf{B}\|_2. \tag{9}$$

The basic idea of our methods is extracting the weak useful signals based on the local correlation difference between signal and noise. The local correlation is calculated by two different denoising results, therefore the selection of denoising methods is the key to our method. We select the denoising methods based on the LPR CH-1 data and previous researches [1]. From Figure 3, the observable useful signals indicate that the terrain of the LPR coverage is relatively flat, so the target signal is a horizontal signal with a certain continuity. To highlight the different characteristics of CH-1 data (horizontal and continuity), we chose the K-L transform (Appendix A) and ST (Appendix B) to process the LPR data. To preserve the deep weak signals, the ability to suppress noise was limited. Since the assumptions of different denoising methods are different, we considered that the noise residuals corresponding to the denoising results were different. Based on the similarity difference, a time–space dependent weighting operator [24] was proposed to extract weak signals and suppress noise residuals:

$$\mathbf{W}(t,x) = \begin{cases} 1 & c_{n,s}(t,x) > v_2 \\ \dfrac{\gamma_{n,s}(t,x) - v_1}{v_2 - v_1} & v_1 \leq c_{n,s}(t,x) \leq v_2 \ , \\ 0 & c_{n,s}(t,x) < v_1 \end{cases} \tag{10}$$

where $c_{n,s}(t,x)$ is the local correlation map, and v_1, v_2 are the thresholds that divide the local correlation map into three parts (Figure 5). v_1 defines the pure noise section, where the weighting operator $\mathbf{W}$ will remove the entire section. v_2 defines the purely useful signal section, where the weighting operator $\mathbf{W}$ will preserve the entire signal. When the local correlation coefficient is in the range of v_1 to v_2, the weak signal and noise are combined in that section, and the threshold function varies in a weighted length manner.

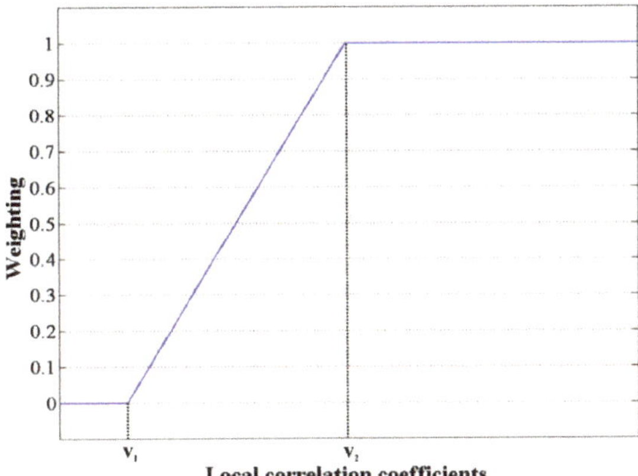

Figure 5. Demonstration of weighting operator.

Note that v_1, v_2 controls the degree of signal extraction. To extract all the detectable useful events and attenuate some noise, we determine v_1, v_2 based on the average of similarity coefficients in pure noise section and purely signal section. The pure signal section covers the first few events, which have large and apparent similarity coefficients. And pure signal section covers the last few detectable events, which have small similarity coefficients.

We utilized the weighting operator on the two denoising results ($\mathbf{D}_{KL}$ and $\mathbf{D}_{ST}$), and then stacked the weighted outputs to obtain the extracted signal $\mathbf{D}$:

$$\mathbf{D} = \frac{\mathbf{W}(\mathbf{D}_{KL}) + \mathbf{W}(\mathbf{D}_{ST})}{2}. \tag{11}$$

The specific workflow was as shown:

1. Use the K-L transform to highlight horizontal signals,
2. Use the ST to highlight continuous signals,
3. Calculate the local correlation map based on the two pre-denoised results, and then construct the weighting operator,
4. Utilize the operator to extract signals with high similarity,
5. Stack the signal sections extracted from the denoising results.

The proposed method extracted weak signals using the similarity difference between signals and noise. Therefore, the choice of denoising parameters was more elegant, which avoided weak signal damage. Moreover, this method could take advantage of different denoising methods to improve the final weak signal extraction results.

3. Simulated Data Test

Figure 6 demonstrates the workflow of the proposed method. As mentioned above, the target weak signal is the horizontal signal with a certain continuity. Therefore, we present a noisy horizontal signal to test the effectiveness of the proposed method. The signal (Figure 6a) contains five events with different slowness, i.e., $(0, 0, 3, -4, -2) \cdot 10^{-5}$ and we add Gaussian random noise to obtain s noisy signal (Figure 6b). Then we calculated the local correlation map (Figure 6e) between the two pre-denoised results (Figure 6c,d) and built the local correlation based weighting operator with $v_1 = 0.1, v_2 = 0.4.$, and Figure 6f was the final result. The weak signal (60 ns) was distorted by noise. Both the K-L transform and ST suppressed some of the noise, but we still observed noise residuals in the denoised section. The local correlation map indicated the similarity difference between noise residuals and weak signals. Noise residuals were well attenuated by the weighting operator. To quantitatively compare the effects of weak signal extraction, we calculated the SNR of the denoising result:

$$\text{SNR} = 10\log_{10}\frac{\|\mathbf{d}_{\text{signal}}\|_2^2}{\|\mathbf{d}_{\text{noisy}} - \mathbf{d}_{\text{denoise}}\|_2^2}, \tag{12}$$

where $\mathbf{d}_{\text{noisy}}$ is the noisy signal, and $\mathbf{d}_{\text{signal}}$, $\mathbf{d}_{\text{denoise}}$ are the signal section and denoised section, respectively.

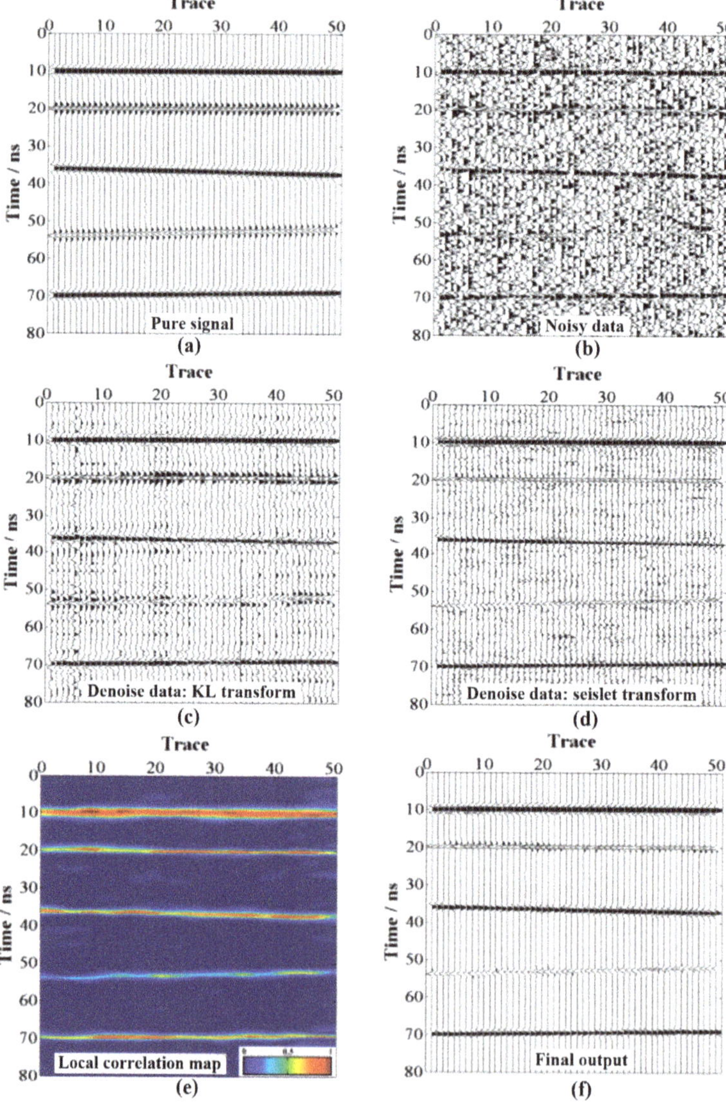

Figure 6. Demonstration of weak signal extraction. (**a**) Signal; (**b**) Noisy signal; (**c**) Pre-denoised result: K-L transform; (**d**) Pre-denoised result: ST; (**e**) Local correlation map of (**c**,**d**); (**f**) Extracted signal using our method.

Table 1 shows the signal-to-noise ratio obtained by the four processing methods. As is well known, a larger SNR value indicates a stronger signal. Therefore, the local correlation algorithm used in this paper can obtain a better signal to noise ratio.

Table 1. Comparison of the signal-to-noise ratio (SNR).

Data	Noisy Data	K-L	ST	Proposed Method
SNR	−9.02	8.42	6.32	15.13

4. Results

In this section, we applied the K-L transform and ST to pre-denoise the LPR CH-1 data. Then we calculated the local correlation map between the two denoised results and built the weighting operator. After thresholding using the weighting operator, we stacked the two processed datasets to highlight the weak signal. Finally, we interpreted the processed LPR data. In the interpretation, we extracted two layers of the paleoregolith which separated the covered basalts in different periods.

Figure 7 shows the pre-denoised results using the K-L transform and ST. From Figure 7, we observe that the residuals in the denoised sections were quite different, which showed the different advantages of denoising methods.

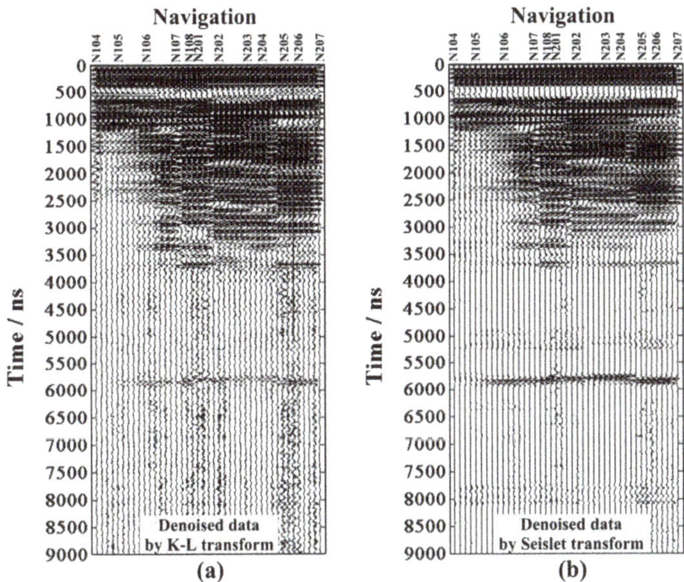

Figure 7. Demonstration of pre-denoised results. (**a**) K-L transform; (**b**) ST.

Figure 8 shows the final extracted signal using our method. Besides the extracted weak signal, the false signals (3700 and 5800 ns) were also extracted. The reason was the large amplitude and high continuity in the noisy section, which showed the effectiveness of signal extraction using local correlation. Note that the selection of denoising methods varies with the target signal. For example, to extract signals from random noise, we can select the *f-x* deconvolution, median filter, bandpass filter, etc.

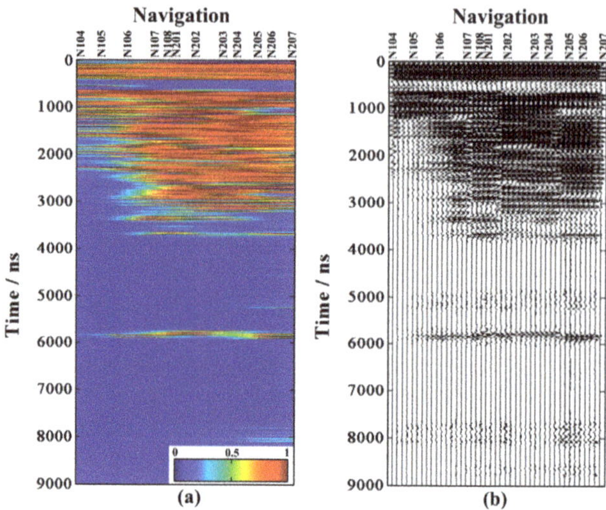

Figure 8. (**a**) Local correlation map and (**b**) final extracted signal.

Furthermore, we also observed some suspected signals in Figure 8a,b, especially in the ranges 4950–5300 ns and 7750–8100 ns. Figure 9 shows the zoomed section of the two suspected signals. The two extracted signals denoted the large similarity in the local correlation map, which meant that the extracted signals had a certain continuity in the horizontal direction.

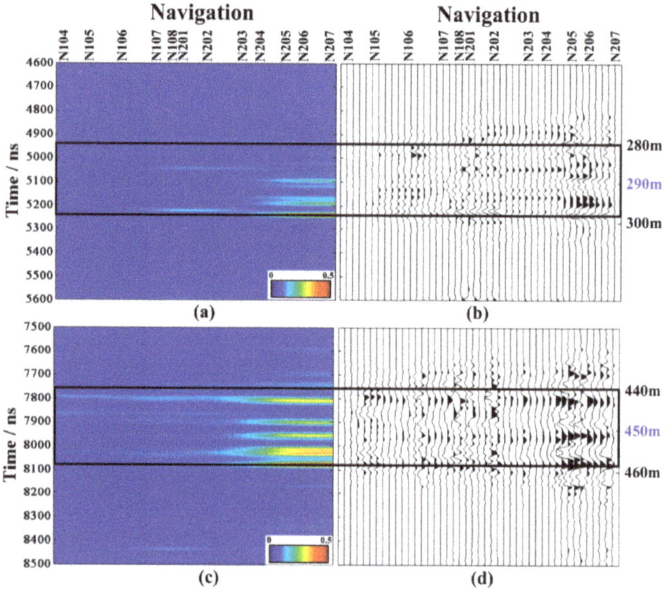

Figure 9. Zoomed local correlation map and zoomed section of the final extracted signal. (**a**) Local correlation map (Near 290m) (**b**) Extracted signal (Near 290m) (**c**) Local correlation map (Near 450m) Local correlation map (**d**) Extracted signal (Near 450m).

After the time–depth conversion, Figure 10 shows the result of the data. Except for the false signals (3700 and 5800 ns), we also extracted two layers of weak reflection at ~290 m and ~450 m. According to the formation mechanism of the mare [25,26], the Mare Imbrium is covered by basalts during different historical periods. After each basalt layer was covered, during that time interval, ancient regolith was formed, due to the impact of various meteorites. Based on the two weakly reflected events that we extracted from the CH-1 data, we explained these two layers (Figure 11). During the formation of Mare Imbrium, two layers of the paleoregolith at 290 m and 450 m were formed. The thickness of both is about 10 m.

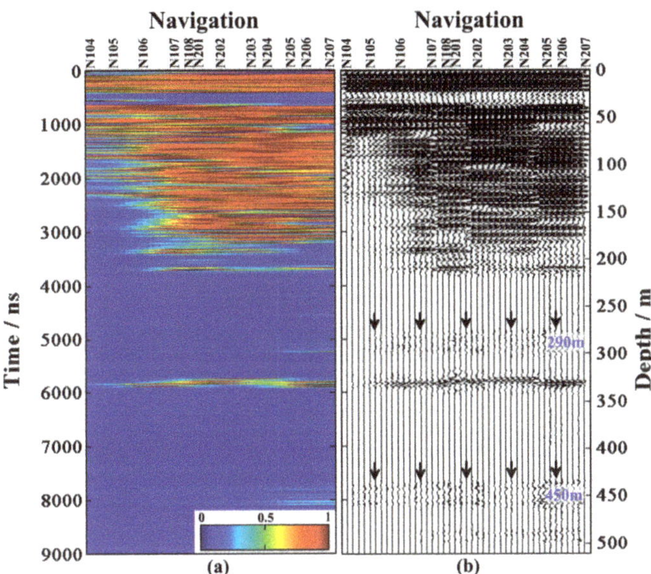

Figure 10. Demonstration of the extracted weak signal (with depth marked). (**a**) Local correlation map (**b**) Extracted signal using our method.

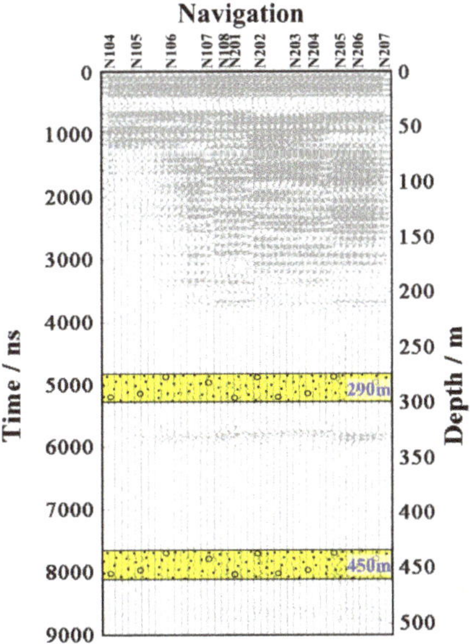

Figure 11. Interpretation of the data.

5. Discussions

Xiao et al. show that at least nine underground horizons can be determined by LPR data and comprehensive geological interpretation, indicating that the area has undergone complex geological processes, since the Imbrian is compositionally distinct from the Apollo and Luna landing site [1]. Zhang et al. reveal the in situ spectral reflectance and elemental analysis of the lunar soil at the landing site. The young basalt can be from the mantle reservoir rich in ilmenite and then assimilated by 10-20% of the last residual melt in the lunar magma ocean [4]. Yuan et al. interpret as different period lava flow sequences deposited on the lunar surface. The most probable directions of these flows were inferred from layer depths, thicknesses, and other geological information. Moreover, the apparent Imbrian paleoregolith homogeneity in the profile supports the suggestion of a quiescent period of lunar surface evolution. Similar subsurface structures are found at the NASA Apollo landing sites, indicating that the cause and time of formation of the imaged phenomena may be similar between the two distant regions [27].

It should be noted that the above-mentioned layer judgment is performed under the condition that the CH-1 data is reliable and true. However, Li et al.'s paper tries to solve this controversy by carefully analyzing and comparing the data collected by the Yutu rover on the moon and the LPR prototype of the CE-3 lunar rover model installed on the ground. This analysis shows that deep radar features previously attributed to lunar shallow stratum are not true reflectors, but rather they may be signal artifacts produced by the system and their electromagnetic interaction with metal rover [11].

Based on the local correlation method, we extracted two layers of weak reflection at ~290 m and ~450 m. According to the formation mechanism of the mare, after each basalt layer was covered, during that time interval, ancient regolith was formed, due to the impact of various meteorites. During the formation of Mare Imbrium, two layers of the paleoregolith at 290 m and 450 m were formed. The thickness of both is about 10 m.

Our result avoids the extracted events mentioned by Li et al. And we propose two new weakly reflective layers after Xiao et al. These two new reflective layers enhance the deep utilization of CH-1 data to some extent.

The proposed method is based on the assumption that the noise residuals by different denoising methods are orthogonal which have a small value of local correlation coefficients, and useful signals are just the opposite. The local correlation-based methods not only can deal with random noise, but also coherent noise [28]. Due to the complex moon acquisition conditions, the LPR data contains various types of noise. Strong noise adaptability makes the proposed method more promising in LPR data processing.

Another advantage of the proposed method is balancing the advantages of different denoising methods, even the conventional methods. And the most important step of our method is the selection of denoising method. The choices of denoising method are determined by the target signal characteristics. For CH-1 data, the denoising method which highlights horizontal and continuous signal is selected. Furthermore, the useful signals in CH-2 data are interfered by rocks caused diffraction noise [7], the dip filtering methods are more applicable, such as a f-k filter.

6. Conclusions

In this paper, we proposed a weak signal extraction method based on local correlation to deal with LPR CH-1 data, and then we studied the structure of the lunar regolith. First, we built a pre-processing workflow to improve the signal-to-noise ratio (SNR). Second, we applied the K-L transform to reserve horizontal signals, and the ST to reserve continuous signals, respectively. Then, the local correlation map was calculated according to the two denoising results, and a time–space dependent weighting operator was constructed to suppress the noise residuals. The weak signal after noise suppression can provide a new reference for subsequent data interpretation. Finally, combining with previous research and the LPR data, we provided some speculative interpretations of the LPR CH-1 data.

Except for the false signals, we extracted two layers of weak reflection at ~290 m and ~450 m. According to the formation mechanism of the mare, the two layers are explained from the two weakly reflected events. During the formation of Mare Imbrium, two layers of paleoregolith at 290 m and 450 m were formed. The thickness both was about 10 m.

These results provide valuable information to understand the reflections of LPR data, and they offer a reference for future lunar missions.

Author Contributions: Conceptualization, methodology, investigation and Writing, Z.J.; Supervision, S.L.; Data curation, L.Z. and B.H.; Formal analysis, J.Z.

Funding: This research was supported by the Major Projects of the National Science and Technology of China (2016YFC0600505) and the National Natural Science Foundation of China under Grant 41574109, 41504085 and 41874136.

Acknowledgments: We thank Madagascar software for the open-source code and the open-source data in http://moon.bao.ac.cn.

Conflicts of Interest: The authors declare no conflict of interest.

Abbreviations

The following abbreviations are used in this manuscript:

LPR	lunar penetrating radar
CH-1	channel 1
CH-2	channel 2
K-L	Karhunen-Loeve
ST	seislet transform
NAOC	National Astronomical Observatories in China
SNR	signal-to-noise ratio
CE-3	Chang'E-3
EM	Electromagnetic
CMP	common-midpoint

Appendix A Review of the K-L Transform

To perform the K-L transform on 2D data $\mathbf{X}$ with N traces and M samples, we first need to find the transformation matrix $\mathbf{L}$. The row vector of $\mathbf{L}$ is composed of the eigenvectors of the covariance matrix of $\mathbf{X}$ [18]. The estimated covariance matrix $\mathbf{U}$ of the data matrix X can be expressed by:

$$\mathbf{U} \approx \mathbf{X}\mathbf{X}^{T}. \tag{A1}$$

Let $\mathbf{L} - (l_1, l_2, \ldots, l_N)$, where $l_1, l_2, \ldots, l_N$ are N eigenvectors of $\mathbf{U}$, then the forward K-L transform can be expressed as:

$$h_j(t) = \sum_{i=1}^{N} l_{ij} x_i(t), \tag{A2}$$

where $x(t), h(t)$ is the vector of input and output and $i, j = 1, 2, \ldots, N$. The K-L forward transform can be written as a matrix form:

$$\psi = \mathbf{L}^T \mathbf{X}, \tag{A3}$$

and the details are shown as

$$\begin{bmatrix} h_{11} & h_{12} & \cdots & h_{1M} \\ h_{21} & h_{22} & \cdots & h_{2M} \\ \vdots & \vdots & \ddots & \vdots \\ h_{N1} & h_{N2} & \cdots & h_{NM} \end{bmatrix} = \begin{bmatrix} l_{11} & l_{12} & \cdots & l_{1N} \\ l_{21} & l_{22} & \cdots & l_{2N} \\ \vdots & \vdots & \ddots & \vdots \\ l_{N1} & l_{N2} & \cdots & l_{NN} \end{bmatrix} \begin{bmatrix} x_{11} & x_{12} & \cdots & x_{1M} \\ x_{21} & x_{22} & \cdots & x_{2M} \\ \vdots & \vdots & \ddots & \vdots \\ x_{N1} & x_{N2} & \cdots & x_{NM} \end{bmatrix}, \tag{A4}$$

where l_{ij} is the element of $\mathbf{L}^T$. Given that $\mathbf{L}^T$ is an orthogonal matrix, the output $h_j(t)$ can be selected as a set of orthogonal basis vectors. $x_i(t)$ can be expressed as

$$x_i(t) = \sum_{j=1}^{N} l_{ji} h_j(t), \tag{A5}$$

where $i, j = 1, 2, \ldots, N$. The matrix form of the K-L reverse transform can be written as

$$\mathbf{X} = \mathbf{L}\psi, \tag{A6}$$

and the details are shown as

$$
\begin{bmatrix}
x_{11} & x_{12} & \cdots & x_{1M} \\
x_{21} & x_{22} & \cdots & x_{2M} \\
\vdots & \vdots & \ddots & \vdots \\
x_{N1} & x_{N2} & \cdots & x_{NM}
\end{bmatrix}
=
\begin{bmatrix}
l_{11} & l_{21} & \cdots & l_{N1} \\
l_{12} & l_{22} & \cdots & l_{N2} \\
\vdots & \vdots & \ddots & \vdots \\
l_{1N} & l_{2N} & \cdots & l_{NN}
\end{bmatrix}
\begin{bmatrix}
h_{11} & h_{12} & \cdots & h_{1M} \\
h_{21} & h_{22} & \cdots & h_{2M} \\
\vdots & \vdots & \ddots & \vdots \\
h_{N1} & h_{N2} & \cdots & h_{NM}
\end{bmatrix}.
\tag{A7}
$$

The reconstructed form of the first m principal components is

$$\tilde{x}_i(t) = \sum_{j=1}^{m} l_{ji} h_j(t); \ \ m < N, \tag{A8}$$

and the matrix form is

$$
\begin{bmatrix}
\tilde{x}_{11} & \tilde{x}_{12} & \cdots & \tilde{x}_{1M} \\
\tilde{x}_{21} & \tilde{x}_{22} & \cdots & \tilde{x}_{2M} \\
\vdots & \vdots & \ddots & \vdots \\
\tilde{x}_{N1} & \tilde{x}_{N2} & \cdots & \tilde{x}_{NM}
\end{bmatrix}
=
\begin{bmatrix}
l_{11} & l_{21} & \cdots & l_{m1} \\
l_{12} & l_{22} & \cdots & l_{m2} \\
\vdots & \vdots & \ddots & \vdots \\
l_{1N} & l_{2N} & \cdots & l_{mN}
\end{bmatrix}
\begin{bmatrix}
h_{11} & h_{12} & \cdots & h_{1M} \\
h_{21} & h_{22} & \cdots & h_{2M} \\
\vdots & \vdots & \ddots & \vdots \\
h_{m1} & h_{m2} & \cdots & h_{mM}
\end{bmatrix}.
\tag{A9}
$$

According to the above equation, the horizontal signal is reconstructed with the first few principal components.

Appendix B Review of the ST

The wavelet-lifting scheme is defined based on the cross-correlation between even elements and odd elements, and it calculates the difference $\mathbf{r}$ between the true odd elements and predicts the difference between even elements [29]. In this scheme, the basic function of the ST is defined and the ST pairs [14] are described as follows:

$$\mathbf{r} = \mathbf{o} - \mathbf{P}(\mathbf{e}), \tag{A10}$$

$$\mathbf{c} = \mathbf{e} + \mathbf{U}(\mathbf{r}), \tag{A11}$$

$$\mathbf{e} = \mathbf{c} - \mathbf{U}(\mathbf{r}), \tag{A12}$$

$$\mathbf{o} = \mathbf{r} + \mathbf{P}(\mathbf{e}), \tag{A13}$$

where $\mathbf{P}$ denotes the prediction operator and $\mathbf{U}$ denotes the update operator. The difference between the true trace and the predicted trace is represented by $\mathbf{r}$, while $\mathbf{c}$ stands for a rough approximation of the dataset. The prediction and update processes are accomplished through local slope estimation as in Reference [30]:

$$\mathbf{P}(\mathbf{e})_k = (\mathbf{P}_k^+(\mathbf{e}_{k-1}) + \mathbf{P}_k^-(\mathbf{e}_k))/2, \tag{A14}$$

$$\mathbf{U}(\mathbf{r})_k = (\mathbf{P}_k^+(\mathbf{r}_{k-1}) + \mathbf{P}_k^-(\mathbf{r}_k))/4, \tag{A15}$$

where $\mathbf{P}_k^+$ and $\mathbf{P}_k^-$ are the event shifting operators based on the local slope for the corresponding trace.

For random noise attenuation based on the seislet transform, it can be achieved by forward transform, hard thresholding, and the inverse transform. The threshold function is shown as

$$T_{\text{hard}}\{\mathbf{X}\}_{ij} = \begin{cases} |\mathbf{X}_{ij}| \cdot \text{sgn}(d_{ij}) & |\mathbf{X}_{ij}| \geq \sigma \\ 0 & |\mathbf{X}_{ij}| < \sigma \end{cases}, \tag{A16}$$

where σ is the threshold parameter.

References

1. Xiao, L.; Zhu, P.; Fang, G.; Xiao, Z.; Zou, Y.; Zhao, J.; Zhao, N.; Yuan, Y.; Qiao, L.; Zhang, X.; Zhang, H.; et al. A young multilayered terrane of the northern Mare Imbrium revealed by Chang'E-3 mission. *Science* **2015**, *347*, 1226–1234. [CrossRef] [PubMed]
2. Fang, G.; Zhou, B.; Ji, Y.; Zhang, Q.; Shen, S.; Li, Y.; Guan, H.; Tang, C.; Gao, Z.; Lu, W.; Ye, S.; et al. Lunar Penetrating Radar onboard the Chang'e-3 mission. *Res. Astron. Astrophys.* **2014**, *14*, 1607–1622. [CrossRef]
3. Su, Y.; Fang, G.; Feng, J.; Xing, S.; Ji, Y.; Zhou, B.; Gao, Y.; Li, H.; Dai, S.; Xiao, Y.; Li, C. Data processing and initial results of Chang'e-3. lunar penetrating radar. *Res. Astron. Astrophys.* **2014**, *14*, 1623–1632. [CrossRef]
4. Zhang, J.; Yang, W.; Hu, S.; Lin, Y.; Fang, G.; Li, C.; Peng, W.; Zhu, S.; He, Z.; Zhou, B.; Lin, H.; et al. Volcanic history of the Imbrium basin: A close-up view from the lunar rover Yutu. *Proc. Natl. Acad. Sci. USA* **2015**, *112*, 5342–5348. [CrossRef] [PubMed]
5. Fa, W.; Zhu, M.; Liu, T.; Plescia, J. Regolith stratigraphy at the Chang'E-3 landing site as seen by lunar penetrating radar. *Geophys. Res. Lett.* **2016**, *42*, 179–187. [CrossRef]
6. Lai, J.; Xu, Y.; Zhang, X.; Tang, Z. Structural analysis of lunar subsurface with Chang'E-3 lunar penetrating radar. *Planet. Space Sci.* **2016**, *120*, 96–102. [CrossRef]
7. Zhang, L.; Zeng, Z.; Li, J.; Huang, L.; Huo, Z.; Zhang, J.; Huai, N. A Story of Regolith Told by Lunar Penetrating Radar. *Icarus* **2019**, *321*, 148–160. [CrossRef]
8. Dong, Z.; Fang, G.; Ji, Y.; Gao, Y.; Wu, C.; Zhang, X. Parameters and structure of lunar regolith in Chang'E-3 landing area from lunar penetrating radar (LPR) data. *Icarus* **2016**, *282*, 40–46. [CrossRef]
9. Zhang, L.; Zeng, Z.; Li, J.; Huang, L.; Huo, Z.; Wang, K.; Zhang, J. Parameter Estimation of Lunar Regolith from Lunar Penetrating Radar Data. *Sensors* **2018**, *18*, 2907. [CrossRef]
10. Gao, Y.; Dong, Z.; Fang, Y.; Ji, Y.; Zhou, B. The Processing and Analysis of Lunar Penetrating Radar Channel-1 Data from Chang'E-3. *J. Radars* **2016**, *4*, 518–526.
11. Li, C.; Xing, S.; Lauro, S.E.; Su, Y.; Dai, S.; Feng, J.; Cosciotti, B.; Di Paolo, F.; Mattei, E.; Xiao, Y.; Ding, C. Pitfalls in GPR Data Interpretation: False Reflectors Detected in Lunar Radar Cross Sections by Chang'e-3. *IEEE Trans. Geosci. Remote Sens.* **2017**, *56*, 99–199.
12. Fomel, S. Local seismic attributes. *Geophysics* **2007**, *72*, 29–33. [CrossRef]
13. Fomel, S. Shaping regularization in geophysical-estimation problems. *Geophysics* **2007**, *24*, 29–36. [CrossRef]
14. Liu, G.; Fomel, S.; Jin, L.; Chen, X. Stacking seismic data using local correlation. *Geophysics* **2009**, *74*, 43–48. [CrossRef]
15. Liu, G.; Fomel, S.; Chen, X. Time-frequency characterization of seismic data using local attributes. *Geophysics* **2009**, *76*, 23–24. [CrossRef]
16. Chen, Y.; Jiao, S.; Ma, J.; Chen, H.; Zhou, Y.; Gan, S. Ground-roll noise attenuation using a simple and effective approach based on local band-limited orthogonalization. *IEEE Geosci. Remote Sens. Lett* **2015**, *12*, 1–5. [CrossRef]
17. Chen, Y.; Fomel, S. Random noise attenuation using local signal-and-noise orthogonalization. *Geophysics* **2015**, *80*, 23–24. [CrossRef]
18. Wang, Z.; Zeng, Z.; Xue, J.; Liu, S. The Application of KL Transform to Remove Horizontal Coherent Noise in GPR Record. *J. Jiling Univ.* **2005**, *35*, 127.
19. Huo, Z.; Wang, M. The Application of KL Transform to Remove Direct Wave in Ground Penetrating Radar Records. In Proceedings of the Fourth International Conference on Image and Graphics (ICIG 2007), Sichuan, China, 22–24 August 2007. [CrossRef]
20. Fomel, S.; Liu, Y. Seislet transform and seislet frame. *Geophysics* **2010**, *75*, 25–38. [CrossRef]
21. Liu, Y.; Fomel, S.; Liu, C.; Wang, D.; Liu, L.; Feng, X. High-order seislet transform and its application of random noise attenuation. *Chin. J. Geophys.* **2009**, *52*, 2142–2151.
22. Chen, Y.; Fomel, S.; Hu, J. Iterative deblending of simultaneous-source seismic data using seislet-domain shaping regularization. *Geophysics* **2013**, *79*, 179–189. [CrossRef]
23. Liu, C.; Li, P.; Liu, Y.; Wang, D.; Feng, X.; Liu, D. Iterative data interpolation beyond aliasing using seislet transform. *Chinese J. Geophys.* **2013**, *56*, 1619–1627.
24. Chen, Y.; Zhang, L.; Mo, L. Seismic data interpolation using nonlinear shaping regularization. *J. Seism Explor* **2015**, *24*, 327–342.

25. Morgan, G.A.; Campbell, B.A.; Campbell, D.B.; Hawke, B.R. Investigating the stratigraphy of Mare Imbrium flow emplacement with Earth-based radar. *J. Geophys. Res.-Planets* **2016**, *121*, 1498–1513. [CrossRef]

26. Heiken, G.H.; Vaniman, D.T.; French, B.M. *Lunar Source Book: A Users Guide to the Moon*; Cambridge University Press: Cambridge, UK, 1991; p. 753.

27. Yuan, Y.; Zhu, P.; Zhao, N.; Xiao, L.; Garnero, E.; Xiao, Z.; Zhao, J.; Qiao, L. The 3D geological model around Chang'E-3 landing site based on lunar penetrating radar Channel-1 data: 3D Geological model of CE-3 landing site. *Geophys. Res. Lett* **2017**, *44*, 13. [CrossRef]

28. Chen, Y.; Fomel, S. Random noise attenuation using local similarity. In *SEG Technical Program Expanded Abstracts 2014*; Society of Exploration Geophysicists: Tulsa, OK, USA, 2014; pp. 4360–4365. [CrossRef]

29. Sweldens, W. The lifting scheme: A custom-design construction of biorthogonal wavelets. *Appl. Comput. Harmon. Anal.* **1996**, *3*, 186–200. [CrossRef]

30. Fomel, S. Applications of plane-wave destruction filters. *Geophysics* **2002**, *67*, 1946–1960. [CrossRef]

 *electronics*

Article

Calculation and Interpretation of Ground Penetrating Radar for Temperature and Relative Water Content of Seasonal Permafrost in Qinghai-Tibet Plateau

Qing Wang [1] and Yupeng Shen [2,*]

[1] School of Information and Communication Engineering, Beijing Information Science and Technology University, 100192 Beijing, China

[2] School of Civil Engineering, Beijing Jiaotong University, 100044 Beijing, China

* Correspondence: ypshen@bjtu.edu.cn; Tel.: +86-1381-087-3345

Received: 1 May 2019; Accepted: 25 June 2019; Published: 27 June 2019

Abstract: Due to the seasonal permafrost thawing, the Qinghai–Tibet Highway has a depression and instability of the roadbed. In order to obtain the ablation interface and water content characteristics of seasonal permafrost areas, non-destructive ground penetrating radars using electromagnetic wave detection methods are widely used. Regarding the imaging of the ablation interface in permafrost regions, this paper proposes a high-precision procedure for seasonal permafrost media using waveform difference analysis, electromagnetic wave attenuation attribute calculation and relative wave impedance conversion. It improves the resolution and division accuracy of the imaging. In addition, the study demonstrates a method to calculate the temperature and water content of the ablation zone by mining attenuation attribute, relative wave impedance attribute, absolute instantaneous amplitude attribute and the weighted average frequency attribute parameters under the constraints of the measured data. This method has high accuracy and high efficiency and can be used in the rapid calculation of temperature and water content of seasonal permafrost on the Qinghai–Tibet Highway.

Keywords: GPR; seasonal permafrost; electromagnetic wave attribute; relative water content

---

1. Introduction

Various rocks and soils with temperatures below 0 °C and containing ice are often referred to as frozen soils. According to the time when the soil is in a state of continuous freezing, the frozen soil is generally divided into short-term frozen soil (hours to half-months), seasonal permafrost (half months to several months), and permafrost (several years to tens of thousands of years). China's permafrost is mainly distributed in the Qinghai–Tibet Plateau, the northeastern Daxing, Xiaoxing'anling, Songnen Plain and some high mountains in the west, with a total area of about 2.07×106 km^2, accounting for 21.5% of China's land area. Permafrost in the Qinghai–Tibet Plateau is the representative of high-altitude permafrost in the low latitudes of the world, with a distribution area of about 1.5×106 km^2, accounting for 70% of the total permafrost area in China [1].

Permafrost is extremely sensitive to temperature changes, especially high-altitude permafrost, and slight changes in temperature have a significant impact on the temperature and stability of Permafrost. [2,3]. According to the temperature data of the central hole along the Qinghai–Tibet Highway, a thawing nucleus with a thickness of 2–7 m has been formed under the Qinghai–Tibet Highway, and tens of kilometers of a "thawing channel" has been formed under the roadbed along the longitudinal direction of the road. Most of the thawing nucleus stores water for many years, resulting in strong uneven subsidence and road surface damage [4,5]. Therefore, detecting the interface of seasonal permafrost thawing, the formation temperature and the water content in the soil are of great significance for judging the stability and safety of the road project. In addition, it also provides data for global warming research.

As the seasons and temperature change, the dielectric constant and the conductivity of soil in seasonal permafrost soils change significantly. The dielectric constant of the seasonal permafrost soil area can vary from 3 to 30. These changes will be directly reflected in the geophysical wave field [6]. The ground penetrating radar (GPR) using electromagnetic wave detection method has been widely used since the 1990s because of its advantages of being fast, of high efficiency and non-destructive [7]. Arcone et al. [8,9] of the US Army Engineering Cold Research and Engineering Laboratory (CRREL) successfully detected groundwater below the frozen soil in permafrost and rock masses that exist within the permanent frozen layer in Alaska using GPR of 100 MHz and 50 MHz. De Pascale et al. (2008) used a combination of GPR and capacitively coupled resistivity profilometry to examine the distribution of ice bodies in multi-year frozen layers on several sections of Richards Island in the western Canadian Arctic. Xiao and Liu [10] used dual-frequency (100 MHz and 400 MHz) ground penetrating radar data to study the important layered interfaces such as the seasonal active layer and the top surface of the frozen soil. These studies mainly focus on two aspects: 1) the size, shape and distribution of ice bodies in the permafrost; 2) the layered structure of the thermal state. In the study of permafrost soil's water content study using GPR data, scholars have proposed many theoretical and empirical model formulas. The more famous formulas are: Topp formula, Malicki formula, Roth formula and Alharathi formula [11,12]. Du and Sperl proposed a method for measuring soil water content using GPR common offset data [13]. Huisman [14] et al. used the WARR method to apply 225 MHz and 450 MHz antennas respectively to determine the water content of the soil based on the slope of the ground wave. Shen et al. [15] improved the Topp formula using the GPR instantaneous quality factor attribute to study the relative water content of frozen soil. However, these studies also have the following problems. Firstly, in the study of seasonal permafrost soil in the Qinghai–Tibet Plateau, the division of different temperature layers and thawing layers is not elaborate enough, and only the in-phase reflection of GPR data is utilized. No more GPR data attributes have been analyzed. Secondly, there are few studies on using the GPR data to calculate the water content in the frozen soil area. The method of using only a single attribute for water content calculation is not precise enough and lacks constraints.

In order to solve these problems, this study comprehensively analyzed the characteristics of GPR data attributes in seasonal permafrost soil areas. Accurate imaging and stratigraphic division of seasonal frozen soil on the Qinghai–Tibet Plateau are carried out in terms of waveform variation, wavefield attenuation and wave impedance variation. In addition, under the constraint of water content and temperature data, the water content measurement model of frozen soil area was reconstructed by comprehensively utilizing various GPR data sensitive attributes. The water content and formation temperature of the measurement area were calculated.

2. Data Acquisition

2.1. Test Site

The field site is located on the Qinghai–Tibet highway, which is close to Tuotuo River, Hoh Xil plateau region, center of Tibetan Plateau, China (N 34°10′, E 92°23′). The data collection site is on the Qinghai–Tibet Highway, 40 km south of Tanggula Town (Figure 1). The range of temperature change with the seasons is large [16,17]. Affected by high altitude, the average surface temperature in the area is 10–14 °C lower than other areas in the same latitude. However, the free temperature in summer is 5–7 °C higher than the temperature in the same latitude plain. The annual average temperature in the study area is below 0 °C, and the lowest is −7.0 °C. The average temperature in June–September is positive during the year, and the temperature is negative from October to May. The average temperature in July is the highest, about 5.0–6.0 °C, and the lowest in January is about −20.0 to −16.0 °C [7]. In addition, road construction in permafrost areas affects the thermal regime of frozen soils which results in permafrost degradation and road damage [18]. Due to long-term man-made construction activities and vehicle traffic, the seasonal thawing permafrost on the Qinghai–Tibet Highway is very obvious. Due to the heterogeneity of the permafrost medium on the Qinghai–Tibet Highway, seasonal ablation caused many road collapses and undulations [7].

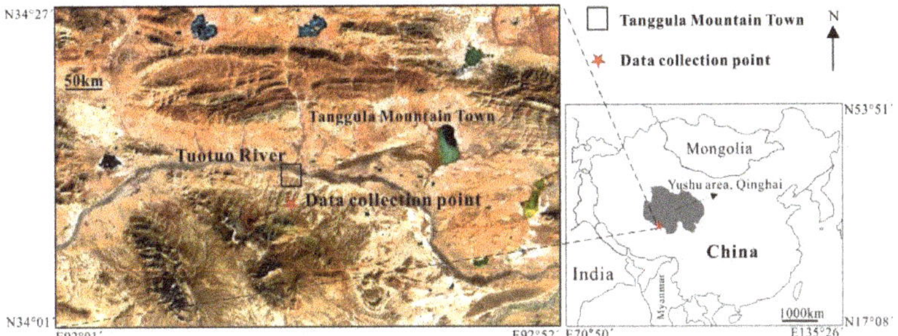

Figure 1. Location map of test site.

2.2. Test Experiment and Data Acquisition

The SIR-20 GPR (GSSI, Inc.) system equipped with 200 MHz central frequency antennas was used to perform wheel survey acquisition in the test site. The measurement line layout is shown in Figure 2. Lines 1 to 6 are arranged along the direction of the road, with a line spacing of 3 meters; lines 7 to 12 are arranged perpendicular to the road, with a line spacing of 2 meters. Due to the width of the road, line 1 to 6 have 181 traces, line 7 to 12 have 101 traces. The time window of each trace is 200 ns with 512 samples. Using the measuring wheel, we use the zero offset distance measurement mode. Figure 3 shows the spatial distribution of 12 lines of data.

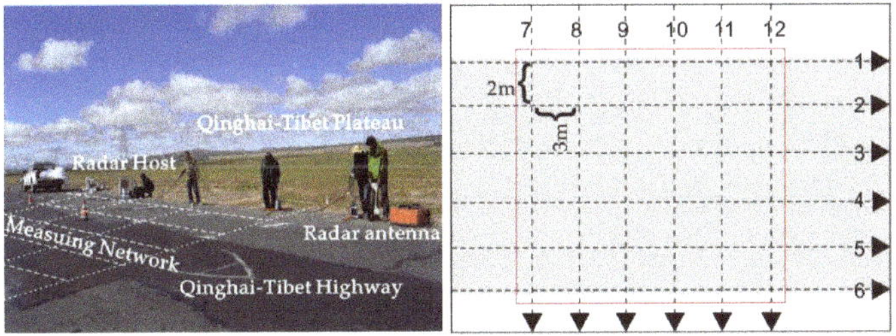

Figure 2. Test experiment and measurement line layout.

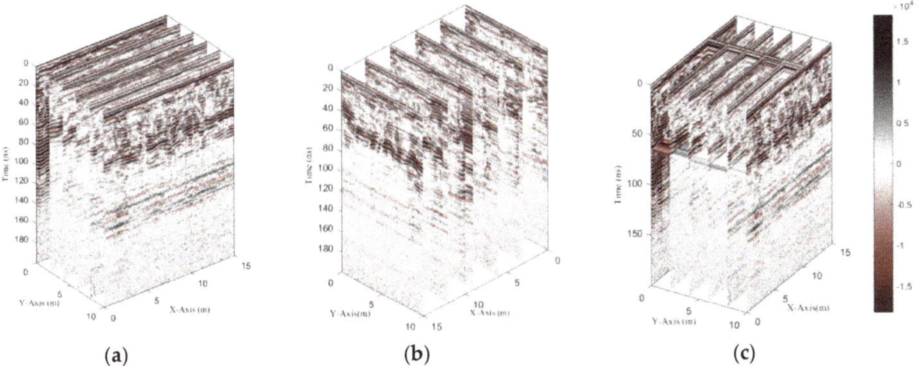

Figure 3. The spatial distribution of 12 lines of data. (**a**) Lines 1,2,3,4,5,6; (**b**) lines 7,8,9,10,11,12; (**c**) Cross and slice display.

2.3. Experimental Data Processing

In order to ensure the quality of the data, we performed spectrum analysis and imaging capability analysis on these 12 GPR data. The data is centered at 200 MHz. In the frequency range from 0 to 500 MHz, the data energy is strong, ensuring the spectral integrity of its signal. The original data collected by the experiment uses a distance sampling mode in which the shielded antenna approximates zero offset. The amplitude data of the original data is distributed between 0–800 MHz and carries background noise pollution (Figure 4a). After noise suppression, the amplitude energy is mainly concentrated in the 50–400 MHz interval. This result is consistent with the signal characteristics of the antenna with a center frequency of 200 MHz (Figure 4b). In addition, due to the small influence from external noise and terrain interference, in order to better serve GPR data interpretation, the processing steps can be summarized as follows: correct position, background removal, frequency filtering, horizontal scaling and deconvolution.

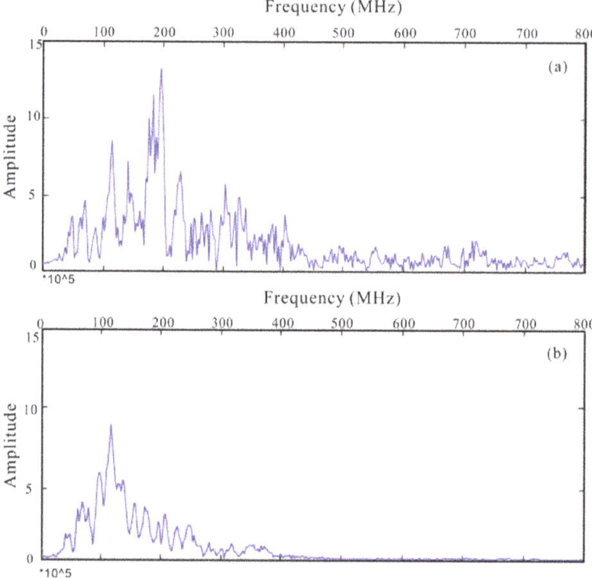

Figure 4. Data spectrum analysis. (**a**) Raw data; (**b**) processed data.

Figure 5 is a lithology and stratigraphic column of the shoulder-drilled formation in the data acquisition block. The surface layer is artificial gravel and soil filling with a thickness of 1 m. When the depth is within 7 meters, the formation lithology is mainly clay and sandstone. The lithological changes within 7 meters are mainly reflected in the size of the grain. In formations above 3.7 m depth, the melting of the frozen soil results in water in the medium. When the depth is greater than 3.7, the water content in the frozen soil is extremely low, and the water is stored in the soil in the form of ice.

Stratigraphic histogram	Lithology	Depth (m)	Lithological properties and description of frozen soil
Gravel and soil		1.00	Subgrade fill, brown gravel and soil, wet, dense.
Turf clay		0.35	Turf clay, hard plasticity
Gravel bearing sand		0.70	Gravel-containing clay, brown, watery.
Gravel sand		0.35	Yellowish brown with crushed stone sub-clay, plastic
Gravel bearing sand		0.80	Yellow-brown pebbled sub-clay, watery
Subclay		0.20	Gray clay, soft plasticity
Mud stone		0.30	See frozen soil, gray-white, fully weathered mudstone and shale
Silt		0.30	Black green silt with ice, massive structure
Mudstone		2.60	Gray soily ice layer, rich frozen soil at the bottom

Figure 5. Lithology and stratigraphic column.

3. Data Analysis and Discussion

3.1. Frozen Soil Layered Description

The continuity of the GPR data in phase and the strong impedance interface are often used to divide the various interfaces of the frozen soil [19]. Figure 6 shows amplitude of the 2 and 5 line which are parallel to the road's alignment and the 8 and 12 lines which are perpendicular to the road alignment. Due to the change of the dielectric constant, a relatively reflective interface appears in the vicinity of 20 ns, 50 ns and 100 ns of the data.

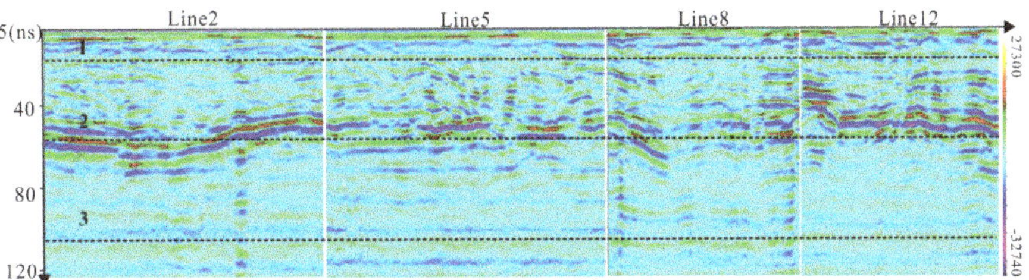

Figure 6. Amplitude of ground penetrating radar (GPR) data: the black dashed lines 1, 2, and 3 respectively notes 20 ns, 50 ns and 100 ns.

In order to more accurately determine the location of the interface, we performed waveform view and attenuation attributes analysis (Figure 7) on the data. In Figure 7, the attenuation attributes of the GPR signal are calculated. That represents the instantaneous power of electromagnetic waves over time in different media [20]. Mean and median attenuation respectively represent two calculation methods. In the calculation process, they use the mean and median instantaneous amplitude of all traces. The mean is the arithmetic average of a set of numbers, or distribution. It is the most commonly used measure of central tendency of a set of numbers. The median is described as the numeric value separating the higher half of a sample, a population, or a probability distribution, from the lower half. Two statistical algorithms can be used to avoid the incompleteness of any one algorithm, making the results more credible. This also determines best fitting models for power-law and exponential attenuation based on the median attenuation data. Among the four physical layers, the attenuation of electromagnetic waves also exhibits differences due to various values in dielectric constant, density, and water content. At time in 20 ns, 50 ns, and 100 ns, there is a significant inflection point in the instantaneous power.

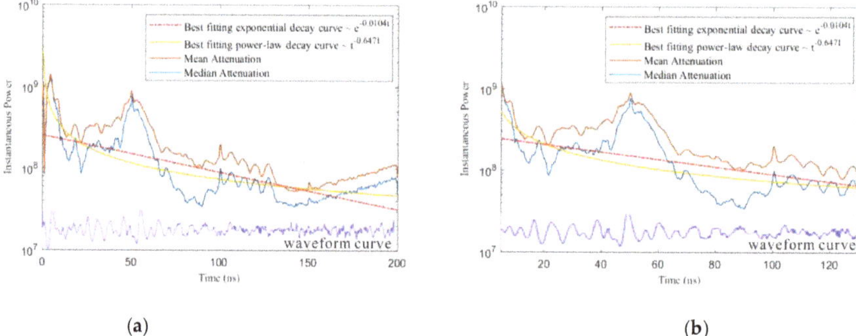

Figure 7. Electromagnetic waveform view and wave attenuation analysis. (**a**) Raw data; (**b**) processed data.

For calculating the attenuation attribute, we use the classic equation as follows:

$$\beta = 20 \times \log_{10} \frac{A_2}{A_1} \tag{1}$$

where A_2 and A_1 are the amplitude values of two adjacent peaks. A change in the dielectric constant will produce impedance of the electromagnetic wave. Relative to the amplitude attribute, it can better reflect the changing characteristics of the interface. Due to the lack of accurate dielectric constant and physical property information of the buried media, it is difficult to establish the initial model needed to construct absolute wave impedance conversion. Fortunately, relative wave impedance conversion (RAI) does not rely on an initial model. Relative wave impedance is a simplified conversion. It provides a quick and easy way to view data without subjecting it to an conversion process. It highlights areas where there are changes in reflectivity illuminating specific subsurface medium properties [21]. Due to its ability to identify thin layers, it is ideal for radar data interpretation that requires high resolution. The magnitude of reflection coefficient γ_i indicates the intensity of the reflection between two thawing layers according to the classic Equation (2) [22]:

$$\gamma_i = \frac{\sqrt{\varepsilon_i} - \sqrt{\varepsilon_{i+1}}}{\sqrt{\varepsilon_i} + \sqrt{\varepsilon_{i+1}}} \tag{2}$$

where ε_i and ε_{i+1} are the dielectric constant of permafrost thawing layers *i* and *i+1* respectively. Continue to simplify and derive Equation (3):

$$\lambda = \int \gamma dt \approx \frac{1}{2}\int \frac{\Delta \varepsilon}{2\varepsilon} = \frac{1}{2}\ln \varepsilon \tag{3}$$

where ε is the average dielectric constant value of ε_i and ε_{i+1}. $\Delta \varepsilon$ is the difference between ε_i and ε_{i+1}. λ is relative wave impedance. It is proportional to the natural logarithm of the permittivity values.

Figure 8 shows the relative wave impedance conversion profiles for lines 2, 5, 8, and 12, respectively. Relative to the amplitude attribute, the relative wave impedance data is clearer and the resolution is higher. The black dotted lines 1, 2, and 3 respectively show the stratified positions of the frozen soil physical interface.

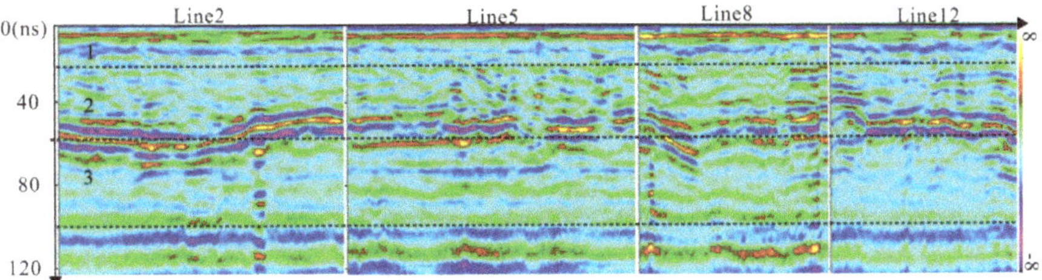

Figure 8. Relative wave impedance conversion profiles.

Based on the above analysis, we divide the seasonal permafrost into four parts. The 0–20 ns interval is the complete thawing layer; the 20–50 ns is the thawing layer, wherein the 50 ns depth is the upper interface of the seasonal permafrost ablation; the 50 ns is the frozen soil unablated layer, of which 50–100 ns is the partial ablation of the early frozen soil. The dielectric property change layer, below 100 ns is a completely frozen soil layer. The reason for the early partial ablation at 50–100 ns was that the depth of 50 ns was the surface before the Qinghai–Tibet Highway was built. Combined with the drilling measurement data, when the dielectric constant of the medium is 11, the GPR data has the best correspondence with the formation horizon.

3.2. Relative Water Content and Temperature Calculation

The water content calculation of seasonal permafrost is an important part of ground penetrating radar data. Temperature is the main factor affecting the water content of frozen soil. At the same time, due to the change of water content, the physical property parameter information will be directly reflected in the electromagnetic wave data of ground penetrating radar. By mining the multi-attribute information of GPR data, the correspondence between the GPR data attribute and the measured water content and temperature is the basis for the water content calculation using GPR data. When the correlation between the attribute value and the water content is more than 80%, it is judged as a valuable attribute for calculating the water content. By calculation, we get the valuable attribute: The absolute instantaneous amplitude and the weighted average frequency attribute. The weighted average frequency is an improvement to the instantaneous frequency. Instantaneous frequency $w(t)$ is defined as the derivative of the phase of the signal. Signal $z(t)$ can be written as the sum of N items of the index of signal:

$$z(t) = \sum_{n=0}^{N=+\infty} a_n(t)e^{j\phi_n(t)} = a(t)e^{j\phi(t)} \tag{4}$$

where $a_n(t)$ is a constant parameter. Because instantaneous frequency is prone to generating spikes and noise for GPR data. To solve this problem, weighted instantaneous frequency is introduced. Weighted instantaneous frequency $\omega(t)$ is defined as follows:

$$\omega(t) = \frac{\sum\limits_{n=0}^{N=+\infty} a_n^2(t)\phi_n'(t)}{\sum\limits_{n=0}^{N=+\infty} a_n^2(t)} \tag{5}$$

The weighted average frequency attribute is rarely affected by short wavelength and noise [23]. In Figure 9a,b, there is a high correlation between absolute instantaneous amplitude (Ai), the weighted average frequency attribute ω and the measured water content. The high red absolute instantaneous amplitude and weighted average frequency attribute corresponds well to the water content of the black display. On the mining of valuable attributes representing temperature, we use the correlation value greater than 65% as the reference value for attribute mining. Due to differences in media properties and particles in the surface roadbed, the correlation between temperature and water content in the surface layer is not high. The weighted average frequency attribute can basically reflect the temperature distribution characteristics of a fully melted layer with water and an incompletely melted ice layer. However, because the temperature change value is not large, the accuracy cannot reach the standard of water content correlation. So, we use this attribute to show the trend of temperature (Figure 9b). It is mainly used as a control factor for the calculation of water content. Figure 10 shows the weighted average frequency attribute reflecting the formation of temperature. Figure 10 indicates the reflecting effect of temperature in the water-ice contact. The temperature of the fully melted layer is higher than 0 °C and the temperature of the ice-containing layer is lower than 0 °C.

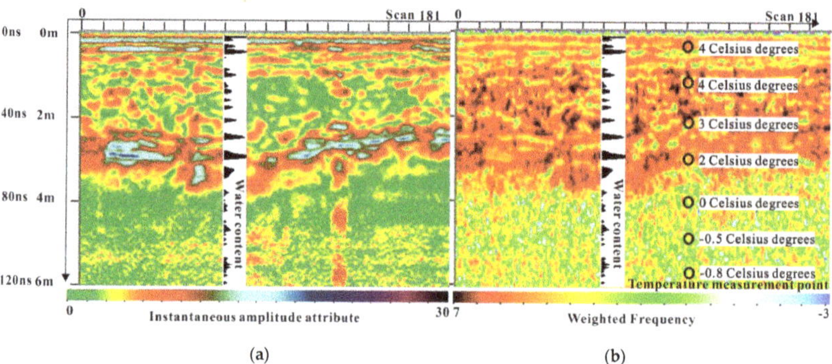

Figure 9. Correspondence diagram between electromagnetic wave attribute and measured data. (**a**) Absolute instantaneous amplitude attribute; (**b**) weighted average frequency attribute; water content is inserted in (**a**) and (**b**); black circle notes temperature in (**b**). The dielectric constant of the medium is 11.

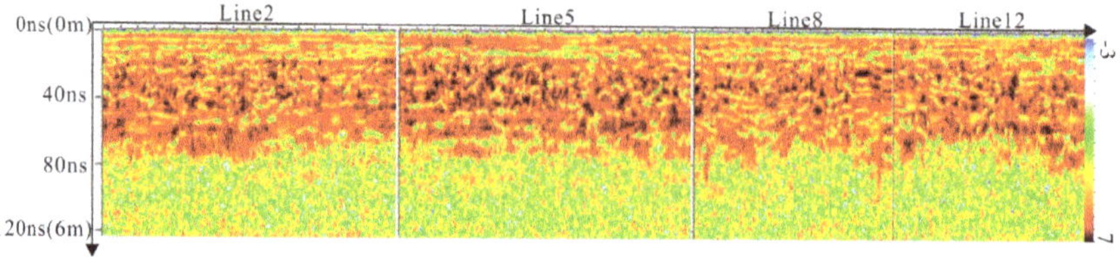

Figure 10. The weighted average frequency attribute.

The relationship between measured data and electromagnetic wave attributes mining and screening studies provides attenuation attribute, relative wave impedance, absolute instantaneous amplitude attribute and the weighted average frequency attribute for water content conversion. Values of water content in the medium lead to differences in dielectric constant. The change of dielectric constant of the medium causes the change of the electromagnetic wave impedance value. Because of the positive correlation among them, the electromagnetic wave impedance parameter can be added to the calculation of the relative water content. The water content calculation operator can be expressed as:

$$\delta = \frac{\beta + \lambda + Ai + \omega}{4}$$
$$\beta = 20 \times \log_{10} \frac{A_2}{A_1}$$
$$\lambda = \int \gamma dt = \tfrac{1}{2} \ln \varepsilon$$
$$(6)$$

where attenuation attribute β, relative wave impedance λ, absolute instantaneous amplitude attribute (Ai) and the weighted average frequency attribute ω are the normalized values parameters. δ is water content conversion operator. The Topp model is a formula for accurately calculating the relationship between soil water content and dielectric constant. Since GPR data cannot directly obtain accurate absolute dielectric constants, substitute Equation 4 into the Topp model formula [24]:

$$\theta = -0.53 \times 10^{-2} + 2.92 \times 10^{-2}\delta - 5.5 \times 10^{-4}\delta^2 + 4.3 \times 10^{-6}\delta^3 \tag{7}$$

where θ is relative water content. Figure 12 shows the working methods and processes of the entire research work. The relative water content information is mined from the GPR data by calculating the value of the sensitive attribute and the relative wave impedance (Figure 11).

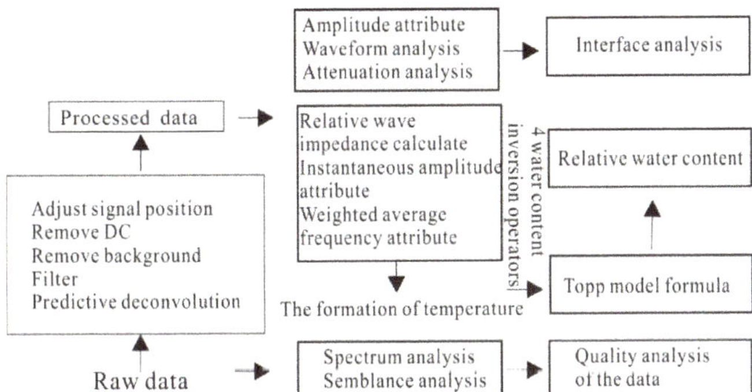

Figure 11. Method flow diagram.

We use Equation 5 for the calculation of relative water content. Figure 12 shows the water content profile for the conversion of different line data. Because in the research of frozen soil water content calculation, the main concern area is the complete melting zone of frozen soil nodule between 20~55 ($\pm$10) ns. The measured sample data is in good agreement with the water content value calculated by the GPR data in this region. In the 50 ($\pm$10) ~100 ns and 100–120 ns areas, it is not the area of concern. The soil in this area mainly contains ice, and the measured water content is inferior to the calculated value. In addition, due to the low quality of deep data, deep data carries less physical information. We calculate and present the data within 65 ns. Figure 13 shows the planar variation characteristics of water content at different depths of time. Red represents high water content and green represents low water content. Before 10 ns, due to the roadbed medium, the surface medium was relatively dry and the water content was very low. As the depth increases, the water content in the thawing layer increases. As the depth continues to increase, the water content becomes smaller. These characteristics and values are consistent with the characteristics and measured data of seasonal permafrost ablation.

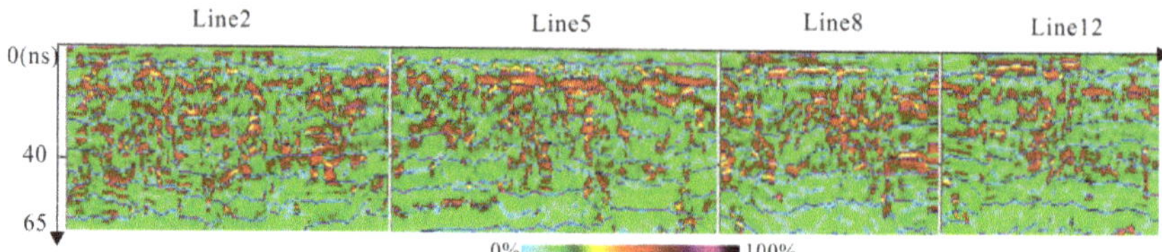

Figure 12. Relative water content calculation for the set of lines 2, 5, 8 and 12.

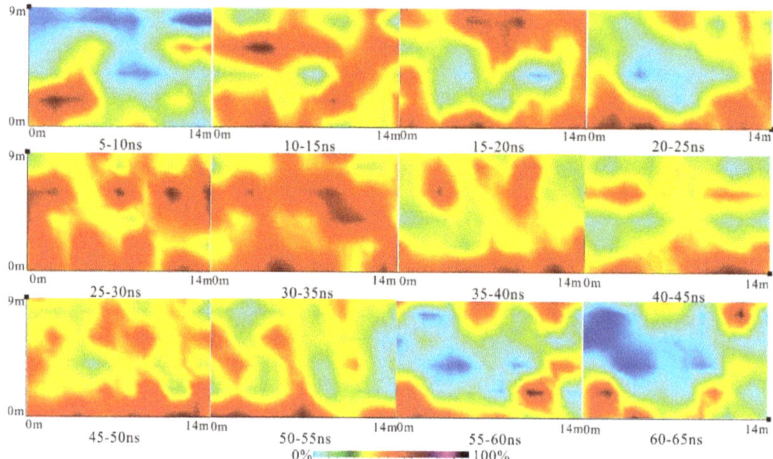

Figure 13. The planar variation characteristics of relative water content at different depths of time.

4. Conclusions

We provide an interface identification method for using ground penetrating radar data to carry out seasonal permafrost areas on the Tibetan Plateau. By contrast with the traditional low-resolution method, only the amplitude data change is used to distinguish the interface. In this study, amplitude analysis, the waveform analysis and electromagnetic wave attenuation attribute are used to make a more detailed imaging division of the seasonal permafrost area. The medium in the seasonal permafrost area is divided into a finer four-layer structure. They are: subgrade, frozen soil ablation layer, frozen soil semi-ablative layer and permafrost layer. Through the calibration of the drilling data, dielectric constant of time-depth conversion in the Qinghai–Tibet highway, which is close to Tuotuo River, and the Hoh Xil plateau region, the center of Tibetan Plateau, is near 11.

In addition, when the correlation between the attribute value and the water content is more than 80%, it is judged as a valuable attribute for calculating the water content. Absolute instantaneous amplitude and the weighted average frequency attribute can react well to changes in water content. Using GPR data is more difficult for temperature calculation. The weighted average frequency attribute can basically reflect temperature and change, but its accuracy does not reach the standard of water content correlation.

Since the GPR data cannot directly obtain the dielectric constant throughout the medium, we replace the dielectric constant with a calculation factor that combines the attenuation attribute, relative wave impedance attribute, absolute instantaneous amplitude attribute, and the weighted average frequency attribute to calculate the relative moisture content. The calculation results are in good agreement with the actual sample measurements in the frozen soil ablation zone. When the ice-containing state occurs in the soil, the error of this calculation method becomes large.

Author Contributions: Q.W. and Y.S. jointly collected data needed for research on the Qinghai-Tibet Plateau. Q.W. studied data processing algorithms; Y.S. studied frozen soil theory and results verification. They completed the research and article together.

Funding: This was financially supported by National 973 Project of China (NO.2012CB026104), the National Natural Science Foundation (NSFC) under grants (No.51578053, No. 41772330 and No. 41171064), Beijing Information Science and Technology University Research Fund project (182501), the State Key Laboratory of Frozen Soils Engineering(Grant Nos.SKLFSE201808), the Scientific Research Project of Beijing Educational Committee (KM201811232010), Beijing Science and Technology Innovation Service Capacity Building-Fundamental Research Fund(PXM2018_014224_000032) and the Project for Promoting Connotative Development of Universities-"Information +" project.

Acknowledgments: The authors would like to thank the professors of Beijing Jiaotong University for suggestions on the theory of seasonal permafrost background. The authors would like to thank to the National 973 Project of China for support. The authors are also grateful to the reviewers and editor for their valuable comments and remarks.

Conflicts of Interest: The authors declare no conflict of interest. The funders had no role in the design of the study; in the collection, analyses, or interpretation of data; in the writing of the manuscript; or in the decision to publish the results.

References

1. Zhou, Y.W.; Guo, D.X.; Qiu, G.Q. *Geocryology in China*; Beijing Science Press: Beijing, China, 2000; pp. 5–35.
2. Niu, F.J.; Lin, Z.H.; Liu, H.; Xu, Z.Y. Characteristics of Roadbed Settlement in Embankment-Bridge Transition Section along the Qinghai–Tibet Railway in Permafrost Regions. *Cold Reg. Sci. Technol.* **2001**, *65*, 437–445. [CrossRef]
3. Wang, S.L. Discussion on the Degradation of Frozen Soil and the Problem of Frozen Soil Environment in the Qinghai-Tibet Plateau. In Proceedings of the Fifth National Glacier and Frozen Soil Conference, Lanzhou, China, 1996; pp. 11–17.
4. Wu, Q.B.; Dong, X.F.; Liu, Y.Z. Response Analysis of Permafrost along the Qinghai-Tibet Highway to Climate Change and Engineering Impact. *Glacier* **2005**, *27*, 50–54.
5. Jin, H.J.; Zhao, L.; Wang, S.L.; Jin, R. Characteristics and Degradation Modes of Frozen Soil Along the Qinghai-Tibet Highway. *Chin. Sci. Bull. D Earth Sci.* **2006**, *11*, 1009–1019.
6. Davis, J.L.; Annan, A.P. Ground Penetrating Radar for High Resolution Mapping of Soil and Rock Stratigraphy. *Geophys. Prospect.* **1989**, *37*, 531–551. [CrossRef]
7. Luo, J. Study on instability and susceptibility evaluation of frozen soil slope in Qinghai-Tibet Engineering Corrido. PhD Thesis, Chinese Academy of Sciences, Lanzhou, China, 2015.
8. Arcone, S.A.; Lawson, D.E.; Delaney, A.J. Ground-Penetrating Radar Reflection Profiling of Groundwater and Bedrock in An Area of Discontinuous Permafrost. *Geophysics* **1998**, *63*, 1573–1584. [CrossRef]
9. Arcone, S.A.; Prentice, M.L.; Delaney, A.J. Stratigraphic Profiling with Ground-Penetrating Radar in Permafrost: A Review of Possible Analogs for Mars. *J. Geophys. Res.* **2002**, *107*, 1–14. [CrossRef]
10. Xiao, J.; Liu, L. Signal Fusion Using Extrapolation with Deterministic Deconvolution on Multi-Frequency Qinghai-Tibet Railway GPR Data for Permafrost Subgrade Detection. In Proceedings of the 15th International Conference On Ground Penetrating, Radar, Brussels, Belgium, 30 June—4 July 2014; pp. 586–589.
11. Robert, C. Time—Domain Reflectometry Method and Its Application for Measuring Moisture Content in Po Rous. *Mater. Meas.* **2009**, *42*, 329–336.
12. Weiler, K.W.; Teenhuis, T.S.; Kung, K.S. Comparison of Ground Penetrating Radar and Time Domain Reflectometry as Soil Water Sensors. *Soil Sci.* **1998**, *62*, 1237–1239. [CrossRef]
13. Grote, K.H.; Ubbard, S.; Rubin, Y. Field -Scale Estimation of Volumetric Water Content Using Ground Penetrating Radar Ground Wave Techniques. *Water Resour. Res.* **2003**, *39*, 1–13. [CrossRef]
14. Huisman, J.A.; Sperl, C.; Bouten, W. Soil Water Content Measurements at Different Scales: Accuracy of Time Domain Reflectometry and Ground Penetrating Radar. *J. Hydrol.* **2001**, *245*, 48–58. [CrossRef]
15. Shen, Y.P.; Zuo, R.; Liu, J.K. Characterization and Evaluation of Permafrost Thawing Using GPR Attributes in The Qinghai-Tibet Plateau. *Cold Reg. Sci. Technol.* **2018**, *151*, 302–313. [CrossRef]
16. Zhao, L.; Cheng, G.; Li, S. Thawing and Freezing Processes of Active Layer in Wudaoliang Region of Tibetan Plateau. *Chin. Sci. Bull.* **2000**, *45*, 2181–2186. [CrossRef]

17. Zhao, W.K.; Forte, E.; Pipan, M.; Tian, G. Ground Penetrating Radar (GPR) Attribute Analysis for Archaeological Prospection. *J. Appl. Geophys.* **2014**, *97*, 107–117. [CrossRef]

18. Hinkel, K.M.; Doolittle, J.A.; Bockheim, J.G.; Nelson, F.E.; Paetzold, R.; Kimble, J.M.; Travis, R. Detection of Subsurface Permafrost Features with Ground-Penetrating Radar, Barrow, Alaska. *Permafr. Periglac. Process.* **2001**, *12*, 179–190. [CrossRef]

19. Liu, L.B.; Qian, R.Y. Ground Penetrating Radar: A Critical Tool in Near-Surface Geophysics. *Chin. J. Geophys.* **2015**, *58*, 2606–2617.

20. Tzanis, A. matGPR Release 2: A Freeware MATLAB Package for The Analysis & Interpretation of Common and Single Offset GPR Data. *Fast Times* **2010**, *15*, 17–43.

21. Rahmani, A.; Belmokhtar, A.; Murineddu, A.; Khazanehdari, J.; English, J.; Roumane, H.; Godfrey, B. The Art of Seismic Conversion-An Example from Erg Chouiref Algeria. *SEG Tech. Program Expand. Abstr.* **2006**, *25*, 264–268.

22. Maser, K.R. *Ground Penetrating Radar Surveys to Characterize Pavement Layer Thickness Variations at GPS Sites*; Report; Strategic Highway Research Program National Research Council: Washington, DC, USA, 1994; p. 397.

23. Loughlin, P.J. Comments on the interpretation of instantaneous frequency. *IEEE Signal Process Lett.* **1997**, *4*, 123–125. [CrossRef]

24. Topp, G.C.; Davis, J.L.; Annan, A.P. Electromagnetic Determination of Soil Water Content: Measurements in Coaxial Transmission Lines. *Water Resour. Res.* **1980**, *16*, 574–582. [CrossRef]

Article

Evaluation of a Straight-Ray Forward Model for Bayesian Inversion of Crosshole Ground Penetrating Radar Data

Hui Qin [1,2] , **Xiongyao Xie** [2,3] **and Yu Tang** [4,*]

1   School of Civil Engineering, Dalian University of Technology, Dalian 116024, China; hqin@dlut.edu.cn
2   Key Laboratory of Geotechnical and Underground Engineering of Ministry of Education, Tongji University, Shanghai 200092, China; xiexiongyao@tongji.edu.cn
3   Department of Geotechnical Engineering, Tongji University, Shanghai 200092, China
4   School of Hydraulic Engineering, Dalian University of Technology, Dalian 116024, China
*   Correspondence: ytang@dlut.edu.cn; Tel.: +86-411-84707232

Received: 29 April 2019; Accepted: 31 May 2019; Published: 4 June 2019

Abstract: Bayesian inversion of crosshole ground penetrating radar (GPR) data is capable of characterizing the subsurface dielectric properties and qualifying the associated uncertainties. Markov chain Monte Carlo (MCMC) simulations within the Bayesian inversion usually require thousands to millions of forward model evaluations for the parameters to hit their posterior distributions. Therefore, the CPU cost of the forward model is a key issue that influences the efficiency of the Bayesian inversion method. In this paper we implement a widely used straight-ray forward model within our Bayesian inversion framework. Based on a synthetic unit square relative permittivity model, we simulate the crosshole GPR first-arrival traveltime data using the finite-difference time-domain (FDTD) and straight-ray solver, respectively, and find that the straight-ray simulator runs 450 times faster than its FDTD counterpart, yet suffers from a modeling error that is more than 7 times larger. We also perform a series of numerical experiments to evaluate the performance of the straight-ray model within the Bayesian inversion framework. With modeling error disregarded, the inverted posterior models fit the measurement data nicely, yet converge to the wrong set of parameters at the expense of unreasonably large number of iterations. When the modeling error is accounted for, with a quarter of the computational burden, the main features of the true model can be identified from the posterior realizations although there still exist some unwanted artifacts. Finally, a smooth constraint on the model structure improves the inversion results considerably, to the extent that it enhances the inversion accuracy approximating to those of the FDTD model, and further reduces the CPU demand. Our results demonstrate that the use of the straight-ray forward model in the Bayesian inversion saves computational cost tremendously, and the modeling error correction together with the model structure constraint are the necessary amendments that ensure that the model parameters converge correctly.

Keywords: crosshole ground penetrating radar (GPR); Bayesian inversion; Markov chain Monte Carlo (MCMC); forward model; modeling error; discrete cosine transform (DCT)

---

1. Introduction

The crosshole ground penetrating radar (GPR) is an effective tool to map the subsurface properties, and found widespread application in soil moisture estimation [1,2], hydraulic parameter qualification [3,4], geological investigation [5–7], and civil structure inspection [8,9]. This method uses a transmitting antenna in one borehole to emit high-frequency (10 MHz to 1 GHz) electromagnetic (EM) waves and a receiving antenna in an adjacent borehole to receive them. By analyzing the acquired crosshole GPR data (first-arrival traveltimes, first-cycle amplitudes, or waveforms), the spatial distribution of the dielectric properties (the dielectric permittivity, ε and electrical conductivity, σ) in-between the two boreholes can be characterized for a better understanding of the subsurface features that are sensitive to those properties [10].

To derive the EM properties from crosshole GPR data, a variety of inversion methods have been developed. Perhaps the most popular methods are the ray-based tomographic algorithms that simplify the EM wave propagation to a straight or bending ray from the transmitter to receiver [11,12]. These approaches that use the information of first-arrival traveltimes and maximum first-cycle amplitudes solve iteratively for the EM wave velocity and attenuation fields [13–16]. Ray tomography is usually computationally efficient but the resolution is limited to the scale of the first Fresnel zone due to the high-frequency approximation [17,18]. In contrast, the waveform-based inversion techniques that make the best of the information of the full-waveforms can reach a sub-wavelength resolution [19–22]. In the process of full-waveform inversions, the forward computations need numerical solutions of the Maxwell's equations in either the time or frequency domain and place a heavy burden on computational resources. Recently a neural network-based model is proposed for forward computing [23]. This method uses a trained neural network to calculate crosshole GPR traveltimes, which is fast to evaluate and more accurate than the ray-based model. Yet the main challenges lie in generating a large training data and training the neural network.

The commonly used inversion methods that adopt gradient-based approaches to search for a group of "optimal" model parameters (permittivity or conductivity values) to fit the measurement data are referred to as deterministic inversion methods. They provide only a single realization and are not able to quantify the result uncertainties. On the contrary, probabilistic inversion methods treat different sources of error explicitly and provide a set of solutions drawn from the posterior distribution of model parameters [24–28]. In previous work we have developed a Bayesian inversion method to determine the relative permittivity fields, ε_r ($\varepsilon_r = \varepsilon/\varepsilon_0$, where ε_0 signifies the dielectric permittivity in free space) from crosshole GPR waveform data [29]. This method treats the grid values of the discretized ε_r model as unknown parameters and uses the discrete cosine transform (DCT) to reduce the parameter dimensionality [30]. We employ a 2D finite-difference time-domain (FDTD) solver of the Maxwell's equations for forward computing [31,32], and resort to the Markov-chain Monte Carlo (MCMC) simulation with the DREAM$_{(ZS)}$ algorithm to explore the posterior distribution of model parameters (DCT-coefficients) [33–35]. The usefulness of the proposed inversion method was demonstrated on both numerical and real-world applications [28,29]. However, we find the use of the CPU-intensive FDTD forward modeling leads to the inversion work a daunting task as the MCMC iteration requires thousands to millions of model evaluations. In order to improve the computational efficiency, ray-based models could be possible replacements for the forward calculations. Yet due to the ray assumption, any scattering effects of EM waves are neglected and notable modeling errors might be produced and bias the inversion results. Before the ray-based model can be implemented successfully within the Bayesian inversion framework, the modeling errors should be carefully considered [36–38].

In this paper, we focus our attention on the applicability of the widely used straight-ray forward model in our Bayesian inversion framework. We first briefly summarize the basic idea of the Bayesian inversion method and formulation of the straight-ray model, then evaluate the modeling error and computational efficiency of the forward simulator, followed by a detailed analysis of the impact of the modeling error and model structure constraint on the inversion results, and conclude with a summary of the main findings.

2. Methodology

The crosshole GPR measurement can be described by the following equation

$$\tilde{\mathbf{y}} = f(\mathbf{m}) + \mathbf{e}, \tag{1}$$

where the forward operator $f(\cdot)$ simulates the physical relation between model parameters, $\mathbf{m}$ and measurement crosshole GPR data, $\tilde{\mathbf{y}}$ and $\mathbf{e}$ contains varies sources of errors including measurement error and modeling deficiencies. In this study the model parameters signify the 2D discretized relative permittivity values of the subsurface media. The measurement data are either GPR waveforms or first-arrival traveltimes. The forward functions $f(\cdot)$ that we use are a FDTD solver of Maxwell's equations that simulates GPR waveforms and a straight-ray forward kernel that simulates first-arrival traveltimes.

In probabilistic inversion, the model parameters, $\mathbf{m}$ can be derived from measurement GPR data, $\tilde{\mathbf{y}}$ using Bayes theorem

$$p(\mathbf{m}|\tilde{\mathbf{y}}) = kp(\mathbf{m})L(\mathbf{m}|\tilde{\mathbf{y}}), \tag{2}$$

where $p(\mathbf{m}|\tilde{\mathbf{y}})$ denotes the posterior distribution of model parameters, $p(\mathbf{m})$ describes prior knowledge of $\mathbf{m}$ before carrying out crosshole GPR measurement, and $L(\mathbf{m}|\tilde{\mathbf{y}})$ is the likelihood function that summarizes the distance between simulated and measurement data. The normalization constant k ensures the posterior distribution integrates to unity.

In the absence of detailed information about the model structure, a uniform distribution is often used as non-informative prior. Another strategy is to include a smooth constraint in the prior distribution to reduce model complexity and stabilize the solution [39]. Following the study of Rosas-Carbajal et al. [40], the model constraint can be defined using a normal distribution

$$p(\mathbf{m}) = \frac{1}{(2\pi\lambda^2)^{R_x+R_z}} \exp\left(\frac{1}{2\lambda^2}\left(\mathbf{m}^{\mathrm{T}}\left(\mathbf{W}_x^{\mathrm{T}}\mathbf{W}_x + \mathbf{W}_z^{\mathrm{T}}\mathbf{W}_z\right)\mathbf{m}\right)\right), \tag{3}$$

where $\mathbf{W}_x$ and $\mathbf{W}_z$ denote the first difference operators in the horizontal and vertical directions with rank $\mathbf{R}_x$ and $\mathbf{R}_z$, respectively, and λ the standard deviations of model gradients in the horizontal and vertical directions. For numerical stability we work with the following logarithmic form

$$\log\left(p(\mathbf{m})\right) = -(R_x + R_z)\log\left(2\pi\lambda^2\right) - \frac{1}{2\lambda^2}\left(\mathbf{m}^{\mathrm{T}}\left(\mathbf{W}_x^{\mathrm{T}}\mathbf{W}_x + \mathbf{W}_z^{\mathrm{T}}\mathbf{W}_z\right)\mathbf{m}\right). \tag{4}$$

If we assume the measurement errors to be independent and identically distributed following a normal distribution with mean zero and standard deviation σ, the likelihood function takes the form

$$L(\mathbf{m}|\mathbf{d}) = \frac{1}{(\sqrt{2\pi}\sigma)^N} \exp\left(-\frac{1}{2}\sum_{i=1}^{N}\frac{(f_i(\mathbf{m}) - \mathbf{d}_i)^2}{\sigma^2}\right), \tag{5}$$

where N signifies the total number of crosshole GPR measurements, and i denotes the i-th measurement. The log-likelihood function can then be given by

$$l(\mathbf{m}|\mathbf{d}) = -\frac{N}{2}\log(2\pi) - \frac{N}{2}\log(\sigma^2) - \frac{1}{2}\sigma^{-2}\sum_{i=1}^{N}(f_i(\mathbf{m}) - \mathbf{d}_i)^2. \tag{6}$$

The third term on the right-hand side of Equation (6) measures the distance between simulated and observed GPR data. Thus the value of the log-likelihood function evaluates how well the forward model fits the observed data given a set of model parameters. The Gaussian likelihood function allows for homoscedastic and heteroscedastic measurement errors, and is widely used when the error residuals are normally distributed [2,29,41,42]. When it comes up with non-Gaussian error residual distributions, other forms of likelihood functions need to be constructed. For example the generalized likelihood function (GLF) proposed by Schoups and Vrugt allows for the treatment of nontraditional error residual distributions [43].

In forward calculations the ε_r model is discretized in 2D Cartesian space and each grid value defines a model parameter. This Cartesian parameterization would involve the inference of many thousands of unknowns (2500 in this work), resulting in the inversion being a time-consuming task [44]. We therefore resort to a much more efficient parameterization strategy using the discrete cosine transform (DCT) [30] defined by

$$\mathbf{B}(p,q) = \alpha_p \alpha_q \sum_{p=0}^{P-1} \sum_{q=0}^{Q-1} \mathbf{A}(i,j) \cos\left(\frac{\pi(2i+1)p}{2P}\right) \cos\left(\frac{\pi(2j+1)q}{2Q}\right), \tag{7}$$

where $\mathbf{A}$ and $\mathbf{B}$ are the uniformly discretized model and its DCT-coefficient matrix with P rows and Q columns. The counter i, j and p, q denote the row and column index of $\mathbf{A}$ and $\mathbf{B}$, and coefficients α_p and α_p are given by

$$\alpha_p = \begin{cases} \frac{1}{\sqrt{P}}, & p = 0 \\ \sqrt{\frac{2}{P}}, & 1 \le p \le P-1, \end{cases} \tag{8}$$

and

$$\alpha_q = \begin{cases} \frac{1}{\sqrt{Q}}, & q = 0 \\ \sqrt{\frac{2}{Q}}, & 1 \le q \le Q-1. \end{cases} \tag{9}$$

The DCT approach has the advantage that it concentrates most spatial information of $\mathbf{A}$ into the upper-left corner of $\mathbf{B}$. Thus we can retain only a few lower-order DCT-coefficients without losing significant information. Estimating the retained DCT-coefficients reduces the parameter dimensionality dramatically and improves the computational efficiency.

Once the prior distribution and likelihood function have been defined, the main work left is to derive the posterior distribution of model parameters. As the inverse problem is high-dimensional and non-linear, it is practically very difficult to derive the posterior distribution analytically. We therefore resort to MCMC simulation with the DREAM$_{(ZS)}$ algorithm to generate samples from the posterior target distribution. The basic idea of MCMC simulation is a Markov chain that generates a trail move from the current state $\mathbf{m}_{t-1}$ to a new state $\mathbf{m}_t$. This candidate point is accepted with probability known as the Metropolis ratio [45]

$$p_{\text{acc}}(\mathbf{m}_{t-1} \rightarrow \mathbf{m}_t) = \min\left[1, \frac{p(\mathbf{m}_t)}{p(\mathbf{m}_{t-1})}\right], \tag{10}$$

where $p(\cdot)$ denotes the posterior probability. If the probability of the proposed model, $p(\mathbf{m}_t)$ is greater than that of the current state, $p(\mathbf{m}_{t-1})$, the chain moves to the new state. Otherwise it remains at its current location. After many iterations, samples generated with the Markov chain are distributed to the posterior target distribution. The DREAM$_{(ZS)}$ algorithm, which is an adaptive MCMC algorithm, was designed to accelerate convergence for high-dimensional problems and details of this algorithm can be found in [33,35,42,46].

3. Forward Modeling

The crosshole GPR method uses a transmitting and receiving antenna in two adjacent boreholes and measures the EM properties in-between the two boreholes. The propagation of EM waves through subsurface medium is governed by the Maxwell's equations. For wave propagation in the (x, z) plane, the Maxwell's equations can be written in transverse electric (TE) mode

$$\frac{\partial E_x}{\partial t} = \frac{1}{\varepsilon}\left(\frac{\partial H_y}{\partial z} - \sigma E_x\right) \tag{11}$$

$$\frac{\partial E_z}{\partial t} = \frac{1}{\varepsilon}\left(\frac{\partial H_y}{\partial x} - \sigma E_z\right) \tag{12}$$

$$\frac{\partial H_y}{\partial t} = \frac{1}{\mu}\left(\frac{\partial E_z}{\partial x} - \frac{\partial E_x}{\partial z}\right), \tag{13}$$

where E_x and E_z are the x and z components of the electric field, and H_y is the y component of the magnetic field. ε represents the dielectric permittivity, σ denotes the conductivity, and μ signifies the magnetic permeability.

In most cases Equations (11)–(13) cannot be solved analytically. Alternatively, we can use the FDTD approach that discretizes the partial derivatives of Maxwell's equations in space and time using central differencing to provide numerical solutions. By FDTD modeling of crosshole GPR measurement, full waveform as well as first-arrival traveltime data can be obtained as simulated data in a inverse problem. This method generates GPR data with high precision, yet the FDTD calculation is very time consuming, especially when the model is discretized with fine grid size.

To seek a more efficient forward simulator for crosshole GPR data, we turn our attention to the most widely used straight-ray forward model [11]. This model simplifies the EM wave between the source and receiver to a straight ray, l and calculates the first-arrival traveltime, t through the raypath by

$$t = \int_l s(l)\, dl, \tag{14}$$

where s denotes the slowness along the raypath. Under the low loss condition, s can be derived by $s = \sqrt{\varepsilon_r}/c$, and c is the EM wave velocity in free space. To perform calculations the slowness filed can be discretized into $P \times Q$ grid cells, and Equation (14) can be written as

$$t = \sum_{i=1}^{P \times Q} s_i l_i. \tag{15}$$

To put N measured first-arrival traveltimes in the vector **d** and slowness in **m**, a series of equations can be built in terms of matrix multiplication.

$$\mathbf{d} = \mathbf{Gm}, \tag{16}$$

where **G** is a sparse matrix with N rows and $P \times Q$ columns, also called the forward kernel. Giving a set of model parameters (slownesses), the first-arrival traveltimes can be calculated straightforward using Equation (16).

The straight-ray model presented above has much higher computational efficiency compared with the FDTD approach. However, because it simplifies EM wave propagation into a straight ray that any scattering effects are neglected, the ray approximation may produce considerable modeling error that bias inversion results.

We now use a synthetic example to evaluate the performance of the straight-ray forward model compared with its FDTD counterpart. As illustrated in Figure 1a, we create a 1.0 m $\times$ 1.0 m ε_r field as the reference model, in which a 0.2 m $\times$ 0.2 m square-shaped target is simulated using $\varepsilon_r = 12$, higher than $\varepsilon_r = 9$ for the surrounding medium. To simulate crosshole GPR measurements, the transmitting and receiving antennas are placed on the left and right side of the model, marked with red dots and black crosses, respectively. Multi-offset gathering (MOG) is used to collect data with step length of 0.02 m for both transmitting and receiving antennas. For each position of the transmitting antenna, GPR data are recorded at all receiving antenna locations. This results in 51 $\times$ 51 transmitting-receiving antenna pairs and a total number of 2601 crosshole GPR data.

We first discretize the reference model (Figure 1a) with grid size of 0.005 m $\times$ 0.005 m (fine-grid model) and implement the 2D-FDTD solver (gprMax-2D developed by Giannopoulos [32]) to simulate the crosshole GPR experiment and extract first-arrival traveltimes for the 51 $\times$ 51 transmitting-receiving antenna pairs. Other setups of the FDTD modeling include Ricker source wavelet with central frequency of 500 MHz, time window of 20 ns, and perfectly matched layer (PML) boundary condition on each side of the model. We take the first-arrival traveltimes calculated by the fine-grid FDTD modeling as real data (plotted with red lines in Figure 2, assuming that no modeling error is accounted for in this simulation.

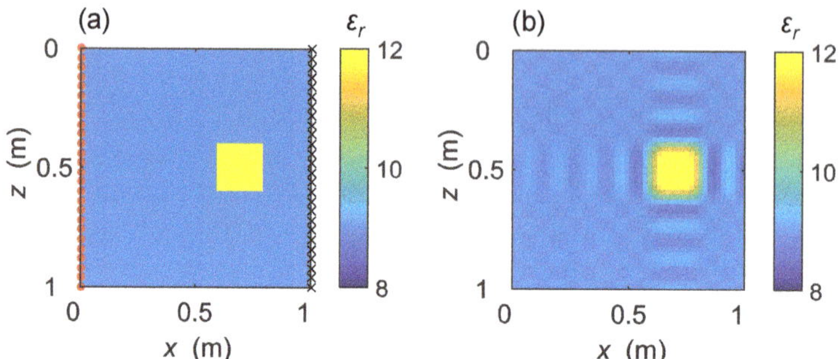

Figure 1. (**a**) A synthetic relative permittivity model, and (**b**) reconstructed model using 256 (16 × 16) discrete cosine transform (DCT)-coefficients. The red dots and black crosses mark the positions of transmitting and receiving ground penetrating radar (GPR) antennas, respectively.

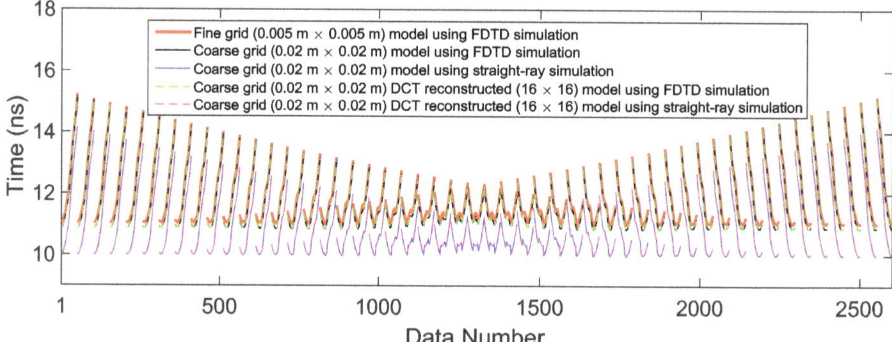

Figure 2. Simulated crosshole GPR first-arrival traveltimes using finite-difference time-domain (FDTD) and straight-ray modeling.

We next investigate the modeling errors of the FDTD and straight-ray models with grid size of 0.02 m × 0.02 m (coarse-grid model), which is practical and computational affordable in Bayesian inversion. The same settings except for grid size are used for the FDTD calculations, and the simulated first-arrival traveltimes are depicted with black lines in Figure 2. Meanwhile, we use the straight-ray forward model to generate first-arrival traveltimes, and show the data in Figure 2 with blue lines. It is obvious that the crosshole GPR data simulated by the FDTD modeling (black lines) are much closer to the real data (red lines) than those (blue lines) calculated by the straight-ray modeling. Thus the straight-ray model suffers from bigger modeling errors compared with the FDTD counterpart.

We also consider the modeling errors caused by the DCT parameterization approach. In order to do so, we create a DCT representation (Figure 1b) of the coarse-grid model (2500 grid cells) using 256 (16 × 16) lower order DCT-coefficients. Here we use the peak signal-to-noise ratio (PSNR) to quantify the quality of parameter reduction, which is a common used tool in image compression [47]. PSNR is defined as

$$
\mathrm{PSNR} = 10 \log \left(\frac{\max{(\mathbf{A})}^2}{\frac{1}{PQ} \sum_{i=1}^{P} \sum_{j=1}^{Q} \left(A(i,j) - B(i,j) \right)^2} \right), \tag{17}
$$

where $\mathbf{A}$ and $\mathbf{B}$ are the ε_r matrix and its DCT realization, respectively. A larger PSNR value stands for the smaller distortion of image reconstruction. For the above case, we use 256 DCT-coefficients to recover the matrix with 2500 grid cells and derive the PSNR value of 35.12, well above the common threshold in image compression, meaning that the reconstructed model preserves the structural details with high fidelity [41].

The green and pink dotted lines in Figure 2 show the first-arrival traveltimes calculated with the DCT reconstructed model using the FDTD and straight-ray solvers, respectively. The lines almost overlap with the data simulated using the full-parameter model, providing us with confidence that the DCT approach can reduce the parameter dimensionality considerably without causing noteworthy modeling errors.

We now summarize the performances of FDTD and straight-ray forward models in Table 1. As a benchmark, the error-free first-arrival traveltimes are created using the FDTD modeling with fine-grid (0.005 m $\times$ 0.005 m) parameterization. With the coarse-grid (0.02 m $\times$ 0.02 m) parameterization, the data simulated by the straight-ray modeling show a more than 7 times larger root mean squared-error (RMSE) than that simulated by the FDTD modeling, indicating that the straight-ray model may cause considerable modeling errors that bias the inversion result. With forward simulations using the DCT reconstructed model, neglectable difference can be observed between the reconstructed and full-parameter model, or, to be precise, the difference is less than 1% for both FDTD and straight-ray models. This is an encouraging result that the use of DCT for dimensionality reduction causes no further modeling errors. We also resort our attention to the computational requirements of the two forward models. For a single forward simulation of the 51 $\times$ 51 first-arrival traveltimes, the FDTD modeling takes about 9.5790 s while the straight-ray modeling uses only 0.0214 seconds, which is roughly 450 times faster for one run (The computing time is tested on a computer with an i7-7700 CPU and 16 GB RAM). Therefore, the straight-ray forward model is computationally efficient thus can help relieve CPU burden of the inversion work, whereas the side effect that it brings relatively high modeling errors to data should be considered and treated carefully, and this will be addressed in the next section.

Table 1. Performances of finite-difference time-domain (FDTD) and straight-ray forward simulators.

Forward Model	Grid Size (m)	Parameterization	Number of Parameters	RMSE (ns)	Computing Time (s)
FDTD	0.005 $\times$ 0.005	Grid value	40,000 (200 $\times$ 200)	0	90.7318
FDTD	0.02 $\times$ 0.02	Grid value	2500 (50 $\times$ 50)	0.1373	9.5790
FDTD	0.02 $\times$ 0.02	Discrete cosine transform (DCT)	256 (16 $\times$ 16)	0.1364	9.5790
Straight-ray	0.02 $\times$ 0.02	Grid value	2500 (50 $\times$ 50)	1.0181	0.0214
Straight-ray	0.02 $\times$ 0.02	DCT	256 (16 $\times$ 16)	1.0172	0.0214

4. Inversion Results

In this section we investigate the use of the straight-ray forward model in our Bayesian inversion framework. We first present the inversion results using the FDTD forward modeling with waveform data, followed by a detailed analysis of the impact of the modeling error and model structure constraint of the the straight-ray model on the inversion results, and conclude with a summary of the model performances.

We take the synthetic relative permittivity model of Figure 1a as the reference model. Measurement data (GPR waveforms and first-arrival traveltimes) are created by the fine-grid (0.005 m $\times$ 0.005 m) FDTD modeling of 51 $\times$ 51 transmitting-receiving antenna pairs, and a total number of 2601 crosshole GPR observations are obtained. The simulated data are contaminated with artificial white noise with standard deviations of 3% of the simulated data, which are 2.06 (in amplitude) for waveform data and 0.24 ns for first-arrival traveltime data. These data serve as our measurement data set, and are used to infer the relative permittivity values in the following inversion cases.

MCMC simulation with the DREAM$_{(ZS)}$ algorithm is employed to explore the posterior distributions of model parameters (relative permittivity values). We use a Jeffreys prior (uniform distribution in the log-transformed space) [48] for the relative permittivity values with the lower and upper bound of 6 and 15, respectively, and implement the likelihood function in the form of Equation (6). To maximize the computational efficiency, we run the DREAM$_{(ZS)}$ algorithm in parallel by evaluating 4 Markov chains on different CPU-cores. We also set the number of crossover values to

20, and scaling factor of the jump rate 75% lower to raise the acceptance rate of proposals, while keep all other settings the default values of the algorithm.

4.1. FDTD Model with Waveform Data

The first inversion starts with the FDTD forward model and waveform data. A coarse-grid of 0.02 m × 0.02 m is used to discrete the model, resulting in a total of $50 \times 50 = 2500$ relative permittivity values that need to be estimated. The DCT approach is used for dimensionality reduction and different number of DCT-coefficients are tested. Figure 3 illustrates the inversion results using 64, 144 and 256 DCT-coefficients. For each case, the reconstructed posterior mean (left column) and maximum a-posterior (MAP) estimations (middle column) of the relative permittivity fields appear visually identical and both pinpoint correctly the higher ε_r area (target). With more DCT-coefficients, the inversion resolution increases as the square-shape and sharp-boundary of the target become more clear. Also, the RMSE values of the posterior models (right column) decrease as the number of the DCT-coefficients increases, which means the larger number of the DCT-coefficients used the better the posterior models fit the data. However, due to the modeling error caused by the coarse-grid FDTD, the posterior RMSE values are greater than the measurement error (2.06 in amplitude) for all three cases.

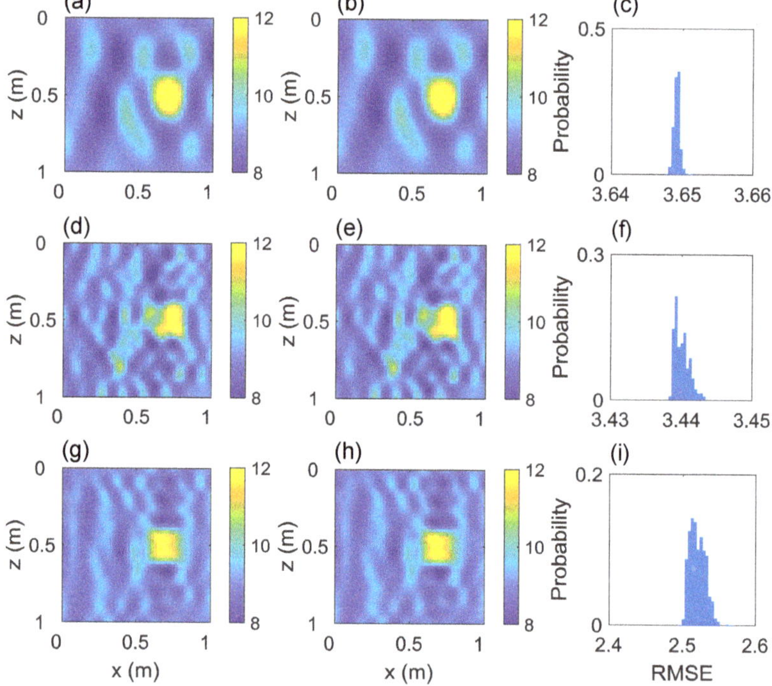

Figure 3. Bayesian inversion using FDTD forward modeling with waveform data. The three rows from top to bottom involve the inversion results using (**a**–**c**) 64, (**d**–**f**) 144, and (**g**–**i**) 256 DCT-coefficients. The left and middle columns display the posterior mean and maximum a-posterior (MAP) density solutions of ε_r field, whereas the right column plots the histograms of the root mean squared-error (RMSE) values of the posterior solutions.

Figure 4 quantifies the impact of the number of DCT-coefficients on the inversion results. By investing more DCT-coefficients, the RMSE values (Figure 4a) calculated using the posterior mean (blue lines with triangle markers) and MAP (red lines with square markers) models decrease from

3.65 to 2.54 that are in good agreement with the plots in Figure 3c,f,i, indicating the improvement of data fit. The PSNR values (Figure 4b) of the posterior mean and MAP models increase from 27.1 to 29.8 with the number of DCT-coefficients growing from 64 to 256, confirming the perspective stated in the visual inspection that better reconstructed ε_r fields are obtained with more DCT-coefficients. Besides, all these PSNR values that are greater than 27 demonstrate that high fidelity of the reconstructed ε_r fields is achieved using the FDTD forward model and waveform data. We also monitor the required number of model evaluations (ME) in Figure 4c for the DREAM$_{(ZS)}$ algorithm to declare convergence and a total of 26,000, 68,000 and 136,000 model evaluations are needed for the three cases. An approximately linearly relation can be observed between the computational requirements and the number of DCT-coefficients.

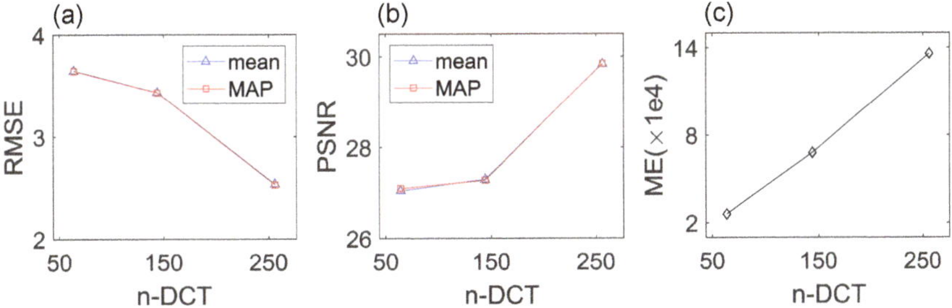

Figure 4. Metrics of the Bayesian FDTD waveform inversion results as a function of the number of discrete cosine transform (DCT)-coefficients (n-DCT): (**a**) Root mean squared-error (RMSE), and (**b**) PSNR of the posterior mean and maximum a-posterior (MAP) model simulations, and (**c**) required number of FDTD model evaluations (ME) to reach convergence.

4.2. Straight-Ray Model without Modeling Error Corrected

We now evaluate the use of the uncorrected straight-ray model in Bayesian inversion of crosshole GPR first-arrival traveltime data with the same settings of model discretization and dimensionality reduction as those of the FDTD model. The inversion results of using 64, 144 and 256 DCT-coefficients are considered and displayed in the top (a–d), middle (e–h) and bottom (i–l) rows in Figure 5. It can be seen from the posterior mean and MAP realizations (the first and second columns) that the use of more DCT-coefficients deteriorates the inversion results. For the latter two cases using 144 and 256 DCT-coefficients, the posterior realizations exhibit significant variations that the target can no longer be resolved correctly. We plot in the third column the simulated first-arrival traveltimes of posterior models and the measurement data, and find the posterior models of the three cases fit the measurement data equally well. Note that although the RMSE values of the posterior solutions (the right column) are distributed very close to the measurement error (0.24 ns), the correct model parameters are not found. That is, without modeling error accounted for, there exists a set of incorrect parameters that can better match the measurement data than the true parameters do.

For a closer investigation of the inversion results, we see in Figure 6a,b that the RMSE values are well around the measurement error, whereas the PSNR values are much smaller than those of the FDTD model. These PSNR values are under the common threshold that the reconstructed ε_r fields hardly resemble the reference model. With more DCT-coefficients, the RMSE values of the posterior mean and MAP models increase while the PSNR values decrease, which means the use of more DCT-coefficients cannot improve the inversion results. Also, the required number of model evaluations grows with the number of DCT-coefficients. Compared with the FDTD model, ten times more model evaluations are needed when using the uncorrected straight-ray model. Therefore, the modeling error caused by the straight-ray model is remarkable and biases the inversion results considerably. Without modeling error corrected, even if the measurement data can be fit nicely, the inverted ε_r fields differ greatly from the true model.

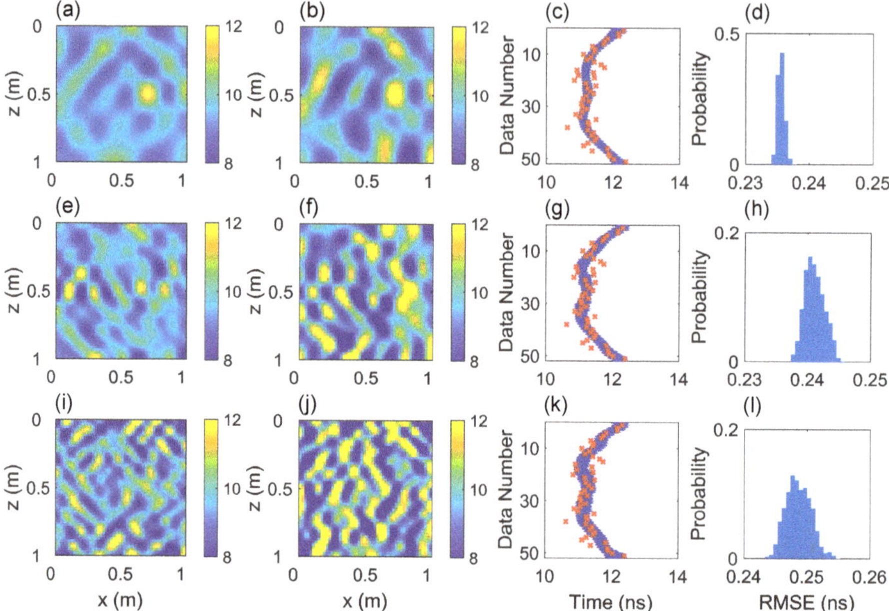

Figure 5. Bayesian inversion using straight-ray forward modeling without modeling error corrected. The three rows from top to bottom involve the inversion results using (**a–d**) 64, (**e–h**) 144, and (**i–l**) 256 DCT-coefficients. The first and second columns display the posterior mean and MAP density solutions of ε_r field. The third column plots the measurement and posterior simulated first-arrival traveltimes with the transmitting antenna located at 0.5 m depth at the left side of the model, and 51 receiving antenna positions distributed equally from 0 to 1 m depth at the right side. The measurement and simulated data are marked with red crosses ($\times$) and blue stars ($*$), respectively. The right column depicts the histograms of the RMSE values of the posterior solutions.

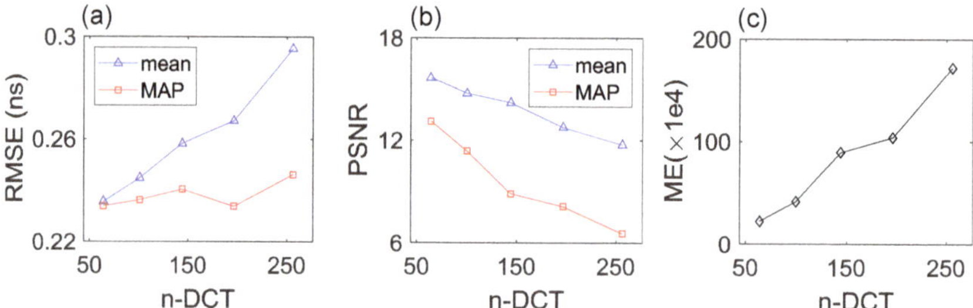

Figure 6. Metrics of the Bayesian inversion results as a function of the number of DCT-coefficients. The straight-ray forward model is used without modeling error corrected. (**a**) RMSE, and (**b**) peak signal-to-noise ratio (PSNR) of the posterior mean and MAP model simulations, and (**c**) required number of model evaluations to reach convergence.

4.3. Straight-Ray Model with Modeling Error Corrected

To take into consideration the modeling error of the straight-ray simulator, we adopt the idea in [38] that we first calculate an orthonormal basis for the modeling error using the fine-grid FDTD and coarse-grid straight-ray solvers prior to the inversion work. Then in MCMC, we use this basis to calculate the modeling error and subtract it from the residual the between simulated and measurement data before calculating the likelihood function.In this way the modeling error is isolated outside the Bayesian inversion and no longer biases the inversion results.

Figure 7 plots the inversion results using the straight-ray forward model with modeling error corrected. From the posterior mean and MAP realizations (the first and second columns) we can see improvements compared with the previous cases using the uncorrected forward model that the target can be identified in all three cases with 64, 144 and 256 DCT-coefficients. But like the uncorrected model example, with more DCT-coefficients, the reconstructed posterior mean and MAP models suffer from larger spatial variation that worsens the inversion results. It can be also seen that after modeling error is corrected, the simulated data do not closely fit the observations (the third column) and the RMSE values of the posterior solutions are much larger than the measurement error. This is because the simulated data incorporate the measurement and modeling errors. With the modeling error accounted for, the residuals between the simulated and measurement data need not to be minimized.

Figure 8 reveals similar patterns to those found in the uncorrected model example. The RMSE values increase and the PSNR values decrease with the number of DCT-coefficients, indicating that the use of more DCT-coefficients deteriorates the inversion results. Yet as a whole, the PSNR values derived from corrected models (Figure 8b) are much larger than those of uncorrected models (Figure 6b), demonstrating that the modeling error correction improves the inversion results considerably. Besides, after modeling error correction, only less than one quarter of model evaluations are needed for the parameters to reach convergence (Figure 8c), which improves the computational efficiency obviously.

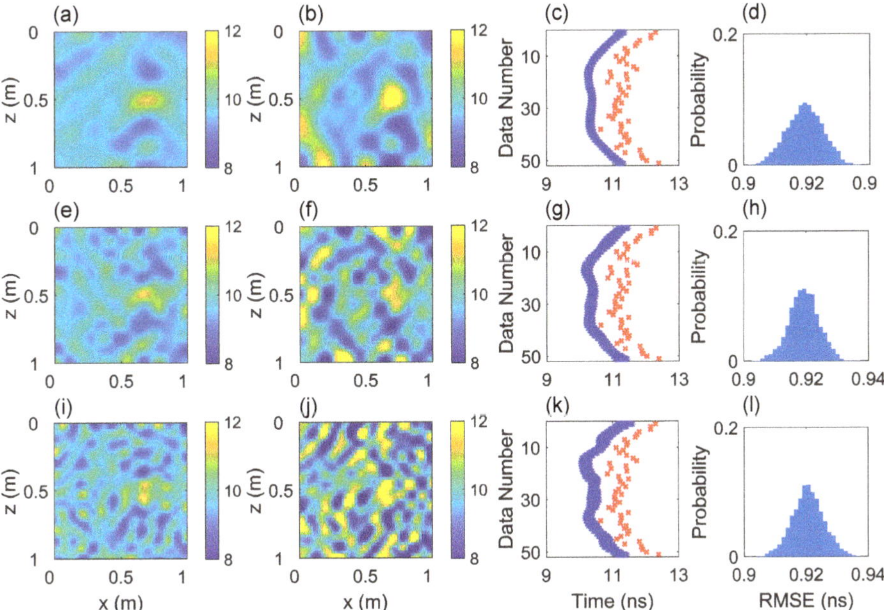

Figure 7. Bayesian inversion using straight-ray forward modeling with modeling error corrected. The three rows from top to bottom involve the inversion results using (**a–d**) 64, (**e–h**) 144, and (**i–l**) 256 DCT-coefficients. The first and second columns display the posterior mean and MAP density solutions of ε_r field. The third column plots the measurement and posterior simulated first-arrival traveltimes with the transmitting antenna located at 0.5 m depth at the left side of the model, and 51 receiving antenna positions distributed equally from 0 to 1 m depth at the right side. The measurement and simulated data are marked with red crosses ($\times$) and blue stars ($*$), respectively. The right column depicts the histograms of the RMSE values of the posterior solutions.

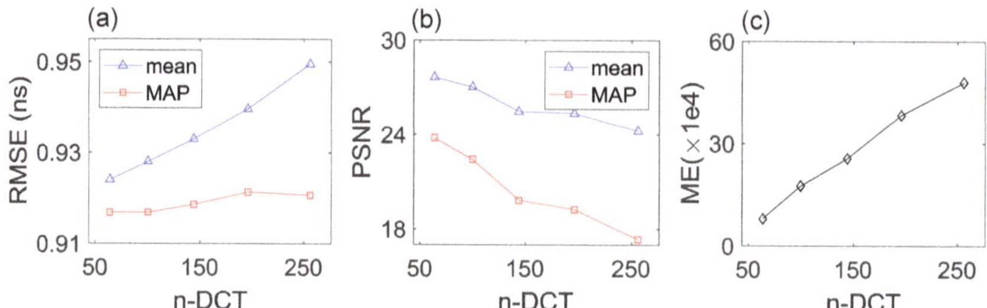

Figure 8. Metrics of the Bayesian inversion results as a function of the number of DCT-coefficients. The straight-ray forward model is used and modeling error is corrected. (**a**) RMSE, and (**b**) peak signal-to-noise ratio (PSNR) of the posterior mean and MAP model simulations, and (**c**) required number of model evaluations to reach convergence.

We now turn our attention to the impact of the number of DCT-coefficients. In the FDTD model example, the use of more DCT-coefficients improves the inversion results, whereas in the straight-ray model examples (both corrected and uncorrected), to increase the number of DCT-coefficients makes the inversion results even worse. We know that the DCT approach reduces the model parameters to a few DCT-coefficients. The more DCT-coefficients we use the higher the inversion resolution we can achieve, meanwhile the more unknowns we need to infer from data. Consider the FDTD model and the corrected straight-ray model, assuming that they both have no modeling errors, then the difference between the two examples is the data they use. In this work the FDTD model works with waveform data, while the straight-ray model works with first-arrival traveltime data. Although the 51×51 antenna pairs generate the same amount of waveform traces and first-arrival traveltimes, the number of data used for the FDTD inversion is far more than that for the straight-ray inversion as each trace contains hundreds of data points. In this work each trace contains 212 samples and a total of 551,412 data points are used for the FDTD inversion, yet for the straight-ray inversion the number of measurement first-arrival traveltimes is only 2601. Therefore, with the same amount of crosshole GPR measurements, the waveform data are much more informative to estimate more DCT-coefficients for a better description of the model structure, whereas the first-arrival traveltimes are insufficient and only a limited number of DCT-coefficients can be estimated correctly.

The choice of the number of DCT-coefficients is a trade-off between spatial resolution and uncertainty in the inversion results. In previous work we have introduced a method to determine the number of DCT-coefficients by examining the similarity between the reference model and its reduced order DCT representations [28]. If some structural features of the subsurface can be used as prior information, a set of realizations can be generated from the training image (TI) and the DCT-coefficients with the occurrence probabilities larger than the threshold are considered as unknown model parameters in the inversion [49]. One should be also aware that when using the first-arrival traveltimes as measurement data, the amount of data is always insufficient to infer a large number of unknown model parameters, thus a relatively small number of DCT-coefficients should be used for inversion.

4.4. Model Constraint

As a limited number of the first-arrival traveltime data is incapable of providing enough knowledge about the relatively large number of DCT-coefficients, while to increase considerably the amount of data is practically very expensive, we here learn the experience of deterministic inversion that to use a regularization term to decrease the ill-posedness of the inversion problem. In the view of the probabilistic inversion, the regularization term serves as the prior information that refines the range of model parameters [40]. In this work, we impose a smooth constraint on the model structure by including Equation (4) in the prior distribution. The regularization weight, λ scales how much weight is assigned to the regularization term, and here we simply use the optimal value ($\lambda_{\text{opt}} = 0.0984$) calculated by maximizing the regularization term given the true model parameter values. Note that λ can also be estimated together with model parameters in Bayesian inversion.

The inversion results using the corrected straight-ray forward model and constrained model structure are shown in Figure 9. For all three cases using 64, 144 and 256 DCT-coefficients, the higher permittivity area is clearly depicted in the posterior mean and MAP realizations, and they appear no significant difference between each other. With modeling error accounted for, the simulated first-arrival traveltimes no longer closely fit the measurement data and the RMSE values of the posterior solutions are much larger than the measurement error. Note that with model constraint, the variation of simulated data becomes more modest than that of the previous example (Figure 7). This reflects a smooth variation of the model structure when the model constraint is applied in the inversion.

As depicted in Figure 10, the RMSE and PSNR values almost remain unchanged with the number of DCT-coefficients. This is because the regularization term reduces the model structure complexity and stabilizes the solution of the inverse problem. It can be also seen in Figure 10b that the PSNR values of the cases using different number of DCT-coefficients stay around 27.8, indicating relatively high quality of the reconstructed models. The model constraint also decreases the required number of model evaluations (Figure 10c) for the parameters to reach convergence to 60% of those using an unconstrained model (Figure 8c).

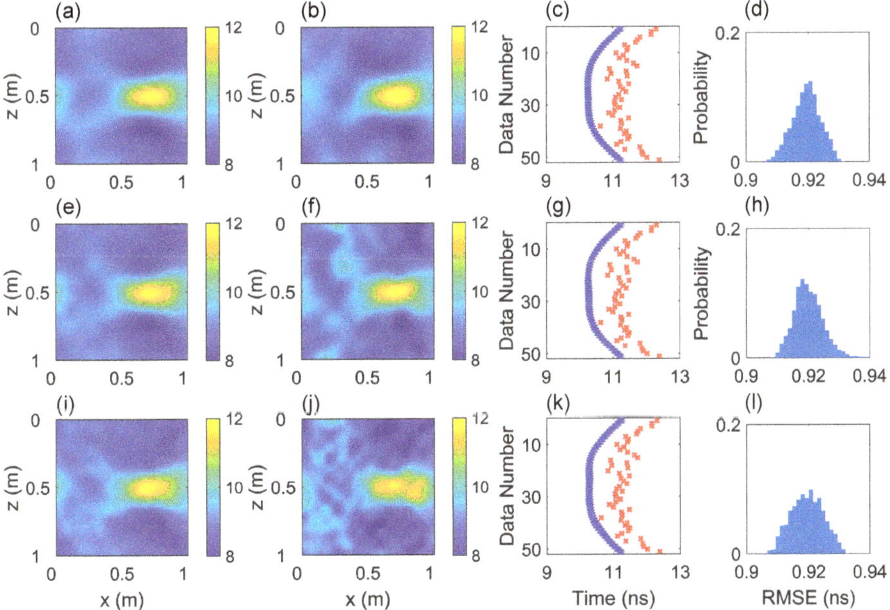

Figure 9. Bayesian inversion using straight-ray forward model with modeling error corrected and smooth model structure constrained. The three rows from top to bottom involve the inversion results using (**a–d**) 64, (**e–h**) 144, and (**i–l**) 256 DCT-coefficients. The first and second columns display the posterior mean and MAP density solutions of ε_r field. The third column plots the measurement and posterior simulated first-arrival traveltimes with the transmitting antenna located at 0.5 m depth at the left side of the model, and 51 receiving antenna positions distributed equally from 0 to 1 m depth at the right side. The measurement and simulated data are marked with red crosses ($\times$) and blue stars ($\ast$), respectively. The right column depicts the histograms of the RMSE values of the posterior solutions.

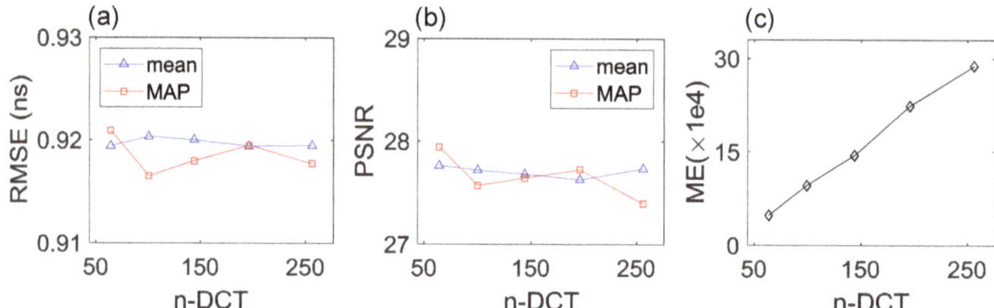

Figure 10. Metrics of the Bayesian inversion results as a function of the number of DCT-coefficients. The straight-ray forward model is used with modeling error corrected and smooth model structure constrained. (**a**) RMSE, and (**b**) PSNR of the posterior mean and MAP model simulations, and (**c**) required number of model evaluations to reach convergence.

We conclude our numerical examples in Figure 11 with a bar chart that summarizes the performance of different forward models in Bayesian inversion of crosshole GPR data. We use the PSNR values (blue bars) of posterior mean models to evaluate the inversion accuracy, and the number of model evaluations (orange bars) and computing time (gray bars) to measure the computational efficiency. The FDTD model with waveform data has the merit of the highest inversion accuracy and requires the least number of model evaluations among the examples, yet suffers from significantly long computing time (90.47 h in this study) due to huge CPU demand of the FDTD modeling. The straight-ray forward model, on the other hand, has the advantage of higher computational efficiency. Although the required number of model evaluations are much larger than that of the FDTD model, the inversion work takes far less computing time. However, due to modeling error, the straight-ray model cannot reconstruct the ε_r field correctly and leads to a low PSNR value (11.76). By correcting the modeling error and constraining the model structure, the PSNR value is increased to 27.74 and the computing time is reduced to 0.43 h, which improves the inversion accuracy and computational efficiency considerably.

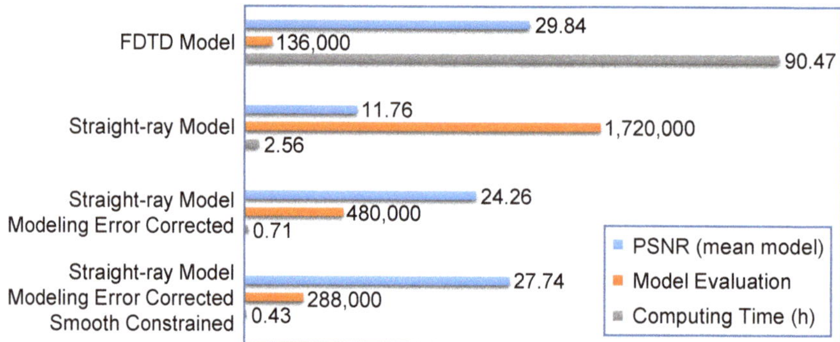

Figure 11. Bar chart that summaries the goodness of reconstruction and computational efficiency of different forward models in Bayesian inversion. 256 DCT-coefficients are used for inversion and the PSNR values are calculated using the posterior mean models.

5. Conclusions

In Bayesian inversion of crosshole GPR data, the use of FDTD forward model with waveform data can generate posterior realizations with high accuracy, yet thousands to millions of model evaluations required by MCMC iterations are always computationally infeasible. We therefore in this paper turned our attention to a computational efficient straight-ray forward model and evaluated the applicability in the Bayesian inversion framework.

Based on a reference relative permittivity model, we calculated the first-arrival traveltime data using the FDTD and straight-ray solvers, respectively. With the same model grid size and measurement setup, the straight-ray simulator ran 450 times faster than the FDTD simulator, whereas suffered from more than 7 times larger modeling error. The DCT approach for dimensionality reduction was also tested in the forward simulations. We found that the dimensionality reduction using DCT contributed less than 1% to the modeling error.

We then performed Bayesian inversion of crosshole GPR first-arrival traveltime data with the straight-ray forward model using a set of synthetic examples. First the modeling error was disregarded and the straight-ray solver was used directly for forward calculations. In this case the measurement data were nicely fitted, yet the posterior realizations were not able to resemble the main features of the true model, and using more DCT-coefficients led to more unwanted artifacts in the posterior realizations. With modeling error accounted for, although the simulated first-arrival traveltimes no longer fit closely the measurement data, there were less unwanted features in the posterior realizations and the main features of the true model can be identified correctly. However, again, the inversion results became worse with an increased number of DCT-coefficients. Finally, we imposed a smooth constraint on the model structure and improved the inversion results considerably. No apparent unwanted features were observed in the posterior realizations and the number of DCT-coefficients had no significant impact on the inversion results. By investigating the computing time and PSNR values of the synthetic examples, we conclude that the use of the straight-ray forward model reduces the computational burden remarkably, meanwhile, with the modeling error corrected and model structure constrained, the inversion results are approximate to those of the FDTD model.

It should also be noted that the applicability of our inversion method is based on the evaluation of the synthetic example involving a moderate degree of nonlinearity. When the inverse problem is of a higher degree of nonlinearity, such as higher relative permittivity differences, larger object dimensions or in the presence of multiple objects, the straight-ray model we use, which is a linear model, might bias the inversion results. Future work will incorporate bending-ray based or artificial neural network (ANN) based forward kernels in our inversion framework to improve the capability of dealing with highly nonlinear inverse problems.

Author Contributions: H.Q., X.X. and Y.T. conceived the main idea of this work; H.Q. prepared the manuscript and performed simulations; Y.T. analyzed data; X.X. and Y.T. reviewed and revised the paper.

Funding: This research was funded by the Open Fund of Key Laboratory of Geotechnical and Underground Engineering of Ministry of Education at Tongji University (Grant No. KLE-TJGE-B1804), and the Fundamental Research Funds for the Central Universities (Grant No. DUT19JC23).

Acknowledgments: The authors would like to thank Professor Jasper A. Vrugt for providing the DREAM$_{(ZS)}$ package.

References

1. Huisman, J.A.; Hubbard, S.S.; Redman, J.D.; Annan, A.P. Measuring soil water content with ground penetrating radar: A review. *Vadose Zone J.* **2003**, *2*, 476–491. [CrossRef]
2. Linde, N.; Vrugt, J.A. Distributed soil moisture from crosshole ground-penetrating radar travel times using stochastic inversion. *Vadose Zone J.* **2013**, *12*, 1–16. [CrossRef]
3. Irving, J.; Singha, K. Stochastic inversion of tracer test and electrical geophysical data to estimate hydraulic conductivities. *Water Resour. Res.* **2010**, *46*, W11514. [CrossRef]
4. Yang, X.; Klotzsche, A.; Meles, G.; Vereecken, H.; van der Kruk, J. Improvements in crosshole GPR full-waveform inversion and application on data measured at the Boise Hydrogeophysics Research Site. *J. Appl. Geophys.* **2013**, *99*, 114–124. [CrossRef]
5. Dorn, C.; Linde, N.; Doetsch, J.; Le Borgne, T.; Bour, O. Fracture imaging within a granitic rock aquifer using multiple-offset single-hole and cross-hole GPR reflection data. *J. Appl. Geophys.* **2012**, *78*, 123–132. [CrossRef]

6.   Keskinen, J.; Klotzsche, A.; Looms, M.C.; Moreau, J.; van der Kruk, J.; Holliger, K.; Stemmerik, L.; Nielsen, L. Full-waveforminversion of crosshole GPR data: Implications for porosity estimation in chalk. *J. Appl. Geophys.* **2017**, *140*, 102–116. [CrossRef]

7.   Liu, S.; Liu, X.; Meng, X.; Fu, L.; Lu, Q.; Deng, L. Application of time-domain full waveform inversion to cross-hole radar data measured at Xiuyan jade mine, China. *Sensors* **2018**, *18*, 3114. [CrossRef]

8.   Hu, S.F.; Zhao, Y.H.; Rao, C.F.; Qin, T.; An, C.; Ge, S.C. GPR tomography based on regularization method for concrete defect detection. In Proceedings of the 16th International Conference on Ground Penetrating Radar (GPR), Hong Kong, China, 13–16 June 2016; pp. 1–6. [CrossRef]

9.   Qin, H.; Xie, X.; Tang, Y.; Wang, Z. Detection of diaphragm wall defects using crosshole GPR. In Proceedings of the 17th International Conference on Ground Penetrating Radar, Rapperswil, Switzerland, 18–21 June 2018; pp. 1–4. [CrossRef]

10.   Lai, W.W.L.; Dérobert, X.; Annan, P. A review of ground penetrating radar application in civil engineering: A 30-year journey from locating and testing to imaging and diagnosis. *NDT E Int.* **2018**, *96*, 58–78. [CrossRef]

11.   Dines, K.A.; Lytle, R.J. Computerized geophysical tomography. *Proc. IEEE* **1979**, *67*, 1065–1073. [CrossRef]

12.   Witten, A.J.; Molyneux, J.E.; Nyquist, J.E. Ground penetrating radar tomography: Algorithms and case studies. *IEEE Trans. Geosci. Remote Sens.* **1994**, *32*, 461–467. [CrossRef]

13.   Holliger, K.; Musil, M.; Maurer, H.R. Ray-based amplitude tomography for crosshole georadar data: A numerical assessment. *J. Appl. Geophys.* **2001**, *47*, 285–298. [CrossRef]

14.   Hanafy, S.; al Hagrey, S.A. Ground-penetrating radar tomography for soil-moisture heterogeneity. *Geophysics* **2005**, *71*, K9–K18. [CrossRef]

15.   Giroux, B.; Gloaguen, E.; Chouteau, M. bh_tomo—A Matlab borehole georadar 2D tomography package. *Comput. Geosci.* **2007**, *33*, 126–137. [CrossRef]

16.   Balkaya, C.; Akcıg, Z.; Gokturkler, G. A comparison of two travel-time tomography schemes for crosshole radar data: Eikonal-equation-based inversion versus ray-based inversion. *J. Environ. Eng. Geophys.* **2010**, *15*, 203–218. [CrossRef]

17.   Williamson, P.R.; Worthington, M.H. Resolution limits in ray tomography due to wave behavior: Numerical experiments. *Geophysics* **1993**, *58*, 727–735. [CrossRef]

18.   Chang, P.Y.; Alumbaugh, D. An analysis of the cross-borehole GPR tomography for imaging the development of the infiltrated fluid plume. *J. Geophys. Eng.* **2011**, *8*, 294–307. [CrossRef]

19.   Ernst, J.R.; Maurer, H.; Green, A.G.; Holliger, K. Full-waveform inversion of crosshole radar data based on 2-D finite-difference time-domain solutions of Maxwell's equations. *IEEE Trans. Geosci. Remote Sens.* **2007**, *45*, 2807–2828. [CrossRef]

20.   Ernst, J.R.; Green, A.G.; Maurer, H.; Holliger, K. Application of a new 2D time-domain full-waveform inversion scheme to crosshole radar data. *Geophysics* **2007**, *72*, J53–J64. [CrossRef]

21.   Meng, X.; Liu, S.X. Source-independent time-domain waveform inversion of cross-hole GPR data. In Proceedings of the 16th International Conference on Ground Penetrating Radar (GPR), Hong Kong, China, 13–16 June 2016; pp. 1–4. [CrossRef]

22.   Van der Kruk, J.; Liu, T.; Mozaffari, A.; Gueting, N.; Klotzsche, A.; Vereecken, H.; Warren, C.; Giannopoulos, A. GPR full-waveform inversion, recent developments, and future opportunities. In Proceedings of the 17th International Conference on Ground Penetrating Radar (GPR), Rapperswil, Switzerland, 18–21 June 2018; pp. 1–6. [CrossRef]

23.   Hansen, T.M.; Cordua, K.S. Efficient Monte Carlo sampling of inverse problems using a neural network-based forward-applied to GPR crosshole traveltime inversion. *Geophys. J. Int.* **2017**, *211*, 1524–1533. [CrossRef]

24.   Bikowski, J.; Huisman, J.A.; Vrugt, J.A.; Vereecken, H.; van der Kruk, J. Integrated analysis of waveguide dispersed GPR pulses using deterministic and Bayesian inversion methods. *Near Surf. Geophys.* **2012**, *10*, 641–652. [CrossRef]

25.   Scholer, M.; Irving, J.; Looms, M.C.; Nielsen, L.; Holliger, K. Bayesian Markov-chain-Monte-Carlo inversion of time-lapse crosshole GPR data to characterize the vadose zone at the Arrenaes site, Denmark. *Vadose Zone J.* **2012**, *11*, 1–19. [CrossRef]

26.   Dafflon, B.; Barrash, W. Three-dimensional stochastic estimation of porosity distribution: Benefits of using ground-penetrating radar velocity tomograms in simulated-annealing-based or Bayesian sequential simulation approaches. *Water Resour. Res.* **2012**, *48*, 1–13. [CrossRef]

27. Hunziker, J.; Laloy, E.; Linde, N. Inference of multi-Gaussian relative permittivity fields by probabilistic inversion of crosshole ground-penetrating radar data. *Geophysics* **2017**, *82*, H25–H40. [CrossRef]

28. Qin, H.; Vrugt, J.A.; Xie, X.; Zhou, Y. Improved characterization of underground structure defects from two-stage Bayesian inversion using crosshole GPR data. *Autom. Constr.* **2018**, *95*, 233–244. [CrossRef]

29. Qin, H.; Xie, X.; Vrugt, J.A.; Zeng, K.; Hong, G. Underground structure defect detection and reconstruction using crosshole GPR and Bayesian waveform inversion. *Autom. Constr.* **2016**, *68*, 156–169. [CrossRef]

30. Ahmed, N.; Natarajan, T.; Rao, K.R. Discrete cosine transform. *IEEE Trans. Comput.* **1974**, *100*, 90–93. [CrossRef]

31. Yee, K.S. Numerical solution of initial boundary value problems involving Maxwell's equations in isotropic media. *IEEE Trans. Antennas Propag.* **1966**, *AP-14*, 302–307. [CrossRef]

32. Giannopoulos, A. Modelling ground penetrating radar by gprmax. *Constr. Build. Mater.* **2005**, *19*, 755–762. [CrossRef]

33. ter Braak, C.J.F.; Vrugt, J.A. Differential evolution Markov chain with snooker updater and fewer chains. *Stat. Comput.* **2008**, *18*, 435–446. [CrossRef]

34. Vrugt, J.A.; Stauffer, P.H.; Wöhling, T.; Robinson, B.A.; Vesselinov, V.V. Inverse modeling of subsurface flow and transport properties: A review with new developments. *Vadose Zone J.* **2008**, *7*, 843–864. [CrossRef]

35. Vrugt, J.A.; ter Braak, C.J.F.; Diks, C.G.H.; Robinson, B.A.; Hyman, J.M.; Higdon, D. Accelerating Markov chain Monte Carlo simulation by differential evolution with self-adaptive randomized subspace sampling. *Int. J. Nonlinear Sci. Numer. Simul.* **2009**, *10*, 271–288. [CrossRef]

36. Hansen, T.M.; Cordua, K.S.; Jacobsen, B.H.; Mosegaard, K. Accounting for imperfect forward modeling in geophysical inverse problems Exemplified for crosshole tomography. *Geophysics* **2014**, *79*, H1–H21. [CrossRef]

37. Köpke, C.; Irving, J.; Elsheikh, A.H. Accounting for model error in Bayesian solutions to hydrogeophysical inverse problems using a local basis approach. *Adv. Water Resour.* **2018**, *116*, 195–207. [CrossRef]

38. Köpke, C.; Irving, J.; Roubinet, D. Stochastic inversion for soil hydraulic parameters in the presence of model error: An example involving ground-penetrating radar monitoring of infiltration. *J. Hydrol.* **2019**, *569*, 829–843. [CrossRef]

39. deGroot Hdlin, C.; Constable, S. Occam's inversion to generate smooth, two-dimensional models from magnetotelluric data. *Geophysics* **1990**, *55*, 1613–1624. [CrossRef]

40. Rosas-Carbajal, M.; Linde, N.; Kalscheuer, T.; Vrugt, J.A. Two-dimensional probabilistic inversion of plane-wave electromagnetic data: methodology, model constraints and joint inversion with electrical resistivity data. *Geophys. J. Int.* **2014**, *193*, 1508–1524. [CrossRef]

41. Lochbühler, T.; Vrugt, J.A.; Sadegh, M.; Linde, N. Summary statistics from training images as prior information in probabilistic inversion. *Geophys. J. Int.* **2015**, *201*, 157–171. [CrossRef]

42. Vrugt, J.A. Markov chain Monte Carlo simulation using the DREAM software package: Theory, concepts, and MATLAB implementation. *Environ. Model. Softw.* **2015**, *75*, 273–316. [CrossRef]

43. Schoups, G.; Vrugt, J.A. A formal likelihood function for parameter and predictive inference of hydrologic models with correlated, heteroscedastic, and non-Gaussian errors. *Water Resour. Res.* **2010**, *46*, W10531. [CrossRef]

44. Belhadj, J.; Romary, T.; Gesret, A.; Noble, M.; Figliuzzi, B. New parameterizations for Bayesian seismic tomography. *Inverse Probl.* **2018**, *34*, 065007. [CrossRef]

45. Metropolis, N.; Rosenbluth, A.W.; Rosenbluth, M.N.; Teller, A.H.; Teller, E. Equation of state calculations by fast computing machines. *J. Chem. Phys.* **1953**, *21*, 1087–1092. [CrossRef]

46. Shockley, E.M.; Vrugt, J.A.; Lopez, C.F. PyDREAM: High-dimensional parameter inference for biological models in python. *Bioinformatics* **2018**, *34*, 695–697. [CrossRef] [PubMed]

47. Huynh-Thu, Q.; Ghanbari, M. Scope of validity of PSNR in image/video quality assessment. *Electron. Lett.* **2008**, *44*, 800–801. [CrossRef]

48. Jeffreys, H. An invariant form for the prior probability in estimation problems. *Proc. R. Soc. Lond. Ser. A Math. Phys. Sci.* **1946**, *186*, 453–461.

49. Moghadas, D. Probabilistic inversion of multiconfiguration electromagnetic induction data using dimensionality reduction technique: A numerical study. *Vadose Zone J.* **2019**, *18*, 1–16. [CrossRef]

Article

Multipath Ghost Suppression Based on Generative Adversarial Nets in Through-Wall Radar Imaging

Yong Jia [1,*], Ruiyuan Song [1], Shengyi Chen [1], Gang Wang [1], Yong Guo [1], Xiaoling Zhong [1] and Guolong Cui [2]

[1] College of Information Science & Technology, Chengdu University of Technology, Chengdu 610059, China; songruiyuan915@gmail.com (R.S.); cdut_chenshengyi@outlook.com (S.C.); wangganglim@gmail.com (G.W.); guoy@cdut.edu.cn (Y.G.); zhongxiaoling@cdut.cn (X.Z.)

[2] School of Information and Communication Engineering, University of Electronic Science and Technology of China, Chengdu 611731, China; cuiguolong@uestc.edu.cn

* Correspondence: jiayong2014@cdut.edu.cn; Tel.: +86-189-0822-7416

Received: 21 April 2019; Accepted: 31 May 2019; Published: 3 June 2019

Abstract: In this paper, we propose an approach that uses generative adversarial nets (GAN) to eliminate multipath ghosts with respect to through-wall radar imaging (TWRI). The applied GAN is composed of two adversarial networks, namely generator G and discriminator D. Generator G learns the spatial characteristics of an input radar image to construct a mapping from an input to output image with suppressed ghosts. Discriminator D evaluates the difference (namely, the residual multipath ghosts) between the output image and the ground-truth image without multipath ghosts. On the one hand, by training G, the image difference is gradually diminished. In other words, multipath ghosts are increasingly suppressed in the output image of G. On the other hand, D is trained to improve in evaluating the diminishing difference accompanied with multipath ghosts as much as possible. These two networks, G and D, fight with each other until G eliminates the multipath ghosts. The simulation results demonstrate that GAN can effectively eliminate multipath ghosts in TWRI. A comparison of different methods demonstrates the superiority of the proposed method, such as the exemption of prior wall information, no target images with degradation, and robustness for different scenes.

Keywords: generative adversarial nets; through-wall radar imaging; multipath ghost suppression; generator and discriminator

---

1. Introduction

For through-wall radar imaging (TWRI), the presence of furniture and walls, floors, and ceilings makes electromagnetic waves have strong reflections between the targets and them, which brings multipath returns to the received radar signal. Based on imaging algorithms, such as the back-projection algorithm [1–3], target-like images called multipath ghosts are produced at nontarget locations, which makes the performance of detection and recognition significantly worse.

To solve this problem, a group of methods was designed via the multipath model based on prior information about the walls' locations and antennas. Specifically, in References [4,5], first-order multipath ghosts were mapped back to the positions of associated targets, while target images that overlapped with multipath ghosts were mistakenly removed from true positions. To preserve the overlapped target images, multipath echoes were removed form the raw radar data in Reference [6]. In addition, in Reference [7], multiple estimated images gained by two different kinds of imaging dictionaries were fused to obtain an image without multipath ghosts.

Nevertheless, prior information of accurate walls' locations is difficult to gain in an actual detection scene. To achieve multipath-ghost suppression without walls' locations, the aspect-dependence (AD)

feature of multipath ghosts is utilized to develop suppression algorithms. In Reference [8], two aspects of subaperture images were multiplied to an image without multipath ghosts. However, this method has a poor performance in suppressing multipath ghosts of the back wall, as they appear close together in both subaperture images. In Reference [9], multiple images with different array rotation angles were fused to yield an image without multipath ghosts. However, the two methods based on the AD feature both needed complicated parameter deployment. In other words, the subaperture method should find suitable subapertures, and the array rotating method should find appropriate rotating angles.

Depending on the ratio of coherent power to the pixel with the incoherent power, a coherence factor, a phase coherence factor (PCF), and a sign coherence factor were designed to weigh images for suppressing multipath ghosts [10–12]. These methods have poor suppression performances for well-focused multipath ghosts in the case of synthetic aperture imaging. Moreover, the methods based on these coherence factors and the aforementioned AD feature enlarge the energy differences between target images, which makes it difficult to identify degraded targets with a low signal-to-multipath-clutter ratio (SMCR).

Considering that generative adversarial nets (GAN) [13,14] is classified as a structured learning network that is applied to construct spatial-structure mapping from input images to output images and multipath-ghost suppression is a typical process of spatial-structure mapping, in this paper, GAN, including a generator G and a discriminator D, is introduced to suppress multipath ghosts in through-wall radar imaging. With regard to an input radar image with multipath ghosts, generator G exploits spatial characteristics to generate an output image with reduced multipath ghosts and adversarial discriminator D recognizes the difference between the output image and the ground truth image. The recognized difference is sent to G to improve the generative ability. Through training, G and D alternate and recur until the end, G generates a desired image without multipath ghosts, and D loses effectiveness. The simulation results verify the feasibility of the proposed method. The comparison of different methods demonstrates the superiorities of the proposed method, which are that the proposed method

- has robustness in finishing multipath-ghost suppression without accurate walls' locations;
- preserves the target images even if they are overlapped with multipath ghosts;
- finishes multipath ghost suppression without the use of complicated tuning parameters in different detection scenes; and
- prevents the energy difference of target images from enlarging, which is beneficial in identifying all targets.

The remainder of this paper is organized as follows. Section 2 briefly describes the first-order multipath model. Section 3 analyzes the details of the generative adversarial model. Section 4 indicates the detailed structure of the proposed networks. Simulations on different datasets are presented in Section 5. Section 6 concludes this paper.

2. Multipath Model

Assume that a single-channel radar is monitoring an enclosed room with four separate homogeneous walls and a synthesized array centered at the origin is placed against the wall surface at $R_1, R_2, ..., R_N$, as shown in Figure 1. The front-wall surface is located along the x-axis, and the back wall is parallel to the x-axis with a length D_x. The left- and right-side walls are symmetric about the y-axis with a length D_y. We consider the direct path from target P to antenna R_n as path A and three first-order multipaths as paths B, C, and E. The refraction points of the first-order multipaths on the back wall and the left- and right-side walls are B_r, C_r, and E_r, respectively.

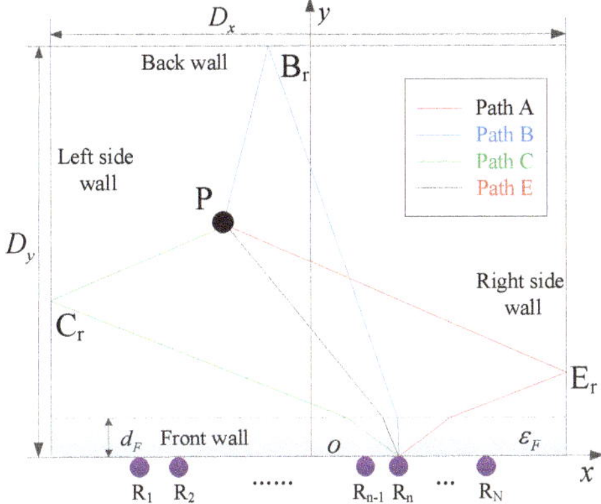

Figure 1. An illustration of the multipath model.

The one-way propagation delays of these four paths are denoted as $\tau_p^{(n)}$, $p \in \{A, B, C, E\}$ and, of which the numerical solutions were obtained in References [4,7]. Therefore, the radar echo with a direct path and first-order multipaths is given by

$$r_n(t) = T_{An}^2 s\left(t - 2\tau_A^{(n)}\right) + \sum_{q \in \{B,C,E\}} T_{An} T_{qn} s\left(t - \tau_A^{(n)} - \tau_q^{(n)}\right), \tag{1}$$

where $s(\cdot)$ is the transmitting signal and T_{An} and T_{qn} are the complex amplitude associated with reflection and transmission coefficients. Based on the back-projection algorithm, multipaths are transformed into multipath ghosts in the formed image.

3. Generative Adversarial Model

GAN is a novel way to train a generative model, which consists of two adversarial nets, namely, a generator G and a discriminator D. In order to make the generator have a wide range of generalization abilities, generator G establishes mapping from a predefined noise distribution p_z to a predefined data distribution p_{data} in the initial GAN [13]. As a result, it can output a high-quality image rather than an image that is full of noise with any input. p_g is defined to represent the output distribution of G. Discriminator D outputs a score to evaluate that x is from p_{data} rather than from p_g. G and D are alternately trained to achieve $p_g \approx p_{data}$. Optimizing the parameter of G is to minimize $\log(1 - D(G(z)))$, where $D(G(z))$ indicates the output of D with the input $G(z)$ and $G(z)$ denotes the output of G with input z. The feedback of discriminator D improves the generative ability (to make discriminator D unable to distinguish whether the data are from p_{data} or p_g). Optimizing the parameters of D is to increase the correct label of the training sample and generating sample, which means to improve the discriminating ability by trying to make $D(x) = 1$ and $D(G(z)) = 0$, where $D(x)$ denotes the output of D with input x. The whole process is just like two players playing a game, where one adjusts G to minimize the objective function $L_{GAN}(D, G)$ and where another adjusts D to maximize it, namely,

$$\min_G \max_D L_{GAN}(D,G) = E_x[\log D(x)] + E_z[\log(1 - D(G(z)))], \tag{2}$$

where $E[\cdot]$ denotes the mean value. In order to enhance the controllability of G in Equation (2), an additional message y was introduced in Reference [14] that can be accomplished by simultaneously introducing y into G and D. The objective function in Equation (2) can be modified as follows:

$$\min_{G} \max_{D} L_{cGAN}(D,G) = E_{x,y}[\log D(x,y)] + E_{z,y}[\log(1 - D(G(z,y),y))], \qquad (3)$$

where $D(x,y)$ indicates the output of D with the inputs of x and y. $G(z,y)$ indicates the output of G with the inputs of z and y. In this paper, a radar image with multipath ghosts is used as a conditional input. The training mechanism is shown in Figure 2. The entire training is an iterative dynamic process of alternately training G and D. First, the parameters of G are fixed, and only D is trained to distinguish the output of G and ground truth as much as possible by labeling the output of G as fake and the ground truth as real. Then, the parameters of D are fixed to train G. By optimizing the parameters of G, a realistic output can be judged as real by D. Eventually, through the process of iterative training, discriminator D is unable to differentiate between the two distributions p_g and p_{data}. Theoretically, there exists a unique solution of $D(x,y) = D(G(z,y),y) = 0.5$ that can be used as a sign of the end of training.

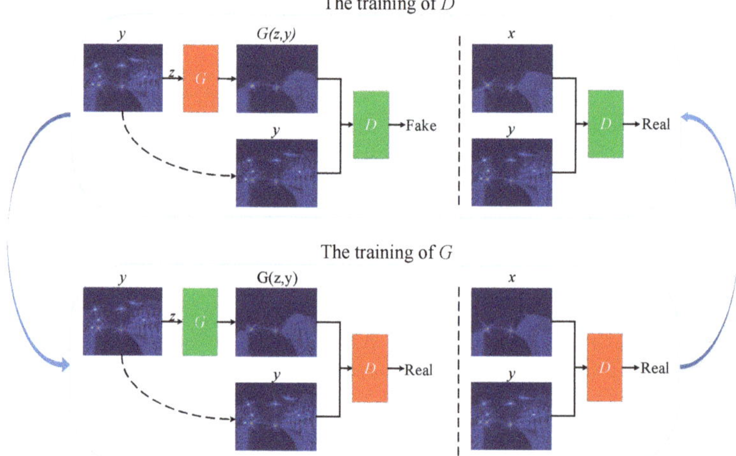

Figure 2. The training mechanism of conditional generative adversarial nets (GAN): The red rectangles indicate that the parameters of this network are fixed. The green rectangles indicate that the parameters of this network are trainable. y is the input as the control condition. x is the ground-truth image.

For the objective function in Equation (3), a better output can be generated by combining the $L1$ distance [15], expressed as follows:

$$L_{L1}(G) = E_{x,y,z}[\|x - G(z,y)\|_1]. \qquad (4)$$

Therefore, in this paper, we apply the objective function as follows:

$$Loss = \min_{G} \max_{D} L_{cGAN}(D,G) + \lambda \cdot L_{L1}(G), \qquad (5)$$

where λ is a parameter to limit the difference between the output and ground truth.

4. Network Architecture

In this section, the structures of generator G and discriminator D are described in detail. In this paper, G makes use of the type of U-net [16] and D adopts the discriminator of PatchGAN [17].

4.1. Generator G

In the original GAN generator [13,18], they mainly adopt a decoder structure to map a vector to an image. Conditional GAN [19,20] almost continues this tradition by an encoder-decoder structure [21], as shown in Figure 3a, which has two drawbacks of information loss and a high training complexity because all information flows through the whole network. However, in the radar image, there is some shared information between the input and output, such as the edges and positions of target images. In order to solve information sharing, in this paper, a skip connection is adopted to enable some information to bypass the middle layer and its specific structure is U-net network. The specific method is to connect the i layer with the $n - i + 1$ layer, as shown in Figure 3b.

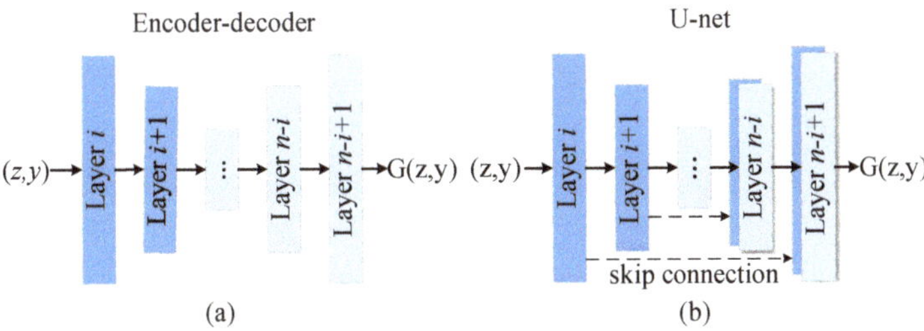

Figure 3. The structures of two generators. (**a**) Encoder-decoder. (**b**) U-net.

4.2. Discriminator D

With respect to the applied objective function in Equation (5), $L1$ loss concerns the global information of the input radar image to generate the mean of all possible images. If the effect of $L_{cGAN}(D, G)$ is ignored, the output image is blurred [15]. This means that $L1$ loss determines the low-frequency information of the output image. For this reason, $L_{cGAN}(D, G)$ only needs to generate high-frequency information. In order to force discriminator D to pay more attention to high-frequency information, a superior way is to focus on the locality of the image and to narrow the receptive field. This type of discriminator D is a PatchGAN discriminator [17], which is used to discriminate whether each $N \times N$ block is real or fake. Let D convolve across the entire image to obtain all output values and average them as the final output. The structure of a PatchGAN discriminator is shown in Figure 4.

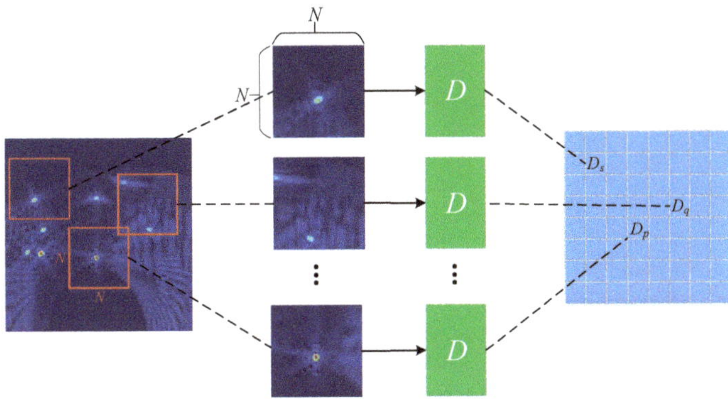

Figure 4. A PatchGAN discriminator where the receptive field of discriminator is $N \times N$.

4.3. Detailed Architectures of G and D

Based on the aforementioned description, generator G adopts the U-net network and discriminator D adopts the full convolution network with a receptive field of 70×70. For simplification, CBR_k is used to represent a Convolution–BatchNorm–ReLU layer with k filters, and $CBDR_k$ denotes a Convolution–BatchNorm–Dropout–ReLU layer with k filters and a dropout rate of 0.5. All convolutional layers adopt a filter with a size of 4×4 and a stride of 2. ReLUs are leaky with a slope of 0.2. In this paper, the detailed architectures of G and D are used as follows.

- Generator architecture
  encoder: $CBR_{64} - CBR_{128} - CBR_{256} - CBR_{512} - CBR_{512} - CBR_{512} - CBR_{512} - CBR_{512}$

  decoder: $CBDR_{512} - CBDR_{1024} - CBDR_{1024} - CBR_{1024} - CBR_{1024} - CBR_{512} - CBR_{256} - CBR_{128} - CBR_3(tanh)$
- Discriminator architecture
  $CBR_{64} - CBR_{128} - CBR_{256} - CBR_{512}(stride:1) - CBR_1(stride:1)(sigmoid)$

stride: 1 indicates the stride in this layer is 1. *tanh* and *sigmoid* denote the activation functions using tanh or sigmoid in this layer, and the others adopt default parameters.

5. Simulation and Discussion

5.1. Data Preparation

Two groups of data are generated with MATLAB to verify the potential of the method, as shown in Figure 1. A synthesized array with 31 single-channel radars monitors an enclosed room. The transmitting signal is a stepped-frequency continuous-wave signal with a carrier frequency of 1.5 GHz and a bandwidth of 1 GHz. The synthesized array is equidistantly placed with a spacing of 0.1 m. The lengths of the back wall and the side walls are both a random number from 5 to 7 m, namely the scenes are changeable. For simplification, the front wall is removed to avoid a penetration effect. The reflection and transmission coefficients T_{qn} and T_{An} are set to 0.5. All point targets are set at random locations inside the enclosed room. Based on Equation (1), echoes with first-order multipaths are obtained to form the input image of generator G. Echoes without first-order multipaths are obtained to form the ground-truth image of discriminator D. Specifically, a back-projection algorithm is used to form these images. The size of the input images and the ground-truth images is set to 256×256.

- Dataset 1
  The number of targets is set to a random number ranging from one to four; 1000 samples and 100 samples of data are respectively used as a training set and a validation set.
- Dataset 2
  The number of targets is increased to a random number ranging from ten to twenty; 2000 samples and 200 samples of data are respectively used as a training set and a validation set.

5.2. Training Details

For the convenience of practical training, minimizing $\log(1 - D(G(z,y),y))$ in objective function Equation (5) is replaced by maximizing $\log D(G(z,y),y)$. Minibatch SGD (stochastic gradient descent) and the Adam optimizer are adopted with momentum parameters $\beta_1 = 0.5$ and $\beta_2 = 1.00$. The batch size is set to 1. Moreover, λ in Equation (5) is set to 100. All training is run on a single GeForce GTX1080Ti GPU (with 11 GB memory). The loading process of a G network requires 54.414 MB of memory, and the loading process of a D network requires 2.769 MB of memory. As a result, the training of GAN requires at least 57.183 MB of memory, and the testing of GAN requires at least 54.414 MB. The weights of all filters are initialized from a Gaussian distribution with a mean of 0 and a standard deviation of 0.02. In the training of Dataset 1, 50 epochs are trained and each epoch consumes an average of 144 s. In the training of Dataset 2, 150 epochs were trained and each epoch consumes an average of 298 s. The learning rate of the first 100 epochs is 0.0002, and the learning rate of the last 50 epochs is reduced by 0.000002 each time.

5.3. Result Analysis

After 50 epochs of training of Dataset 1, generator G could correctly eliminate the multipath ghosts. The training-loss curve is shown in Figure 5. Specifically, the curve of $L_{cGAN}(D,G)$ has an undulating trend, since one of generator G and discriminator D is always in a dominant position during the adversarial process. There is a slight downward trend in the curve of $L_{L_1}(G)$, which indicates that the similarity between the image generated by G and the ground truth is slightly improved. The initial stage of D_{real} is almost greater than D_{fake}, which means that generator D can completely distinguish whether the sample is from G or the ground truth. However, both of them later begin to approach each other, which indicates that generator G can correctly eliminate multipath ghosts so that D hardly distinguishes the radar image from G or ground truth. The performance of GAN changing over the iterations is shown in Figure 6, which indicates that multipath ghosts are gradually suppressed but target images are gradually formed.

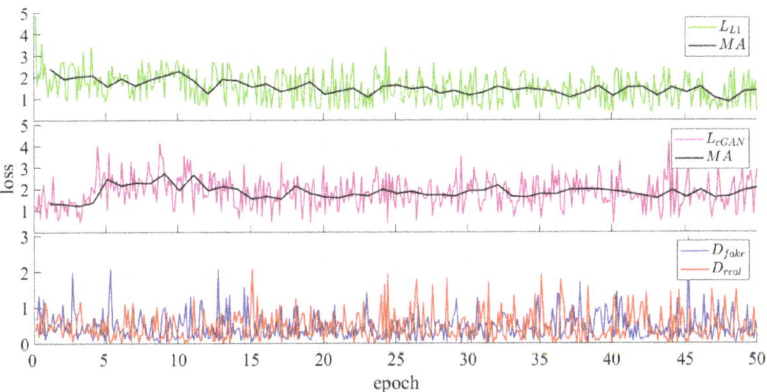

Figure 5. The loss of dataset 1 over time. MA means the moving average curve with the cycle of one epoch. D_{real} indicates the curve of $D(x,y)$, and D_{fake} represents the curve of $D(G(z,y),y)$.

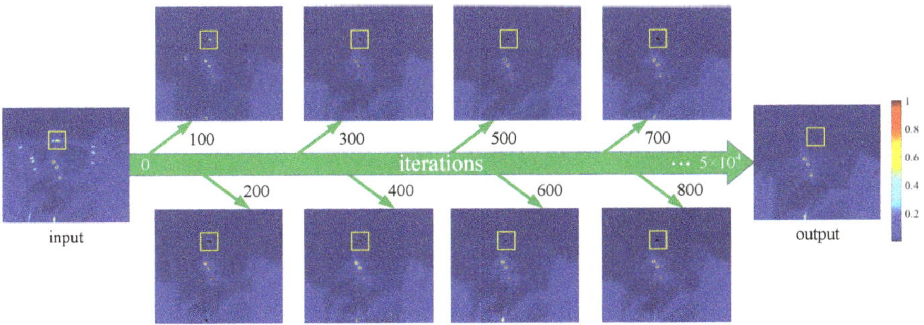

Figure 6. The output of GAN varies from the iterations of Dataset 1. The yellow rectangles indicate a part of the multipath ghosts' positions.

The results of generator network G are shown in Figure 7, which indicates that multipath ghosts are correctly eliminated. The differences between the output image and the ground truth image are only the grating lobes and side lobes marked with a red oval which can be learned by continuing training. However, as this paper mainly focuses on multipath-ghost suppression, it can be reasonably considered that training is completed. It is worth noting that marks, axis, and color bars are absent in the training samples.

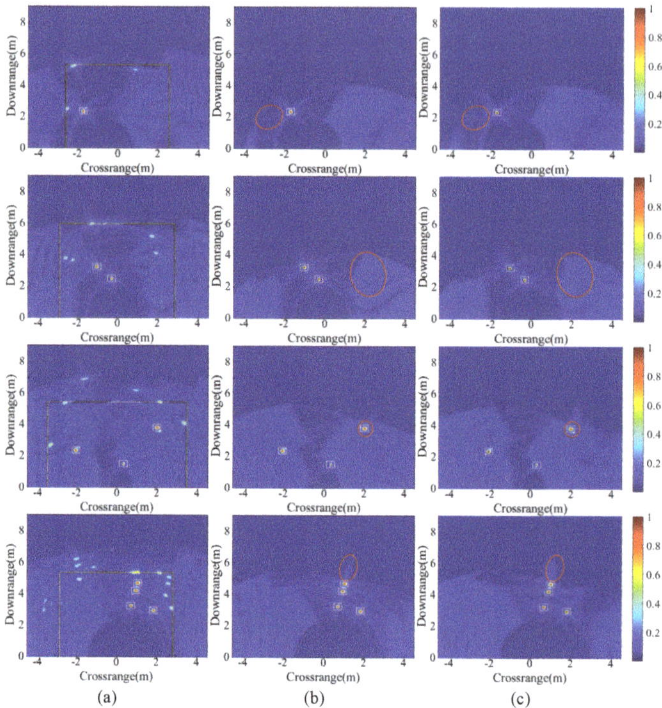

Figure 7. The results of Dataset 1. (**a**) Input images. (**b**) Ground-truth images. (**c**) Output images. The red ellipses mark the differences between the output images and the ground-truth images. The yellow lines mark the walls' locations. The white rectangles mark the targets' positions.

After 150 epochs of training of Dataset 2, the curve of loss is shown in Figure 8. Compared with Figure 5, the L_{cGAN} has a different (ascent) trend due to the mismatch of evolution speed between G and D in the early stages. In a complex situation with a large number of multipath ghosts, the reason for a mismatch could be summarized into two conflict points. On the one hand, complex multipath ghosts bring convenience to D to identify the radar image from G or ground truth. On the other hand, it makes it difficult for G to eliminate multipath ghosts. The mismatch increases the training time, appearing as the ascent trend of L_{cGAN} in Figure 8. The performance of GAN, varying from the iterations, is shown in Figure 9, which indicates that multipath ghosts are gradually suppressed but target images are gradually formed.

The result of the final training is shown in Figure 10. Although the situation is more complicated, multipath ghosts are still correctly eliminated. It is worth noting that true target images can be preserved even if targets are overlapped with multipath ghosts. For example, the overlapped target images marked with red rectangles in the input images are clearly preserved in the output image.

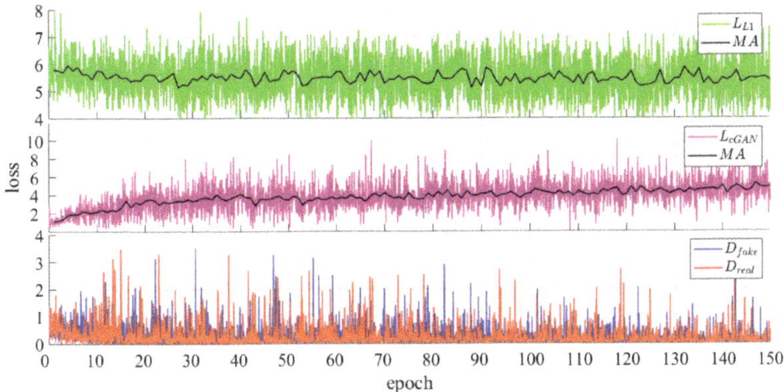

Figure 8. The loss of Dataset 2 over time. MA means the moving average curves with the cycle of one epoch. D_{real} indicates the curve of $D(x,y)$, and D_{fake} represents the curve of $D(G(z,y),y)$.

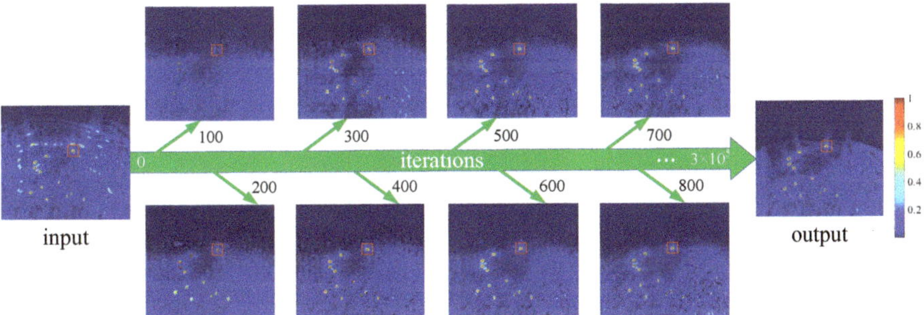

Figure 9. The output of GAN varies from the iterations of Dataset 2. The red rectangles indicate a part of the targets' positions.

In addition, the performance of elimination is quantitatively measured, and the results are shown in Table 1. The above two networks are separately tested with 200 new test samples. The rate of one error and two or more errors are separately counted, where an error indicates a residual multipath ghost or a lost target image. The statistical results demonstrate the proposed method can effectively eliminate multipath ghosts.

Table 1. Accuracy.

	One Error	Two Errors or More
GAN trained by Dataset 1	0.5%	0%
GAN trained by Dataset 2	2.5%	0.5%

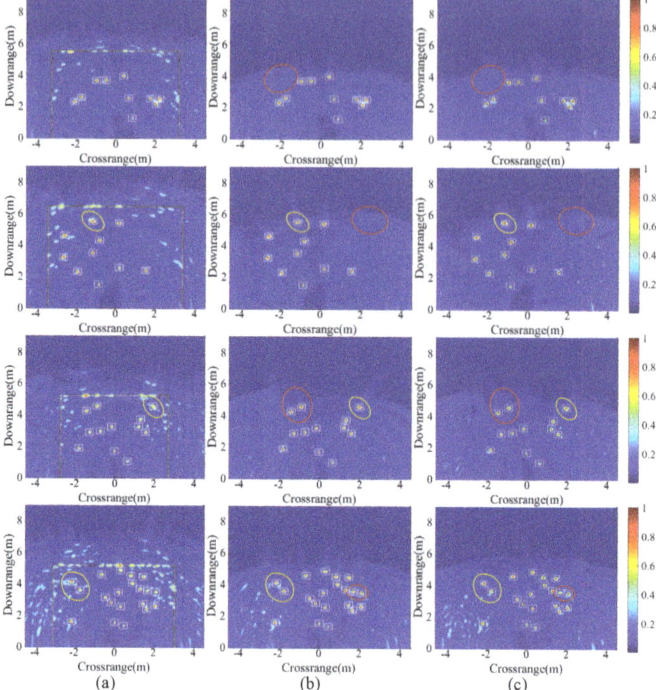

Figure 10. The results of dataset 2. (**a**) Input images. (**b**) Ground-truth images. (**c**) Output images. The red ellipses mark the differences between the output images and the ground-truth images. The yellow lines mark the walls' locations. The white rectangles mark the targets' positions. Moreover, the yellow ellipses mark the target images that are overlapped with multipath ghosts.

5.4. Comparison of Different Methods

In this section, the proposed method is compared with the PCF method [11], the subaperture-fusion method [8], and the imaging-dictionary-based method [7]. The results are shown in Figure 11. Specifically, the walls' locations need to be known in advance by the imaging-dictionary-based method. Table 2 illustrates the averaging computation time of 100 trials for each multipath-suppression method. The PCF method, subaperture-fusion method, and imaging-dictionary-based method run on Matlab 2017a, while the proposed method runs on Python. All methods adopt a workstation including a Intel 2.60 GHz Core(TM) i7-6700HQ CPU processor (with 8 GB of memory) and a NVIDIA GeForce GTX1080Ti GPU (with 11 GB of memory) with CUDA (compute unified device architecture) acceleration. The comparison results in Table 2 demonstrate that the proposed method and subaperture-fusion method have similar time consumptions that are superior to the PCF and imaging-dictionary-based methods.

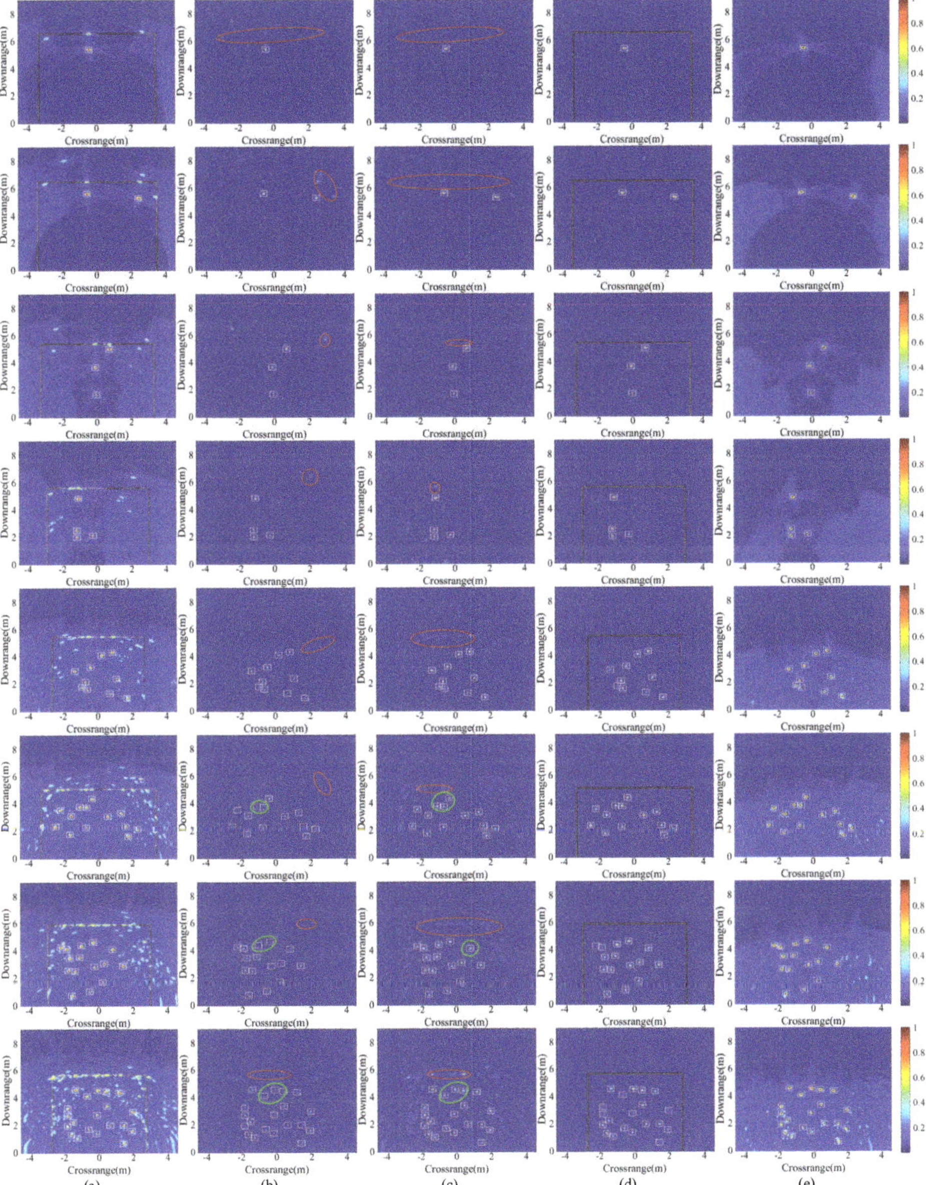

Figure 11. The comparison results of different methods. (**a**) The original images. (**b**) The PCF method [11]. (**c**) The subaperture-fusion method [8]. (**d**) The imaging-dictionary-based method [7]. (**e**) The proposed method. The yellow lines mark the walls' locations, which indicate that the imaging-dictionary-based method requires prior wall location information. The red ellipses mark a part of the multipath ghost locations. The green ellipses mark a part of the degraded target images. The white rectangles mark the targets' positions.

Table 2. The averaging computation time of 100 trials for four different multipath suppression methods.

Method	Computation Times
PCF	1.09
Subaperture-fusion	0.62
Imaging-dictionary	1.65
The proposed method	0.65

As shown in Figure 11b, the PCF method is unable to eliminate well-focused multipath ghosts, marked by red ellipses in the case of synthetic aperture imaging (SAI). In Figure 11c, a part of the multipath ghosts, marked by the red ellipses especially about the back wall, still exist with the subaperture-fusion method, as they appear at the close positions in both subaperture images. Figure 11d demonstrates that the imaging-dictionary-based method has an excellent performance in suppressing multipath ghosts while it needs the prior walls' locations. In Figure 11b,c, the PCF method and the subaperture-fusion method enlarge the energy differences between the target images, which makes it difficult to identify degraded targets such as target images marked by green ellipses with a low SMCR that is a ratio between the peak of the target image and its multipath ghost [10]. As a comparison, as shown in Figure 11e, the proposed method achieves an excellent multipath suppression without walls' locations. Moreover, the proposed method prevents the difference of target images from enlarging, which is beneficial to identifying all targets.

To strengthen the point of the proposed method being well-suited for the application, the advantages and disadvantages of each method are summarized in Table 3. Furthermore, the SMCR is applied to quantitatively evaluate the performances of multipath-ghost suppression for different methods. Specifically, as shown in Figure 12, the scene with two targets in Figure 11 is chosen as a sample. As shown in Table 4, the proposed method has a much lower SMCR than the PCF method (by about 20–50 dB) and the subaperture-fusion method (by about 15–35 dB). The imaging-dictionary-based method also has the highest SMCR. As a result, both the proposed method and the imaging-dictionary-based method have an excellent multipath suppression while the imaging-dictionary-based method needs prior walls' locations.

Table 3. The advantages and disadvantages of different methods.

Method	Advantages	Disadvantages
PCF	1. Does not require prior walls' locations. 2. Does not require complicated tuning parameters.	1. Poor multipath suppression for SAI. 2. A part of target images are degraded with low SMCR.
Subaperture-fusion	Does not require prior walls' locations.	1. Poor suppression for the multipath ghosts of back wall. 2. A part of target images are degraded with low SMCR. 3. Requires complicated tuning parameters.
Imaging-dictionary	1. Excellent multipath suppression. 2. Does not require complicated tuning parameters.	Requires prior walls' locations.
The proposed method	1. Excellent multipath suppression. 2. Does not require prior walls' locations. 3. No target images with degradation. 4. Does not require complicated tuning parameters.	Requires a dataset with labels.

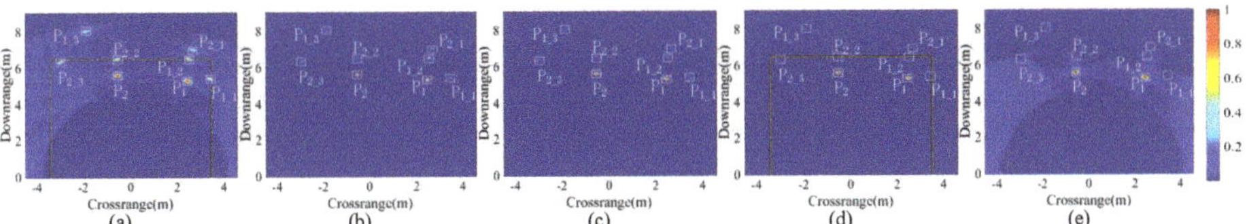

Figure 12. The selected sample for quantitatively evaluating multipath-ghost suppression for different methods. (**a**) The original images. (**b**) The PCF method. (**c**) The subaperture-fusion method. (**d**) The imaging-dictionary-based method. (**e**) The proposed method. The white rectangles mark the regions of target images and their multipath ghosts.

Table 4. Ratios between the peaks of the target images and their multipath ghosts (dB).

Method	$P_1 VS.P_{1\_1}$	$P_1 VS.P_{1\_2}$	$P_1 VS.P_{1\_3}$	$P_2 VS.P_{2\_1}$	$P_2 VS.P_{2\_2}$	$P_2 VS.P_{2\_3}$
Original image	5.89	5.90	8.48	6.37	5.74	6.00
PCF	13.09	6.99	35.11	10.47	12.81	12.53
Subaperture-fusion	17.41	17.15	24.93	20.89	17.41	18.42
Imaging-dictionary	103.56	119.37	172.02	112.93	75.06	123.06
The proposed method	32.45	54.49	56.49	43.49	31.27	27.06

It is worth noting that the side/grating lobes are preserved in ground-truth images and output images. On the one hand, GAN is classified as a structured learning network that is applied to construct spatial structure mapping from input images to output images. The preservation of side/grating lobes is equivalent to preserving the spatial structure, which promotes the elimination of multipath ghosts and the preservation of target images. On the other hand, side/grating lobes in output images could be eliminated by threshold detection thanks to a high signal-to-noise ratio. In other words, side/grating lobes are effective information for multipath suppression and have no effect on the target-detection performance.

6. Conclusions

A GAN-based multipath-ghost suppression algorithm is presented in this paper. Based on Matlab simulation datasets, the generator of GAN is trained to be able to efficiently reduce multipath ghosts, along with fighting with the discriminator. It is demonstrated that GAN has the potential for multipath elimination in TWRI. In a future work, we will research the modification of GAN and use complicated simulation datasets (such as changeable radar parameters) and practical measured datasets to outline the potential of GAN.

Author Contributions: Y.J., R.S., and S.C. provided ideas and wrote the paper. R.S. finished the training of GAN. S.C. generated the simulation data. Y.G. provided the funding acquisition and resources with X.Z. G.C. reviewed and edited the paper with G.W.

Funding: This work was funded by the National Natural Science Foundation of China No. 41574136 and 61501062, the 2018 National Students' Innovation, Entrepreneurship Training Program No. 201810616086 and 201810616118, and the Key Research and Development Project of Sichuan Science and Technology Program of China under Grants 2019YFG0097 and 2018GZ0454.

Acknowledgments: The authors would like to thank Chao Yan and Junyou Xiao for their useful suggestions and anonymous reviewers.

Conflicts of Interest: The authors declare no conflict of interest.

References

1. Jia, Y.; Zhong, X.; Liu, J.; Guo, Y. Single-side two-location spotlight imaging for building based on MIMO through-wall-radar. *Sensors* **2016**, *16*, 1441. [CrossRef] [PubMed]
2. Halman, J.I.; Shubert, K.A.; Ruck, G.T. SAR processing of ground-penetrating radar data for buried UXO detection: results from a surface-based system. *IEEE Trans. Antennas Propag.* **1998**, *46*, 1023–1027. [CrossRef]
3. Lei, C.; Ouyang, S. A time-domain beamformer for UWB through-wall imaging. In Proceedings of the IEEE Region 10 Conference, Taipei, Taiwan, 30 October–2 November 2007; Volume 10, pp. 1–4.
4. Setlur, P.; Amin, M.; Ahmad, F. Multipath model and exploitation in through-the-wall and urban radar sensing. *IEEE Trans. Geosci. Remote Sens.* **2011**, *49*, 4021–4034. [CrossRef]
5. Setlur, P.; Alli, G.; Nuzzo, L. Multipath exploitation in through-wall radar imaging via point spread functions. *IEEE Trans. Image Process.* **2013**, *22*, 4571–4586. [CrossRef] [PubMed]
6. Chen, X.; Chen, W. Multipath ghost elimination for through-wall radar imaging. *IET Radar Sonar Navig.* **2016**, *10*, 299–310. [CrossRef]
7. Guo, S.; Cui, G.; Kong, L.; Yang, X. An Imaging Dictionary Based Multipath Suppression Algorithm for Through-Wall Radar Imaging. *IEEE Trans. Aerosp. Electron. Syst.* **2018**, *54*, 269–283. [CrossRef]
8. Li, Z.; Kong, L.; Jia, Y.; Zhao, Z.; Lan, F. A novel approach of multi-path suppression based on sub-aperture imaging in through-wall-radar imaging. In Proceedings of the IEEE Radar Conference, Ottawa, ON, Canada, 29 April–3 May 2013; pp. 1–4.
9. Guo, S.; Yang, X.; Cui, G.; Song, Y.; Kong, L. Multipath Ghost Suppression for Through-the-Wall Imaging Radar via Array Rotating. *IEEE Geosci. Remote Sens. Lett.* **2018**, *15*, 868–872. [CrossRef]
10. Liu, J.; Jia, Y.; Kong, L.; Yang, X.; Liu, Q.H. Sign-coherence-factor-based suppression for grating lobes in through-wall radar imaging. *IEEE Geosci. Remote Sens. Lett.* **2016**, *13*, 1681–1685. [CrossRef]
11. Lu, B.; Sun, X.; Zhao, Y.; Zhou, Z. Phase coherence factor for mitigation of sidelobe artifacts in through-the-wall radar imaging. *J. Electromagn. Waves Appl.* **2013**, *27*, 716–725. [CrossRef]
12. Burkholder, R.J.; Browne, K.E. Coherence factor enhancement of through-wall radar images. *IEEE Antennas Wirel. Propag. Lett.* **2010**, *9*, 842–845. [CrossRef]
13. Goodfellow, I.; Pouget-Abadie, J.; Mirza, M.; Xu, B.; Warde-Farley, D.; Ozair, S.; Courville, A.; Bengio, Y. Generative Adversarial Nets. In Proceeding of the Advances in Neural Information Processing Systems 27, Montreal, QC, Canada, 8–13 December 2014; pp. 2672–2680.
14. Mirza, M.; Osindero, S. Conditional generative adversarial nets. *arXiv* **2014**, arXiv:1411.1784.
15. Isola, P.; Zhu, J.Y.; Zhou, T.; Efros, A.A. Image-to-image translation with conditional adversarial networks. *arXiv* **2017**, arXiv:1611.07004.
16. Ronneberger, O.; Fischer, P.; Brox, T. U-net: Convolutional networks for biomedical image segmentation. In Proceedings of the International Conference on Medical image computing and computer-assisted intervention, Munich, Germany, 5–9 October 2015; pp. 234–241.
17. Li, C.; Wand, M. Precomputed real-time texture synthesis with markovian generative adversarial networks. In Proceedings of the European Conference on Computer Vision ECCV, Amsterdam, The Netherlands, 8–16 October 2016; pp. 702–716.
18. Radford, A.; Metz, L.; Chintala, S. Unsupervised representation learning with deep convolutional generative adversarial networks. *arXiv* **2015**, arXiv:1511.06434.
19. Pathak, D.; Krahenbuhl, P.; Donahue, J.; Darrell, T.; Efros, A.A. Context encoders: Feature learning by inpainting. In Proceedings of the IEEE Conference on Computer Vision and Pattern Recognition, Las Vegas, NV, USA, 27–30 June 2016, pp. 2536–2544.
20. Wang, X.; Gupta, A. Generative image modeling using style and structure adversarial networks. In Proceedings of the European Conference on Computer Vision, Amsterdam, The Netherlands, 11–14 October 2016; pp. 318–335.
21. Hinton, G.E.; Salakhutdinov, R.R. Reducing the dimensionality of data with neural networks. *Science* **2006**, *313*, 504–507. [CrossRef] [PubMed]

 *electronics*

Article

Entropy-Based Low-Rank Approximation for Contrast Dielectric Target Detection with Through Wall Imaging System

Mandar Bivalkar [1], Dharmendra Singh [1,*] and Hirokazu Kobayashi [2]

[1]   Indian Institute of Technology Roorkee, Roorkee, Uttarakhand 247667, India; mbivalkar@somaiya.edu
[2]   Osaka Institute of Technology Osaka, Osaka Prefecture 535-8585, Japan; hirokazu.kobayashi@oit.ac.jp
*   Correspondence: dharmfec@gmail.com

Received: 29 April 2019; Accepted: 30 May 2019; Published: 5 June 2019

Abstract: In through wall imaging, clutter plays an important role in the detection of objects behind the wall. In the literature, extensive studies have been carried out to eliminate clutter in the case of targets with the same dielectric. Existing clutter reduction techniques, such as the sub-space approach, differential approach, entropy-based time gating, etc., are able to detect a single target or two targets with the same dielectric behind the wall. In a real-time scenario, it is not necessary that targets with the same dielectric will be present behind the wall. Very few studies are available for the detection of targets with different dielectrics; here we termed it "contrast target detection" in the same scene. Recently, low-rank approximation (LRA) was proposed to reduce random noise in the data. In this paper, a novel method based on entropy thresholding for low-rank approximation is introduced for contrast target detection. It was observed that our proposed method gives satisfactory results.

Keywords: through-wall imaging; contrast target detection; clutter reduction; entropy thresholding; low-rank approximation

---

1. Introduction

Through wall imaging (TWI) is emerging as an important technology for surveillance, security and rescue missions. The main aim of TWI is seeing through a wall with the help of electromagnetic waves. In any radar system, the signal-to-clutter ratio (SCR) plays an important role in the improvement of the detection of the objects. The SCR can be improved either by classical or statistical methods [1]. Classical methods use different classical digital filters, while statistical methods exploit the statistical nature of the received signal to separate the clutter from the signal.

Digital filtering technique [2] uses frequency analysis of the clutter geometrical model and the signal geometrical model, while in [3], the coupled iterative procedure was used to reduce the ground reflections for the application of ground penetrating radar (GPR). The classical clutter reduction algorithm (CCRA) was proposed in [4,5], but this method does not optimize the coefficients for the representation of noise and the target. Kalman filtering uses the background component model in [6], but designing the Kalman filter is computationally intensive. Parametric clutter reduction was proposed in [3] by modeling the variations in the received signal, but it requires a reference signal, which cannot practically be made available.

In the literature, various statistical clutter reduction techniques have been proposed. Different statistical reduction techniques were compared in [7]. Verma et al. [1] applied various statistical techniques for clutter reduction such as singular value decomposition (SVD), principle component analysis (PCA), factor analysis (FA), and independent component analysis (ICA), and concluded that ICA performs better for the detection of low dielectric material behind a wall.

The major contribution to the clutter in TWI is due to reflections from the wall. Different wall removal techniques are proposed in the literature, such as the sub-space projection approach that is used in [8] for wall removal. SVD frequently using the sub-space projection approach for clutter reduction, and it has previously been used to enhance the signal-to-clutter ratio for the application of ground penetrating radar (GPR) [9] and TWI in [10]. In SVD, Eigen-images of the B-scan are determined and used to identify wall clutter and target subspace. In [11], it is stated that the first two Eigen-values correspond to the wall and target, respectively, but [8] shows that wall clutter is spread along with the high dimensional subspace and weak wall singular components interleave with target subcarriers. Recently, the empirical low- rank approximation method was proposed in [12] for seismic data, where all Eigen-values corresponding to the noise subspace were considered to identify the weak signals. If we consider all the Eigen-values along with the Eigen-values corresponding to the signal, then noise is also get added in the signal.

Compared to other imaging systems such as GPR and biomedical, TWI has to deal with more severe problems like changes in the propagation environment and sensor positioning [13,14]. Another problem in TWI is the propagation medium, where multiple unknowns and either homogenous or non-homogenous walls are involved [15]. In a real-time scenario, it may be possible that targets with different dielectrics will be present behind the wall. It is challenging to detect low dielectric targets such as wood ($\approx$4) in the presence of metal ($\approx\infty$) behind the wall because in TWI images, low dielectric targets are obscured in the presence of high noise. We termed the target detection and imaging problem in which targets with different dielectrics are present as "contrast target detection" and "contrast target imaging", respectively. The contribution of this paper is that first we propose a novel method to detect contrast targets using a sub-space projection approach based on low rank approximation (LRA) and modify it using entropy in the Eigen-values to reduce the clutter from the useful signal. Second, we solve the inherent problem of considering a large rank for the signal recovery in LRA by introducing an entropy-based threshold. The critical analysis of existing clutter reduction techniques shows that they cannot detect contrast targets in the scene. The performance of the proposed method is compared with other traditional techniques such as an average trace subtraction, subspace projection, entropy based time gating, SVD, ICA and the differential approach for the targets having contrast dielectrics.

The remainder of this paper is organized as follows. Geometry for the TWI imaging is presented in Section 2. Different clutter reduction techniques are presented along with the results in Section 3. Novel methods for entropy-based low-rank approximation for contrast imaging is proposed in Section 4. Section 5 concludes the paper.

2. TWI Data Collection and Beamforming

2.1. Data Collection

Data is collected by placing different dielectric materials such as metal or wood behind the wall at different distances in our experimental work. Synthetic aperture radar (SAR) in the multi-static mode in which an array of antennas are used to scan the whole wall at M different locations, and the reflection coefficient (S11) is measured for P- targets in the scene using system parameters given in Table 1 and then set-up, which is shown in Figure 1. Received frequency domain data is converted to time domain [16]

Step (1): Transformation from Time domain to spatial domain

Time domain signal is converted to spatial domain to determine the range profile by using $z = c * t/2$ where c is the speed of light and t is a delay.

Step (2): External calibration

The metallic plate is placed in front of the antenna [17] to find the delay due to the antenna system, which will be subtracted from the observed data. The range profile is corrected using the difference in the delay.

Step (3): Velocity correction

As the antenna is placed at the standoff distance from the wall, the signal propagates through the air, then the wall and again through the air up to the target. The presence of the wall scattered the signal, and shifting of the target position took place. This shifting was compensated using the method of velocity correction. The mathematical equation for this is given next in the paper.

Table 1. System parameters.

Sr. No.	Parameters	Value
01	Radar type	SFCW
02	Frequency range	1 GHz–3 GHz
03	Transmitted power	3 dBm
04	Number of frequency points	201
05	Bandwidth	2 GHz
06	Cross-range resolution	15 cm
07	Down-range resolution	7.5 cm
08	Polarization	VV
09	Antenna type	Horn
10	Gain of Antenna	20 dB
11	Beam-width	15.92° and 17.02°

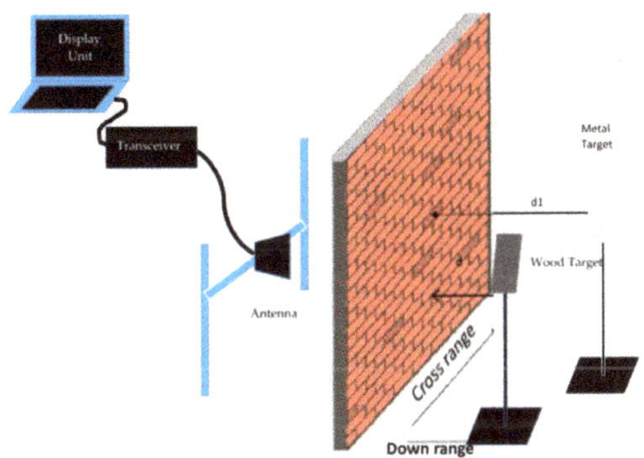

Figure 1. Through Wall Imaging (TWI) set- up.

2.2. Beamforming

The antenna is placed in front of the wall at a fixed standoff distance and data is collected for M different locations that received the signal, represented as Equation (1).

$$x(n,t) = \sum_{p=0}^{P-1} \sigma_p s(t - \tau_{n,p}) \tag{1}$$

where $s(t)$ is the transmitted signal convolved with the transfer function of the wall [14] for the SFCW radar. σ_p is the reflection coefficient, and $\tau_{n,p}$ is the two-way delay—the time between the nth antenna position and the target P. When the signal propagates between the nth antenna positions and p^{th} target, the two-way delay-time is given as Equation (2)

$$\tau_{n,p} = \frac{2}{c} \sqrt{(x_p - x_n)^2 + (y_p - y_n)^2} \tag{2}$$

where c is the speed of light.

In this paper, the focused image has been developed using DS (delay-and-sum) beamforming for collected data. The ith pixel value in the DS image is given by Equation (3)

$$b(i, j) = \frac{1}{N} \sum_{n=0}^{N-1} x(n, t + \tau_n(i, j)) \tag{3}$$

where $\tau_n(i, j)$ is the propagation delay on both sides of the wall. The SFCW radar, for which stepped size depends upon the selection of frequency bins, requires trade-off between a number of frequency bins and scanning time. The SFCW radar waveform consists of Q- narrowband signals defined as Equation (4)

$$b(i, j) = \sum_{n=0}^{N-1} \sum_{q=0}^{Q-1} x(n, f_q) \tag{4}$$

where $x(n, f_q)$ is the signal received at the nth antenna position for the frequency q.

S11 data is collected either in the time domain or frequency domain [18]. We collected data for three targets with different dielectrics by arranging them at different positions behind the wall (refer to Appendix A). Our method was tested on collected data for illustration purposes, with a few results given in Sections 3 and 4. The geometry for the TWI is shown in Figure 2, if we consider the point target at X_p, then developing the image transformation is required from the time to the spatial domain.

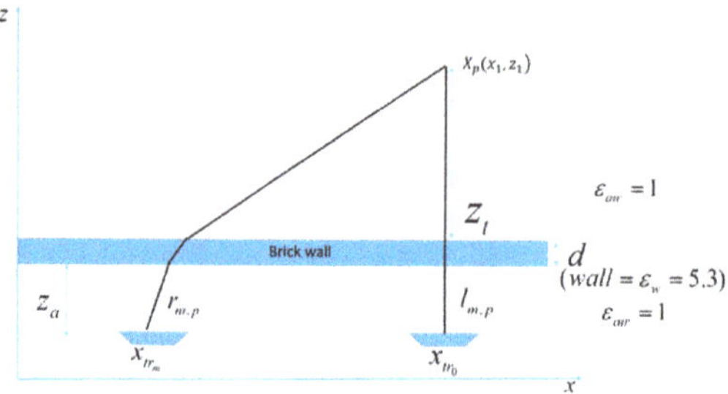

Figure 2. Through Wall Imaging (TWI) Geometry.

Delay-sum-beamforming is the most popular and least complex imaging algorithm, interested readers may refer to [19] for more details. The signal received at the antenna location is $z[m, q]$ of the frequency f_n with a delay $\tau_{p,m}$, then $z[m, q]$ can be represented as Equation (5)

$$z[m, q] = \sum_{p=0}^{P-1} \sigma_p \exp\left\{-j2\pi f_n \tau_{p,m}\right\} \tag{5}$$

where m and q represents the spatial index and frequency index, respectively. In our experimental set-up, we considered a homogenous wall of thickness $d = 15$ cm and relative permittivity of the wall of $\varepsilon_w = 5.3$. The dielectric constant of the wall is measured as described in [20]. The distance from the antenna to the wall is (z_a) and from a wall to the target is (z_t). The velocity correction [21] for geometry shown in Figure 2 is given by Equation (6)

$$d_v = z_a + d \sqrt{\varepsilon_w} + z_t \tag{6}$$

where d_v is the actual distance between the antenna and the target after velocity correction. $\tau_{n,p}$ can be estimated by putting Equation (6) into Equation (2)

$$\tau_{n,p} = \sqrt{(x_{tr_0} - x_{tr_n})^2 + (d_v + X_p)^2} \tag{7}$$

We can recover the image $s[k, l]$ by DS–beamforming using Equation (8)

$$s[x_k, z_l] = \frac{1}{MQ} \sum_{m=0}^{M-1} \sum_{q=0}^{Q-1} z[m, q] \exp\{j2\pi f_n \tau_{n,p}\}$$ (8)

where k and l are the number of pixels in the image.

3. Clutter Reduction Techniques

Clutter is the unwanted reflections due to other objects in the room. Clutter overwhelms the target, and so clutter reduction techniques can be used to separate clutter from the target. The data collected for the nth observation can be denoted as

$$s(t) = s_a(t) + s_w(t) + s_p(t)$$ (9)

where $s_a(t)$ are the reflections due to an antenna mismatch, $s_w(t)$ are the reflections due to the wall and $s_p(t)$ are the contribution due to p (the number of targets behind the wall). The discrete form of collected data for M antenna locations at N different instances can be arranged in an $M \times N$ data matrix.

$$s = [s_0, s_1, \ldots, s_{M-1}]$$ (10)

We try to separate the $s_p(t)$ signal from $s_a(t)$ and $s_w(t)$ using clutter reduction methods [22].

Imaging for raw data was done with the help of DS beamforming as discussed in Section 2. Raw normalized images for different targets after pre-processing are shown in Figure 3. It can be seen from the DS images for different targets, along with the target, clutter due to the wall and other objects also dominate, which may obscure the target. The reflections from the low dielectric targets are generally weak compared to clutter from the interior and exterior of the wall, making detection of such targets difficult. Efficient clutter reduction technique is required to remove the clutter from the B-scan image to detect low dielectric targets [23]. After pre-processing, different commonly used clutter reduction techniques are implemented and the results of this are discussed here. Their performance is given in terms of the PSNR (peak signal-to-noise ratio) [24].

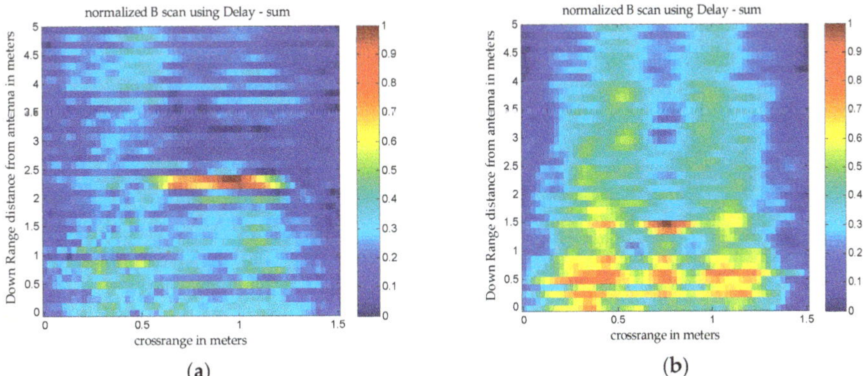

Figure 3. *Cont.*

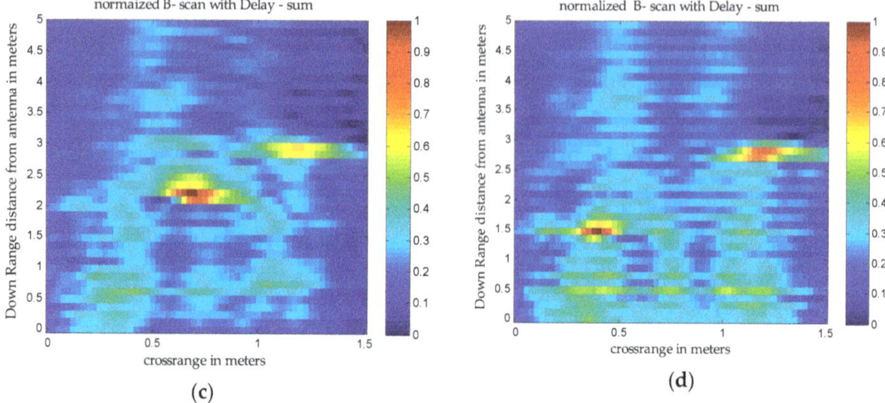

Figure 3. Delay and Sum B-scan images (**a**) Target ID 01: Metal target, (**b**)Target ID 02: Wood target, (**c**) Target ID 05: Two metal targets, (**d**)Target ID 07: Metal and wood targets where color bar represents normalized intensity value.

3.1. Average Trace Subtraction

In general, clutter remains constant with respect to target reflections for data collected [25], hence we can consider that clutter will be constant for a homogenous wall. We can separate a constant signal from the non-constant signal using spatial filtering [26] in the time domain, which can be represented as

$$s_{av}(n,m) = s(n,m) - \bar{s}(m) \quad m = 0 \ldots M-1$$
$$\text{Where } \bar{s}(m) = \frac{1}{N}\sum_{n=0}^{N-1} s(n,m) \quad m = 0 \ldots M-1 \tag{11}$$

where $s(n,m)$ is the data matrix element and $\bar{s}(m)$ is the average of the data matrix. The Fourier transform for (11) can be given as

$$\widehat{s}_{av}(k_x,m) = \sum_{n=0}^{N-1} s_{av}(n,m)\exp(-ik_x n\Delta x)\Delta x$$
$$= \widehat{s}(k_x,m) - \bar{s}(m)\frac{\sin(N\Delta x k_x/2)}{\sin(\Delta x k_x/2)} * \exp[-i\Delta x k_x(N-1)/2] \tag{12}$$

where k_x represents spatial frequency ($\Delta x \leq 2\pi/k_0$), and k_0 is the frequency wave number. This condition cover $[-k_0,k_0]$ and no filtering is introduced by the grating lobe. Due to the Dirichlet condition appearing in Equation (12), the low-frequency spatial spectrum of target signal s_n will also get rejected, and due to this target which is placed near the wall, cannot be detected.

3.2. Differential Approach

In this approach the clutter is removed by subtracting the adjacent two traces [27], for the time domain data this is represented as [28]

$$s_{DA}(n,m) = s(n+1,m) - s(n,m) \quad m = 0 \ldots \ldots M-1, \; n = 0 \ldots \ldots N-2 \tag{13}$$

The corresponding spatial spectrum can be given as

$$\widehat{s}_{DA}(k_x,m) \simeq 2j\widehat{s}(k_x,m)\left(\sin(\frac{k_x\Delta x}{2})\right) \tag{14}$$

The observation point corresponds to $x_n = (n+1/2)\Delta x \; n = 0, \ldots, N-1$ and $n = 0, \ldots, N-2$ in the first and second term. It can be seen from Equation (14) that de-cluttering can be achieved locally and that not all traces are exploited simultaneously. If Δx is very small, then the **sin** term rises very slowly, and if it is very high, then artifacts corrupt the reconstruction.

3.3. Subspace Projection Approach

Subspace approaches are used to separate out complementary subspaces, called the target and clutter, in order to increase SCR. In the literature, many methods for clutter reduction using the subspace projection approach are given. Here we will restrict our discussion to SVD and ICA.

3.3.1. Singular Value Decomposition (SVD)

SVD is the most efficient technique from linear algebra for clutter reduction. SVD decomposition of the B-scan matrix (S) of dimension $M \times N$ is given by

$$S = UDV^T \tag{15}$$

where $U = [U_1, U_2, \ldots, U_M]$ and $V = [V_1, V_2, \ldots, V_N]$ are the left and right singular matrix whose column are Eigen-vectors. The D matrix is the diagonal matrix for which singular values are arranged in decreasing order. The B-scan matrix for SVD of S is given by

$$S = \sum_{i=1}^{N} w_i u_i v_i^T \tag{16}$$

where $u_i v_i^T$ are the Eigen-component and w_i is the Eigen-value for the ith component. The first Eigen-value represents the strong reflections, which are generally from the wall in the case of TWI, the remaining values represent other reflections from the target and noise. We can categorize Eigen-space into target sub-space and noise sub-space. Let $E_i = u_i v_i^T$ then

$$E = \left[E_{1 \to k} \middle| E_{k+1 \to p} \middle| E_{p+1 \to N} \right] \tag{17}$$

where $E_{1 \to k}$ represents strong reflections, $E_{k+1 \to p}$ represents reflections from the target and $E_{p+1 \to N}$ represents noise.

3.3.2. Independent Component Analysis (ICA)

ICA divides the data into statistically independent components. Statistical independence considers higher order moments for data matrix S. ICA takes a linear combination of S_x such that

$$I_x = \sum_{j=1}^{N} a_{ij} s_j \quad j = 1, 2, \ldots NI = SA \tag{18}$$

where A is the matrix holding N independent source components. The output signal matrix Y is for the input matrix, matrix I can be determined with the help of the full rank matrix (W), such that

$$Y = WI \tag{19}$$

where W is the matrix which makes I as independent as possible for dependent sensor signals S.

3.4. Entropy-based Time Gating

Recently in [25], entropy-based time gating was proposed for clutter reduction, and it is shown that this method is efficient for clutter reduction compared to earlier methods in the literature. In this method behavior of the clutter, which is similar over each time trace, is used to exploit the entropy.

Clutter signal gives higher entropy compared to the target. The threshold set in this paper is as Equation (20)

$$W(m) = 0 \text{ if entropy } \geq \alpha \log (N-1) \text{ and} W(m) = 1 \text{ } elsewhere \tag{20}$$

where $\alpha < 1$ is the tolerance for the threshold and N is the number of scanning points

The time trace after incorporating the threshold is given as

$$e_w(n,m) = W(m)e(n,m) \tag{21}$$

To illustrate the results using the above methods, we used the data collected with a single metal target and two targets of metal and wood for contrast imaging as described in Section 2. The PSNR for each method in both cases is given in Tables 2 and 3.

Table 2. PSNR in dB for Metal target.

Sr. No.	Clutter Reduction Method	PSNR in dB
1.	Average trace subtraction	10.7504
2.	Singular value decomposition	7.6220
3.	Differential approach	10.1494
4.	Independent component analysis	12.6255

Table 3. PSNR in dB for Metal and wood.

Sr. No.	Clutter Reduction Method	PSNR in dB
1.	Average trace subtraction	9.9608
2.	Singular value decomposition	9.8052
3.	Differential approach	9.3390
4.	Independent component analysis	12.8549

It can be seen from Figure 4 that both targets are visible along with some clutter by using average trace subtraction, but they are not visible by other methods. It is necessary to develop an efficient clutter removal technique to detect a weak target in the presence of the strong clutter.

PSNR is the ratio to analyze the distortion in the final image with respect to the input low-resolution image is given by

$$PSNR = 10 \log_{10} \frac{1}{MSE} \tag{22}$$

$$MSE = \frac{(O.I. - F.I.)^2}{(V.P. * H.P.)} \tag{23}$$

where

MSE—Mean square error

O.I.—Original normalized image

F.I.—Final image

V.P.—Number of vertical scanning points

H.P.—Number of horizontal scanning points

It can be observed from Tables 1 and 2, the PSNR for ICA is high among all other clutter reduction techniques. Even if ICA is efficient for the detection of the low dielectric target it cannot detect a wood target when it is placed at a different down range position compared to the metal target.

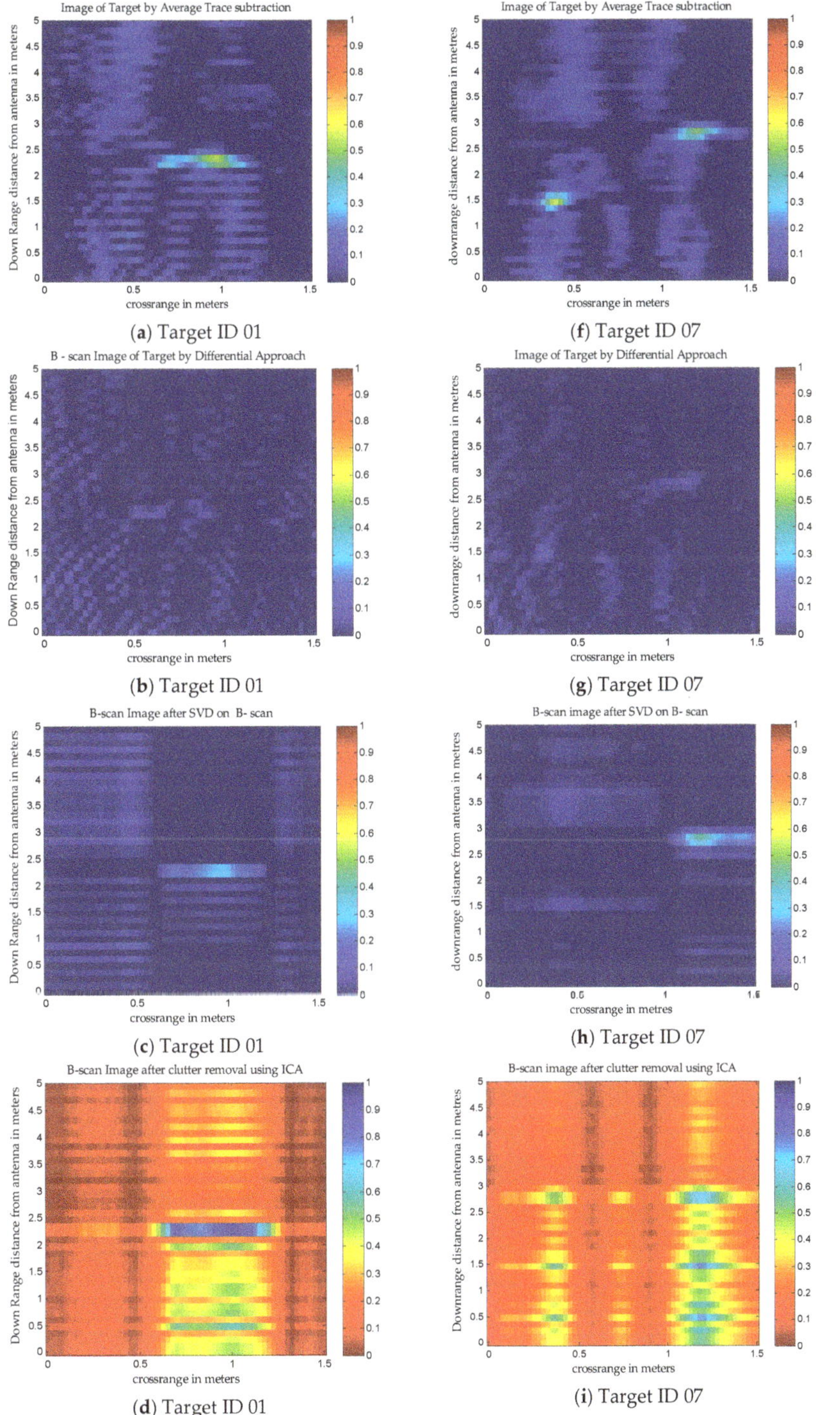

(**a**) Target ID 01

(**f**) Target ID 07

(**b**) Target ID 01

(**g**) Target ID 07

(**c**) Target ID 01

(**h**) Target ID 07

(**d**) Target ID 01

(**i**) Target ID 07

Figure 4. *Cont.*

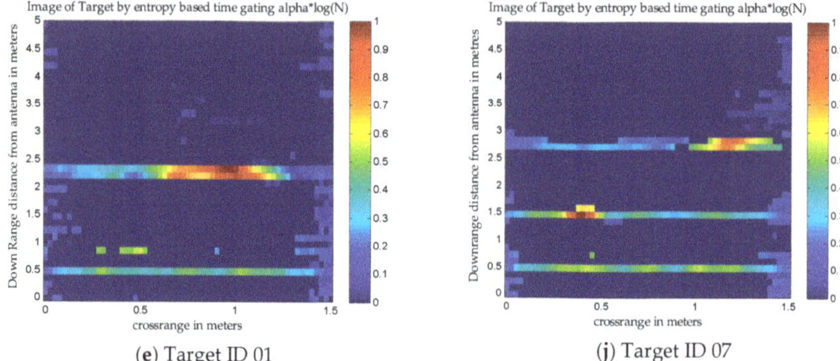

(e) Target ID 01   (j) Target ID 07

Figure 4. B-scan images. (**a**) Average trace subtraction for metal target; (**b**) Differential approach for metal target; (**c**) SVD for metal target; (**d**) ICA for metal target; (**e**) Entropy-based time gating for metal target; (**f**) Average trace subtraction for metal and wood targets; (**g**) Differential approach for metal and wood targets; (**h**) SVD for metal and wood targets; (**i**) ICA for metal and wood targets; (**j**) Entropy-based time gating for metal and wood targets where color bar represents normalized intensity values.

4. A Proposed Novel Method for Contrast Imaging

Low-rank approximation (LRA) has been used recently for seismic data [12]. LRA is efficient compared to SVD and ICA as it exploits the noise space by considering the large rank, this motivates us to use LRA for the detection of contrast targets. LRA is a rank reduction technique, the principle requirement of the LRA is that the data should be low rank. In TWI imaging, the number of targets are less than the number of scanning points ($P < MN$), hence the collected data is inherently sparse and low rank. Steps for traditional LRA are

Step (1): Calculate the SVD for data matrix S.

$$S = U * D * V^T \tag{24}$$

Step (2): Select n largest diagonal singular values from the matrix D and set other values to zero.

$$\widehat{D} = D(1:n, 1:n) \tag{25}$$

Step (3): Calculate the LRA matrix.

$$\widehat{S} = U\widehat{D}V^T \tag{26}$$

While selecting the n largest singular or Eigen-values from matrix D, we ignore the first Eigen-value, which represents the strongest reflections from the wall [1]. The modified LRA for TWI can be given as

Step (4): Ignore first Eigen-value corresponding to wall reflections, hence the matrix D is given as

$$\widehat{D} = D(2:n, 2:n) \tag{27}$$

To satisfy the principle of the algorithm, LRA works in the local windows, where deciding the optimum rank is difficult. Since LRA cannot estimate the optimum rank in the local window, it will consider the large rank to preserve the useful energy in the signal. In the attempt to preserve the useful energy using the large rank, unwanted clutter is also added to the signal. This problem can be solved by using an optimum threshold while selecting Eigen-values during the reconstruction of the useful signal.

Step (5): Select the optimum threshold for Eigen-values from LRA using the entropy-based criterion.

We used the entropy-based criterion [25] for selecting the threshold for the Eigen-values in the LRA. The idea for entropy-based thresholding was adapted to discriminate between target and clutter signals. To select the optimum threshold, we consider the criteria that entropy is maximum for clutter and minimum for the target. First, we construct the normalized time traces as

$$S_{Nz}(n,m) = \frac{S(n,m)^2}{\sum\limits_{l=0}^{N-1} S(l,m)^2} \quad m = 0,\ 1,\dots,\ M-1 \tag{28}$$

$$\text{Now } S_{Nz}(n,m) \geq 0 \text{ and } \sum\limits_{n=0}^{n-1} S_{Nz}(n,m) = 1 \text{ for all } m \tag{29}$$

At each instant, normalized data is considered to be a probability density function (PDF) [29]. Introducing PDF allows us to adopt the entropy-based criterion to determine the threshold and entropy measure, which is given as

$$E_S(m) = -\sum\limits_{n=0}^{N-1} S_{Nz}(n,m)\log(S_{Nz}(n,m)) \tag{30}$$

The entropy of the clutter signals gives large values and the clutter in the observations is generally constant, hence the average value for the Eigen-values can be the optimum threshold. The flow chart for the proposed method is given in Figure 5.

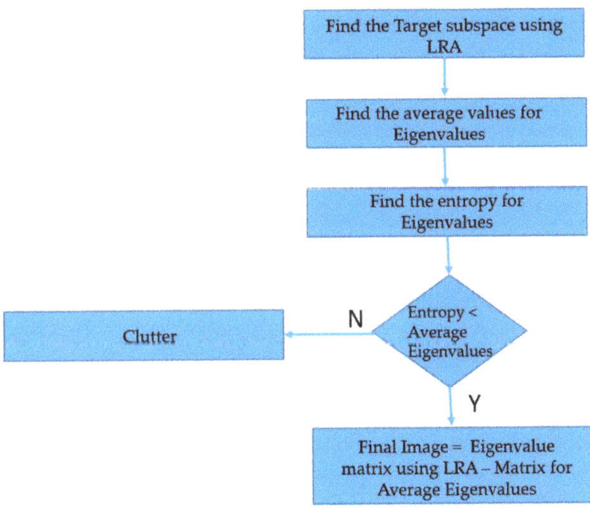

Figure 5. Flow chart for the proposed method.

For subspace approaches, such as SVD and ICA, only a few dominant Eigen-values are considered as the target and lower Eigen-values are considered as the noise, as a result, weak targets are considered as noise. In the proposed developed method, we consider all the Eigen-values and set the optimum threshold to eliminate the noise–space, hence we are able to detect weak targets such as wood along with a strong target such as metal.

The data is processed with different targets to check the capability of the method. For contrast imaging, two targets with different dielectrics are chosen behind the wall, i.e., metal and wood. The results for which are shown in Figure 6.

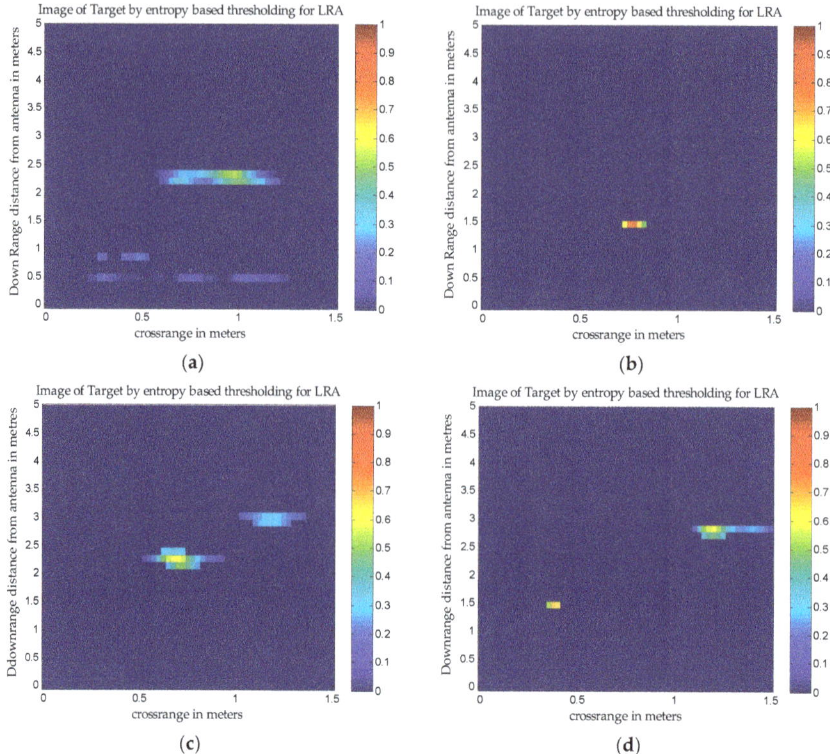

Figure 6. B- scan images with a proposed method with (**a**) Target ID 01: single metal target, (**b**) Target ID 02: single wood target, (**c**) Target ID 05: Two metal targets, (**d**) Target ID 07: Metal and wood targets where color bar represents normalized intensity values.

5. Conclusions

In this paper, the problem of contrast imaging is addressed and it is shown that our developed method is able to detect a weak target in the presence of a strong target. The inherent problem of considering the large rank in LRA is solved by setting the optimum threshold using the entropy-based criterion. The entropy-based LRA method was compared to other methods, such as *average trace subtraction, entropy-based time gating, SVD, ICA* and *DA*, and found to be very effective for different types and arrangements of the target.

Another advantage of the proposed method is that it also avoids the filtering of low spatial targets and therefore this method allows a better reconstruction of low as well as high dielectric targets in the scene. In our future work, we are going to work on rank optimization problem of LRA.

Author Contributions: Conceptualization, Experimental details, D.S.; Supervision, D.S.; Writing—original draft, Experimental work, M.B.; Writing—review & editing, D.S. and H.K.

Funding: This research received no external funding.

Conflicts of Interest: The authors declare no conflict of interest.

Appendix A

We collected 45 datum points for different arrangements of a target position behind the wall, target size and target thickness. The distance of the target is measured from the antenna mouth. Table 3 gives target types while Figures A1–A3 gives A-scan, B-scan and C-scan images using the proposed method as described in Section 4, which are given for validation.

Table A1. Types of targets.

Target ID	Number of Targets	Target Type	The Distance of the Targets from the Antenna Mouth	Target Size/Thickness
01	01	Metal	2.3 m	17.5 cm × 14.5 cm/1 cm
02	01	Wood	1.5 m	Thick wood: 50 cm × 30 cm/2 cm Thin wood: 30 cm × 30 cm/1 cm
03	01	Teflon	1.5 m	50 cm × 40 cm/1 cm
04	02	Metal-Metal	3 m	17.5 cm × 14.5 cm/1 cm
05	02	Metal-Metal	2.3 m and 3 m	17.5 cm × 14.5 cm/1 cm
06	02	Metal-Wood	1.73 m	17.5 cm × 14.5 cm/1 cm, Thick wood: 50 cm × 30 cm/2 cm
07	02	Metal-Wood	2.3 m and 1.5 m	17.5 cm × 14.5 cm/1 cm, Thick wood: 50 cm × 30 cm/2 cm
08	02	Metal-Teflon	2.3 m	17.5 cm × 14.5 cm/1 cm, 50 cm × 40 cm/1 cm
09	02	Metal-Teflon	2.3 m and 1 m	17.5 cm × 14.5 cm/1 cm, 50 cm × 40 cm/1 cm
10	02	Wood (thick)–Wood (thin)	1.73 m	Thick wood: 50 cm × 30 cm/2 cm Thin wood: 30 cm × 30 cm/ 1 cm
11	02	Wood (thick)–Wood (thin)	3.5 m and 2.5 m	Thick wood: 50 cm × 30 cm/2 cm Thin wood: 30 cm × 30 cm/1 cm
12	03	Metal Wood (Thick)-Wood (thin)	1.5 m	17.5 cm × 14.5 cm/ 1 cm, Thick wood: 50 cm × 30 cm/2 cm Thin wood: 30 cm × 30 cm/ 1 cm
13	03	Metal-Wood (Thick)-Wood (thin)	3.5 m, 2.5 m, 1.5 m	17.5 cm × 14.5 cm/ 1 cm, Thick wood: 50 cm × 30 cm/2 cm Thin wood: 30 cm × 30 cm/1 cm

TWI scanning methods—In TWI, three types of scanning are done for target detection and shape identification. The A-Scan or range profile is a dimensional plot, which provides information about the presence of a target along with the approximate location. The B-Scan gives information about a number of targets present in the down-range and the C-scan gives information about shape, height, and width. A-Scan plots for different target ID are shown in Figure A1. B-Scan and C-Scan images are developed using 30 horizontal scans and 15 vertical scans.

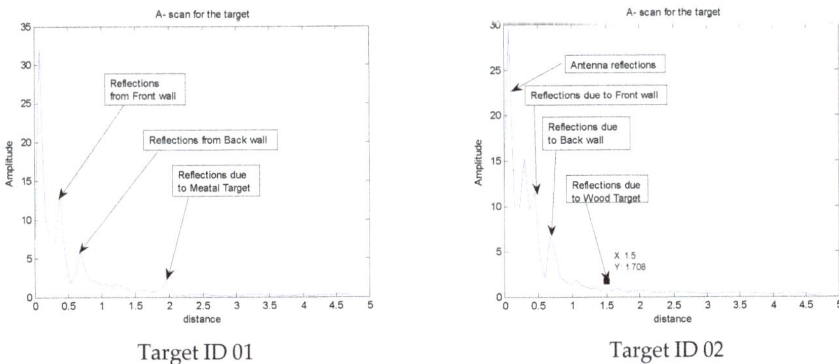

Target ID 01   Target ID 02

Figure A1. *Cont.*

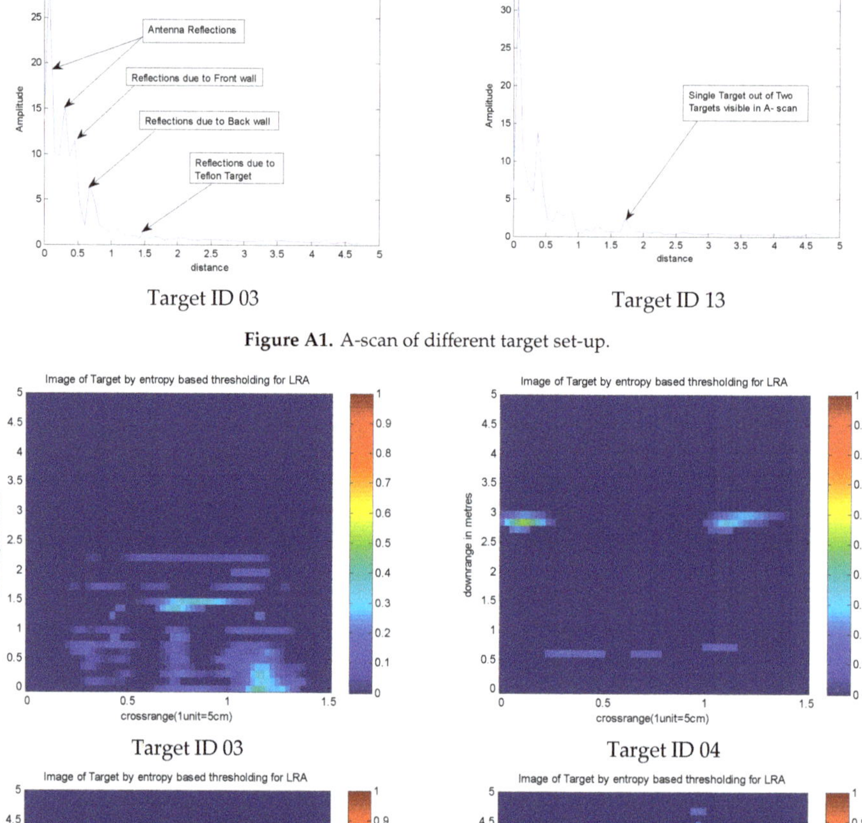

Figure A1. A-scan of different target set-up.

Figure A2. *Cont.*

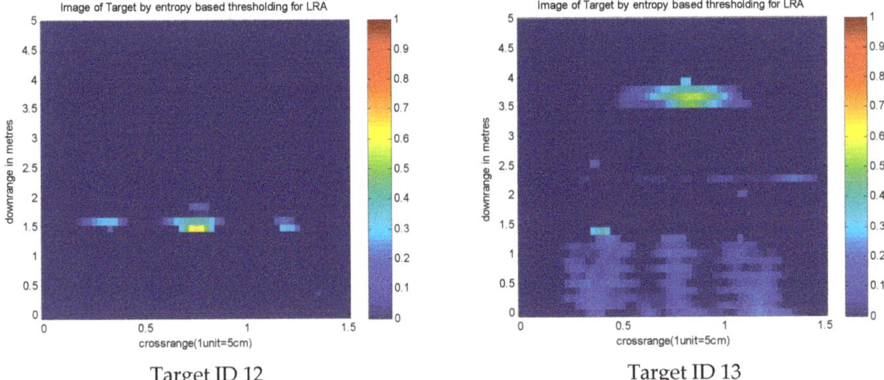

Target ID 12   Target ID 13

Figure A2. B-scan images for different targets.

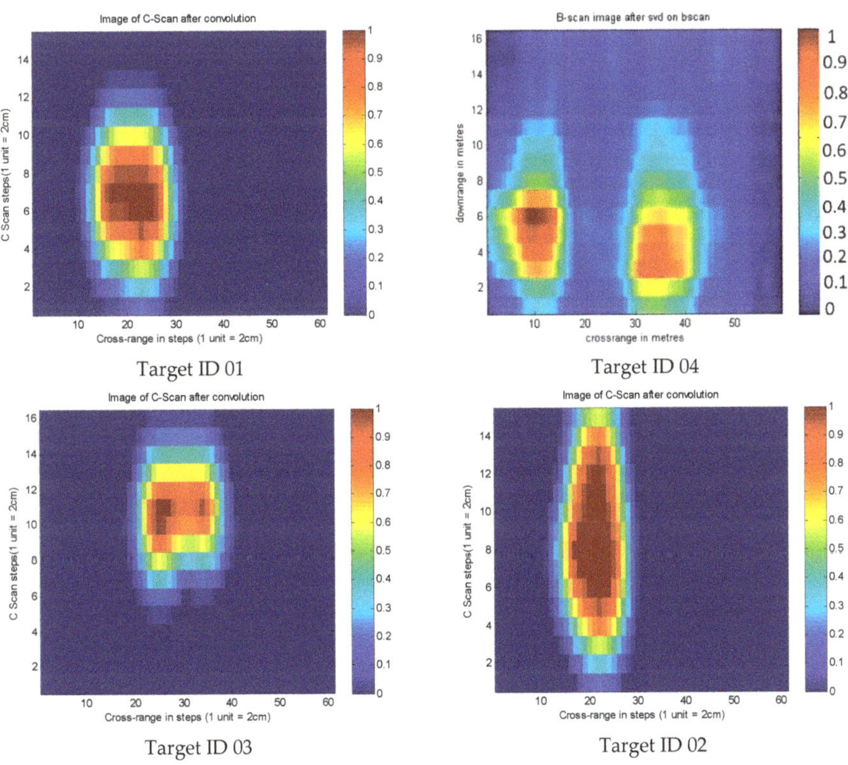

Target ID 01   Target ID 04

Target ID 03   Target ID 02

Figure A3. C-scan images for different targets where the color bar represents normalized intensity values.

References

1. Verma, P.K.; Gaikwad, A.N.; Singh, D.; Nigam, M.J. Analysis of Clutter Reduction Techniques for through Wall Imaging in UWB Range. *Prog. Electromagn. Res.* **2009**, *17*, 29–48. [CrossRef]
2. Potin, D.; Duflos, E.; Vanheeghe, P. Landmines Ground-Penetrating Radar Signal Enhancement by Digital Filtering. *IEEE Trans. Geosci. Remote Sens.* **2006**, *44*, 2393–2406. [CrossRef]
3. van der Merwe, A.; Gupta, I.J. A novel signal processing technique for clutter reduction in GPR measurements of small, shallow land mines. *IEEE Trans. Geosci. Remote Sens.* **2000**, *38*, 2627–2637.

4.   Daniels, D.J. A review of landmine detection using GPR. In Proceedings of the 2008 European Radar Conference, Amsterdam, Netherlands, 30–31 October 2008; pp. 280–283.

5.   van Kempen, L.; Sahli, H. Signal processing techniques for clutter parameters estimation and clutter removal in GPR data for landmine detection. In Proceedings of the 11th IEEE Signal Processing Workshop on Statistical Signal Processing (Cat. No.01TH8563), Singapore, 8 August 2001; pp. 158–161.

6.   Zoubir, A.M.; Chant, I.J.; Brown, C.L.; Barkat, B.; Abeynayake, C. Signal processing techniques for landmine detection using impulse ground penetrating radar. *IEEE Sens. J.* **2002**, *2*, 41–51. [CrossRef]

7.   Gaikwad, A.N.; Singh, D.; Nigam, M.J. Application of clutter reduction techniques for detection of metallic and low dielectric target behind the brick wall by stepped frequency continuous wave radar in ultra-wideband range. *IET Radar Sonar Navig.* **2011**, *5*, 416–425. [CrossRef]

8.   Tivive, F.H.C.; Bouzerdoum, A.; Amin, M.G. A Subspace Projection Approach for Wall Clutter Mitigation in Through-the-Wall Radar Imaging. *IEEE Trans. Geosci. Remote Sens.* **2015**, *53*, 2108–2122. [CrossRef]

9.   Tebchrany, E.; Sagnard, F.; Baltazart, V.; Tarel, J.; Dérobert, X. Assessment of statistical-based clutter reduction techniques on ground-coupled GPR data for the detection of buried objects in soils. In Proceedings of the 15th International Conference on Ground Penetrating Radar, Brussels, Belgium, 30 June–4 July 2014; pp. 604–609.

10.   Tivive, F.H.C.; Bouzerdoum, A.; Amin, M.G. An SVD-based approach for mitigating wall reflections in through-the-wall radar imaging. In Proceedings of the 2011 IEEE RadarCon (RADAR), Kansas City, MO, USA, 23–27 May 2011; pp. 519–524.

11.   Tivive, F.H.C.; Amin, M.G.; Bouzerdoum, A. Wall clutter mitigation based on eigen-analysis in through-the-wall radar imaging. In Proceedings of the 2011 17th International Conference on Digital Signal Processing (DSP), Corfu, Greece, 6–8 July 2011; pp. 1–8.

12.   Chen, Y.; Zhou, Y.; Chen, W.; Zu, S.; Huang, W.; Zhang, D. Empirical Low-Rank Approximation for Seismic Noise Attenuation. *IEEE Trans. Geosci. Remote Sens.* **2017**, *55*, 4696–4711. [CrossRef]

13.   Mohsin Riaz, M.; Ghafoor, A. Through-Wall Image Enhancement Based on Singular Value Decomposition. Available online: https://www.hindawi.com/journals/ijap/2012/961829/ (accessed on 29 May 2018).

14.   Zhang, Y.; Xia, T. In-Wall Clutter Suppression Based on Low-Rank and Sparse Representation for Through-the-Wall Radar. *IEEE Geosci. Remote Sens. Lett.* **2016**, *13*, 671–675. [CrossRef]

15.   Wang, G.; Amin, M. A new approach for target location of through the wall radar imaging in the presence of wall ambiguities. In Proceedings of the Fourth IEEE International Symposium on Signal Processing and Information Technology, Rome, Italy, 18–21 December 2004; pp. 183–186.

16.   Boje, E. Fast discrete Fourier transform with exponentially spaced points. *IEEE Trans. Signal Process.* **1995**, *43*, 3033–3035. [CrossRef]

17.   Nicolaescu, I.; Genderen, P. van Procedures to improve the performances of a Sfcw radar used for landmine detection. In Proceedings of the 2008 Microwaves, Radar and Remote Sensing Symposium, Kiev, Ukraine, 22–24 September 2008; pp. 250–255.

18.   Mikhnev, V.A.; Vainikainen, P. Single-reference near-field calibration procedure for step-frequency ground penetrating radar. *IEEE Trans. Geosci. Remote Sens.* **2003**, *41*, 75–80. [CrossRef]

19.   Ahmad, F.; Amin, M.G.; Kassam, S.A. A beamforming approach to stepped-frequency synthetic aperture through-the-wall radar imaging. In Proceedings of the 1st IEEE International Workshop on Computational Advances in Multi-Sensor Adaptive Processing, Puerto Vallarta, Mexico, 13–15 December 2005; pp. 24–27.

20.   Kaushal, S.; Kumar, B.; Singh, D. An autofocusing method for imaging the targets for TWI radar systems with correction of thickness and dielectric constant of wall. *Int. J. Microw. Wirel. Technol.* **2019**, *11*, 15–21. [CrossRef]

21.   Yao, Q.; Qifu, W. Kirchoff Migration Algorithm for Ground Penetrating Radar Data. In Proceedings of the 2012 International Conference on Computer Science and Electronics Engineering, Hangzhou, China, 23–25 March 2012; Volume 2, pp. 396–398.

22.   Yektakhah, B.; Dehmollaian, M. A Method for Cancellation of Clutter Due to an Object in Transceiver Side of a Wall for Through-Wall Sensing Applications. *IEEE Geosci. Remote Sens. Lett.* **2012**, *9*, 559–563. [CrossRef]

23.   Kumar, B.; Upadhyay, R.; Singh, D. Development of an Adaptive Approach for Identification of Targets (Match Box, Pocket Diary and Cigarette Box) under the Cloth with MMW Imaging System. *Prog. Electromagn. Res.* **2017**, *77*, 37–55. [CrossRef]

24.   Kumar, B.; Sharma, P.; Singh, D. Development of an efficient approach for MMW imaging system to identify concealed targets inside the book. *Microw. Opt. Technol. Lett.* **2017**, *59*, 2982–2990. [CrossRef]

25. Solimene, R.; Cuccaro, A. Front Wall Clutter Rejection Methods in TWI. *IEEE Geosci. Remote Sens. Lett.* **2014**, *11*, 1158–1162. [CrossRef]
26. Yoon, Y.S.; Amin, M.G. Spatial Filtering for Wall-Clutter Mitigation in Through-the-Wall Radar Imaging. *IEEE Trans. Geosci. Remote Sens.* **2009**, *47*, 3192–3208. [CrossRef]
27. Dehmollaian, M.; Thiel, M.; Sarabandi, K. Through-the-Wall Imaging Using Differential SAR. *IEEE Trans. Geosci. Remote Sens.* **2009**, *47*, 1289–1296. [CrossRef]
28. Hyvarinen, A. Fast and robust fixed-point algorithms for independent component analysis. *IEEE Trans. Neural Netw.* **1999**, *10*, 626–634. [CrossRef] [PubMed]
29. Stickley, G.F.; Noon, D.A.; Cherniakov, M.; Longstaff, I.D. Gated stepped-frequency ground penetrating radar. *J. Appl. Geophys.* **2000**, *43*, 259–269. [CrossRef]

Article

Coherent Integration for Radar High-Speed Maneuvering Target Based on Frequency-Domain Second-Order Phase Difference

Ke Jin [1,*], Tao Lai [1], Yubing Wang [2], Gongquan Li [1] and Yongjun Zhao [1]

[1] National Digital Switching System Engineering and Technological Research Center, Zhengzhou 450000, China; ltnudt@126.com (T.L.); lgq1225@126.com (G.L.); zhaoyjzz@163.com (Y.Z.)
[2] Air Traffic Control and Navigation College, Air Force Engineering University, Xi'an 710000, China; wangyubing_AFU@126.com
* Correspondence: jinke_xd@outlook.com

Received: 18 January 2019; Accepted: 27 February 2019; Published: 4 March 2019

Abstract: In recent years, target detection has drawn increasing attention in the field of radar signal processing. In this paper, we address the problem of coherent integration for detecting high-speed maneuvering targets, involving range migration (RM), quadratic RM (QRM), and Doppler frequency migration (DFM) within the coherent processing interval. We propose a novel coherent integration algorithm based on the frequency-domain second-order phase difference (FD-SoPD) approach. First, we use the FD-SoPD operation to reduce the signal from three to two dimensions and simultaneously eliminate the effects of QRM and DFM, which leads to signal-to-noise ratio improvement in the velocity-acceleration domain. Next, we estimate the target motion parameters from the peak position without the need for a search procedure. We show that this algorithm can be easily implemented by using complex multiplications combined with fast Fourier transform (FFT) and inverse FFT (IFFT) operations. We perform comparisons with several representative algorithms and show that the proposed technique can be used to achieve a good trade-off between computational complexity and detection performance. We present both simulated and experimental data to demonstrate the effectiveness of the proposed method.

Keywords: maneuvering target detection; coherent integration; motion parameter estimation; second-order phase difference (SoPD); time-frequency analysis

---

1. Introduction

With the increasing requirements for space target detection and high-resolution imaging, radar high-speed maneuvering target detection has drawn growing attention [1–11]. Normally, a low-speed target is located in the same range cell during the short observation time, and the conventional moving target detection (MTD) algorithm [12] can complete coherent integration by using fast Fourier transform (FFT). It is well known that in order to improve the detection ability in far-range and low radar cross section (RCS) targets, a long-term coherent integration is always required [13]. In this case, for high-speed maneuvering targets, the linear range migration (LRM), quadratic range migration (QRM), and Doppler frequency migration (DFM) effects will inevitably occur, thereby deteriorating integration performance.

As for radar coherent integration, many successful detection algorithms have been proposed, such as the keystone transform (KT) [14,15], scaled inverse Fourier transform (SCIFT) [16,17], frequency-domain deramp-keystone transform (FDDKT) [18], modified location rotation transform (MLRT) [19], and Radon Fourier transform (RFT) [20]. For a moving target with linear range migration, these algorithms achieve satisfactory antinoise performance, parameter estimation accuracy, and detection ability with reasonable computational cost. Nevertheless, they may suffer from integration performance degradation due to ignoring the effects of QRM and DFM caused by the target's acceleration.

To address these issues, many advanced methods have been recently proposed. They can be roughly divided into three categories.

(a)   Radon transform-based algorithms, such as generalized Radon Fourier transform (GRFT) [21], Radon-fractional Fourier transform (RFRFT) [22], and Radon-Lv's Distribution (RLVD) [23–26]. These kinds of algorithms implement phase compensation and parameter estimation by searching the maneuvering target motion trajectory. Although they can obtain coherent integration under a low signal-to-noise ratio (SNR), the huge computational load seriously limits their practical application.

(b)   KT based algorithms, such as second-order keystone transform (SoKT) [27], Doppler keystone transform (DKT) [28], keystone-Lv's distribution (KT-LVD) [29], and so on. The KT is used to correct the QRM blindly, which reduces the calculation cost to a certain extent, but it still needs to use parameter searching to eliminate the Doppler ambiguity.

(c)   Correlation-based algorithms: The representative adjacent cross-correlation function and Lv's distribution (ACCF-LVD) algorithm proposed in References [30–32] reduces the migration order by ACCF and quickly estimates the motion parameters without any searching procedure, which greatly reduces the computational burden and benefits practical applications. Unfortunately, this method is only effective when the input SNR is high [33]. The three-dimensional scaled transform (TDST) method was then presented to realize coherent integration and motion parameters estimations for maneuvering targets under a low SNR background [34]. This method eliminates the coupling effectively among spatial frequency, slow time, and time delay. However, the complex three-dimensional transform is usually less suitable for realistic applications.

Aiming to realize the coherent integration of radar high-speed maneuvering targets with low computational complexity, we propose a novel frequency-domain second-order phase difference (FD-SoPD) algorithm in this paper. First, the SoPD is used in the spatial frequency domain to eliminate the impact of acceleration. Then, we can simultaneously estimate the velocity and acceleration from the peak position, followed by phase compensation and coherent integration. The proposed technique has the following contributions: (a) It reduces the signal from three to two dimensions, thus avoiding the complex operation of TDST; (b) the target motion parameters can be easily estimated by FFT without any searching process; (c) the phase difference eliminates the Doppler ambiguity, thus the high speed of target can be accurately estimated; (d) it achieves a good balance between the computational cost and detection ability. Finally, we present both simulated and experimental data to demonstrate the effectiveness of the proposed method.

The remainder of this paper is organized as follows. In Section 2, the signal model for the maneuvering target is established. In Section 3, we deduce the principle of FD-SoPD in detail and discuss the situations of single target and multi-targets, respectively. Section 4 analyses the computational burden. In Section 5, we evaluate the performance via several numerical experiments. Finally, conclusions are drawn in Section 6.

2. Signal Model and Problem Formulation

Suppose the radar transmits a linear frequency modulated (LFM) signal, which can be expressed as:

$$s_t(\hat{t}) = \text{rect}\left(\frac{\hat{t}}{T_p}\right) \exp\left(j2\pi f_c \hat{t} + j\pi\gamma\hat{t}^2\right) \tag{1}$$

where,

$$\text{rect}\left(\frac{\hat{t}}{T_p}\right) = \begin{cases} 1, & |\hat{t}| < T_p/2 \\ 0, & |\hat{t}| > T_p/2 \end{cases} \tag{2}$$

is the rectangular window function, and $\hat{t}$ is the fast time. T_p, f_c and γ indicate the pulse width, carrier frequency, and frequency modulation rate, respectively. Assume that there are K targets in the scene of radar observation. During the accumulation time, the distance between the maneuvering target and radar can be approximated as second order, i.e.,

$$R_i(t_m) = r_i + v_i t_m + a_i t_m^2/2 \tag{3}$$

where $t_m = m/PRF$ is the slow time, m and PRF denote the transmitted pulse number index and the pulse repetition frequency (PRF). r_i, v_i, and a_i are respectively the initial slant range, radial velocity, and acceleration of the ith target.

Ignoring the influence of noise, the received signal of K targets after down conversion can be expressed as [35]:

$$s_r\left(\hat{t}, t_m\right) = \sum_{i=1}^{K} A_{r,i} \text{rect}\left(\frac{\hat{t} - 2R_i(t_m)/c}{T_p}\right) \exp\left\{-j\frac{4\pi f_c R_i(t_m)}{c}\right\} \exp\left\{j\pi\gamma\left(\hat{t} - \frac{2R_i(t_m)}{c}\right)^2\right\} \tag{4}$$

where $A_{r,i}$ is the target reflectivity, and c is the light speed.

After pulse compression, the radar echoes are written as:

$$s_c\left(\hat{t}, t_m\right) = \sum_{i=1}^{K} A_{c,i} \text{sinc}\left[B\left(\hat{t} - \frac{2R_i(t_m)}{c}\right)\right] \exp\left\{-j\frac{4\pi f_c R_i(t_m)}{c}\right\} \tag{5}$$

where $A_{c,i}$ denotes the amplitude after compression and $B = \gamma T_p$ is the bandwidth of transmitted signal.

Substituting Equation (3) into Equation (5), we obtain:

$$s_c\left(\hat{t}, t_m\right) = \sum_{i=1}^{K} A_{c,i} \text{sinc}\left[B\left(\hat{t} - \frac{2\left(r_i + v_i t_m + a_i t_m^2/2\right)}{c}\right)\right] \exp\left\{-j\frac{4\pi f_c\left(r_i + v_i t_m + a_i t_m^2/2\right)}{c}\right\} \tag{6}$$

As can be seen from Equation (6), the signal envelope indicates the target range, which changes nonlinearly with the slow time. When the offset exceeds the range resolution $\Delta r = c/2B$, the LRM will occur. If the target has a large acceleration, the QRM can be seen in the envelope. In this case, the conventional MTD is invalid. In addition, the phase in Equation (6) indicates a linear change in the Doppler frequency of the ith target, i.e.,

$$f_{d,i} = \frac{2}{\lambda}\frac{\text{d}\left(r_i + v_i t_m + a_i t_m^2/2\right)}{\text{d}t_m} = \frac{2v_i + 2a_i t_m}{\lambda} \tag{7}$$

where $\lambda = c/f_c$ is the wave length. Similarly, when the offset exceeds a Doppler resolution, DFM would occur and defocus the target energy in the Doppler domain. Moreover, for high-speed targets, we often have $f_{d,i} \gg PRF$, which induces the Doppler ambiguity and makes it hard to estimate the target's velocity. Therefore, the coherent accumulation of high-speed maneuvering targets can only be achieved by effectively eliminating the LRM, QRM, and DFM.

3. The Principle of the FD-SoPD

3.1. FD-SoPD with Mono-Target

According to Equation (6), the compressed signal of the ith target is

$$s_c(\hat{t}, t_m) = A_{c,i}\text{sinc}\left[B\left(\hat{t} - \frac{2\left(r_i + v_i t_m + a_i t_m^2/2\right)}{c}\right)\right]\exp\left\{-j\frac{4\pi f_c\left(r_i + v_i t_m + a_i t_m^2/2\right)}{c}\right\} \tag{8}$$

Performing the Fourier transform (FT) along the $\hat{t}$ axis, we obtain the spatial spectrum of the signal, i.e.,

$$\begin{aligned}S(f_r, t_m) &= A_{f_r,i}\text{rect}\left(\tfrac{f_r}{B}\right)\exp\left(-j\tfrac{4\pi(f_r+f_c)R_i(t_m)}{c}\right)\\&= A_{f_r,i}\text{rect}\left(\tfrac{f_r}{B}\right)\exp\left\{-j4\pi\tfrac{(f_r+f_c)}{c}\left(r_i + v_i t_m + a_i t_m^2/2\right)\right\}\end{aligned} \tag{9}$$

where f_r is the frequency of the spatial harmonic from the spatial spectrum of the fast time signal record, and $A_{f_r,i}$ is the amplitude of the spatial harmonic obtained by FT of the fast time signal record.

As shown in Equation (9), the coupling between f_r and t_m (or t_m^2) is the fundamental cause of LRM or QRM. Moreover, the existence of t_m^2 broadens the Doppler spectrum and makes the signal energy defocused. If the velocity v_i and acceleration a_i are accurately estimated, it is easy to perform phase compensation and coherent integration.

The proposed FD-SoPD is defined as:

$$R_{SoPD}(t_m, \tau, \tau_1; f_r) = S\left(f_r; t_m + \tfrac{\tau}{2}\right)S^*\left(f_r; t_m - \tfrac{\tau}{2}\right)\left[S\left(f_r; t_m + \tfrac{\tau_1}{2}\right)S^*\left(f_r; t_m - \tfrac{\tau_1}{2}\right)\right]^* \tag{10}$$

where τ and τ_1 are two lag time variables. Substituting Equation (9) into Equation (10) yields:

$$\begin{aligned}R_{SoPD}(f_r, t_m, \tau; \tau_0) &= \left|A_{f_r,i}\right|^4\text{rect}\left(\tfrac{f_r}{B}\right)\exp\left[j4\pi\tfrac{f_r+f_c}{c}v_i(\tau_1 - \tau)\right]\\&\quad\times\exp\left[j4\pi\tfrac{f_r+f_c}{c}a_i t_m(\tau_1 - \tau)\right]\end{aligned} \tag{11}$$

When τ and τ_1 have fixed nonzero lag time difference, the coupling between τ and τ_1 will be eliminated, i.e.,

$$2\tau_0 = \tau_1 - \tau \tag{12}$$

where τ_0 is the fixed lag time. Equation (11) can be further expressed as:

$$\begin{aligned}R_{SoPD}(f_r, t_m, \tau; \tau_0) &= \left|A_{f_r,i}\right|^4\text{rect}\left(\tfrac{f_r}{B}\right)\exp\left(j8\pi\tfrac{f_r+f_c}{c}v_i\tau_0\right)\\&\quad\times\exp\left(j8\pi\tfrac{f_r+f_c}{c}a_i\tau_0 t_m\right)\end{aligned} \tag{13}$$

As shown in Equation (13), three axes, f_r, t_m and τ, exist in $R_{SoPD}(f_r, t_m, \tau; \tau_0)$. However, the signal energy is constant along the τ axis, and thus can be accumulated coherently by the addition operation as follows:

$$\begin{aligned}R_A(f_r, t_m; \tau_0) &= \text{ADD}_\tau[R_{SoPD}(f_r, t_m, \tau; \tau_0)]\\&= G_m\left|A_{f_r,i}\right|^4\text{rect}\left(\tfrac{f_r}{B}\right)\exp\left(j8\pi\tfrac{f_r+f_c}{c}v_i\tau_0\right)\exp\left(j8\pi\tfrac{f_r+f_c}{c}a_i\tau_0 t_m\right)\end{aligned} \tag{14}$$

where $\text{ADD}_\tau(\cdot)$ is the addition function along the τ axis and G_m denotes the corresponding integration gain.

Remark 1. *From Equation (14), we may find three features of the FD-SoPD. (a) The signal is reduced from three to two dimensions, which avoids the multidimensional scaled transform in TDST algorithm. (b) The QRM and*

DRM are simultaneously eliminated. (c) Equation (14) is equivalent to a uniform motion model, whose velocity is $-2a_i\tau_0$. Thus, the Doppler ambiguity is eliminated. If the envelope migration caused by the velocity $-2a_i\tau_0$ exceeds a range cell, the KT is needed, i.e.,

$$(f_r + f_c)t_m = f_c\tau_m \tag{15}$$

where τ_m denotes the scaled slow-time variable.

After performing the KT on Equation (14), we have:

$$\begin{aligned}
R_A(f_r, \tau_m; \tau_0) = G_m\left|A_{f_r,i}\right|^4 rect\left(\tfrac{f_r}{B}\right) \exp\left(j8\pi\tfrac{f_c}{c}v_i\tau_0\right) \\
\times \exp\left(j8\pi\tfrac{f_r}{c}v_i\tau_0\right) \exp\left(j8\pi\tfrac{f_c}{c}a_i\tau_0\tau_m\right)
\end{aligned} \tag{16}$$

Applying the FT with respect to f_r and τ_m, we get:

$$\begin{aligned}
S_F(\hat{t}, f_{\tau m}) &= FT_{\tau_m}\left\{FT_{f_r}[R_A(f_r, \tau_m; \tau_0)]\right\} \\
&= A_{F,i}\exp\left(j\tfrac{8\pi f_c v_i\tau_0}{c}\right)\mathrm{sinc}\left[B\left(\hat{t} + \tfrac{4v_i\tau_0}{c}\right)\right]\mathrm{sinc}\left[CI\left(f_{\tau m} - \tfrac{4a_i\tau_0}{\lambda}\right)\right]
\end{aligned} \tag{17}$$

where $A_{F,i}$ is amplitude after two-dimensional FT, $f_{\tau m}$ is the frequency with respect to τ_m, and CI denotes the coherent integration time.

From Equation (17), we can simultaneously estimate the velocity v_i and acceleration a_i of the ith target, i.e.,

$$\left(\hat{v}_i = \frac{-c\hat{t}_{max}}{4\tau_0}, \hat{a}_i = \frac{\lambda f_{\tau m,max}}{4\tau_0}\right) \tag{18}$$

Notice that the fixed lag time constant τ_0 is important in the implementation of the SoPD. A large fixed lag time τ_0 will improve the parameter estimation accuracy, whereas spectrum aliasing may occur. Therefore, the compromise consideration usually chooses $\tau_0 < T_a/4$, where T_a is the observation time.

Utilizing the estimated parameters, we can construct the phase compensation function to compensate the LRM, QRM, and DFM in Equation (9)

$$H_{com}(f_r, t_m) = \exp\left(j4\pi f_r\frac{v_i t_m + a_i t_m^2/2}{c}\right)\exp\left(j4\pi f_c\frac{a_i t_m^2/2}{c}\right) \tag{19}$$

Finally, the signal energy will be integrated by the IFT and the FT operations,

$$\begin{aligned}
E(\hat{t}, f_d) &= FT_{t_m}\left\{IFT_{f_r}[S(f_r, t_m)H_{com}(f_r, t_m)]\right\} \\
&= A_{E,i}\mathrm{sinc}\left[B\left(\hat{t} - \tfrac{2r_i}{c}\right)\right]\delta(f_d + f_{d0,i})
\end{aligned} \tag{20}$$

where f_d is the Doppler frequency with respect to t_m, and $f_{d0,i}$ is the Doppler frequency of the target.

In Equation (20), the signal energy of a high-speed maneuvering target is integrated into a single peak in the range-Doppler domain. The peak position is $(2r_i/c, -f_{d0,i})$ and the peak value is $|E(2r_i/c, -f_{d0,i})|$. Here, the constant false alarm rate (CFAR) [36] technique can be used for the target detection, i.e.,

$$|E(2r_i/c, -f_{d0,i})| \mathop{\gtrless}\limits_{H_0}^{H_1} \eta \tag{21}$$

where η is the CFAR threshold. If $|E(2r_i/c, -f_{d0,i})|$ is larger than the threshold, there will be a moving target. Otherwise, no target is detected.

Remark 2. *Different from the Radon transform based algorithms in [13,21–23], the proposed FD-SoPD method avoids the brute-force searching procedure of unknown motion parameters. In addition, it can be easily implemented by FFT and IFFT, which significantly reduces the computational complexity.*

In the following, we will give an example to demonstrate how the FD-SoPD works to accomplish target motion parameter estimation and coherent integration.

Example 1. *We use an ideal maneuvering point target in this example. The parameters of frequency-modulated continuous-wave (FMCW) radar are set as: The carrier frequency $f_c = 1$ GHz, the bandwidth $B = 100$ MHz, the sampling frequency $f_s = 2$ MHz, pulse repetition frequency PRF $= 128$ Hz, and the number of integration pulses $M = 256$ and $\tau_0 = T_a/5$. The motion parameters of the maneuvering point target are: $A_{r,i} = 1$, $r_i = 3km$, $v_i = 15\,m/s$, $a_i = 1m/s^2$. Simulation results are shown in Figure 1.*

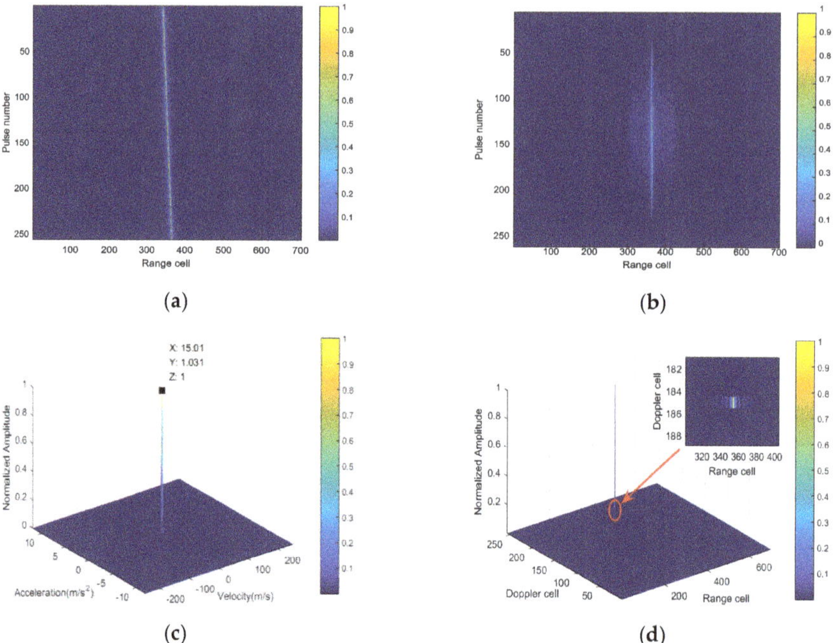

Figure 1. Simulation results of Example 1. (**a**) The result of pulse compression; (**b**) the result of the frequency-domain second-order phase difference (FD-SoPD); (**c**) velocity–acceleration distribution; (**d**) the result of coherent integration.

Figure 1a shows the result of pulse compression, where serious LRM occurs due to the target's high speed and the radar's high resolution. Figure 1b shows the result of FD-SoPD transform. It is obvious that, after the FD-SoPD with respect to slow time, the envelope migration is effectively eliminated, and the target energy is located in the same range cell. Performing the FT with respect to τ_m, we can get the velocity-acceleration distribution, as shown in Figure 1c. The target energy is well accumulated as one peak, and we can estimate $\hat{v}_i = 15.01\,\text{m/s}$ and $\hat{a}_i = 1.031\text{m/s}^2$ from its position. Finally, coherent integration is easily accomplished with FFT, and the result is given in Figure 1d.

3.2. FD-SoPD with Multi-Targets

In this subsection, we will analyze the performance of FD-SoPD under multi-targets in detail. Assume that there are K maneuvering targets in the scene of radar observation. The compressed signal can be expressed as Equation (6). Accordingly, the signal in the spatial frequency domain is:

$$
\begin{aligned}
S(f_r, t_m) &= \sum_{i=1}^{K} A_{f_r,i} rect\left(\frac{f_r}{B}\right) \exp\left(-j\frac{4\pi(f_r+f_c)R_i(t_m)}{c}\right) \\
&= \sum_{i=1}^{K} A_{f_r,i} rect\left(\frac{f_r}{B}\right) \exp\left\{-j4\pi\frac{(f_r+f_c)}{c}\left(r_i + v_i t_m + a_i t_m^2/2\right)\right\}
\end{aligned}
\tag{22}
$$

Substituting Equation (22) into Equation (10) along the slow time, we obtain the FD-SoPD of $S(f_r, t_m)$

$$
R_{SoPD}(f_r, t_m, \tau; \tau_0) = R_{auto}(f_r, t_m, \tau; \tau_0) + R_{cross}(f_r, t_m, \tau; \tau_0)
\tag{23}
$$

where $R_{auto}(f_r, t_m, \tau; \tau_0)$ and $R_{cross}(f_r, t_m, \tau; \tau_0)$ denote the auto-terms and cross terms, and can be written as

$$
\begin{aligned}
R_{auto}(f_r, t_m, \tau; \tau_0) &= \sum_{i=1}^{K} \left|A_{f_r,i}\right|^4 rect\left(\frac{f_r}{B}\right) \exp\left(j8\pi\frac{f_r+f_c}{c}v_i\tau_0\right) \\
&\times \exp\left(j8\pi\frac{f_r+f_c}{c}a_i\tau_0 t_m\right)
\end{aligned}
\tag{24}
$$

$$
R_{cross}(f_r, t_m, \tau; \tau_0) = R_2 + R_3 + R_4 + R_5
\tag{25}
$$

The summation R_5 can be further expanded as the following three parts:

$$
R_5 = R_6 + R_7 + R_8
\tag{26}
$$

The detailed expressions of $R_i (i = 2, 3 \cdots, 8)$ are given in the Appendix A.

After the addition, KT and two-dimensional FT, the velocity and acceleration of target will be estimated simultaneously, i.e.,

$$
S_F(\hat{t}, f_{\tau m}) = FT_{\tau m}\left\{FT_{f_r}\{KT\{ADD_\tau[R_{SoPD}(f_r, t_m, \tau; \tau_0)]\}\}\right\}
\tag{27}
$$

According to the specific motion of the maneuvering target, we consider the cross-terms resulting from the following two cases.

Case 1. *The acceleration of any two targets is different, i.e., $a_i \neq a_j, \forall i, j = 1, 2, \ldots, K, i \neq j$. In this case, R_{auto} has a similar form with Equation (13), which can be integrated after FFT. R_2 has the linear term of τ and the coupling term between t_m and τ, which cannot be accumulated in the addition operation and FFT of Equation (27). R_3, R_4, R_7 and R_8 have the quadratic term τ^2 and a coupling term between t_m and τ, and thus cannot be accumulated as well. It is known from Equation (34) that R_6 has a symmetric property about τ, which will become a sinusoidal oscillation term after the addition in Equation (27). Therefore, the energy of R_6 will be smeared after performing two-dimensional FFT. In summary, the cross terms can be ignored compared to the auto-terms. Here, we give an example to illustrate the discussion of Case 1.*

Example 2. *In this example, we use two maneuvering targets designated as Tr1 and Tr2, respectively. Radar parameters are the same as those in Example 1. Target motion parameters are set as: $A_{r,1} = 1, r_1 = 3km$, $v_1 = 15\,m/s, a_1 = 1m/s^2$ for target Tr1; $A_{r,2} = 1, r_2 = 3.2km, v_2 = -12\,m/s, a_2 = -0.6m/s^2$ for target Tr2.*

Figure 2a is the result of pulse compression. Figure 2b gives the result of FD-SoPD. It is obvious that the auto-terms are corrected into beelines, while the cross-terms cannot be corrected. Thus, after the addition and two-dimensional FT, only the auto-terms are accumulated into two peaks, as shown in Figure 2c. We can estimate the velocity and acceleration of targets as $\hat{v}_1 = 15.01\,m/s$, $\hat{a}_1 = 1.067m/s^2$,

$\hat{v}_2 = -12\,\text{m/s}$, $\hat{a}_2 = -0.5822\text{m/s}^2$. After compensating the RM, QRM, and DFM with the estimated motion parameters, these two targets are coherently integrated, as shown in Figure 2d,e.

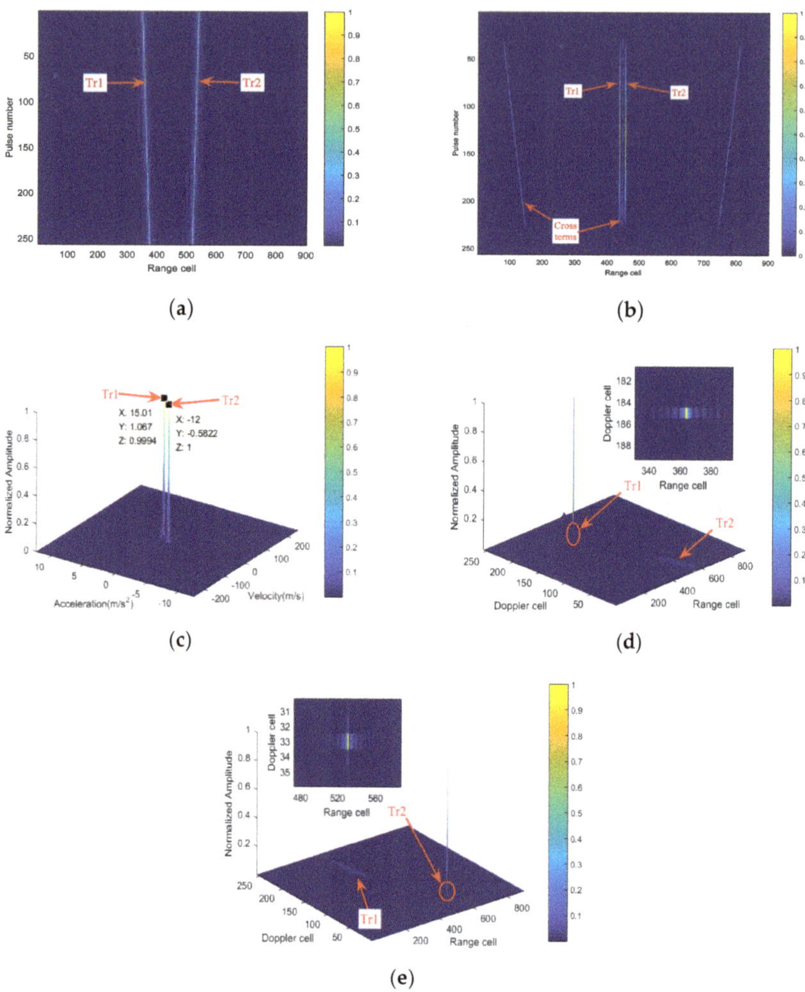

Figure 2. Simulation results of Example 2. (**a**) The result of pulse compression; (**b**) the result of FD-SoPD; (**c**) the velocity-Acceleration distribution; (**d**) coherent integration result of Tr1; (**e**) coherent integration result of Tr2.

Case 2. *Some of the accelerations coincide, i.e., $a_i = a_j$ or $c_{i2} = c_{j2}$, $\exists i, j = 1, 2, \ldots, K$, $i \neq j$. In this case, the coupling term between t_m and τ in R_2 is eliminated, but there are still linear terms of τ. The energy of R_2 will be accumulated only when $c_{i1} = c_{j1}$, which means the two targets have the same velocity and acceleration. i.e., $R_1 = R_2$. R_3, R_4 and R_8 have the quadratic term τ^2 and coupling term between t_m and τ, and thus cannot be accumulated. R_7 has linear terms of τ and a random initial phase regarding target reflectivity and the initial range, which defocuses the target energy.*

As for R_6, when $c_{i2} = c_{j2}$, it can be simplified as:

$$R_6 = \sum_{\substack{i=1 \\ i \neq j}}^{K} \sum_{j=1}^{K} |A_i|^2 |A_j|^2 \exp\left\{ j\frac{4\pi}{\varepsilon}\left[(c_{i1} + c_{j1})\tau_0 + 4c_{i2}\tau_0 t_m \right] \right\} \tag{28}$$

Substituting Equation (28) into Equation (27), we can see that R_6 can achieve energy accumulation, and the peak position is in the middle of the auto-terms, that is, the acceleration is the same as the real value, while the velocity is estimated as the average of the two targets.

Example 3. *In this example, two maneuvering targets designated as Tr1 and Tr2 have the same acceleration. Radar parameters are the same as those in Example 1. Target motion parameters are set as: $A_{r,1} = 1$, $r_1 = 2.9km$, $v_1 = 15\,m/s$, $a_1 = 1m/s^2$ for Target Tr1; $A_{r,2} = 1$, $r_2 = 3.1km$, $v_2 = -15\,m/s$, $a_2 = 1m/s^2$ for target Tr2.*

Figure 3a shows the target trajectories after pulse compression. Figure 3b is the result of FD-SoPD. It is obvious that, in addition to the auto-terms, the cross term R_6 is also corrected as a beeline, which locates in the middle of them. Thus, the energy of R_6 is accumulated into Peak 1 in Figure 3c, and the auto-terms form Peak 2 and Peak 3. Moreover, the motion parameters of Peak 1 also confirm the theoretical analysis in Case 2.

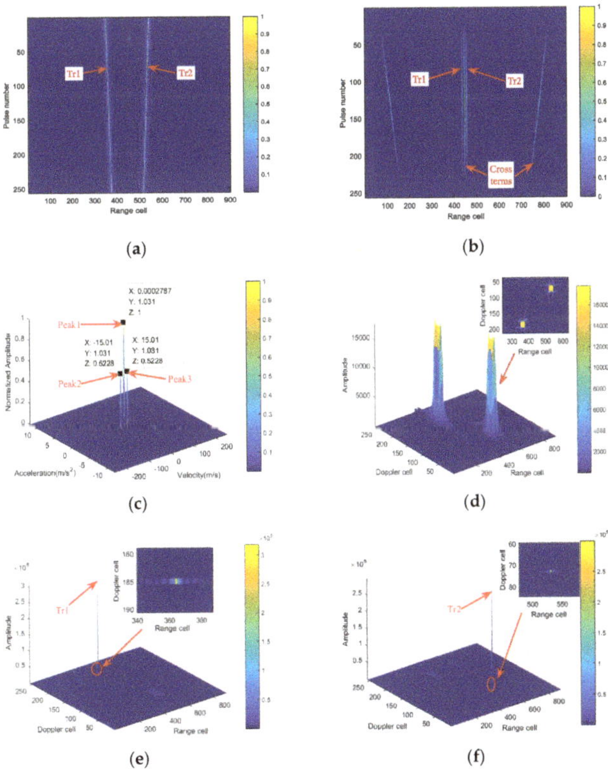

Figure 3. Simulation results of Example 3; (**a**) the result of pulse compression; (**b**) the result of the FD-SoPD; (**c**) the velocity-Acceleration distribution; (**d**) the coherent integration result with peak 1; (**e**) Coherent integration result with peak 2; (**f**) the coherent integration result with peak 3.

The next step is to determine whether all of these peaks are real maneuvering targets. After phase compensation with Peak 1, the coherent integration result is shown in Figure 3d, where no target will be detected by CFAR detection. Therefore, Peak 1 belongs to a cross-term peak. In contrast, integration with Peak 2 or Peak 3 can both produce a single sharp peak in the range-Doppler domain. Thus, Peak 2 and Peak 3 belong to the auto-term peaks. This also provides us with a method for estimating the target motion parameters combined with CFAR detection. The detailed flowchart of the FD-SoPD algorithm is given in Figure 4.

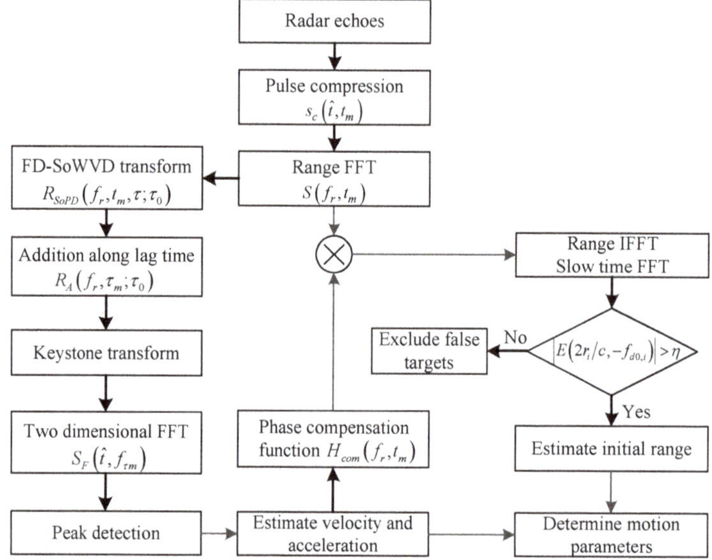

Figure 4. Flowchart of the proposed coherent detection algorithm.

4. Computational Burden Analysis of the FD-SoPD Algorithm

In this section, we will analyze the computational burden of the algorithm. The SCIFT [16], TDST [34], and ACCF-LVD [30] are selected for comparisons.

We denote the number of range cells and pulses by N_r and M. For SCIFT, its main procedures include the symmetric autocorrelation function ($O(2N_r M \log_2 M)$), the chirp-z based SCIFT ($O(3MN_r \log N_r)$), and FFT along the lag time axis ($O(N_r M \log_2 M)$). Therefore, the computational complexity is about $O(3MN_r(\log_2 M + \log N_r))$.

For TDST, to complete the two steps of scaled Fourier transform (SCFT), the computational complexities are $O(3N_r M^2 \log_2 M)$ and $O(3M^2 N_r \log_2 N_r)$, respectively. Thus, the total computational complexity is in the order of $O(3N_r M^2 \log_2 N_r M)$.

For ACCF-LVD, its main procedures include ACCF operation ($O(2MN_r \log_2 N_r)$) and chirp-z based LVD algorithm ($O(3M^2 \log_2 M)$). Therefore, its computational cost is about $O(2MN_r \log_2 N_r + 3M^2 \log_2 M)$.

The implementation of the proposed algorithm needs the calculation of $R_{SoPD}(f_r, t_m, \tau; \tau_0)$ ($O(2M^2 N_r)$), chirp-z based KT ($O(3N_r M \log_2 M)$), and two dimensional FFT ($O(MN_r(\log_2 N_r + \log_2 M))$). Therefore, the overall computational cost of the proposed method is in the order of $O(2M^2 N_r)$.

The computational complexities are listed in Table 1. Under the assumption of $N_r = M$, Figure 5 shows the computational complexities of the above four methods. Obviously, the TDST takes too much time and is not suitable for real-time processing. Table 1 also gives the detailed values of computational resources. The TDST takes up much more memory to store the three-dimensional matrix [34]. In comparison, the SCIFT, ACCF-LVD, and FD-SoPD show great advantages in this aspect. Therefore, we could conclude that the proposed FD-SoPD cost moderates computational time and resources, which helps practical applications.

Table 1. The computational burden comparisons of different algorithms.

Method	Computational Complexity	Time Cost (s) [1]	Computational Resources
SCIFT	$O(3MN_r(\log_2 M + \log N_r))$	6.35	$O(2MN_r)$
TDST	$O(3N_r M^2 \log_2 N_r M)$	335.43	$O(2M^2 N_r)$
ACCF-LVD	$O(2MN_r \log_2 N_r + 3M^2 \log_2 M)$	6.06	$O(MN_r)$
FD-SoPD	$O(2M^2 N_r)$	12.24	$O(MN_r)$

[1] The main configuration of the computer. CPU: Intel Core i7-6700HQ 2.60 GHz; RAM: 16.00G; Operating System: Windows 7; Software: Matlab 2015a.

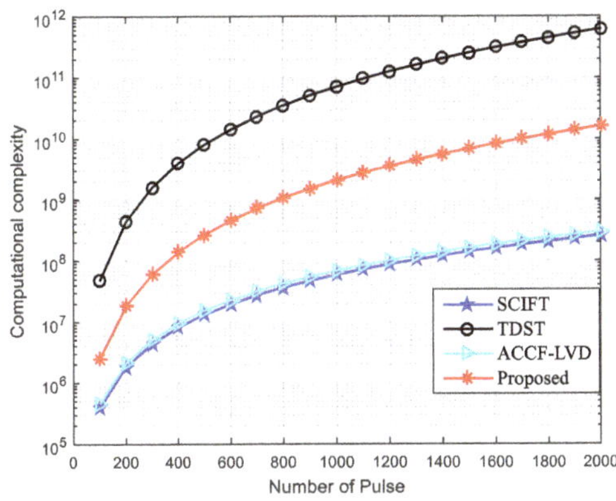

Figure 5. Computational complexity comparison.

5. Numerical Results

In the section, we will give several numerical experiments to demonstrate the effectiveness of the proposed algorithm. The simulation parameters are given in Table 2.

Table 2. Simulation parameters for the radar and target.

Parameters	Value	Parameters	Value
Carrier frequency	1 GHz	Bandwidth	100 MHz
Sample frequency	2 MHz	PRF	128 Hz
Pulse duration	2 ms	Pulse number	256
Initial slant range	3 km	Radial velocity	15 m/s
Radial acceleration	1 m/s^2	-	-

5.1. Coherent Integration Performance

In this part, the coherent integration performance of the proposed method for a maneuvering target is evaluated. We choose the representative MTD, SCIFT, TDST, and ACCF-LVD algorithms as references. Complex zero-mean white Gaussian noise is added to radar echoes, and the SNR is set to be 5dB after compression. Figure 6a shows the target trajectory, and the result of MTD is given in Figure 6b. It is obvious that the MTD cannot integrate the target energy due to ignoring the LRM, QRM, and DFM. The velocity estimation and coherent integration results of the SCIFT are shown in Figure 6c,d, respectively. Unfortunately, the SCIFT is also invalid due to ignoring the target's acceleration.

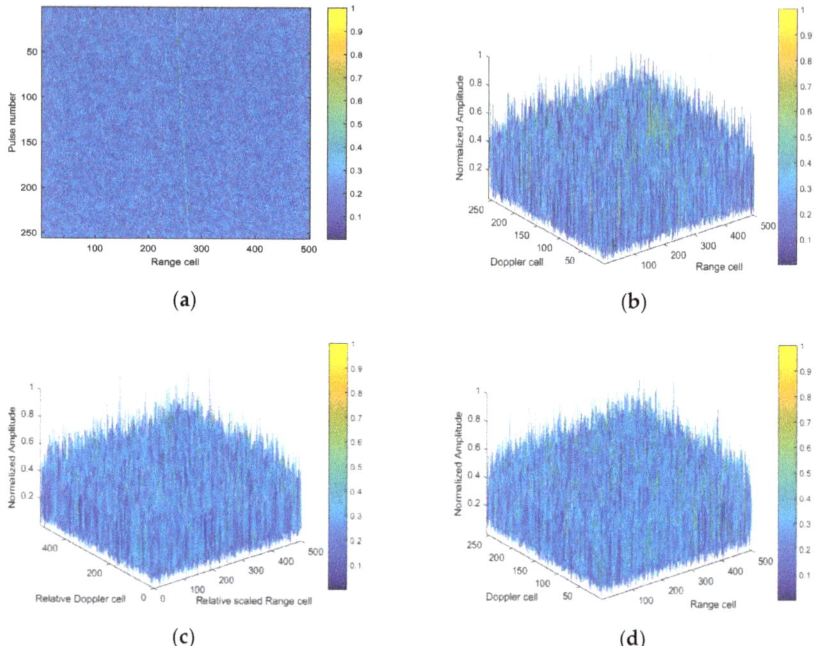

Figure 6. Coherent integration for a maneuvering target. (**a**) The result after pulse compression; (**b**) the integration result of moving target detection (MTD); (**c**) the velocity estimation result of the scaled inverse Fourier transform (SCIFT); (**d**) the integration result of the SCIFT.

Figure 7a shows the integration result of LVD, where no significant peak can be found. Thus, the ACCF-LVD cannot integrate target energy in such a low SNR, as shown in Figure 7b. Figure 7c–f give respectively the parameter estimation and integration results of the FD-SoPD and TDST. Although both algorithms can accurately estimate the target motion parameters and perform coherent accumulation, the proposed algorithm has much more advantages in computational efficiency and resources.

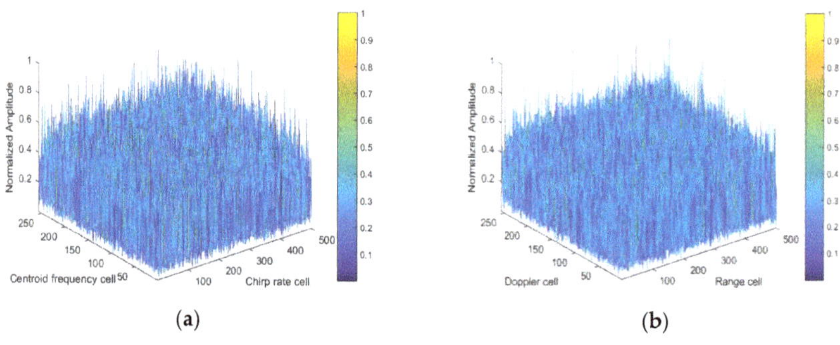

Figure 6. *Cont.*

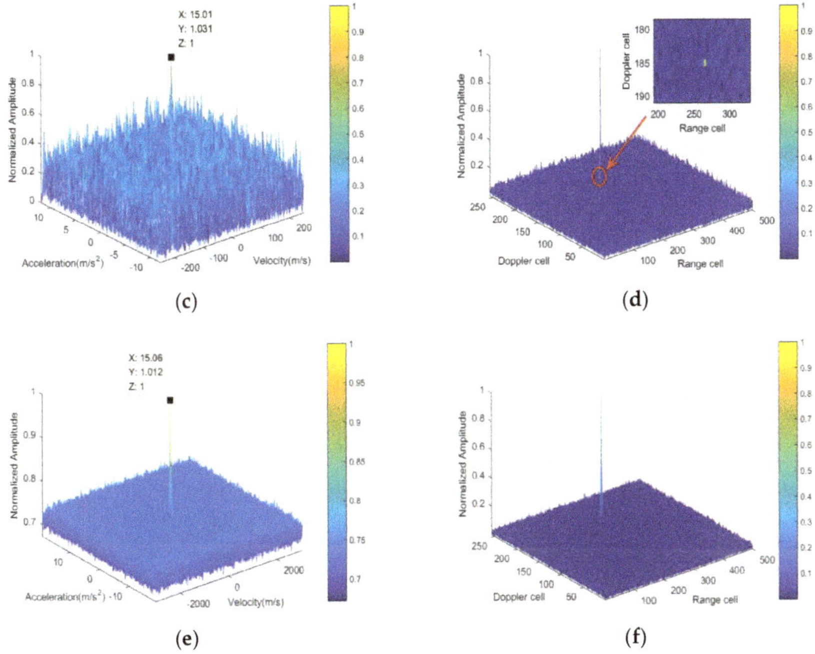

Figure 7. Integration performance comparison. (**a**) Parameter estimation result of LVD; (**b**) integration result of the ACCF-LVD; (**c**) parameter estimation result of the proposed algorithm; (**d**) integration result of the proposed algorithm; (**e**) parameter estimation result of the TDST; (**f**) integration result of the TDST.

Detailed results of parameter estimation and target detection are given in Table 3.

Table 3. Comparisons of simulated parameter estimation and target detection results.

	Initial Range (km)	Velocity (m/s)	Acceleration (m/s^2)	Detection Result
MTD	2.95	−1.65	-	No target
SCIFT	3.07	655.72	-	No target
ACCF-LVD	2.88	1043.84	11.386	No target
Proposed	3.00	15.01	1.031	Detected
TDST	3.00	15.06	1.012	Detected

5.2. Detection Performance

The detection ability of the above five algorithms is evaluated combined with the CFAR detector. Assume the radar data is contaminated by the zero-mean white Gaussian noise and input SNRs after pulse compression are set as [-20:1:20] dB. 200 trials are done for each SNR value. The false alarm rate is set as $P_{fa} = 10^{-6}$. Figure 8 shows the simulation result, where one can see that the MTD and SCIFT have the poorest detection probability due to ignoring the QRM or DFM. The adjacent cross-correlation function suffers more energy loss than the SoPD in the slow time domain [7]. Thus, the required SNR of FD-SoPD is about 4 dB less than ACCF-LVD. However, compared with TDST, the proposed algorithm suffers about 7 dB loss due to two-order bilinear transformation in SoPD. Considering the advantages of FD-SoPD, we can conclude that the proposed coherent detection algorithm achieves a good balance between the computational burden and detection ability.

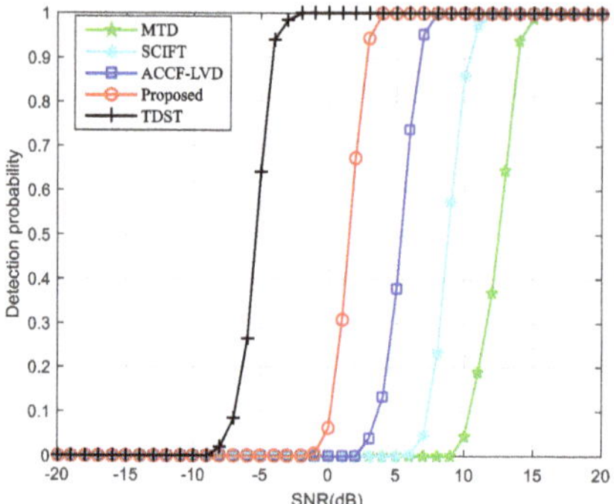

Figure 8. Detection probability of five algorithms.

5.3. Parameter Estimation Performance

We also evaluate the motion parameters estimation performance of FD-SoPD. The SNR after range compression varies from -15dB–20dB. The parameters for the radar and target are given in Table 2. two-hundred Monte Carlo simulations are performed for each SNR value. The root mean square error (RMSE) is utilized as a benchmark. The ACCF-LVD and TDST, which can estimate the velocity and acceleration of target, are selected for comparisons. Figure 9a,b show the RMSEs of the estimated velocity and acceleration. It can be seen that the TDST has the best estimation performance at the cost of huge computational burden. The performance of the proposed method is about 4dB better than those of ACCF-LVD on the input SNR threshold. However, compared with TDST, the FD-SoPD suffers from about 8dB SNR loss due to the constant delay in Equation (12). Overall, the proposed technique strikes a better balance between parameter estimation performance and computational cost.

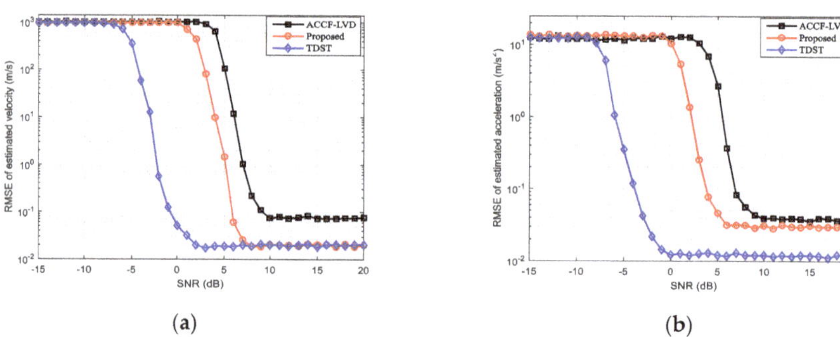

(a)   (b)

Figure 9. Motion parameters estimation performance of the three methods. (**a**) Estimation of the root mean square error (RMSE) of velocity; (**b**) estimation RMSE of acceleration.

5.4. Experimental Data Processing

In this subsection, we adopt the measured data of a DJI Phantom 3 commercial UAV to demonstrate the proposed FD-SoPD method. The data was collected in March 2017 by the National University of Defense Technology, Hunan, China. Figure 10a–c show the experimental scene, FMCW radar system, and radar antennas, respectively. Radar parameters are given in Table 4. Figure 10d shows the target trajectory after pulse compression, where the UAV moves across 7 range cells during the observation time. Figure 10e gives the parameter estimation result, where we could read the velocity and acceleration of the UAV, i.e., $\hat{v} = 1.217\,m/s$ and $\hat{a} = 0.2145m/s^2$. Finally, coherent integration of FD-SoPD can be obtained with the estimated velocity and acceleration, as shown in Figure 10f. At the same time, the integration results of SCIFT and MTD are also given in Figure 10g,h. Due to ignoring the LRM, QRM, or DFM, the target energy is distributed in the range-Doppler domain. However, the proposed method can estimate the acceleration of the target accurately. Thus, a well-focused peak is obtained, which is beneficial to target detection. Detailed results of parameter estimation and coherent integration are given in Table 5.

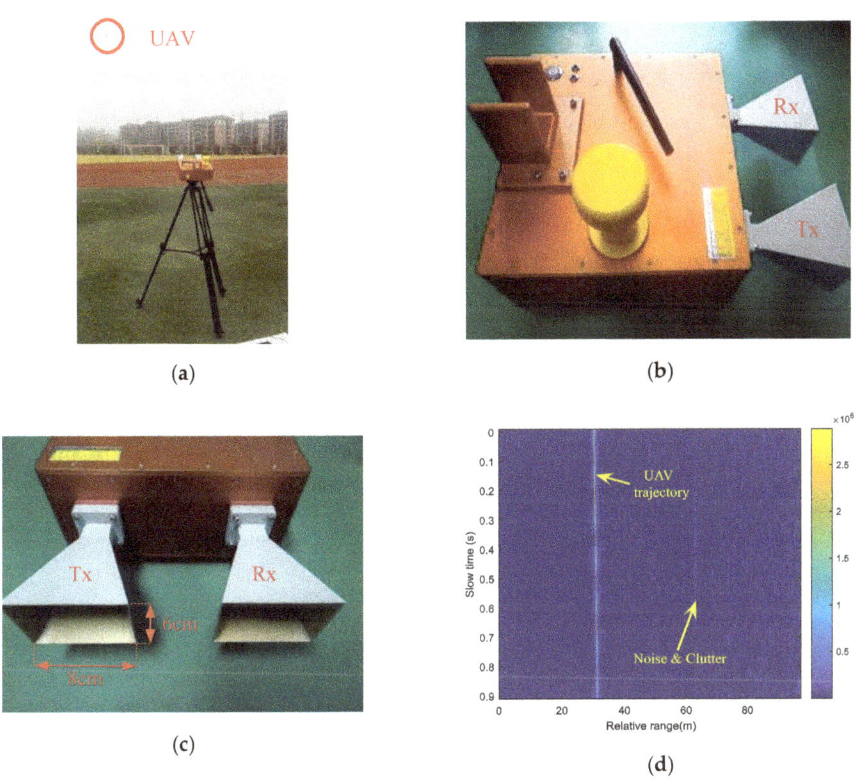

Figure 9. *Cont.*

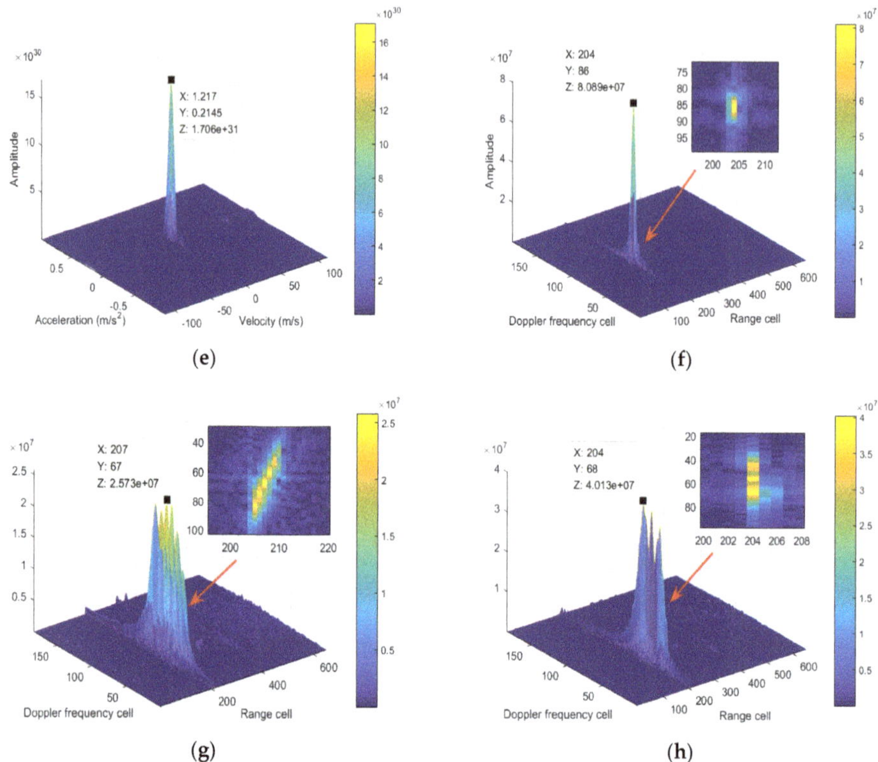

Figure 10. Experimental data processing results. (**a**) The experimental scene; (**b**) the FMCW radar system; (**c**) the radar antennas; (**d**) moving trajectory of the UAV; (**e**) parameter estimation result of FD-SoPD; (**f**) the coherent integration result of FD-SoPD; (**g**) the coherent integration result of the MTD; (**h**) the coherent integration result of the SCIFT.

Table 4. Frequency-modulated continuous-wave (FMCW) radar parameters.

Radar Parameter	Value	Radar Parameter	Value
Carrier frequency	9.5 GHz	PRF	50 Hz
Bandwidth	1 GHz	Sampling frequency	1 MHz
Pulse width	0.0102 s	Coherent time	0.92 s
Transmit power	25 dbm	Weight	7 kg
Radar length	35 cm	Radar width	24 cm
Radar height	20 cm	-	-

Table 5. Comparisons of experimental parameter estimation results.

	Range Cell	Velocity (m/s)	Acceleration (m/s^2)	Peak Value
MTD	207	−0.279	-	2.573×10^7
SCIFT	204	−0.283	-	4.013×10^7
FD-SoPD	204	1.217	0.2145	8.089×10^7

6. Conclusions

A frequency-domain second-order phase difference method is proposed to achieve coherent integration and parameter estimation. First, the FD-SoPD is performed to eliminate the QRM, DFM, and Doppler ambiguity simultaneously. After that, parameter estimation and coherent integration are accomplished. Compared with ACCF-LVD, the FD-SoPD could obtain better detection performance with moderate computation complexity. Simulations and experimental data processing results demonstrate the effectiveness of the proposed algorithm.

Author Contributions: K.J., T.L. and Y.Z. conceived and designed the experiments; G.L. performed the experiments; Y.W. analyzed the data and contributed analysis tools; K.J. and Y.W. wrote the paper.

Funding: This research received no external funding.

Acknowledgments: The authors would like to thank Tian Jin for his useful suggestions as well as the help of the real radar data processing.

Conflicts of Interest: The authors declare no conflict of interest.

Appendix A

In this appendix, we will give the expressions of cross terms R_2-R_8. In order to simplify the expression form, we define

$$
\begin{aligned}
\varepsilon &= c/(f_r + f_c) \\
c_{i1} &= v_i \\
c_{i2} &= a_i/2 \\
A_i &= A_{f_r,i} \mathrm{rect}\left(\tfrac{f_r}{B}\right) \exp\left(-j4\pi \tfrac{(f_r+f_c)}{c} r_i\right)
\end{aligned}
\tag{A1}
$$

Then the cross terms R_2-R_8 can be written as follows.

$$
R_2 = \sum_{\substack{i=1\\i\neq j}}^{K} \sum_{j=1}^{K} |A_i|^2 |A_j|^2 \exp\left\{ j\frac{4\pi}{\varepsilon}\left[(c_{i1} - c_{j1})\tau + 2c_{i1}\tau_0 + 2(c_{i2} - c_{j2})t_m\tau + 4c_{i2}\tau_0 t_m\right] \right\}
\tag{A2}
$$

$$
\begin{aligned}
R_3 &= \sum_{p=1}^{K} |A_p|^2 \exp\left[-j\tfrac{4\pi}{\varepsilon}(c_{p1}\tau + 2c_{p2}t_m\tau)\right] \times \sum_{\substack{k=1\\k\neq l}}^{K} \sum_{l=1}^{K} A_k^* A_l \\
&\times \exp\left\{ j\tfrac{4\pi}{\varepsilon}\left[c_{k1}\left(t_m + \tfrac{\tau}{2} + \tau_0\right) + c_{k2}\left(t_m + \tfrac{\tau}{2} + \tau_0\right)^2 - c_{l1}\left(t_m - \tfrac{\tau}{2} - \tau_0\right) + c_{l2}\left(t_m - \tfrac{\tau}{2} - \tau_0\right)^2 \right] \right\}
\end{aligned}
\tag{A3}
$$

$$
\begin{aligned}
R_4 &= \sum_{q=1}^{K} |A_q|^2 \exp\left[j\tfrac{4\pi}{\varepsilon}(c_{q1}\tau + 2c_{q2}t_m\tau + 4c_{q2}t_m\tau_0)\right] \times \sum_{\substack{i=1\\i\neq j}}^{K} \sum_{j=1}^{K} A_i A_j^* \\
&\exp\left\{ -j\tfrac{4\pi}{\varepsilon}\left[c_{i1}\left(t_m + \tfrac{\tau}{2}\right) + c_{i2}\left(t_m + \tfrac{\tau}{2}\right)^2 - c_{j1}\left(t_m - \tfrac{\tau}{2}\right) - c_{j2}\left(t_m - \tfrac{\tau}{2}\right)^2 \right] \right\}
\end{aligned}
\tag{A4}
$$

$$
\begin{aligned}
R_5 &= \sum_{\substack{i=1\\i\neq j}}^{K} \sum_{j=1}^{K} A_i A_j^* \exp\left\{ -j\tfrac{4\pi}{\varepsilon}\left[c_{i1}\left(t_m + \tfrac{\tau}{2}\right) + c_{i2}(t_m+)^2 \right] \right\} \\
&\times \exp\left\{ -j\tfrac{4\pi}{\varepsilon}\left[c_{j1}\left(t_m - \tfrac{\tau}{2}\right) + c_{j2}\left(t_m - \tfrac{\tau}{2}\right)^2 \right] \right\} \\
&\times \sum_{\substack{k=1\\k\neq l}}^{K} \sum_{l=1}^{K} A_k^* A_l \exp\left\{ j\tfrac{4\pi}{\varepsilon}\left[c_{k1}\left(t_m + \tfrac{\tau}{2} + \tau_0\right) + c_{k2}\left(t_m + \tfrac{\tau}{2} + \tau_0\right)^2 \right] \right\} \\
&\times \exp\left\{ -j\tfrac{4\pi}{\varepsilon}\left[c_{l1}\left(t_m - \tfrac{\tau}{2} - \tau_0\right) + c_{l2}\left(t_m - \tfrac{\tau}{2} - \tau_0\right)^2 \right] \right\}
\end{aligned}
\tag{A5}
$$

$$
\begin{aligned}
R_6 &= \sum_{\substack{i=1\\i\neq j}}^{K} \sum_{j=1}^{K} |A_i|^2 |A_j|^2 \exp\left\{ j\tfrac{4\pi}{\varepsilon}\left[(c_{i1} + c_{j1})\tau_0 + (c_{i2} - c_{j2})\tau_0^2 \right] \right\} \\
&\times \exp\left\{ j\tfrac{4\pi}{\varepsilon}\left[2(c_{i2} + c_{j2})\tau_0 t_m + (c_{i2} - c_{j2})\tau_0\tau \right] \right\}
\end{aligned}
\tag{A6}
$$

$$R_7 = \sum_{\substack{i=1 \\ i \neq j}}^{K} \sum_{j=1}^{K} A_i A_j^* \exp\left\{-j\tfrac{4\pi}{\varepsilon}\left[c_{i1}\left(t_m + \tfrac{\tau}{2}\right) + c_{i2}\left(t_m + \tfrac{\tau}{2}\right)^2\right]\right\}$$

$$\times \exp\left\{j\tfrac{4\pi}{\varepsilon}\left[c_{j1}\left(t_m - \tfrac{\tau}{2}\right) + c_{j2}\left(t_m - \tfrac{\tau}{2}\right)^2\right]\right\}$$

$$\times \sum_{\substack{k=1 \\ k \neq i, l \neq j, k \neq l}}^{K} \sum_{l=1}^{K} A_k^* A_l \exp\left\{j\tfrac{4\pi}{\varepsilon}\left[c_{k1}\left(t_m + \tfrac{\tau}{2} + \tau_0\right) + c_{k2}\left(t_m + \tfrac{\tau}{2} + \tau_0\right)^2\right]\right\} \tag{A7}$$

$$\times \exp\left\{-j\tfrac{4\pi}{\varepsilon}\left[c_{l1}\left(t_m - \tfrac{\tau}{2} - \tau_0\right) + c_{l2}\left(t_m - \tfrac{\tau}{2} - \tau_0\right)^2\right]\right\}$$

$$R_8 = \sum_{\substack{i=1 \\ i \neq j}}^{K} \sum_{j=1}^{K} A_i A_j^* \exp\left\{-j\tfrac{4\pi}{\varepsilon}\left[c_{i1}\left(t_m + \tfrac{\tau}{2}\right) + c_{i2}\left(t_m + \tfrac{\tau}{2}\right)^2 - c_{j1}\left(t_m - \tfrac{\tau}{2}\right) - c_{j2}\left(t_m - \tfrac{\tau}{2}\right)^2\right]\right\}$$

$$\times \sum_{\substack{k=1 \\ k=i, l \neq j, k \neq l}}^{K} \sum_{l=1}^{K} A_k^* A_l \exp\left\{j\tfrac{4\pi}{\varepsilon}\left[c_{k1}\left(t_m + \tfrac{\tau}{2} + \tau_0\right) + c_{k2}\left(t_m + \tfrac{\tau}{2} + \tau_0\right)^2 - c_{l1}\left(t_m - \tfrac{\tau}{2} - \tau_0\right) - c_{l2}\left(t_m - \tfrac{\tau}{2} - \tau_0\right)^2\right]\right\}$$

$$+ \sum_{\substack{i=1 \\ i \neq j}}^{K} \sum_{j=1}^{K} A_i A_j^* \exp\left\{-j\tfrac{4\pi}{\varepsilon}\left[c_{i1}\left(t_m + \tfrac{\tau}{2}\right) + c_{i2}\left(t_m + \tfrac{\tau}{2}\right)^2 - c_{j1}\left(t_m - \tfrac{\tau}{2}\right) - c_{j2}\left(t_m - \tfrac{\tau}{2}\right)^2\right]\right\} \tag{A8}$$

$$\times \sum_{\substack{k=1 \\ k \neq i, l = j, k \neq l}}^{K} \sum_{l=1}^{K} A_k^* A_l \exp\left\{j\tfrac{4\pi}{\varepsilon}\left[c_{k1}\left(t_m + \tfrac{\tau}{2} + \tau_0\right) + c_{k2}\left(t_m + \tfrac{\tau}{2} + \tau_0\right)^2 - c_{l1}\left(t_m - \tfrac{\tau}{2} - \tau_0\right) - c_{l2}\left(t_m - \tfrac{\tau}{2} - \tau_0\right)^2\right]\right\}$$

References

1. Huang, P.H.; Liao, G.S.; Yang, Z.Z.; Xia, X.G.; Ma, J.T.; Ma, J.T. Long-Time Coherent Integration for Weak Maneuvering Target Detection and High-Order Motion Parameter Estimation Based on Keystone Transform. *IEEE Trans. Signal Process.* **2016**, *64*, 4013–4026. [CrossRef]

2. Suo, P.C.; Tao, S.; Tao, R.; Nan, Z. Detection of high-speed and accelerated target based on the linear frequency modulation radar. *IET Radar Sonar Navig.* **2014**, *8*, 37–47. [CrossRef]

3. Zhu, S.Q.; Liao, G.S.; Yang, D.; Tao, H.H. A New Method for Radar High-Speed Maneuvering Weak Target Detection and Imaging. *IEEE Geosci. Remote Sens. Lett.* **2014**, *11*, 1175–1179.

4. Li, X.L.; Sun, Z.; Yi, W.; Cui, G.L.; Kong, L.J.; Yang, X.B. Computationally efficient coherent detection and parameter estimation algorithm for maneuvering target. *Signal Process.* **2018**, *155*, 130–142. [CrossRef]

5. Li, X.L.; Cui, G.L.; Yi, W.; Kong, L.J. Fast coherent integration for maneuvering target with high-order range migration via TRT-SKT-LVD. *IEEE Trans. Aerospace Electr. Syst.* **2016**, *52*, 2803–2814. [CrossRef]

6. Wu, W.; Wang, G.H.; Sun, J.P. Polynomial Radon-Polynomial Fourier Transform for Near Space Hypersonic Maneuvering Target Detection. *IEEE Trans. Aerospace Electr. Syst.* **2018**, *54*, 1306–1322. [CrossRef]

7. Zhang, J.C.; Su, T.; Zheng, J.B.; He, X.H. Novel Fast Coherent Detection Algorithm for Radar Maneuvering Target with Jerk Motion. *IEEE J. Sel. Top. Appl. Earth Observ. Remote Sens.* **2017**, *10*, 1792–1803. [CrossRef]

8. Lao, G.C.; Yin, C.B.; Ye, W.; Sun, Y.; Li, G.J. An SAR-ISAR Hybrid Imaging Method for Ship Targets Based on FDE-AJTF Decomposition. *Electronics* **2018**, *7*, 46. [CrossRef]

9. Lv, Y.K.; Wu, Y.H.; Wang, H.Y.; Qiu, L.; Jiang, J.W.; Sun, Y. An Inverse Synthetic Aperture Ladar Imaging Algorithm of Maneuvering Target Based on Integral Cubic Phase Function-Fractional Fourier Transform. *Electronics* **2018**, *7*, 148. [CrossRef]

10. Lazarov, A.; Minchev, C. ISAR Geometry, Signal Model, and Image Processing Algorithms. *IET Radar Sonar Navig.* **2017**, *11*, 1425–1434. [CrossRef]

11. Li, X.L.; Kong, L.J.; Cui, G.L.; Yi, W. A low complexity coherent integration method for maneuvering target detection. *Dig. Signal Process.* **2016**, *49*, 137–147. [CrossRef]

12. Rao, X.; Tao, H.H.; Su, J.; Xie, J.; Zhang, X.Y. Detection of Constant Radial Acceleration Weak Target via IAR-FRFT. *IEEE Trans. Aerospace Electr. Syst.* **2015**, *51*, 3242–3253. [CrossRef]

13. Xu, J.; Yu, J.; Peng, Y.N.; Xia, X.G. Radon-Fourier Transform for Radar Target Detection, I: Generalized Doppler Filter Bank. *IEEE Trans. Aerospace Electr. Syst.* **2011**, *47*, 1186–1202. [CrossRef]

14. Zhu, D.Y.; Li, Y.; Zhu, Z.D. A Keystone Transform Without Interpolation for SAR Ground Moving-Target Imaging. *IEEE Geosci. Remote Sens. Lett.* **2007**, *4*, 18–22. [CrossRef]

15. Sun, G.C.; Xing, M.D.; Xia, X.G.; Wu, Y.R.; Bao, Z. Robust Ground Moving-Target Imaging Using Deramp–Keystone Processing. *IEEE Trans. Geosci. Remote Sens.* **2013**, *51*, 966–982. [CrossRef]
16. Zheng, J.B.; Su, T.; Zhu, W.T.; He, X.H.; Liu, Q.H. Radar High-Speed Target Detection Based on the Scaled Inverse Fourier Transform. *IEEE J. Sel. Top. Appl. Earth Observ. Remote Sens.* **2015**, *8*, 1108–1119. [CrossRef]
17. Niu, Z.Y.; Zheng, J.B.; Su, T.; Zhang, J.C. Fast implementation of scaled inverse Fourier transform for high-speed radar target detection. *Electr. Lett.* **2017**, *53*, 1142–1144. [CrossRef]
18. Zheng, J.B.; Su, T.; Liu, H.W.; Liao, G.S.; Liu, Z.; Liu, Q.H. Radar High-Speed Target Detection Based on the Frequency-Domain Deramp-Keystone Transform. *IEEE J. Sel. Top. Appl. Earth Observ. Remote Sens.* **2016**, *9*, 285–294. [CrossRef]
19. Sun, Z.; Li, X.L.; Yi, W.; Cui, G.L.; Kong, L.J. A Coherent Detection and Velocity Estimation Algorithm for the High-Speed Target Based on the Modified Location Rotation Transform. *IEEE J. Sel. Top. Appl. Earth Observ. Remote Sens.* **2018**, *11*, 2346–2361. [CrossRef]
20. Xu, J.; Yu, J.; Peng, Y.N.; Xia, X.G. Radon-Fourier Transform for Radar Target Detection (II): Blind Speed Sidelobe Suppression. *IEEE Trans. Aerospace Electr. Syst.* **2011**, *47*, 2473–2489. [CrossRef]
21. Xu, J.; Xia, X.G.; Peng, S.B.; Yu, J.; Peng, Y.N.; Qian, L.C. Radar Maneuvering Target Motion Estimation Based on Generalized Radon-Fourier Transform. *IEEE Trans. Signal Process.* **2012**, *60*, 6190–6201.
22. Chen, X.L.; Guan, J.; Liu, N.B.; He, Y. Maneuvering Target Detection via Radon-Fractional Fourier Transform-Based Long-Time Coherent Integration. *IEEE Trans. Signal Process.* **2014**, *62*, 939–953. [CrossRef]
23. Li, X.L.; Cui, G.L.; Yi, W.; Kong, L.J. Coherent Integration for Maneuvering Target Detection Based on Radon-Lv's Distribution. *IEEE Signal Process. Lett.* **2015**, *22*, 1467–1471. [CrossRef]
24. Lv, X.L.; Bi, G.A.; Wan, C.R.; Xing, M.D. Lv's Distribution: Principle, Implementation, Properties, and Performance. *IEEE Trans. Signal Process.* **2011**, *59*, 3576–3591. [CrossRef]
25. Lv, X.L.; Xing, M.D.; Zhang, S.H.; Bao, Z. Keystone transformation of the Wigner-Ville distribution for analysis of multicomponent LFM signals. *Signal Process.* **2009**, *59*, 791–806. [CrossRef]
26. Zheng, J.B.; Liu, H.W.; Liu, Q.H. Parameterized Centroid Frequency-Chirp Rate Distribution for LFM Signal Analysis and Mechanisms of Constant Delay Introduction. *IEEE Trans. Signal Process.* **2017**, *65*, 6435–6447. [CrossRef]
27. Kirkland, D. Imaging moving targets using the second-order keystone transform. *IET Radar Sonar Navig.* **2011**, *5*, 902–910. [CrossRef]
28. Li, G.; Xia, X.G.; Peng, Y.N. Doppler Keystone Transform: An Approach Suitable for Parallel Implementation of SAR Moving Target Imaging. *IEEE Geosci. Remote Sens. Lett.* **2008**, *5*, 573–577. [CrossRef]
29. Li, X.L.; Cui, G.L.; Yi, W.; Kong, L.J. Manoeuvring target detection based on keystone transform and Lv's distribution. *IET Radar Sonar Navig.* **2016**, *10*, 1234–1242. [CrossRef]
30. Li, X.L.; Cui, G.L.; Yi, W.; Kong, L.J. A Fast Maneuvering Target Motion Parameters Estimation Algorithm Based on ACCF. *IEEE Signal Process. Lett.* **2015**, *22*, 270–274. [CrossRef]
31. Li, X.L.; Cui, G.L.; Kong, L.J.; Yi, W. Fast Non-Searching Method for Maneuvering Target Detection and Motion Parameters Estimation. *IEEE Trans. Signal Process.* **2016**, *64*, 2232–2244. [CrossRef]
32. Li, X.L.; Cui, G.L.; Yi, W.; Kong, L.J. Radar Maneuvering Target Detection and Motion Parameter Estimation Based on TRT-SGRFT. *Signal Process.* **2017**, *133*, 107–116. [CrossRef]
33. He, X.P.; Liao, G.S.; Zhu, S.Q.; Xu, J.W.; Guo, Y.F.; Wei, J.Q. Fast Non-Searching Method for Ground Moving Target Refocusing and Motion Parameters Estimation. *Digital Signal Process.* **2018**, *79*, 152–163. [CrossRef]
34. Zheng, J.B.; Liu, H.W.; Liu, J.; Du, X.L.; Liu, Q.H. Radar High-Speed Maneuvering Target Detection Based on Three-Dimensional Scaled Transform. *IEEE J. Sel. Top. Appl. Earth Observ. Remote Sens.* **2018**, *11*, 2821–2833. [CrossRef]
35. Zheng, J.B.; Su, T.; Zhang, L.; Zhu, W.T.; Liu, Q.H. ISAR Imaging of Targets with Complex Motion Based on the Chirp Rate–Quadratic Chirp Rate Distribution. *IEEE Trans. Geosci. Remote Sens.* **2014**, *52*, 7276–7289. [CrossRef]
36. Guida, M.; Longo, M.; Lops, M. Biparametric CFAR procedures for lognormal clutter. *IEEE Trans. Aerospace Electr. Syst.* **1993**, *29*, 798–809. [CrossRef]

 *electronics*

Article

Feasibility Study of Passive Bistatic Radar Based on Phased Array Radar Signals

Jiameng Pan [1,*], Panhe Hu [1], Qian Zhu [1], Qinglong Bao [1] and Zengping Chen [2]

[1] National Key Laboratory of Science and Technology on ATR, National University of Defense Technology, Changsha 410073, China
[2] School of Electronics and Communication Engineering, Sun Yat-sen University, Guangzhou 510275, China
* Correspondence: 3090100472@zju.edu.cn

Received: 5 June 2019; Accepted: 20 June 2019; Published: 26 June 2019

Abstract: This paper presents the concept of a passive bistatic radar (PBR) system using existing phased array radar (PAR) as the source of illumination. Different from PBR based on common civil illuminators of opportunity, we develop an experimental PBR system using an high-power air surveillance PAR with abundant signal modulation forms as the transmitter. After the introduction of the PBR system and PAR signals, it can be concluded that the agility of the waveform parameters of PAR signal brings two problems to the signal processing of the PBR systems, which are not discussed in conventional PBR systems. The first problem is the time and frequency synchronization of the system, so we propose a direct wave parameter estimation method based on template matching to estimate the parameters of the transmitted signal in real time to achieve time and frequency synchronization of the system. The second problem is the coherent integration for moving target detection and weak target detection, so we propose a coherent integration method based on Radon–Nonuniform Fast Fourier Transform (Radon-NUFFT) to deal with the problems introduced by the agile waveform parameters. Preliminary results from the field experiment demonstrate the feasibility of the PBR system based on PAR signals, and the effectiveness of the proposed methods is verified.

Keywords: passive bistatic radar; phased array radar; parameter estimation; coherent integration; aircraft surveillance

---

1. Introduction

A passive bistatic radar (PBR) system performs target detection and localization by exploiting noncooperative illuminators of opportunity. PBR offers many advantages over conventional monostatic radar systems, including lower cost, harder to detect, higher immunity to electronic countermeasures, no requirement for frequency allocation, and the ability to counter stealth targets due to the bistatic configuration of PBR systems. Therefore, many kinds of civil illuminators have been exploited by PBR systems, such as FM radio [1], Integrated Services Digital Broadcasting-Terrestrial (ISDB-T) [2], Digital Video Broadcasting-Terrestrial (DVB-T) [3–5], Long-Term Evolution (LTE) [6], Wireless Fidelity (Wi-Fi) [7], and Global Navigation Satellite System (GNSS) [8]. However, those communications signals are not designed for use in radar, which may cause ambiguities due to the signal structure [9], and the maximum detectable range may be small due to the limited transmitting power.

Compared with civil illuminators, a dedicated radar transmitter usually transmits signals with a more ideal ambiguity function and higher transmitting power. German passive radar system Klein Heidelberg [10] was the first PBR system based on existing hostile radar, and in recent years, some PBR systems which utilize existing radar as their transmitter have been proven to perform target detection successfully [11–13]. Furthermore, phased array radar (PAR) has the advantages of stronger transmitting power and higher reliability. However, the complexity of PAR signals poses unique challenges to the signal processing of a PBR system, so PBR systems based on PAR signals still remains a broad area of research.

The exploited illuminator of our PBR system is a high-power air surveillance PAR, which is defined as cooperative but non-dedicated. It is cooperative in the sense that the information about signal waveform is available but is considered non-dedicated as its operations is solely for its own monostatic radar purposes and that no changes are applied to enhance PBR capabilities. The PAR adopts agile radar parameters to improve the anti-jamming ability. Liner frequency modulated (LFM) waveform is used as a transmitting signal. The PAR transmits a group of signals with the same carrier frequency (CF) and bandwidth (BW) for seconds; the CF and BW change randomly among different groups, while the pulse width (PW) and pulse repetition interval (PRI) are different among different pulses in each pulse group. The agility of the waveform parameters of transmitting signal brings two problems to the signal processing of our PBR system. The first problem is the time and frequency synchronization of the system, and the second problem is the coherent integration for moving target detection and weak target detection.

For the first problem, we use a reference channel to receive the direct wave signal and to estimate the parameters of the transmitted signal in real time to achieve time and frequency synchronization of the system. The parameters we need to estimate include BW, PW, CF, and time of arrival (TOA). To solve the problem of LFM signal detection and parameter estimation, many literatures have proposed many algorithms for different applications, such as Wigner–Hough transform [14], Lv's distribution [15], and Ensemble Empirical Mode Decomposition-Fractional Fourier Transform (EEMD-FRFT) [16]. Qian proposed a method based on generalized Radon Fourier transform for the parameter estimation of direct wave signal [17], but it requires a lot of searching and calculation, which cannot be realized in real time. Since the PAR transmits a deterministic signal and the parameter template library of the signal waveform has been established in advance, we propose a direct wave parameter estimation method based on template matching to realize the time and frequency synchronization of the PBR system.

For the second problem, since the duration of a group of pulses with the same CF and BW is generally longer than a second and the coherent processing interval (CPI) for coherent integration will not exceed one second, we perform coherent integration for signals with the same CF and BW, while the PRI is random and PW is staggered. In recent years, many methods have been proposed for coherent integration, such as Keystone transform [18], Radon Fourier transform (RFT) [19], axis rotation moving target detection [20], and scaled inverse Fourier transform [21], which are only applicable for conventional radar. As for radar signals with random PRI and staggered PW, it can be concluded that the random PRI will introduce the problems of irregular range cell migration (RCM) and nonuniform phase fluctuations among different pulses and that the staggered PW will introduce the problem of the irregular range-Doppler coupling effect. The problem of nonuniform phase fluctuations among different pulses can be converted to the problem of spectral analysis of nonuniformly sampled complex-valued data. As for spectral analysis of nonuniformly sampled data, a method based on interpolation [22,23] cannot be applied when the Doppler frequency of the moving target is higher than the pulse repetition frequency of the radar. Li proposed a method based on nonuniform Fast Fourier Transform (NUFFT) for random PRI PD radar [24], but the compensation of the irregular RCM is not covered in this paper. Tian proposed a method based on Radon nonuniform Fractional Fourier Transform (Radon-NUFrFT) for a random PRI radar [25], but it is only applicable for periodic nonuniformly sampled signals, which cannot solve the non-periodic, nonuniform phase fluctuations and the irregular range-Doppler coupling effect. Pan proposed a method based on the Radon-iterative adaptive approach (Radon-IAA) [26] to solve the problems introduced by the random PRI, but it requires a lot of iterative calculations, which is difficult to apply directly to engineering practice. Hence, we propose a method based on Radon-NUFFT to deal with the problems in the coherent integration for a radar signal with random PRI and staggered PW.

The remainder of this paper is organized as follows. Section 2 introduces the PBR system in general, including the geometry, hardware, and signal processing flow of the PBR system. Section 3 establishes the mathematical model of direct wave signal and describes the parameter estimation method based on template matching. Section 4 establishes the mathematical model of echo signal of a moving target firstly; then, the coherent integration method based on Radon-NUFFT is described, and simulation results are performed. Section 5 presents the experimental results to verify the performance of the PBR system. Section 6 concludes the paper.

2. Overview of Experimental PBR System

2.1. Geometry of the PBR System

The experimental PBR system consists of a reference antenna, a surveillance antenna, and a receiver. As shown in Figure 1, the reference antenna points in the direction of transmitter and receives the direct wave signal and the surveillance antenna covers the surveillance area to receive the scattered signal simultaneously.

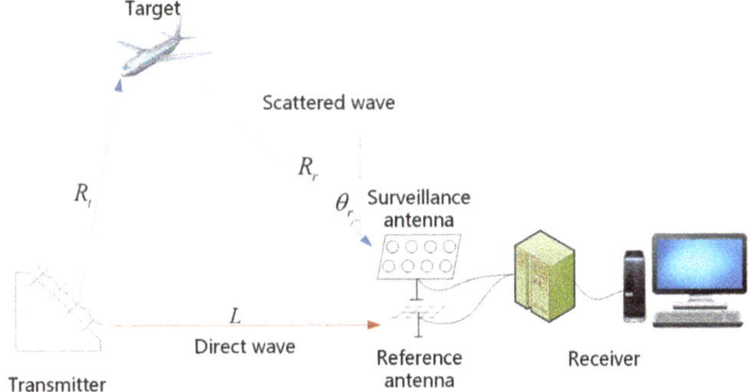

Figure 1. Geometry of the passive bistatic radar (PBR) system.

As shown in Figure 1, the transmitter and receiver are separated by a distance noted as L. Suppose that there is an aircraft target in the air, the distance between the target and transmitter is R_t and the distance between the target and receiver is R_r. The azimuth of the target relative to the surveillance antenna is θ_r.

When scattered off a target, the combined distance traveled by the signal can be calculated as

$$R_t + R_r = L + c\Delta t \tag{1}$$

where c is the velocity of light and Δt is the time delay between the target echo and the direct wave signal. After the range sum $(R_t + R_r)$ is calculated, according to the triangular geometry, the distance between the target and receiver can be calculated as follows

$$R_r = \frac{(R_t + R_r)^2 - L^2}{2(R_t + R_r + L\sin\theta_r)} = \frac{(L + c\Delta t)^2 - L^2}{2(L + c\Delta t + L\sin\theta_r)} \tag{2}$$

Therefore, the target positioning can be realized by the PBR system as long as we get the time delay Δt and the azimuth of the target relative to the surveillance antenna θ_r.

2.2. Hardware of the PBR System

The receiving antenna we designed consists of a reference antenna and a surveillance antenna. Figure 2 shows the photographs of the reference antenna. A Yagi antenna with 30 elements is chosen as the reference antenna due to its good directivity and simplicity, which is used to receive the direct wave signal to estimate the detailed parameters of the transmitted signal. The Yagi antenna has a gain of 13.5 dBi and a half-power beam width (HPBW) of 14°. The surveillance antenna is used to receive the scattered wave signal. Since the scattered wave is extremely weak relative to the direct wave, it requires a high gain and the ability to cover a wide range of airspace. Therefore, we use a phased array antenna, which has a total of 64 array elements with 8 rows and 8 columns in triangle arrangement; each array element has a gain of 9 dBi and is circularly polarized.

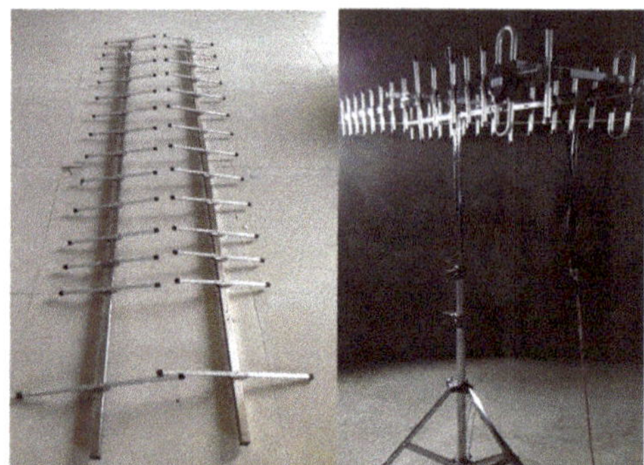

Figure 2. Photographs of the reference antenna.

In order to sample the signal from reference antenna and surveillance antenna simultaneously, a high performance Analog-to-Digital Converter (ADC) chip EV10AQ190 of E2V company was applied in the sampling module. The highest sampling frequency of the chip is 5 GHz, and the 3 dB bandwidth is 3 GHz. A high performance signal processing platform based on VPX is utilized to deal with large amounts of real-time data. Figure 3 shows the photographs of the high performance sampling module and signal processing platform based on VPX.

Figure 3. High performance sampling module and signal processing platform.

2.3. Signal Processing Method

The flow chart of the signal processing method is shown in Figure 4. Respectively, for the reference channel signal and surveillance channel signal, the brief process is described as follows.

Firstly, for the reference channel signal, the CF, BW, PW, and TOA of the direct wave pulse are obtained by the parameter estimation method based on template matching, which is a key step in the signal processing flow, and we will discuss it in detail in Section 3. The TOA is used to distinguish different pulse signals, and then each pulse signal is down-converted to baseband according to the estimated CF. The estimated PW and BW are used to reconstruct the baseband reference signal to perform matched filtering on the basedband scattered wave signal.

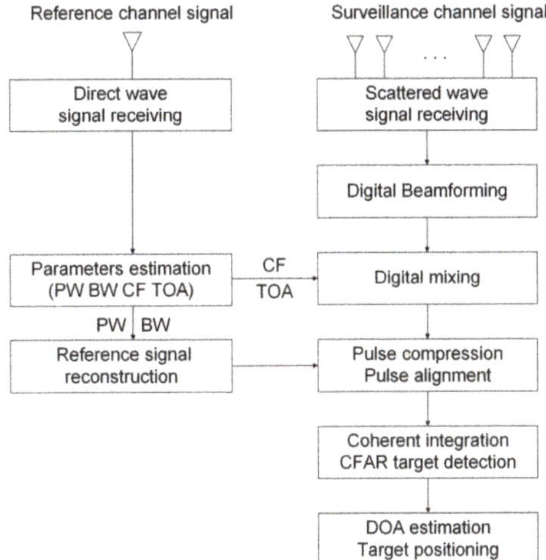

Figure 4. Flow chart of the signal processing method.

As for the surveillance channel signal, the simultaneous multi-beam forming method is adopted to cover the surveillance airspace. In addition, since the CF of transmitted signal hops in a broad frequency range, wideband beamforming algorithm must be adopted. The PBR system forms seven frequency invariant beams to cover the surveillance airspace, as shown in Figure 5.

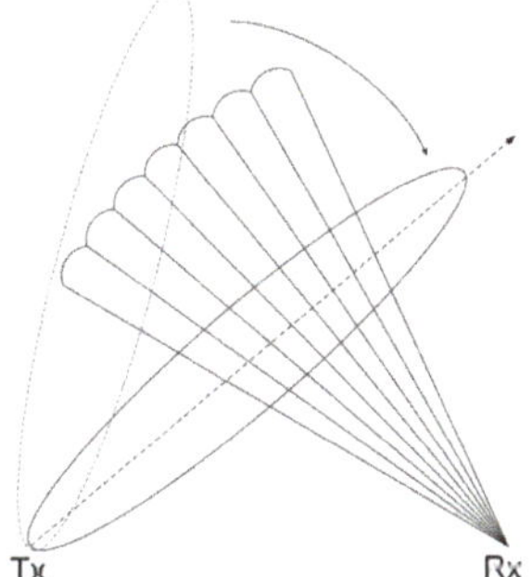

Figure 5. Sketch map of simultaneously multi-beamforming.

After digital beamforming, the receiving data are transformed from array element space to beamspace. We use the estimated CF to down-convert the signal to baseband to reduce the amount of data to be processed subsequently. Then, pulse compression is performed and combined with the reconstructed reference signal, which concentrates the energy of target echo and improves the range resolution. Then, the pulses are realigned according to the peak position of the direct wave signal after pulse compression.

After pulse alignment, we divide the pulses into different groups according to the CF of pulses and, then, perform coherent integration according to the CPI we set for the signals in each group, which can be used to filter out fixed clutter and to improve moving target detection performance. The coherent integration method based on Radon-NUFFT will be introduced in Section 4. Then, the constant false alarm rate (CFAR) detection method is used to detect targets.

Finally, after target detection, an amplitude comparison direction measurement is implemented by taking advantage of the beams that may contain echo of targets. Combined with the azimuth and bistatic range of the target, the actual position of the target can be obtained.

3. Direct Wave Signal Parameter Estimation Method Based on Template Matching

Different from conventional PBR based on a civil illuminator that directly use the direct wave as the reference signal for cross correlation with scattered wave, we first estimate the parameters of the direct wave signal to achieve the time and frequency synchronization of PBR system. Therefore, the parameter estimation of the direct wave signal is a very important step in the whole signal processing flow and should be accurately completed in real time. Therefore, combined with the parameter template library of PAR signal, we propose a direct wave parameter estimation method based on template matching.

3.1. Signal Model

The PAR transmits the LFM signal, which can be expressed as

$$s_t\left(t_m, \hat{t}\right) = A \operatorname{Rect}\left(\frac{\hat{t}}{\tau_m}\right) \exp\left[j2\pi f_m\left(t_m + \hat{t}\right)\right] \exp\left(j\pi\mu_m\hat{t}^2\right) \tag{3}$$

where A is the amplitude of transmitted signal, t_m is the slow time, m indicates the pulse number, $\hat{t}$ is the fast time, $\operatorname{Rect}(\cdot)$ is the window function, τ_m is the pulse width, f_m is the carrier frequency, and $\mu_m = B_m/\tau_m$ is the chirp rate with bandwidth B_m. Then, the direct wave signal received by the reference antenna can be represented as

$$s_d\left(t_m, \hat{t}\right) = \sigma_d A \operatorname{Rect}\left(\frac{\hat{t} - L/c}{\tau_m}\right) \exp\left[j2\pi f_m\left(t_m + \hat{t} - L/c\right)\right] \exp\left[j\pi\mu_m(\hat{t} - L/c)^2\right] \tag{4}$$

where σ_d is the attenuation coefficient and L is the distance between the PAR transmitter and reference antenna. The parameters to be estimated are f_m, τ_m, B_m, and the time of arrival.

Based on the a priori knowledge of the waveform of PAR, we can build a template library of radar parameters, including CF template library noted as $\left\{f^{(k)} | \mathrm{k} = 1, 2, \ldots, N_f\right\}$, PW template library noted as $\left\{\tau^{(k)} | \mathrm{k} = 1, 2, \ldots, N_\tau\right\}$, and BW template library noted as $\left\{B^{(k)} | \mathrm{k} = 1, 2, \ldots, N_B\right\}$, where $f^{(k)}$, $\tau^{(k)}$, and $B^{(k)}$ are the specific templates in the signal parameter template library and N_f, N_τ, and N_B are the number of parameters of CF, PW, and BW respectively. Based on these prior information, we can use a method based on template matching to estimate the parameters of direct wave signals in real time.

3.2. Proposed Method

Figure 6 shows the flow chart of the direct wave parameter estimation method. Firstly, the double threshold detection method is used to detect the LFM signal and to roughly estimate the PW and TOA; then, by matching with the PW template, the precise estimation of the PW can be obtained. Secondly, the BW is accurately estimated by the dechirp method, and the CF is roughly estimated, then the precise estimation of CF is obtained by matching with the CF template. Finally, the accurate PW, BW, and CF are used to reconstruct a reference signal so as to accurately estimate the TOA by means of matched filtering. Next, we will introduce the process of direct wave parameter estimation step by step.

Step 1. In order to detect the LFM signal, it is necessary to determine the detection threshold. Since the signal-to-noise ratio (SNR) of a direct wave signal is generally high, after calculating the noise energy σ^2, set a threshold V_T that is slightly larger than σ^2. Using the double threshold detection method, it is considered that the pulse has arrived only when the signal amplitude continuously exceeds the threshold for p times. Similarly, the pulse is considered to be ended only when the amplitude is continuously below the threshold for q times, wherein the values of p and q are determined by the SNR of the direct wave signal. After that, the preliminary estimated arrival time $\hat{T}_s$, end time $\hat{T}_e$, and preliminary estimated PW $\hat{\tau}$ are obtained.

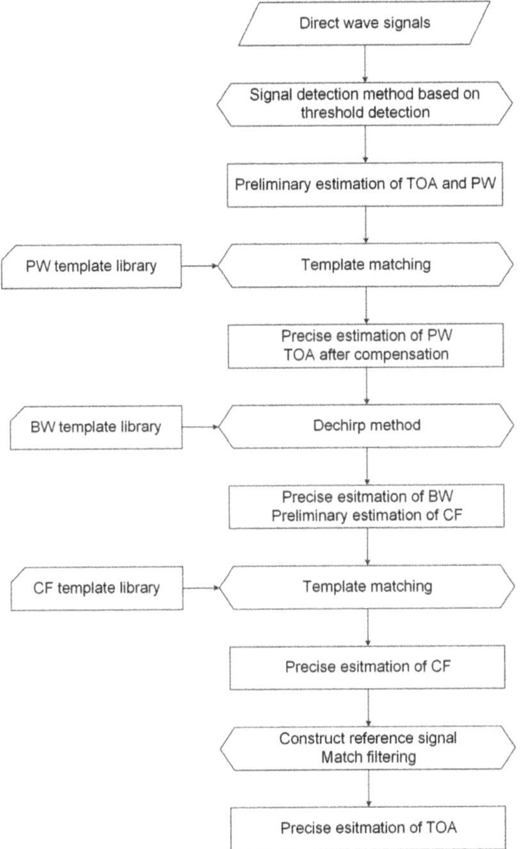

Figure 6. Flow chart of direct wave parameter estimation method.

Step 2. Comparing $\hat{\tau}$ with the PW template library $\left\{ \tau^{(k)} | k = 1, 2, \dots, N_\tau \right\}$, choose τ_0 which is closest to $\hat{\tau}$ in the PW template library; then, τ_0 is the precise estimation of PW. Thus, the PW estimation error is $\varepsilon = \hat{\tau} - \tau_0$, and the arrival time and end time are compensated, which can be expressed as

$$\hat{T}_{start} = \hat{T}_s + \varepsilon/2 \, , \ \hat{T}_{end} = \hat{T}_e + \varepsilon/2 \tag{5}$$

Step 3. Combined with the BW template library, the dechirp method is used to accurately estimate the BW and to preliminarily estimate the CF of the direct wave signal. Firstly, intercept the signal s_p of interval $\left[\hat{T}_{start}, \hat{T}_{end} \right]$; s_p contains a total of $M = \tau_0 \cdot F_s$ points, where F_s is the sampling rate of the signal. As the BW template library is $\left\{ B^{(k)} | k = 1, 2, \dots, N_B \right\}$, the possible chirp rate of the signal can be

$$k_1 = \frac{B^{(1)}}{\tau_0}, k_2 = \frac{B^{(2)}}{\tau_0}, \cdots, k_{N_B} = \frac{B^{(N_B)}}{\tau_0} \tag{6}$$

Then, construct reference signals with different chirp rates, which can be expressed as

$$s_{ref1} = e^{-j\pi k_1 (i/F_s)^2}, i = 1, 2, ..., M$$

$$s_{ref2} = e^{-j\pi k_2 (i/F_s)^2}, i = 1, 2, ..., M$$

$$\cdots$$

$$s_{refN_B} = e^{-j\pi k_{N_B} (i/F_s)^2}, i = 1, 2, ..., M \tag{7}$$

Multiply s_p with different reference signals $s_{ref1}, s_{ref2}, \cdots, s_{refN_B}$ to obtain the dechirped signal $s_{dc1}, s_{dc2}, ..., s_{dcN_B}$, and then, perform a fixed-point FFT on the signal to obtain the spectrum $S_{dc1}, S_{dc2}, ..., S_{dcN_B}$. Measuring and comparing the peak value of $S_{dc1}, S_{dc2}, ..., S_{dcN_B}$, the chirp rate $k_{\max}$ corresponding to the spectrum with the largest peak value is the chirp rate of the signal, and then, the accurate BW can be calculated as $B_0 = k_{\max} \cdot \tau_0$. At the same time, the frequency position corresponding to the peak value in the spectrum is the preliminary estimation of CF of the signal, noted as $\hat{f}_c$. Comparing $\hat{f}_c$ with the CF template library $\left\{ f^{(k)} | k = 1, 2, \ldots, N_f \right\}$, choose f_0 which is closest to $\hat{f}_c$ in the CF template library; then, f_0 is the precise estimation of CF.

Step 4. Combining with the estimated parameters of the signal, including B_0, τ_0 and f_0, reconstruct a reference signal noted as

$$s_{mf} = e^{j\left[2\pi f_0 \frac{i}{F_s} + \pi \frac{B_0}{\tau_0} \left(\frac{i}{F_s}\right)^2\right]}, i = 1, 2, ..., M \tag{8}$$

Using the reference signal s_{mf} to perform matched filtering on the whole signal; then, the peak position of the result is the accurate TOA of the direct wave pulse.

4. Coherent Integration Method Based on Radon-NUFFT

Since the radar signal we want to perform coherent integration on has random PRI and staggerd PW, after analyzing the signal model of the echo of moving target, a coherent integration method based on Radon-NUFFT is proposed to solve the problems caused by the random PRI and staggered PW.

4.1. Signal Model

Suppose the bistatic velocity of the target is v_0, the initial distance between the target and transmitter is R_{t0}, and the initial distance between the target and receiver is R_{r0}; then, the bistatic range sum of the target can be expressed as

$$R\left(t_m, \hat{t}\right) = R_{t0} + R_{r0} - v_0 \left(t_m + \hat{t}\right) = R_0 - v_0 \left(t_m + \hat{t}\right) \tag{9}$$

where $R_0 = R_{t0} + R_{r0}$ and R_0 is the initial range sum.

Then the time delay between the target echo and the transmitted signal is $\Delta t = R\left(t_m, \hat{t}\right)/c$. As the transmitted signal is noted as Equation (3), the echo signal from the target can be represented as

$$s_r\left(t_m, \hat{t}\right) = \sigma_r A \, \text{Rect}\left(\frac{\hat{t} - \Delta t}{\tau_m}\right) \exp\left[j\pi \frac{B}{\tau_m}(\hat{t} - \Delta t)^2\right] \exp\left[j2\pi f_c\left(t_m + \hat{t} - \Delta t\right)\right] \tag{10}$$

where σ_r is the scattering coefficient of target and the signals to be analyzed have the same CF noted as f_c and BW noted as B.

The baseband signal of the target echo after down-conversion can be given as

$$s_b\left(t_m, \hat{t}\right) = \sigma_r A \, \text{Rect}\left(\frac{\hat{t} - \Delta t}{\tau_m}\right) \exp\left[j\pi \frac{B}{\tau_m}(\hat{t} - \Delta t)^2\right] \exp\left(-j2\pi f_c \Delta t\right) \tag{11}$$

After the reconstruction of reference signal, pulse compression is performed and the result can be represented as

$$s_{PC}\left(t_m, \hat{t}\right) = \sigma_r A \sqrt{B\tau_m} \operatorname{sinc}\left[\pi\left(B + f_d\right)\left(\hat{t} - \frac{2\left(R_0 - v_0 t_m\right)}{c} + \frac{f_d}{\mu_m}\right)\right] \exp\left(-j\pi\frac{f_d^2}{\mu_m}\right)$$
$$\cdot \exp\left[-j4\pi\frac{\left(f_c - f_d\right)\left(R_0 - v_0 t_m\right)}{c}\right] \exp\left[j\pi f_d\left(\hat{t} - \frac{2\left(R_0 - v_0 t_m\right)}{c} + \frac{f_d}{\mu_m}\right)\right] \tag{12}$$

where $f_d = 2v_0 f_c / c$ denotes the Doppler frequency and $\mu_m = B/\tau_m$ is the chirp rate. Based on the assumption that the radar signal is narrowband signal and $v_0 \ll c$, the above formula can be simplified as follows:

$$s_{PC}\left(t_m, \hat{t}\right) = A_m \operatorname{sinc}\left[\pi B\left(\hat{t} - \left(\frac{2\left(R_0 - v_0 t_m\right)}{c} - \frac{f_d}{\mu_m}\right)\right)\right] \exp\left(j2\pi f_d t_m\right) \tag{13}$$

where A_m is the complex amplitude of the signal.

It can be observed from Equation (13) that the amplitude of the signal is proportional to $\sqrt{B\tau_m}$; since the signal pulse width is agile, the amplitudes of different pulses will be different. Also, the signal envelope varies with $[2\left(R_0 - v_0 t_m\right)/c - f_d/\mu_m]$, which can be divided into two terms for analysis. The former term $[2\left(R_0 - v_0 t_m\right)/c]$ is the RCM caused by target motion; the latter term f_d/μ_m is caused by the range-Doppler coupling effect. Since the slow time is nonuniform, which will cause nonuniform RCM and since μ_m is different for different pulses, it is necessary to compensate the RCM and range-Doppler coupling of different pulses accordingly. In addition, the phase of the signal with respect to the slow time is $\exp\left(j2\pi f_d t_m\right)$, so the nonuniform slow time will lead to the nonuniform phase fluctuations among different pulses and traditional Fourier transform among slow time dimension cannot be applied.

4.2. Radon-NUFFT Method

Since the irregular RCM $2\left(R_0 - v_0 t_m\right)/c$ is correlated with R_0, $v0$, and t_m and the irregular range-Doppler coupling effect $f_d/\mu_m = 2v_0 B/\lambda\tau_m$ is correlated with v_0 and τ_m, we can compensate them by searching through the motion parameters R_0 and v_0 for each pulse respectively.

As for the nonuniform phase fluctuations $\exp\left(j2\pi f_d t_m\right)$ among different pulses, which can be converted to the problem of spectral analysis of nonuniformly sampled complex-valued data, we use the nonuniform discrete Fourier transform (NUDFT) of nonuniform sampling data, which is given as follows:

$$y(f_k) = \sum_{i=0}^{N-1} x_i e^{-j\frac{2\pi}{N} f_k t_i} \tag{14}$$

where x_i is the nonuniformly sampled data at nonuniform sampling time t_i, N is the total number of sampling points, and f_k is the frequency point we want to analyze. However, the NUDFT algorithm requires a large amount of computation, so we use a NUFFT method based on a class of regular Fourier matrices [27], which is much faster than NUDFT. For the nonuniformly sampled data sequence $\{x(t_i)\}_{i=1}^{N}$, the spectral estimation at f_k can be denoted as $y(f_k) = \mathrm{NUFFT}[x(t_i), f_k]$.

Figure 7 shows the flowchart of the proposed method. The coherent integration time T_n, the number of pulses N_p, the range search scope $[r_{min}, r_{max}]$, and interval Δr are predetermined according to the dwell time of antenna and radar parameters. The searching scope of the initial velocity and interval are preset based on prior information such as moving status of targets to be detected. Therefore, the number of range searching parameters is $N_r = \lceil\left(r_{max} - r_{min}\right)/\Delta r\rceil$, where $\lceil\cdot\rceil$ denotes the round up to an integer operation and the number of velocity searching parameters is $N_v = \lceil 2v_{max}/\Delta v\rceil$. Therefore, the moving trajectories of the target determined by the searching parameters can be given as follows:

$$r(t_m) = r_i - v_j t_m \tag{15}$$

where $m = 1, 2, \cdots, N_p$, $r_i = r_{min} + (i-1)\Delta r$, $i = 1, 2, \cdots, N_r$, $v_j = -v_{max} + (j-1)\Delta v$, and $j = 1, 2, \cdots, N_v$. Then, compensate the irregular range-Doppler coupling $2v_j\tau_m/B\lambda$ for each pulse.

For the signal shown in Equation (13), extract the N_p dimension data vector for coherent integration.

$$\mathbf{X}_{N_p} = s_{PC}\left[t_m, \frac{2(r_i - v_j t_m)}{c} - \frac{2v_j\tau_m}{B\lambda}\right] \tag{16}$$

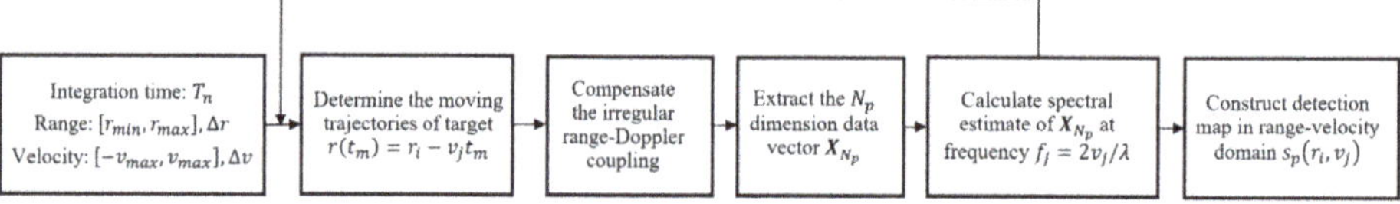

Figure 7. Flowchart of the proposed method.

At last, calculate the spectral estimate of $\mathbf{X}_{N_p}$ at frequency $f_j = 2v_j/\lambda$ to compensate the irregular RCM and range-Doppler coupling and nonuniform phase fluctuations among different pulses simultaneously, which can be expressed as

$$s_p(r_i, v_j) = \text{NUFFT}\left\{ s_{PC}\left[t_m, \frac{2(r_i - v_j t_m)}{c} - \frac{2v_j\tau_m}{B\lambda}\right], \frac{2v_j}{\lambda}\right\} \tag{17}$$

where $s_p(r_i, v_j)$ is the coherent integration result of the target with the initial bistatic range sum r_i and constant radial velocity v_j. Go through all the searching parameters of range and velocity; then a two-dimensional result defined in the (r, v) plane can be formed, which is the result of coherent integration based on the proposed method.

According to the abovementioned analysis, the sampled data are accumulated coherently along the slow time dimension via a general Doppler filter bank, so when the searching parameters are equal to the real motion parameters, the irregular RCM and range-Doppler coupling can be compensated and the energy of target can be accumulated completely.

4.3. Simulation Results

In order to validate the performances of the proposed method, simulation experiments are conducted. The simulated parameters ofthe radar and moving target are listed in Table 1, which is based on the monostatic radar model.

Table 1. Simulated parameters.

System Parameters (Unit)	Values
Carrier frequency (MHz)	680
Bandwidth (MHz)	2
Sampling frequency (MHz)	5
Pulse width template library (us)	{40 80 120 160 200}
Average PRI (us)	1500
Variation range of PRI (us)	[1000 2000]
Number of coherently integrated pulses	128
Initial distance of target (km)	90
Radial velocity of target (m/s)	800

Suppose the SNR of the received target echo before pulse compression is 10 dB. Figure 8 shows the range–time map of signal after pulse compression. It can be seen that the trajectory of target is not a straight line due to the irregular RCM and range-Doppler coupling. In addition, the amplitude of different pulse is different because the amplitude of signal after pulse compression is proportional to $\sqrt{B\tau_m}$, and τ_m is staggered.

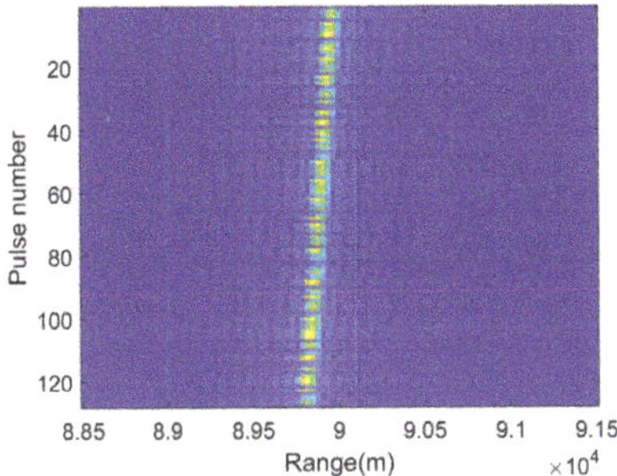

Figure 8. Range-time map after pulse compression.

Figure 9a shows the result of the signal after compensation for irregular RCM via Radon algorithm by using the accurate velocity of target. It can be seen that the trajectory of the target is still not a vertical straight line due to the irregular range-Doppler coupling. Figure 9b is the result of Figure 9a after compensating for range-Doppler coupling. It can be seen that the envelope of the target is aligned on the same vertical line.

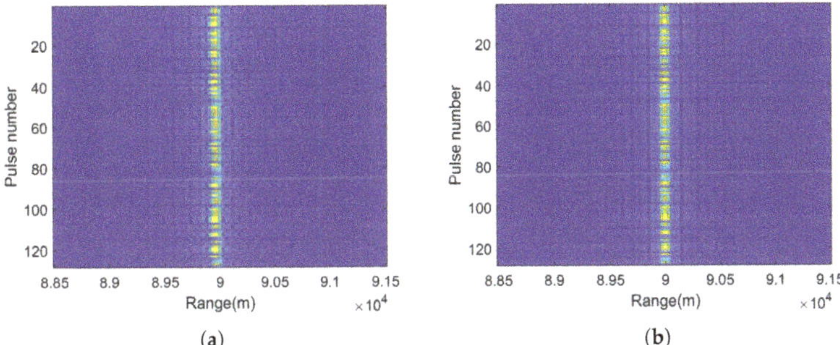

(a)   (b)

Figure 9. Results of signal after compensation. (a) Result after compensation of irregular RCM; (b) Result after compensation of irregular RCM and range-Doppler coupling.

Figure 10 compares the results of coherent integration via four methods. Figure 10a is the result of moving target detection (MTD), i.e., perform FFT directly in the slow time dimension without compensation, which cannot obtain any apparent peak in the range–velocity plane. Figure 10b shows the result of RFT, i.e., perform FFT in the slow time dimension after compensation of RCM and range-Doppler coupling, which cannot obtain any apparent peak either. Therefore, we can get the conclusion that FFT cannot solve the nonuniform phase fluctuations among different pulses, which has a very serious influence on the coherent integration.

Figure 10c shows the result of using NUFFT among a slow time dimension without compensation of RCM and range-Doppler coupling, and Figure 10d shows the result of Radon-NUFFT. It can be seen that both of them show a peak in the range–velocity plane, which reveals the coherently integrated echo signals. However, in Figure 10c, the estimated initial range and velocity of the target indicated by the peak location in the range–velocity domain is inaccurate, since the irregular RCM and range-Doppler coupling are not compensated. Moreover, it can be seen that the peak value in Figure 10d is higher than that in Figure 10c, and the initial range and velocity indicated by the peak position in Figure 10d are both accurate, which proves the effectiveness of the proposed method.

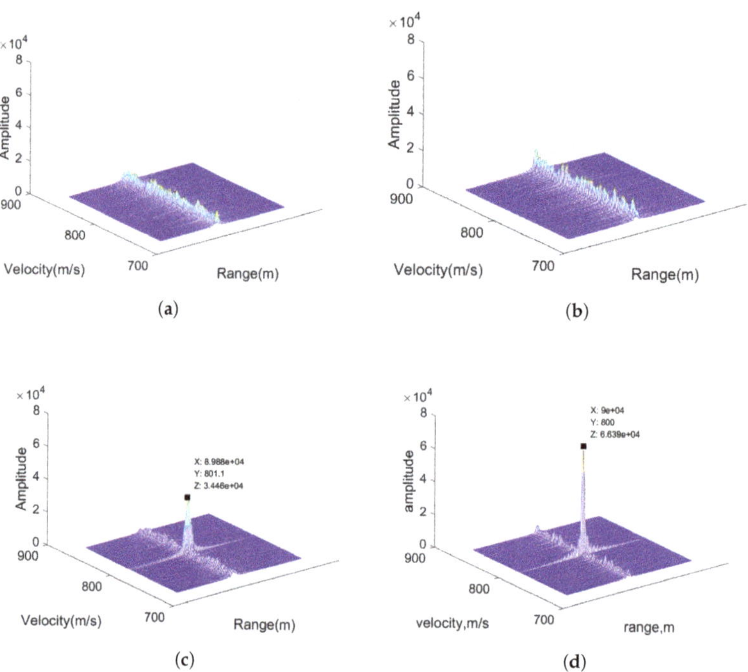

Figure 10. Coherent integration via four method: (**a**) MTD, (**b**) RFT, (**c**) NUFFT among slow time, (**d**) Radon-NUFFT.

5. Experimental Results

In this section, some of the field experimental results are given. Figure 11 shows the experimental scenario geometry. The PAR transmitter and the PBR system are located at different locations, the distance between them is 35 km. There is a civil airport in the north of the PBR system with a distance of 48 km, thus ensuring the presence of aircraft targets in the air, which can be used to verify the target surveillance capability of the PBR system.

Based on a priori knowledge of the waveform of PAR signals, it can be concluded that there are 81 carrier frequency points of the transmitted signals, which vary from 558 MHz to 643 MHz. Therefore, a supersonic heterodyne receiver down-converts a high radio frequency (RF) signal from the reference antenna into intermediate frequencies (IF) signal by mixing the RF signal with a 600-MHz signal from a Local Oscillators (LO). Figure 12 shows the imaginary part of the time domain signal and the spectrogram of direct wave IF signal.

It can be seen from Figure 12 that there are two groups of signals with different CF and BW and that the PRI and PW of different pulses in the same group are different. For direct wave signal, the CF, BW, and PW of the signal are obtained by the parameter estimation method described in Section 3 and the scattered wave signal received by the surveillance antenna is down-converted to baseband combined with the estimated CF. Figure 13 shows the imaginary part of the time domain signal and the spectrogram of the scattered wave baseband signal.

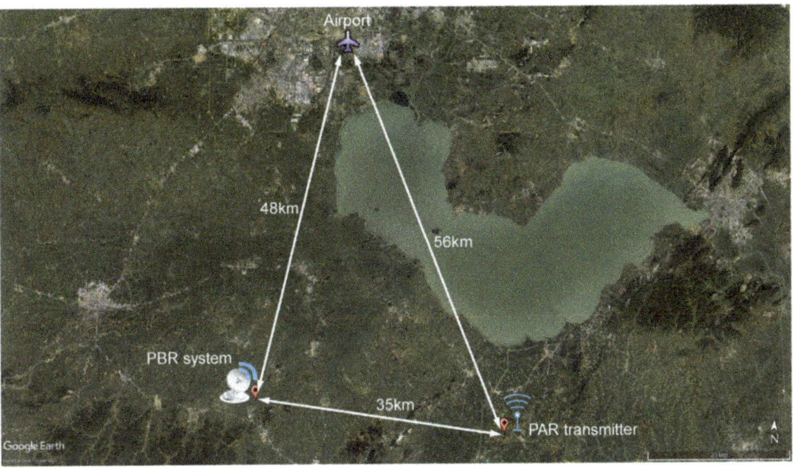

Figure 11. The experimental scenario geometry.

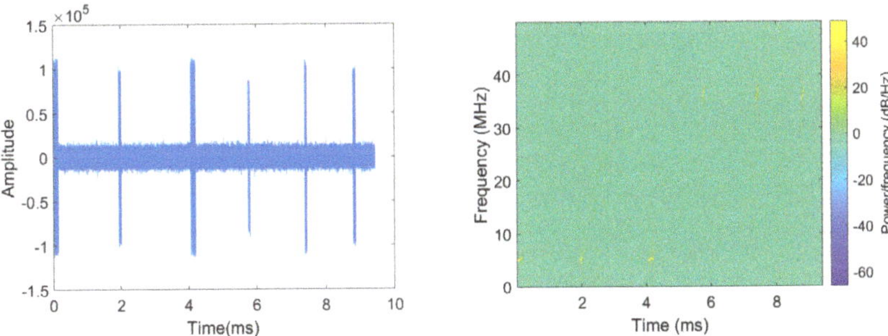

Figure 12. Time domain waveform and spectrogram of direct wave IF signal.

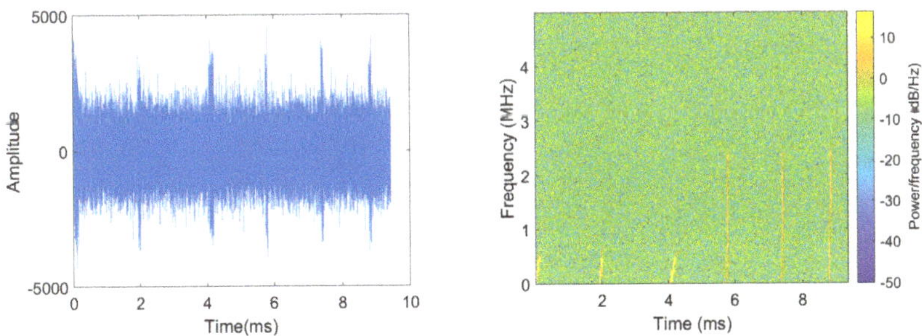

Figure 13. Time domain waveform and spectrogram of scattered wave baseband signal.

It can be seen that, although the surveillance antenna is not directed to the PAR transmitter, there are still direct wave signals in the surveillance channel, since the surveillance antenna can receive the sidelobe energy from the PAR emitter. However, it is obvious that the SNR of the direct wave in Figure 13 is much lower than that in Figure 12.

To verify the performance of matched filtering based on reconstructed reference signal, for the baseband signal in Figure 13, the last direct wave pulse signal is intercepted and the time-domain and frequency-domain signals are shown in Figure 14a, while the time-domain and frequency-domain plots of the reconstructed reference signal based on the parameters of the direct wave signal estimated in the reference channel are shown in Figure 14b, it can be seen that the SNR of the reconstructed reference signal is much higher.

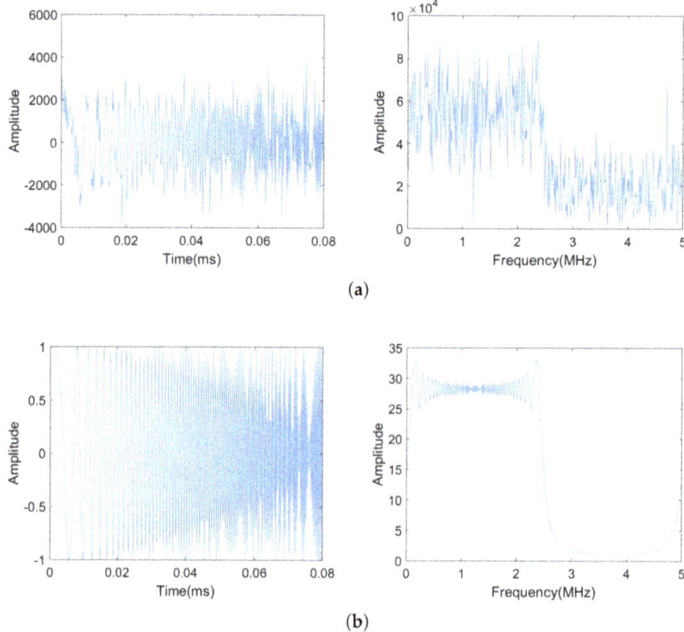

Figure 14. Time-domain and frequency-domain plots of the intercepted direct wave signal and reconstructed reference signal. (**a**) Intercepted direct wave signal; (**b**) Reconstructed reference signal

Then, the intercepted direct wave signal and the reconstructed reference signal are used for matched filtering of the whole echo signal, and the results are shown in Figure 15; it can be seen that the target echo has a higher SNR after matched filtering using the reconstructed reference signal.

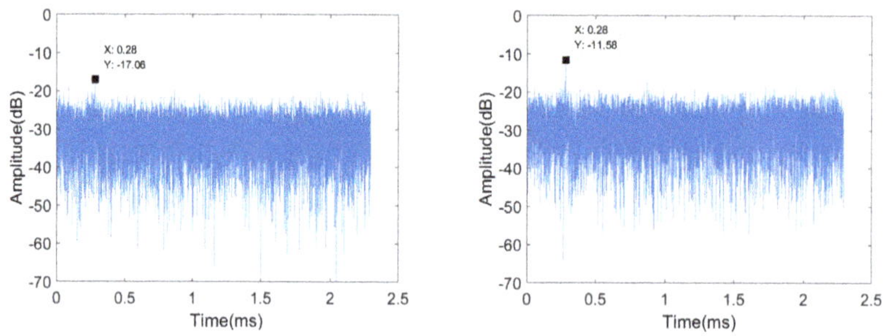

Figure 15. Results of matched filtering using intercepted direct wave signal and reconstructed reference signal.

Since the CF and BW of the PAR signal are agile for different groups and a group of signal can last for a few seconds, we set 0.2 s as the CPI for the coherent integration of scattered wave signals. A total of 128 pulses in a CPI of the 4th beam are matched filtered and realigned according to the peak position of the direct wave signal, and the result is shown in Figure 16.

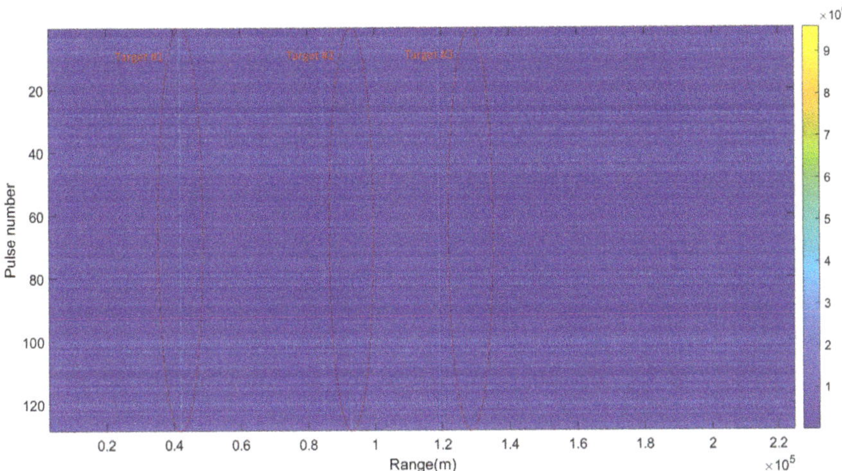

Figure 16. Time-range map after pulse compression and realignment.

In Figure 16, three trajectories of target can be faintly seen; then, they are intercepted and enlarged, which are shown in Figure 17.

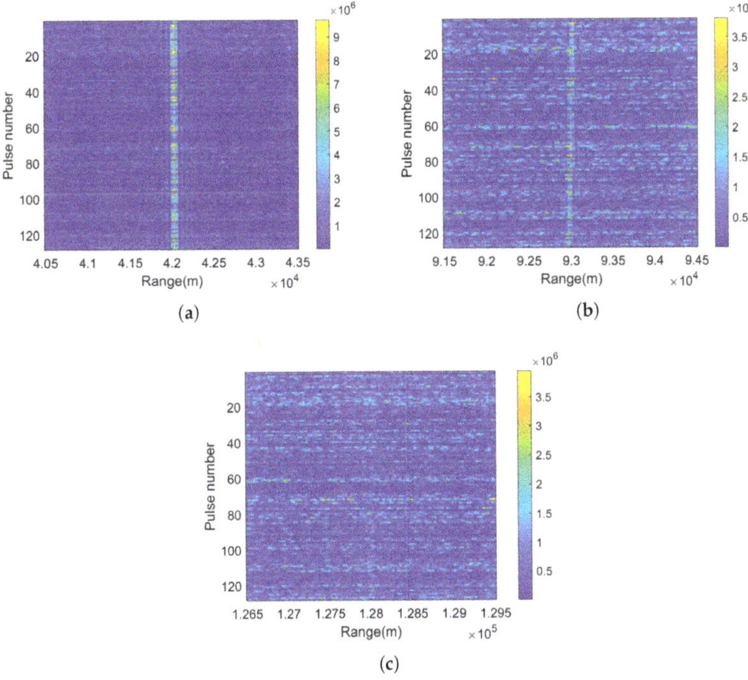

Figure 17. Time-range map of three targets.

It can be seen that the energy of the target echo is inversely proportional to its range, and since the PW of the signal is staggered, the energy of different echo signals after pulse compression are different. In addition, it is obvious that target #1 in Figure 17a is not moving, target #2 is a moving target since

RCM occurs in Figure 17b, while the SNR of the target #3 is too low to judge whether it is moving. Therefore, the energy of the target echo should be accumulated by using the coherent integration method proposed in Section 4 so as to improve the detection probability of the target. Also, the velocity of targets can be estimated, thereby filtering out the fixed clutter.

Figure 18 shows the results of coherent integration of the target echo in Figure 17 using the Radon-NUFFT method. According to Figure 18, the distance and velocity information of the three targets can be obtained as shown in Table 2.

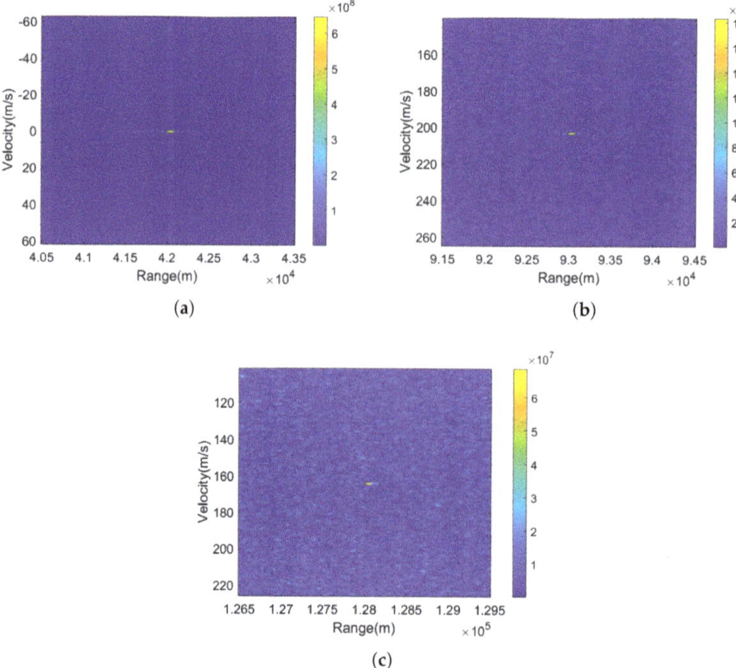

Figure 18. Coherent integration results of three targets.

Table 2. Motion parameters of the targets.

Target Number	Distance (km)	Velocity (m/s)
#1	42.03	0.98
#2	93.06	203
#3	128.03	164

Therefore, the motion parameters of the aircraft target can be effectively obtained by using the coherent integration method proposed in Section 4. After coherent integration and filtering of fixed clutter, CFAR detection is carried out for signals of different beams with a false alarm rate of 10^{-4}. After obtaining the target detection result, the target azimuth is obtained by amplitude comparison direction measurement. Finally, the distance between the target and the PBR system can be calculated according to Equation (2) combined with the bistatic range and the azimuth of the target, so the target positioning can be achieved afterwards.

The target positioning results processed by PBR system for 120 s and target tracks converted from the real-time location information of flights provided by Automatic Dependent Surveillance-Broadcast

(ADS-B) at the same time are drawn on the same figure to verify the target detection performance of the PBR system, which is shown in Figure 19.

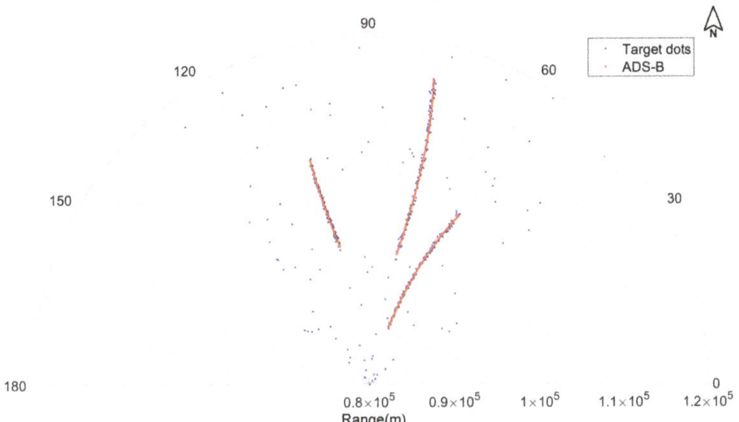

Figure 19. Result of target positioning and Automatic Dependent Surveillance-Broadcast (ADS-B) tracks.

It can be seen that the PBR system achieves the detection and localization of three moving aircraft targets effectively. There are some deviations between the target dots and the tracks provided by ADS-B since the error of target's monostatic range depends on the target's bistatic range and azimuth measurement errors at the same time. Moreover, it can be seen that there are some erroneous points far from the actual target tracks, which are false alarm points caused by clutter and noise; since these erroneous points are not continuous in time, they can be filtered out by the target tracking method, which will be explored in our future work. To validate the accuracy of the PBR system, we compare the monostatic distance and azimuth of the target dots near the three target tracks in 120 s with an actual target position, and the root mean square error (RMSE) of the distance and azimuth of the target dots can be calculated to be 984 m and 2.7°, respectively. It can be concluded that the PBR system based on PAR signals is able to locate the aircraft targets in the surveillance airspace.

6. Conclusions

In this paper, we conducted an experimental study to verify the feasibility of PBR based on PAR signals for aircraft target detection and localization. First, the experimental PBR system was described in general; after the analysis of the complex modulation form of PAR signals, two problems to be solved in signal processing were proposed, including the time and frequency synchronization of the PBR system and the coherent integration of radar signal with random PRI and staggered PW. Then, aiming at these two problems, the paper first proposed a direct wave parameter estimation method based on template matching to achieve a time and frequency synchronization of the PBR system and, then, proposes a coherent integration method based on Radon-NUFFT to deal with the irregular RCM and nonuniform phase fluctuations among different pulses caused by random PRI and the irregular range-Doppler coupling caused by staggered PW. Finally, the results of field experiments were presented. The performance of the matched filtering method based on reconstructed reference signal was verified by comparing the matched filtering results using different reference signals. The effectiveness of the coherent integration method based on Radon-NUFFT was verified by the coherent integration results of three targets. Finally, by comparing the target positioning results with the ADS-B tracks, the performance of target detection and localization of the PBR system based on PAR signals was verified.

Author Contributions: J.P. proposed and implemented the methods, Q.B. designed the experiment, J.P. and Q.Z. processed the data, and P.H. and Z.C. revised the manuscript. All authors of the article provided substantive comments.

Funding: This research received no external funding.

Conflicts of Interest: The authors declare no conflict of interest.

References

1. Howland, P.; Maksimiuk, D.; Reitsma, G. FM radio based bistatic radar. *IEE Proc. Radar Sonar Navig.* **2005**, *152*, 107. [CrossRef]
2. Honda, J.; Otsuyama, T. Feasibility Study on Aircraft Positioning by Using ISDB-T Signal Delay. *IEEE Antennas Wireless Propag. Lett.* **2016**, *15*, 1787–1790. [CrossRef]
3. Langellotti, D.; Colone, F.; Lombardo, P.; Sedehi, M.; Tilli, E. DVB-T based Passive Bistatic Radar for maritime surveillance. In Proceedings of the IEEE Radar Conference, Cincinnati, OH, USA, 19–23 May 2014; pp. 1197–1202. [CrossRef]
4. Bolvardi, H.; Derakhtian, M.; Sheikhi, A. Dynamic Clutter Suppression and Multitarget Detection in a DVB-T-Based Passive Radar. *IEEE Trans. Aerosp. Electron. Syst.* **2017**, *53*, 1812–1825. [CrossRef]
5. Feng, W.; Friedt, J.M.; Cherniak, G.; Sato, M. Passive bistatic radar using digital video broadcasting–terrestrial receivers as general-purpose software-defined radio receivers. *Rev. Sci. Instrum.* **2018**, *89*, 104701. [CrossRef] [PubMed]
6. Salah, A.; Abdullah, R.R.; Aziz, N.A.; Ismail, A.; Hashim, F. Experimental study of LTE signals as illuminators of opportunity for passive bistatic radar applications. *Electron. Lett.* **2014**, *50*, 545–547. [CrossRef]
7. Falcone, P.; Colone, F.; Bongioanni, C.; Lombardo, P. Experimental results for OFDM WiFi-based passive bistatic radar. In Proceedings of the IEEE Radar Conference, Washington, DC, USA, 10–14 May 2010; pp. 516–521. [CrossRef]
8. Clemente, C.; Soraghan, J.J. GNSS-Based Passive Bistatic Radar for Micro-Doppler Analysis of Helicopter Rotor Blades. *IEEE Trans. Aerosp. Electron. Syst.* **2014**, *50*, 491–500. [CrossRef]
9. Harms, H.A.; Davis, L.M.; Palmer, J. Understanding the signal structure in DVB-T signals for passive radar detection. In Proceedings of the IEEE Radar Conference, Washington, DC, USA, 10–14 May 2010; pp. 532–537. [CrossRef]
10. Griffiths, H.; Willis, N. Klein Heidelberg—The First Modern Bistatic Radar System. *IEEE Trans. Aerosp. Electron. Syst.* **2010**, *46*, 1571–1588. [CrossRef]
11. Ito, T.; Takahashi, R.; Morita, S.; Hirata, K. Experimental result of passive bistatic radar with unknown transmitting radar pulse. In Proceedings of the European Microwave Conference, Nuremberg, Germany, 6–10 October 2013; pp. 1767–1770. [CrossRef]
12. Samczynski, P.; Krysik, P.; Kulpa, K. Passive radars utilizing pulse radars as illuminators of opportunity. In Proceedings of the IEEE Radar Conference, Johannesburg, South Africa, 27–30 October 2015; pp. 168–173. [CrossRef]
13. Wang, Y.; Bao, Q.; Wang, D.; Chen, Z. An Experimental Study of Passive Bistatic Radar Using Uncooperative Radar as a Transmitter. *IEEE Geosci. Remote Sens. Lett.* **2015**, *12*, 1868–1872. [CrossRef]
14. Barbarossa, S. Analysis of multicomponent LFM signals by a combined Wigner-Hough transform. *IEEE Trans. Signal Process.* **1995**, *43*, 1511–1515. [CrossRef]
15. Lv, X.; Bi, G.; Wan, C.; Xing, M. Lv's Distribution: Principle, Implementation, Properties, and Performance. *IEEE Trans. Signal Process.* **2011**, *59*, 3576–3591. [CrossRef]
16. Hao, H. Multi component LFM signal detection and parameter estimation based on EEMD–FRFT. *Optik* **2013**, *124*, 6093–6096. [CrossRef]
17. Qian, L.; Xu, J.; Meng, C.; Li, J.; Zhou, X.; Long, T. Multi-waveform parameter estimation of external illuminator for passive bistatic radar. In Proceedings of the CIE International Conference on Radar (RADAR), Guangzhou, China, 10–13 October 2016; pp. 1–5. [CrossRef]
18. Perry, R.P.; DiPietro, R.C.; Fante, R.L. Coherent Integration With Range Migration Using Keystone Formatting. In Proceedings of the IEEE Radar Conference, Boston, MA, USA, 17–20 April 2007; pp. 863–868. [CrossRef]
19. Xu, J.; Yu, J.; Peng, Y.N.; Xia, X.G. Radon-Fourier Transform for Radar Target Detection, I: Generalized Doppler Filter Bank. *IEEE Trans. Aerosp. Electron. Syst.* **2011**, *47*, 1186–1202. [CrossRef]

20. Rao, X.; Tao, H.; Su, J.; Guo, X.; Zhang, J. Axis rotation MTD algorithm for weak target detection. *Digit. Signal Process.* **2014**, *26*, 81–86. [CrossRef]

21. Zheng, J.; Su, T.; Zhu, W.; He, X.; Liu, Q.H. Radar High-Speed Target Detection Based on the Scaled Inverse Fourier Transform. *IEEE J. Sel. Topics Appl. Earth Observ. Remote Sens.* **2014**, *8*, 1108–1119. [CrossRef]

22. Babu, P.; Stoica, P. Spectral analysis of nonuniformly sampled data—A review. *Digit. Signal Process.* **2010**, *20*, 359–378. [CrossRef]

23. Wu, Y.; Sepehri, N. Interpolation of bandlimited signals from uniform or non-uniform integral samples. *Electron. Lett.* **2011**, *47*, 53. [CrossRef]

24. Li, J.; Chen, Z. Research on random PRI PD radar target velocity estimate based on NUFFT. In Proceedings of the 2011 IEEE CIE International Conference on Radar, Chengdu, China, 24–27 October 2011; Volume 2, pp. 1801–1803. [CrossRef]

25. Tian, J.; Xia, X.G.; Cui, W.; Yang, G.; Wu, S.L. A Coherent Integration Method via Radon-NUFrFT for Random PRI Radar. *IEEE Trans. Aerosp. Electron. Syst.* **2017**, *53*, 2101–2109. [CrossRef]

26. Pan, J.; Li, J.; Hu, P.; Bao, Q. Coherent integration method based on Radon-iterative adaptive approach for irregular pulse repetition interval radar. *J. Appl. Remote Sens.* **2019**, *13*, 016521. [CrossRef]

27. Liu, Q.H.; Nguyen, N. An accurate algorithm for nonuniform fast Fourier transforms (NUFFT's). *IEEE Microw. Guid. Wave Lett.* **1998**, *8*, 18–20. [CrossRef]

Article

Improved ISRJ-Based Radar Target Echo Cancellation Using Frequency Shifting Modulation

Qihua Wu *, Feng Zhao *, Junjie Wang, Xiaobin Liu and Shunping Xiao

State Key Laboratory of Complex Electromagnetic Environment Effects on Electronics and Information System, National University of Defense Technology, Changsha 410073, China; wangjunjienudt@163.com (J.W.); highge@126.com (X.L.); xiaoshunping_nudt@163.com (S.X.)

* Correspondence: wuqihua13@nudt.edu.cn (Q.W.); zhaofeng321@nudt.edu.cn (F.Z.);
   Tel.: +86-731-8700-3528 (Q.W.); +86-731-8457-6240 (F.Z.)

Received: 10 December 2018; Accepted: 28 December 2018; Published: 1 January 2019

Abstract: Target echo cancellation is an ingenious method that protects the target of interest (TOI) from being detected by radar. Interrupted-sampling repeater jamming (ISRJ) is a novel deception jamming method for linear frequency modulation (LFM) radar countermeasures, which has been applied in target echo cancellation recently. Compared with the conventional cancellation method, not only can the target echo be successfully cancelled at radar receiver, but a train of false targets is also produced and forms deception jamming by applying the ISRJ technique. In this paper, an improved radar target echo cancellation method based on ISRJ is proposed that utilizes an extra frequency shifting modulation on the intercepted LFM radar signal. The jammer power is more efficiently utilized by the proposed method. Moreover, more flexible multi-false-target deception jamming can be obtained by adjusting the interrupted sampling frequency. The real target remains effectively protected by the false preceding target in the presence of amplitude mismatch of cancellation signal and target echo. Numerical simulations and measured data experiments are conducted to demonstrate the effectiveness of the proposed method.

Keywords: radar echo cancellation; frequency shifting modulation; interrupted sampling; radar jamming; deception jamming

---

1. Introduction

Radar plays an important role in both civil and military fields as its all-weather and day-night capacities superior to the optical sensors [1–4]. To protect the targets of interest (TOI) from being detected by the radar, radar jamming techniques have been widely studied over the past few decades, including blanket jamming and deception jamming. Interrupted-sampling repeater jamming (ISRJ) is a novel radar deception jamming technique proposed in 2006 [5,6]. By sampling and repeating the radar signal at sub-Nyquist rates, a train of false targets is produced after radar matched filtering (MF) processing. Thus ISRJ is widely applied in radar jamming including synthetic aperture radar (SAR) [7–10] and inverse synthetic aperture radar (ISAR) [11,12]. On the other hand, the anti-ISRJ technique also develops rapidly in the past decades [13].

Radar target echo cancellation is an ingenious jamming method that cancels the target echo at the radar receiver [14–22]. The core idea lies in transmitting a synthesized replica of the target echo except for its being 180° out of phase to the radar by an active source. For linear frequency modulation (LFM) pulse compression radar, ISRJ can produce a train of false targets with controllable amplitudes and phases. Based on this phenomenon, a radar target echo cancellation method using self-protection ISRJ is proposed [23,24]. By designing the interrupted sampling frequency, the repeater time-delay and the jammer power, the ISRJ signal not only ideally cancels radar target echo with −1 order false target, but also produces a train of false targets. Thus a better cancellation performance is obtained compared with the conventional cancellation methods due to the multi-false-target deception jamming. In our previous work, the cancellation method based on nonperiodic ISRJ has been further proposed, considering the unavoidable amplitude mismatch of the cancellation signal and the target echo [25]. However, on the one hand, the energy of -1 order false target is lower than 0 order false target in ISRJ. Hence a relatively large transmitting power is needed, which may lead to hostile anti-radiation weapons attack. If 0 order false target can be used to cancel the target echo, the jammer power can be reduced efficiently. This protects the jammer equipment effectively. On the other hand, interrupted sampling frequency should be precisely designed according to radar signal parameters as noted in [23], so the position of the false targets produced by ISRJ remains fixed in the radar MF output, which makes it easier to be countered [26].

Range-Doppler coupling is a unique property for LFM signal, which causes the peak of the compressed pulse to shift in time by an amount proportional to the Doppler frequency [27]. By utilizing this property, some effective methods against LFM radar such as frequency-shifting deception jamming have been proposed [28,29]. Inspired by this, the main contribution of this paper is to propose an ISRJ-based radar target echo cancellation method with an extra frequency-shifting modulation. By the frequency shifting modulation on the ISRJ signal, the radar target echo can be cancelled by the 0 order false target. Thus the jammer power is more efficiently utilized, which will also protect the jammer due to the smaller radiation energy. Moreover, the interrupted sampling frequency can be flexibly adjusted to change the position of false targets in the proposed method, thus better deception jamming is performed. Last but not least, the preceded -1 order false target will shield the TOI more effectively in the presence of amplitude mismatch of the cancellation signal and the target echo.

The remainder of the paper is organized as follows. In Section 2, the existing radar target echo cancellation method based on ISRJ is reviewed, and the shortage is analyzed. In Section 3, the improved cancellation method is proposed. In Section 4, numerical simulations and measured SAR data experiments are conducted to illustrate the validity of the proposed method. Finally, conclusions are drawn in Section 5.

2. Review of Radar Target Echo Cancellation Based on ISRJ

2.1. Amplitude and Phase Characteristics of the ISRJ Signal

As shown in Figure 1a, assume the interrupted sampling function is a rectangular envelope pulse train denoted as $p(t)$, then

$$p(t) = rect(\frac{t}{T_p}) \otimes \sum_{n=-\infty}^{+\infty} \delta(t - nT_s) \tag{1}$$

where t denotes the time variable, T_p is the pulse width, T_s is the pulse repetition interval (PRI), $\delta(\cdot)$ is the impulse function, n is the pulse number, $\otimes$ represents the convolution operation, and $rect(t/T_p) = \begin{cases} 1 & |t/T_p| < 0.5 \\ 0 & others \end{cases}$. The sampling duty ratio is defined as T_p/T_s.

Via Fourier transform, the spectrum of $p(t)$ is given as

$$P(f) = T_p f_s \sum_{n=-\infty}^{+\infty} sinc(nf_s T_p)\delta(f - nf_s) \tag{2}$$

where $sinc(x) = sin(\pi x)/(\pi x)$, and $f_s = 1/T_s$ represents the interrupted sampling frequency.

The radar transmits LFM pulse signal, denoted as

$$x(t) = \frac{1}{\sqrt{T}} rect(\frac{t}{T}) exp(j\pi kt^2) \tag{3}$$

where T is the pulse width, k is the chirp rate and the bandwidth is $B = kT$.

Figure 1b presents the interrupted sampling processing of LFM signal, then the ISRJ signal can be expressed as

$$s(t) = p(t)x(t) \tag{4}$$

As demonstrated in [5], the MF output of ISRJ signal at radar receiver is formed by a train of false targets, given by

$$y(t) = \sum_{n=-\infty}^{+\infty} a_n y_{sn}(t) \tag{5}$$

where

$$a_n = T_p f_s sinc(nf_s T_p) \tag{6}$$

is the amplitude coefficient of n order false target.

$$y_{sn}(t) = sinc[(kt + nf_s)(T - |t|)](1 - \frac{|t|}{T})exp(j\pi nf_s t) \tag{7}$$

represents the pulse compression output of n order false target.

From Equation (7), the peak of n order false target lies in

$$t_n = -nf_s/k \tag{8}$$

The phase of n order false target is

$$\varphi_n = \pi n f_s t|_{t=-nf_s/k} = -(nf_s)^2 \pi/k \tag{9}$$

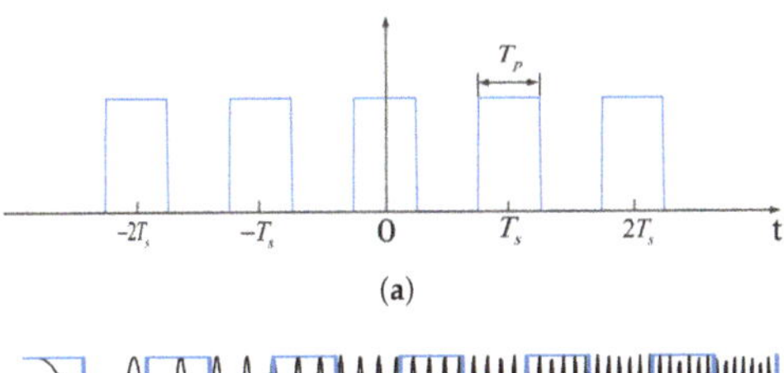

(a)

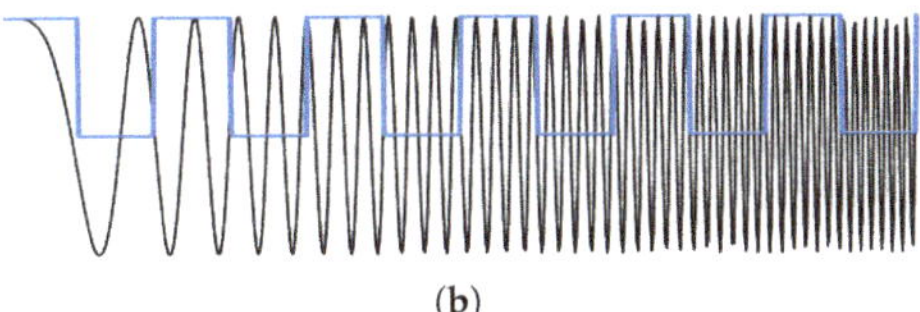

(b)

Figure 1. Signal model of interrupted sampling. (**a**) The interrupted sampling function. (**b**) Interrupted sampling of linear frequency modulation (LFM) signal.

2.2. Target Echo Cancellation Using the ISRJ Signal

For radar target echo cancellation, three conditions should be satisfied including range synchronization, phase coherent and amplitude match. The phase of the radar target echo consists of two components including the propagation phase and the signature of the radar target itself. In self-protection jamming, the target and jammer echoes travel the same distance to the radar, therefore the propagation phase difference is zero. On the other hand, it is assumed, for simplicity, that the target cross-section amplitude equals 1 and the phase equals zero.

For self-protection jamming, the 0 order target lags behind the target echo due to the unavoidable repeater time-delay. Thus parameters of -1 order false target are designed to cancel the target echo [23].

The repeater time-delay satisfies

$$\tau_d = -t_{-1} = f_s/k \tag{10}$$

to guarantee the MF output of the -1 order false target synchronizes with the target echo in time domain.

The interrupted sampling frequency satisfies

$$f_s = \sqrt{k} \tag{11}$$

to guarantee $\varphi_{-1} = -\pi$, which makes the phase of -1 order false target opposites the phase of the real target.

The amplitude modulation coefficient of the jammer satisfies

$$A_J = \frac{1}{a_{-1}} = \frac{1}{T_p f_s sinc(f_s T_p)} \tag{12}$$

to guarantee the amplitudes of two echoes equal.

Further, the radiant power of the jammer should satisfy

$$ERP_J = \frac{PG_t\sigma}{4\pi R^2} A_J^2 \tag{13}$$

where P is the peak power of radar, G_t is the receiving antenna gain, σ is the target RCS, and R is the distance between the jammer and radar.

When Equations (10)–(12) are satisfied, ideal target echo cancellation can be realized in radar receiver. Thus the amplitude and phase modulation of the ISRJ signal is subtly utilized to cancel the target echo by parameter designs.

As analyzed above, the -1 order false target of ISRJ signal is used as the cancellation source, not the 0 order false target which has the strongest amplitude response. It is because the 0 order false target lags behind the real target echo due to the repeater time-delay of self-protection ISRJ. Besides, Equation (9) indicates that the phase of 0 order target is 0, equal to the phase of real target, which may strengthen the real target echo at radar receiver on the contrary. It may lead to hostile anti-radiation weapons attack due to the larger tranmitting power for the existing method. If the 0 order false target can be utilized as the cancellation source by some special modulations, the required jammer power will reduce comparatively. Besides, interrupted sampling frequency must be accurately set by Equation (11). Hence the false target will locate at the fixed position in radar MF output according to Equation (8), which greatly affects the jamming performance.

3. Improved ISRJ-Based Cancellation Method Using Frequency Shifting Modulation

In this section, an improved radar target echo cancellation method using ISRJ is proposed to overcome the shortage of the existing method proposed in Section 2.

As known, the LFM waveform exhibits range-Doppler coupling which causes the peak of the compressed pulse to shift in time by an amount proportional to the Doppler frequency. In particular, the peak occurs earlier in time for a positive LFM slope, compared with the peak response for a stationary target [27]. Inspired by this unique property, a frequency shift f_d can be added to the ISRJ signal to make the 0 order false target synchronize with the real target echo in time domain.

After adding frequency shift, the ISRJ signal can be expressed as

$$s'(t) = s(t)exp(j2\pi f_d t) = p(t)x(t)exp(j2\pi f_d t) \tag{14}$$

Let $x'(t) = x(t)exp(j2\pi f_d t)$ be the complete LFM signal after adding frequency shift. The MF output at radar receiver is

$$s_{mf}(t) = sinc[(kt + f_d)(T - |t|)](1 - \frac{|t|}{T})exp(j\pi f_d t) \tag{15}$$

Similar to Equations (5) and (15), the MF output of ISRJ signal with frequency shift will be

$$y(t) = \sum_{n=-\infty}^{+\infty} a_n y'_{sn}(t) \tag{16}$$

where a_n is the amplitude coefficient given by Equation (6).

$$y'_{sn}(t) = sinc[(kt + nf_s + f_d)(T - |t|)](1 - \frac{|t|}{T})$$
$$\cdot exp(j\pi(f_d + nf_s)t) \tag{17}$$

represents the pulse compression output of the n order false target.

According to Equation (17), after frequency modulation, the peak of the n order false target locates at

$$t'_n = -f_d/k - nf_s/k \tag{18}$$

The phase is

$$\varphi'_n = \pi(f_d + nf_s)t|_{t=-(f_d+nf_s)/k} = \frac{-(nf_s + f_d)^2}{k}\pi \tag{19}$$

From Equations (18) and (19), both the position and phase of 0 order false target are modulated by the frequency shift. Then the 0 order false target can be utilized to cancel the real target echo. Similarly, three conditions proposed in Section 2 must be satisfied to realize the cancellation.

To guarantee the phase of 0 order false target opposites the phase of the real target, $\varphi_0 = -\frac{f_d^2}{k}\pi = -(2m + 1)\pi$, where m is an integer. Then the frequency shift should satisfy

$$f_d = \sqrt{(2m + 1)k} \tag{20}$$

To guarantee the range synchronization in time domain, the repeater time-delay should satisfy

$$\tau_d = -t_0 = f_d/k = \sqrt{(2m + 1)/k} \tag{21}$$

Assume that the minimum required analyzing time for the canceller is t_{min}. Obviously τ_d should be no smaller than t_{min}, then

$$m \geq (k \cdot t_{min}^2 - 1)/2 \tag{22}$$

The minimum m is $m_0 = \lceil (k \cdot t_{min}^2 - 1)/2 \rceil$, then $\tau_{dmin} = \sqrt{(2m_0 + 1)/k}$, $f_{dmin} = \sqrt{(2m_0 + 1)k}$. It is necessary to point out that when the required processing time t_{min} is larger than the time-delay t_{-1} determined by Equation (10) for the method proposed in Section 2, the echo cancellation will fail. Hence the improved method is more robust from this point of view.

Finally, the required jammer power is calculated. Due to the utilization of 0 order false target, the amplitude modulation coefficient changes to

$$A'_j = \frac{1}{a_0} = \frac{1}{T_p f_s} \tag{23}$$

Compared with the method proposed in Section 2, the radiant power of the jammer will be

$$ERP'_J = (\frac{A'_J}{A_J})^2 ERP_J = sinc^2(f_s T_p) ERP_J \tag{24}$$

From Equation (24), by joint frequency shift and repeater time-delay modulation given by Equations (20) and (21) respectively, the required jammer power reduces to $sinc^2(f_s T_p)$ times compared with that of the method proposed in Section 2. Besides, the interrupted sampling frequency is not necessary to be designed particularly. It means that the position of false targets can be flexibly designed by adjusting the interrupted sampling frequency to accord with the demands of electronic countermeasures (ECM). Thus more effective jamming performance can be obtained. Since the cancellation signal is transmitted back to the radar along with the target echo, it is effective against both real-time processing radars and off-line processing radars such as SAR.

However, it is necessary to point out that the tramsmitting power given by Equation (24) is difficult to precisely set due to the unavoidable parameter estimation error. Then the amplitude match of the cancellation signal and target echo cannot be guaranteed. In this case, the target echo cannot be completely erased out at the radar receiver. That is exactly the reason why the nonperiodic interrupted sampling modulation is adopted to protect the target residual with the continuous jamming strip in [25]. In this paper, the false targets generated by the periodic ISRJ are expected to accomplish the target protection. Hence the acceptable intensity of the effective echo cancellation should be analyzed. Since the false targets are expected to protect the target residual, it is reasonable to define that the amplitude of the false target should be larger than that of the suppressed real target for effective cancellation. Hence to achieve the acceptable suppression level, we have

$$A_1 \geq |A_C - A_T| \tag{25}$$

where A_1 is the amplitude of the 1 order false target, A_C is the amplitude of the cancellation signal, and A_T is the amplitude of the target echo. According to the interrupted sampling theory, the amplitude of the 1 order false target and the 0 order false target can be given by

$$A_1 = A_C sinc(f_s T_p) \tag{26}$$

Substitute (26) into (25), then

$$\frac{|A_C - A_T|}{A_C} \leq sinc(f_s T_p) \tag{27}$$

Define the amplitude mismatch degree γ as

$$\gamma = \frac{|A_C - A_T|}{A_T} \tag{28}$$

The amplitude mismatch includes two cases, one is the amplitude of the cancellation signal is larger than that of the target echo e.g. $A_C > A_T$, the other is the opposite. Here the former case is mainly discussed. Then the mismatch degree should satisfy

$$\gamma|1 - \gamma| \leq sinc(f_s T_p) \tag{29}$$

The block diagram of the canceller can be concluded in Figure 2.

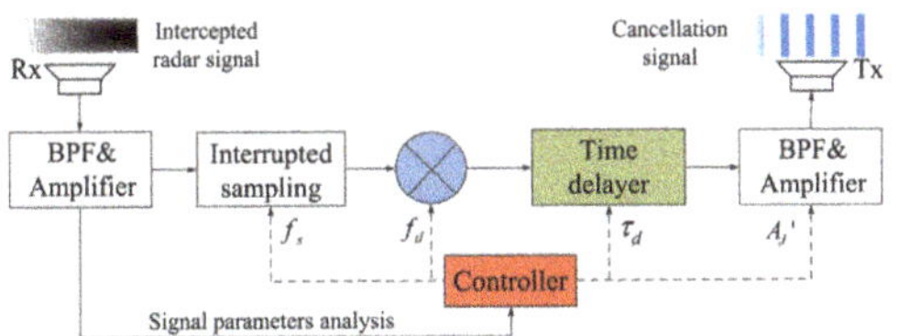

Figure 2. Block diagram of the canceller.

As shown in Figure 2, the block diagram of the canceller is similar to that proposed in [25]. The main difference is the interrupted sampling modulation here is periodic. Then the multi-false-target deception jamming is obtained by the proposed method instead of the blanket jamming in [25]. The unavoidable limitations of the canceller are the signal parameters should be precisely measured similar to some other types of deception jammings such as the frequency shifting jamming. The processing procedure can be listed as follows.

Step 1: Intercept the radar signal and analyze signal parameters including chirp rate k, pulse width T.

Step 2: Execute interrupted sampling processing of the intercepted signal with sampling frequency f_s and sampling pulse width T_p, then modulate the cancellation signal by frequency shift f_d and time-delay τ_d given by Equations (20) and (21) respectively.

Step 3: Calculate the transmitting power given by Equation (24), then retransmit the cancellation signal to radar.

4. Simulation Results and Analysis

In this section, the performance of the proposed method is analyzed by simulations. For convenience, the existing method proposed in [23] is named as Method 1, and our improved method proposed in Section 3 is named as Method 2.

4.1. Comparison of Required Jammer Power

Firstly, the required jammer power of two methods is compared. The main parameters are listed in Table 1.

Table 1. Main parameters of radar and target.

Parameters	Value
radar antenna gain	60 dB
peak power	810 kW
pulse width	100 us
chirp rate	4×10^{10} Hz/s
target RCS	0.1 m^2

The required jammer power of Method 1 and the proposed Method 2 can be calculated according to Equations (13) and (24) respectively. Figure 3 presents the comparison result of the required jammer power, where the sampling duty ratio of the ISRJ is set as 20% and 50% respectively.

As shown in Figure 3, the required jammer power of Method 2 is lower than that of Method 1 when the duty ratio is equal, and the reduction is more obvious with higher duty ratio. When the target distance is 1000 km and the duty ratio of ISRJ is 50%, the required jammer power reduces by 5 dBW by applying the proposed Method 2. It demonstrates the proposed method can efficiently reduce the required jammer power as expected. Thus a more efficient utilization of the jammer power can be obtained. Then the jammer can be effectively protected due to the smaller radiation energy.

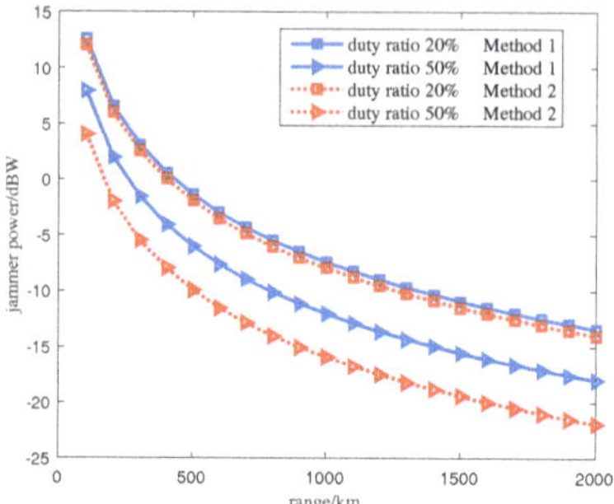

Figure 3. Required jammer power of two methods.

Besides, the required jammer power reduces with the increase of duty ratio, because higher duty ratio means that the ISRJ signal acquires more energy from the complete radar signal. As shown in Figure 3, the required jammer power of Method 2 with the duty ratio 20% is larger than that of Method 1 with the duty ratio 50%. It indicates higher duty ratio is also very significant for reducing the required jammer power in actual applications.

4.2. Cancellation Performance Analysis by Numerical Simulations

In this sub-section, the target echo cancellation performance of the two methods is compared. Assume that the peak output of the target echo after MF locates at $t = 0$. For Method 1, the interrupted sampling frequency $f_s = 200$ kHz, and the time-delay $\tau_d = 5$ us. For Method 2, the frequency shift $f_d = 200$ kHz, the repeater time-delay $\tau_d = 5$ us, and the interrupted sampling frequency is set as $f_s = 200$ kHz and $f_s = 100$ kHz respectively. The distance between the jammer and the radar R = 1000 km. The duty ratio is 50% for both methods. The jammer power for the two methods is set according to Equations (12) and (24). Then the amplitude of the target echo and the cancellation can be equal at radar receiver for ideal cancellation. However, it is a great challenge to estimate the parameters accurately given by Equations (12) and (24) as analyzed in Section 3. Thus it is quite difficult to guarantee the amplitude match of the cancellation signal and target echo in actual applications. Hence the simulations are divided into two parts, one is the cancellation performance with ideal amplitude match, the other is the cancellation performance with amplitude mismatch.

4.2.1. Cancellation Performance with Ideal Amplitude Match

Firstly, the assumption of amplitude match is guaranteed, which means that the corresponding parameters in Equations (12) and (24) are accurately estimated. Figure 4 present the target echo cancellation results of the ISRJ signal with the parameters given by Method 1 and Method 2 respectively, where the amplitude of the target echo is assumed to be 1 and the normalization of the amplitude is done with respect to that of the target echo.

From Figure 4a,b, the MF output of ISRJ signal at radar receiver is formed by a train of false targets spreading symmetrically in the range domain, and the power of the false targets decreases rapidly from the center to the edge, which is consistent with the characteristics of the ISRJ [5]. Figure 4c,d indicate that both Method 1 and Method 2 can ideally cancel the real target echo after MF at radar receiver with the designed parameters. For Method 1, the -1 order false target is utilized to cancel the target echo. Zero and 1 order false targets remain in the output of MF, and forms multi-false-target deception jamming as revealed in Figure 4c. For Method 2, the 0 order false target is utilized to cancel the target echo. The ±1 order false targets form multi-false-target deception jamming as revealed in Figure 4d. Thus the effectiveness of the proposed method for target echo cancellation with a smaller jammer power is demonstrated.

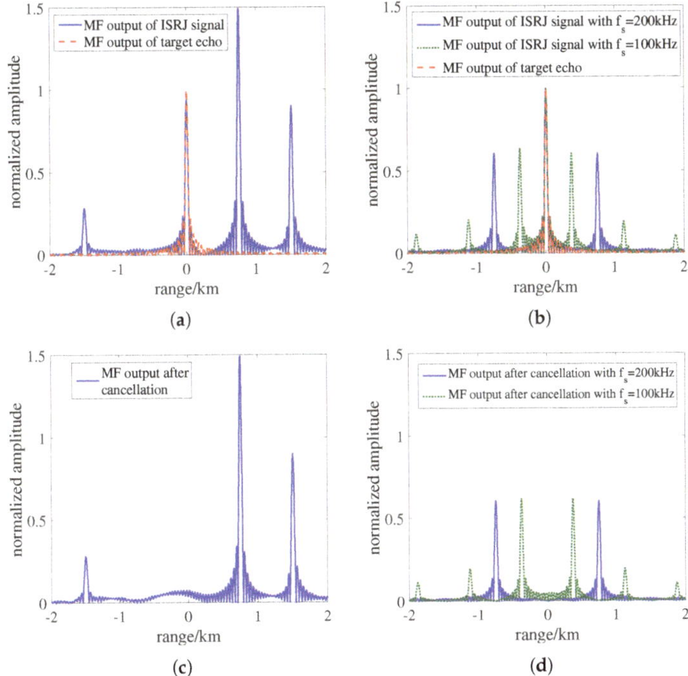

Figure 4. Target echo cancellation result with amplitude match. (**a**) matched filtering (MF) output of the interrupted-sampling repeater jamming (ISRJ) signal and target echo of Method 1. (**b**) MF output of the ISRJ signal and target echo of Method 2. (**c**) MF output after echo cancellation of Method 1. (**d**) MF output after echo cancellation of Method 2.

For Method 2, the cancellation result with different interrupted sampling frequencies is shown in Figure 4d. When the interrupted sampling frequency is set as 200 kHz and 100 kHz, the ± 1 order false targets appear at ± 750 m and ± 375 m respectively. It indicates that the position of false targets after MF can be flexibly adjusted by designing the interrupted sampling frequency. Thus more flexible jamming performance is obtained.

4.2.2. Cancellation Performance with Amplitude Mismatch

Cancellation results with amplitude mismatch for both methods are presented in Figure 5 considering parameter estimation errors. Assume the mismatch degree $\gamma = 50\%$ and $\gamma = 80\%$ respectively. Similarly, the amplitude of the target echo is assumed to be 1 and the normalization of the amplitude is done with respect to that of the target echo.

From Figure 5, the echo cancellation fails for both methods with amplitude mismatch. However, the ISRJ cancellation signal still contributes to the protection of real target. On the one hand, the amplitude of the target echo greatly reduces due to the cancellation signal, which makes the real target less visible by the radar. On the other hand, the generated false targets can confuse the radar and effectively protect the real target. When comparing the two methods, although the false target near the true target is with higher power for Method 1, the main 0 and 1 order false targets both lag off the real target for Method 1 as shown in Figure 5a,c. While the main -1 order false target will be ahead of the real target in time domain for Method 2 as revealed in Figure 5b,d. Thus the real target can be more effectively protected by the preceded false target in the improved Method 2.

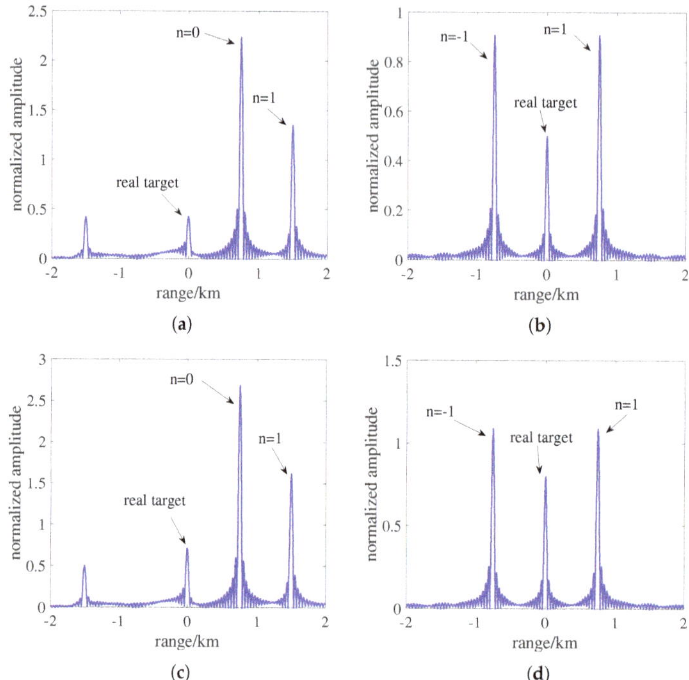

Figure 5. Target echo cancellation result with amplitude mismatch (f_s = 200 kHz). (**a**) Method 1 (γ = 50%). (**b**) Method 2 (γ = 50%). (**c**) Method 1 (γ = 80%). (**d**) Method 2 (γ = 80%).

To evaluate the deception jamming performance with amplitude mismatch for the proposed Method 2 quantitatively, Figure 6 presents the amplitude ratio of the preceded −1 order false target to the real target after cancellation when mismatch degree γ varies from 0.1 to 0.8.

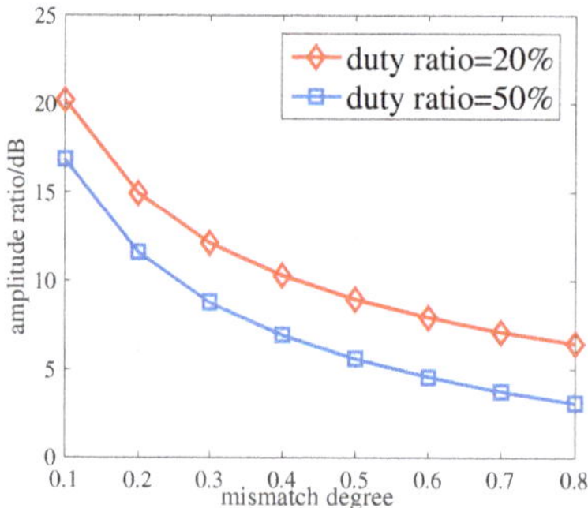

Figure 6. Amplitude ratio of −1 order false target to real target after cancellation for Method 2.

As revealed in Figure 6, the amplitude ratio reduces with the increase of mismatch degree, which is reasonable because cancellation performance gets worse when amplitude mismatch degree increases. However, the amplitude of −1 order false target is still much larger than that of real target even when the mismatch degree γ reaches 0.8, e.g. 6.4 dB for duty ratio =20% and 3.1 dB for duty ratio =50%

respectively. Hence the real target can be effectively protected even with large amplitude mismatch for our proposed method.

In conclusion, both of the methods can ideally cancel the target echo at radar receiver by accurate parameter designs. Besides, they can also contribute to target protection by multi-false-target deception jamming even when the amplitude of the cancellation signal mismatches with that of the target echo. The proposed Method 2 is superior to Method 1 by exploring the following advantages. Firstly, the jammer power is more efficiently utilized. Secondly, more flexible deception jamming performance can be obtained by adjusting the interrupted sampling frequency. Last but not least, the generated preceded false target can protect the real target more effectively.

4.3. Measured SAR Data Experiments Verification

To further present the performance of our proposed Method 2, the measured mini SAR complex imagery provided by the Sandia National Laboratories is used [30]. Since the comparison between the two methods has been made in the previous section and the performance with the SAR data when applying Method 1 has been presented in [24], here only the peformance of the proposed Method 2 is presented in this section. Table 2 presents the main parameters of the SAR imaging scene.

Table 2. Main parameters of the synthetic aperture radar (SAR) imaging scene.

Parameters	Value
Center frequency	9 GHz
Resolution(range and azimuth)	0.5 m $\times$ 0.5 m
Platform velocity	180 m/s
Odd number	1001
Chirp rate	1.5×10^{14} Hz/s

Figure 7a presents the original SAR image by the Range Doppler Algorithm. A dotted circle is added to indicate the position of the plane target that needs to be cancelled. The raw signal of the target is inverted using the algorithm proposed in [31], then the target echo is simulated as shown in Figure 7b.

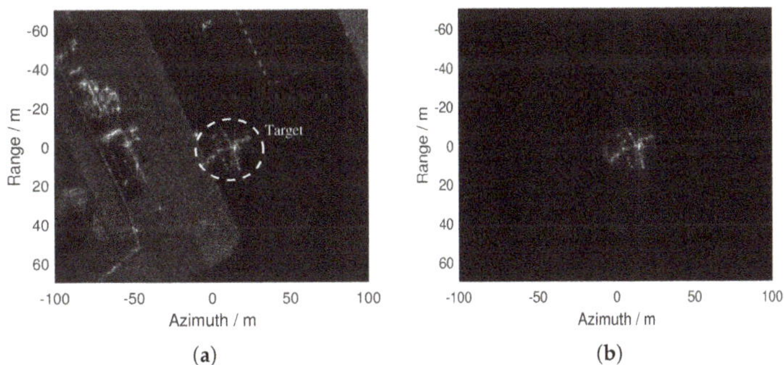

Figure 7. Original synthetic aperture radar (SAR) scene. (**a**) The original image. (**b**) The target.

4.3.1. Cancellation Results

By calculation, the frequency shift $f_d = 12.2$ MHz, the time-delay $\tau_d = 0.082$ us. Firstly assume the jammer power is ideally set to guarantee the amplitude match at radar receiver. Figure 8a,b present the cancellation results with interrupted sampling frequencies set as $f_s = 20$ MHz and $f_s = 40$ MHz respectively, and the sampling duty ratio is 50%.

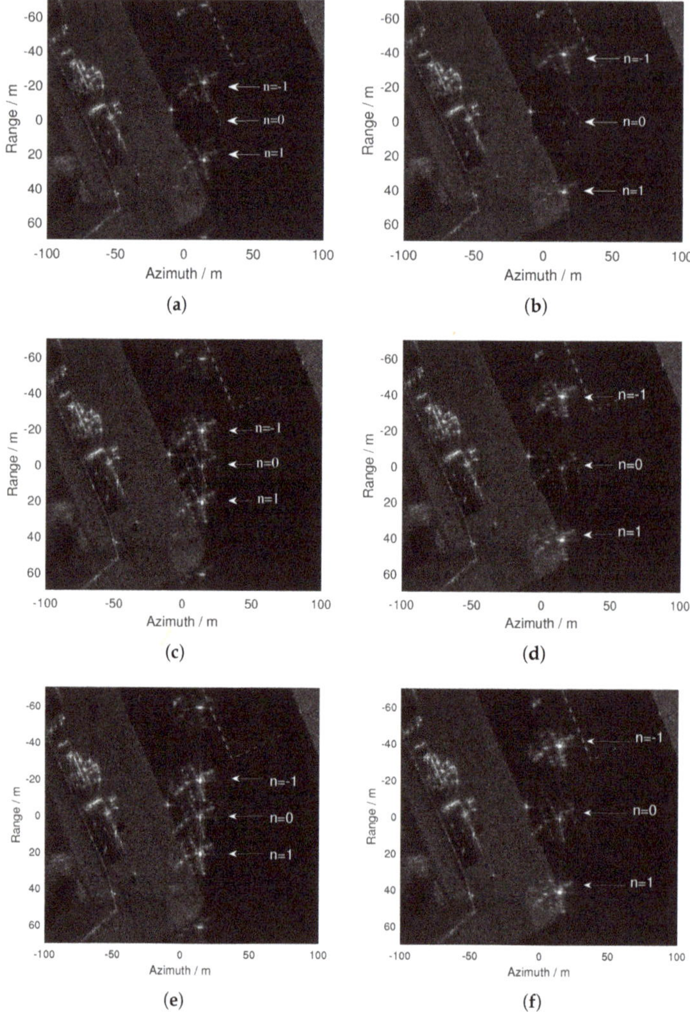

Figure 8. Results of measured SAR data with the proposed method. (**a**) Ideal amplitude match (f_s = 20 MHz). (**b**) Ideal amplitude match (f_s = 40 MHz). (**c**) Amplitude mismatch (γ = 50%, f_s = 20 MHz). (**d**) Amplitude mismatch (γ = 50%, f_s = 40 MHz). (**e**) Amplitude mismatch (γ = 80%, f_s = 20 MHz). (**f**) Amplitude mismatch (γ = 80%, f_s = 40 MHz).

As shown in Figure 8a,b, the target echo is successfully cancelled by the proposed method. By calculation, the position of the ± 1 order false targets should be ± 20 m and ± 40 m for f_s = 20 MHz and f_s = 40 MHz respectively, which is consistent with the results shown in Figure 8a,b. It indicates that the position of false targets is precisely adjusted by designs of the interrupted sampling frequency as expected.

Similarly, the cancellation results with amplitude mismatch are presented in Figure 8c–f, where the amplitude mismatch degree is set as γ = 50% and γ = 80% respectively. As revealed in Figure 8c–f, the target echo cannot be ideally cancelled with amplitude mismatch. The target becomes more and more obvious with the increase of the amplitude mismatch. However, the amplitude of target greatly reduces and the target is less visible in SAR image due to the cancellation compared with original image. Besides, the false targets can effectively shield the real target.

4.3.2. Evaluation of the Cancellation Performance

To access the performance of cancellation, Figure 9 presents the range cuts of Figure 8 in azimuth unit 20 m, where the dotted blue lines represent the range cuts of the original SAR image, and the red lines represent the range cuts after cancellation.

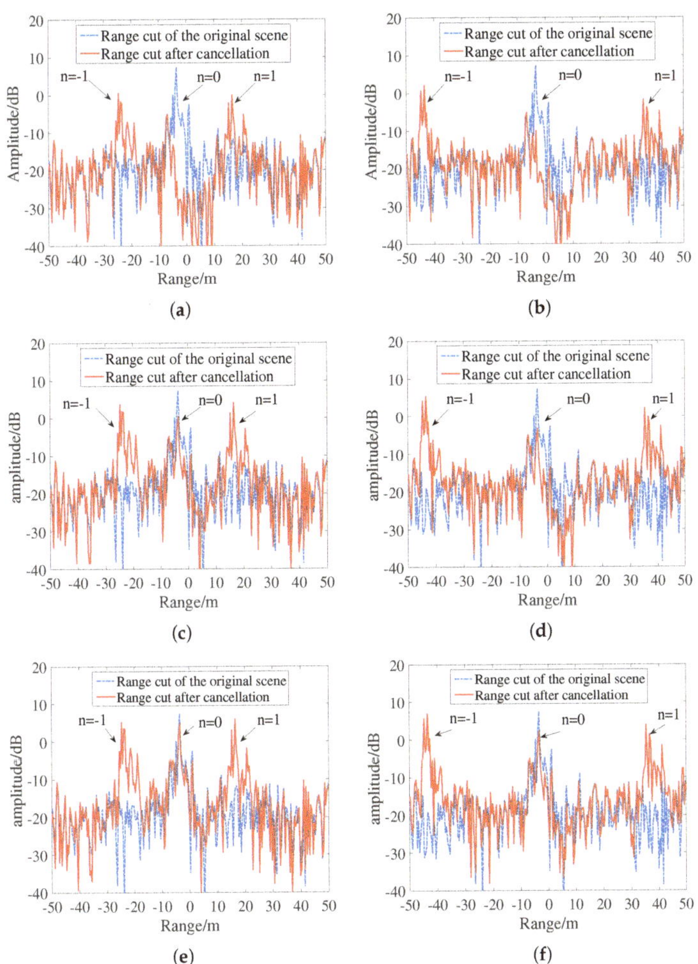

Figure 9. Range cut at azimuth unit 20 m. (**a**) Ideal amplitude match (f_s = 20 MHz). (**b**) Ideal amplitude match (f_s = 40 MHz). (**c**) Amplitude mismatch (γ = 50%, f_s = 20 MHz). (**d**) Amplitude mismatch (γ = 50%, f_s = 40 MHz). (**e**) Amplitude mismatch (γ = 80%, f_s = 20 MHz). (**f**) Amplitude mismatch (γ = 80%, f_s = 40 MHz).

As revealed in Figure 9a,b, in terms of the absort amplitude of the imaging result, there is more than a 20 dB drop in the vicinity of target due to the presence of cancellation signal with ideal cancellation assumptions. After cancellation, the amplitude of the false target is much larger than that of the real target, which makes the radar difficult to distinguish the real target. As revealed in

Figure 9c–f, the amplitude of the real target reduces by 10 dB and 5 dB approximately after cancellation in the presence of 50% and 80% amplitude mismatch, respectively. Thus the cancellation performance gets worse in the presence of amplitude mismatch compared with the ideal cancellation conditions. However, the amplitude of generated false target is about 5 dB larger ($\gamma = 50\%$) and 3 dB larger ($\gamma = 80\%$) than the amplitude of real target, thus the false targets can still effectively protect the real target by stronger amplitude response.

5. Conclusions

In this paper, an improved radar target echo cancellation method that uses frequency shifting modulation is proposed. The proposal uses the frequency shifting modulation to cancel the radar target echo using 0 order false targets produced by ISRJ. Compared with the existing method, the improved method utilizes the jammer power more efficiently, and more flexible jamming performance can be obtained. Besides, the real target can still be effectively protected by the preceded false target even with amplitude mismatch. Simulation results and measured SAR data experiments demonstrate its effectiveness. This work provides good information towards future jammer designs. In futhre work, further investigations on the influence of measuring accuracies of the radar signal parameters will be conducted. Besides, the hardware system designs of the canceller will be another important topic.

Author Contributions: Data curation, F.Z. and J.W.; Formal analysis, Q.W.; Funding acquisition, F.Z.; Investigation, X.L. and S.X.; Validation, Q.W., X.L. and S.X.; Writing—original draft, Q.W. and F.Z.; Writing—review & editing, Q.W. and F.Z.

Funding: This work was supported by the National Natural Science Foundation of China (Grant No. 61890542).

Conflicts of Interest: The authors declare no conflict of interest.

References

1. Skolnik. *Introduction to Radar Systems*; McGraw-Hill: Boston, MA, USA, 2001.
2. Liu, Y.; Wang, W.; Pan, X.; Dai, D.; Feng, D. A frequency-domain three-stage algorithm for active deception jamming against synthetic aperture radar. *IET Radar Sonar Navigat.* **2014**, *8*, 639–646. [CrossRef]
3. Xu, L.; Feng, D.; Wang, X. Improved synthetic aperture radar micro-Doppler jamming method based on phase-switched screen. *IET Radar Sonar Navigat.* **2016**, *10*, 525–534. [CrossRef]
4. Zhang, R.; Cao, S. 3D imaging millimeter wave circular synthetic aperture radar. *Sensors* **2017**, *17*, 1419–1440. [CrossRef] [PubMed]
5. Wang, X.; Liu, J.; Zhang, W.; Fu, Q.; Liu, Z.; Xie, X. Mathematic principles of interrupted-sampling repeater jamming (ISRJ). *Sci. China-Inf. Sci.* **2006**, *50*, 891–901. [CrossRef]
6. Feng, D.; Tao, H.; Yang, Y.; Liu, Z. Jamming dechirping radar using interrupted-sampling repeater. *Sci. China-Inf. Sci.* **2011**, *54*, 2138–2146. [CrossRef]
7. Almslmany, A.; Cao, Q.; Wang, C. A new airborne self-protection jammer for countering ground radars based on sub-Nyquist. *IEICE Electron. Express* **2015**, *12*, 2138–2149. [CrossRef]
8. Feng, D.; Xu, L.; Pan, X.; Wang, X. Jamming wideband radar using interrupted-sampling repeater. *IEEE Trans. Aerosp. Electron. Syst.* **2017**, *53*, 1341–1354. [CrossRef]
9. Tai, N.; Cui, K.; Wang, C.; Yuan, N. The design of a novel coherent noise jammer against LFM radar. *IEICE Electron. Express* **2016**, *13*, 2138–2149. [CrossRef]
10. Wu, X.; Wang, X.; Lu, H. Study of intermittent-sampling repeater jamming to SAR. *J. Astronaut.* **2009**, *30*, 2043–2048.
11. Wang, W.; Pan, X.; Liu, Y.; Feng, D.J.; Fu, Q.X. Sub-Nyquist sampling jamming against ISAR with compressed sensing. *IEEE Sensors J.* **2014**, *14*, 3131–3136. [CrossRef]
12. Pan, X.; Wang, W.; Feng, D.; Liu, Y.; Fu, Q.; Wang, G. On deception jamming for countering ISAR based on sub-Nyquist sampling. *IET Radar Sonar Navigat.* **2014**, *8*, 173–179. [CrossRef]
13. Gong, S.; Wei, X.; Li, X. ECCM scheme against interrupted sampling repeater jammer based on time-frequency analysis. *J. Syst. Eng. Electron.* **2014**, *25*, 996–1003. [CrossRef]

14. Meller, M. Cheap Cancellation of Strong Echoes for Digital Passive and Noise Radars. *IEEE Trans. Signal Process* **2012**, *60*, 2654–2659. [CrossRef]
15. Lu, Y.; Fowler, R.; Tian, W.; Thompson, L. Enhancing Echo Cancellation via Estimation of Delay. *IEEE Trans. Signal Process* **2005**, *53*, 4159–4168.
16. Wang, Y.; Zhao, G.; Wang, H. Echo cancelling algorithm for the LFM radar. *J. Xidian Univ.* **2008**, *6*, 1031–1035.
17. Ufimtsev, P.Y. Comment on Diffraction Principles and Limitations of RCS Reduction Techniques. *Proc. IEEE* **1996**, *85*, 1830–1851. [CrossRef]
18. Xu, S.; Xu, Y. Simulation analysis of an active cancellation stealth system. *Optik* **2014**, *125*, 5273–5277.
19. Yang, X.; Zhao, W.; Huang, L. Calculation of RCS of targets and statistical analysis of cancellation effect. *Chin. J. Radio Sci.* **2002**, *17*, 88–92.
20. Xiang, Y.; Qu, C.; Su, F.; Yang, M. Active Cancellation Stealth Analysis of Warship for LFM Radar. *Proc. ICSP* **2010**, *17*, 2109–2112.
21. Xiang, Y.; Qu, C.; Li, B.; Hou, H. Simulation Research on Cancellation Stealth of Warship Based on Its Radar Scattering Properties. *J. Syst. Simul.* **2013**, *25*, 104–110.
22. Wei, Y.; Zhang, J.; Li, Z. A novel successive cancellation method to retrieve sea wave components form spatio-temporal remote sensing image sequences. *Remote Sens.* **2016**, *8*, 606–626. [CrossRef]
23. Feng, D.; Xu, L.; Wang, W.; Yang, H. Radar target echo cancellation using interrupted-sampling repeater. *IEICE Electron. Express* **2014**, *11*, 1–6. [CrossRef]
24. Xu, L.; Feng, D.; Dai, D.; Pan, X.; Wang, X. A Dual-Antenna Active-Echo-Cancellation Method for Synthetic Aperture Radar. In Proceedings of the 10th European Conference on Antennas and Propagation (EuCAP), Davos, Switzerland, 10–15 April 2016; pp. 1–5.
25. Wu, Q.; Liu, J.; Wang, J.; Zhao, F.; Xiao, S. Improved Active Echo Cancellation Against Synthetic Aperture Radar Based on Non periodic Interrupted Sampling Modulation. *IEEE Sensors J.* **2018**, *18*, 4453–4461. [CrossRef]
26. Hanbali, S.; Kastantin, R. Technique to counter active echo cancellation of self-protection ISRJ. *IET Electron. Lett.* **2017**, *53*, 680–681. [CrossRef]
27. Xu, L.; Feng, D.; Liu, Y.; Pan, X.; Wang, X. A three-stage active cancellation method against synthetic aperture radar. *IEEE Sensors J.* **2015**, *15*, 6173–6178. [CrossRef]
28. Hanbali, S.; Kastantin, R. Countering a self-protection frequency-shifting jamming against LFM pulse compression radars. *INTL J. Electron. Telecommun.* **2017**, *63*, 145–150. [CrossRef]
29. Yang, Y.; Zhang, W.M.; Yang, J.H. Study on frequency-shifting jamming to linear frequency modulation pulse compression radars. In Proceedings of the 2009 International Conference on Wireless Communications & Signal Processing, Nanjing, China, 13–15 November 2009; pp. 1–5.
30. The Sandia Complex Image. 2006. Available online: https://www.sandia.gov/radar/complex-data/ (accessed on 10 December 2018).
31. Franceschetti, G.; Guida, R.; Iodice, A., Riccio, D.; Ruello, G. Efficient simulation of hybrid stripmap/spotlight SAR raw signals from extended scenes. *IEEE Trans. Geosci. Remote Sens.* **2004**, *42*, 2385–2396. [CrossRef]

Article

PBR Clutter Suppression Algorithm Based on Dilation Morphology of Non-Uniform Grid

Qian Zhu [1], Tao Li [2], Jiameng Pan [1] and Qinglong Bao [1,*]

[1] Science and Technology on Automatic Target Recognition Laboratory, National University of Defense Technology, Changsha 410072, China; zq1142006@126.com (Q.Z.); 3090100472@zju.edu.cn (J.P.)
[2] Artificial Intelligence Research Center (AIRC), National Innovation Institute of Defense Technology (NIIDT), Beijing 100071, China; litao_nudt@163.com
* Correspondence: cbpest@163.com; Tel.: +86-132-7241-5200

Received: 15 May 2019; Accepted: 20 June 2019; Published: 22 June 2019

Abstract: Many new challenges are faced by the PBR (passive bi-static radar) employing non-cooperative radar illuminators. After the CFAR (constant false alarm) processor, the appearance of the amount of false alarm clutter points impacts the following tracing performance. To enhance the PBR tracing performance, we consider to reduce these clutter points before target tracing as soon as possible. In this paper, we propose a PBR clutter suppression algorithm based on dilation morphology of non-uniform grid. Firstly, we construct the non-uniform polar grid based on the acquisition geometry of PBR. Then, with the help of the grid platform, we separate the false alarm clutter points based on the dilation morphology. To efficiently operate the algorithm, we build up its parallel iteration scheme. To verify the performance of the proposed algorithm, we utilize both simulated data and field data to do the experiment. Experimental results show that the algorithm can effectively suppress most of the clutter points. Besides, we respectively combine the proposed suppression algorithm with two typical tracking algorithms to test the performance. Experimental results reveal that the compound tracing algorithm outperforms the traditional one. It can enhance the PBR tracing performance, reduce the occurrence probability of false tracks and meanwhile save time.

Keywords: PBR (passive bistatic radar); clutter suppression; non-uniform grid; dilation morphology

---

1. Introduction

Passive radar, employing non-cooperative illuminators, has attracted increasing interests in recent years [1–12]. It has many obvious advantages over traditional active radars, such as low-cost, feasibility of various illuminators and immune to the anti-radiation missiles [1]. In existing literature, illuminators of opportunity for passive radar are generally categorized into four groups: broadcast signals (DVB-T, FM, DAB, etc.) [3,4], mobile communication signals (Wi-Fi, GSM (Global System for Mobile communication), LTE (Long Term Evolution), etc.) [6,7], geolocalization signals (GNSS (Global Navigation Satellite System), GPS (Global Positioning System)) [8–10] and radar signals [11,12]. Most of researches focus on passive systems employing the first three types of illuminators, while the literature on passive systems employing non-cooperative radar signal as illuminator is rare because of the difficulties in signal processing. In this paper, we explore the research based on the PBR (passive bi-static radar) employing non-cooperative radar illuminators.

The operation geometry of the PBR system is illustrated in Figure 1. The system consists of two channels: echo channel and reference channel. The former is designed for receiving scattered wave, and the latter is for direct wave. The uncooperative transmitter is generally equipped with phased array. Compared to the traditional mechanical scanning radar, the phased array radar has flexible multi-beam scanning and various beam dwell times. It can achieve search and tracing simultaneously. Nowadays, many modern radars are equipped with phased array for detection. It is meaningful to exploit the phased array radar signal as the illuminator. The beam scan of the uncooperative illuminator is agile and flexible with unknown purpose. Since it is hard to predict and track its rapid changing beam steering, we choose to adopt multi-beam forming simultaneously covering the surveillance range [11] to realize space synchronization. Besides, to enhance the ability of anti-jamming and the detection probability, the uncooperative radar usually transmits the signal that is agile in frequency, PW (pulse width), BW (band width) and PRI (pulse recurrence interval). Based on the characteristics of PBR above, it faces many new challenges.

- The space synchronization accuracy is not as good as the traditional radar, resulting in the decreased SNR (signal noise ration) and the poor location precision.
- Simultaneous multi-beam forming leads to the redundant data being increased.
- The reference channel is not ideally compatible to the echo channel due to the multipath and the minor difference of antenna performance. The performance of the following pulse compression degrades.
- Due to the agility of the illuminator parameters, the number of the pulses utilized for detection is less. Besides, the scattered wave of the target depends on the opportunity of the beam steering. Thus, the valid data rate is decreased.
- Since the illuminator parameters are agile pulse by pulse, it is hard to adopt coherent integration to suppress clutter like traditional radar.
- Low SNR calls for low threshold during CFAR (constant false alarm), that is to increase the detection rate, whereas the false-alarm rate increases correspondingly.

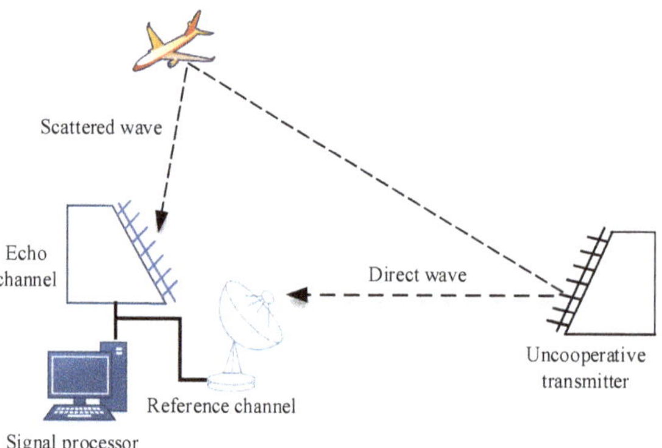

Figure 1. Geometry of PBR utilizing uncooperative radar signal as transmitter.

Thus, after pulse compression and CFAR processor, the difficulties during target tracking can be concluded into four points. That is breaking tracks, amounts of false-alarm clutter data, random interval between adjacent tracing points and huge computation.

To enhance the PBR tracking performance, we consider to reduce the false-alarm clutter points before tracking as soon as possible. The existing clutter suppression algorithms for passive radar are aimed at the direct wave interference and ground clutter. In addition, they mainly focus on the spatial domain, the temporal domain and the sub-carrier domain. In spatial domain, there is ABF (adaptive beamforming) and its extension version [13,14]. In temporal domain, researchers propose many adaptive filter algorithms applied on PBR, such as LMS (least mean square) [15], fast-block LMS [16], GANF (generalized adaptive notch filter) [17] and so on. Adaptive filter is of low convergence speed; however, ECA [18] (extensive cancellation algorithm) covers its deficiency. In recent years, many algorithms around ECA have been proposed, such as ECA-S (ECA-sliding) [19], ECA-ES (ECA-expectation simplified) [20] and so on. In sub-carrier domain, algorithms only work in orthogonal frequency-division multiplexing-based PBR, such as RLS-C (recursive least square by sub-carrier) [21], ECA-C (ECA by carrier) [22], ECA-CD (ECA by carrier and doppler shift) [23] and so on.

In addition to the direct wave interference and ground clutter, the radar illuminator-based PBR is also influenced by the false-alarm clutter during processing, as analyzed above. Rare literatures discuss the false-alarm clutter suppression algorithm in spatial–temporal domain before tracing. In this paper, we aim to put forward a PBR false-alarm clutter suppression algorithm. To make a low budget solution, we resort to the grid-based method so that we avoid calculating point-to-point Euclidean distance. In [24], grid-based DBSCAN is proposed for clustering objects in radar data. The method is not specially designed for PBR and its model is simple. In [25], ENM (ellipsoid norm method) is proposed to promise optimal result in passive multi-static location. It focuses on finding the nearest grid point in grid-based method. However, it only works in the noise-free scenario and is not suitable for dense clutter environment. Thus, based on the acquisition geometry of PBR, we firstly propose a non-uniform polar grid construction method. In addition, with the help of the grid platform, a false-alarm clutter separation method is proposed based on dilation morphology. The combination of these two steps is the whole algorithm we proposed in this paper.

The remaining part is organized as follows. Section 2 illustrates the geometrical relationship of PBR and describes the proposed non-uniform polar grid construction method. Section 3 describes the grid-based clutter suppression method and its parallel iteration scheme. Section 4 describes both simulated data experiment results and field data experiment results. Section 5 is the conclusion.

2. Non-Uniform Polar Grid Construction for PBR

According to the bi-static radar position principle, Figure 2 demonstrates the geometrical relationship between target, transmitter and receiver. R_r stands for the range of the target, that is the distance between target and receiver. R_t is the distance between target and transmitter. L stands for the baseline range between transmitter and receiver. β stands for the bi-static angle, that is the intersection angle between the line from receiver to target and the line from transmitter to target. θ_r stands for the azimuth angle, that is the supplementary angle between the line from receiver to transmitter and the line from receiver to target.

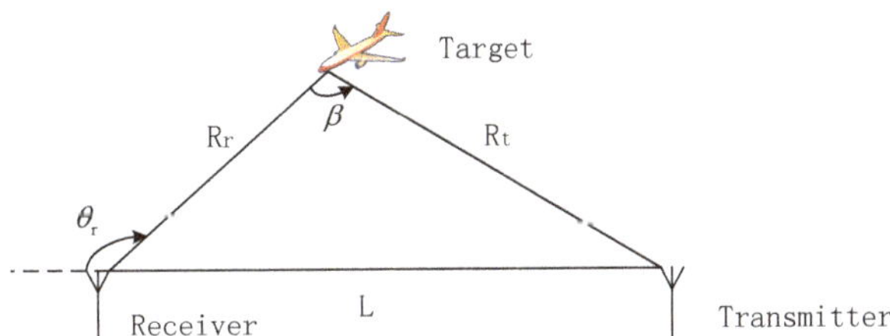

Figure 2. The geometrical relationship of PBR.

To locate the target, we should find its azimuth and range. The azimuth can be directly measured through the passive bi-static radar system. However, the range can only be calculated through the bi-static range sum. Suppose R_S is the bi-static range sum. Thus,

$$R_S = R_t + R_r \tag{1}$$

Referring to the cosine law, the relationships between R_r and R_s can be derived into Equations (2) and (3).

$$R_r = (R_s^2 - L^2)/(2(R_s + Lcos\theta_r)) \tag{2}$$

$$R_s = R_r + \sqrt{R_r^2 + L^2 + 2R_r Lcos(\theta_r)} \tag{3}$$

From Equation (2), we can find that the mono-static range R_r is a non-linear function of the bi-static range sum R_S and the azimuth angle θ_r.

In passive bi-static radar system, firstly, achieve the time synchronization of the echo channel with the help of the reference channel. Then, measure the difference between R_S and L through the time delay of echo. Since L is fixed, R_S can be directly measured from echo channel. As R_S and θ_r are direct measurements, the measuring and detecting error transfer to R_r by Equation (2) at last. Due to non-linear relationship between R_S and R_r, it is necessary for us to construct a non-uniform grid for the latter clutter suppression processing.

At first, we derive the first order Taylor series expansion of Equation (2) at an arbitrary position (R_{s0}, θ_0), shown in Equation (4).

$$
\begin{aligned}
&R_r(R_{s0} + \Delta r, \theta_0 + \Delta\theta) \\
&= R_r(R_{s0}, \theta_0) + \Delta r \frac{\partial R_r(R_s,\theta_0)}{\partial R_s}\Big|_{R_s=R_{s0}} + \Delta\theta \frac{\partial R_r(R_{s0},\theta)}{\partial\theta}\Big|_{\theta=\theta_0} + o(\Delta r, \Delta\theta)
\end{aligned}
\tag{4}
$$

Thus, the transfer error of the R_r caused by R_S and θ at the position $M(R_{s0}, \theta_0)$ is derived below.

$$
\begin{aligned}
\Delta R_r &= R_r(R_{s0} + \Delta r, \theta_0 + \Delta\theta) - R_r(R_{s0}, \theta_0) \\
&= \Delta r \frac{\partial R_r(R_s,\theta)}{\partial R_s}\Big|_{R_s=R_{s0},\theta=\theta_0} + \Delta\theta \frac{\partial R_r(R_s,\theta)}{\partial\theta}\Big|_{R_s=R_{s0},\theta=\theta_0} + o(\Delta r, \Delta\theta) \\
&= \Delta r \frac{R_{s0}^2 + L^2 + 2R_{s0}Lcos(\theta_0)}{2(R_{s0} + Lcos(\theta_0))^2} + \Delta\theta \frac{(R_{s0}^2 - L^2)Lsin(\theta_0)}{2(R_{s0} + Lcos(\theta_0))^2} + o(\Delta r, \Delta\theta)
\end{aligned}
\tag{5}
$$

Assume $\rho_1(R_{s0}, \theta_0) = \frac{R_{s0}^2 + L^2 + 2R_{s0}Lcos(\theta_0)}{2(R_{s0} + Lcos(\theta_0))^2}$, $\rho_2(R_{s0}, \theta_0) = \frac{(R_{s0}^2 - L^2)Lsin(\theta_0)}{2(R_{s0} + Lcos(\theta_0))^2}$.

Then, omit the Peano remainder term $o(\Delta r, \Delta\theta)$.

$$
\Delta R_r(\Delta r, \Delta\theta)|_{R_s=R_{s0},\theta=\theta_0} \approx \rho_1(R_{s0}, \theta_0)\Delta r + \rho_2(R_{s0}, \theta_0)\Delta\theta
\tag{6}
$$

From Equation (6), we can find that when Δr and $\Delta\theta$ are fixed, the transfer error changes with the position (R_{s0}, θ_0). To facilitate following operation, we divide the detection coverage into grids based on the transfer error expansion and project the processing data into grids.

Assume the bi-static range error Δr and the azimuth angle error $\Delta\theta$ obey the Gaussian distribution with zero-mean, as shown below.

$$
\Delta r \sim N(0, \sigma_r^2); \Delta\theta \sim N(0, \sigma_\theta^2)
\tag{7}
$$

In general, σ_r, the standard deviation of Δr, mostly relates to the range resolution. In addition σ_θ, the standard deviation of $\Delta\theta$, mostly relates to the space synchronization accuracy and the array error. We suppose them as known constant. Further discussion about them is not included in this paper.

Since ΔR_r is the linear function of Δr and $\Delta\theta$, ΔR_r is also obey the Gaussian distribution with zero-mean, as shown below.

$$
\Delta R_r \sim N(0, \rho_1^2(R_{s0}, \theta_0)\sigma_r^2 + \rho_2^2(R_{s0}, \theta_0)\sigma_\theta^2)
\tag{8}
$$

There is a 3σ principle of Gaussian distribution in the probability theory. In Gaussian distribution, the probability of the data distributing in the range $(\mu - 3\sigma, \mu + 3\sigma)$ is 99.74%. Where μ is the mean value of sample set. σ is the standard deviation of database.

Based on the analysis above, the grid spacing is designed shown in Equations (9) and (10).

$$
\delta_r(R_{s0}, \theta_0) = 3\sqrt{\rho_1^2(R_{s0}, \theta_0)\sigma_r^2 + \rho_2^2(R_{s0}, \theta_0)\sigma_\theta^2}
\tag{9}
$$

$$
\delta_\theta(R_{s0}, \theta_0) = 3\sigma_\theta
\tag{10}
$$

where $\delta_r(R_{s0}, \theta_0)$ and $\delta_\theta(R_{s0}, \theta_0)$ are the grid spacing at the position (R_{s0}, θ_0) in range dimension and angular dimension respectively.

As the δ_θ is uncorrelated with the position (R_{s0}, θ_0), the grid is divided evenly in the angular dimension. However, in range dimension, the pace will be iteratively calculated based on the present position. Assume the observing scope is from θ_0 to θ_{max} and from r_0 to r_{max}. The polar mesh grid calculation step is shown in Table 1.

Table 1. The polar mesh grid calculation steps.

---

1. Calculate angular coordinate

The angular coordinate set is $\Theta = \left\{ \theta_k | \theta_{k+1} - \theta_k = 3\sigma_\theta, k = 0 \ldots N, N = \left[\frac{\theta_{max} - \theta_0}{3\sigma_\theta} \right] \right\}$. Where $[\cdot]$ is the symbol of round down, and N is the mesh counts in angular dimension.

---

2. Aiming at each angular coordinate θ_k in Θ, iteratively calculate the grid division in range dimension.

For $\theta_k, k = 0 \ldots N$
Initialization: $i = 0, R_i = r_0$;
Iteration: $Rs_i = R_i + \sqrt{R_i^2 + L^2 + 2R_i L cos(\theta_k)}$;
$\rho_1(R_{si}, \theta_k) = \frac{R_{si}^2 + L^2 + 2R_{si} L cos(\theta_k)}{2(R_{si} + L cos(\theta_k))^2}, \rho_2(R_{si}, \theta_k) = \frac{(R_{si}^2 - L^2) L sin(\theta_k)}{2(R_{si} + L cos(\theta_k))^2}$.
$R_{i+1} = R_i + 3\sqrt{\rho_1^2(R_{si}, \theta_k)\sigma_r^2 + \rho_2^2(R_{si}, \theta_k)\sigma_\theta^2}$;
$i = i + 1$;
Terminate when $R_i > r_{max}$.
$N_k = i$. N_k is the mesh counts in range dimension for θ_k.
The range coordinate set is $\Lambda = \{R_{i,\theta_k} | i = 0 \ldots N_k, \theta_k \in \Theta\}$.

Referring to the proposed polar grid calculation algorithm, we calculate and make an example map of the polar grid in Figure 3. Set the observing scope 100° to 160° in azimuth and 20 km to 100 km in range. Blue lines denote the grid division in angular dimension, while red lines stand for the grid division in range dimension. It is obvious that closer to the baseline angle 180°, the grid size is bigger.

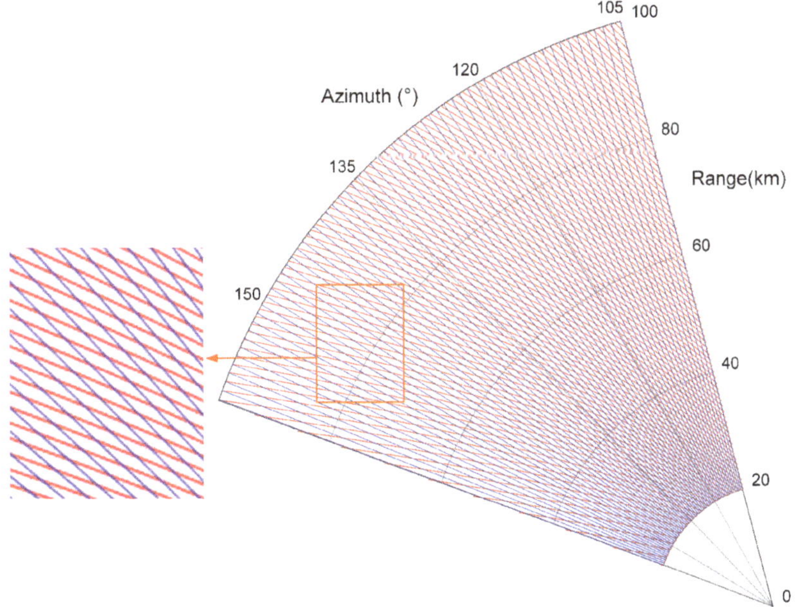

Figure 3. An example map of the proposed polar grid.

3. Separate False Alarm Clutter from Data

We utilize the location dependency of target track within consecutive frames to separate false alarm clutter from data. To reduce the computation load, the method in this section is proposed based on the grid constructed in Section 2. More specifically, in this section, we resort to morphology, usually applied in image processing, to achieve clutter separation.

The process is divided into three steps. Firstly, mark the point on the polar grid according to its position measured. Secondly, separate false alarm clutter point from data based on morphology algorithm. Thirdly, iteratively complete the operation frame by frame.

3.1. Mark the Point on Grid

After the CFAR processor and the Direction of Angle (DoA) estimation, we obtain the point data carrying with its own location information, namely range and azimuth angle. The first step of the separation is to mark those points on the grid constructed in Section 2 based on their locations. As the grid is uniform in angular dimension, to facilitate the calculation, we operate in order from azimuth angle to range. Assume n_a, n_r as the grid index of the point in angular dimension and range dimension respectively. r and α refer to its range and azimuth angle. The calculation method is described in Equations (11) and (12).

$$n_a = \left[\frac{\alpha}{\Delta\theta} \right] + 1 \tag{11}$$

$$n_r = \{ind|net(ind, n_a) \leq r < net(ind + 1, n_a)\} \tag{12}$$

where $[\cdot]$ is the symbol of round down. $\Delta\theta$ is the angular spacing of the grid. *net* is the grid coordinate matrix in accord with the range coordinate set Λ which is calculated in Section 2. The number of columns in *net* is the same as the number of elements in the angular coordinate set Θ where each column vector is the range division corresponding to each angular value. According to the range and angular index calculated above, mark the point on a matrix A with the same size of *net*.

3.2. Separate False Alarm Clutter from Data Based on the Dilation Morphology

In general, the true points of the target track have the feature of location dependency among several consecutive frames, whereas the false alarm clutter points are relatively isolated. Assume F_n is the object frame. $\Psi = \{F_{n-k} \ldots F_{n-1}, F_{n+1} \ldots F_{n+k}\}$ is the group of the reference frames, before and after several frames of F_n. Where k is the half number of the reference frames. Firstly, mark the points of the frame in group Ψ one by one. Obtain the mark-matrix $A_{n-k} \ldots A_{n-1}, A_{n+1} \ldots A_{n+k}$ respectively, and compose them into a new mark-matrix Γ_n, shown in Equation (13). Meanwhile, mark the points of the object frame on the matrix A_n.

$$\Gamma_n = A_{n-k} \cup \ldots \cup A_{n-1} \cup A_{n+1} \cup \ldots \cup A_{n+k} \tag{13}$$

To facilitate following iteration calculation, we replace (13) by another more specific operation, described in Equation (14). Where $\mathfrak{Bm}(\cdot)$ is the symbol of binarization.

$$\Gamma_n = \mathfrak{Bm}(A_{n-k} + \cdots + A_{n-1} + A_{n+1} + \cdots + A_{n+k}) \tag{14}$$

Next, to ensure the target points in the neighborhood of the points from reference frames, we do morphological dilation on Γ_n with a rectangular structural element B. The size of B depends on the coarse estimation of the target's move range. The dilation result matrix marks the neighborhood area of the points from reference frames. M_n in Equation (15) is the dot product of the dilation result and the object mark-matrix A_n.

$$M_n = (\Gamma_n \oplus B) \cdot {*} A_n \tag{15}$$

where $\oplus$ is the symbol of dilation. $*$ is the symbol of the dot product. The matrix M_n stands for the final marked area of screened data for the frame F_n. To facilitate realization, the dilation of binary matrix can be expressed by binarization after convolution. So Equation (15) can be rewritten as

$$M_n = \mathfrak{Bm}(\Gamma_n \otimes B) \cdot {*}A_n \tag{16}$$

where $\otimes$ is the symbol of convolution.

The processing progress is illustrated in Figure 4.

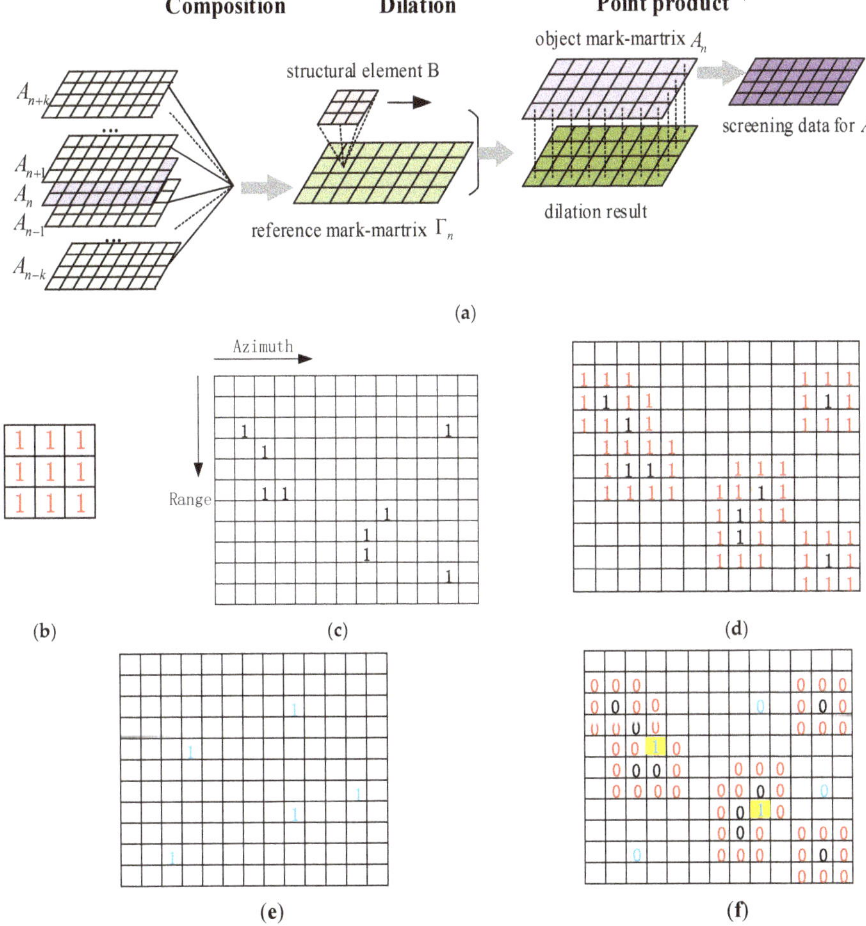

Figure 4. The sketch map of processing progress: (**a**) general view; (**b**) structural element B; (**c**) reference mark-matrix Γ_n; (**d**) dilation result; (**e**) object mark-matrix A_n; (**f**) screened data for A_n.

We choose a simple case to clearly illustrate how the proposed algorithm operates. Figure 4a indicates the general view of the operation. Figure 4b–f shows the processing result in each step. After the composition of reference frames, dilation and the point product with object frame, we can obtain the screened data, where the clutter points are filtered out. In this case, Structural element B is a 3×3 square matrix, as shown in Figure 4b. Figure 4c shows the reference mark-matrix Γ_n. Figure 4d shows the dilation result of the B and Γ_n. Figure 4e is the object mark-matrix A_n with five suspected areas. Figure 4f shows the result, the screened data for A_n. Obviously, the points in two areas of yellow background are retained, yet rest of them are suspected as clutters and filtered out.

It is noteworthy that since the transfer error changes with the point location, the real size and shape of the structural element for dilation is not fixed in fact. It changes along with the location of the suspected point. However, as the processing data has been abstracted through the non-uniform polar grid, the change of the structural element does not involve in this section. The left part of Figure 5 is the partial enlarged map of Figure 3. There are two structural elements, corresponding to point A and B, with different shape and size. Both are projected into the same element for convenience during processing.

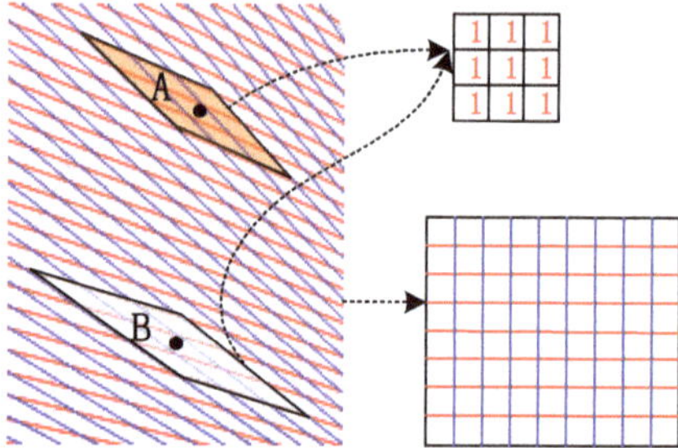

Figure 5. The sketch map of the grid projection.

3.3. Iteratively Calculation Frame by Frame

To efficiently operate the algorithm, we further explore its iterative and parallel calculating scheme. Referring to Equation (14), we can derive the reference mark-matrix Γ_{n+1}:

$$
\begin{aligned}
\Gamma_{n+1} &= \mathfrak{Bm}(A_{n+1-k} + \cdots + A_n + A_{n+2} + \cdots + A_{n+1+k}) \\
&= \Gamma_n + \mathfrak{Bm}(A_n + A_{n+1+k} - A_{n+1} - A_{n-k}) \\
&= \Gamma_n + \mathfrak{Bm}(A_n - A_{n-k}) + \mathfrak{Bm}(A_{n+1+k} - A_{n+1})
\end{aligned}
\tag{17}
$$

Assume

$$
\Psi A_n = \begin{cases} \mathfrak{Bm}(A_n - A_{n-k}) & n > k \\ \mathfrak{Bm}(A_n) & n \leq k \end{cases}
\tag{18}
$$

Thus, Equation (17) will be derived into Equation (19).

$$
\Gamma_{n+1} = \Gamma_n + \Psi A_n + \Psi A_{n+1+k}
\tag{19}
$$

where k is the half number of the reference frames. When $n = 1$, $\Gamma_1 = \mathfrak{Bm}(A_2 + \cdots + A_{1+k})$.

It can be observed that Equations (11), (12), (16), (18) and (19) are relatively independent of calculation. To promote the efficiency of the algorithm, we split the whole process into two parts for parallel calculation and build up an intermediate database to link them together. One is frame processing part, and the other is interframe processing part. Both are designed to operate in parallel. Based on the above analysis, the clutter separation algorithm is described in below chart, shown in Figure 6.

The data frame F_i flown from the CFAR processor is input in this system. i is the index of the current frame flowing in. When $i > k$, the interframe processing part starts operating. n is the index of the current processing frame, and is independent of the index i. In frame processing part, mark the data of F_i on the grids and calculate ΨA_i. Save A_i and ΨA_i into the intermediate database called by clutter separation in interframe processing part. Finally, iteratively compute and output the mark matrix M_n in the interframe processing part.

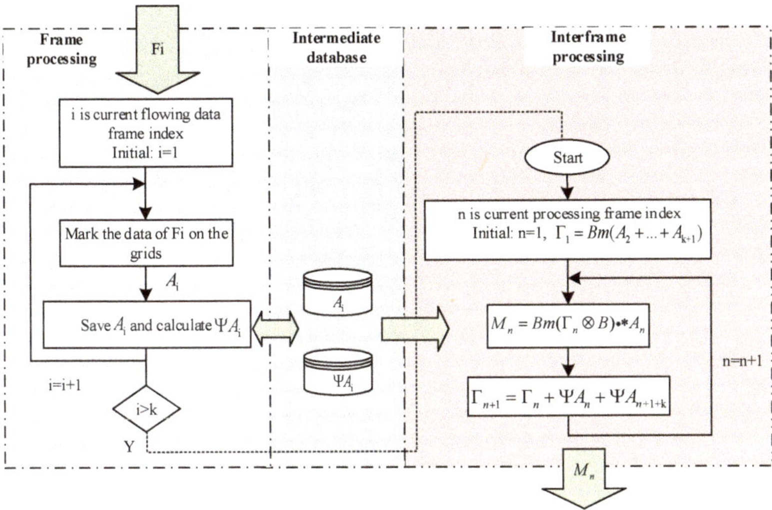

Figure 6. The flow chart of the clutter separation algorithm.

4. Experiment result and Analysis

In this section, the proposed algorithm is tested on a simulated scenario in Section 4.1 and a real scenario in Section 4.2. The computer configuration for experiment: Inter(R) Core(TM) i7-4790 CPU @ 3.60 GHz. RAM: 16.0 GB. All operations in this section run on Matlab R2018a.

4.1. Testing by Simulated Data

4.1.1. Scenario for Simulation

The surveillance scope is set from 60° to 170°. There are five targets moving with constant velocity in the scope. Table 2 lists the track settings of five targets. Plot these target tracks in polar coordinates, as Figure 7a shows.

Table 2. The track settings of five targets.

	Start Position (km, degree) in Polar Coordinates	Start Position (km) in Cartesian Coordinates	Track Slope	Track Intercept (km)
Target 1	(86.023,144.5)	(−70, 50)	5	60
Target 2	(70.456,96.5)	(−8, 70)	10	100
Target 3	(80.623,82.9)	(10, 80)	−3	10
Target 4	(76.158,113.2)	(−30, 70)	−10	80
Target 5	(70.711,135)	(−50, 50)	30	55

Since the parameter of the illuminator is agile and various, PBR utilizes only part of pulses with specific aims for detection. To make closer to reality, the time interval between pulses utilized is not constant. The whole time length of simulated data is 50 s. The number of valid pulses is set as 667. The time interval between adjacent valid pulses is allocated randomly. Figure 8 shows the pulse interval allocation of simulated data. So, the detection result from the ideal echo of the target is not uniformly continues. Figure 7b demonstrates the target tracks points based on the pulse interval allocation with measurement error.

Besides, due to the flicker of target's RCS in PBR, target can only be detected from part of valid pulses. Figure 7c demonstrates the real target points detected. The signal to clutter ratio in this experiment is defined in Equation (20).

$$SCR = \log\left(\frac{N_{sig}}{N_{clu}}\right) \tag{20}$$

where N_{sig} represents the mean number of the valid target detection points in each frame. N_{clu} represents the mean number of the false alarm clutter points in each frame. Set SCR as -1.26 dB. Figure 7d shows the final detection result from CFAR, which is the simulated data for following experiment.

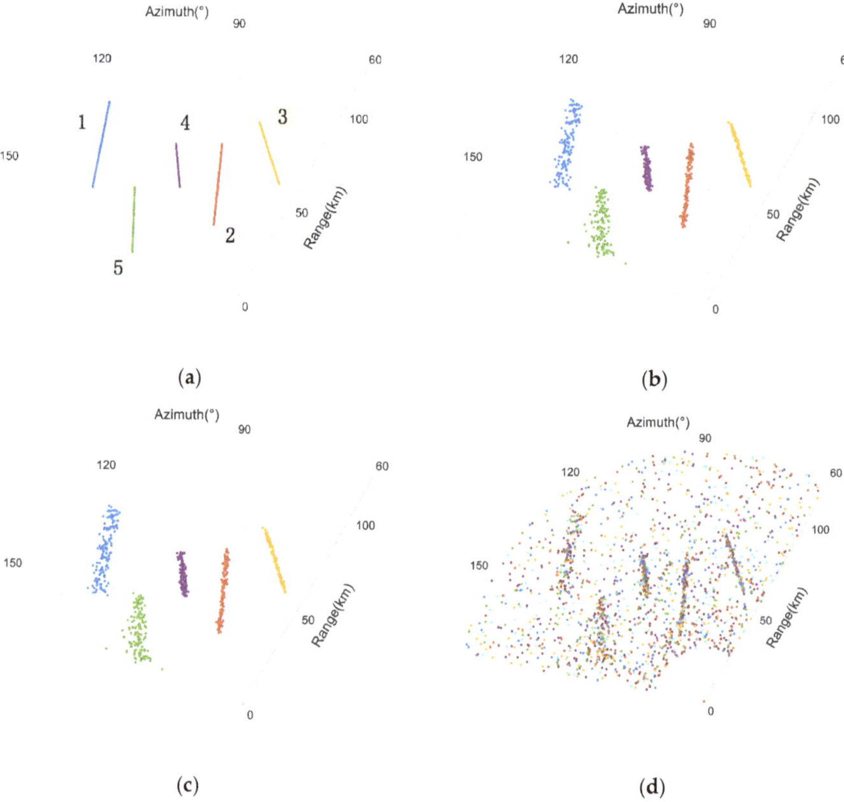

Figure 7. Production of the simulated data: (**a**) true tracks of five targets; (**b**) target tracks with measurement error; (**c**) target tracks detected; (**d**) detection results from CFAR.

Figure 8. Pulse interval allocation.

4.1.2. The Clutter Suppression Performance Analysis

For grid construction, the standard deviation of range is set as 300 m according to the bandwidth of the illuminator. The standard deviation of angle is set as 0.3 according to the number of beams simultaneously covering the surveillance range. The range is from 20 km to 120 km. The angle scope is from 60° to 170°. The baseline range is 400 km. For frame operation, we categorize the simulated data into frame data by every 0.5 seconds. The structural element size is 3 × 3. The half number of the reference frames is set as 3. Set SCR as −1.26 dB. Figure 9 demonstrates the contrast before and after the suppression. In addition Table 3 shows three performance indexes of the suppression algorithm. The detection accuracy rate is the ratio of the number of correct target points extracted to the whole number of the correct target points. The false alarm decline rate is the ratio of the number of the false points extracted to the whole number of the clutter points set before. The miss detection rate is the ratio of the number of missing target points to the whole number of correct target points. From Table 3 and Figure 9, we can conclude that near 90% of clutter points are suppressed, while 97.45% of target points retain. To illustrate the algorithm performance comprehensively, change the SCR of the simulation scenario from −3 dB to 1 dB. In each SCR scenario, do Mont-Carlo experiment for 50 times and calculate the mean value of the performance indexes. We obtain following results, as Figure 10 shows. Red line stands for the correct detection rate. Blue line stands for the CFAR decline rate. Green line stands for the missing rate. Over 90% of target points can be extracted correctly in this algorithm. In addition when SCR is up to −2 dB, the number of false alarm points can be decline to the 30% of the original number. When SCR is up to −1 dB, over 90% of clutter points can be suppressed.

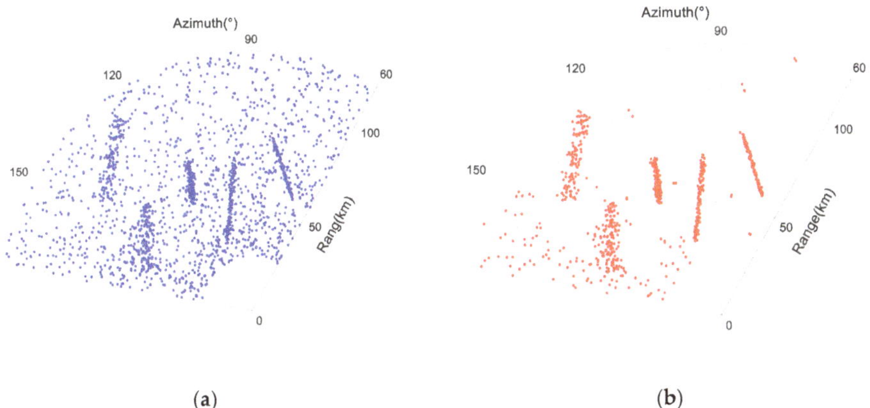

(a)   (b)

Figure 9. Contrast of before and after the suppression: (**a**) simulated data before suppression; (**b**) clutter suppression result.

Table 3. The performance indexes of the suppression algorithm.

Detection Accuracy Rate	False Alarm Decline Rate	Miss Detection Rate
97.45%	10.24%	2.55%

4.1.3. Computation Analysis

When the uncooperative illuminator and the receiver of the PBR are located at fix sites, the non-uniform polar grid is fixed according to the acquisition geometry. The calculation of the grid is done in preprocess only once. So, the computation analysis of the grid construction is not involved in this section. In frame processing, assume the grid size is $m_1 * m_2$. The calculation of A for one processing point concludes $m_1 + m_2$ additions. The calculation of ΨA concludes $m_1 m_2$

additions. In interframe processing, assume the structural element size is $s_1 * s_2$. The calculation of Γ concludes $2m_1 * m_2$ additions. The calculation of M needs $(s_1 * s_2 + 1) \cdot (m_1 * m_2)$ multiplications and $(s_1 * s_2 - 1) \cdot (m_1 * m_2)$ additions.

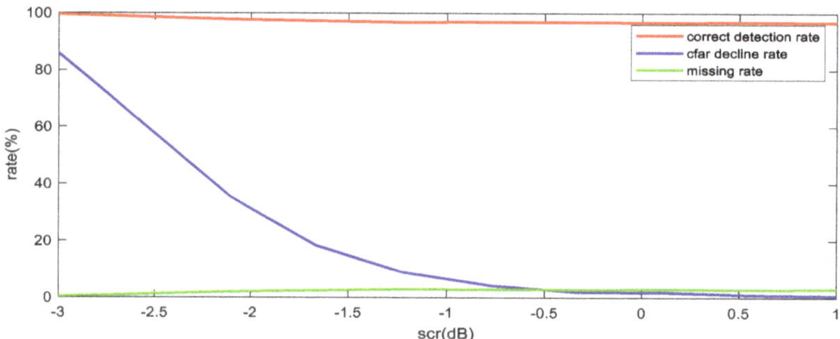

Figure 10. The change of the performance indexes along with SCR.

Suppose the frame number is k and the points number in each frame is n. So the whole process needs $k[n(m_1 + m_2) + (2 + s_1 * s_2) \cdot (m_1 * m_2)]$ additions and $k(s_1 * s_2 - 1) \cdot (m_1 * m_2)$ multiplications. That is $o(k * n)$ additions and $o(k)$ multiplications. Therefore, the addition times depends on the total number of points. In addition, the multiplication times only depends on the number of frames.

During tracing, the calculation amount relating to the number of clutter points is mostly caused by the Euclidean distance calculation between points. Assume M is the total number of points for tracing in each frame. In two consecutive frames, there are M^2 point pairs for processing. For each point pair, the calculation of the Euclidean distance needs 3 additions and 3 multiplications. Thus, the calculation amount of the Euclidean distance is $o(k * M^2)$ additions and $o(k * M^2)$ multiplications. Where k is the total number of frames.

Suppose our algorithm can suppress 90% clutter points, and this suppression process only consumes $o(k * M)$ additions and $o(k)$ multiplications. After suppression, the calculation amount of the Euclidean distance will descend to the 1% of the original. Thus, we pay low calculation amounts for reducing much more computation amounts of tracing.

4.1.4. Test the Performance Combining with Tracking Algorithm

We combine the proposed suppression algorithm with two typical tracking algorithms to test the performance. One is traditional NN TO-MHT algorithm (Nearest Neighbor Track Oriented-Multiple Hypothesis Tracking) [26], abbreviated as NN-MHT in this paper. The other is SNN-Kalman tracking algorithm (Suboptimal Nearest Neighbor - Kalman) [27], which is proposed aiming at multi-target tracking in non-cooperative passive system. To explicitly name the proposed algorithm, we name its abbreviation as MCSNG (Multi-frame Clutter Suppression based on Non-uniform Grid). Combine the clutter suppression with two tracking algorithms mentioned above respectively. For each frame, the tracking process follows with the clutter suppression in pipeline operation. Based on the difference of the tracking process, we name these compound algorithms as MCSNG-NN and MCSNG-SNN-K respectively. To make comparison, we utilize the original data without clutter suppression for tracing. Choose four indexes (total number of traces, mean trace length, maximum trace length, time consuming) to evaluate the tracing performance. Set the maximum velocity and the maximum accelerated velocity as 1000 m/s and 200 m/s^2 respectively. In Kalman filtering process, the maximum time period of blind prediction is set as 15 s.

Figure 11a,b illustrates the tracing result of NN-MHT and MCSNG-NN, respectively. In addition Table 4 lists the tracing result indexes of both. Utilize different colors to distinguish different tracks.

The number of false tracks produced from NN-MHT is 17, which is much more than MCSNG-NN. Besides, the time consuming of the MCSNG-NN is 76% less than the one of the NN-MHT.

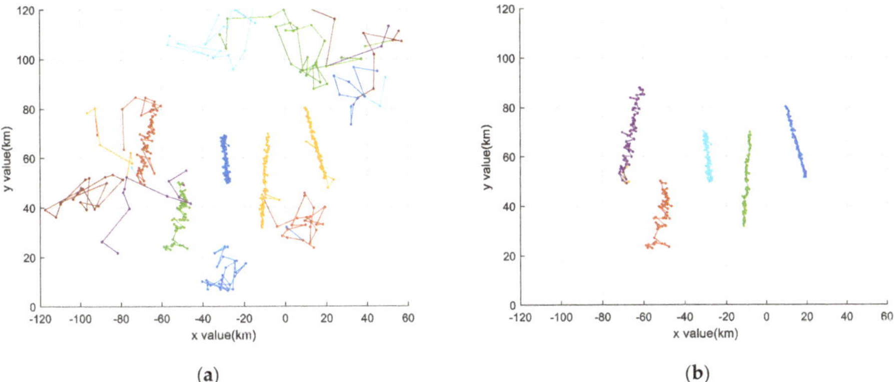

(**a**) (**b**)

Figure 11. Tracing results comparison of both algorithms: (**a**) NN-MHT tracing result; (**b**) MCSNG-NN tracing result.

Table 4. The performance indexes of NN-MHT and MCSNG-NN.

	Total Number of Traces	Mean Trace Length	Max Trace Length	Time Consuming (s)
NN-MHT	22	39.13	89	2.99
MCSNG-NN	6	78.83	89	0.7

Figure 12a,b illustrates the valid tracing result of SNN-Kalman and MCSNG-SNN-K respectively. Comparing with five target tracks set before, we calculate the mean trace error of the valid tracing results from both algorithms, as Table 5 list. It is obvious that the tracking precision of MCSNG-SNN-K is higher than the one of SNN-Kalman. Since the time interval is not constant, the velocity estimation accuracy will be affected. However, the velocity estimation of MCSNG-SNN-K is closer to the true value than SNN-Kalman. Table 6 shows the index of the tracing result. It can be found that MCSNG-SNN-K is more efficiency and produces less false tracks than SNN-Kalman.

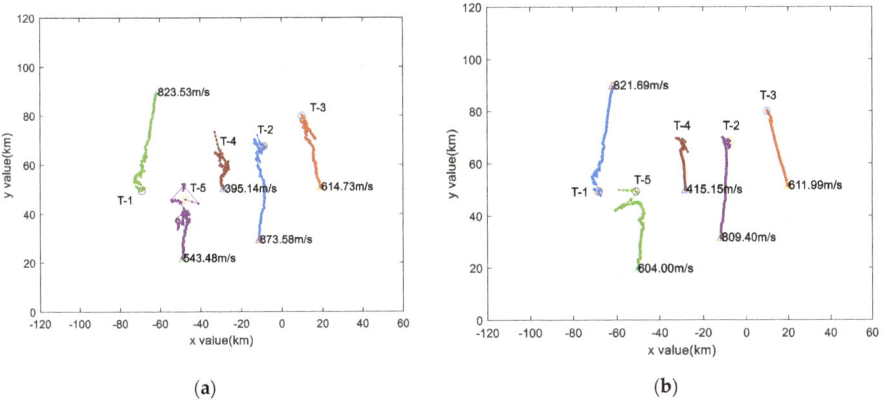

(**a**) (**b**)

Figure 12. Valid tracing results comparison of both algorithms: (**a**) SNN-Kalman tracing result; (**b**) MCSNG-SNN-K tracing result.

Table 5. The comparison of five targets tracing results between SNN-Kalman and MCSNG-SNN-K.

Track NO.	Mean Trace Error (m)		Velocity (m/s)		
	SNN-Kalman	MCSNG-SNN-K	SNN-Kalman	MCSNG-SNN-K	True Value
1	548.87	434.56	823.5	821.7	800
2	1596.55	389.63	873.6	809.4	750
3	644.06	141.85	614.7	612.0	600
4	1286.59	283.97	395.1	415.2	400
5	2721.07	1840.33	543.5	604.0	600

Table 6. The performance indexes of SNN-Kalman and MCSNG-SNN-K.

	Total Number of Traces	Mean Trace Length	Max Trace Length	Time Consuming (s)
SNN-Kalman	104	64.89	654	0.6715
MCSNG-SNN-K	23	124.47	474	0.6081

4.2. Testing by the Field Data

In this part, we utilize the field data to test the performance of the proposed algorithm. The PBR field experiment is done with an uncooperative radar with frequency, PW and PRI agile, and it aims to detecting the air-flights. We make validations with the ADS-B (Automatic dependent surveillance-broadcast) dataset. The detection scope is from 80° to 170°. The detection range is from 50 km to 200 km. Due to unknown parameters of illuminator, the performance of pulse compression among several pulses degrades. To increase the detection probability, we reduce the CFAR rate to 10^{-2}. After CFAR detection, we adopt the multi-beam amplitude comparison direction measurements. Figure 13a illustrates the final detection point map of the field data with 390 s duration. Two directions of jamming are located at 144° and 148°. Suppress the jamming in two directions and adopt the proposed clutter suppression algorithm. The structural element size is 5×5. The half number of the reference frames is set as 2. The suppression result is shown in Figure 13b. It is obvious that most of points are filtered out. Instead, points in three suspected track areas are retained. Referring to the ADS-B dataset, we plot the real-time civil flight information in Figure 13c, which is selected with the same duration and detection scope as the field data. The line with different colors stands for different flight track. There are three flights in the detection scope. Comparing with the ADS-B data, we can find that the proposed algorithm can effectively suppress the clutters and retain most of the target information.

Like the operations in Section 4.1.4, we test two compound tracking algorithms (MCSNG-NN and MCSNG-SNN-K) by the field data. The maximum velocity is set as 1200 m/s. The maximum accelerate velocity is set as 200 m/s^2. In Kalman filtering process, the maximum time period of blind prediction is set as 15 s. Each frame consists of the clustered point data with 0.5 s period. Figure 14 and Table 7 illustrates the tracing results and the performance indexes of NN-MHT and MCSNG-NN. Figure 15 and Table 8 illustrates the tracing results and the performance indexes of SNN-Kalman and MCSNG-SNN-K. Comparing with the ADS-B dataset, we marked the true tracks by the ellipses with dotted line. It is obvious that MCSNG-NN reduces the occurrence probability of false tracks relative to NN-MHT. In addition its time-consumption drops to 28.16% of the original NN-MHT time-consumption. Besides, similar conclusions are suitable to SNN-Kalman and MCSNG-SNN-K. We can find that MCSNG-SNN-K reduces the occurrence probability of false tracks and saves time relative to traditional SNN-Kalman.

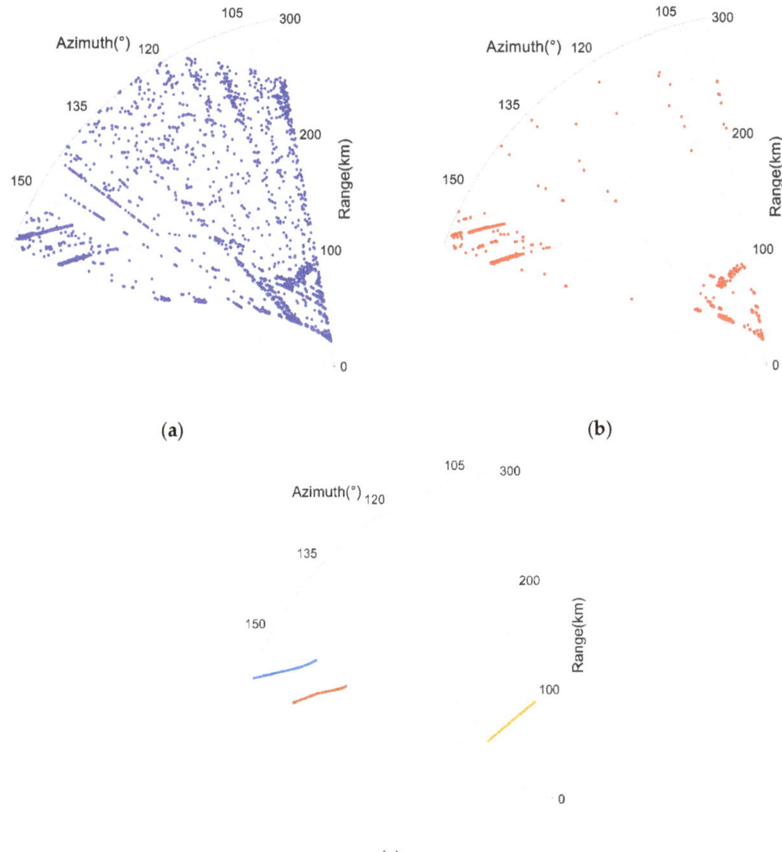

Figure 13. The contrast of before and after the suppression and the comparison with ADS-B dataset: (**a**) field data before suppression; (**b**) clutter suppression result; (**c**) real-time fights information from ADS-B dataset.

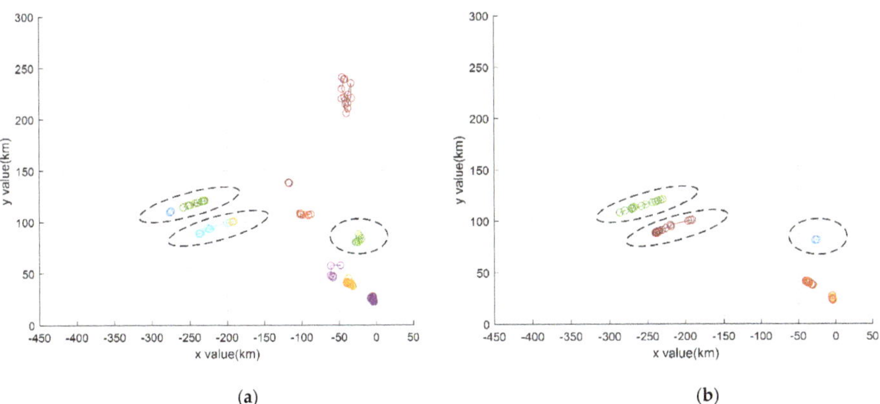

Figure 14. Tracing results of NN-MHT and MCSNG-NN: (**a**) NN-MHT tracing result; (**b**) MCSNG-NN tracing result.

Table 7. The performance indexes of NN-MHT and MCSNG-NN.

	Total Number of Traces	Mean Trace Length	Max Trace Length	Time Consuming (s)
NN-MHT	31	38.7	135	4.83
MCSNG-NN	9	39.3	120	1.36

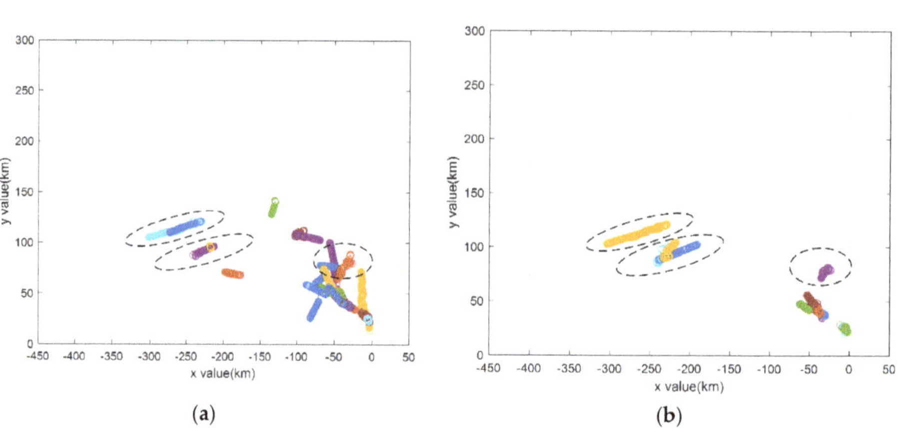

(a)　　　　(b)

Figure 15. Tracing results of SNN-Kalman and MCSNG-SNN-K: (**a**) SNN-Kalman tracing result; (**b**) MCSNG-SNN-K tracing result.

Table 8. The performance indexes of SNN-Kalman and MCSNG-SNN-K.

	Total Number of Traces	Mean Trace Length	Max Trace Length	Time Consuming (s)
SNN-Kalman	23	254.5	734	0.8536
MCSNG-SNN-K	13	196.6	424	0.6467

5. Conclusions

For PBR detection, especially for those illuminators with frequency, PRI and PW agile, it brings many challenges in following target tracing due to heavy clutters. Thus, combining with the features of PBR, a preprocessing operation is introduced before target tracing. In this paper, we propose a PBR cluttering suppression algorithm based on dilation morphology of non-uniform grid. According to the acquisition geometry of PBR, the nonuniform grid construction method is proposed at first. Then, iteratively separate false-alarm clutters from the point data based on dilation morphology. We perform experiments utilizing both simulated data and field data. Experiment results show that the proposed algorithm can effectively filter most false alarm clutters. Besides, combining with the tracing algorithm, it can enhance the PBR tracing performance, reduce the occurrence probability of false tracks and meanwhile save time. Furthermore, the theory of the proposed algorithm is also applicable for 3-D passive tracking, if the non-uniform grid for dilation is modified into cube. In current algorithm, the non-uniform grid is calculated through the first order Taylor expansion. Its reminder term is larger compared to the one of the higher order Taylor expansion. To balance the computation cost and the model accuracy, it is meaningful to exploit the maximum acceptable magnitude of measurement error in different positions. Future researches will focus on building up more specific non-uniform grid for clutter suppression combined with the target tracing, especially for the grids close to the baseline direction.

Author Contributions: Q.Z. and T.L. implemented the methods and designed the experiment. Q.Z. and J.P. performed the experiments and analyzed the data. Q.B. supervised the research. Q.Z. wrote the paper. All authors of the article provided substantive comments.

Funding: This research received no external funding.

Conflicts of Interest: The authors declare no conflict of interest.

References

1. Kuschel, H.; Cristallini, D.; Olsen, K.E. Tutorial: Passive radar tutorial. *IEEE Aerosp. Electron. Syst. Mag.* **2019**, *34*, 2–19. [CrossRef]
2. Zhou, X.; Wang, H.; Cheng, Y.; Qin, Y. Radar coincidence imaging by exploiting the continuity of extended target. *IET Radar Sonar Navig.* **2017**, *11*, 60–69. [CrossRef]
3. Malanowski, M.; Kulpa, K.; Kulpa, J.; Samczynski, P.; Misiurewicz, J. Analysis of detection range of FM-based passive radar. *IET Radar Sonar Navig.* **2014**, *8*, 153–159. [CrossRef]
4. Palmer, J.E.; Harms, H.A.; Searle, S.J.; Davis, L. DVB-T Passive Radar Signal Processing. *IEEE Trans. Signal Process.* **2013**, *61*, 2116–2126. [CrossRef]
5. Pastina, D.; Colone, F.; Martelli, T.; Falcone, P. Parasitic exploitation of wi-fi signals for indoor radar surveillance. *IEEE Trans. Veh. Technol* **2015**, *64*, 1401–1415. [CrossRef]
6. Samczynski, P.; Kulpa, K.; Malanowski, M.; Krysik, P.; Maślikowski, Ł. A concept of GSM-based passive radar for vehicle traffic monitoring. In Proceedings of the Microwaves, Radar and Remote Sensing Symposium, Kiev, Ukraine, 25–27 August 2011; pp. 271–274.
7. Raja, R.A.; Noor, A.A.; Nur, A.R.; Asem, A.S.; Fazirulhisyam, H. Analysis on target detection and classification in lte based passive forward scattering radar. *Sensors* **2016**, *16*, 1607. [CrossRef] [PubMed]
8. Hong-Cheng, Z.; Jie, C.; Peng-Bo, W.; Wei, Y.; Wei, L. 2-d coherent integration processing and detecting of aircrafts using gnss-based passive radar. *Remote Sens.* **2018**, *10*, 1164. [CrossRef]
9. Ma, H.; Antoniou, M.; Pastina, D.; Santi, F.; Pieralice, F.; Bucciarelli, M.; Cherniakov, M. Maritime Moving Target Indication Using Passive GNSS-based Bistatic Radar. *IEEE Trans. Aerosp. Electron. Syst.* **2018**, *54*, 115–130. [CrossRef]
10. Suberviola, I.; Mayordomo, I.; Mendizabal, J. Experimental results of air target detection with a gps forward-scattering radar. *IEEE Geosci. Remote Sens. Lett.* **2012**, *9*, 47–51. [CrossRef]
11. Wang, Y.; Bao, Q.; Wang, D.; Chen, Z. An experimental study of passive bistatic radar using uncooperative radar as a transmitter. *IEEE Geosci. Remote Sens. Lett.* **2015**, *12*, 1–5. [CrossRef]
12. Zhu, Q.; Bao, Q.; Hu, P.; Chen, Z. Experimental study of aircraft detection by PBR exploiting uncooperative radar as illuminator. In Proceedings of the International Conference on Information, Electronic and Communication Engineering, Beijing, China, 28–29 October 2018; pp. 185–190.
13. Wang, Y.; Bao, Q.; Chen, Z. Robust adaptive beamforming using IAA-based interference-plus-noise covariance matrix reconstruction. *Electron. Lett.* **2016**, *52*, 1185–1186. [CrossRef]
14. Yang, X.; Xie, J.; Li, H.; He, Z. Robust adaptive beamforming of coherent signals in the presence of the unknown mutual coupling. *IET Commun.* **2018**, *12*, 75–81. [CrossRef]
15. Meller, M. Cheap Cancellation of Strong Echoes for Digital Passive and Noise Radars. *IEEE Trans. Signal Process.* **2012**, *60*, 2654–2659. [CrossRef]
16. Dwivedi, S.; Aggarwal, P.; Jagannatham, A.K. Fast block lms and rls-based parameter estimation and two-dimensional imaging in monostatic mimo radar systems with multiple mobile targets. *IEEE Trans. Signal Process.* **2018**, *66*, 1775–1790. [CrossRef]
17. Guan, X.; Hu, D.H.; Zhong, L.H.; Ding, C.B. Strong echo cancellation based on adaptive block notch filter in passive radar. *IEEE Geosci. Remote Sens. Lett.* **2015**, *12*, 339–343. [CrossRef]
18. Colone, F.; O'Hagan, D.W.; Lombardo, P.; Baker, C.J. A multistage processing algorithm for disturbance removal and target detection in passive bistatic radar. *IEEE Trans. Aerosp. Electron. Syst.* **2009**, *45*, 698–722. [CrossRef]
19. Colone, F.; Palmarini, C.; Martelli, T.; Tilli, E. Sliding extensive cancellation algorithm for disturbance removal in passive radar. *IEEE Trans. Aerosp. Electron. Syst.* **2016**, *52*, 1309–1326. [CrossRef]
20. Fu, Y.; Wan, X.; Zhang, X.; Yi, J.; Zhang, J. Parallel processing algorithm for multipath clutter cancellation in passive radar. *IET Radar Sonar Navig.* **2018**, *12*, 121–129. [CrossRef]
21. Zhao, Z.; Zhou, X.; Zhu, S.; Hong, S. Reduced complexity multipath clutter rejection approach for drm-based hf passive bistatic radar. *IEEE Access* **2017**, *5*, 20228–20234. [CrossRef]
22. Zhao, Z.; Wan, X.; Shao, Q.; Gong, Z.; Cheng, F. Multipath clutter rejection for digital radio mondiale-based HF passive bistatic radar with OFDM waveform. *IET Radar Sonar Navig.* **2012**, *6*, 867–872. [CrossRef]

23. Chabriel, G.; Barrère, J.; Gassier, G.; Briolle, F. Passive Covert Radars using CP-OFDM signals: A new efficient method to extract targets echoes. In Proceedings of the IEEE International Radar Conference, Lille, France, 13–17 October 2014; pp. 1–6.

24. Kellner, D.; Klappstein, J.; Dietmayer, K. Grid-based dbscan for clustering extended objects in radar data. In Proceedings of the IEEE Intelligent Vehicles Symposium, Madrid, Spain, 3–7 June 2012; pp. 365–370.

25. Zhang, T.; Mao, X.; Zhao, C.; Liu, J. A novel grid selection method for sky-wave time difference of arrival localization. *IET Radar Sonar Navig.* **2019**, *13*, 538–549. [CrossRef]

26. Guo, J.; Zhang, R. Efficient radar data processing algorithm for dense cluttered environment. In Proceedings of the IEEE Cie International Conference on Radar, Chengdu, China, 24–27 October 2011; pp. 1692–1695.

27. Pan, S.S. Research and Engineering Realization of Multi-Target Tracking Technology for Non-Cooperative Passive Detection System. Master's Thesis, National University of Defense Technology, Changsha, China, 2017.

electronics

Article

Saliency Preprocessing Locality-Constrained Linear Coding for Remote Sensing Scene Classification

Lipeng Ji [1,*], Xiaohui Hu [2] and Mingye Wang [1]

[1] School of Automation Science & Electrical Engineering, Beihang University, Beijing 100083, China; marcel0829@126.com

[2] Institute of Software Chinese Academy of Science, Beijing 100190, China; hxh@iscas.ac.cn

[*] Correspondence: jlp_1987@buaa.edu.cn; Tel.: +86-010-8231-7751

Received: 30 July 2018; Accepted: 26 August 2018; Published: 30 August 2018

Abstract: Locality-constrained Linear Coding (LLC) shows superior image classification performance due to its underlying properties of local smooth sparsity and good construction. It encodes the visual features in remote sensing images and realizes the process of modeling human visual perception of an image through a computer. However, it ignores the consideration of saliency preprocessing in the human visual system. Saliency detection preprocessing can effectively enhance a computer's perception of remote sensing images. To better implement the task of remote sensing image scene classification, this paper proposes a new approach by combining saliency detection preprocessing and LLC. This saliency detection preprocessing approach is realized using spatial pyramid Gaussian kernel density estimation. Experiments show that the proposed method achieved a better performance for remote sensing scene classification tasks.

Keywords: saliency preprocessing LLC; saliency detection; image processing; scene classification

---

1. Introduction

Over recent decades, an overwhelming amount of high-resolution (HR) remote sensing images have become available. Since remote sensing images have abundant structural patterns and spatial information that are difficult to be fully applied directly, we need to correctly classify them by senses before further processing. Therefore, remote sensing scene classification is a central issue in remote sensing applications [1,2].

Various methods have been proposed to classify remote sensing scenes over the years. Bag-of-Features (BoF) [3,4] is a classical method in whole-image categorization tasks. This method first forms a histogram based on a remote sensing image's local features, and then uses the histogram to represent the remote sensing image. However, this method lacks consideration of the spatial layout information of features in remote sensing images. There are some improvements based on the BoF method, such as those shown in References [5,6], and the Spatial Pyramid Matching (SPM) method [7], which is a successful method. The SPM method divides a remote sensing image into different scale spatial sub-regions. Then, histograms of local features from each sub-region are computed. Usually, $2^l \times 2^l$ sub-regions (where $l = 0, 1, 2$) are used. SPM has shown a better performance than BoF in most image classification tasks; however, the traditional SPM approach also has limitations. It requires nonlinear classifiers to complete classification. To improve SPM, a new coding algorithm, named Locality-constrained Linear Coding (LLC), was proposed [8]. The LLC method is widely applied in image classification tasks. It considers both the locality constraints and global sparsity when coding remote sensing image. LLC shows a state-of-the-art classification accuracy. In this paper, the proposed remote sensing scene classification approach is based on LLC. The proposed approach will improve the codebook technology in traditional LLC by combing saliency detection technology, which is good for remote sensing scene classification.

LLC has achieved state-of-the-art performances in several image classification tasks; however, it is still based on modeling the cognitive mechanism of human vision to complete the image classification task. Physiological and psychophysical evidence indicates that the human's visual system has evolved a specialized focus processing treatment called "attentive" mode, which is directed to particular locations in the visual field [9]. Based on the "attentive" mode, researchers have proposed several visual saliency detection methods [10]. LLC does not take into account visual saliency detection. Thus, we can improve LLC by combining it with visual saliency detection.

Saliency detection was introduced into the field of computer vision in the late 1990s. According to the computational view, saliency detection can be grouped into different algorithms. One category is the center-surround thought. This type of algorithm assumes that a local window exists, which can divide an image into a center containing an object and a surround [10–13]. Another category of saliency detection algorithms is based on frequency domain methods [14–16]. The other category is relying on information theory concepts [17–19]. In Reference [13], Fast and Efficient Saliency Detection (FESD) was proposed, which belongs to the first saliency detection algorithm category. In the FESD method, a saliency map is built by computing the kernel density estimation, which has a faster computing performance compared with several other density estimations [20,21]; however, most remote sensing images, in which saliency maps are obtained using FESD, are not always the best. This is because remote sensing images usually have a wide field of view and the size of the kernel used for FESD, chosen by experience, is not necessarily optimal. However, we found that FESD can be improved using a simple technique. The technique is named spatial pyramid Gaussian kernel density estimation (SPGKDE), which is suitable for saliency detection of remote sensing images. In our research, SPGKDE is used for the preprocessing of remote sensing images. Then, by combining LLC encoding, the classification accuracy of remote sensing images can be improved.

The main contribution of this paper is to propose a new kind of remote sensing scene classification method by combing SPGKDE saliency detection preprocessing and LLC to improve remote sensing scene classification accuracy. This paper is structured as follows. Section 2 describes the remote sensing image scene classification proposed method in detail. In particular, this section outlines SPGKDE saliency detection preprocessing, and explains how and why remote sensing scene classification accuracy can be improved by SPGKDE preprocessing. Section 3 shows a comparison of the experimental results based on traditional LLC classification and the proposed method, followed by discussions. Section 4 concludes the paper.

2. Methodology

As mentioned above, this paper proposes a saliency preprocessing locality-constrained linear coding method for remote sensing scene classification. Figure 1 shows the realization of this method. The core technologies are SPGKDE and saliency preprocessing LLC. More specifically, descriptions of these two techniques are as follows.

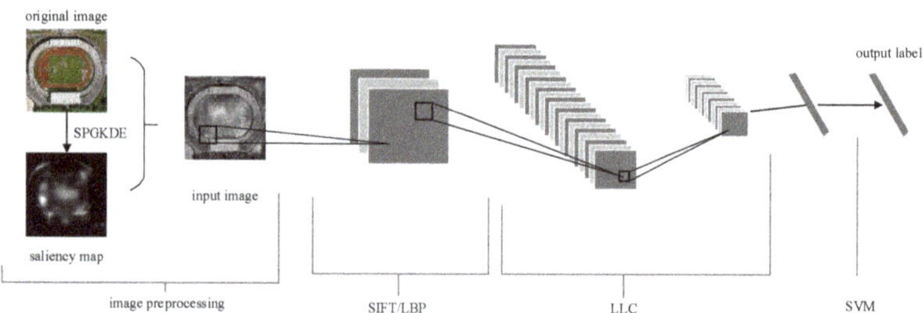

Figure 1. The flow chart of the proposed method.

2.1. Spatial Pyramid Gaussian Kernel Density Estimation Saliency Detection Preprocessing

FESD builds a saliency map of a remote sensing image using Gaussian kernel density estimation [13]. This method also implicitly considers sparse sampling and center bias. Since the human eye is more focused on the center of an image and is accustomed to taking a photo with the subject in the center, this method works well for most images. However, this method is not universal for remote sensing images. In some remote sensing images, the scene occupies the entire image. Center bias can lead to the loss of some salient information, which is detrimental to the classification of remote sensing images. We therefore need to improve FESD and propose SPGKDE. Figure 2 shows the weakness of the FESD method, which is the loss of salient information.

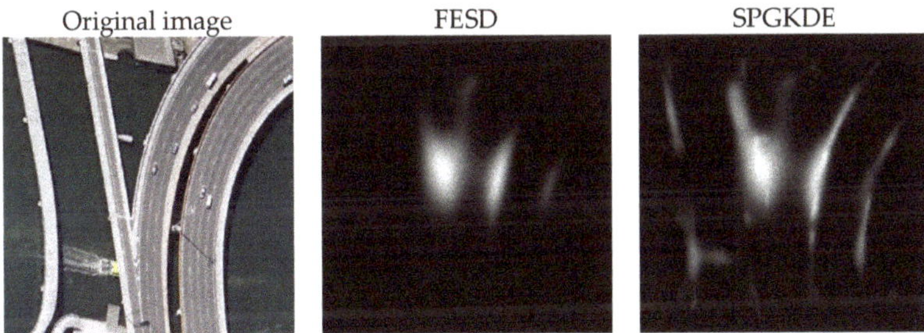

Figure 2. Saliency maps comparison between FESD (fast and efficient saliency detection method, which is proposed by reference [13]) and SPGKDE (spatial pyramid Gaussian kernel density estimation).

In this paper, a spatial pyramid Gaussian kernel density estimation (SPGKDE) saliency detection based on FESD is proposed. It requires minor changes but offers a great improvement on FESD. It is proposed for obtaining a remote sensing image's saliency map more effectively, which is used for saliency preprocessing locality-constrained linear coding.

The spatial pyramid method (SPM [7]) is a simple and practical method in computer vision. This method divides a remote sensing image, from coarse to fine, by level. Then, local features in each level are aggregates later. Usually, $2^l \times 2^l$ sub-regions (where $l = 0, 1, 2$) are used. The proposed method, SPGKDE, is based on this thought. Figure 3 shows the realization of SPGKDE.

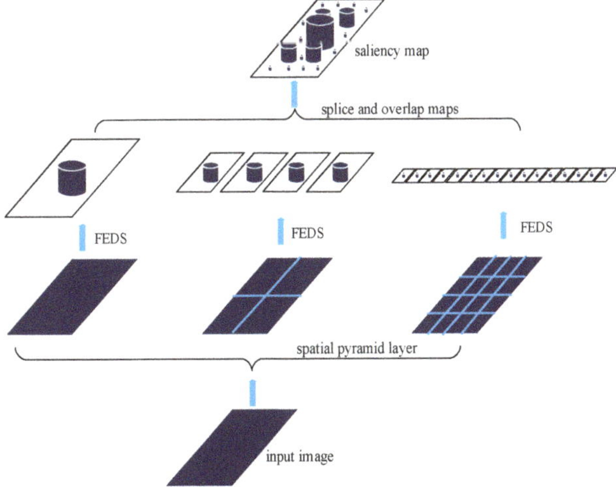

Figure 3. The flow chart of the SPGKDE.

Assume that there is a remote sensing image I and R_i^l is one of its sub-regions. For each sub-region image R_i^l, each pixel exists as $x = (\overline{x}, f)$, where f is a feature vector extracted from R_i^l, and $\overline{x}$ denotes the coordinate of pixel x in R_i^l. For each sub-region R_i^l, we can get its corresponding saliency map S_i^l using the FESD method proposed in Reference [13].

$$S_i^l(x) = \mathfrak{A} * \left[P_r^n(1|f, \overline{x}) \right]^\alpha \tag{1}$$

where $*$ is convolution operator, $\mathfrak{A}$ is a circular averaging filter, $P_r^n(1|f, \overline{x})$ is a calculated probability for characterizing the pixels in the saliency areas, and $\alpha \geq 1$ is a factor that affects high probability areas.

Then, the saliency map S of remote sensing image I can be calculated as follows:

$$S = d_0 S_1^0 + d_1 \sum_{i=1}^{4} S_i^1 + d_2 \sum_{i=1}^{16} S_i^2 = \sum_{l=0}^{2} \sum_{i=1}^{4^l} d_l S_i^l \tag{2}$$

where $d_l = (l = 0, 1, 2)$ means a weight, and it can be defined as $d_l = \frac{1}{2^l}$, and S_i^l is acquired by Equation (1). When the level l is increasing, the weight d_l is decaying to prevent the collection of too much useless salient information.

Finally, a preprocessed image is obtained by:

$$I' = I + \xi \cdot S \tag{3}$$

where I' is the preprocessed image, and $\xi \in (0, 1]$ is used in order to avoid adding invalid details caused by the saliency map.

SPGKDE has a stronger nature of saliency detection than FESD because of its re-aggregation of salient information from different image space scales. After saliency detection preprocessing, the salient information can increase the inter-class variations between different remote sensing scenes. This is beneficial for improving the accuracy of classification tasks. In this paper, the Gaussian kernel is fixed as a 9×9 size, and the proportional control coefficient ξ is fixed at 0.5.

2.2. Saliency Preprocessing Locality-Constrained Linear Coding

Different kinds of coding algorithms have been proposed in the past few decades [5,8], most of which usually consist of feature extraction and feature coding. Experimental results have shown that, with a certain visual codebook, utilizing different coding schemes will directly affect the remote sensing scene classification accuracies [22,23]. Meanwhile, sparsity is less essential than locality under certain assumptions, as pointed out in Reference [24]. Therefore, Locality-constrained Linear Coding (LLC) was proposed [8].

LLC is widely applied in image classification tasks. It adds local restrictions to achieve global sparsity. Based on a given visual codebook, LLC provides analytical solutions. This coding method also has a fast coding speed. In this paper, LLC is selected as the basic method for remote sensing image feature coding.

Let F be a set of L-dimensional local descriptors extracted from the remote sensing image, i.e., $F = [f_1, f_2, ..., f_P] \in R^{L \times P}$. Given a codebook $B = [b_1, b_2, ..., b_Q] \in R^{L \times Q}$ with Q entries, LLC obeys the following criteria:

$$\min_C \sum_{i=1}^{P} \|f_i - Bc_i\|^2 + \lambda \|d_i \odot c_i\|^2, \quad s.t. \mathbf{1}^T c_i = 1, \forall i \tag{4}$$

where $d_i \in R^Q$ is the locality adaptor, and $\odot$ is the element-wise multiplication. Usually, we have:

$$d_i = \exp\left(\frac{dist(f_i, B)}{\sigma}\right) \tag{5}$$

where $dist(f_i, B) = [dist(f_i, b_1), ..., dist(f_i, b_q)]^{\mathrm{T}}$, and $dist(f_i, b_i)$ represent the distance between f_i and b_i. σ is the weight used for adjusting the locality adaptor decay speed. Further, we can get the following solutions:

$$\tilde{c}_i = (C_i + \lambda diag^2(d_i))\mathbf{1} \tag{6}$$

$$c_i = \tilde{c}_i / \mathbf{1}^{\mathrm{T}} \tilde{c}_i \tag{7}$$

$$C_i = (B^{\mathrm{T}} - \mathbf{1} f_i^{\mathrm{T}})(B^{\mathrm{T}} - \mathbf{1} f_i^{\mathrm{T}})^{\mathrm{T}} \tag{8}$$

Unlike most coding methods, LLC can provide analytical solutions. It is of great benefit to computation. It also can be seen from Equation (8) that the quality of the given codebook B directly affects the coding results in LLC, and, further, indirectly affects the classification accuracy of remote sensing image scene categories. To better complete the task of remote sensing scene classification, we make full use of the advantages of the saliency detection information. Thus, we propose saliency preprocessing LLC.

For each remote sensing image I, we can obtain its preprocessed image I' by the method proposed in Section 2.1. We use image I and I' to get the corresponding codebook B and B'. Then, the given codebook B in Equation (8) can be replaced by B_{final}, which uses the following formula:

$$B_{final} = \frac{1}{2}(B + B') \tag{9}$$

where codebook B represents the original remote sensing images features codebook, and B' comes from the computations of corresponding saliency preprocessed remote sensing images. Figure 4 shows the differences between the traditional LLC codebook and the proposed method codebook. Obviously, the features used to generate the new codebook B_{final} are more prominent than those of codebook B constructed from traditional LLC features, and this new codebook B_{final} will improve the performance of remote sensing scene classification.

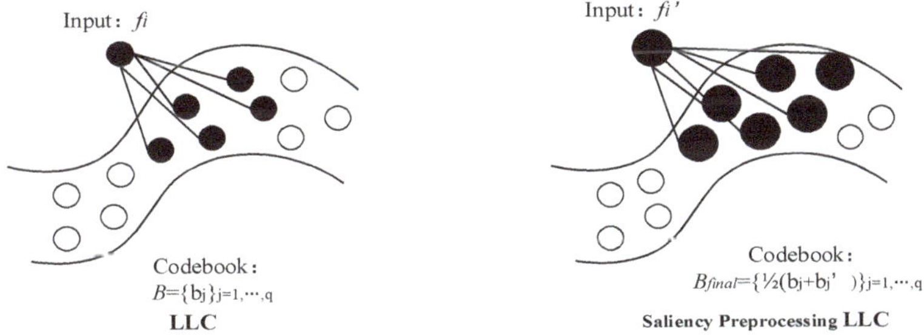

Figure 4. Comparison between LLC (Locality-constrained Linear Coding) and proposed method.

Saliency preprocessing LLC only solves the problem of feature coding. In this paper, Scale Invariant Feature Transform (SIFT) [25] and Local Binary Pattern (LBP) [26] are used to extract image features and a support vector machine (SVM) [27,28] is used as the training classifier. Of course, to prove the effectiveness of the proposed method, this paper uses a public 19-class remote sensing scene dataset to conduct experiments [29,30]. The dataset is proposed by Dengxin Dai and Wen Yang, and named as WHU-RS dataset. The images in this dataset are a fixed size of 600 × 600 pixels. All the images are collected from Google Earth. In the early days, the dataset contained 12 categories of physical scenes in the satellite imagery [29]; later, it was expanded to 19 classes [30]. The 19 classes of the dataset are airport, beach, bridge, river, forest, farmland, meadow, mountain, pond, parking, port, park, viaduct, desert, football field, railway station, residential area, industrial area and commercial area.

The dataset has higher intra-class variations and smaller inter-class dissimilarities. The experimental results and discussions are given in the next section.

3. Experiments and Discussion

In this section, we will focus on experiments based on a WHU-RS dataset, demonstrate the advantage of SPGKDE preprocessing for remote sensing images, and then show the performance of both the traditional LLC and the proposed method. Thereafter, the results will be briefly discussed.

3.1. SPGKDE Preprocessing

Figure 5 shows several sample images of the WHU-RS dataset and the corresponding saliency detection results computed by SPGKDE (proposed in this paper).

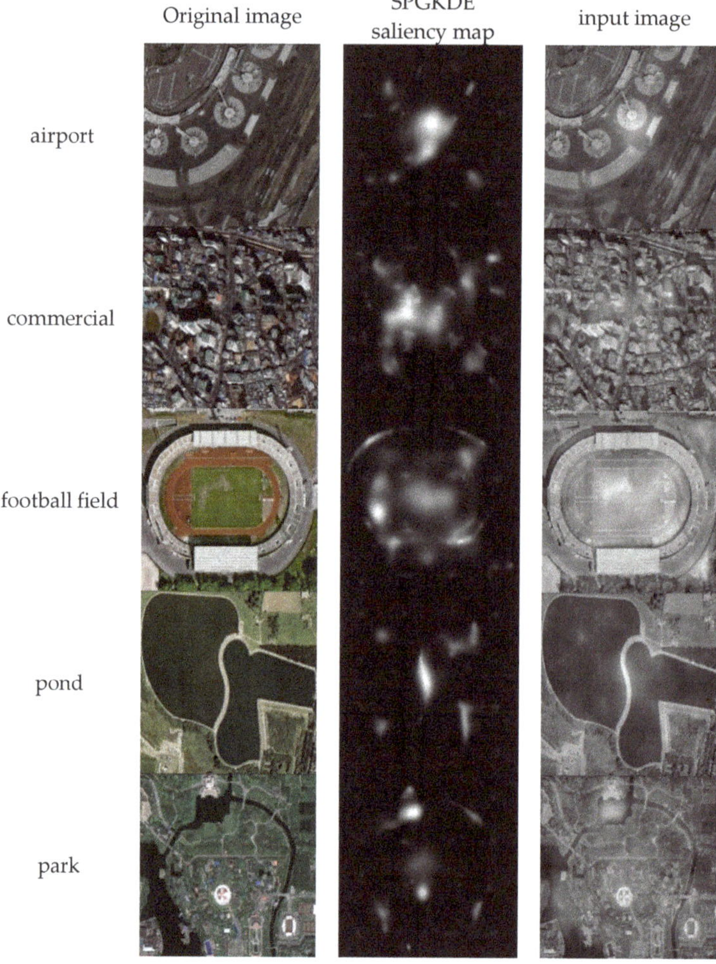

Figure 5. *Cont.*

Figure 5. Samples of a 19-class dataset, SPGKDE saliency detection, and the final preprocessed images for LLC input.

As shown in Figure 5, most of the images can obtain useful and expected saliency maps via the proposed SPGKDE saliency detection, and after detection preprocessing, key parts of the images can be made more prominent. The preprocessing technique can increase inter-class variations and reduce inter-class dissimilarities. Though, inevitably, there are some unsatisfactory saliency maps for images that may lead to classification confusion; for instance, the pond scene image in Figure 5, where the features of the pond are vague, whereas the bridge becomes rather prominent after saliency detection preprocessing and would be wrongly classified into the park scene class, though this phenomenon is very rare. In fact, a very high percentage of images can correctly receive saliency detection preprocessing.

3.2. Performance of Traditional LLC and Proposed Method

In this paper, two kinds of low-level feature are used to form a codebook for LLC separately; namely, Scale Invariant Feature Transform (SIFT) and Local Binary Pattern (LBP).

3.2.1. Performance Based on SIFT Feature

The SIFT vector has a dimension of 128. Half of the dataset is used as the training set and the other half is used for the test. The accuracy of classification of traditional LLC and the proposed method based on SIFT is shown in Table 1, where the BoF performance is also shown as a baseline method.

Table 1. Classification accuracies based on Scale Invariant Feature Transform (SIFT).

Descriptors Methods	Traditional BoF	Traditional LLC	Proposed Method
SIFT	72.87%	73.27%	79.01%

Figure 6 shows the confusion matrices generated by the traditional LLC and the proposed method based on SIFT. It is benefit for observing more differences in detail between the two methods. We can observe that more scene image categories, such as airport, bridge and commercial area were correctly classified by the proposed method. Although the classification accuracies of meadow, parking and residential area categories were reduced, especially the classification accuracy of meadow, which dropped sharply by 12%, the method improved the classification accuracy of the entire public dataset by about 6%.

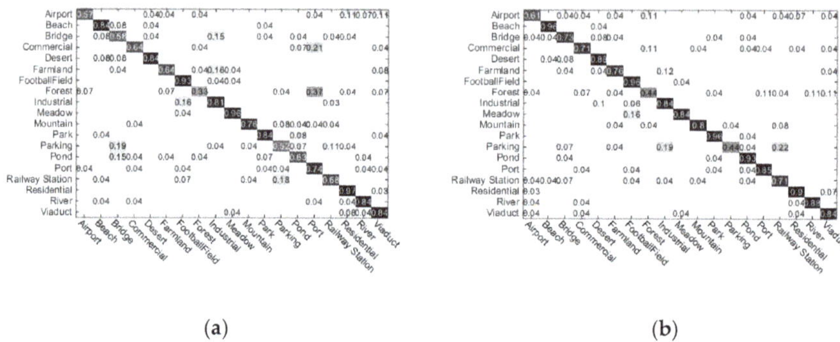

(a)   (b)

Figure 6. Confusion matrices based on SIFT; (**a**) Traditional LLC (SIFT); (**b**) Proposed method (SIFT).

To investigate the impact of saliency detection preprocessing, different proportions of training and test are used for the experiments. The training proportion is ranged from 20% to 80%. The experimental results are shown in Figure 7. The classification accuracies are clearly improved with our proposed method.

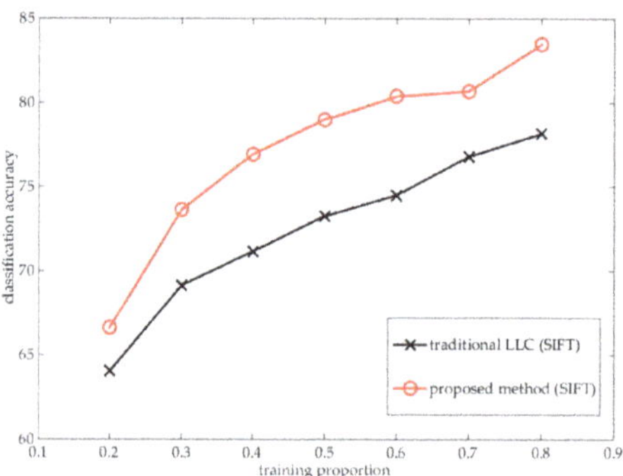

Figure 7. Different ratios of training set experiments based on SIFT.

3.2.2. Performance Based on LBP Feature

LBP vector is a 256-dimension feature vector. Half of the dataset is used as the training set and the other half is used for the test. The accuracy of classification of the traditional LLC and the proposed method based on LBP is shown in Table 2, where the BoF performance is shown as a baseline method as well.

Table 2. Classification accuracies based on Local Binary Pattern (LBP).

Methods Descriptors	Traditional BoF	Traditional LLC	Proposed Method
LBP	68.71%	72.87%	78.22%

To observe more classification details of the proposed method, we gave the confusion matrices of traditional LLC and proposed method by Figure 8. The confusion matrices are generated by the traditional LLC and the proposed method based on LBP. We can see that the classification accuracies of the four categories is reduced, namely football field, industrial, meadow and mountain, accounting for 21% of all categories, whereas the classification accuracy of the entire public dataset is improved by approximately 5.4%.

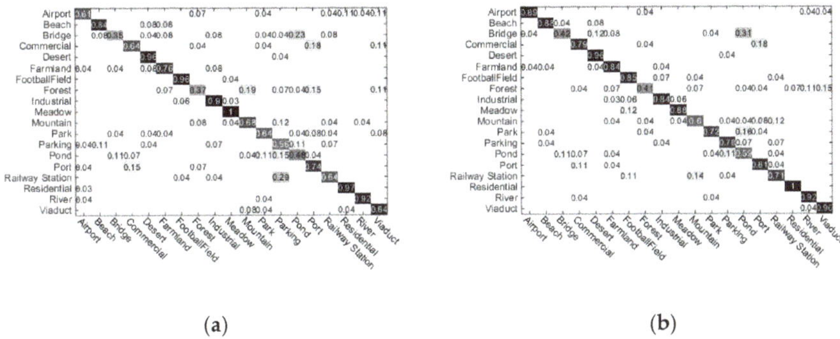

(a)   (b)

Figure 8. Confusion matrices based on LBP. (**a**) Traditional LLC (LBP); (**b**) Proposed method (LBP).

To investigate the impact of saliency detection preprocessing, different proportions of training and test are used for the experiments. The training proportion ranged from 20% to 80%. The experimental results are shown in Figure 9. As can be seen, the classification accuracies are improved with our proposed method.

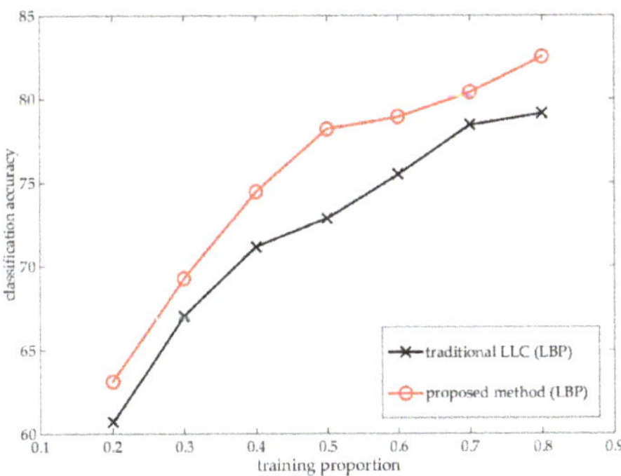

Figure 9. Different ratios of training set experiments based on LBP.

Through the above experiments, we can see that the proposed method can improve the scene classification accuracies of remote sensing images, regardless of the features extracted by the SIFT or LBP methods. However, the classification accuracies of the meadow category are reduced using the proposed method. To determine why the classification of the meadow scene images is more difficult to identify using the proposed method, we reviewed the entire process of processing meadow category images. We found that almost all the meadow-category images in the WHU-RS dataset did not have an obvious saliency region, as shown in Figure 10. The most striking feature of the meadow category images was the color. When we used SPGKDE to preprocess meadow category images, not only did we get little color information, but there was also additional untrue saliency region information. Furthermore, weakening the color information and forcibly mining saliency region information, can lead to generating the wrong codebook for LLC, which is detrimental for classification based on LLC. This may explain why the proposed method reduced the classification accuracies of the meadow category. Fortunately, there are saliency regions in most remote sensing images and this happens only for meadow category images.

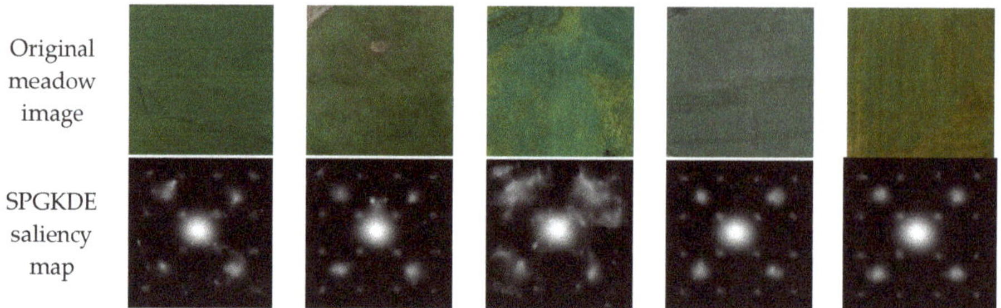

Figure 10. Meadow category images and corresponding saliency maps.

From the above experimental results, we have good reason to believe that the proposed method can improve remote sensing scene classification performance.

4. Conclusions

In this paper, an improvement saliency detection method named SPGKDE is proposed based on the existing saliency detection FESD for remote sensing image preprocessing. The new saliency detection method re-aggregates salient information from different image space scales and is obviously more applicable to remote sensing images. SPGKDE has a wider field of view than FESD.

Thus, a new kind of remote sensing scene classification method that combines SPGKDE saliency detection preprocessing and LLC is proposed. The method is easy to operate. Visual saliency detection plays an important role in remote sensing image analysis and the traditional LLC classification technology ignores this technology, resulting in a limited accuracy for image classification. This paper proposes integrating an improved saliency detection method—SPGKDE— into the LLC classification. This method can increase inter-class variations and reduce inter-class dissimilarities. In fact, the preprocessing method improves the codebook technology in traditional LLC. It is a core technology that directly determines the classification result. This method achieves a better simulation of the human visual system than traditional LLC. In this paper, both SIFT and LBP features were used for experiments. The experiments show that the proposed method is useful and can improve remote sensing scene classification accuracy.

Author Contributions: Conceptualization, L.J.; Data curation, L.J. and M.W.; Formal analysis, L.J.; Funding acquisition, X.H.; Investigation, M.W.; Methodology, L.J. and X.H.; Software, L.J. and M.W.; Writing—original draft, L.J.

Funding: This research was funded by the National Natural Science Foundation of China (U1435220).

Acknowledgments: The authors acknowledge the National Natural Science Foundation of China (U1435220).

References

1. Cheng, G.; Han, J.; Guo, L.; Liu, Z.; Bu, S.; Ren, J. Effective and efficient midlevel visual elements-oriented land-use classification using VHR remote sensing images. *IEEE Trans. Geosci. Remote Sens.* **2015**, *53*, 4238–4249. [CrossRef]
2. Anwer, R.M.; Khan, F.S.; van de Weijer, J.; Molinier, M.; Laaksonen, J. Binary patterns encoded convolutional neural networks for texture recognition and remote sensing scene classification. *ISPRS J. Photogramm. Remote Sens.* **2018**, *138*, 74–85. [CrossRef]
3. Sivic, J.; Zisserman, A. Video google: A text retrieval approach to object matching in videos. In Proceedings of the IEEE International Conference on Computer Vision, Nice, France, 13–16 October 2003; Volume 2, pp. 1470–1477.
4. Csurka, G.; Dance, C.R.; Fan, L.; Bray, C.; Csurka, G. Visual categorization with bags of keypoints. *Workshop Statist. Learn. Comput. Vis. Eccv.* **2004**, *44*, 1–22.
5. Bosch, A.; Zisserman, A.; Muñoz, X. Scene classification using a hybrid generative/discriminative approach. *IEEE Trans. Pattern Anal. Mach. Intell.* **2008**, *30*, 712–727. [CrossRef] [PubMed]
6. Yang, L.; Jin, R.; Sukthankar, R.; Jurie, F. Unifying discriminative visual codebook generation with classifier training for object category recognition. In Proceedings of the 26th IEEE Conference on Computer Vision and Pattern Recognition, Anchorage, AK, USA, 23–28 June 2008.
7. Lazebnik, S.; Schmid, C.; Ponce, J. Beyond bags of features: spatial pyramid matching for recognizing natural scene categories. In Proceedings of the Conference on Computer Vision and Pattern Recognition, New York, NY, USA, 17–22 June 2006; pp. 2169–2178.
8. Wang, J.; Yang, J.; Yu, K.; Lv, F.; Huang, T.; Gong, Y. Locality-constrained linear coding for image classification. In Proceedings of the Computer Vision and Pattern Recognition, San Francisco, CA, USA, 13–18 June 2010; pp. 3360–3367.
9. Koch, C.; Ullman, S. Shifts in selective visual attention: Towards the underlying neural circuitry. *Human Neurobiol.* **1985**, *4*, 219–227.
10. Achanta, R.; Estrada, F.; Wils, P.; Süsstrunk, S. Salient region detection and segmentation. In *Computer Vision Systems*; Springer: Berlin, Germany, 2008; pp. 66–75.
11. Seo, H.J.; Milanfar, P. Training-free, generic object detection using locally adaptive regression kernels. *IEEE Trans. Pattern Anal. Mach. Intell.* **2009**, *32*, 1688–1704.
12. Rahtu, E.; Heikkilä, J. A Simple and efficient saliency detector for background subtraction. In Proceedings of the IEEE International Conference on Computer Vision Workshops, Kyoto, Japan, 27 September–4 October 2009; pp. 1137–1144.
13. Tavakoli, H.R.; Rahtu, E. Fast and efficient saliency detection using sparse sampling and kernel density estimation. In *Image Analysis*; Springer-Verlag: Berlin, Germany, 2011; pp. 666–675.
14. Guo, C.; Ma, Q.; Zhang, L. Spatio-temporal saliency detection using phase spectrum of quaternion fourier transform. In Proceedings of the 26th IEEE Conference on Computer Vision and Pattern Recognition, Anchorage, AK, USA, 23–28 June 2008.
15. Hou, X.; Zhang, L. Saliency detection: A spectral residual approach. In Proceedings of the IEEE Computer Society Conference on Computer Vision and Pattern Recognition, Minneapolis, MN, USA, 17–22 June 2007.
16. Achanta, R.; Hemami, S.; Estrada, F.; Susstrunk, S. Frequency-tuned salient region detection. In Proceedings of the 2009 IEEE Conference on Computer Vision and Pattern Recognition, Miami, FL, USA, 20–25 June 2009; pp. 1597–1604.
17. Bruce, N.D.B.; Tsotsos, J.K. Saliency based on information maximization. In Proceedings of the NIPS'05 Proceedings of the 18th International Conference on Neural Information Processing Systems, Cambridge, MA, USA, 5–8 December 2005; pp. 155–162.
18. Lin, Y.; Fang, B.; Tang, Y. A computational model for saliency maps by using local entropy. In Proceedings of the National Conference on Artificial Intelligence, Atlanta, GA, USA, 11–15 July 2010; Volume 2, pp. 967–973.
19. Mahadevan, V.; Vasconcelos, N. Spatiotemporal saliency in dynamic scenes. *IEEE Trans. Pattern Anal. Mach. Intell.* **2009**, *32*, 171–177. [CrossRef] [PubMed]
20. Aiazzi, B.; Alparone, L.; Baronti, S.; Garzelli, A.; Zoppetti, C. Nonparametric change detection in multitemporal SAR images based on mean-shift clustering. *IEEE Trans. Geosci. Remote Sens.* **2013**, *51*, 2022–2031. [CrossRef]

21. Liu, J.; Tang, Z.; Cui, Y.; Wu, G. Local competition-based superpixel segmentation algorithm in remote sensing. *Sensors* **2017**, *17*, 1364. [CrossRef] [PubMed]
22. Chen, J.; Li, Q.; Peng, Q.; Wong, K.H. CSIFT based locality-constrained linear coding for image classification. *Pattern Anal. Appl.* **2015**, *18*, 441–450. [CrossRef]
23. Yu, K.; Zhang, T.; Gong, Y. Nonlinear learning using local coordinate coding. In Proceedings of the Advances in Neural Information Processing Systems 22, Vancouver, BC, Canada, 7–10 December 2009; pp. 2223–2231.
24. Wang, S.; Ding, Z.; Fu, Y. Marginalized denoising dictionary learning with locality constraint. *IEEE Trans. Image Process.* **2018**, *27*, 500–510. [CrossRef] [PubMed]
25. Lowe, D.G.; Lowe, D.G. Distinctive image features from scale-invariant keypoints. *Int. J. Comput. Vis.* **2004**, *60*, 91–110. [CrossRef]
26. Ojala, T.; Pietikainen, M.; Maenpaa, T. Multiresolution gray-scale and rotation invariant texture classification with local binary patterns. *Pattern Anal. Mach. Intell. IEEE Trans.* **2002**, *24*, 971–987. [CrossRef]
27. Boser, B.E.; Guyon, I.M.; Vapnik, V.N. Training algorithm for optimal margin classifiers. In Proceedings of the Fifth Annual ACM Workshop on Computational Learning Theory, New York, NY, USA, 27–29 July 1992; pp. 144–152.
28. Fan, R.E.; Chang, K.W.; Hsieh, C.J.; Wang, X.-R.; Lin, C.-J. LIBLINEAR: A library for large linear classification. *J. Mach. Learn. Res.* **2008**, *9*, 1871–1874.
29. Dai, D.; Yang, W. Satellite image classification via two-layer sparse coding with biased image representation. *IEEE Geosci. Remote Sens. Lett.* **2011**, *8*, 173–176. [CrossRef]
30. Sheng, G.; Yang, W.; Xu, T.; Sun, H. High-resolution satellite scene classification using a sparse coding based multiple feature combination. *Int. J. Remote Sens.* **2012**, *33*, 2395–2412. [CrossRef]

Article

Multi-Sensor Satellite Data Processing for Marine Traffic Understanding

Marco Reggiannini * and Luigi Bedini

Institute of Information Science and Technologies, National Research Council of Italy, 56124 Pisa, Italy; luigi.bedini@isti.cnr.it
* Correspondence: marco.reggiannini@isti.cnr.it; Tel.: +39-050-621-3469

Received: 12 December 2018; Accepted: 26 January 2019; Published: 1 February 2019

Abstract: The work described in this document concerns the estimation of the kinematics of a navigating vessel. This task can be accomplished through the exploitation of satellite-borne systems for Earth observation. Indeed, Synthetic Aperture Radar (SAR) and optical sensors installed aboard satellites (European Space Agency Sentinel, ImageSat International Earth Remote Observation System, Italian Space Agency Constellation of Small Satellites for Mediterranean basin Observation) return multi-resolution maps providing information about the marine surface. A moving ship represented through satellite imaging results in a bright oblong object, with a peculiar wake pattern generated by the ship's passage throughout the water. By employing specifically tailored computer vision methods, these vessel features can be identified and individually analyzed for what concerns geometrical and radiometric properties, backscatterers spatial distribution and the spectral content of the wake components. This paper proposes a method for the automatic detection of the vessel's motion-related features and their exploitation to provide an estimation of the vessel velocity vector. In particular, the ship's related wake pattern is considered as a crucial target of interest for the purposes mentioned. The corresponding wake detection module has been implemented adopting a novel approach, i.e., by introducing a specifically tailored gradient estimator in the early processing stages. This results in the enhancement of the turbulent wake detection performance. The resulting overall procedure may also be included in marine surveillance systems in charge of detecting illegal maritime traffic, combating unauthorized fishing, irregular migration and related smuggling activities.

Keywords: remote sensing; SAR; radon transform; speckle noise filtering; maritime traffic monitoring; wake detection and analysis

---

1. Introduction

Monitoring the maritime surface represents a crucial task to authorities and institutions. It provides meaningful information to develop suitable policies regulating a number of human activities and it helps to detect critical circumstances early and eventually counteract or mitigate related consequences. Within this scenario, maritime traffic represents a critical activity which deserves uninterrupted supervision. Approximately 600,000 vessels navigate daily on the world seas, with purposes related to fishing, travel, tourism, military or mercantile business. Nowadays, several types of maritime traffic are under observation, such as migrant flow, unauthorized fishing and environmental pollution. Existing monitoring tools (Automatic Identification System [1]) are based on the active collaboration from navigating vessels. They turn out to be intrinsically unreliable because the data can be easily counterfeited by malicious users. As an alternative, satellite missions rotating around earth-centered orbits provide remote sensing information in the form of multi-sensor multi-resolution data, captured on a daily basis. These data carry relevant information about moving vessels and their attributes. The observed vessel's features may relate to morphological-geometrical properties (center of mass location, length, width), to radiometric properties (backscattering intensity statistics) as well as to pertubations generated in the water medium by the ship motion. Examples of high-resolution data can be provided by currently orbiting satellites such as European Space Agency (ESA) Copernicus Sentinels, ImageSat International Earth Remote Observation System (ISI EROS) and Italian Space Agency Constellation of Small Satellites for Mediterranean basin Observation (ASI COSMO-SkyMed).

The automatic monitoring of the maritime traffic is currently a relevant research topic. For this purpose, Information and Communication Technology (ICT) provide powerful tools that can be exploited in the implementation stage. The present scientific literature provides examples of software platforms, based on a computer vision approach, which are primarily dedicated to the observation and classification of marine traffic [2–8].

The ship detector applied to an SAR/optical map typically returns a small patch cut out of the entire original map (an example in Figure 1, left side), centered on a candidate vessel, supplied with relevant information, such as a coarse estimate of the ship centroid coordinates and a first approximated ship geometry. The detector returns an output patch that includes the vessel body and, if available, the surrounding motion's related features.

A typical approach (Figure 2) is based on the cascade pipeline of multiple machine learning and image processing steps. The most relevant modules are represented by the ship detector, the ship segmentation/classification module and the ship kinematics estimation module.

Figure 1. Example of the crop image of a navigating vessel (**left**), cut out from a larger SAR map. (COSMO-SkyMed Product—© ASI 2016 processed under license from ASI—Agenzia Spaziale Italiana, all rights reserved, distributed by e-GEOS). To enhance the visualization of the input data, the original crop has been processed by an histogram correction algorithm. In the enhanced image (**right**), the corresponding wake is clearly visible in the vessel's nearby. Additionally the Azimuth shift effect is also observable.

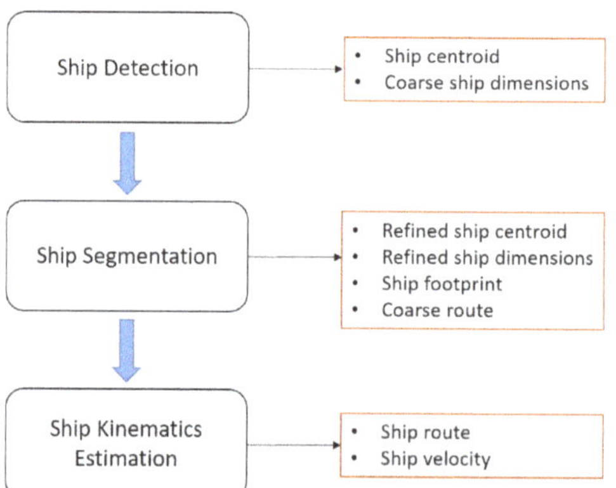

Figure 2. Block diagram illustrating the processing sequence.

The following segmentation step provides accurate estimates of the vessel centroid positioning, a binary model of the vessel shape representing the area occupied by the ship, the hull main dimensions and an estimate of the vessel orientation, which is related to the minimum inertia axis of the candidate target (see [9]). For this reason, the vessel course is provided with a 180° ambiguity, since at this stage it is not yet possible to univocally identify the target's fore and aft.

The geometrical and morphological information provided by the previous steps can be enriched by inspecting the water surface surrounding the detected ship position. Indeed, it is known that the ship's kinematics is directly related to the peculiar wake pattern generated by the motion of the ship itself through the water surface (see [10,11]). In particular, by detecting the linear envelopes of the main wake components and performing a proper analysis of the detected signals, it is possible to estimate respectively the ship's heading (univocally) and the ship's velocity module (see, for example, [12–14]).

The present work focuses on the latter topic, by discussing the conception and development of a suite of algorithms dedicated to the estimation of a vessel's route and velocity values through the detection and analysis of the vessel's motion related features. Interesting features for the mentioned purposes are (i) the displacement, exclusively observable in SAR imagery, between the vessel target center of mass and the corresponding wake pattern tip and (ii) the spatial wavelength of the plane wave oscillation located at the edge sector of the wake envelope (*Kelvin wake*). The extraction of these features is carried out following a novel approach based on the computation of the gradient of the input signal, according to a criterion of robustness w.r.t. noise. Integrating this preliminary stage of signal manipulation in the processing pipeline yields an enhancement in the wake pattern detection results, as suggested by the discussion reported in the following sections. The presented procedure has been implemented in the Octave environment [15] and tested on a SAR imagery dataset captured by ESA Sentinel-I and provided through the Copernicus Open Access Hub. Octave represented an attractive option because of the available basic image processing functionalities usually featuring high computational performances, and thus suitable for real-time application purposes.

The rest of the paper is arranged as follows: Section 2 concerns a detailed discussion of the signal features that are related to the vessel motion and that may be employed to estimate the vessel kinematics. Section 3 concerns the preliminary processing stages of the proposed pipeline, which are in charge of enhancing the Signal-to-Noise Ratio of the captured imagery. Section 4 concerns the description of the algorithms implemented for the purposes of this work and the related results. Section 5 concludes the paper by presenting a summary of the main results and discussing future developments.

2. Motion's Related Features

For the purpose of the work described in this document, the analysis will converge on those peculiar features that refer more or less directly to the motion of the vessel itself. The problem of estimating the ship velocity from an image can be separated in the estimation of the direction and the magnitude of the velocity vector.

For what concerns the ship orientation, we make the hypothesis that each moving vessel has a velocity versor that coincides with the principal axis of the hull, identified by the stern-bow oriented segment. A first attempt to provide information about the vessel kinematics consists of performing a refined segmentation of the candidate target and in estimating the 2D principal inertia axis of the target. Eventually, the ship's main axis is identified as the minimum inertia axis [9]. Furthermore, previous literature [16] has proven that an object moving on the water surface at constant heading and speed generates a wake pattern made up of divergent and transverse wave components (the diagram in Figure 3 shows the crests profiles for a generic wake pattern). We assume that all the wakes observed in the captured data share common morphological features, such as the angular aperture between the different wake components, and differ for what concerns the spatial frequencies of the generated oscillations. This hypothesis can be considered valid for a certain range of velocities (from a few up to tens of km/h, see, for example, [17]), and for a variety of vessels' typologies with different hull dimensions.

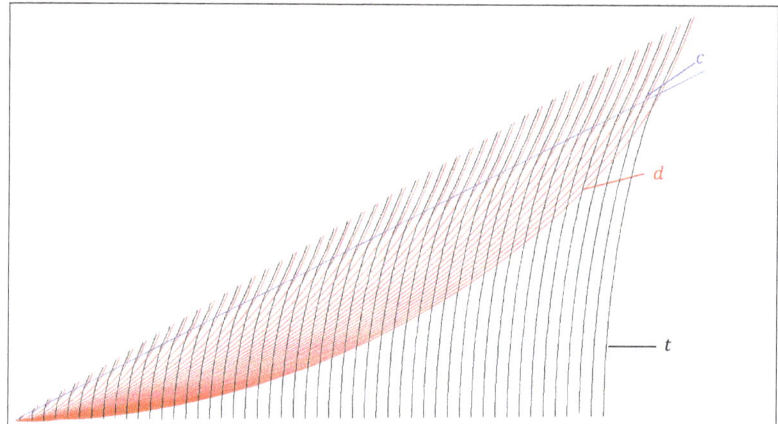

Figure 3. Wake crests diagram. Divergent (*d*) crests are in red, transverse (*t*) crests are in black. The external boundary (*c*) of the wake pattern corresponds to the cusp waves.

Generally speaking, a wake pattern in a SAR map is usually observed as the composition of three macro-structures that develop along linear directions: (i) a central turbulent strip, usually appearing as two collinear lines, a bright one next to a dark one, behind the vessel, with the same orientation of the vessel's heading and (ii) two linear bright stripes, located approximately at the wake boundaries, about $\theta = \pm 19.47°$ with respect to the central turbulent wake, in the so-called cusp wave regions of the wake. The wake formation results from the combination of multiple oscillatory components whose summation exhibits a V-shaped pattern centered on the ship route axis. The angular aperture of this V usually features a constant value of, approximately, 39°. Exploiting these observable phenomena, the route direction can be estimated by first detecting the V pattern (Figure 4) through a radon-transform-inspired linear detector [18], and by later identifying the wake center axis.

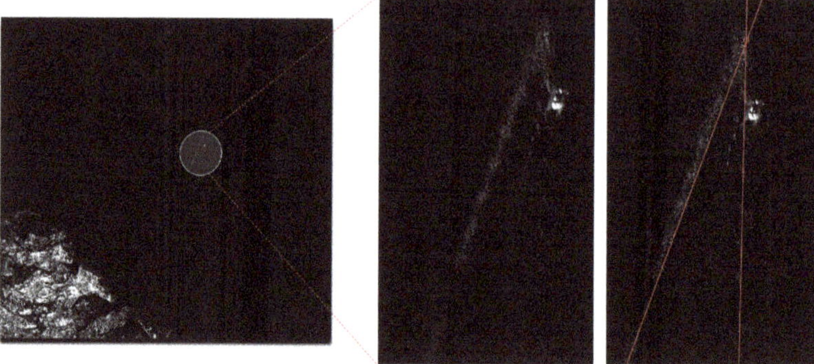

Figure 4. Wake pattern detection. (COSMO-SkyMed Product—© ASI 2016 processed under license from ASI—Agenzia Spaziale Italiana, all rights reserved, distributed by e-GEOS).

The wake pattern carries information about the vessel speed—for example, the oscillatory components observed in the external boundaries of the wake and feature wavelength values that relate to the velocity of the ship itself. Hence, provided the image resolution is large enough to observe these specific wake details, a frequency analysis of the external wake components is performed, followed by the computation of the dominant wavelength λ.

The ship's velocity v is finally estimated by means of the following expression (see [19]):

$$v = \sqrt{\frac{\sqrt{3}g\lambda}{4\pi}},\tag{1}$$

where $g = 9.81 \text{ m/s}^2$.

A second method for estimating the vessel's speed exploits the *Azimuth shift* effect, a distortion which affects SAR remote sensing, causing an artificial separation, observed in the resulting map, between the moving ship and its wake. The separation length Δ_{as} is proportional to the vessel speed according to the following expression (see [20]):

$$v = \frac{V_{sat} \cdot \Delta_{as}}{R_{sr} \cos \beta},\tag{2}$$

where V_{sat} is the satellite speed, R_{sr} is the slant range from satellite to the target and β is the angle between the vessel's velocity vector and the radar beam. Hence, the vessel speed can be obtained by measuring the separation length directly on the SAR map. Figure 5 illustrates examples of the image processing methods mentioned.

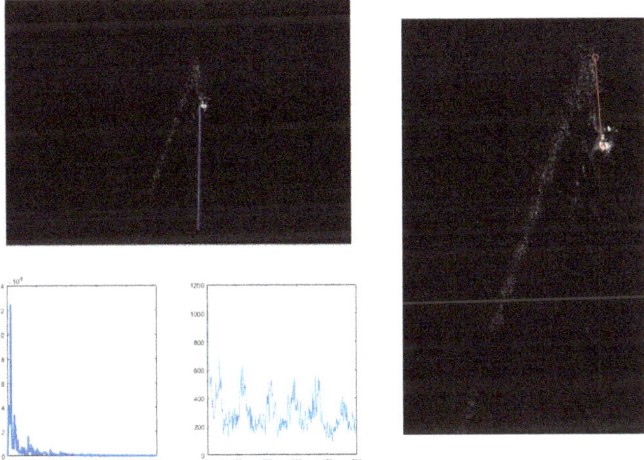

Figure 5. External wake component detection and processing for kinematics estimation purposes. In the upper left, an SAR image example, representing a moving vessel and its related wake, with a Kelvin cusp wave highlighted in color. On the right side, the identified Azimuth shift displacement is highlighted in color. On the lower left side, the results of the frequency analysis on the Kelvin cusp wave signal are shown (spectrum on the left side, the sampled signal on the right side). (COSMO-SkyMed Product—© ASI 2016 processed under license from ASI—Agenzia Spaziale Italiana, all rights reserved, distributed by e-GEOS).

The work described in the following will focus on (i) the detection of the linear envelopes of the wake, in order to estimate the ship heading, (ii) the estimation of the ship velocity by exploiting the causality with the azimuth shift effect occurring in SAR mapping and, in the unlikely circumstance that the internal wake components are represented at a sufficiently large resolution, the spectral analysis of the cusp waves, whose oscillation properties are also related to the kinematics of the ship.

Indeed, ship wakes are not usually visible or too faint to be detected. The most common type of detectable wakes are the turbulent wakes. Typically, due to very low Signal-to-Noise Ratio (*SNR*) values, the detection of a wake is a task with very small probabilty of success. Thus, different approaches have been proposed (see, e.g., [21,22]) in order to improve the poor results obtained through classical methods, e.g., based on the straightforward application of the radon/Hough transform on intensity images. In Section 4.3, a method inspired by [20] is described while, in the following Section 4.4, a novel method exploiting the image gradient is proposed. The latter shows promising results concerning the detection of the turbulent component of a wake. Accordingly, a comparison between the two mentioned approaches is proposed.

The remaining parts of this section concern detailed descriptions of each processing stage that contributes to the fulfillment of the aforementioned tasks.

3. Signal Pre-Processing for Multiplicative Noise Reduction

Due to the coherent nature of the SAR imaging system, the data capture is affected by a multiplicative noise, called *speckle*. It results from the combination of echoes coming from different scatterers, coherently generating a non-null backscatter value.

A consequence of the combination of speckle noise with the signal backscattered by homogeneous areas is that the pixels' statistical properties will vary spatially. This effect can be partially reduced by integrating different captured images (*looks*) of the same scene. The latter operation entails a decrease of the signal variance and the subsequent enhancement of the signal-to-noise ratio.

A notable amount of previous literature has concerned speckle filtering issues in SAR imagery (see, e.g., [23,24]). Given a point (x, y) in the image plane, a popular model for the speckled image formation is:

$$I(x,y) = R(x,y)u(x,y), \tag{3}$$

where I is the image intensity, R is the radar reflectivity and u is the multiplicative speckle noise term. The approach adopted within this work consists of describing the imaged scene and the speckle noise through their related probability density functions (*pdf*). According to the Bayes theorem, the conditional probability of R given I takes the following form:

$$p(R|I) = \frac{p(I|R)p(R)}{p(I)}, \tag{4}$$

where $p(R)$ represents the a priori knowledge about the radar reflectivity. $p(R|I)$ is the a posteriori information about reflectivity, information that is gathered during the measurement process. In this framework, the unknown quantity R is estimated by adopting a Maximum A Posteriori (MAP) approach, i.e., by computing the mode of the posterior probability in Equation (4).

In accordance with [25], we assume the following statistical properties:

- $P_R(R)$ is modeled as a Gamma (The generic expression for a Gamma *pdf* with parameters κ and λ is given by

$$p_x(x; \kappa, \lambda) = \frac{\lambda^\kappa}{\Gamma(\kappa)} e^{-\lambda x} x^{\kappa-1},$$

  where $\Gamma(\kappa) = 1 \cdot 2 \cdots (\kappa - 1)$, $\mu_x = \kappa/\lambda$ and $\sigma_x^2 = \kappa/\lambda^2$). distribution with $\kappa = \alpha$ (α is called the *heterogeneity coefficient* and is locally estimated according to [25]) and, accordingly, $\lambda = \alpha/E[R]$;
- speckle u is modeled as a Gamma distributed variable, with parameters $\mu_u = 1$ and $\sigma_u^2 = 1/L$, where L is the equivalent number of looks.

According to Equation (3), $I|R$ also follows a Gamma distribution, with a form that is proven [24] to be

$$P_{I|R}(I|R) = \frac{L^L}{\Gamma(L)} \left(\frac{I}{R}\right)^{L-1} e^{-\frac{LI}{R}} \frac{1}{R}.$$

By substituting in Equation (4), taking the logarithm and computing the derivative with respect to R, we get:

$$\alpha R^2 + (1 + L - \alpha) <I> R - L <I> I = 0, \tag{5}$$

which, upon selecting the positive solution, gives the restored signal

$$R = \frac{(\alpha - L - 1) <I> + \sqrt{(\alpha - L - 1)^2 <I>^2 + 4\alpha L I <I>}}{2\alpha}.$$

(6)

Each of the considered input maps have been preliminary processed by applying the signal restoration defined by Equation (6) (see an output example in Figure 6).

Figure 6. Example of Sentinel I SAR map including a vessel and its corresponding wake pattern (contains modified Copernicus Sentinel data, 2018). The signal has been processed by a MAP filtering procedure and represented on a logarithmic scale.

4. Wake Features Extraction and Analysis

The following sections concern detailed descriptions of the algorithms that have been developed to perform the estimation of a vessel kinematics. The corresponding processing pipeline is illustrated in Figure 7.

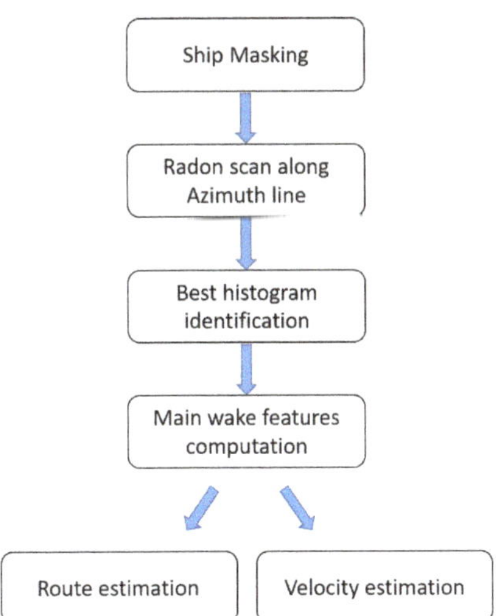

Figure 7. Conceptual diagram of the vessel kinematics estimation.

4.1. Ship Masking and Data Preprocessing

For the purpose of properly identifying the signals of interest, the captured data must be preliminary filtered to limit false positive detections, due to noise or known spurious signals that have to be neglected a priori. In case of SAR data input, the wake detector performance can be critically affected by the signal backscattered by the ship's body, which usually features pixel values that are several orders of magnitude larger than the backscattering amplitudes related to the wake pattern. We assume that pixels associated to the main vessel's body can be filtered out provided that a ship footprint is available from preliminary estimations, obtained through the methods discussed in Section 1. This way, a binary mask can be generated, based on the mentioned footprint, and exploited to substitute the ship's body pixels with proper intensity values, such as the image mean value (Figure 8).

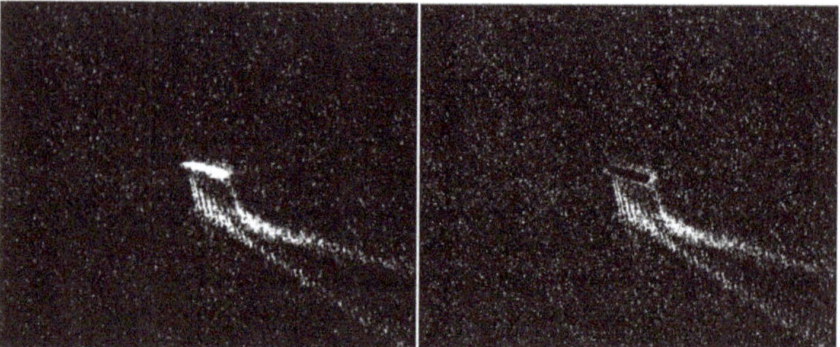

Figure 8. Exploitation of the ship footprint returned by a ship segmentation module.

As previously mentioned, in case of SAR remote sensing of moving objects, the surface wake generated by the ship's motion through the water will be represented in the map as displaced, with respect to the ship centroid, by a certain amount of pixels along the Azimuth direction. This can be exploited for limiting the wake pattern search to those points that lie along the Azimuth line (Figure 9). The criteria adopted to implement the mentioned wake search are presented in the following section.

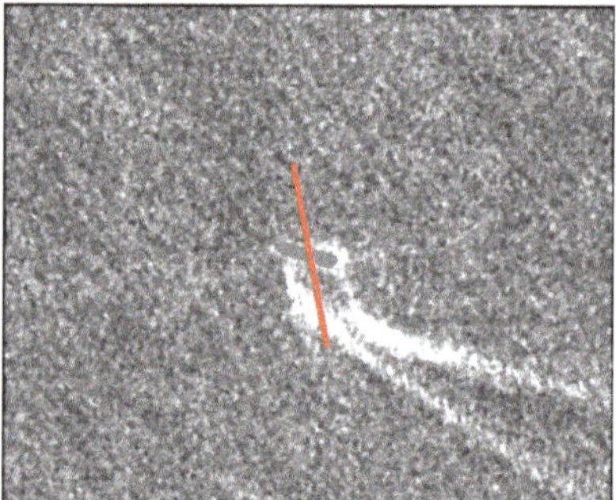

Figure 9. Azimuth line direction. The wake search is performed on the red linear subset. (COSMO-SkyMed Product—© ASI 2016 processed under license from ASI—Agenzia Spaziale Italiana, all rights reserved, distributed by e-GEOS).

4.2. Azimuth Line Scan

The wake search is performed along the line passing through the ship centroid, with orientation given by the satellite heading (Azimuth). This approach, graphically explained in Figure 10, is inspired by the work [20].

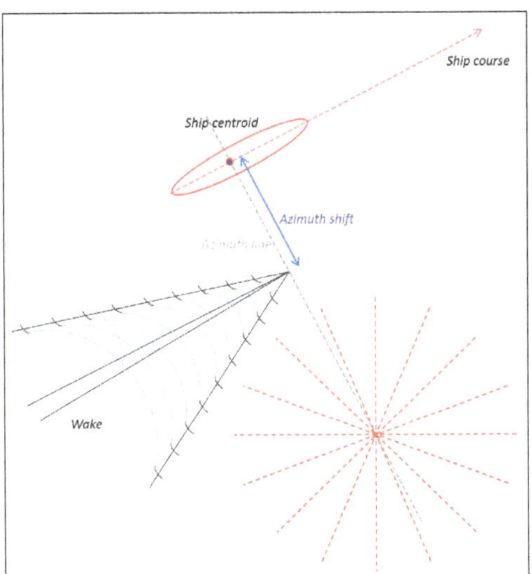

Figure 10. Wake Pattern formation and representation in SAR remote sensing and punctual radon transform.

It is assumed that the wake tip is located on the mentioned line. For each point p_j on that line, consider the family of straight lines having that point as the origin. In case a $1°$ angular pace is set, we will have a family of 360 straight lines for each point. For each of the 360 angle values, the integral summation of the pixel intensities along each orientation is computed as follows:

$$h_{p_j}(\phi_i) = \sum_{p_k}^{n_{p_k}} I(p_k)/n_{p_k}, \quad \forall\, p_k \in I : \arctan\left[\frac{(p_k - p_j)_y}{(p_k - p_j)_x}\right] = \phi_i. \tag{7}$$

Hence, each point on the azimuth line has an associated angular histogram, describing the average image value along 360 angular directions. In case a linear pattern crosses a point located on the azimuth line, the integral summation process will generate a peak for that particular direction. As soon as the scan on the azimuth-oriented line gets to the wake tip, the corresponding histogram will exhibit a number of peaks as large as the number of observed wake arms. Hence, a first detection criterion consists of selecting the point p_j on the Azimuth line (Figure 9) such that the corresponding angular histogram features the largest peak:

$$p_j : \max_j h_{p_j}(\phi). \tag{8}$$

Performing this operation on the data represented in Figure 1 returns the histogram function in Figure 11.

4.3. Wake Detection

As a consequence of the previous stage, in every crop, one single point lying on the Azimuth line has been identified as the potential tip of a wake trace. Between all the points that lie on the Azimuth line, the selected one is the most probable, since its associated angular histogram function, i.e., the distribution of the pixel mean values computed along 360 directions, features the largest peak amplitude.

In order to decide whether an angular histogram is generated by a meaningful linear object (e.g., a wake pattern arm) or it is due to noise/spurious signals, the histogram undergoes a dedicated detection stage, following the ideas described in [20].

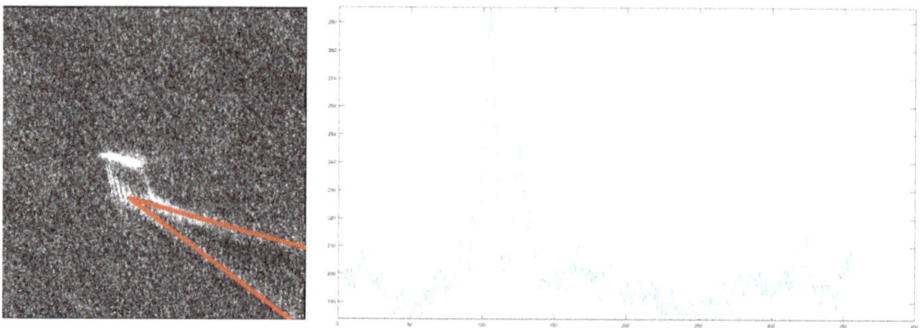

Figure 11. Example of wake linear components detection resulting from the identification of the histogram with larger peak response. (COSMO-SkyMed Product—© ASI 2016 processed under license from ASI—Agenzia Spaziale Italiana, all rights reserved, distributed by e-GEOS).

Given a point p_j on the Azimuth line, the histogram function $h_{p_j}(\phi_i)$ can be approximated, according to [26], by a linear combination $g(\phi_i)$ of n Chebyshev polynomials, where n is the order of the fit and ϕ_i is defined in the angular domain $[1°, 360°]$.

The Chebyshev orthogonal polynomials are defined as:

$$T_k(\phi) = \cos[k \arccos(\phi)] \tag{9}$$

with $k = 1, 2, 3, \ldots$ and $\phi = 1°, \ldots, 360°$.

Thus, $g(\phi_i)$ can be expressed as:

$$g(\phi_i) = \sum_{k=0}^{n} c_k T_k(\phi_i). \tag{10}$$

The coefficients c_k are estimated adopting a least squares approach, i.e., they are computed by minimizing the following error:

$$\epsilon = \sum_{\phi_i=1}^{360} [g(\phi_i) - h(\phi_i)]^2. \tag{11}$$

If the fit order value is chosen as sufficiently small with respect to the number of histogram bins, e.g., taking the first few Chebyshev polynomials to approximate g, the fit operation returns a continuous curve that smoothly follows the input signal (red dotted curve in Figure 12).

Each point in the fit curve has an associated uncertainty which depends on the statistical hypothesis adopted to model the signal capture process.

According to the assumed hypothesis, the pixel values are samples of the related random distribution, hence, for every point on the fit curve, the statistical deviation from the mean value can be expressed by a numerical value σ_i, as in [20].

The computed deviation σ_i is exploited to decide whether a histogram value lies within a regular range or if it represents an anomaly.

Indeed, a histogram h_{p_j} is associated with a positive detection in case one or more of its points overshoot a given threshold. The threshold is defined point by point, since each value of the histogram is computed as the average of a varying number of pixels. For the i-th angle, the threshold has been placed at p times the corresponding standard deviation σ_i above the fit curve value. To obtain the purple and yellow curves in Figure 12, we adopted the same approach specified in [20] with $p = 3.5$. It has to be remarked that the choice of the optimal detection threshold represents a delicate decision in the processing chain. Indeed, it is not usually possible to have an a priori knowledge of the noise statistical properties, since it is typically modeled as a non-stationary space-varying speckle process.

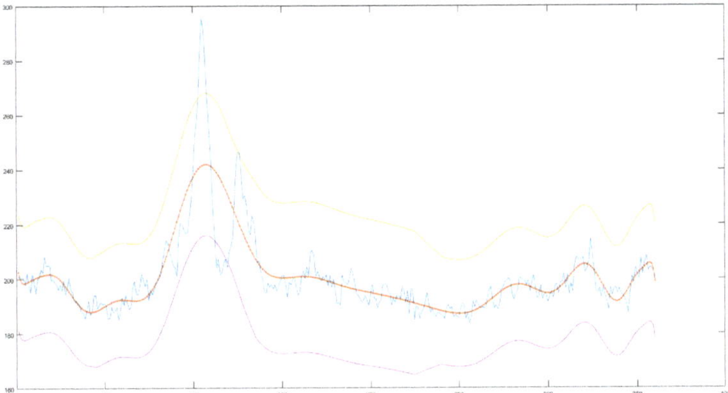

Figure 12. Results of the Chebyshev polynomial fitting applied to the histogram in Figure 11. The fit order n has been set empirically to 30.

Some exception rules have been implemented in order to avoid spurious signals from passing the detection stage. In case of SAR remote sensing, the backscatter signal captured by strong reflectors, such as the vessel metallic hull, is usually affected by typical cross-shaped artifacts (Figure 13, upper right). These artifacts typically feature low intensity values, but they become relevant in case the signal undergoes a preprocessing stage to enlarge the intensity range, e.g., by applying a gamma correction algorithm (see, for example, [18]). In these circumstances, spurious signals typically generate relevant peaks in the angular histogram, causing false positive detections.

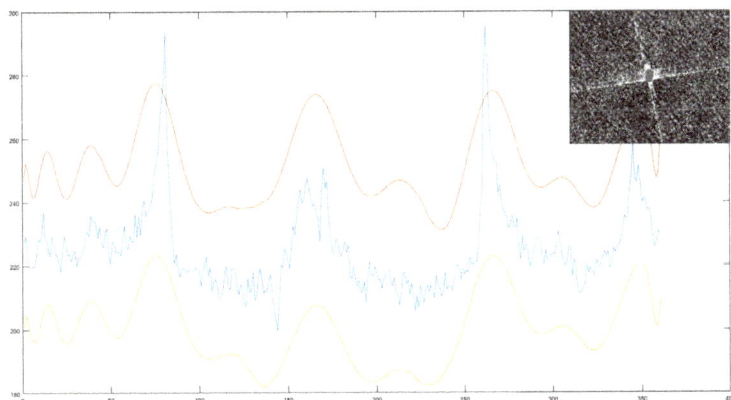

Figure 13. Angular histogram computed on the SAR crop in the upper right.

In order to avoid misleading pitfalls, the algorithm is enhanced by introducing dedicated rules, such as the exclusion of those candidates whose histograms exhibit peaks separated by specific angular distances that may relate to the presence of cross-shaped artifacts (e.g., multiples of 90°). An example of an angular histogram returned by the application of the wake detection pipeline to a signal affected by a star-shaped artifact is represented in Figure 13.

4.4. Gradient Based Wake Detection

In SAR imaging of moving vessels, a typical turbulent wake appears as a central dark line aligned with the ship longitudinal axis. Two bright linear features (narrow V-wakes) can appear within a half-angle of 1.5°–4°. In this section, we focus on the detection of the turbulent band by first computing the signal gradient through an approach specifically suited for speckle-affected signals, followed by the detection of the wake central component based on the methods discussed in Sections 4.2 and 4.3.

Indeed, we expect that the local gradient exhibits relevant variations along the direction orthogonal to the wake center axis (Figure 14). However, it is a proven fact (see [27]) that estimating the gradient through differences of standard masks (e.g., $[-1, 0, 1]$ or $[-1, 0, 1]^T$) results in low performances on speckle-affected data.

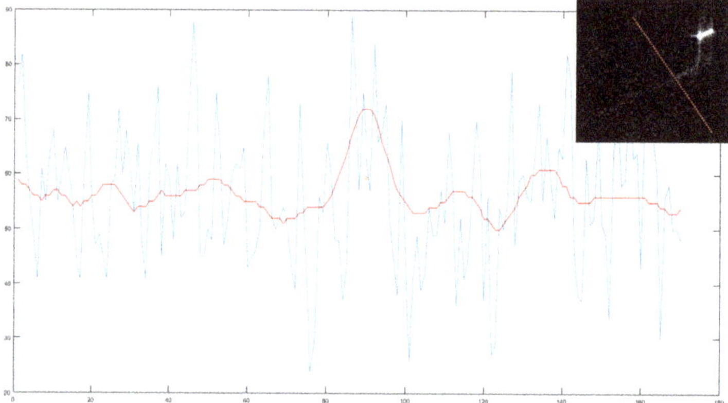

Figure 14. SAR signal sampled along a direction orthogonal to the turbulent component of the wake, faintly visible in the small crop. The original signal is represented in blue color (sampled along the dotted line in the upper right patch) while the red curve represents the result of a Gaussian lowpass filtering, applied to reduce speckle noise. The turbulent wake corresponds to the bell-shaped signal in the range $[80 \div 100]$.

In order to robustly estimate the signal gradient, we adopted the approach described in [28]. In the article mentioned, the signal derivative at a given point (x, y) is computed through the ratio of average (*roa*) estimator (see Figure 15), which is defined, for the horizontal and vertical cases, as

$$roa_h(x, y) = \frac{< I_L >}{< I_R >}, \quad roa_v(x, y) = \frac{< I_U >}{< I_D >}. \tag{12}$$

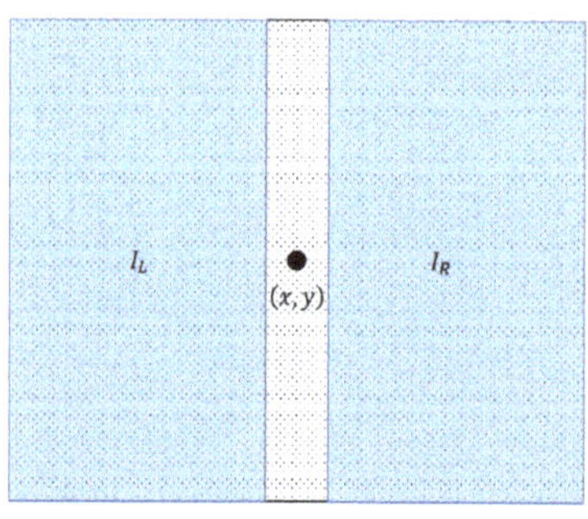

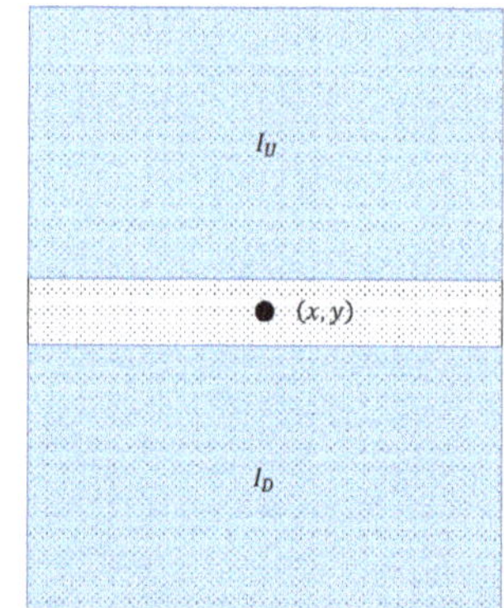

Figure 15. Ratio of average computation along the horizontal and vertical directions.

The horizontal and vertical *roa* correspond to the horizontal and vertical gradient components obtained as:

$$G_H(x,y) = log(roa_h(x,y)), \quad G_V(x,y) = log(roa_v(x,y)) \tag{13}$$

and the related magnitude and phase values

$$M_G(x,y) = \sqrt{G_H(x,y)^2 + G_V(x,y)^2}, \quad \theta_G(x,y) = \arctan G_V(x,y)/G_H(x,y). \tag{14}$$

Hence, gradient phase and amplitude are estimated, for every pixel in the image, by means of Equation (13). Eventually, in order to capture and isolate the maximum variation of the signal in the wake neighborhood, the estimated gradient is projected onto the direction orthogonal to the previously estimated main axis of the ship. The resulting gradient is finally processed by the wake detection method discussed in Section 4.2.

The proposed turbulent wake detector has been applied to the dataset in Figure 16. The detection of the turbulent band can be visually observed in Figure 17 while numerical validation is illustrated in Figures 18 and 19, where the estimated Azimuth displacement and turbulent wake orientation have been plotted versus the corresponding true values. In particular, the first set of diagrams represents the comparison between the results obtained by applying the proposed pipeline directly on the intensity maps (standard approach following [20], Figure 18), while the second set (Figure 19) accounts for the results obtained by preprocessing the intensity map by the described gradient estimator. It is relevant to consider the data dispersion around the red line (the $y = x$ line) in Figure 19, and observe the better performance of the proposed method w.r.t. the standard method (Figure 18).

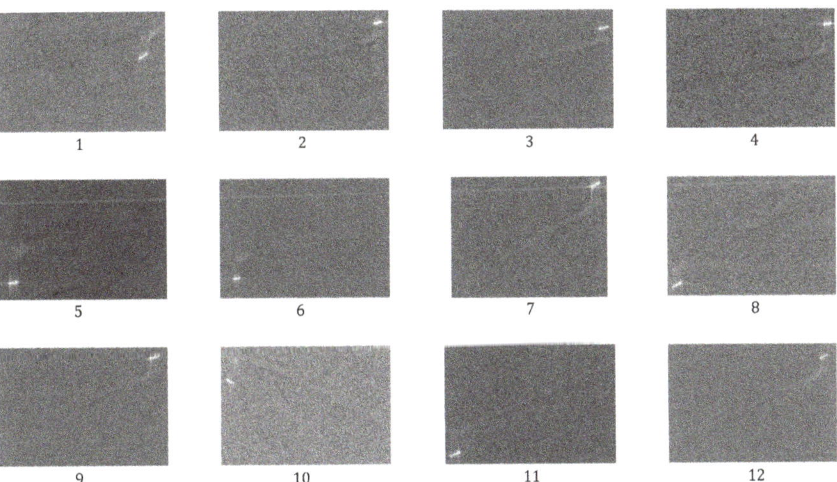

Figure 16. Dataset employed to test the turbulent wake detector (contains modified Copernicus Sentinel data, 2018).

4.5. Wake Analysis

In case a crop candidate passes all the detection phases, it is further processed to extract all the information that is considered of interest. First of all, the peaks that overshoot the $p\sigma$ threshold are considered wake components. As already stated in [2], in the most favourable case, the observable wake components are the central turbulent band, directed as the vessel route, and two external envelopes placed symmetrically at $\pm 19.47°$ w.r.t. the central component.

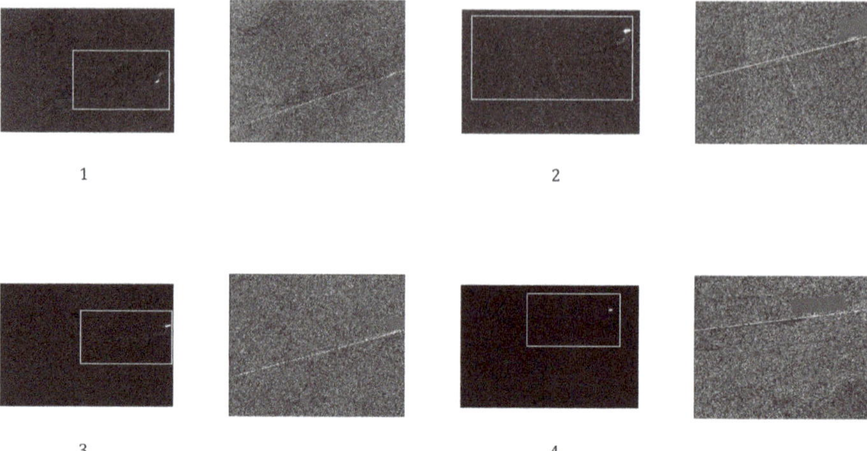

1   2

3   4

Figure 17. Results of the turbulent wake detector applied on the first four wakes from the dataset represented in Figure 16. Input images are identified by their ID number. Processing results of relevant areas (identified by white bounding rectangles in the input maps) are represented next to input images.

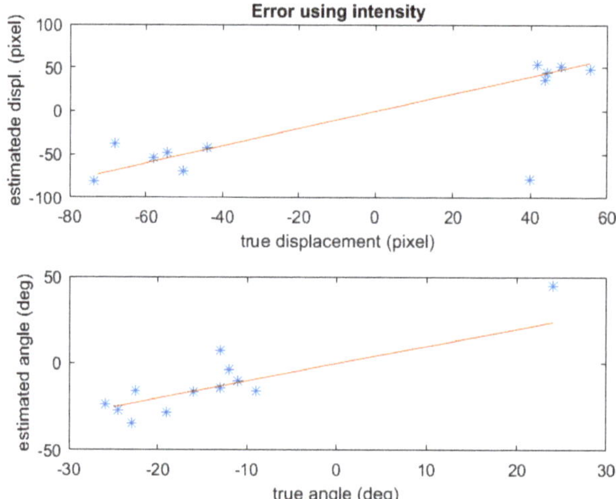

Figure 18. Standard method error diagram.

Since the vessel route coincides with the center band of the wake pattern, recognizing the turbulent component represents a crucial goal for the wake analysis task. In case the number of observed components is at a maximum, the route is defined as the bisector between the most external components, while, in case a single line is detected, this one will be automatically identified as the turbulent component. In the most ambiguous circumstance, i.e., when the algorithm detects two different lines, the adopted approach consists of selecting the component that exhibits the largest mean value. Since the central turbulent band of a wake usually features the largest backscattering signal, the largest peak in the angular histogram is finally labeled as the wake central orientation.

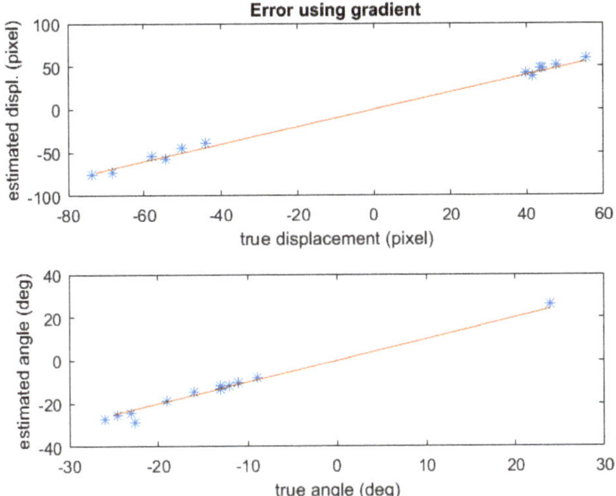

Figure 19. Proposed method error diagram.

After the vessel route computation, the wake pattern is considered for estimating the ship's speed. Indeed, the kinematics information intrinsically included in the wake pattern geometry can be exploited to estimate the vessel velocity according to the methods discussed in Section 2.

The analysis of the wake pattern has been performed on the ship target represented in Figure 1. For that candidate target, it has been possible to get reliable ground truth data from a commercial Automatic Identification System (AIS) provider. The analysis performed on the detected wake pattern provided the results in Table 1. Velocity estimates are provided according to the two implemented methods, i.e., the method exploiting the Azimuth shift (A.S.) and the one exploiting the spectral analysis on the external Kelvin wake (F.A.). Concerning the latter method, the Fourier analysis has been performed on three lines starting from the wake tip (Figure 20, left side). The result of the Fourier analysis provided three periodogram curves (Figure 20, right side). The blue curve clearly exhibits a peak corresponding to the wavelength of the periodical component.

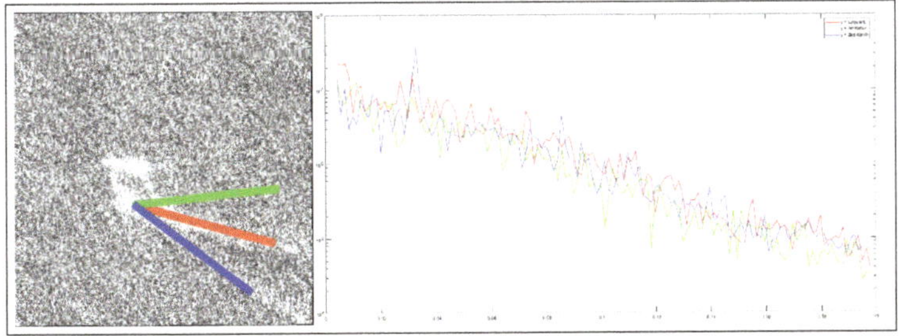

Figure 20. Linear sampling on the wake signal presented in Figure 1. The observed peak in the blue curve corresponds to the wavelength of the periodical wake component (COSMO-SkyMed Product—© ASI 2016 processed under license from ASI—Agenzia Spaziale Italiana, all rights reserved, distributed by e-GEOS)

As mentioned in Section 1, Kelvin wakes are rarely observed in SAR data and they typically feature low SNR values. Nevertheless, when they are successfully detected, it is possible, provided that the signal resolution is sufficiently large, to perform accurate estimations (see Table 1).

Table 1. Kinematics estimation output. R. stands for Route and V. for Velocity. The presented velocity results correspond respectively to the Azimuth Shift (A.S.V.) method and to the Fourier Analysis method (F.A.V.).

AIS R. (deg)	Estimated R. (deg)	AIS V. (m/s)	A.S.V. (m/s)	F.A.V. (m/s)
290	285	6.0	5.89	6.22

5. Conclusions

This document presents the implementation and the results related to a software procedure dedicated to the ship kinematics estimation task. This procedure takes as input remote sensing imagery and returns the estimated values of the vessel route and the vessel speed. This is primarily conceived to process radar imagery but can also be applied to optical data, provided some proper preliminary processing stages are introduced in order to enhance the wake pattern traces.

Wake patterns are hardly detectable in SAR maps, hence future developments will also be devoted to the refinement of the wake recognition process, based on the exploitation of additional information, such as the fine estimate of the vessel position as well as the constraints of this peculiar hydrodynamics problem, e.g., the theoretically expected wake angular aperture. A novel method for the detection of the wake has been introduced. Promising results (see Figure 17) obtained by processing the dataset in Figure 16 suggest that employing the proposed gradient-based approach may enhance the accuracy concerning the estimation of the wake motion's related features (see charts in Figures 18 and 19).

Taking inspiration from cutting edge literature of machine learning, novel prospects will also be devised, especially concerning the development of procedures for wake detection by means of neural networks based methods, also benefitting from the availability of large amounts of open access data (e.g., the ESA Copernicus Open Access Hub). For what concerns the estimation of the vessel's speed, novel methods are currently being investigated to evaluate their potential in terms of kinematics information extraction. In particular, along-track-interferometry techniques represent interesting tools for the purpose of estimating the line-of-sight velocity value through the analysis of single-look-complex SAR data. Moreover, the Doppler centroid of the SAR signal varies according to the kinematics of the backscatterer. Estimating the variation between the Doppler centroid of a moving object w.r.t. a stationary one provides an additional velocity estimation method, which sounds worthy of being further studied.

The presented platform is being currently tested within the framework of the OSIRIS (Optical/SAR data and system Integration for Rush Identification of Ship models), an ESA project with the main goal of developing a platform dedicated to sea surveillance, capable of detecting and identifying illegal maritime traffic. The main goal of this platform is to detect and identify target vessels within a given sea surface area, which is remotely supervised by orbiting satellites such as Sentinel 1/2, CosmoSKy-Med and EROS missions. Radar and optical images represent the main input data for the platform described. These are processed by a suite of algorithms which are sequentially applied to the data returning information about (i) the ship positioning within the inspected area, (ii) the main ship geometrical attributes, such as length overall, beam overall and heading, and (iii) the ship kinematics status represented by its velocity vector. Future developments will also be devoted to improving the overall performance of the platform by enhancing the accuracy of each individual stage in the processing pipeline.

OSIRIS will represent a new tool for combating unauthorized fishing, irregular migration and related smuggling activities.

Author Contributions: M.R. contributed to the design and development of the wake detector, to the ship kinematics estimation module and to the preparation of the paper, and L.B. contributed to the design and development of the presented methods through brilliant discussions and insightful suggestions.

Funding: This research has also been made possible thanks to the OSIRIS project, funded by ESA, grant number ESA-IPL-POE-SBo-sp-RFP-1008-2015.

Acknowledgments: This work has been partly possible thanks to the OSIRIS project (Optical/SAR data and system Integration for Rush Identification of Ship models), funded by the European Space Agency (ESA). OSIRIS is carried out within the ESA General Support Technology funding Programme (GSTP).

Conflicts of Interest: The authors declare no conflict of interest.

References

1. Automatic Identification System. Available online: http://www.imo.org/en/OurWork/Safety/Navigation/Pages/AIS.aspx (accessed on 12 December 2018).
2. Reggiannini, M.; Bedini, L. Synthetic Aperture Radar Processing for Vessel Kinematics Estimation. In Proceedings of the International Workshop on Computational Intelligence for Multimedia Understanding, IWCIM 2017, Kos, Greece, 2 September 2017. Available online: https://tinyurl.com/y79l38o7 (accessed on 12 December 2018).
3. Reggiannini, M. *OSIRIS—Satellite SAR Imagery Processing for Vessel Kinematics Estimation*; Project Report, 2017-PR-013; CNR Internal Report: Pisa, Italy, 2017.
4. Lombardo, P.; Sciotti, M. Segmentation-based technique for ship detection in SAR images. *IEEE Proc. Radar Sonar Navig.* **2001**, *148*, 147–159. [CrossRef]
5. Beaulieu, J.M.; Touzi, R. Segmentation of textured polarimetric SAR scenes by likelihood approximation. *IEEE Trans. Geosci. Remote Sens.* **2004**, *42*, 2063–2072. [CrossRef]
6. Crisp, D.J. *The State-of-the-Art in Ship Detection in Synthetic Aperture Radar Imagery*; Report DSTO-RR-0272; Australian Government, Department of Defence: Canberra, Australia, 2004.
7. Allard, Y.; Germain, M.; Bonneau, O. *Harbour Protection through Data Fusion Technologies*; Springer Science + Business Media B.V.: Berlin, Germany, 2009; pp. 243–250.
8. Askari, F.; Zerr, B. *Automatic Approach to Ship Detection in Spaceborne Synthetic Aperture Radar Imagery: An Assessment of Ship Detection Capability Using RADARSAT*; Technical Report SACLANTCEN-SR-338; SACLANT Undersea Research Centre: La Spezia, Italy, 2000.
9. Bedini, L.; Righi, M.; Salerno, E. Size and Heading of SAR-Detected Ships through the Inertia Tensor. *Proceedings* **2018**, *2*, 97. [CrossRef]
10. Tuck, E.O.; Collins, S.I.; Wells, W.H. On Ship Wave Patterns and Their Spectra. *J. Ship Res.* **1991**, *15*, 11–21.
11. Crawford, F.S. Elementary Derivation of the Wake Pattern of a Boat. *Am. J. Phys.* **1984**, *51*, 782–785. [CrossRef]
12. Tunaley, J.K.E. The Estimation of Ship Velocity from SAR Imagery. In Proceedings of the 2003 IEEE International Geoscience and Remote Sensing Symposium, Toulouse, France, 21–25 July 2003; Volume 1, pp. 191–193.
13. Scherbakov, A.; Hanssen, R.; Vosselman, G.; Feron, R. Ship wake detection using Radon transforms of filtered SAR imagery. *Proc. SPIE* **1996**, *2958*, 96–106.
14. Graziano, M.D.; D'Errico, M.; Rufino, G. Ship Heading and Velocity Analysis by Wake Detection in SAR Images. *Acta Astronaut.* **2016**, *128*, 72–82. [CrossRef]
15. Eaton, J.W.; Bateman, D.; Hauberg, S.; Wehbring, R. GNU Octave Version 4.4.1 Manual: A High-Level Interactive Language for Numerical Computations. Available online: https://www.gnu.org/software/octave/doc/v4.4.1/ (accessed on 12 December 2018).
16. Thomson, W. On the Waves Produced by a Single Impulse in Water of Any Depth, or in a Dispersive Medium. *Proc. R. Soc. Lond.* **1887**, *42*, 80–83. [CrossRef]
17. Rabaud, M.; Moisy, F. Ship wakes: Kelvin or Mach angle? *Phys. Rev. Lett.* **2013**, *110*, 214503. [CrossRef] [PubMed]
18. Gonzalez, R.C.; Woods, R.E. *Digital Image Processing*; Pearson Prentice Hall: Upper Saddle River, NJ, USA, 2008.
19. Zilman, G.; Zapolski, A.; Marom, M. The speed and beam of a ship from its wake's SAR images. *IEEE Trans. Geosci. Remote Sens.* **2004**, *42*, 2335–2343. [CrossRef]
20. Eldhuset, K. An automatic ship and ship wake detection system for spaceborne SAR images in coastal regions. *IEEE Trans. Geosc. Remote Sens.* **1996**, *34*, 1010–1019. [CrossRef]
21. Krishnaveni, M.; Thakur, S.K.; Subashini, P. An Optimal Method for Wake Detection in SAR Images Using Radon Transformation Combined with Wavelet Filters. *Int. J. Comput. Sci. Inf. Secur.* **2009**, *6*, 066–069.

22. Chaillan, F.; Courmontagne, P. On the Use of the Stochastic Matched Filter for Ship Wake Detection in SAR Image. In Proceedings of the OCEANS 2006, Boston, MA, USA, 18–21 September 2006; doi:10.1109/OCEANS.2006.307122.

23. Argenti, F.; Lapini, A.; Bianchi, T.; Alparone, L. A Tutorial on Speckle Reduction in Synthetic Aperture Radar Images. *IEEE Geosci. Remote Sens. Mag.* **2013**, *1*, 6–35. [CrossRef]

24. Aubert, G.; Aujol, J.F. A Variational Approach to Remove Multiplicative Noise. *SIAM J. Appl. Math.* **2008**, *68*, 925–946. [CrossRef]

25. Nezry, E. Adaptive Speckle Filtering in Radar Imagery. In *Land Applications of Radar Remote Sensing*; Closson, D., Ed.; InTech: London, UK, 2014; doi:10.5772/58593. Available online: https://www.intechopen. com/books/land-applications-of-radar-remote-sensing/adaptive-speckle-filtering-in-radar-imagery (accessed on 12 December 2018).

26. Cheney, W.; Kincaid, D. *Numerical Mathematics and Computing*; Brooks/Col: Monterey, CA, USA, 1980.

27. Dellinger, F.; Delon, J.; Gousseau, Y.; Michel, J.; Tupin, F. SAR-SIFT: A SIFT-like Algorithm for SAR Images. *IEEE Trans. Geosci. Remote Sens.* **2015**, *1*, 453–466. [CrossRef]

28. Song, S.; Xu, B.; Yang, J. SAR Target Recognition via Supervised Discriminative Dictionary Learning and Sparse Representation of the SAR-HOG Feature. *Remote Sens.* **2016**, *8*, 683. [CrossRef]

Article

Study of Algorithms for Wind Direction Retrieval from X-Band Marine Radar Images

Hui Wang [1], Haiyang Qiu [1,*], Pengfei Zhi [1], Lei Wang [2], Wei Chen [1], Rizwan Akhtar [1] and Muhammad Asif Zahoor Raja [3]

[1] School of Electronics and Information, Jiangsu University of Science and Technology, Zhenjiang 212003, China
[2] State Key Laboratory of Information Engineering in Surveying, Mapping and Remote Sensing, Wuhan University, Wuhan 430079, China
[3] Department of Electrical and Computer Engineering, COMSATS University Islamabad, 43600 Attock Campus, Pakistan
* Correspondence: hy.qiu@just.edu.cn; Tel.: +86-1889-665-8699

Received: 29 April 2019; Accepted: 5 July 2019; Published: 8 July 2019

Abstract: After decades of research, X-band marine radars have been broadly used for wind measurement. For retrieving the wind direction based on the wind-induced streaks, a lot of effort has been expended on three celebrated approaches—the local gradient method (LGM), the adaptive reduced method (ARM), and the energy spectrum method (ESM). This paper presents a scientific study of these methods. The contrast of retrieving the real measured marine radar images and vane measured results is evaluated, in perspective of the error statistics and algorithm operation efficiency. Interference factors, such as the historical information of the measured area, reference wind speed, and sea condition showing in the monitoring equipment are also concerned. The tentative results showed that LGM is robust, which can be implemented in most radar images, because it allows for a lower selection of requirements compared with the other two methods. For ARM, the better retrieval performance is a tradeoff with extra computation, which is expensive. ESM is superior to the other two algorithms in terms of accuracy and computation load; however, this algorithm is sensitive in rain-contaminated radar images, meaning it is a good choice for data post-processing in the lab.

Keywords: marine radar; wind direction retrieval; small wind streak; local gradient method; adaptive reduced method; energy spectrum method

---

1. Introduction

A near-surface wind field is formed by the horizontal movement of air relative to the sea surface, which is an important driving force for marine dynamics. Wind field parameters are imperative for the safety of navigation and marine engineering, necessary for military activities such as vessels' or ships' voyages. Conventional wind direction measurement methods are classified into two categories—site-based measurement and remote sensing. Site-based measurement can obtain ocean wind direction information through a wind vane [1], however this method may be influenced by the surrounding environment, such as platform movement or vile weather. Additionally, it is difficult to mount a vane in a remote ocean area, and to have a wind field error of up to 10% [2]. At present, remote sensing mainly depends on spaceborne/airborne microwave scatterometers [3], SARs [4], and so on. These devices can capture a large area of ocean wind field information, however, as they are restricted to a low spatial resolution, it is difficult for them to consistently monitor sea surface wind, and they are vulnerable to weather or other defects at the same time [5].

In order to make up for the deficiency existing in remote sensing measurement, low-cost marine X-band radars, with the advantage of high resolution and timely feedback, have been broadly used in many monitoring scenarios. X-band radar image sequences have previously been used to measure two-dimensional wave spectra in order to retrieve sea wave components [6,7]. The image sequences of the ocean surface can be utilized for target identification [8], detection [9], mean near-surface current measurement [10], and current field measurement [11]. Specially, X-band radar image sequence is becoming the state-of-the-art for wind field retrieving [12].

By operating at a low grazing angle, the transmitting signal can be reflected by the sea surface to form backscatter, which is mainly caused by sea surface roughness. Both a long surface gravity and local wind can modulate the roughness, thus generating radar images [13]. Particularly, small-scale roughness is highly dependent on local wind; hence, the wind direction can be retrieved accordingly, based on collecting marine radar images.

From a graphic perspective, the normalized radar cross section (NRCS), indicating an intensity level of radar echo, keeps a high correlation with the wind speed and direction. NRCS has only one peak in the upwind direction, and a minimum in the crosswind direction—simultaneously, it is an exponential function of wind speed [14]. According to these characters, two key approaches are posed for wind direction retrieval from X -band radar image sequences [15].

The first one is based on the relationship of NRCS and the radar look direction—this approach necessitates the azimuth being equal or greater than 180°, which is difficult to realize for shore-based radars because of the sheltering of land [16]. Recently, Lund et al. [17] and Vicen Bueno et al. [18] utilized the same principle to retrieve wind field information with independent platform movement. However, the abovementioned two methods neglect the following crucial factors: in cases of low sea states' circumstances—the obstruction of the radar field and the appearance of islands—the wind measurement errors will be increased [19]. Although the latter method has been improved by Ying Liu et al. [19], it cannot be applied to local X-band radar images or to a sheltering of land greater than 180° [20].

The second approach is based on small-scale streaks modulated by the wind in radar images—the determined wind directions as the orientations of the wind streaks that are approximately aligned with the mean surface wind direction. A large amount of literature has proven that this method has the advantage of high precision, fast operation, and is robust in a variety of circumstances, and it has thus gradually dominated X-band sea environmental monitoring [21–23]. There are quite a few algorithms for small-scale streak retrieving, including, but not limit to, the local gradient method (LGM), the adaptive reduced method (ARM), the energy spectrum method (ESM), and the optical flow-based motion (OFM). Because a wind field is in the presence of static signals as a result of the radar image sequence, the outcome of the OFM algorithm has a lower accuracy that will not meet the demand of an actual sea monitoring project [24]; thus, we dropped this approach for this paper. LGM needs to iteratively smooth and subsample the wind streaks' images three times in order to obtain a so-called Gaussian pyramid. From each of these images, the local wind directions are computed using the standard local gradient [25]. ARM puts forward an adaptive reduction operator, which means that the algorithm can automatically select the reduction operator with a different step size according to the judgment result, and the local wind directions are computed using the adaptive local gradient [26]. ESM is proposed for extracting the wind direction through wind streaks' image energy spectrum—the wind direction gained directly from the energy spectrum image based on the energy concentrated areas perpendicular to the wind streaks [27]. To the authors' knowledge, there is limited literature making an explicit comparison of the three algorithms using real-time collected radar data. The main contribution of this paper lies in that we, in detail, analyze and compare the three algorithms' performance error statistics and operating efficiency, and simultaneously consider the influence factors, which are helpful and vital for engineering realization problems.

The data for analyzing and researching is gathered by the HEU wave monitoring system, which is developed by the HEU marine monitoring research group [28,29]. The remainder of the paper is organized as follows. Section 2 presents a preliminary introduction of wind feature extraction from a small-scale wind streak. In Section 3, we give a review and contrast of LGM, ARM, and ESM.

These methods are analyzed and discussed with the testing data in Section 4. Finally, we draw a conclusion in Section 5.

2. Data and System Model Overview

The above-mentioned algorithms cannot be directly used for the collection of raw radar data, as the mapped objective is a small-scale wind streak. This section will give an overview of the data collection, and will introduce the data pre-processing in terms of rain recognition and image filtering, in order to extract small-scale wind streaks initially.

2.1. Data Overview

The data used for testing was collected from the Pingtan General Testing Ground, an institution in the south of Haitan Island, the Fujian Province of China. Haitan Island is the fifth largest Chinese sea island, located between the Taiwan Strait and Haitan Strait. The average depth of the radar-measured sea area is about 28 m; this area often happens in high-sea conditions due to terrain and strong storms. The measured tidal level difference is up to 5 m, accompanied by an average wind speed of about 6.9 m/s. The average strong-wind (over seven level) days are nearly 125 days a year. Strong storms happen about 6.3 times a year, which indicates that the ground is one of the strongest wind areas in Fujian Province. The test site can provide abundant data for algorithm analyzing. Figure 1 shows how the HEU wave monitoring system collects and records data at this site, in which a standard RM-1290 marine X-band radar is mounted on top of a platform tower.

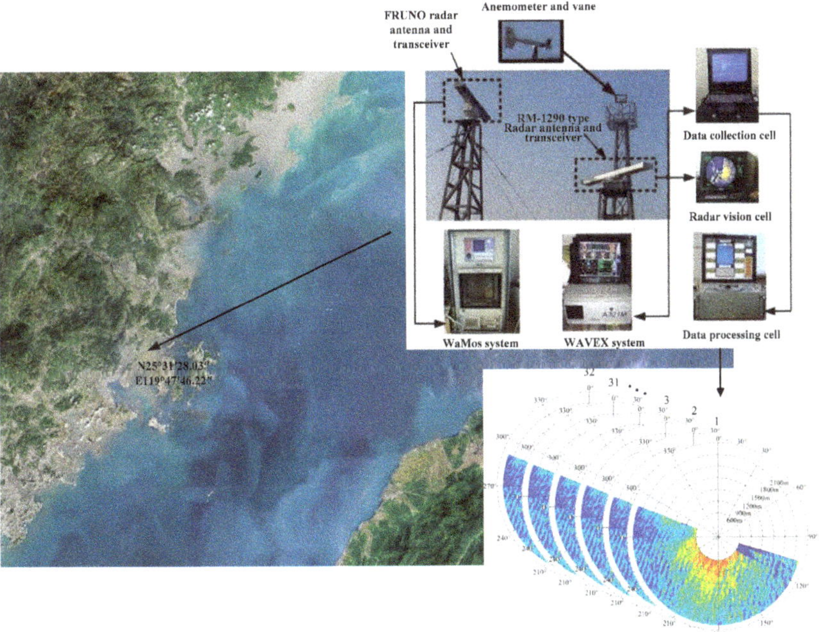

Figure 1. The testing site as well as the monitoring equipment.

The radar transceiver signal is replicated into triplet form—the replicated signal is sent to the data acquisition unit, radar display unit, and data processing unit, respectively. For reference, another piece of sophisticated equipment, WAVEX, works simultaneously, so as to measure the wave height and sea current direction. The X-band marine radar operates at a low grazing incidence with horizontal polarization in the transmitting and receiving process—it belongs to a sort of shot pulsed radar with a pulse width of 0.5–10 µs. The radar operates at 9.4 GHZ, with a sampling frequency of 20 MHz The specifications detail of the RM-1290 marine radar are shown in Table 1.

Table 1. Specifications of the RM-1290 marine radar. HH—horizontally; RF—radiofrequency; IF—infrared.

Antenna Rotating Speed	26^{+2}_{-6} **rpm**
Gain	28.5 dB
Polarization	HH
Horizontal beam width	≤1.3° (antenna of 1.8 m)
Vertical beam width	23° ± 2
Radio frequency	9410 ± 30 MHz
RF pulse envelope width	Short pulse: $0.05^{+0.03}_{-0.02}$ µs
Pulse repetition frequency	Short pulse: 1300 (± 10%) Hz
RF pulse peak power	Short pulse: ≥16 KW
Receiver IF bandwidth	Short pulse: 20 ± 3 MHz
Distance resolution	≤23 m
Azimuth resolution	≤1.3°

The radar antenna rotatory period is 2.5 s (24 rpm), with a pulse repetition frequency of 1.3 kHz. Each of the radar-image sequences collected here consists of 32 images, with a time interval of 80 s. Using the azimuth-range bin and scales data, the monitoring system digitizes the radar backscatter intensity into a 14-bit, that is, digitized back-scatter intensity, in the range of 0 to 8192 ($2^{(14-1)}$), as shown at the right bottom corner in Figure 1.

2.2. Data Pre-Processing

Before using the radar data to retrieve the wind direction, a basic data pre-processing quality control procedure should be conducted. The procedure includes the rain recognition and image filter.

2.2.1. Rain Recognition

All of the methods work only for rain-free radar images, which have to been automatically selected depending on the rain recognition data pre-processing. Lund et al. [17] proposed using the zero-pixel percentage (ZPP) in the whole radar image in order to identify the rainfall radar image. Considering the influence of sea conditions on the statistical results, this paper will extend a method for rainfall image recognition based on the statistical results of ZPP in the occlusion area of a radar image. The ZPP is as follows:

$$P = \frac{f_0}{f} \tag{1}$$

where f is the total number of pixels in the radar image, and f_0 is the number of normalized zero intensity pixels.

No matter whether using a ship-based or shore-based navigation radar image, when the propagation route of the radar electromagnetic wave is blocked because of obstacles such as a chimney, mast, island, and so on, a large number of invalid signals will be generated in the radar image, resulting in a fanned shadow area (shielding area) in the radar image. When there is no rain, the shielding area is almost blue, and there is no radar echo. When there is rainfall, the raindrop can reflect the radar electromagnetic wave signal and generate uniform backscattering, so that the blocking area appears as flaky rain echo. As the zero-intensity percentage of the statistical occlusion area will not be affected by the change of sea conditions, it will be more effective and accurate to detect whether the radar image is affected by rainfall.

Figure 2 shows the occlusion zero pixel proportion (OZPP) distribution of 200 radar images, in which the selected radar region is 80–600 points in the radial direction, and 50–90 degrees in the azimuth direction. The rectangular area in the figure is the data of the rainfall in the radar image.

As can be seen from the figure, when there is no rain, the OZPP basically maintains a stable change range, very close to 1, because there is almost no radar echo at this time. When rainfall occurs, the percentage of zero-intensity in the shielding area decreases immediately by a large margin. Finally, according to the statistical results, 0.94 is selected as the threshold to detect whether the radar image is affected by rainfall. When the OZPP is greater than 0.94, it indicates that there is no rainfall in the detection area; when the OZPP is less than 0.94, it can be determined that there is rainfall in the detection area.

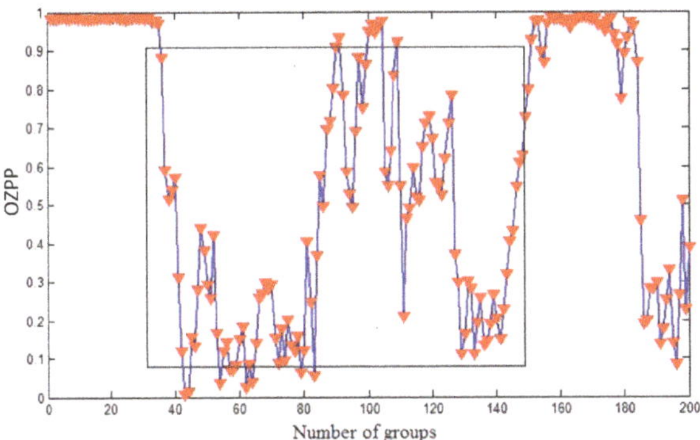

Figure 2. The OZPP distribution of 200 radar images.

2.2.2. Image Filter

In order to effectively remove the influence of the same frequency, a small amount of rainfall, and a small target on the radar image, it is necessary to filter the radar images to the extract the sea wind direction. While a linear filter will alter the intrinsic features of the radar images, a nonlinear median filter, which not only owns a superior denoising performance, but also has a protective effect on the image detail features, is more preferable in this paper. Median filtering is implemented by overlapping the slide window with the image; the gray value of the pixel in the window is arranged according to a monotonic rise or monotonic fall. A median filtered image is obtained by traversing the slide window over the entire image. It can be written as follows:

$$g(x, y) = median\{f(x - i, y - j), (i, j) \in W\} \tag{2}$$

where $g(x, y)$ represents the filtered image, $f(x - i, y - j)$ is the input image, and W represents the slide window. According to the resolution of the input data, the size of W is determined to be 3×3.

2.3. Small Scale Wind Streak Characteristics Extraction

Some streaks can form, evolve, and decay over relatively short lifetimes of only dozens of minutes, and then rapidly regenerate—so, it is named "small-scale wind streak" [30,31]. The light and dark stripes' characteristics modulated by wind fields exist in marine radar images. The experimental results show that the change of roughness caused by the turbulence of wind shear stress can be detected by marine radar, and these results can be reflected on the radar image in a stripe distribution [32]. The relationship between the near-surface wind vector and NRCS can be described by a geophysical model function (GMF), as described in [14]:

$$\sigma^0 = A(\theta)u^{\gamma(\theta)}(1 + B(u, \theta)\cos\Phi + C(u, \theta)\cos 2\Phi) \tag{3}$$

where σ^0 and u represent the NRCS and wind speed, respectively; Φ means the angle between the radar look and wind direction; and θ is the angle of grazing incidence. A, B, C, and γ, typically determined by the radar frequency and polarization mode, are empirical parameters. It is well known that NRCS represents the level of near-surface roughness, and the GMF model shows that there is a high correlation with the wind field. Thus, wind streaks possessing same distribution characteristics with near-surface roughness will exist in the X-band marine radar image. In other words, the wind streaks characteristics are consistent with the stripe-like turbulence characteristics, and exist in the marine radar image sequence, shown as light-, dark-, and small-scale forms. The visible streaks in the radar images hover around 200 m to 500 m; these streaks have been shown to be well aligned with the mean surface wind direction, with their frequency being close to static.

The small-scale wind streaks cannot be observed in a single radar image, because of the single observation lasting for only a few seconds, whereas the small-scale wind streaks have static features. The following formulation shows how to separate wind from other signatures according to the small-scale wind streak evolving period. Firstly, integrate a set of radar-image sequences (typically, 32 images representing 80 s duration of data) over time, and then construct an image space low-pass filter to average the radar image at the same position's pixel points, as follows [33]:

$$f(\theta, r) = \frac{\sum_{t=0}^{N_t} f(\theta, r, t)}{N_t} \tag{4}$$

where $f(\theta, r)$ represents the sea surface static feature image, including the numbers of the small scale wind streaks, which we are interested in; $f(\theta, r, t)$ is the single radar image at time t; θ and r are the radar image azimuth angle and radial radius; and N_t is the sum of the sequence. In the following, we will compare three kinds of sea surface wind retrieval algorithms based on $f(\theta, r)$.

3. Review and Illustration of LGM, ARM, and ESM

In this section, we firstly give a review of three different wind-direction retrieval methods of the ocean surface based on small-scale wind streaks, and then a complexity comparison of the three algorithms is discussed.

3.1. The Traditional LGM

Under the assumption that small-scale streaks are supposed to be approximately parallel with the mean wind direction, LGM can retrieve the wind direction by using a space domain iterative to reduce and smooth; the flow chart of LGM is shown in Figure 3.

The reduce process in Step 4 can be computed, depending on the Gaussian pyramid (GP), as follows:

$$F_{G1} = C_{(\downarrow 2)} F \tag{5}$$

$$C_{(\downarrow 2)} = \frac{1}{4} \begin{bmatrix} 1 & 1 \\ 1 & 1 \end{bmatrix} \tag{6}$$

where $C_{(\downarrow 2)}$ is the GP reduce operator, and the subscript means that the sampling rate is 2. F denotes the original image, and F_{G1} is the first-time reduced image. F_{G1} needs to be smoothed in order to filter out the high-frequency signals, this process uses both fourth- and second-order Yang hui triangular filter matrixes, as follows:

$$H^4 = \frac{1}{256} \begin{bmatrix} 1 & 4 & 6 & 4 & 1 \\ 4 & 16 & 24 & 16 & 4 \\ 6 & 24 & 36 & 24 & 6 \\ 4 & 16 & 24 & 16 & 4 \\ 1 & 4 & 16 & 4 & 1 \end{bmatrix} H^2 = \frac{1}{16} \begin{bmatrix} 1 & 2 & 1 \\ 2 & 4 & 2 \\ 1 & 2 & 1 \end{bmatrix} \tag{7}$$

to formulate an operator for $R = H^2 C_{(\downarrow K)} H^4$, the image including the wind streaks (sea surface static feature image) has firstly been denoted to be as follows: $F_{R1} = RF_{G1}$; the smooth process does not change the image resolution. For an equivalent of the wind steak scale, the upper entire process, including reducing and smoothing, has to repeated three times, and finally, we obtain F_{R3}.

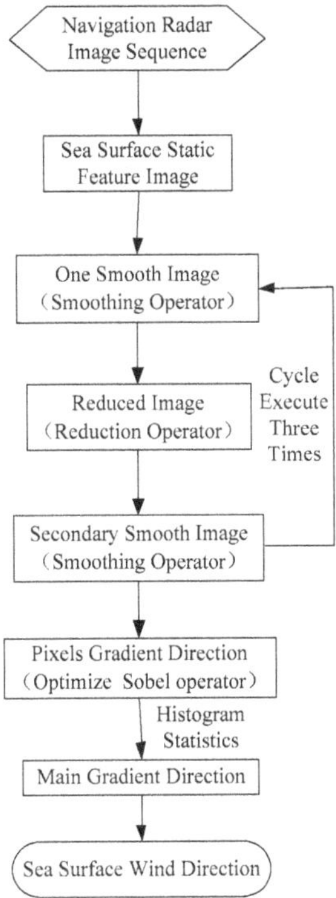

Figure 3. The flow chart of the local gradient method (LGM).

With F_{R3}, the gradient direction of the wind streaks image can be calculated by applying the optimized Sobel operator along the x and y axes. Then, we have $W_X = D_X F_{R3}$ and $W_Y = D_Y F_{R3}$, and the optimized Sobel operator is as follows:

$$\begin{cases} D_x = \frac{1}{32}\begin{bmatrix} 3 & 0 & -3 \\ 10 & 0 & -10 \\ 3 & 0 & -3 \end{bmatrix} \\ D_y = D_x^T \end{cases} \tag{8}$$

Hence, the main sea wind direction can be obtained by the normal main gradient direction. Danker [21] has verified the effectiveness of this method, but he did not consider whether the subsampled image resolution could march with the streaks scale or not. This may lead to a lower accuracy, with some data being unavailable.

3.2. The Improved ARM

On the basis of the LGM algorithm, ARM uses an adaptive reduction operator, and this means that ARM can automatically select the reduction operator based on the loop index of K, and then determine the dimension size of the slide window. The flow chart of ARM is shown in Figure 4. The adaptive reduction operator is shown, as follows:

$$C_{\downarrow K} = \frac{1}{K}\left.\begin{bmatrix} 1 \\ 1 \\ \vdots \\ 1 \end{bmatrix}\right\} K \times \frac{1}{K}\underbrace{[\; 1 \quad 1 \quad \cdots \quad 1\;]}_{K} = \frac{1}{K^2}\left.\begin{bmatrix} 1 & 1 & \cdots & 1 \\ 1 & 1 & \cdots & 1 \\ 1 & 1 & \ddots & \vdots \\ 1 & 1 & \cdots & 1 \end{bmatrix}\right\} K \tag{9}$$

where $C_{(\downarrow K)}$ is the reduction operator of the reduced rate K; according to the marine radar resolution, the rate can be determined as $K = 2, 3, \ldots 8$, in which scope maximum value is 8. The flow chart of ARM is shown in Figure 4. The reduce process using $C_{(\downarrow K)}$ can be similarly computed, depending on the Gaussian pyramid (GP), as follows:

$$F_{GK} = C_{(\downarrow K)}F \tag{10}$$

The image uses H^4 to smooth before and after the reduction so as to filter out the high frequency noise, and then formulate an operator for $R = H^2 C_{(\downarrow K)}H^4$; the image including the wind streaks has been denoted to be $F_{RK} = RF_{GK}$, and the smoothing process does not change the resolution of the image. For a match of the wind steak scale, the upper entire process, including the reducing and smoothing, has to repeat N times, and then we obtain F_{RK}. The gradient of each pixel along the x and y axes can be written as follows:

$$\begin{cases} W_x = F_{RK} * D_x \\ W_y = F_{RK} * D_y \end{cases} \tag{11}$$

Thus, the gradient direction W_θ of each pixel is obtained, as follows:

$$W_\theta = \arctan\frac{W_y}{W_x} \tag{12}$$

Histogram statistics were performed for all of the pixel gradient directions. From the probability distribution map of the gradient direction, we could get the direction with the maximum frequency $W_{\theta_{max}}$. In order to calculate the stability factor, we selected all of the gradient directions in the range of $\lambda W_{\theta_{max}} \sim W_{\theta_{max}}$, where λ is a scale parameter, and can be chosen depending on the practical situation. In this paper, we choose $\lambda = 0.7$, and the array of direction consisted of $W_P = \left[W_1, W_2, \cdots W_p\right]$.

ARM uses an optimal reduction rate and stability coefficient to determine when the image reduction process should stop. It is upon this that the reduced image resolution and scale of the small-scale wind streaks can achieve the best ratio. The stability coefficient (η) can be calculated, as follows:

$$\eta = \frac{\sqrt{\frac{1}{p-1}\sum_{i=1}^{p}\left(W_i - \frac{1}{p}\sum_{i=1}^{p}W_i\right)^2}}{\frac{1}{p}\sum_{i=1}^{p}W_i} \times 100\% \tag{13}$$

where, W_i is i-th pixel gradient value of the reduced image, $i = 1, 2, 3, \cdots p$ and p is the amount of pixels. The stability coefficient (η) means the aggregation density of the gradient direction W_i; if η is small, the distribution of gradient direction is dense, and vice versa. A dense gradient direction can better reflect the wind streak information, while a sparse one indicates an implicit interference of another signal. Figure 5 shows the reduction rate for the $K = 2$ and $K = 7$ gradient direction probability distribution on a set of collected radar data. The maximum probability of the gradient direction is 0.012 when the reduction rate is $K = 2$. The maximum probability of the gradient direction is 0.025 when the reduction rate is $K = 7$. So, $K = 7$ is the optimal reduction rate in ARM. At the optimal reduction rate, the reduced image resolution more closely matches the small-scale wind streak. From the data, it is shown that the ARM result for the wind direction is about 37°, and the anemometer result for the wind direction about 35°.

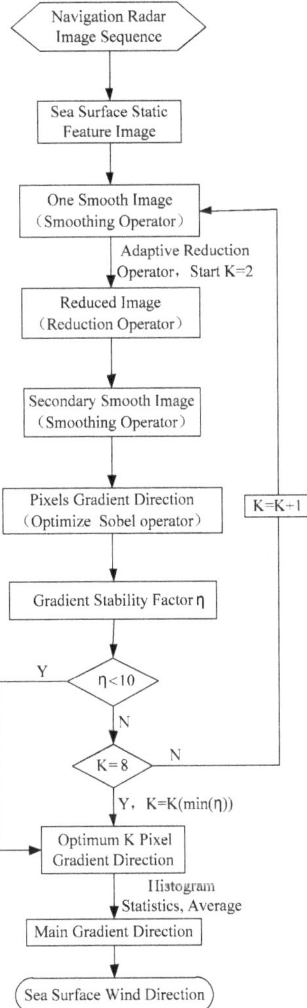

Figure 4. The flow chart of the adaptive reduced method (ARM).

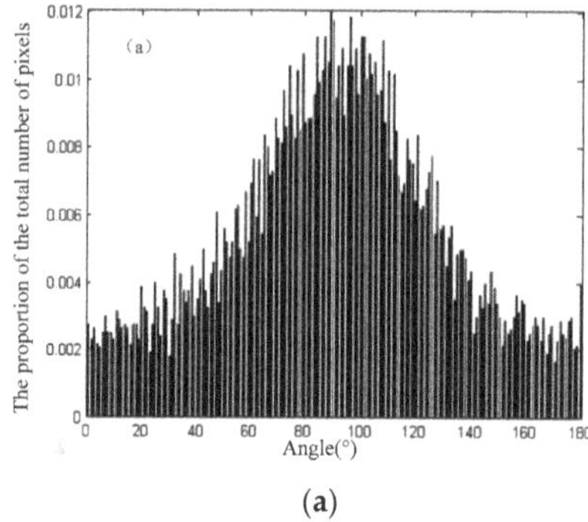
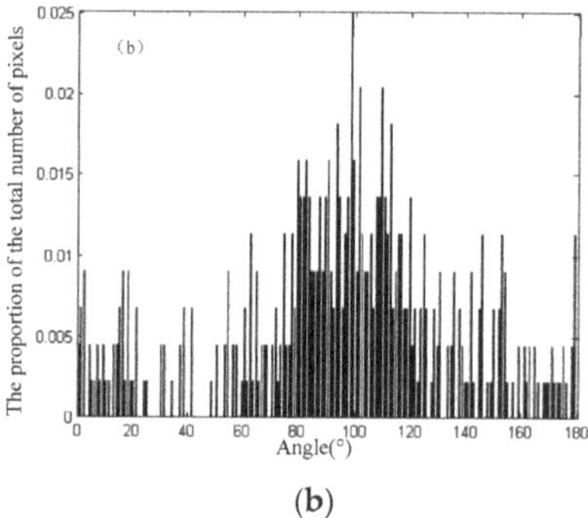

(a)

(b)

Figure 5. (a) $K = 2$: gradient direction probability distribution. (b) $K = 7$: gradient direction probability distribution.

3.3. The New ESM

Unlike retrieving the wind direction in the spatial domain, the ESM algorithm is posed in order to extract the wind direction through the SAR image energy spectrum. The flow chart of the ESM is shown in Figure 6.

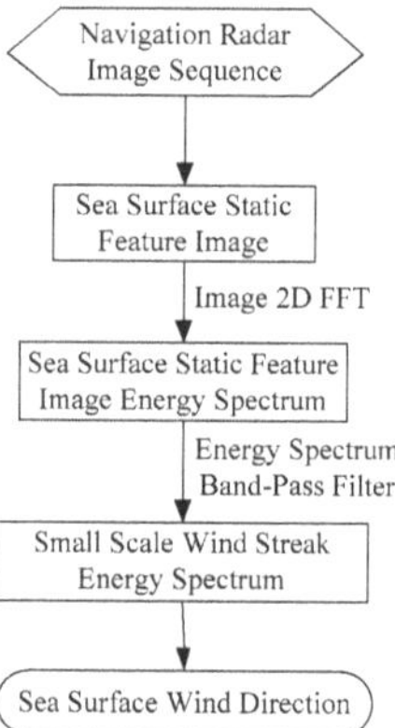

Figure 6. The flow chart of the energy spectrum method (ESM).

The two-dimensional image is composed of sins and cosines signals with different frequencies, amplitudes, and phases. Thus, the energy spectrum distribution of the 2D image can be obtained by 2D FFT. The mathematical model of the 2D FFT for the two-dimensional sea surface static feature images, $g(x, y)$, in Cartesian coordinates, is as follows:

$$F(k_x, k_y) = \sum_{i=1}^{N_x} \sum_{j=1}^{N_y} g(x_i, y_j) \exp\left[-i * 2\pi\left(\frac{x_i * k_x}{\max(x_i)} + \frac{y_j * k_y}{\max(y_j)}\right)\right] \tag{14}$$

where, $F(k_x, k_y)$ is the Fourier coefficient of $g(x, y)$, (k_x, k_y) is the frequency domain coordinates, and N_x and N_y are the pixel elements of $g(x, y)$ in the space domain, respectively. In the formula, the complex exponential term can be expanded as follows:

$$\exp\left[-i * 2\pi\left(\frac{x_i * k_x}{\max(x_i)} + \frac{y_j * k_y}{\max(y_j)}\right)\right] = \cos 2\pi\left(\frac{x_i * k_x}{\max(x_i)} + \frac{y_j * k_y}{\max(y_j)}\right) + i \sin 2\pi\left(\frac{x_i * k_x}{\max(x_i)} + \frac{y_j * k_y}{\max(y_j)}\right) \tag{15}$$

with the relationship of the wavenumber and wavelength ($k_x = \frac{2\pi}{\max(x_i) * d_x}$ and $k_y = \frac{2\pi}{\max(y_j) * d_y}$, respectively), d_x and d_y are the radial and amplitude resolution of the radar image, respectively. As $F(k_x, k_y)$ has both an imaginary and real part, using Equation (14), we can obtain the energy spectrum $A(k_x, k_y)$ of the static characteristic image, $g(x, y)$, as follows:

$$A(k_x, k_y) = \sqrt{\left[\mathrm{Re}(F(k_x, k_y))\right]^2 + \left[\mathrm{Im}(F(k_x, k_y))\right]^2} \tag{16}$$

where $\mathrm{Re}(F(k_x, k_y))$ is the real part of $F(k_x, k_y)$, and $\mathrm{Im}(F(k_x, k_y))$ is the imaginary part of $F(k_x, k_y)$. ESM uses a small-scale wind streak band-pass filter to extract the small-scale streak energy spectrum. This band-pass filter can be shown as follows:

$$I(k_x, k_y) = \begin{cases} A(k_x, k_y)(\sqrt{k_x^2 + k_y^2} \in [|k_d|, |k_t|]), \\ 0(\text{else}) \end{cases} \tag{17}$$

where $I(k_x, k_y)$ is the energy spectrum of the wind streaks, and $k_d = \frac{2\pi}{L_t}$ is the wind streak energy spectrum wave number low-limit, where L_t is the scale upper-limit. Similarly, $k_t = \frac{2\pi}{L_d}$ is the upper-limit, where L_d is the scale low-limit. The small-scale wind streaks wave number energy spectrum is shown in Figure 7.

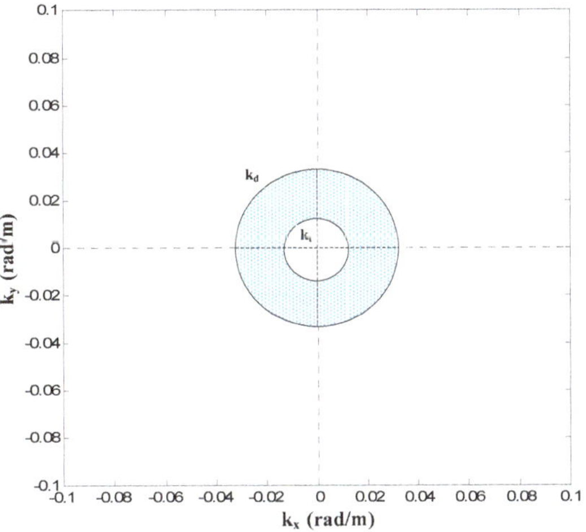

Figure 7. Small-scale wind streak energy spectrum.

As mentioned above, wind direction is retrieved based on the feature that small-scale streaks are assumed to be approximately parallel with the mean wind direction. We can find that the connection line of two energy concentrated areas is perpendicular to the wind streak, and then perpendicular to the wind direction. Figure 8 shows the filtered wind streak energy spectrum from the previously mentioned data set. The inner dotted circle represents the upper-limit, correspondingly, the outer circle represents the low-limit—both are labeled in this figure. The wind direction is perpendicular to the connection line of two energy concentrated areas. From this calculation, it is possible to obtain a vertical line of 36° and 144°, with respect to the north direction, and the 180° ambiguity problem was removed according to the quadrant comparison method. Therefore, this is an example of taking a radar image sequence and applying ESM to obtain the wind direction of 36°. It can be seen that the ESM result is far more agreeable with the anemometer data.

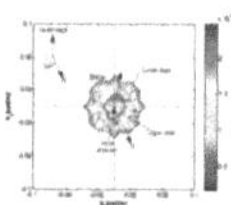

Figure 8. Wind direction retrieved from the filtered wind streak ES.

3.4. Complexity Analysis

The ARM is an improved variant of the traditional LGM. It utilizes the stability coefficient to judge the algorithm loop; recycles the adaptive reduction operator to reduce the image; and carries out two rounds of smoothing, before and after reduction. LGM employs a fixed number of judgments (three times); similarly, it applies a fixed two-order reduction operator to reduce the images, as well as applying the same smoothing process. The time complexity of the LGM is $5O(1)$(for the image processing) plus $O(n)$ (the gradient derivation), so the total algorithm complexity is $O(n)$. As a result of the increase in the calculation and cycle adaptive judgment, ARM takes longer than LGM; similarly, the time complexity is calculated as $kO(1)$ (k is random, with a maximum value of 8) plus $O(n)$, and as the ARM complexity equals LGM, both are $O(n)$.

The new ESM is used on the frequency domain characteristics to retrieve the wind direction. Compared with the other two methods, the computation time of ESM mainly lies in the two-dimensional discrete FFT transformation; however, there is no need to calculate the smoothing and shrinking process for each pixel in the image. At the same time, the small-scale wind band pass filter can be designed to completely separate the small-scale wind streak from the static feature image of the sea surface. Therefore, the information about the sea surface wind direction can be extracted more efficiently by using the feature of a small-scale wind streak parallel to the sea surface wind direction. However, time complexity of 2D FFT is $O(MN \log MN)$; in our algorithm, we made $M = N = 128$ (input image's resolution), so the entire time complexity comes down to $O(N^2 \log N)$.

Because the pixel gradients calculated by the two spatial algorithms (LGM and ARM) do not necessarily reflect the gradient directions of small-scale wind streaks, the inversion accuracy is not as good as the ESM algorithm. Nevertheless, ESM has a higher requirement for the size of the experimental area, and excessive measurement areas contain a lot of noise, which could increase the complexity of the wind streaks, while small measurement areas lead to incomplete wind streaks that will destroy the frequency domain characteristics of the wind steaks. In the application of ESM, experiments should be applied so as to select the appropriate research area, which increases the complexity of the algorithm. Additionally, the time complexity of EMS is a higher order than ARM and LGM, so this algorithm computation would be sensitive to the input image, meaning it will increase sharply with the growing input image resolution.

4. Testing and Results

4.1. Mathematical Metric for the Comparison

In order to measure the performance of the three inversion algorithms, this paper employs a metric of data error to analyze the errors between the inversion and reference wind direction. Let X_i be the time series of the inversion wind direction results, while Y_i refers to the reference wind direction time series, $i = 1, 2 \cdots N$ is the number of navigation radar image sequences used for the inversion of sea surface wind direction. We evaluated the performance of these algorithms by calculating the correlation coefficient, standard deviation, and bias between the inversion and reference wind direction. The calculation formula of these parameters are as follows:

$$A_i = Y_i - X_i \qquad\qquad \text{Error between the single retrieval and reference wind direction}$$

$$Bias = \overline{A} = \frac{1}{N}\sum_{i=1}^{N} A_i \qquad\qquad \text{Average deviation of two sequences}$$

$$Var = \frac{1}{N-1}\left(\sum_{i=1}^{N} A_i^2 - \frac{Bias^2}{N}\right) \qquad\qquad \text{Mean variance of two sequences}$$

$$\sigma = \sqrt{Var} = \sqrt{\frac{1}{N-1}\left(\sum_{i=1}^{N} A_i^2 - \frac{Bias^2}{N}\right)} \qquad\qquad \text{Standard deviation of two sequences}$$

$$r = \frac{\sum_{i=1}^{N}(X_i-\overline{X})(Y_i-\overline{Y})}{\sqrt{\sum_{i=1}^{N}(X_i-\overline{X})^2 \cdot \sum_{i=1}^{N}(Y_i-\overline{Y})^2}} \qquad\qquad \text{Correlation coefficients of two sequences}$$

where $\overline{X} = \frac{1}{N}\sum_{i=1}^{N} X_i$ is the average value of the inversion wind direction, and $\overline{Y} = \frac{1}{N}\sum_{i=1}^{N} Y_i$ refers to the average wind direction. Compared with the other parameters, r is much more powerful for demonstrating the similarity between two sets of data, so in this paper, we choose r as the uniform perform evaluation metric, while *Bias* and σ are simultaneously listed in following tables as the contrasts of the different algorithms.

The comparison of the operating times of the three algorithms is the average inversion time obtained by applying each algorithm to all of the data in the experiment. Because we focused on the post data processing, all of the data were tested using MATLAB software. Suppose t_i is one of the operating time series, then average operating time is as follows:

$$T = \frac{\sum_{i=1}^{N} t_i}{N} \tag{18}$$

4.2. Inversion Results

Figure 9 shows the comparison of the LGM and ARM wind direction from 180 sets of marine radar image-sequences with reference wind directions. Because the two algorithms are both operating in a space-domain, it would be feasible to put them together. The gray bars in the background are the wind speed measured by the anemometer Model-05103. We found wind speed ranges between 1.2–19.4 m/s. The results of ARM are more consistent with the reference wind direction, as shown in Figure 6, while the LGM curve shows a large jitter related to the reference wind direction, especially at a lower wind speed case, as shown from sampling points 29 to 64 in Figure 8.

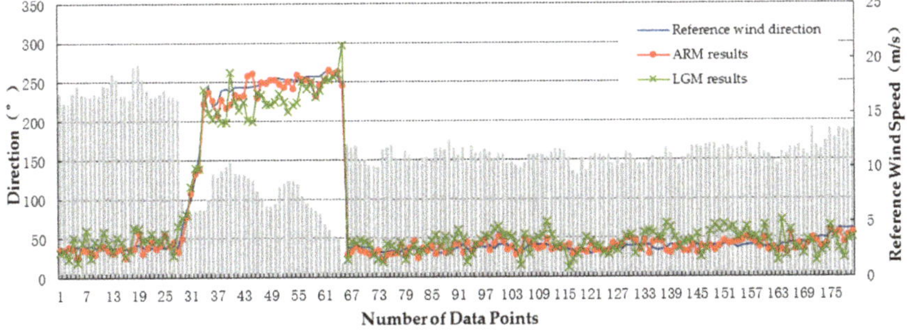

Figure 9. Results of the LGM and ARM algorithms compared with the reference wind directions.

It should be noticed that the discrepancy between LGM and the reference is larger than ARM. LGM's direction is always smaller than the reference in the numerical value, however, the results of LGM maintain a good consistency with the reference wind direction. Although there is a little jitter in the curve, it can quickly track the wind speed to get the ideal results. The comparison of LGM and ESM is in Figure 10.

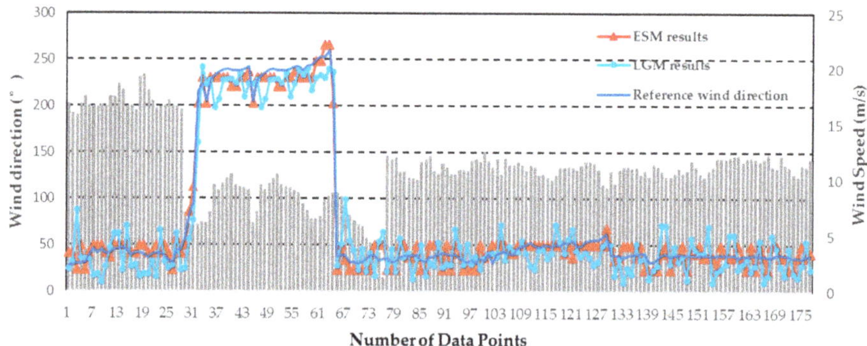

Figure 10. Results of the LGM and ESM algorithms compared with the reference wind directions.

It can be found that ESM contributes a similar performance to LGM, keeping an acceptable fluctuation trend as with LGM, and it can also track the reference values.

Table 2 shows the statistics of the error and running time between the retrieved and reference wind directions. From this information, in terms of the correlation coefficient, standard deviation, and deviation relative to the reference wind direction, we found that ARM is better than LGM and ESM. The correlation coefficient between ARM and the reference wind direction is 0.9956, which illustrates that it has a good consistency with the vane wind direction. The LGM coefficient is slightly inferior to ARM, while EMS establishes only 0.9523. At the same time, the mean differences and standard deviation of ARM are better than for LGM and ESM. the inversion accuracy of LGM and ESM are nearly the same. The last column of Table 2 shows that the average running times for processing a set of 32 radar images are 38.3, 27.1, and 20.5 s, respectively. EMS has the fastest operation time, nearly half that of ARM. From these results, we can conclude that ARM has a better retrieving accuracy compared with LGM and EMS, at the cost of a longer running time. EMS demonstrates a time efficient performance as a result of its spectrum character. LGM contributes a medium performance in the two aspects.

Table 2. Error and running time statistics between the retrieved and reference wind directions. ARM—adaptive reduced method; LGM—local gradient method; ESM—energy spectrum method.

Algorithm	Reference	Correlation Coefficient r	Standard Deviation σ (°)	Deviation Bias (°)	Average Running Time (s)
ARM results	Vane measured Wind direction	0.9956	7.62	−1.04	38.3
LGM results	Vane measured Wind direction	0.9832	17.33	1.18	27.1
ESM results	Vane measured Wind direction	0.9523	18.54	1.21	20.5

As ESM is sensitive to the measured area, for a deeper comparison of ARM and ESM, we picked 1494 sets of radar image sequences, and used both ARM and ESM to retrieve the wind direction, and to optimize an appropriate measurement area. Figure 11 shows the results of ARM and ESM compared with the reference wind directions. It can be found that both algorithms can provide satisfactory retrieving results in most cases, but a dramatic change exists with the reference wind directions between points 500 to 630, and the wind direction using even ARM cannot accurately reflect the wind direction fluctuation. In general, the ESM results are more identical to the reference, although there was always some jitter present in the green curve. By contrast, ARM maintained the up- and down-trend, especially when it comes upon an adverse wind direction, leading ineffective sampling points.

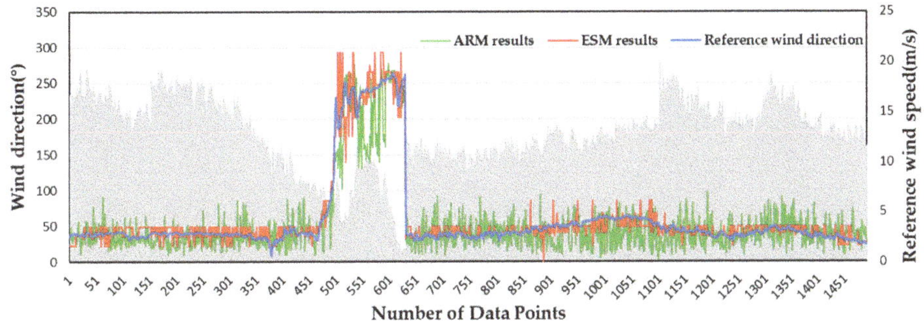

Figure 11. Results of the ARM and ESM algorithms compared with the reference wind directions.

Figure 12 shows the retrieving error distribution of the ARM (subplot-a) and ESM (subplot-b). Both of the retrieval algorithms have greater errors at a reference wind direction about 150°, because of the influence of typhoons and rain, but the ARM wind direction error is superior to that of ESM. The ESM wind direction errors range between −50° to 50°, whereas they are only −20° to 20° for the ARM errors. It can be inferred from the results that ESM is sensitive to rain-contaminated data, and the outlier points enlarge the retrieval range error for the data considered here.

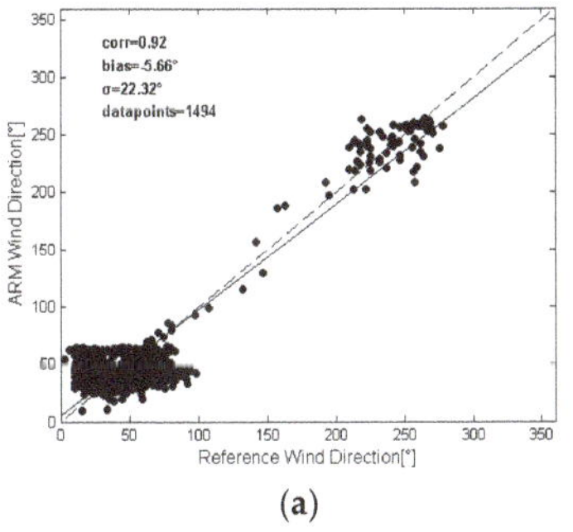

(a)

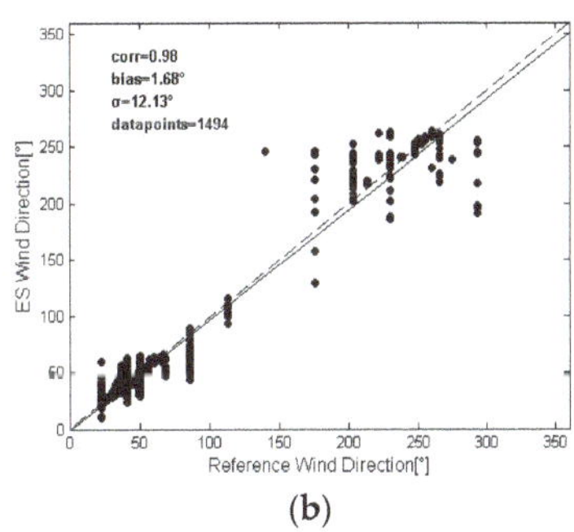

(b)

Figure 12. The scatterplot of ARM and ESM: (**a**) the scatterplot of ARM; (**b**) the scatterplot of ESM.

The statistics of the wind direction results' error and average execution times for the 1494 datasets utilizing the ARM and ESM algorithms are shown in Table 3. It is shown that with an appropriate measured area, the results of ESM are superior to ARM in correlation, bias, and standard deviation, although it appears to have many more outlier points in rain-contaminated circumstance compared with ARM. The average running time of ESM is about 7 s less than that of ARM.

Table 3. Main statistical parameters and time of the ocean wind direction error.

Results	Correlation Coefficient r	Deviation *Bias* (°)	Standard Deviation σ (°)	Average Running Time (s)
ARM results	0.92	−5.66	22.32	25.2
ESM results	0.98	1.68	12.13	18.4

4.3. The Influence Factors

It can be found that ARM performs better than LGM in most aspects. Meanwhile, the preceding experimental results show that ARM and ESM are all in line with the engineering requirements for accuracy and running time. In the following, we will discuss the key factors than can influence algorithm performance, particularly for spatial-domain ARM and frequency-domain ESM, when using real-time data.

4.3.1. The Influence of the Measurement Area Size

For testing the influence of the measurement area size for ARM and ESM, we choose data with almost the same reference wind directions, wind speeds, and sea conditions, in order to make the experiment immune to other factors of interference. WE selected 227 sets of radar image sequences lasting 14 hours on 22 October 2010. In these data, the reference wind direction range was about 31°–43°, the reference wind speed range was about 13.3–19.6 m/s, the average wave height was 3.5 m, and the average wave direction equaled 93°.

The object region area had a radius ranging from 600 to 2100 m, and the azimuth direction ranged from 106°–69° anticlockwise. The measurement area was selected so as to be a square, with a side length lower bound at 500 m and upper bound at 2100 m, because the wind streak scale was 200–500 m. Finally, we took three typical measurement areas of 1485 m × 1485 m, 960 m × 960 m, and 720 m × 720 m, in order to apply the real measured area data to test the influence of the measurement area size.

The influence of the measurement area size on ARM and ESM is shown in Table 4. Here, ARM shows the highest precision in the measurement area, at 1485 m × 1485 m, because ARM retrieves the wind direction based on the small-scale wind streak spatial domain characteristics. The larger the measured area, the more wind streaks, that will lead to the measurement areas have the higher probability of the gradient direction. It demonstrates that if a large measurement area is available, the ARM would be more preferable. It shows that ESM has the highest precision in the measurement area of 960 m × 960 m from the table. Excessive measurement areas contain a lot of noise, which could increase the complexity of the wind streaks, while small measurement areas lead to incomplete wind streaks that will destroy the frequency domain characteristics of the wind steaks. Therefore, the ESM algorithm requires the measurement area to be validated and preprocessed in order to guarantee its performance. However, the average running time is less than that of ARM by about 6 s.

Table 4. Statistical parameters of wind direction considering different areas.

Measurement Area		1485 m × 1485 m	960 m × 960 m	720 m × 720 m
ARM results (°)	Deviation(°)	3.1	13.3	−20.7
	standard deviation (°)	13.2	15.1	27.7
	Correlation coefficient	0.99	0.95	−0.95
	Average running time (s)	24.4	24.3	23.7
ESM results (°)	Deviation (°)	−4.4	−2.2	−4.9
	Standard deviation (°)	9.03	5.6	9.6
	Correlation coefficient	0.99	0.99	0.99
	Average running time (s)	18.8	18.2	16.8

4.3.2. The Influence of Sea Surface Wind Speed

Next, we began to investigate the influence of the sea surface wind speed on retrieving the wind direction. By comparing the error of the wind direction results obtained from the 1494 sampling points, the results of ARM and ESM are shown in Figure 13a,b, respectively.

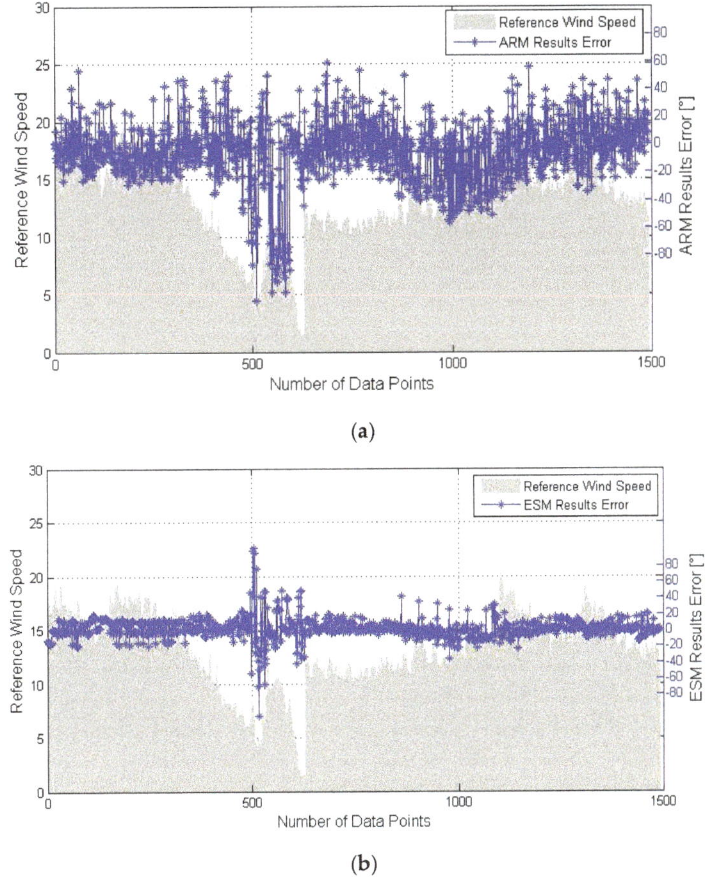

Figure 13. The influence of wind speeds on the error of ARM and ESM. (**a**) ARM; (**b**) ESM.

It can be found that both algorithms suffer large errors at about 500 sampling points, when the reference wind speed decreased sharply from 15 to 5 m/s. The errors became bigger because the small-scale wind streak was collapsed by the increasing ocean surface swell, due to the dramatic changes in wind speed. However, the ESM is much more robust against the uncertain interference. The main reason for this is that the ESM algorithm obtains the wind direction information from the wind streak energy spectrum. The energy spectrum band pass filter removes other streaks' features, which helps to eliminate the influence of the increased wind speed. The ARM algorithm obtains the wind direction information from the processing of small-scale wind streaks in the spatial domain, and the scale of the wind streak increases with the wind speed growth; hence, the characteristics of the reduced small-scale wind streak in the spatial domain will be lost as a result of the wind streak scale increasing, so ARM has a low accuracy in this case.

4.3.3. The Influence of Significant Wave Height

To test the influence of the wave height on ARM and ESM, we compared the wind direction under the data of the significant wave heights from 172 20-min long aligned sampling points of WAVEX. As shown in Figure 14, the average significant wave height in the data was in the range of 1–5 m. It can be found that the wind direction errors of both algorithms increased as the wave height increased.

When the significant wave height was less than 2 m, the wind direction error of the ARM was greater than ESM, but the acquisition point in this condition was too small to study. The wind direction error of ARM keeps on increasing when the significant wave height is more than 2 m, while the wind direction error of ESM is more stable in this case. The wind direction error of ESM has a slow growth, only when the significant wave height is more than 4.5 m.

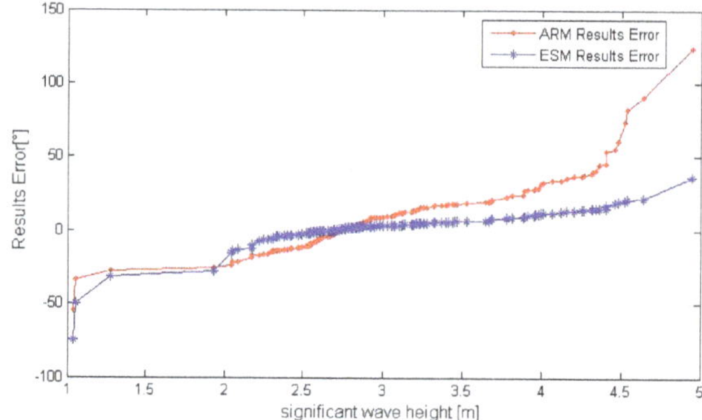

Figure 14. The influence of the significant wave height on the errors of ARM and ESM.

In general, the sea surface is almost swelling when the effective wave height is lower, and the wind field is relatively weak under this condition. The ESM algorithm seems to be more robust because swelling cannot break the small-scale wind streaks of the sea-surface wind field. In contrast, when the significant wave height is high, the sea surface is full of wind waves. It is easy to influence the ESM algorithm, because the small-scale wind streaks are not clear. However, for ARM, the characteristic image of the small-scale wind streak is obtained by the spatial domain reduction and filtering, once the significant wave height is high with increasing winds, it can be inferred that the wind direction error will increase, as the small-scale wind streak is collapsed.

4.3.4. The Influence of Near-Surface Current

To test the influence of the near-surface current on ARM and ESM, the near-surface current and reference wind direction from 172 20-min aligned acquisition points of WAVEX are shown in Figure 15, Figure 15a shows the relationship between the wind direction retrieving error and near-surface current, while Figure 15b gives the retrieving error and reference wind directions. In the following analysis, we considered the above-mentioned two factors at the same time. It should be noted that there were no data for the shored-based near-surface current at 110°–170°, as well as no data for the reference wind directions at 140°–200°.

It has been observed that the wind direction result errors have the most growth in the current at 180° to 200°, and in the reference wind direction at 0° to 50°, and the lower errors appear in the current at 60° to 120°, and in the reference wind direction at 200° to 240°. It is also shown that the error becomes the largest when the current direction is identical to the reference wind direction and is smallest in the opposite direction. For horizontally-polarized (HH-polarized) radars at grazing incidence, the NRCS has only one peak in the upwind direction, and has minimum in the crosswind direction. The influence of the sea-surface wind streaks' characteristics is relatively strong when the NRCS is a large value, further leading a severe wind retrieving error. From Figure 14, we find that the sea-surface current has almost no impact on the two wind direction algorithms. Both algorithms can retrieve a real-time wind direction, even in the varying near-surface current circumstances.

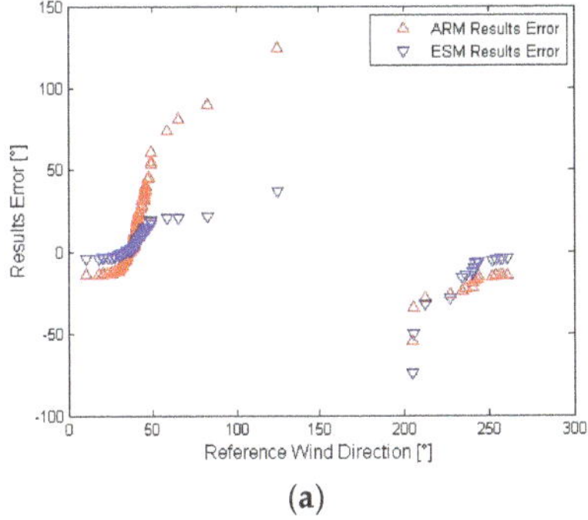
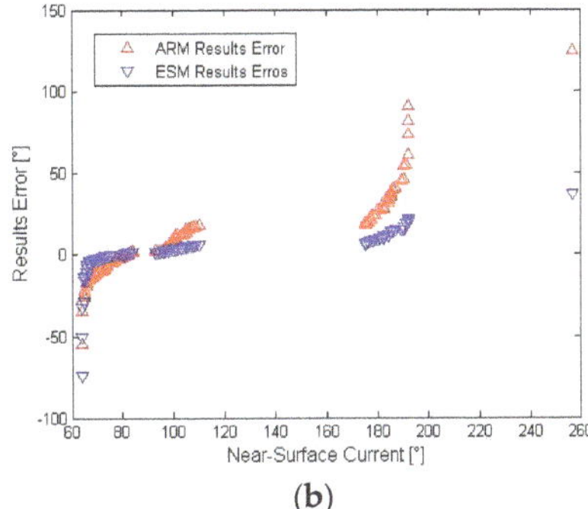

(a)   (b)

Figure 15. The influence of the near-surface current on the errors of ARM and ESM. (**a**) the relationship between the wind direction retrieving error and near-surface current; (**b**) the retrieving error and reference wind directions.

5. Conclusions

In this paper, we studied three celebrated wind direction retrieving methods—LGM, ARM, and ESM—based on small wind streaks extracted from X-band marine radar images. The data for the analyses were collected by the HEU wave monitoring system, and an anemometer on the Chinese South Coast, at Pingtan General Testing Ground. The contributions of this paper are that we give a straightforward comparison of the three methods, and their performances with real-world data are compared, with the additional analysis of the interference factors. The experiment results show that the ARM and ESM algorithms are better than LGM in terms of retrieving accuracy. However, LGM has the shortest running time, and it can be implemented for coarse-retrieving cases.

The ARM algorithm contributes a breakthrough, by retrieving the wind direction from the ocean surface in the spatial domain, and effectively suppresses the drawback of LGM. However, the ARM algorithm is computationally expensive, because it has more recursive cycles and an additional threshold judgment. Both LGM and ARM may overwhelm some characteristics of small-scale wind streaks, simultaneously losing part of the small-scale wind streak images. So, the performances of the spatial domain approaches cannot be guaranteed when the small-scale wind streaks' characteristics are not obvious. As ESM can automatically adapt to small-scale streak scale changes, it is robust with different data for retrieval, and with applying the band-bass filter. It can guarantee the quality of the separated small-scale wind streaks energy spectrum. Therefore, this method can provide an acceptable retrieving accuracy based on the data from X-band marine radars.

It has been shown that although ARM is inferior to ESM in terms of accuracy, it can cover most of the real-time requirements, and provide an acceptable performance. Furthermore, it requires only a single measurement area, compared with ESM. Taking the interference factors into account, ARM is susceptible to sea-surface wind speeds and significant wave heights, but is robust against sea-surface currents. ESM is almost robust against all of the interference factors, except for in a rain-contaminate environment; its wind direction results error increases only if the significant wave height is higher than 4.5 m. We draw a conclusion that ARM is better for real-time wind retrieving, while ESM is much more adaptable for post-data processing, with a superior retrieving performance. Future work will focus on the real-time implementation of the above-mentioned methods in real engineering areas with various circumstances.

Author Contributions: This paper is a collaborative work by all of the authors. H.W. proposed the idea, implemented the system, performed the experiments, analyzed the data, and wrote the manuscript. H.Q. and P.Z. helped to propose the idea, give suggestions, and revise the rough draft. L.W. helped to process the data from the raw sensors. W.C. helped for visualization work. R.A. and M.A.Z.R. helped to write the review and with editing, as well as doing some proofreading work.

Funding: The project was supported by the Natural Science Foundation of Jiangsu Province of China, under grant no. BK20180988; the National Natural Science Foundation of China (61801196); and the Natural Science Foundation of Higher Education Institutions of Jiangsu Province (17KJB510014).

Acknowledgments: We would like to thank the editor and reviewers for the efforts made in processing this submission, and we are particularly grateful to the reviewers for their constructive comments and suggestions, which helped us improve the quality of this paper.

References

1. Lund, B. Development and Evaluation of New Algorithms for the Retrieval of Wind and Internal Wave Parameters from Shipborne Marine Radar Data. Ph.D. Thesis, University of Miami, Miami, FL, USA, 2012.
2. Moat, B.I.; Yelland, M.J.; Molland, A.F. *The Effect of Ship Shape and Anemometer Location on Wind Speed Measurements Obtained from Ships*; University of Southampton: Southampton, UK, 2005.
3. Risien, C.M.; Chelton, D.B. A global climatology of surface wind and wind stress fields from eight years of QuikSCAT scatterometer data. *J. Phys. Oceanogr.* **2008**, *38*, 2379–2413. [CrossRef]
4. Lin, H.; Xu, Q.; Zheng, Q. An overview on SAR measurements of sea surface wind. *Prog. Nat. Sci.* **2008**, *18*, 913–919. [CrossRef]
5. Dankert, H.; Horstmann, J.; Rosenthal, W. Wind-and wave-field measurements using marine X-band radar-image sequences. *IEEE J. Ocean. Eng.* **2005**, *30*, 534–542. [CrossRef]
6. Hessner, K.; Borge, J.C.N.; Reichert, K. Estimation of the significant wave height with x-band nautical radars. In Proceedings of the 28th International Conference Offshore Mechanics and Arctic Engineering (OMAE), Madrid, Spain, 17–22 June 1999.
7. Wei, Y.; Zhang, J.-K.; Lu, Z. A novel successive cancellation method to retrieve sea wave components from spatio-temporal remote sensing image sequences. *Remote Sens.* **2016**, *8*, 607. [CrossRef]
8. Acobs, S.P.; O'Sullivan, J.A. Automatic target recognition using sequences of high resolution radar range-profiles. *IEEE Trans. Aerosp. Electron. Syst.* **2000**, *36*, 364–381. [CrossRef]
9. Wei, Y.; Lu, Z.; Yuan, G.; Fang, Z.; Huang, Y. Sparsity adaptive matching pursuit detection algorithm based on compressed sensing for radar signals. *Sensors* **2017**, *17*, 1120. [CrossRef] [PubMed]
10. Senet, C.M.; Seemann, J.; Ziemer, F. The near-surface current velocity determined from image sequences of the sea surface. *IEEE Trans. Geosci. Remote Sens.* **2001**, *39*, 492–505. [CrossRef]
11. Dankert, H. Retrieval of surface-current fields and bathymetries using radar-image sequences. In Proceedings of the Geoscience and Remote Sensing Symposium (IGARSS'03), Toulouse, France, 21–25 July 2003; pp. 2671–2673.
12. Gommenginger, C.P.; Ward, N.P.; Fisher, G.J.; Robinson, I.S.; Boxall, S.R. Quantitative microwave backscatter measurements from the ocean surface using digital marine radar images. *J. Atmos. Ocean. Technol.* **2000**, *17*, 665–678. [CrossRef]
13. Lee, P.H.Y.; Barter, J.D.; Beach, K.L.; Hindman, C.L.; Lake, B.M.; Rungaldier, H.; Shelton, J.C.; Williams, A.B.; Yee, R.; Yuen, H.C. X band microwave backscattering from ocean waves. *J. Geophys. Res. Ocean.* **1995**, *100*, 2591–2611. [CrossRef]
14. Hatten, H.; Seemann, J.; Horstmann, J.; Ziemer, F. Azimuthal dependence of the radar cross section and the spectral background noise of a nautical radar at grazing incidence. In Proceedings of the International Geoscience and Remote Sensing Symposium, Seattle, WA, USA, 6–10 July 1998; pp. 2490–2492.
15. Huang, W.; Liu, X.; Gill, E. Ocean wind and wave measurements using X-band marine radar: A comprehensive review. *Remote Sens.* **2017**, *9*, 1261. [CrossRef]
16. Lu, Z.; Yang, J.; Huang, Y. Sea wind direction extraction algorithm by X-band radar in moving platform. *Syst. Eng. Electron.* **2016**, *38*, 799–803.
17. Lund, B.; Graber, H.C.; Romeiser, R. Wind retrieval from shipborne nautical X-band radar data. *IEEE Trans. Geosci. Remote Sens.* **2012**, *50*, 3800–3811. [CrossRef]
18. Vicen-Bueno, R.; Horstmann, J.; Terril, E.; De Paolo, T.; Dannenberg, J. Real-time ocean wind vector retrieval from marine radar image sequences acquired at grazing angle. *J. Atmos. Ocean. Technol.* **2013**, *30*, 127–139. [CrossRef]

19. Liu, Y.; Huang, W.; Gill, E.W.; Peters, D.K.; Vicen-Bueno, R. Comparison of algorithms for wind parameters extraction from shipborne X-band marine radar images. *IEEE J. Sel. Top. Appl. Earth Observ. Remote Sens.* **2015**, *8*, 896–906. [CrossRef]

20. Forget, P.; Saillard, M.; Guérin, C.A.; Testud, J.; Le Bouar, E. On the use of x-band weather radar for wind field retrieval in coastal zone. *J. Atmos. Ocean. Technol.* **2016**, *33*, 899–917. [CrossRef]

21. Dankert, H.; Horstmann, J.; Koch, W. Ocean wind fields retrieved radar image sequences. In Proceedings of the IEEE International Geoscience Remote Sensing Symposium, Toronto, ON, Canada, 24–28 June 2002; pp. 2150–2152.

22. Dankert, H.; Horstmann, J. A marine radar wind sensor. In Proceedings of the International Geosciences and Remote Sensing Symposium, Denver, CO, USA, 31 July–4 August 2006; pp. 1296–1299.

23. Dankert, H.; Horstmann, J. Wind measurements at FINO-I marine radar image sequences. In Proceedings of the IEEE International Geoscience Remote Sensing Symposium, Seoul, Korea, 25–29 July 2005; pp. 1777–1780.

24. Dankert, H. A marine radar wind sensor. *J. Atmos. Ocean. Technol.* **2007**, *24*, 1629–1642. [CrossRef]

25. Wang, H.; Lu, L. Determination of X-band radar images ocean wind direction using ARM. *J. Huazhong Univ. Sci. Technol. (Nat. Sci. Ed.)* **2015**, *10*, 42–47.

26. Wang, H.; Lu, Z. Wind direction retrieved based on energy spectrum from X-band radar image sequences. *J. Comput. Theor. Nanosci.* **2016**, *13*, 8288–8297. [CrossRef]

27. Dankert, H.; Horstmann, J.; Rosenthal, W. Ocean surface winds retrieved from marine radar image sequences. In Proceedings of the International Geosciences and Remote Sensing Symposium, Anchorage, AL, USA, 20–24 September 2004; pp. 1903–1906.

28. Lu, Z.; Yang, Y.; Huang, Y. Research on the improvement on the algorithm for retrieving wave height from shadow in the marine radar images. *Chi. J. Sci. Instr.* **2017**, *38*, 212–218.

29. Lu, Z.; Yang, Y.; Huang, Y. Research on correlation in spatial domain to eliminate the co-channel interference of the X-band marine radar. *Syst. Eng. Electron.* **2017**, *39*, 758–767.

30. Etling, D.; Brown, R.A. Roll vortices in the planetary boundary layer: A review. *Bound. Layer Meteorol.* **1993**, *65*, 215–248. [CrossRef]

31. Brow, R. A secondary flow model for the planetary boundary layer. *Atoms. Sci.* **1970**, *27*, 742–757. [CrossRef]

32. Gerling, T.W. Structure of the surface wind field from the Seasat SAR. *J. Geophys. Res. Ocean.* **1986**, *91*, 2308–2320. [CrossRef]

33. Wang, H.; Lu, Z. Research on small scale wind streak factor of marine radar-image sequences. In Proceedings of the 8th International Symposium on Computational Intelligence and Design (ISCID), Hangzhou, China, 12–13 December 2015; pp. 335–339.

 *electronics*

Article

Design of a Novel Double Negative Metamaterial Absorber Atom for Ku and K Band Applications

Saif Hannan [1,*], Mohammad Tariqul Islam [1,*], Ahasanul Hoque [1], Mandeep Jit Singh [1] and **Ali F. Almutairi [2,*]**

[1]   Centre of Advanced Electronic and Communication Engineering, Faculty of Engineering and Built Environment, Universiti Kebangsaan Malaysia, Bangi 43600, Selangor, Malaysia

[2]   Electrical Engineering Department, College of Engineering and Petroleum, Kuwait University, Safat 13060, Kuwait

*   Correspondence: p98220@siswa.ukm.edu.my (S.H.); tariqul@ukm.edu.my (M.T.I.); ali.almut@ku.edu.kw (A.F.A.); Tel.: +60-135-228-124 (S.H.); +60-193-666-192 (M.T.I.); +965-99-843-420 (A.F.A.)

Received: 30 June 2019; Accepted: 30 July 2019; Published: 31 July 2019

Abstract: This paper presents a multiband metamaterial (MM) absorber based on a novel spiral resonator with continuous, dual, and opposite P-shape. The full wave analysis shows 80.06% to 99.95% absorption at frequencies range for Ku and K bands for several substrate materials of 100 mm^2 area. The results indicate that the absorption rate remains similar for different polarizing angles in TEM mode with different substrates. With FR4 (Flame Retardant 4) substrate and 64 mm^2 ground plane, the design acts as single negative (SNG) MM absorber in K band resonance frequencies (19.75–21.37 GHz) and acts as double negative (DNG) absorber in Ku band resonance frequencies (15.28–17.07 GHz). However, for Rogers 3035 substrate and 36 mm^2 ground plane, it acts as an SNG absorber for Ku band resonance frequency 14.64 GHz with 83.25% absorption and as a DNG absorber for K band frequencies (18.24–16.15 GHz) with 83.69% to 94.43% absorption. With Rogers 4300 substrate and 36 mm^2 ground plane, it acts as an SNG absorber for Ku band at 15.04 GHz with 89.77% absorption and as DNG absorber for K band frequencies (22.17–26.88 GHz) with 92.87% to 93.72% absorption. The design was fabricated with all three substrates and showed quite similar results as simulation. In comparison with other broadband absorbers, this proposed MM absorber illustrated broad incidence angles in TEM mode.

Keywords: metamaterial absorber; double negative; dual-band

---

1. Introduction

Electromagnetic (EM) absorbers are recent trends in the field of antenna design, sensing, electromagnetic clocking, low cross-section materials for radar, and stealth technologies for military purposes and thermo-photovoltaic applications [1–12]. A perfect metamaterial (MM) absorber [13] is expected to absorb almost the entire EM signal, with very little or none to be reflected back to the source. So, researchers are working hard to design a perfect EM absorber with less scattering and reflection of EM waves from it [14–16]. An absorber is a double negative (DNG), if both permittivity and permeability of it become negative while the EM waves pass through it, as a result, the refractive index will also be negative. If either permittivity or permeability is negative, it acts as a single negative (SNG) absorber, in this case, the refractive index can be either negative or positive.

To increase stealth performances, radar absorbing surfaces are used. Furthermore, polarization angle insensitive properties along with broad band absorption, are the essential objectives of perfect MM absorbers [17,18]. Nowadays, symmetrical structures of EM absorbers are designed to attain polarization-independent EM absorption, like unique geometry of unit cells with circular shapes, slip-ring- cross resonators and array of these unit cells [19–22]. FR4 substrates are the most common dielectric medium used in these designs. However, the bandwidths of these absorbers are still narrow [11,23,24]. Some works were done with a very small size of unit cells, but they compromised the bandwidth [25–27]. Many researchers are working in this particular field to design a perfect MM absorber with broad band and polarization-insensitive features [28] to employ in stealth and radar systems. Most of the absorber atoms are designed to C and X bands, but absorber for Ku and K bands are rare to find. An absorber for Ku and K bands could be useful for applications like remote sensing, data collection for weather forecasts, wildlife survey, vehicular communication through satellites, etc.

In this paper, a novel spiral resonator with continuous, dual, and opposite P-shape as a unit cell is proposed for almost entire Ku and K band frequencies, which is polarization insensitive for wide incidence angle absorption. The cell was designed on CST microwave studio 2017 software (which was installed on a computer of Intel® Core i3-2120 CPU @ 3.30 GHz and 8.00 GB RAM and Windows 10 operating system), as three layers, patch, a dielectric substrate, and a ground plane. A 10×10 mm unit cell was considered with patch engraved on the top. The patch has a continuous and flipped P-shape resonator with a square border. The width of the border and the patch wire is the same. An FR4 substrate with 8×8 mm ground plane was used. The patch and the ground planes are of annealed copper of thickness 0.035 mm. Both normal incidence and oblique incidence of the polarized TEM waves were considered. The cell was replaced later with Rogers RT 3035 and Rogers RT 4003 substrates with the same patch but with a modified ground of 6×6 mm plane and broad band resonance frequencies in Ku and K bands with more than 80% absorption was obtained. The average time for simulation on CST for the cell to get outputs was around 11 min.

2. Design Methodology

The proposed MMA unit cell was designed with FSS (frequency selective surface) patch on three different substrates of different dielectric properties and thickness, backed by a copper ground plane. The mostly used FR4 material was selected as the primary dielectric substrate (dielectric constant $\epsilon_r = 4.6$) with a substrate layer of a thickness of 1.578 mm. The patch and the ground plane are made of copper (annealed and lossy with conductivity 5.8×10^7 S/m) of thickness 0.035 mm. Figure 1 shows the proposed unit cell with top and back geometry. Here, the metallic FSS layer in Figure 1a (with copper) is presented by a yellow color, and the rest shows the dielectric substrate. Figure 1b shows the ground plane made up of copper (here, for FR4 substrate) just below the substrate. The dimension of the unit cell is shown in Figure 2 and detailed in Table 1. The proposed unit cell has a 10×10 mm surface. Considering the top surface lying on the x-y plane, the cell was designed and critically analyzed for different incident polarization angles. Incident EM waves were propagated from positive z-direction keeping perfect electric field along the x-axis and perfect magnetic field along the y-axis.

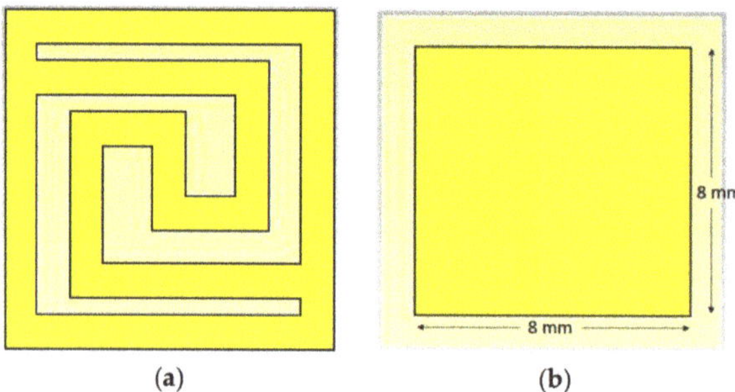

Figure 1. Design of the proposed unit cell (**a**) patch, (**b**) ground (for FR4 (Flame Retardant 4) substrate).

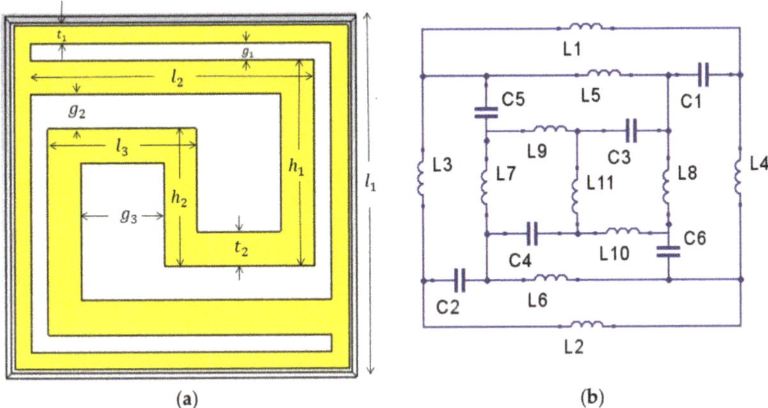

(a)   (b)

Figure 2. (**a**) Unit cell dimension for patch, (**b**) equivalent circuit.

Table 1. Dimension of the continuous, dual, and opposite P-shaped spiral resonator.

Parameter	l_1	l_2	l_3	h_1	h_2	t_1	t_2	g_1	g_2	g_3
Size (mm)	10	8.5	4.5	6	4	1	1	0.5	1	2.5

The patch (with annealed copper) was designed in such a way that, it can absorb mostly at the frequency range of Ku and K bands. The continuous double (and opposite) P-shape acts as the resonator, as shown in Figure 2a. The ground (with same copper layer) acts as a reflector so that no transmission takes place through it. As a result, the S_{11} and S_{21} parameters are found (Figure 3) in the required ranges to ensure maximum absorption of the incident frequencies.

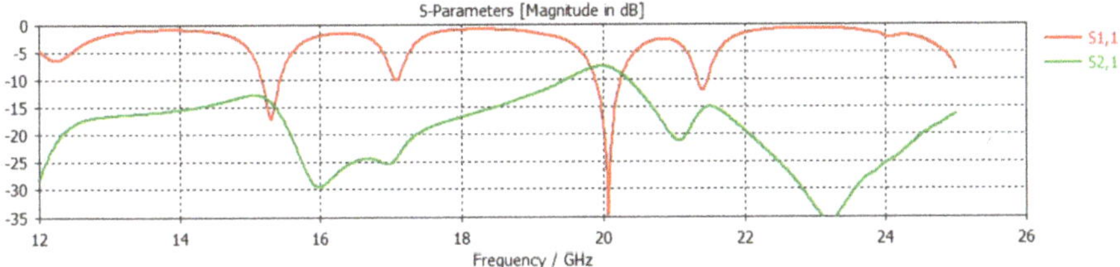

Figure 3. S parameters found after simulation for the model (for FR4 substrate).

3. Results

We know, the Equation for absorption is

$$A(\omega) = 1 - R(\omega) - T(\omega) \tag{1}$$

where $R(\omega)(= S_{11})$ is reflection coefficient and $T(\omega)(= S_{21})$ is transmission coefficient.

A perfect MMA is understood by restraining the transmitted and reflected EM waves to boost the absorption ratio. If no reflection and transmission take place, the MMA will act as a perfect absorber, as per Equation (1). The proposed design was simulated for Ku and K band frequency range (12–26 GHz) with FR4 substrate. After simulation on CST microwave studio, the values of S_{11} and S_{21} parameters were taken with their absolute values from their real and imaginary values and used them on Matlab with proper commands for absorption, permittivity, permeability, and refractive index in both Nicolson-Ross-Weir (NRW) method and direct refractive index (DRI) method. The following

Equations were used in Matlab commands to get permittivity, permeability, and refractive index (in NRW and DRI methods).

Permittivity (relative),

$$\epsilon_r = \frac{2}{\sqrt{-K_\theta d}} \frac{1 - v_1}{1 + v_1} \tag{2}$$

and permeability (relative),

$$\mu_r = \frac{2}{\sqrt{-K_\theta d}} \frac{1 - v_2}{1 + v_2} \tag{3}$$

where, $v_1 = S_{21} + S_{11}$, $v_2 = S_{21} - S_{11}$, $k_\theta = \frac{\omega}{c}$.
where $\omega = 2\pi f$ (f is the frequency of applied EM wave) and c = speed of light, and d = thickness of the substrate.

Refractive index by Nicolson-Ross-Weir (NRW) method

$$= -\text{real}\left(\eta_r\right) \tag{4}$$

and refractive index by direct refractive index (DRI) method

$$= \text{real}\left(\eta\right) \tag{5}$$

where, $\eta_r = \sqrt{\epsilon_r \mu_r}$, and $\eta = \frac{c}{i \pi f d} \sqrt{\frac{(S_{21}-1)^2 - (S_{11})^2}{(S_{21}-1)^2 + (S_{11})^2}}$.

Also, the cell was fabricated and tuned to get S_{11} and S_{21} parameters at the resonance frequencies found in simulation, using Matlab to get the absorption, permittivity, and permeability using Equations (1)–(5). The graphs for permittivity, permeability, and absorption rate with respect to frequencies found from Matlab simulation for both simulated and measured data are shown in Figure 4.

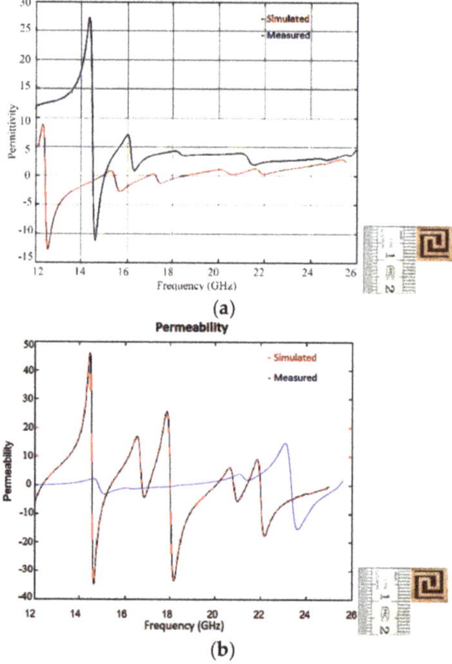

Figure 4. *Cont.*

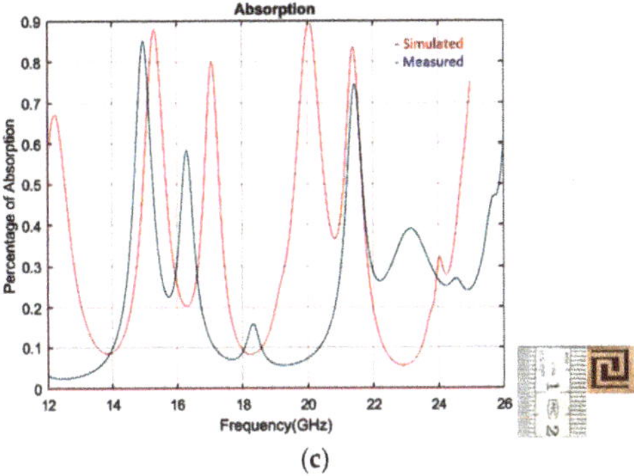

(**c**)

Figure 4. Simulated and measured results for FR4 substrate and 8 × 8 ground (**a**) permittivity vs. frequency graph, (**b**) permeability vs. frequency graph, (**c**) percentage of absorption vs. frequency graph.

The maximum absorptions of 98.04% and 88.93% were found with negative permittivity, permeability, and refractive index (both NRW and DRI methods) for 15.3 GHz and 17.04 GHz frequencies respectively with FR4 substrate in Ku band, which is the DNG absorption property shown by design. In the K band region, SNG property was found with a negative value of either permittivity or permeability and negative value of the refractive index by NRW method at 20.06 GHz and 21.3 GHz with 85.93% and 86.68% absorptions respectively, as shown in Table 2 and Figure 4. So, this design acts as a metamaterial absorber at those frequencies shown in Table 2.

Table 2. Maximum absorptions with double negative (DNG) and single negative (SNG) values.

Band	Ku Band (12–18)		K Band (18–26.5)	
Frequency	15.3 GHz	17.04 GHz	20.06 GHz	21.3 GHz
Permittivity	−0.6466	−0.4076	−0.1	0.8337
Permeability	−1.009	−0.1276	2.252	−1.352
Refractive Index (NRW)	−0.9062	−0.4373	−0.5395	−0.3514
Refractive Index (DRI)	−0.9062	−0.4373	0.5395	0.3514
Absorption	0.9804	0.8893	0.8593	0.8668

Polarization of the incident waves were also considered for the unit cell at different angles (20, 40, 60, and 80 degrees) in TEM mode and found almost same results with insignificant deviations as shown in Table 3. 97.85% to 97.03% absorption were found for DNG properties of EM waves at 15.28 GHz to 15.29 GHz and 80.28% to 89.2% absorption at 16.97 GHz to 17 GHz in Ku band. Whereas in K band, with SNG properties of EM waves, 80.08% to 99.95% absorption was found for 19.75 GHz to 20.02 GHz and 81.41% to 93.52% absorption at 20.23 GHz to 21.37 GHz respectively.

The simulation was done with other substrate materials like, Rogers RT 3035 and Rogers RT 4003 and also fabrication was done using these substrates. It was observed that, with these two substrate materials, we have to change the ground plane significantly to get maximum absorption with DNG properties. The ground was modified with 36 mm² areas of annealed copper of the same thickness, as shown in Figure 5. The simulative and measured results of permittivity, permeability, and absorption rate against frequency range for RT 3035 (dielectric constant = 3.5, thickness = 0.76 mm) substrate is shown in Figure 6.

Table 3. Absorption at different polarizing angles in TEM mode for FR4 substrate.

Polarization Angle φ	Frequency Band	Resonance Frequency (GHz)	Max Absorption (%)	ϵ	μ	η	EM Mode	Substrate	Dielectric Constant (ϵ_r)
0	Ku	15.3	98.04	−0.6466	−1.009	−0.9062			
		17.04	88.93	−0.4076	−0.1276	−0.4373			
	K	20.06	85.93	−0.1	2.252	−0.5395			
		21.3	86.68	0.8337	−1.352	−0.3514			
40	Ku	15.29	97.85	−0.7707	−0.8825	−0.8786	TEM	FR4	4.6
		17	85.34	−0.2505	−0.772	−0.4528			
	K	20.02	99.95	−0.1067	0.206	−0.2			
		21.24	81.76	−0.1156	0.8697	−0.3296			
80	Ku	15.29	97.85	−0.7699	−0.8829	−0.8785			
		16.97	80.28	−0.108	−1.779	−0.4846			
	K	19.75	80.16	0.7407	−1.907	−0.3075			
		21.24	81.81	−0.1162	2.261	−0.3302			

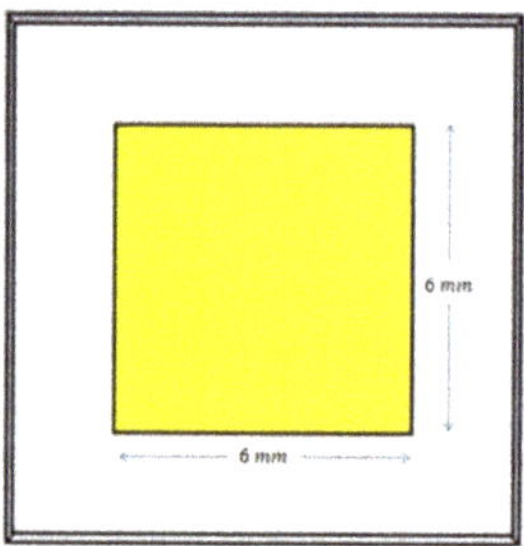

Figure 5. Ground plane for Rogers 3035 and Rogers 4003 substrate.

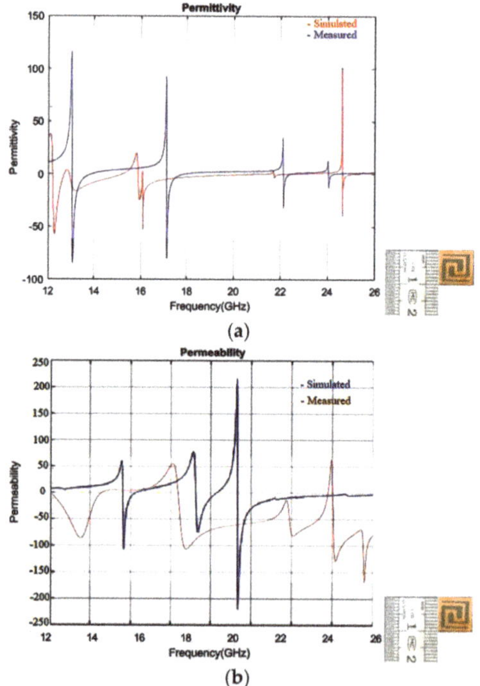

Figure 6. *Cont.*

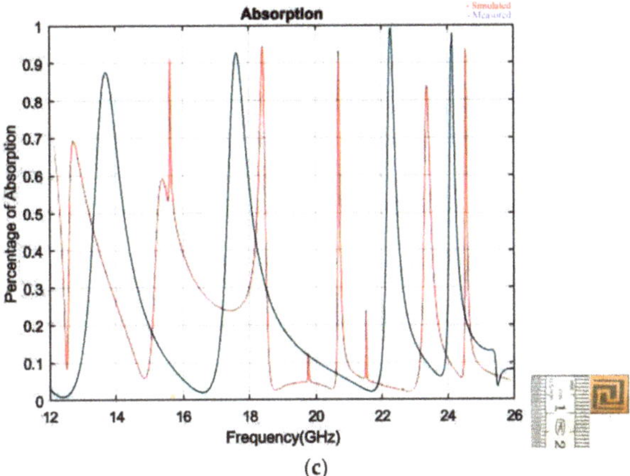

(c)

Figure 6. Simulated and measured results for Rogers RT 3035 and 6 × 6 ground (**a**) permittivity vs. frequency graph, (**b**) permeability vs. frequency graph, (**c**) percentage of absorption vs. frequency graph.

From Table 4 and Figure 6c, it is seen that 83.25% absorption is found with SNG property at 14.64 GHz for all normal incidence and all polarizing angles in TEM mode of operation in the K band. In the K band region, an excellent absorption was found with DNG property at 18.24, 21.2, 24.62, and 26.15 GHz with 94.43%, 92.92%, 83.72%, and 93.55% absorption respectively. So, this design acts as a perfect absorber with DNG MM characteristics in the K band and SNG MM characteristics in the Ku band with RT 3035 substrate.

Table 4. Absorption at different polarizing angles in TEM mode for RT 3035 substrate.

Polarization Angle φ	Frequency Band	Resonance Frequency (GHz)	Max Absorption (%)	ϵ	μ	η	EM Mode	Substrate	Dielectric Constant (ϵ_r)
0	Ku	14.64	83.25	−6.39	5.674	−2.381			
		18.24	94.43	−1.195	−1.064	−1.224			
	K	21.2	92.92	−2.957	−9.058	−4.908	TEM	Rogers 3035	3.5
		24.62	83.71	−4.03	−1.505	−2.983			
		26.15	93.55	−1.002	−3.073	−1.604			
40	Ku	14.64	83.26	−6.39	5.674	−1.383			
		18.24	94.42	−1.195	−1.064	−1.224			
	K.	21.2	92.92	−2.957	−9.058	−2.4			
		24.62	83.72	−4.029	−1.503	−2.982			
		26.15	93.55	−1.003	−3.073	−1.604			
80	Ku	14.64	83.25	−6.39	5.674	−1.382			
		18.24	94.43	−1.195	−1.064	−1.224			
	K	21.2	92.92	−2.957	−9.058	−4.908			
		24.62	83.69	−4.029	−1.503	−2.981			
		26.15	93.54	−1.002	−2.495	−1.604			

With Rogers RT 4003 (dielectric constant = 3.38, thickness = 0.508 mm) substrate, the following results were found, as shown in Figure 7.

With Rogers RT4003 substrate, the results as shown in Figure 7 and Table 5 were achieved. It was observed that the resonance frequencies shifted from those for RT 3035 substrate. Fortunately, the SNG MM characteristics were found in the Ku band and DNG MM characteristics in the K band. 89.78% of absorption was found at 15.04 GHz. 93.28%, 93.72%, and 92.87% absorption were found at 22.17, 25.46, and 26.88 GHz.

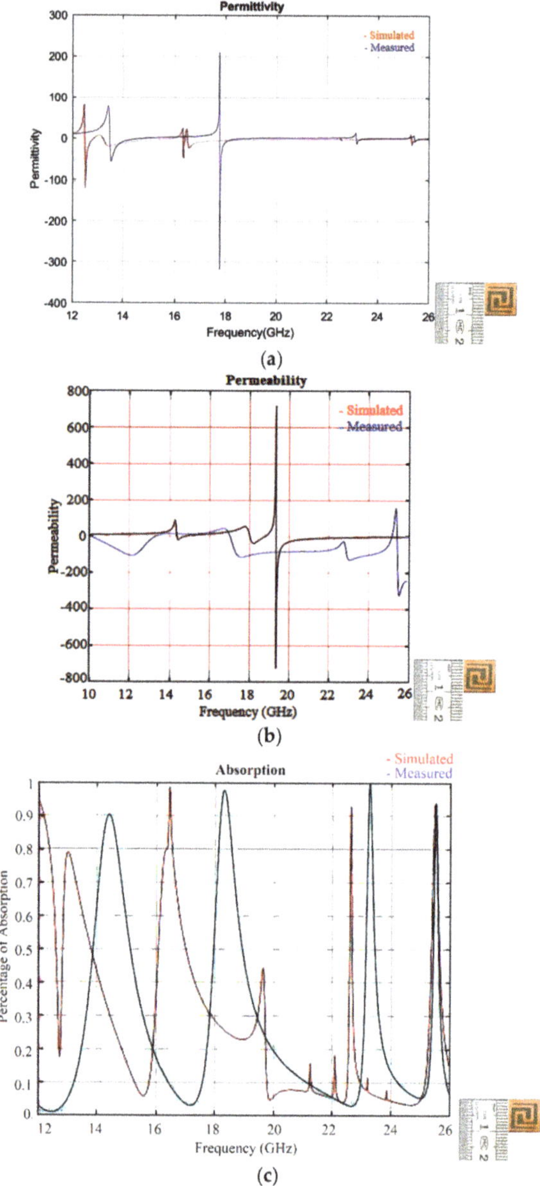

Figure 7. Simulated and measured results for Rogers RT 4003 and 6 × 6 ground (**a**) permittivity vs. frequency graph, (**b**) permeability vs. frequency graph, (**c**) percentage of absorption vs. frequency graph.

Table 5. Absorption at different polarizing angles in TEM mode for RT 4003 substrate.

Polarization Angle φ	Frequency Band	Resonance Frequency (GHz)	Max Absorption (%)	ϵ	μ	η	EM Mode	Substrate	Dielectric Constant (ϵ_r)
0	Ku	15.04	89.77	−5.668	8.465	−2.459			
		22.17	93.28	−1.733	−7.304	−4.914			
	K	25.46	93.72	−2.896	−1.918	−2.659	TEM	Rogers 4003	3.38
		26.88	92.87	−0.7353	−2.112	−1.481			
40	Ku	15.04	89.78	−5.664	8.462	−2.461			
		22.17	93.28	−1.733	−7.304	−4.914			
	K	25.46	93.72	−2.896	−2.078	−2.659			
		26.88	92.87	−0.7353	−2.112	−1.481			
80	Ku	15.04	89.78	−5.663	8.462	−2.461			
		22.17	93.28	−1.733	−7.304	−4.914			
	K	25.46	93.72	−2.896	−2.078	−2.659			
		26.88	92.87	−0.7353	−2.112	−1.481			

4. Discussion

4.1. E-Field, H-Field and Surface Current Analysis

The unit cell was designed with three different substrates (FR4, Rogers RT 3035, and Rogers RT 4003) and as a result, three different types of responses were achieved for the individual substrates from the CST simulations. With FR4 substrate, the E-field, H-field, and surface current distribution were found, as shown in Figure 8. The electric and magnetic field is intense in the regions of bending of the transmission lines inside the patch at 15.3 GHz frequency with 98.04% absorption by it. The values of the permittivity, permeability, and refractive index negative were taken for this case. This is because of the surface current flow in those regions controlled by the transmission lines and the ground placed at the back of the cell. At 17.04 GHz, the electric field becomes more intense in the patch circumference, but magnetic fields shifted towards the center of the patch arms, and hence the value of permeability having much higher value and surface current is less intense than the case for 15.3 GHz, but remains negative with the refractive index and absorption rate reduced to 88.93%. At 20.06 GHz, the electric field is intense at the two opposite arms of the patch only and magnetic field shifts toward the center of the patch, this is because of the surface current distribution is intense at the center and the two opposite arms of the patch. So, permittivity is slightly negative, but permeability becomes positive, and as a result, the refractive index becomes positive by DRI method. However, 85.93% absorption was attained at this frequency. Similarly, at 21.3 GHz, the entire inner circumference of the patch has an intense electric field and associated magnetic field. As a result, the surface current increased significantly, but the electric field became positive, whereas the magnetic field is slightly negative with positive refractive index, and absorption becomes 86.68%. So, with FR4 substrate, the cell acts as a DNG absorber in the Ku band and an SNG absorber in the K band.

For Rogers 3035 substrate, the unit cell shows resonance at 14.64, 18.24, 21.2, 24.62, and 26.15 GHz with 83.25%, 94.43%, 92.92%, 83.71%, and 93.55% absorptions, respectively. The absorptions are high because of the current distributions shown in Figure 9a. The current densities are high around and within the resonator and the ground placed on the opposite side of the substrate. This can be explained with the help of the electric field and magnetic distributions shown in Figure 9b,c.

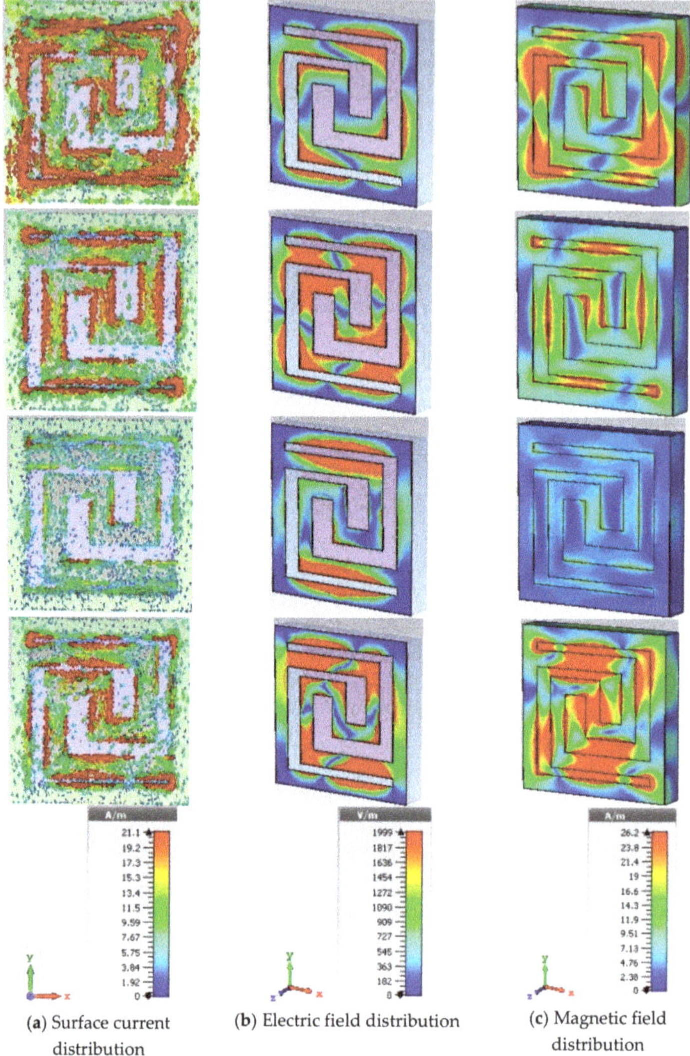

(**a**) Surface current distribution   (**b**) Electric field distribution   (**c**) Magnetic field distribution

Figure 8. Instantaneous distribution of (**a**) surface current, (**b**) electric field, and (**c**) magnetic field at 15.3, 17.4, 20.06, and 21.3 GHz, respectively, for FR4 substrate.

At 14.64 GHz, the electric field distribution shows intense resonance at the circumference of the resonator, but the magnetic field shows resonance on the places where electric fields were less intense. This is why permittivity becomes negative, but permeability became positive. Hence in this frequency, the unit cell acts as an SNG absorber. For the rest of the resonance frequencies, the electric and magnetic fields are intense at the same places of the resonator, and hence both permittivity and permeability are negative, resulting the unit cell to act as a DNG absorber.

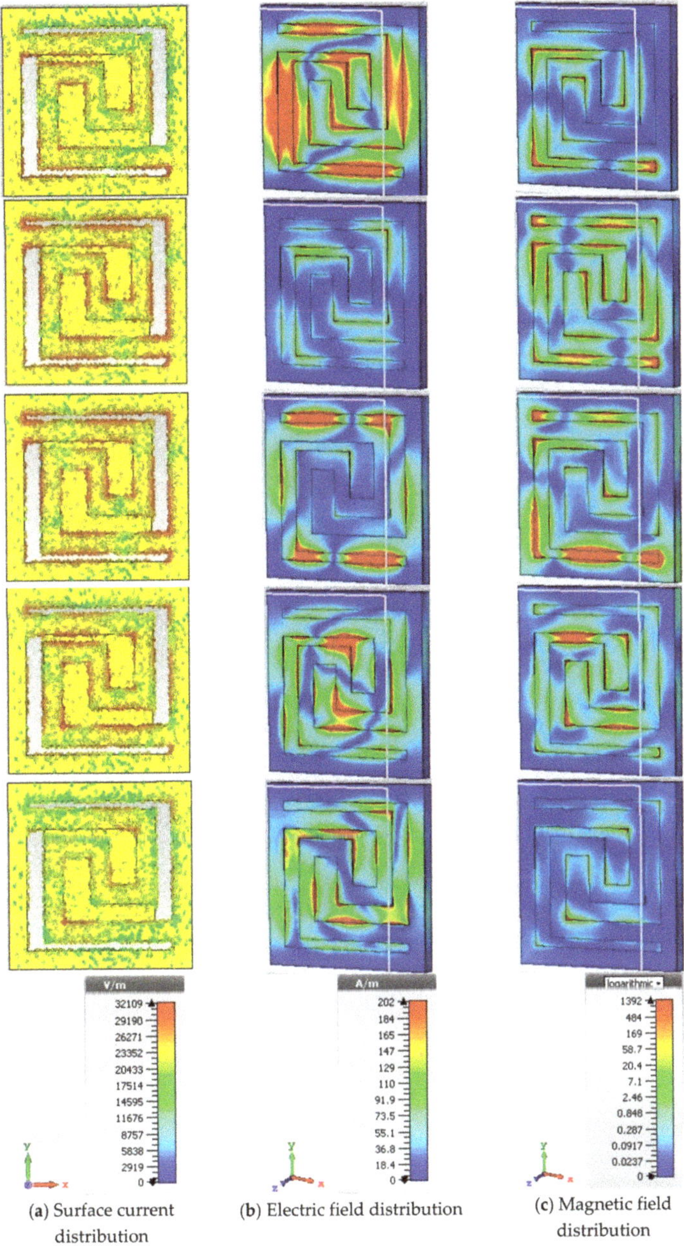

(a) Surface current distribution

(b) Electric field distribution

(c) Magnetic field distribution

Figure 9. Instantaneous distribution of (**a**) surface current, (**b**) electric field, and (**c**) magnetic field at 14.64, 18.24, 21.2, 24.62, and 26.15 GHz, respectively, for Rogers 3035 substrate.

The unit cell with Rogers 4003 substrate showed resonance at 15.04, 22.17, 25.46, and 26.88 GHz frequencies with 89.77%, 93.28%, 93.72%, and 92.87% absorptions respectively. The cell acts as an SNG absorber at the Ku band (15.04 GHz) and a DNG absorber at the K band. The reason can be explained by Figure 10. For 15.04 GHz, the observed electric field and magnetic fields are not showing agitation on the same spots of the resonator, the electric field is intense on the top and right-hand side of the resonator, whereas the magnetic field is intense on the bottom and left-hand side of the resonator. Hence both of them have a high but negative value for permittivity and positive value for permeability, as a consequence of the surface current distribution pattern shown for 15.04 GHz.

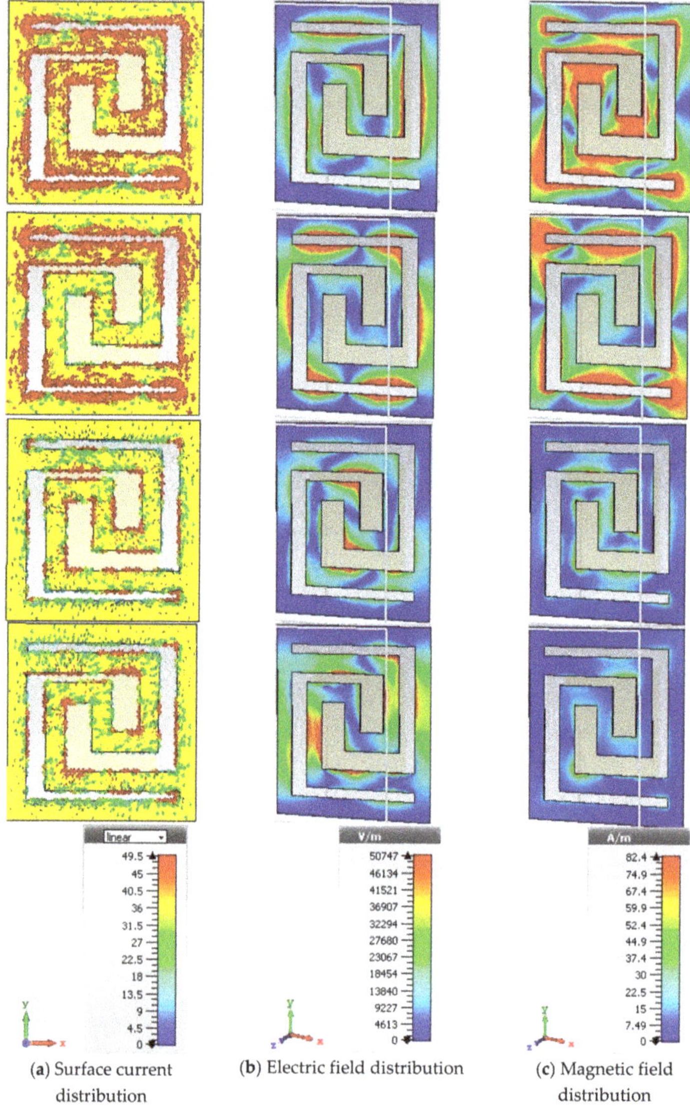

(a) Surface current distribution

(b) Electric field distribution

(c) Magnetic field distribution

Figure 10. Instantaneous distribution of (a) surface current, (b) electric field, and (c) magnetic field at 15.04, 22.17, 25.46, and 26.88 GHz, respectively, for Rogers 4003 substrate.

For the rest of the resonance frequencies, the electric and magnetic field distributions show the same expected patterns because of the surface current distributions. We know that the second Maxwell's Equation relates current density with the magnetic and electric field as

$$\nabla \times H = J + \epsilon \frac{\partial E}{\partial t} \tag{6}$$

Moreover, the relation between the current density and the electric field is

$$J = \sigma E \tag{7}$$

It is evident from Equation (6) that, J depends on E, which is also proved on the Figures of electric fields below, where the high density of electric field relates more current densities on surface current distribution Figures. As J and E are changing with respect to time, Magnetic field also changes along the direction normal to E and J. Hence, Magnetic field is less intense on the areas of the resonator where E is intense. This is because E is maximum where J is maximum, but H changes at right angles to E, so it lags behind E, as a result, shows less intense field on those places. This explanation holds for Figures 8–10 also.

4.2. Absorption Performance Analysis

The unit cell was designed in such a way that, it can be considered for any polarizing angle of the incident EM wave. The resonator was devised of two oppositely connected, continuous P-shape Copper plates connected with the outer square ring, so that it can create essential capacitance and inductances on the resonator with complete control over performance. This unique architecture helped to attain the desired level of absorption for multiple bands.

The Equation of the absorption of EM wave is given by Equation (1). So, the less is reflection and transmission coefficient, the greater is the absorption. Although we have derived the absorption from S parameters found from simulation results, the phenomena can also be explained from physical observation of the resonator from Figures 8–10.

We know from transmission line theory that, reflection coefficient $R(\omega) = \Gamma = \frac{Z_L - Z_0}{Z_L + Z_o}$ and transmission coefficient $T(\omega) = \tau = \frac{2Z_L}{Z_0 + Z_L}$, where $Z_L (= R_L + jX_L)$ is the load impedance and Z_O is the characteristic impedance of the transmission line. The equivalent circuit (Figure 2b) of the resonator shows the load impedance $Z_L \approx 51.99$ to 74.56Ω (calculated) depending on different resonance frequencies and characteristic impedance $Z_O = \sqrt{\frac{L}{C}} \approx 50\Omega$ (this value is predetermined before designing the unit cell and starting the simulation). As transmission coefficient is very small (near to zero due to the ground used (copper is used as ground, hence no transmission took place and EM waves reflected only), it was neglected. So, from Equation (3) it is assumable that, absorption became much higher (more than 80%) due to less reflected wave. Hence the unit cell acted as a perfect absorber at the resonance frequencies.

4.3. Comparison of the Unit Cell with Published Works

The performance of this designed unit cell was compared with some relevant works, which are shown in Table 6, below.

The proposed absorber has maximum absorption and a wide range of resonance frequencies with the smallest area of the unit cell compared to others. Moreover, it has a versatile performance probability with different substrates, which is rare in other works.

Table 6. Comparison of developed multiband metamaterial (MM) absorber with relevant other papers.

Ref. #	Year	Size (mm) [Unit Cell]	Substrate Material	Used Frequency Bands	Max Absorption	Application	Resonance Frequency (GHz)
Sim et. al. [29]	2017	16.8 × 16.8	FR4	X and Ku	>80%	Not specified	11, 12, 13, 14, 15
Madhav et. al. [30]	2018	40 × 40	FR4	Ku, K, and Ka	Not shown	Not specified	1.9, 7.3, 17.8, 25
Khan et. al. [31]	2018	10 × 10	RO4350B	X and Ku	Not shown	Hollow waveguide filter and perfect absorber	7.82, 9.65
Agrawal et. al. [32]	2018	18 × 18	FR4	X and Ku	99.9%	Not specified	7.6, 8.9, 12.3, 12.8
Jafari et. al. [33]	2019	24 × 24	FR4	X and Ku	>84%	Not specified	8.6, 10.2, 11.95
Our proposed work	2019	10 × 10	FR4RT 3035RT 4003	Ku and K	99.95%	Perfect Absorber	14.64–15.3, 17.04–18.24, 20.06–21.3, 24.62–26.88

5. Conclusions

A unique MM absorber was proposed covering the satisfied level of absorption in the Ku and K band region for wide incidence angle EM waves. The design has a spiral resonator with continuous, dual, and opposite P-shape copper patch backed up with a copper ground, to ensure its polarization independence with the horizontally and vertically symmetric structure. The gaps between the spiral arms are also symmetric to keep incidence EM wave angle insensitive. CST 2017 software was used to simulate the design and extract S-parameters to find out absorption, permittivity, permeability, and refractive index by NRW and DRI methods and the cell was fabricated to practically measure these parameters. The measured values are slightly different from simulated values as expected, but still shows absorptions with some negative values of permittivity and permeability. We tried three different substrates for the same design from 10 to 28 GHz and found high absorption rates in Ku and K band regions, with double negative and single negative MM properties. Table 6 shows that our design is much better in comparison with recent works in terms of size of the unit cell, the versatility of using different substrates with the same design, rate of absorption, application, and several resonance frequencies for the highest absorption rate. In all these terms, this design is unique and proves it as a much better absorber, which can be a good candidate for invisibility cloaking, filters, and antennas for satellite communications in Ku and K bands.

Author Contributions: S.H. made significant contributions to this study regarding conception, design and analysis, measurement and writing the manuscript. M.T.I. participated in revising the article critically for remarkable intellectual contents and supervised the whole study. A.H. helped in measurement and A.F.A. and M.J.S. revised the manuscript and provided intellectual suggestions.

Funding: This work is supported by Universiti Kebangsaan Malaysia research grant.

Conflicts of Interest: The authors declare no conflict of interest.

References

1. Sharma, A.; Panwar, R.; Khanna, R. Experimental Validation of a Frequency-Selective-Surface-Loaded Hybrid Metamaterial Absorber with Wide Bandwidth. *IEEE Magn. Lett.* **2019**, *10*, 1–5. [CrossRef]
2. Xin, W.; Binzhen, Z.; Wanjun, W.; Junlin, W.; Junping, D. Design, fabrication, and characterization of a flexible dual-band metamaterial absorber. *IEEE Photonics J.* **2017**, *9*, 1–12. [CrossRef]
3. Ghosh, S.; Lim, S. Perforated Lightweight Broadband Metamaterial Absorber Based on 3-D Printed Honeycomb. *IEEE Antennas Wirel. Propag. Lett.* **2018**, *17*, 2379–2383. [CrossRef]
4. Mishra, N.; Choudhary, D.K.; Chowdhury, R.; Kumari, K.; Chaudhary, R.K. An investigation on compact ultra-thin triple band polarization independent metamaterial absorber for microwave frequency applications. *IEEE Access* **2017**, *5*, 4370–4376. [CrossRef]

5. Zhou, Q.; Yin, X.; Ye, F.; Mo, R.; Tang, Z.; Fan, X.; Cheng, L.; Zhang, L. Optically transparent and flexible broadband microwave metamaterial absorber with sandwich structure. *Appl. Phys. A* **2019**, *125*, 131. [CrossRef]

6. Abbasi, S.; Nourinia, J.; Ghobadi, C.; Karamirad, M.; Mohammadi, B. A sub-wavelength polarization sensitive band-stop FSS with wide angular response for X-and Ku-Bands. *AEU Int. J. Electron. Commun.* **2018**, *89*, 85–91. [CrossRef]

7. Alam, M.J.; Faruque, M.R.I.; Azim, R.; Islam, M.T. Depiction and analysis of a modified H-shaped double-negative meta atom for satellite communication. *Int. J. Microw. Wirel. Technol.* **2018**, *10*, 1155–1165. [CrossRef]

8. Veysi, M.; Kamyab, M.; Moghaddasi, J.; Jafargholi, A. Transmission phase characterizations of metamaterial covers for antenna application. *Prog. Electromagn. Res.* **2011**, *21*, 49–57. [CrossRef]

9. Ullah, M.; Islam, M.; Faruque, M. A near-zero refractive index meta-surface structure for antenna performance improvement. *Materials* **2013**, *6*, 5058–5068. [CrossRef]

10. Islam, M.; Islam, M.T.; Samsuzzaman, M.; Faruque, M.R.I. Compact metamaterial antenna for UWB applications. *Electron. Lett.* **2015**, *51*, 1222–1224. [CrossRef]

11. Hasan, M.; Faruque, M.; Islam, S.; Islam, M. A new compact double-negative miniaturized metamaterial for wideband operation. *Materials* **2016**, *9*, 830. [CrossRef] [PubMed]

12. Hoque, A.; Tariqul Islam, M.; Almutairi, A.; Alam, T.; Jit Singh, M.; Amin, N. A Polarization Independent Quasi-TEM Metamaterial Absorber for X and Ku Band Sensing Applications. *Sensors* **2018**, *18*, 4209. [CrossRef] [PubMed]

13. Landy, N.I.; Sajuyigbe, S.; Mock, J.J.; Smith, D.R.; Padilla, W.J. Perfect metamaterial absorber. *Phys. Rev. Lett.* **2008**, *100*, 207402. [CrossRef] [PubMed]

14. Kollatou, T.; Dimitriadis, A.; Assimonis, S.; Kantartzis, N.; Antonopoulos, C. Multi-band, highly absorbing, microwave metamaterial structures. *Appl. Phys. A* **2014**, *115*, 555–561. [CrossRef]

15. Arjunan, A. Targeted sound attenuation capacity of 3D printed noise cancelling waveguides. *Appl. Acoust.* **2019**, *151*, 30–44. [CrossRef]

16. Arjunan, A. Acoustic absorption of passive destructive interference cavities. *Mater. Today Commun.* **2019**, *19*, 68–75. [CrossRef]

17. Li, H.; Yuan, L.H.; Zhou, B.; Shen, X.P.; Cheng, Q.; Cui, T.J. Ultrathin multiband gigahertz metamaterial absorbers. *J. Appl. Phys.* **2011**, *110*, 014909. [CrossRef]

18. Kollatou, T.M.; Dimitriadis, A.I.; Assimonis, S.; Kantartzis, N.V.; Antonopoulos, C.S. A family of ultra-thin, polarization-insensitive, multi-band, highly absorbing metamaterial structures. *Prog. Electromagn. Res.* **2013**, *136*, 579–594. [CrossRef]

19. Chen, H.; Yang, X.; Wu, S.; Zhang, D.; Xiao, H.; Huang, K.; Zhu, Z.; Yuan, J. Flexible and conformable broadband metamaterial absorber with wide-angle and polarization stability for radar application. *Mater. Res. Express* **2018**, *5*, 015804. [CrossRef]

20. Chen, K.; Luo, X.; Ding, G.; Zhao, J.; Feng, Y.; Jiang, T. Broadband microwave metamaterial absorber with lumped resistor loading. *EPJ Appl. Metamater.* **2019**, *6*, 1. [CrossRef]

21. Li, Y.; Wang, J.; Yang, J.; Wang, J.; Feng, M.; Li, Y.; Zhang, J.; Qu, S. Metamaterial anti-reflection lining for enhancing transmission of high-permittivity plate. *J. Phys. D Appl. Phys.* **2018**, *52*, 03LT01. [CrossRef]

22. Zhi, Y.; Wang, Y.; Nie, Y.; Zhou, R.; Xiong, X.; Wang, X. Design, fabrication and measurement of a broadband polarization-insensitive metamaterial absorber based on lumped elements. *J. Appl. Phys.* **2012**, *111*, 044902.

23. Singh, D.; Srivastava, V.M. Design Implementation of Concentric Loops with Stubs Metamaterial Absorber. *Wirel. Pers. Commun.* **2019**, *104*, 129–148. [CrossRef]

24. Bilotti, F.; Toscano, A.; Alici, K.B.; Ozbay, E.; Vegni, L. Design of miniaturized narrowband absorbers based on resonant-magnetic inclusions. *IEEE Trans. Electromagn. Compat.* **2010**, *53*, 63–72. [CrossRef]

25. Xin, W.; Binzhen, Z.; Wanjun, W.; Junlin, W.; Junping, D. Design and characterization of an ultrabroadband metamaterial microwave absorber. *IEEE Photonics J.* **2017**, *9*, 1–13. [CrossRef]

26. Sen, G.; Islam, S.N.; Banerjee, A.; Das, S. Broadband perfect metamaterial absorber on thin substrate for X-band and Ku-band applications. *Prog. Electromagn. Res.* **2017**, *73*, 9–16. [CrossRef]

27. Kumar, R.; Tripathy, M.R.; Ronnow, D. An Approach to Improve Gain and Bandwidth in Bowtie Antenna Using Frequency Selective Surface. In *Smart Innovations in Communication and Computational Sciences*; Springer: Berlin/Heidelberg, Germany, 2019; pp. 219–227.

28. Lee, J.; Lim, S. Bandwidth-enhanced and polarisation-insensitive metamaterial absorber using double resonance. *Electron. Lett.* **2011**, *47*, 8–9. [CrossRef]

29. Sim, M.S.; You, K.Y.; Esa, F.B.; Chan, Y.L. Broadband metamaterial microwave absorber for X-Ku band using planar split ring-slot structures. In Proceedings of the 2017 Progress in Electromagnetics Research Symposium-Fall (PIERS-FALL), Singapore, 19–22 November 2017; pp. 215–221.

30. Madhav, B.; Krishna, T.R.; Lekha, K.D.S.; Bhavya, D.; Teja, V.D.; Reddy, T.M.; Anilkumar, T. Multiband Semicircular Planar Monopole Antenna with Spiral Artificial Magnetic Conductor. In *Microelectronics, Electromagnetics and Telecommunications*; Springer: Berlin/Heidelberg, Germany, 2018; pp. 599–607.

31. Khan, S.; Eibert, T.F. A Multifunctional Metamaterial-Based Dual-Band Isotropic Frequency-Selective Surface. *IEEE Trans. Antennas Propag.* **2018**, *66*, 4042–4051. [CrossRef]

32. Agrawal, A.; Misra, M.; Singh, A. Wide incidence angle and polarization insensitive dual broad-band metamaterial absorber based on concentric split and continuous rings resonator structure. *Mater. Res. Express* **2018**, *5*, 115801. [CrossRef]

33. Jafari, F.S.; Naderi, M.; Hatami, A.; Zarrabi, F.B. Microwave Jerusalem cross absorber by metamaterial split ring resonator load to obtain polarization independence with triple band application. *AEU Int. J. Electron. Commun.* **2019**, *101*, 138–144. [CrossRef]

 *electronics*

Article

Low Altitude Measurement Accuracy Improvement of the Airborne FMCW Radio Altimeters

Ján Labun [1], Pavol Kurdel [1], Marek Češkovič [1], Alexey Nekrasov [2,3,*] and Ján Gamec [4]

[1] Faculty of Aeronautics, Technical University of Košice, Rampová 7, 041 21 Košice, Slovakia
[2] Department of Radio Engineering Systems, Saint Petersburg Electrotechnical University, Professora Popova 5, Saint Petersburg 197376, Russia
[3] Higher Technical School of Computer Engineering, University of Malaga, 29071 Malaga, Spain
[4] Faculty of Electrical Engineering and Informatics, Technical University of Košice, Letná 9, 042 00 Košice, Slovakia
* Correspondence: alexei-nekrassov@mail.ru; Tel.: +7-8634-360-484

Received: 27 June 2019; Accepted: 8 August 2019; Published: 12 August 2019

Abstract: This manuscript focuses on the analysis of a critical height of radio altimeters that can help for the development of new types of aeronautical radio altimeters with increased accuracy in measuring low altitudes. Altitude measurement accuracy is connected with a form of processing the difference signal of a radio altimeter, which carries information on the measured altitude. The definition of the altitude measurement accuracy is closely linked to the value of a critical height. Modern radio altimeters with digital processing of a difference signal could shift the limit of accuracy towards better values when the basics of the determination of critical height are thoroughly known. The theory results from the analysis and simulation of dynamic formation and the dissolution of the so-called stable and unstable height pulses, which define the range of the critical height and are presented in the paper. The theory is supported by a new method of derivation of the basic equation of a radio altimeter based on a critical height. The article supports the new theory of radio altimeters with the ultra-wide frequency deviation that lead to the increase the accuracy of a low altitude measurement. Complex mathematical analysis of the dynamic formation of critical height and a computer simulation of its course supported by the new form of the derivation of the basic equation of radio altimeter guarantee the correctness of the new findings of the systematic creation of unstable height pulses and the influence of their number on the altitude measurement accuracy. Application of the presented findings to the aviation practice will contribute to increasing the accuracy of the low altitude measurement from an aircraft during its landing and to increasing air traffic safety.

Keywords: FMCW radio altimeter; methodological error; critical height; altitude measurement accuracy; height pulses; ultra-wide frequency deviation

---

1. Introduction

Radio altimeters are being used on board of aircrafts to measure the instant altitude of the flight. Radio altimeters are important from the point of view of flight safety, mainly when approaching landing [1,2]. Due to their specific function of measuring the low altitudes, they use the frequency modulation and continuous transmitted signal. This article focuses on radio altimeters of FMCW type. An FMCW radio altimeter forms information about the flight altitude through the evaluation of the difference in frequency between the transmitted and received high-frequency signals. The instant value of the difference frequency F_d (as a low-frequency information signal of the measured altitude H) is made by the time delay τ of the received signal frequency f_R against the transmitted signal frequency f_T. The delay of the received signal is created by overpassing the height difference on the aircraft–ground–aircraft route [3,4]. Such a radio altimeter possesses its methodological error of height measurement $\pm\Delta H$, which is based on the physical fundamentals of of evaluation of the differential frequency. The methodological error is determined by a so-called critical height ΔH, which is the minimal height range that a radio altimeter can distinguish [4]. The value of the critical height is determined by the following equation:

$$\Delta H = \frac{c}{8\,\Delta f},\tag{1}$$

where c is the speed of light and Δf is the frequency deviation.

The range of the critical height for recent radio altimeters (corresponding with measurement accuracy) has settled on the value of 0.75 m. This manuscript deals with the principle of the formation of critical height and suggests how to decrease its value to increase the altitude measurement accuracy [5,6].

FMCW radio altimeter has a 90-year history of development and improvement for measuring aircraft altitude. At each aviation historical stage, efforts to increase the accuracy of altitude measurement have led to increasing the frequency deviation to higher values of a carrier frequency f_0. At the beginning, the altitude measurement accuracy was ± 2.2 m (at $\Delta f = 17$ MHz and $f_0 = 444$ MHz). Almost 40 years ago, the accuracy of the measurement stabilized at ± 0.75 m (at $\Delta f = 50$ MHz and $f_0 = 4.4$ GHz). However, attempts to increase the measurement accuracy have not stopped even after reaching this limit.

Today, even advanced technologies push the limit a little forward. It is generally known that a further substantial increase of accuracy would be achieved by a significant increase of the values of the frequency deviation and carrier frequency. For instance, $\Delta f = 100$ MHz and $f_0 = 12$ GHz provide an altitude measurement accuracy of ± 0.375 m. However, when flying over various terrains, the use of the very high value of the carrier frequency of 12 GHz is not so advantageous.

Previously, we have proposed some methods to increase the accuracy of altitude measurement by radar altimeters. A new method for measuring the altitude by estimating the period of the differential frequency is presented in [4]. This method does not have a methodological error and provides better measurement accuracy, especially at low altitude, in comparison with the currently used method that is based on the classical calculation of height pulses.

An innovative technique of using the radar altimeter for prediction of terrain collision threats has been presented in [7]. It is based on an atypical way of estimating the Doppler frequency by measuring the ratio of the number of stable and unstable height pulses between the even and odd half-periods of the modulation signal of a radio altimeter. From this point of view, [7] is close to the problem of the current manuscript.

An altitude measurement accuracy improvement with a two-channel method has been considered in [8]. In the two-channel method, the deviation of the carrier frequency of the signal retains its original values.

In this connection, in addition to previously published results, this manuscript supports the new theory of radio altimeters with the ultra-wide frequency deviation leading to an increase in the accuracy of a low altitude measurement, and it justifies that the measurement accuracy is fundamentally influenced by the frequency deviation and not by the carrier frequency itself. For the presented analysis, it is evident that the measurement accuracy determines a radio altimeter methodical error ΔH, which is influenced by the height range of the numbers of the formation and dissolution of unstable height pulses. Using the above mentioned historical and practical experience as well as new theoretical knowledge, the way to increase the accuracy of altitude measurement without the need to increase the carrier frequency is presented in this manuscript with the help of analysis and simulation. The method presented in this manuscript uses a classic single-channel radio altimeter like in [8], but with a doubled value for the carrier frequency deviation.

2. Difference Signal of FMCW Radio Altimeter

In determining the frequency value of the radio altimeter difference frequency signal $u_d(t)$, which carries the information about the measured altitude, based on mutual immediate differences between the frequency-modulated transmitted signal $u_T(t)$ and the time-delayed frequency-modulated received signal $u_R(t)$, it can simply be presented in the following form of $u_d(t) = u_T(t) + u_R(t)$. Technically, this difference between the two high-frequency signals of the radio altimeter is evaluated by a balanced mixer. Since these are two near-frequency signals, mixing the procedure results in a low-frequency difference signal in which the amplitude change in time $U_d(t_1)$ carries the information about the measured altitude. By mathematical analysis of the above-mentioned mixing process, the amplitude of this difference signal is defined as [5,9,10]:

$$U_d(t) = U_T + U_R \cos(\varphi_0 + \varphi_M \cos\Omega_M t), \tag{2}$$

where U_T is the amplitude of the frequency-modulated transmitting signal, U_R is the amplitude of the received signal, ϕ_0 and ϕ_M are the initial and variable phase of the difference signal, Ω_M is the angular frequency modulation of the signal, and t_1 is the time course of the difference signal, $t_1 = (t - \tau/2)$.

This signal allows for the formation of some so-called unstable height pulses N_u. They alternately arise and disappear when the amplitude of the difference signal passes through zero [3,11,12]. Defining the condition of the formation and the dissolution of the unstable height pulses determines the rise H_F and extinction H_E heights of the individual pulses:

$$H_F = \frac{\lambda_0}{8}\frac{2k-1}{1+\xi}, \; H_E = \frac{\lambda_0}{8}\frac{2k-1}{1-\xi}, \tag{3}$$

where λ_0 is the carrier wavelength, k is the sequence number of the height pulse, ξ is the relative value of the frequency deviation, and $\xi = \Delta f/f_0$.

Then, the two heights behind the forming pulses make it possible to define the height range of the formation of individual pulses F, and from the two heights of the following dissolution pulses, Equation (4) is used to define the height range of dissolving pulses E:

$$F = H_{F2} - H_{F1} = \frac{\lambda_0}{4(1+\xi)}, \; E = H_{E2} - H_{E1} = \frac{\lambda_0}{4(1-\xi)}. \tag{4}$$

A height range for the duration of the different unstable height pulse S is determined as the difference of height between the dissolution and formation of any height pulse:

$$S = H_E - H_F = \frac{\lambda_0}{4}(2k-1)\frac{\xi}{1+\xi^2}. \tag{5}$$

Equation (4) shows us that the height range of the gradual formation of the different unstable height pulses is constant and smaller than the value of $\lambda_0/4$ at any measured height. Thus, it is possible to state that $F1 = F2 = F3 = \ldots$. At the same time, it can be seen that the height range of the dissolution of unstable pulses E is also constant but bigger than the value of $\lambda_0/4$ at any measured height. So, it is also possible to state that $E1 = E2 = E3 = \ldots$. Equation (5) also indicates that the height range for the duration of the different unstable height pulses constantly grows with the increase of the consequence number of the pulse k, as a result of the increase of the value of measured altitude. So, it is possible to state that $S1 < S2 < S3 < \ldots$ [3,7].

Gradual increase of the height range for the duration of different unstable pulses together with the growing measured altitude cause more and more of those pulses at the same height to overlap mutually (Figures 1 and 2) [13].

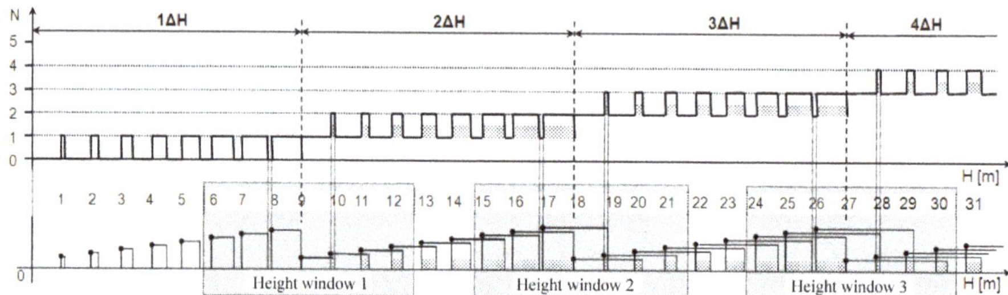

Figure 1. Height dependence of the formation and dissolution of unstable height pulses.

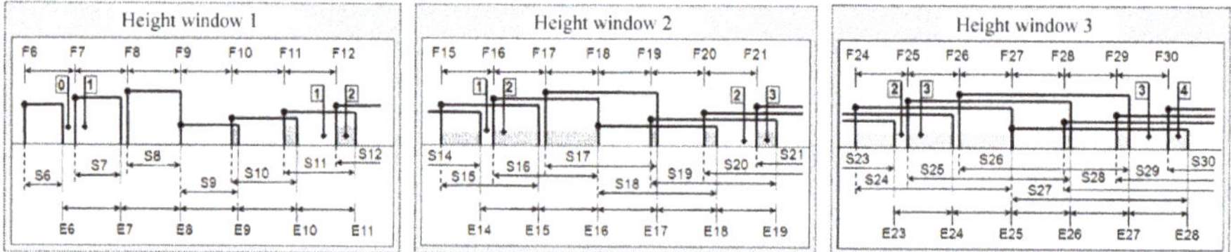

Figure 2. Height ranges and windows in the area of mutual overlapping of unstable height pulses.

3. Simulation of Height Pulses Creation

Based on the above-defined theory, the simulation program for making the stable and unstable height pulses of the radio altimeter has been created. The simulation results at the input parameters equal to the value of the current radio altimeters are shown in Figure 3. The simulation is realized in a height range from 0 m to 2 m, with a mean value of 1 m. In Figure 3, we can see that the first critical height value ends at the 0.75 m height level, and the second one ends at 1.5 m. Obviously, the methodic error of the radio altimeter (0.75 m) is the same as given by the radio altimeter manufacturers.

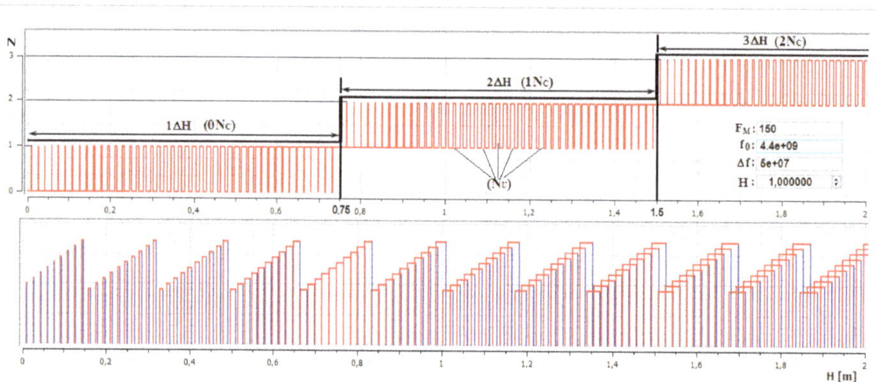

Figure 3. Simulation of the formation of stable and unstable height pulses.

The upper part of Figure 3 represents the height dependence of the creation of numbers of stable (N_C) and unstable (N_U) height pulses. By the term "height pulses," we express the correlation between the number of pulses in the frequency area and the measured altitude expressed in the unit of length. This can be partly compared to the correlation between the harmonic signal in the time domain and the number of periods in the frequency domain, as in Fourier transform. If we use a classical pulse generator, then changing the frequency setting will change the number of output pulses. If we record the generated pulses over time, there will be the time on the horizontal axis and amplitude on the vertical axis. However, if the number of pulses is evaluated in some non-standard but stable time, e.g., per modulation period, then we would plot the frequency on the horizontal axis and the number of pulses per modulation period on the vertical axis.

The radio altimeter is reminiscent of a quasi-pulse generator, which forms the number of pulses that correspond to the flight altitude. The peculiarity is that even with a fluent change in height, the number of pulses does not change fluently, but the formation and dissolution of the pulses are discrete, and their number creates a quantization (stairs) course when changing the altitude (Figure 3). This is due to the fact that the pulses do not occur in a separate low-frequency generator, but are formed from a differential frequency resulting from the mixing of the high frequency and frequency-modulated signals. Thus, when evaluating the number of radio altimeter pulses generated during the modulation period, where the altitude is the control parameter, then the height H is plotted on the horizontal axis, and the number of pulses N for the modulation period is plotted on the vertical axis.

The stable height pulses are those of which the number increases as the altitude increases gradually and discretely with the height range ΔH, that which creates the so-called quantization (stairs) course. In each height range (which is related to the altitude measurement accuracy), the number of these pulses is stable. The frequency value of these pulses is fixed to the altitude of the aircraft, which is expressed by the basic equation of a radio altimeter presented by Equation (12).

Unstable altitude pulses are those of which the number changes discreetly as the height increases, forming on every single stair ΔH a so-called comb course. In each height range ΔH, the formation and dissolution of the unstable pulse means that the total number of pulses within that range ΔH is changed by one pulse. The formation and dissolution of the unstable pulse occurs in a very small height range, so this situation is repeated in the range ΔH tens of times. The course of creation (formation and dissolution) of unstable pulses is the same in every height range ΔH (at each step). So, it is very interesting how many times the unstable pulse formation and dissolution will take place in the height range ΔH, as their number is related to the accuracy of the altitude measurement, which is the main topic of this manuscript.

In the 2 m height range, three critical heights are simulated, which correspond to the two ranges of the existence of the stable pulses $2Nc$. There is no stable pulse in the first range ($0Nc$ in $1\Delta H$ in Figure 3). A larger number of unstable pulses are simulated in the same 2 m height range.

The bottom part of Figure 3 presents the height ranges of the duration of different pulses S. At the critical height of $1\Delta H$, the height ranges of the duration of the unstable height pulses are so small that they do not overlap, and an alternation of zero and the first unstable pulses (0–1) occur at the given critical height. At all the following critical heights, the height ranges of the duration of the unstable height pulses are so big that they mutually overlap. The number of permanently existing unstable pulses is determined by the consequence number of a stable height pulse. At least one unstable pulse exists at each section of the height within the range of $2\Delta H$. It means that one stable pulse exists at the height of alternation of the first and second unstable pulses (1–2) within the given range. Similarly, two unstable pulses exist permanently within the range of $3\Delta H$ in each section of the height. It means that there are two stable pulses.

Closer analysis of this phenomenon can observe that the process of the formation and dissolution of the unstable height of pulses is not chaotic, but has regularity. The regularity of the formation and dissolution of the height pulses is related to the change in the measured height, but their number in the range of critical height corresponds to the technical parameters of the radio altimeter [11,12].

4. Number of Unstable Pulses in Range of Critical Height

It is obvious from Figures 1 and 2, which present the height dependence of the formation of height pulses, as well as from Figure 3, which shows the results of the simulation of the stable and unstable height pulses, that the higher the number of unstable pulses is in the range of critical height, the smaller the precision of height measurement.

For the logical explanation of reasoning and determination of the number of unstable pulses in the range of critical height, it is necessary to move from the frequency area into the area of wavelengths. This thought is based on the generally known facts about the principle of the operation of FMCW radio altimeters presented in Figure 4.

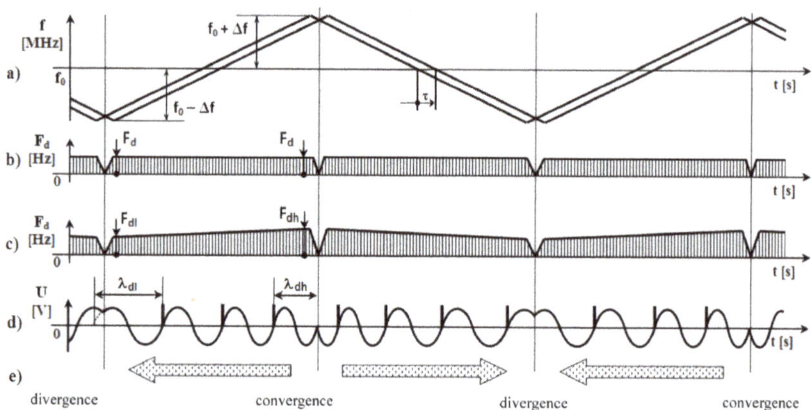

Figure 4. Formation of differential signal for determination of the number of unstable height pulses. (**a**) frequency-modulated transmitted signal, and received signal with the delayed in time by τ; (**b**) theoretical representation of the differential frequency over time; (**c**) actual change of the differential frequency over time; (**d**) change of the period of the differential frequency within the modulation frequency; (**e**) temporary shift of the periods of the differential frequency with increasing the measured altitude.

Based on the principles that if the frequency modulation of the transmitted signal within the whole range of $\pm\Delta f$ is linear (Figure 4a), and that if the time delay of received signal during whole modulation period by value τ is also linear, then the immediate value of differential frequency is constant during whole time of the modulation period (Figure 4b). The areas of the changing of the direction of the frequency deviation in the areas of the maximum $(f_0 + \Delta f)_{max}$ and minimum $(f_0 - \Delta f)_{min}$ transmitted frequency are not considered.

However, the actual situation with the formation of the differential signal is different. The difference in frequency is not formed as a simple mathematical difference, but as interference of transmitted λ_T and received λ_R wavelengths. Then, the wavelength of the carrying signal is minimal λ_{0min} in the case of maximum transmitted frequency $(f_0 + \Delta f)_{max}$ and, vice versa, the wavelength of carrying signal is maximal λ_{0max} at minimum transmitted frequency $(f_0 - \Delta f)_{min}$. With the equivalent time delay τ of the received signal within the whole modulation period but at the different wavelengths λ_{0min} and λ_{0max}, the value of difference frequency (interference) is not formed by the same way. At the wavelength of λ_{0min}, the interference is formed more often and the value of the higher difference frequency F_{dh} is higher and, vice versa, at the wavelength λ_{0max}, the interference is formed less frequently and the value of the lower difference frequency F_{dl} is smaller. Considering the presented phenomenon, the differential frequency is the same within the whole range of the modulation period (Figure 4c).

The above-presented theory is confirmed by the observation of realistic displays of the difference signal of a radio altimeter on an oscilloscope. In the higher frequency area f_{0max} and lower wavelength λ_{0min}, the higher differential frequency with the lower wavelength λ_{dh} will be formed with a higher number of the height pulses N. In the lower frequency area f_{0min} and the higher wavelength λ_{0max}, the lower difference frequency F_{dl} with the higher wavelength λ_{dl} will be formed with a lower number of the height pulses N (Figure 4d).

As a result of such unbalanced formation of differential frequency, and the fluent increase of the measured height, height pulses on the display of the oscilloscope will shift from the areas with the higher number of pulses into the area with the lower number of pulses. The presented shift of the periods of difference frequency and relevant height pulses is shown in Figure 4e.

Determination of the number of unstable pulses within the range of critical height ΔH is considered for a particular case of a radio altimeter with the carrier frequency of f_0 =4400 MHz, the modulation frequency of F_M = 150 Hz, and the frequency deviation of $\pm \Delta f$ = 25 MHz. In this case, we have f_{0max} = 4400 MHz + 25 MHz = 4425 MHz $\Rightarrow \lambda_{0min}$ = 67.797 $\times$ 10^{-3} m, f_0 = 4400 MHz $\Rightarrow \lambda_0$ = 68.182 $\times$ 10^{-3} m, and f_{0min} = 4400 MHz $-$ 25 MHz = 4375 MHz $\Rightarrow \lambda_{0max}$ = 68.571 $\times$ 10^{-3} m.

Then, the difference of wavelengths can be defined $\lambda_{max} - \lambda_{min}$ = 68.571 m $-$ 67.797 m = 0.774 $\times$ 10^{-3} m.

The calculated value expresses the height range of the formation of an unstable height pulse. The half value of the number of the unstable height pulses within half a period of $\lambda_0/2$ of a carrier frequency f_0 presents the number of unstable pulses within the critical height ΔH. Thus, in the case of considered ratio altimeter, the number of unstable pulses is 44.

Figure 5 shows the results of the simulation of the formation of the number of unstable height pulses within the critical height ΔH for the particular type of a radio altimeter.

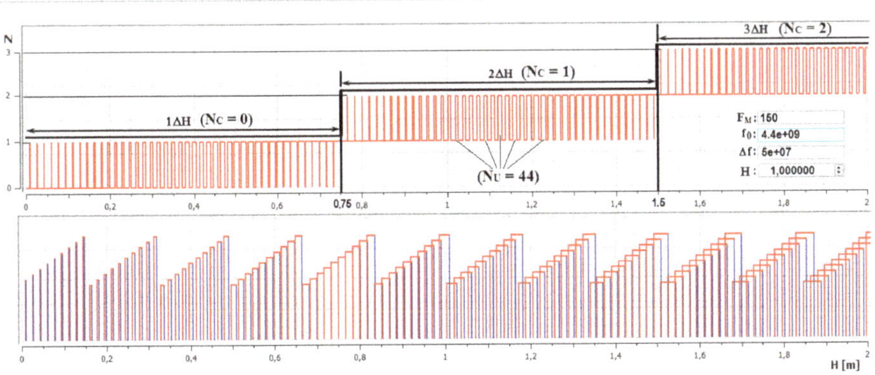

Figure 5. Simulation of number of unstable height pulses for the considering radio altimeter.

5. Number of Stable Pulses in Range of Critical Height

A recent method of increasing the accuracy of altitude measurement by the FMCW radio altimeters is based on the increase of frequency deviation with the simultaneous increase of the carrier signal frequency. However, the recent radio altimeters use the carrier frequency of 4.4 GHz with the frequency deviation of $\pm$25 MHz (overall bandwidth of 50 MHz), and they can provide the measurement accuracy of $\pm$0.75 m.

It is necessary to increase the carrier frequency with the increase of the frequency deviation due to the requirement for balanced frequency transmission characteristics of the high-frequency circuits, as well as active and passive elements of antennas. To increase the accuracy of the altitude measurement two-fold (to the value of $\pm$0.375 m) by the same way, the value of the frequency deviation also should be increased two-fold (to the value of $\pm$50 MHz that, in fact, corresponds to the overall bandwidth of 100 MHz), with the simultaneous increase of the carrier frequency approximately to 10 GHz.

Based on the above-analyzed theory, it is possible to avoid the problem of the necessity of using a higher carrier frequency of 10 GHz through the use of two parallel high-frequency channels at the original carrier frequency of 4.4 GHz. To create a performance signal, the frequency multipliers are commonly used in radio altimeters nowadays.

In such a way, a basic generator forms the frequency modulated signal with the center frequency of 2200 MHz and with the original frequency deviation $\pm\Delta f_1 = \pm25$ MHz. Then, the signal is divided by two frequency band-passes into two individual high-frequency channels. Each channel forms its own frequency deviation; the first channel has $+\Delta f_1$ from 2200 to 2225 MHz, and the second channel has $-\Delta f_1$ from 2175 to 2200 MHz.

Next, the frequency is independently multiplied two times into the value with the extreme frequency deviation of $\pm\Delta f_2 = \pm50$ MHz in each channel; the first channel has $+\Delta f_2$ from 4400 to 4450 MHz, and the second channel has $-\Delta f_2$ from 4350 to 4400 MHz (Figure 6). Each channel would only have a bandwidth that is commonly used, i.e., 50 MHz. At the same time, each channel is tuned in frequency to a different band, but the overall bandwidth is 100 MHz. The particular circuit connection of the radio altimeter with extreme bandwidth, which would comply with the above-presented condition, can be achieved by several solutions [14–17].

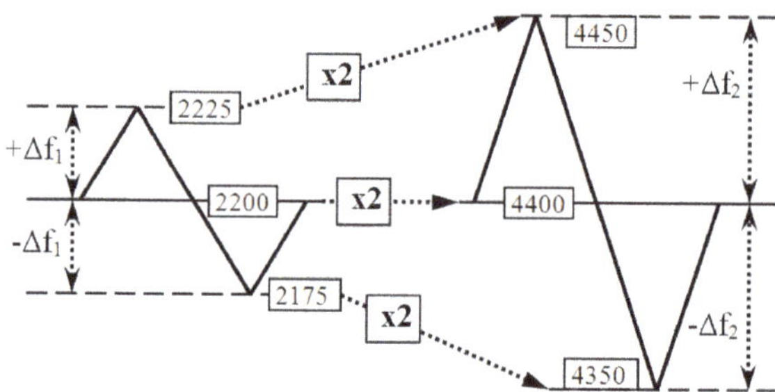

Figure 6. Principle of radio altimeter with extreme bandwidth.

Based on the presented background, parameters of the critical height for a radio altimeter with extreme bandwidth have been analyzed with the help of simulation. The parameters of such a radio altimeter are as the following: the carrier frequency is 4400 MHz, the modulation frequency is 150 Hz, and the frequency deviation is 50 MHz, which in sum gives extreme bandwidth of 100 MHz (Figure 7).

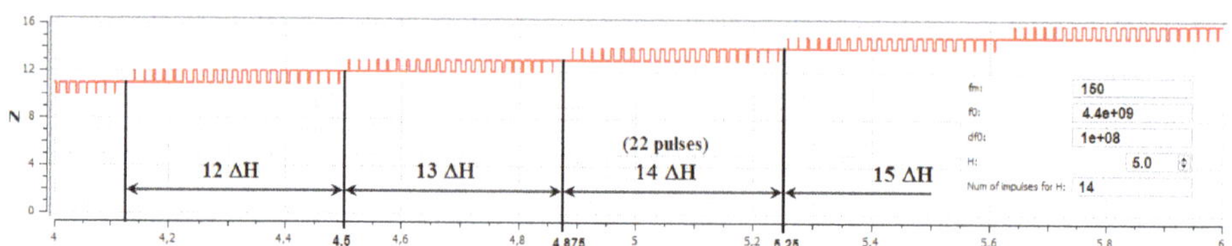

Figure 7. Unstable height pulses of radio altimeter with extreme bandwidth.

From the viewpoint of altitude measurement precision, the results for the radio altimeter with extreme bandwidth are:

(a) critical height has the value of 0.375 m;

(b) methodological error of the radio altimeter is ±0.375 m;

(c) simulated values are at the height of cca 5 m: 4.5 m, 4.875 m, and 5.25 m, respectively, for the bottom, mean, and upper values;

(d) number of simulated unstable pulses subtracted within the range of the critical height is 22;

(e) number of calculated unstable pulses within range of critical height is also 22.

Based on the theory presented in Section 4, the relation for the determination of the number of unstable pulses within the range of critical height has been derived as a part of the research:

$$Nu = \frac{f_0^2 - \Delta f^2}{2\,f_0\,\Delta f}.\tag{6}$$

From the above-presented results, it is possible to conclude that by using the double value of the total frequency deviation of 100 MHz at the carrier frequency of 4.4 GHz (a total frequency deviation of 50 MHz is recently used), it is possible to increase the accuracy of the height measurement two-fold. In this case, the critical height is 0.375 m, and it can be obtained by the formation of 22 unstable height pulses, but only when one stable pulse is formed. The unstable and stable pulses can be supported by the derivation of a basic equation of a radio altimeter from the height pulses formation [8].

The basic equation of a radio altimeter defines the relation between the value of the differential frequency and output information about the measured altitude, which is based on the basic parameters of a radio altimeter. Generally, known derivation comes out of the immediate difference of the frequencies of transmitted and received signals, proportionally to the time delay of the received signal, which depends on the measured altitude. A different method for a definition of a basic equation of a radio altimeter, based on critical height, has been derived during the research connected with the problem presented in the paper.

As it has been presented in this manuscript, a certain number of stable height pulses, shaped during the time of one modulation period T_M, corresponds to the measured altitude:

$$H = \frac{N}{T_M}.\tag{7}$$

The formation of only one stable height pulse during the time of one modulation period corresponds to the height change within the range of critical height:

$$\Delta H = \frac{1}{T_M}.\tag{8}$$

The following proportion can be obtained from Equations (7) and (8):

$$\frac{H}{\Delta H} = \frac{N/T_M}{1/T_M}.\tag{9}$$

Replacing the modulation period by the modulation frequency in Equation (9) we have:

$$\frac{H}{\Delta H} = \frac{N\,F_M}{F_M}.\tag{10}$$

The product of NF_M in Equation (10) presents the difference frequency, which is proportional to the altitude, so:

$$F_d = \frac{F_M}{\Delta H}H.\tag{11}$$

Substitution of the critical height in Equation (11) from Equation (1) produces the basic equation of a radio altimeter presented in a completely different way from the generally known one, with the use of geometric similarity of triangles [18–20]:

$$F_d = \frac{8\,\Delta f\,F_M}{c}H.\tag{12}$$

This fact proves all the above-presented theoretical consideration, which allows for an increase in the accuracy of the altitude measurement by a radio altimeter.

6. Software Application for Simulating the Unstable Pulses Creation

For the needs of the simulation and graphical display of the formation of the unstable pulses, a software application in Qt (C++) environment has been created. The algorithm is based on a theoretical analysis of the formation and dissolution of pulses when measuring altitude by a radio altimeter (Figure 8).

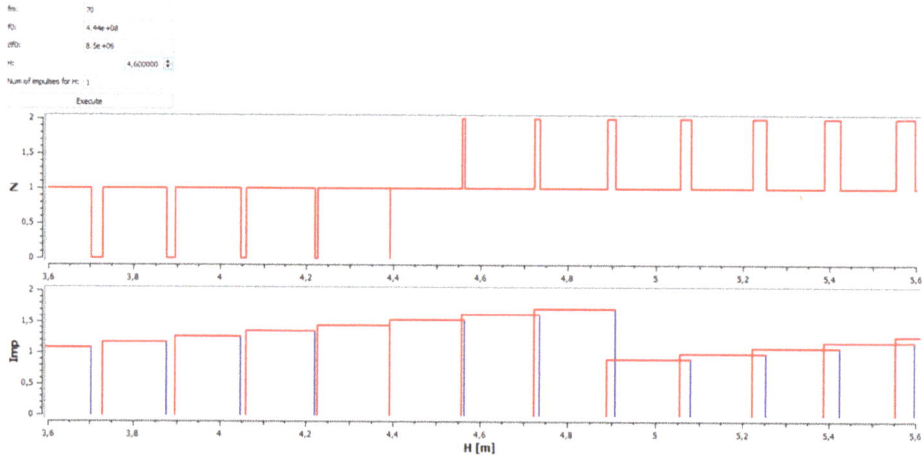

Figure 8. An example of the software application for simulating the unstable pulses creation.

The individual values corresponding to the altitude of the pulse formation and dissolution depend on several parameters: the modulation frequency f_m = 70 MHz, carrier frequency f_0 = 444 MHz, and frequency deviation Δf = 8.5 MHz. One more parameter on the control panel is the choice of the mean value of the flight altitude H, around which the creation of the formation and dissolution of the unstable pulses is displayed. This is the altitude that is interesting for us from a certain point of view. The height range is limited to ±1 m because a wider height range would cause a graphical ambiguity. The last parameter displayed on the control panel represents the calculated number of stable pulses corresponding to the set height of the flight altitude H.

These parameters can be changed with regard to the specific or hypothetical type of radio altimeter and its technical parameters using the application control panel represented in Figure 9.

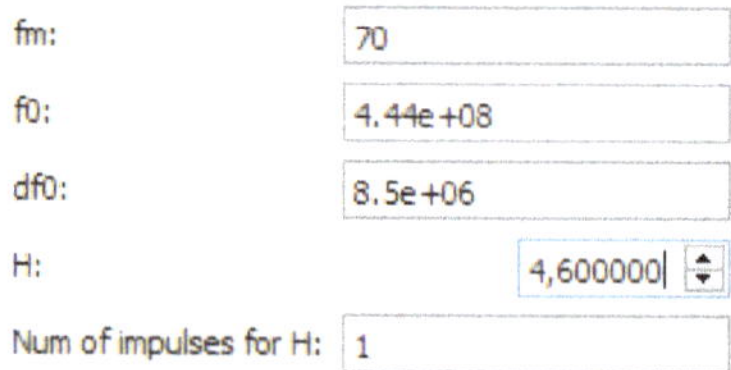

Figure 9. Application control panel.

Figure 8 shows an example of the creation of unstable pulses for a set height of 4.6 m evaluated in the range of 3.6 m to 4.6 m. The parameters of the analyzed radio altimeter are shown in Figure 9.

The bottom part of the graph in Figure 8 shows the process of gradual formation and dissolution of a large number of individual height pulses, with a change in height H. The upper part of the graph presents the result of their mutual overlap by repeated formation and dissolution of one unstable pulse in the ΔH range. The number of repeated formations of a single unstable pulse in the ΔH

range determines the accuracy of the altitude measurement. The resulting quantization (stairs) course of unstable pulses within the ΔH range shapes a number of stable pulses, which correspond to the measured altitude. The software application can be used to compare the height pulse creation and to simulate this process for different types of radio altimeters with various parameters.

7. Discussion and Conclusions

We have performed the analysis of the formation of the range of critical height of a radio altimeter methodological error when measuring the altitude. The analysis has been performed from the viewpoint of the determination of the number of unstable height pulses within the range of critical height. This manuscript focuses on the formation of unstable pulses within the range of critical height and explains the patterns between formation and dissolution of height pulses. As it is generally known, unstable pulses do not carry the information about measured altitude; they are not paid enough attention. Some works even consider the chaotic formation of the number of such pulses. Even though the unstable pulses do not determine measured altitude, and they have a critical influence on the precision of the height measurement.

Precise altitude measuring relates to the method of difference signal processing. The presented analysis supported by simulation has confirmed the applicability of the basic idea for the processing of the differential signal in the form of the creation of stable and unstable pulses.

The process of the determination of the number of unstable pulses within the range of critical height outlines a different view of the basic principle of a radio altimeter (shown in Figure 4c). An original and exact process of the determination of the number of unstable height pulses within the range of critical height for the recent radio altimeters is also presented. This method of calculation logically follows from Equation (6), which has successfully been verified in this manuscript for a different type of a radio altimeter.

The unique solution is based on the decrease of the value of the critical height by the decrease of the number of unstable pulses, which results in the doubled (extreme) increase of the frequency deviation at the original carrier frequency for the recent radio altimeters. In this way, the altitude measurement precision increases from the original value of ± 0.75 m to the two-times lower value of ± 0.375 m, without the need to increase the carrier frequency (two-fold). The correctness of this theory is highlighted by the new way of deriving the basic radio altimeter equation with the use of a critical height parameter. The results presented in the paper can be used for the design of new radio altimeters with the increased accuracy of the low altitude measurement.

Author Contributions: Conceptualization, J.L.; methodology, J.L.; software, M.Č. and P.K.; validation, M.Č., A.N. and J.G.; formal analysis, J.L., A.N. and J.G.; investigation, J.L., A.N. and J.G.; resources, J.L. and P.K.; data curation, M.Č. and P.K.; writing—original draft preparation, J.L.; writing—review and editing, J.L., A.N. and P.K.; visualization, M.Č. and P.K.; supervision, J.L. and A.N.; project administration, J.G.; funding acquisition, J.G.

Funding: Slovak authors J.L., P.K., M.Č., and J.G. has been supported by the Science Grant Agency of the Ministry of Education Science, Research, and Sport of the Slovak Republic, under contract No. 1/0772/17.

Acknowledgments: A.N. wishes to express his sincere appreciation to the University of Malaga for the provided opportunities during his exchange visit.

Conflicts of Interest: The authors declare no conflict of interest.

References

1. Nebylov, A.V. *Aerospace Sensors*; Momentum Press: New York, NY, USA, 2013; p. 348, ISBN 1-60650-059-7. [CrossRef]
2. Kelemen, M.; Szabo, S. *Pedagogical Research of Situational Management in Aviation Education and Forensic Investigation of Air Accidents: Knowledge of Aircraft Operation and Maintenance*; Collegium Humanum: Warsaw, Poland, 2019; p. 144, ISBN 978-83-952951-1-9.
3. Labun, J.; Adamčík, F.; Piľa, J.; Madarász, L. Effect of the measured pulses count on the methodical error of the air radio altimeter. *Acta Polytech. Hung.* **2010**, *7*, 41–49.

4. Soták, M.; Labun, J. The new approach of evaluating differential signal of airborne FMCW radar-altimeter. *Aerosp. Sci. Technol.* **2012**, *17*, 1–6. [CrossRef]

5. Stove, A.G. Linear FMCW radar techniques. *IEE Proc. F (Radar Signal Process.)* **1992**, *139*, 343–350. [CrossRef]

6. Skolnik, M.I. *Introduction to Radar Systems*, 2nd ed.; McGraw Hill Book Company: Singapore, 1981; p. 581, ISBN 0-07-057909-1.

7. Labun, J.; Soták, M.; Kurdel, P. Technical note innovative technique of using the radar altimeter for prediction of terrain collision threats. *J. Am. Helicopter Soc.* **2012**, *57*, 85–87. [CrossRef]

8. Labun, J.; Krchňák, M.; Kurdel, P.; Češkovič, M.; Nekrasov, A.; Gamcová, M. Possibilities of increasing the low altitude measurement precision of airborne radio altimeters. *Electronics* **2018**, *7*, 191. [CrossRef]

9. Taylor, J. Aircraft Operation of Radio Altimeters. In Proceedings of the 24th Meeting of Working Group F, Aeronautical Communications Panel (ACP), Paris, France, 17–21 March 2011.

10. Alivizatos, E.G.; Petsios, M.N.; Uzunoglu, N.K. Architecture of a multistatic FMCW direction-finding radar. *Aerosp. Sci. Technol.* **2008**, *12*, 169–176. [CrossRef]

11. Baskakov, A.I.; Komarov, A.A.; Mikhailov, M.S.; Ruban, A.V. Modeling of the methodical errors of high-precision aircraft radar altimeter operating above the seasurface at low altitudes. In Proceedings of the 2017 Progress in Electromagnetics Research Symposium—Spring (PIERS), St. Petersburg, Russia, 22–25 May 2017; pp. 3236–3240. [CrossRef]

12. Singh, R.; Peterson, R.; Riaz, A.; Hood, C.; Bacchus, R.; Roberson, D. A method for evaluating coexistence of LTE and radar altimeters in the 4.2–4.4 GHz band. In Proceedings of the 2017 Wireless Telecommunications Symposium (WTS), Chicago, IL, USA, 26–28 April 2017; pp. 1–9.

13. Liu, J.; Shen, C.; Wu, S.; Huang, W.; Li, C.; Zhu, Y.; Wang, L. Correction method of the manned spacecraft low altitude ranging based on γ ray. *Aerosp. Sci. Technol.* **2016**, *50*, 71–76. [CrossRef]

14. Labun, J.; Grega, M.; Sopata, M.; Kmec, F. Aircraft Radio Altimeter for Small Heights with Frequency Modulation Design II. Patent 283439; Banská Bystrica, Slovak Republic, 20 June 2003.

15. Grega, M.; Labun, J. Aircraft Radar Altimeter with Frequency Modulation Design. Patent 262626; Prague, Czechoslovak Republic, 16 August 1988.

16. Kayton, M.; Fried, W.R. *Avionics Navigation Systems*; John Wieley & Sons: New York, NY, USA, 1997; p. 773, ISBN 0-471-54795-6.

17. Nekrasov, A. *Foundations for Innovative Application of Airborne Radars: Measuring the Water Surface Backscattering Signature and Wind*; Springer: Dordrecht, The Netherlands, 2014; p. 116, ISBN 978-3-319-00620-8. [CrossRef]

18. Fasano, G.; Accardo, D.; Tirri, A.E.; Moccia, A.; Lellis, E. Radio/electro-optical data fusion for non-cooperative UAS sense and avoid. *Aerosp. Sci. Technol.* **2015**, *46*, 436–450. [CrossRef]

19. Metzger, J.; Meister, O.; Trommer, G.F.; Tumbrägel, F.; Taddiken, B. Adaptations of a comparison technique for terrain navigation. *Aerosp. Sci. Technol.* **2005**, *9*, 553–560. [CrossRef]

20. Estrada, J.; Zurek, P.; Popović, Z. Harvesting of aircraft radar altimeter side lobes for low-power sensors. In Proceedings of the 2018 International Applied Computational Electromagnetics Society Symposium (ACES 2018), Denver, CO, USA, 24–29 March 2018; pp. 1–2. [CrossRef]

Article

Micro-Motion Feature Extraction of a Rotating Target Based on Interrupted Transmitting and Receiving Pulse Signal in an Anechoic Chamber

Feng Zhao, Xiaobin Liu *, Zhiming Xu, Yuan Liu and Xiaofeng Ai

State Key Laboratory of Complex Electromagnetic Environmental Effects on Electronics and Information System, National University of Defense Technology, Changsha 410073, China; zhaofeng321@nudt.edu.cn (F.Z.); zmxu_nudt@163.com (Z.X.); liuyuan17a@nudt.edu.cn (Y.L.); anxifu2001@163.com (X.A.)
* Correspondence: liuxiaobin12@nudt.edu.cn; Tel.: +86-1550-748-6086

Received: 4 September 2019; Accepted: 10 September 2019; Published: 13 September 2019

Abstract: The pulse signal is widely used in micro-motion feature extraction of rapidly rotating targets as its pulse repetition frequency (PRF) can be high. However, when the pulse signal is implemented in a range-limited anechoic chamber for micro-motion feature extraction, the transmitted and reflected pulse signals may be coupled at the receiver. To solve this problem, the interrupted transmitting and receiving (ITR) method is applied to transmit the pulse signal with hundreds of sub-pulses. The target echo can be received when the sub-pulse is not transmitted. Hence, it avoids the coupling effect of transmitted signals and echoes. Then, the whole process of micro-motion feature extraction for rotating target is proposed based on the ITR method. At last, the simulations and experiments verify that the rotating target micro-Doppler can be extracted by the ITR pulse signal.

Keywords: pulse radar; rotating target; micro-motion feature extraction; interrupted transmitting and receiving (ITR)

---

1. Introduction

Radar target feature extraction is widely investigated for automatic target recognition [1,2]. The micro-motion feature is one of the most important features [3–5]. In the past decades, the micro-motion feature measurement in an anechoic chamber has been conducted via the swept frequency signal [6]. Firstly, the target is placed on the turntable at a certain angle. Then, the swept frequency signal is transmitted to the target, and the scattering field data at the specific aspect angle is obtained. The static measurement is conducted with the turntable turning in different aspect angles. Therefore, in order to obtain precise measurement results, the angle interval of the turntable should be small enough which results in a long measurement time. Besides, the pulse repetition frequency (PRF) of the swept frequency signal in [7] is small for micro-motion feature extraction. As a result, the micro-Doppler of the rapidly micro-motional target may be aliasing. Thus, the micro-motion feature cannot be extracted in this case.

Pulse signal is extensively used in target detection and recognition [8]. As a high PRF of the pulse signal can be achieved [9,10], it can be used in the rapid micro-motion feature extraction of micro-motional target. Then, the turntable turns continuously so that the dynamic measurement of micro-motional target can be implemented in the anechoic chamber. For pulse signal, the target has to be located in the unambiguous range so that it can be detected. However, the size of the anechoic chamber is relatively small in comparison with field experiments, and usually smaller than the minimum unambiguous range. Consequently, the pulse signal is difficult to be used in the anechoic chamber measurement directly. The inner reason is that the target echoes are received while the pulse is not completely transmitted. At this time, the transmitted and received pulse signal are coupled at the receiver [11], therefore, the received pulse signal is affected by the transmitted signal. Various wave cancellation methods have been discussed, such as acoustic cancellation [12,13] and radar echo cancellation [14]. However, as the transmitted and received pulse signals are relevant, they are difficult to be separated by the echo cancellation.

In order to conduct the micro-motion feature extraction in anechoic chamber with pulse radar signal, the whole procedure of target measurement is proposed in this paper based on the interrupted transmitting and receiving (ITR) [11,15,16]. Firstly, the pulse radar signal is transmitted and received in sub-microseconds to obtain the target echoes. Then, the target echo property is discussed in detail. The micro-motion extraction procedure based on time-frequency analysis [17,18] is provided and analyzed with experimental results.

The remainder of the paper is organized as follows. Section 2 discusses the principle of ITR for pulse radar signal and the characteristics of ITR echo. In Section 3, the procedure of micro-motion feature extraction for rotating target is provided based on the ITR echo. In Section 4, both simulation and experiment results in anechoic chamber are provided with different ITR control parameters to validate of the proposed rotating feature extraction method. Finally, some conclusions are drawn in Section 5.

2. Characteristics of Interrupted Transmitting and Receiving Echo

2.1. Principle of ITR

As discussed in [11] and [16], ITR means that the antenna is turned on to transmit the pulse signal in sub-microseconds and then interrupted. After that, the other antenna is turned on to receive the echo. Repeating the above procedures for hundreds of times, the whole pulse is fully transmitted. As a result, the complete pulse signal is divided into hundreds of sub-pulses as shown in Figure 1.

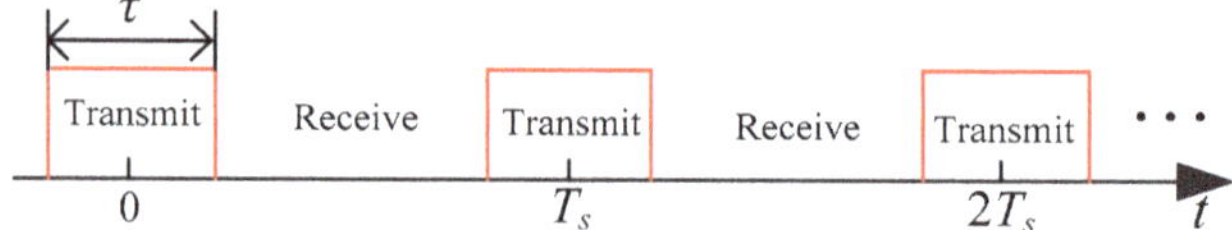

Figure 1. Waveform of p(t).

The ideal signal to control the transmitting and receiving can be expressed as

$$p(t) = \text{rect}(t/\tau) * \sum_{n=-\infty}^{+\infty} \delta(t - nT_s) \tag{1}$$

where $\delta(t)$ is the impulse function, n is the pulse number, T_s is the period, * represents the convolution operation, and $\text{rect}(t/\tau)$ is the rectangular envelope as follows

$$\text{rect}(t/\tau) = \begin{cases} 1, & |t/\tau| < 0.5 \\ 0, & \text{others} \end{cases} \tag{2}$$

where τ is restricted to sub-microsecond to ensure the sub-pulse be fully transmitted before it returns to the receiver in anechoic chamber. In practice, τ should be designed according to the real size of the anechoic chamber at first.

The spectrum of $p(t)$ is

$$P(f) = \tau f_s \sum_{n=-\infty}^{n=+\infty} \sin c(nf_s\tau)\delta(f - nf_s) \tag{3}$$

where $f_s = 1/T_s$, and $\text{sinc}(x) = \sin(\pi x)/(\pi x)$.

2.2. Property of ITR Echo

Assuming the operated pulse signal is linear frequency modulated (LFM) signal and it can be written as

$$s(t) = \text{rect}\left(\frac{t}{T_p}\right) \exp\left[j2\pi\left(f_c t + \frac{1}{2}\mu t^2\right)\right] \tag{4}$$

where f_c is the carrier frequency, T_p is the pulse width, μ is the chirp rate, and the bandwidth is $B = \mu T_p$.

The de-chirp reference signal is

$$s_{ref}(t) = \text{rect}\left(\frac{t - 2R_{ref}/C}{T_{ref}}\right) \exp\left[j2\pi\left(f_c\left(t - \frac{2R_{ref}}{C}\right) + \frac{1}{2}\mu\left(t - \frac{2R_{ref}}{C}\right)^2\right)\right] \tag{5}$$

where R_{ref} is the reference range, T_{ref} is the pulse width of reference signal, and C is the electromagnetic wave velocity.

After de-chirping with the reference signal, the difference-frequency output is

$$s_f(t) = s_r(t)s^*_{ref}(t) \tag{6}$$

where $s_{ref}^*(t)$ is the conjugation of $s_{ref}(t)$, and $s_r(t)$ is the echo which can be expressed as

$$s_r(t) = \sum_{k=1}^{K} \alpha_k s(t - 2R_k/C) \tag{7}$$

where K is the number of scattering centers, α_k is the scattering coefficient, and R_k is the distance between the scattering center k and radar.

Then, the ITR echo is obtained as

$$\begin{aligned}
s_{r_2}(t) &= p_2(t) \cdot s_r(t) \\
&= \left(\text{rect}\left(\frac{t}{\tau}\right) * \sum_{n \to -\infty}^{+\infty} \delta(t - nT_s)\right) s_r(t)
\end{aligned} \tag{8}$$

The high-resolution range profile (HRRP) of ITR echo can be obtained after fast Fourier transform (FFT)

$$S_{f_3}(f) = \tau f_s T_p \sum_{k=1}^{K} \alpha_k \exp\left(-j\frac{4\pi f_c}{C}R_{k,\Delta}\right) \sum_{n \to -\infty}^{+\infty} \left(\sin c(n f_c \tau) \sin c\left(T_p\left(f - n f_s + 2\frac{\mu}{C}R_{k,\Delta}\right)\right)\right) \tag{9}$$

where $R_{k,\Delta} = R_k - R_{ref}$.

From (9), we can find that the amplitude is related to τ, f_s and T_p. The second item sinc($\cdot$) in the summation operation illustrates that the de-chirp output is the accumulation of different sinc($\cdot$) functions. Besides, there are different orders of fake peaks in HRRP with the frequency interval f_s and the distance between the real and fake peaks is $\Delta R = Cf_s/(2\mu)$.

Therefore, the real peaks in HRRP can be fully obtained if $L < \Delta R$, where L is the target length. And then

$$T_s < \frac{C}{2\mu L} \tag{10}$$

3. Micro-Motion Feature Extraction of the Rotating Target

3.1. The Micro-Doppler of the Rotating Target

With the increasing complexity of target structure, many targets contain rotating components, such as aircraft propellers. The distance between the scattering centers and radar varies periodically with time, which produces the particular Doppler shift. Figure 2 shows the typical rotating target structure. Scattering centers A and B are both located at the line apexes and the location of which remains unchanged with the rotation. According to the scattering center classification [19,20], A and B are localized scattering centers which usually correspond to the geometric discontinuities such as corners, edge apexes, cone tips, and so on.

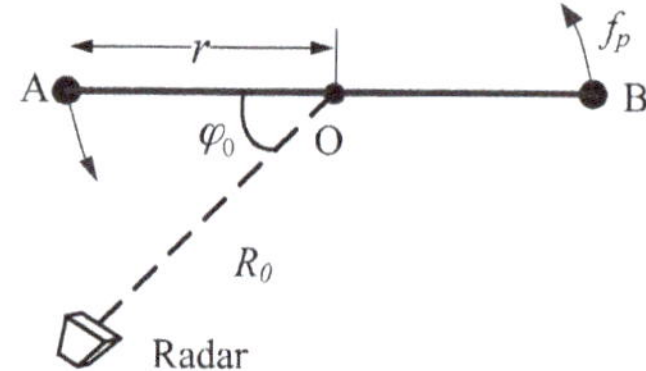

Figure 2. The typical rotating target structure.

Generally, the stop-and-go approximation is adopted, and the position of scattering center is basically unchanged in the fast time. In the slow time t_m, the distance between point A and radar can be expressed as

$$R_A(t_m) = R_0 - r\cos(2\pi f_p t_m + \varphi_0) \tag{11}$$

where R_0 is the distance between reference scattering center O and radar, r is the rotation radius, f_p is the rotation frequency and φ_0 is the initial phase.

According to (9), the HRRP of the ITR echo including micro-motion information can be obtained as

$$\hat{S}_{f_3}(f, t_m) = \tau f_s T_p \sum_{k=1}^{K} \left\{ \alpha_k \exp\left(-j\frac{4\pi f_c}{C} R_{k,\Delta}(t_m)\right) \right.$$
$$\left. \cdot \sum_{n \to -\infty}^{+\infty} \sin c(n f_s \tau) \sin c\left[T_p\left(f - n f_s + 2\frac{\mu}{C} R_{k,\Delta}(t_m)\right)\right] \exp\left(-j\frac{4\pi n f_s}{C} R_{ref}\right) \right\} \tag{12}$$

When $n = 0$, the peak value of HRRP in (9) is

$$\hat{S}'_{f_3}(f, t_m) = \tau f_s T_p \sum_{k=1}^{K} \alpha_k \exp\left(-j\frac{4\pi f_c}{C} R_{k,\Delta}(t_m)\right) \sin c\left[T_p\left(f + 2\frac{\mu}{C} R_{k,\Delta}(t_m)\right)\right] \tag{13}$$

where $R_{k,\Delta}(t_m) = R_k(t_m) - R_{ref}$.

Then, the phase term in (13) is

$$\phi_k(t_m) = -\frac{4\pi}{\lambda} R_{k,\Delta}(t_m) \tag{14}$$

where $\lambda = C/f_c$ is wavelength.

Therefore, the Doppler frequency of point A is obtained as

$$f_{micro,A} = \frac{1}{2\pi}\frac{d\phi_A(t_m)}{dt_m} = \frac{4\pi r f_p}{\lambda}\sin(2\pi f_p t_m + \varphi_0) \tag{15}$$

After time-frequency analysis with (13), the Doppler frequency of point A can be extracted, which is the result of Equation (15).

3.2. Procedure of the Micro-Doppler Extraction with ITR

Based on the above analysis, the whole procedure of micro-motion feature extraction for rotating target is given in Figure 3.

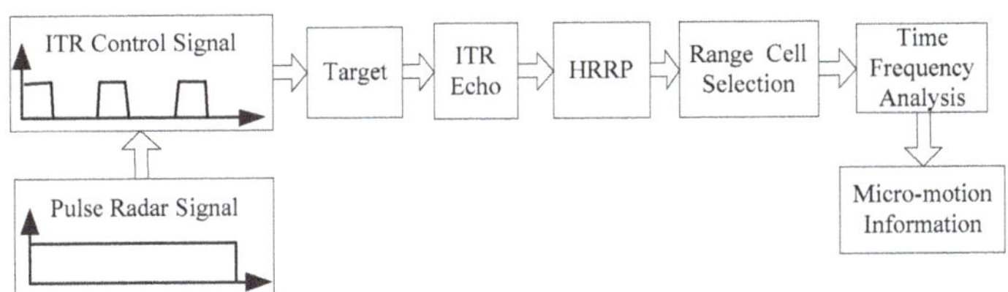

Figure 3. Procedure of micro-motion feature extraction for rotating target.

According to Figure 3, the target echo is received with ITR, and then the target HRRP is obtained by de-chirp processing. After selecting the peak position of target HRRP, the time-frequency analysis can be conducted along the slow time. There are many candidate time-frequency methods such as wavelet transform, Wigner–Ville distribution, short-time Fourier transform (STFT), and so on. Due to simplicity, STFT is chosen in this paper. Finally, the micro-motion information of rotating scattering center can be obtained.

4. Simulation and Experiment Results

4.1. Simulation Results of Rotating Target Micro-Doppler Extraction

The PRF of typical pulse radar signal may vary from hundreds of Hertz to hundreds of thousands of Hertz [9,10]. Therefore, the dynamic measurement of target with high rotation rate can be accomplished by the high PRF of pulse radar signals.

As shown in Figure 2, the typical rotating structure is simplified as two scattering centers, A and B. The distance between the two scattering centers is $2r = 0.2$ m. They are rotating around the midpoint O at a frequency of 8 Hz. The slant range between the midpoint O and radar R_0 is 45 m. The parameters of radar are listed in Table 1.

Table 1. Simulation parameters and values.

Parameter	Value	Parameter	Value
Pulse width T_p	12 μs	Bandwidth B	500 MHz
PRF	1.67 kHz	ITR period T_s	0.6 μs
Wave length λ	0.03 m	ITR pulse width τ	0.2 μs

In Table 1, the bandwidth of the pulse signal and swept frequency signal is 500 MHz. As the pulse width is 12 μs, the distance between the real and fake peaks is 6 m according to $\Delta R = Cf_s/(2\mu)$. Because the target rotation radius is 0.1 m, and the target length is 0.2 m so that the real and fake peaks can be separated. Moreover, the ITR pulse width τ should be smaller than $2R/C$ so that the sub-pulse can be fully transmitted when the reflected signal returns. Because the distance between the target and radar is $R = 45$ m, τ is set as 0.2 μs which is smaller than $2R/C = 0.3$ μs.

The PRF of swept frequency signal is set as 68 Hz for comparison since the equivalent PRF of the swept frequency experimental system is 68 Hz in [7]. The swept frequency interval is 1 MHz and the signal bandwidth is 500 MHz.

Firstly, the typical rotating target echo is simulated with swept frequency signal so that the HRRP can be obtained. According to Equations (7) and (8), the complete pulse echoes and ITR echoes are obtained. Then, the corresponding HRRPs are calculated via FFT. At last, the time-frequency analysis is conducted to extract micro-motion features based on the HRRPs. The micro-Doppler extraction results are shown in Figure 4.

Figure 4a is the range slow-time image obtained by the complete pulse echo. Figure 4b is the micro-Doppler extraction result of the complete pulse echo, in which the micro-Doppler frequency is obtained accurately, and it is the ideal result. Figure 4c is the range slow-time image obtained by the swept frequency signal. Figure 4d is the micro-Doppler extraction result of the swept frequency signal. Because of the PRF of swept frequency signal is small, i.e., $f_{micro,A} > PRF/2$, the extracted micro-Doppler frequency is aliased. Figure 4e is the range slow-time image obtained by the ITR echo. According to (10), the ITR period T_s is smaller than 18 μs, therefore, the real peaks can be extracted by adding a suitable interception window in the HRRP. Then, the time frequency analysis is performed to extract the micro-Doppler. We found that the micro-Doppler information depicted in Figure 4f is basically consistent with that in Figure 4b. In addition, the maximum amplitude values in Figure 4e,f is less than those in Figure 4a,b as depicted in the color bar. It is because the echo energy is reduced after ITR. However, the amplitude reduction can be eliminated by energy compensation.

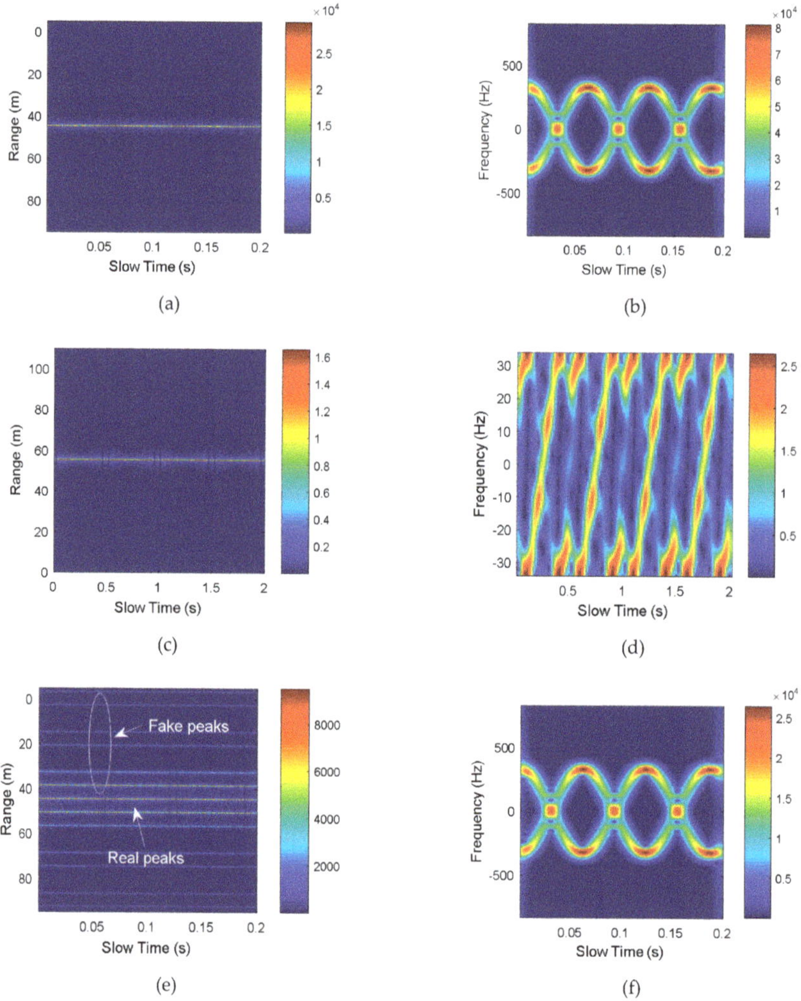

Figure 4. Micro-Doppler extraction results. (**a**) Range slow-time image of the complete pulse echo. (**b**) Time-frequency image of the complete pulse echo. (**c**) Range slow-time image of the swept frequency echo. (**d**) Time-frequency image of the swept-frequency echo. (**e**) Range slow-time image of the ITR echo. (**f**) Time-frequency image of the interrupted transmitting and receiving (ITR) echo.

4.2. Experiment Results in Anechoic Chamber

In this section, the pulse radar signal ITR experimental system is established for rotating target micro-motion extraction. The experimental system and the rotation target are shown in Figure 5.

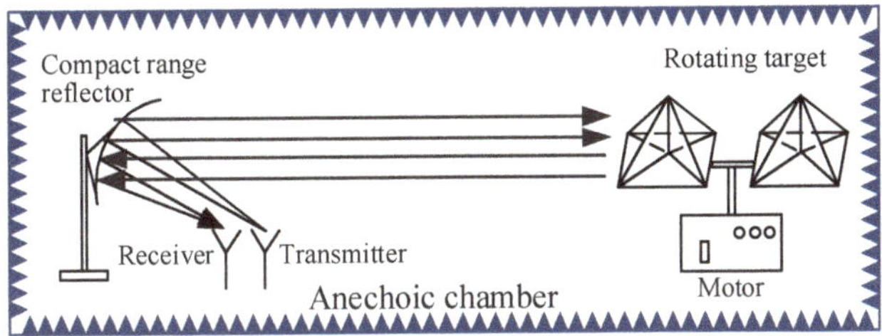

(a) The experiment scenario.

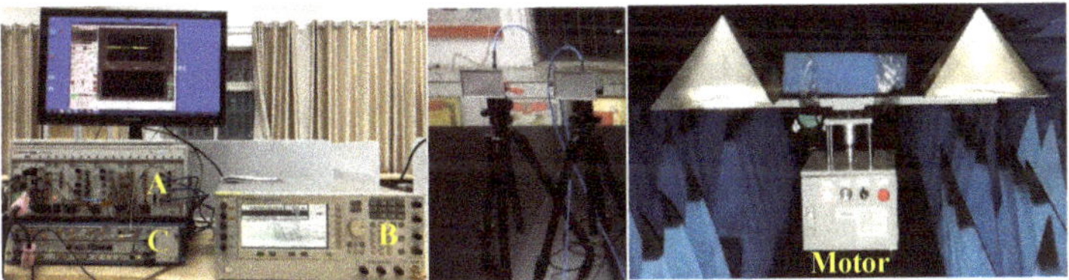

(b) The experimental system. (c) The horn antennas. (d) The rotation target.

Figure 5. Experimental system and rotation target model. (**a**) The experiment scenario. (**b**) The experimental system. (**c**) The horn antennas. (**d**) The rotation target.

The micro-motion feature measurement experimental scenario is presented in Figure 5a. The compact range reflector is used to ensure the plane wave. The width and length of the anechoic chamber are almost 15 m and 20 m, respectively. In this case, the pulse width should be smaller than 0.13 μs at the least to eliminate the coupling effect of the transmitted and reflected signal. Hence, the use of the ITR method is necessary. As shown in Figure 5b, the ITR experimental system is composed of three parts. Part A contains the local oscillator, arbitrary waveform generator (AWG), ITR controller, down-conversion (DC) module and intermediate frequency adjustment module. Part B is the vector signal generator (VSG) to up-convert the intermediate frequency signal. Part C is the intermediate frequency digitizer for data acquisition. Figure 5c. shows the transmitter and receiver horn antennas. The intermediate-frequency LFM signal of 300 MHz is generated by the AWG. Then, the signal is up-converted by the VSG to 9.3 GHz and the antenna radiates the signal to target by ITR antenna controller. Target echo is down-converted by the DC module. After intermediate frequency adjustment, the echo can be stored by the intermediate frequency digitizer module. Finally, the ITR echo is processed on the computer.

In Figure 5d, two corner reflectors are placed at the end of the cross bar to simulate the rotating scattering centers. The rotation is controlled by the motor and the rotation rate of the motor needs to be manual controlled during the experiment. The experimental parameters are listed in Table 2.

According to the procedure in Figure 3, when $B = 10$ MHz and the rotation rate is about 0.85 s, the results of micro-Doppler extraction are drawn as follows. The maximum rotation frequency of the corner reflector is extracted by time frequency distribution–Hough (TFD–Hough) transformation [21,22].

The HRRP of ITR echo is shown in Figure 6b. Then, the range slow-time image can be obtained in Figure 6c. After extracting the target peaks in HRRP, time-frequency analysis can be conducted in the slow time domain to obtain the time-frequency image in Figure 6d. It can be found that two sinusoidal curves with the same period and opposite initial phases are presented in the time-frequency image, which illustrates the micro-Doppler of the two corner reflectors. According to the time-frequency image, the micro-Doppler period is about 0.86 s, which is well consistent with the rotation period 0.85 s set in the experiment. At the same time, when r = 0.38 m, λ = 0.03 m, and f_p = 1/0.85 s = 1.18 Hz, the theoretical maximum micro-Doppler frequency of the corner reflector is 187.26 Hz according to (15). The maximum frequency in the time-frequency image is 185.09 Hz, which is also consistent with the calculation results.

Table 2. Parameter setting of uniform transmitting and receiving period for rotating target.

Parameter	Value	Parameter	Value
Pulse width T_p	12 μs	bandwidth B	10 MHz 300 MHz
PRF	1 kHz	ITR period T_s	0.4 μs
Wave length	0.03 m (10 GHz)	ITR pulse width τ	0.1 μs
Target range	15.5 m	rotation radius	0.38 m

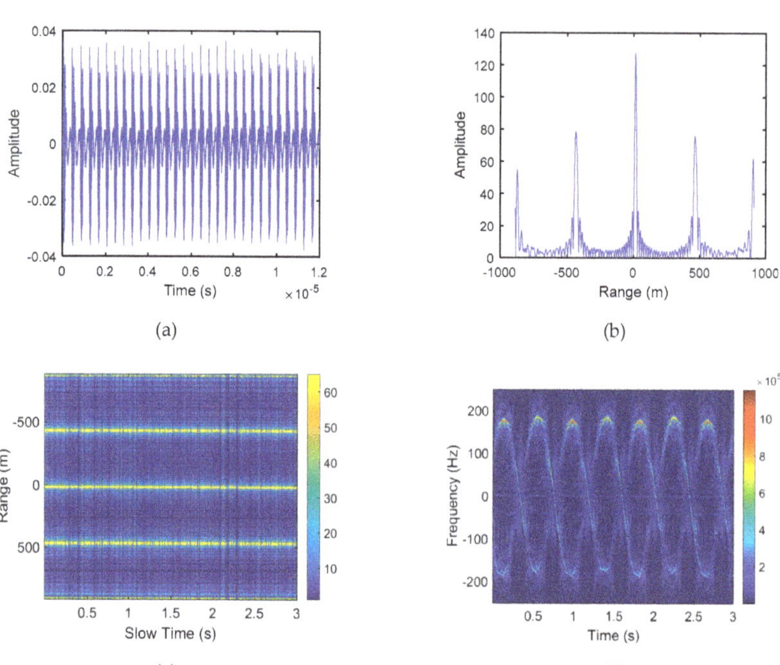

Figure 6. ITR echo and the micro-Doppler extraction results (B = 10 MHz). (**a**) ITR echo. (**b**) HRRP of the ITR echo. (**c**) Range slow-time image. (**d**) Time-frequency image.

When the rotation period is 1.1s and B = 300 MHz, the micro-Doppler extraction results are shown in Figure 7.

The HRRP of ITR echo is shown in Figure 7b. As the signal bandwidth increases, the range resolution is $C/(2B)$ = 0.5 m so that the two reflectors can be distinguished by HRRP in Figure 7c. The range slow-time 2D image is obtained in Figure 7d,e. It can be seen that the periodic characteristic of HRRP is clear in Figure 7e. When the two reflectors are in the same range cell, the scattering points cannot be separated in the HRRP. Time-frequency analysis is conducted by selecting this range cell and the time-frequency image is obtained as presented in Figure 7f. Comparing Figure 7e,f, when two scattering points are in the same range cell, the corresponding micro-Doppler frequency is maximum.

In addition, according to the time-frequency image and range slow-time 2D image, the target rotation period is about 1.122 s and the maximum micro-Doppler frequency is 141.87 Hz. The rotation rate is set as 1.1s, and the maximum micro-Doppler frequency calculated by (15) is 144.7 Hz, which are all consistent with the experiment results.

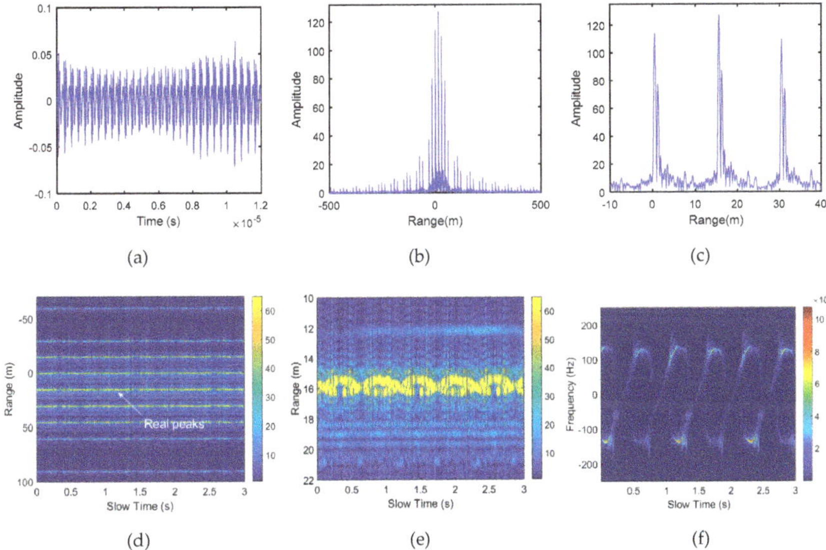

Figure 7. The ITR echo and micro Doppler extraction results ($B = 300$ MHz, $T_s = 0.4$ μs). (**a**) The ITR echo. (**b**) HRRP of ITR echo. (**c**) Magnification of HRRP. (**d**) Range slow-time 2D image. (**e**) Magnification of Range slow-time 2D image. (**f**) Time-frequency image.

5. Conclusions

The whole procedure for the micro-motion feature extraction of rapid micro-motional target is proposed based on the ITR method. The ITR is firstly applied to obtain the target echo in the anechoic chamber. Then, the HRRP of rotating target is deduced with the ITR echo. After that, the target range cell is selected for target micro-Doppler extraction. At last, numerical simulations and experiments in the anechoic chamber are conducted, which verify the validity of the proposed method.

In future, the dynamic measurements of more complicated micro-motional target will be discussed with the proposed method. We will also investigate the echo reconstruction method which makes the ITR echoes approximate the complete pulse echoes as much as possible.

Author Contributions: Methodology, F.Z. and X.L.; software, Z.X.; validation, X.A. and X.L.; formal analysis, F.Z.; investigation, F.Z.; resources, X.A.; data curation, X.L.; writing—original draft preparation, F.Z. and X.L.; writing—review and editing, Z.X. and Y.L.; funding acquisition, F.Z.

Funding: This research was funded by the National Natural Science Foundation of China, grant number 61890542.

References

1. Wang, Y.; Wang, Q.; Chen, D. Research on target recognition based on edge features. In Proceedings of the 2012 5th International Congress on Image and Signal Processing, Chongqing, China, 16–18 October 2012; pp. 1312–1315.

2. Gaikwad, A.N.; Singh, D.; Nigam, M.J. Recognition of target in through wall imaging using shape feature extraction. In Proceedings of the 2011 IEEE International Geoscience and Remote Sensing Symposium, Vancouver, BC, Canada, 24–29 July 2011; pp. 957–960.

3. Chen, V.C.; Li, F.; Ho, S.; Wechsler, H. Micro-Doppler effect in radar: Phenomenon, model, and simulation study. *IEEE Trans. Aerosp. Electron. Syst.* **2006**, *42*, 2–21. [CrossRef]

4. Zhi, X.; Ai, X.; Wu, Q.; Zhao, F.; Xiao, S. Micro-Doppler characteristics of streamlined ballistic target. *Electron. Lett.* **2019**, *55*, 149–157.

5. Ai, X.; Xu, Z.; Wu, Q.; Liu, X.; Xiao, S. Parametric Representation and Application of Micro-Doppler Characteristics for Cone-shaped Space Targets. *IEEE Sens. J.* **2019**. [CrossRef]

6. Liu, J.; Wu, Q.; Ai, X. Experimental study on full-polarization micro-Doppler of space precession target in microwave anechoic chamber. In Proceedings of the IEEE Sensor Signal Processing for Defence, Edinburgh, UK, 22–23 September 2016; pp. 1–5.

7. Liu, J.; Li, G.; Ma, L.; Hu, W.; Li, Y.; Wang, X. Dynamic measurement of micro-motion targets in microwave anechoic chamber. In Proceedings of the IEEE International Radar Conference, Guilin, China, 20–22 April 2009; pp. 1–4.

8. Lin, C.; Tian, R.; Bao, Q.; Chen, Z. A wavelet based denoising method for weak target detection of pulse compression radar. In Proceedings of the Progress in Electromagnetics Research Symposium—Spring (PIERS), St. Petersburg, Russia, 22–25 May 2017; pp. 244–249.

9. Mark, A. *Richards, Fundamentals of Radar Signal Processing*; McGraw-Hill: New York, NY, USA, 2005.

10. Skolnik, M. *Radar Handbook*; McGraw-Hill: New York, NY, USA, 2008.

11. Liu, X.; Liu, J.; Zhao, F.; Ai, X.; Wang, G. An equivalent simulation method for pulse radar measurement in anechoic chamber. *IEEE Geosci. Remote Sens. Lett.* **2017**, *14*, 1081–1085. [CrossRef]

12. Arun, A. Targeted sound attenuation capacity of 3D printed noise cancelling waveguides. *Appl. Acoust.* **2019**, *151*, 30–44.

13. Arun, A. Acoustic absorption of passive destructive interference cavities. *Mater. Today Commun.* **2019**, *19*, 68–75.

14. Huang, Q.; Xu, Y. Active cancellation stealth analysis based on interrupted-sampling and convolution modulation. *Optik Int. J. Light Electron. Opt.* **2016**, *127*, 3499–3503. [CrossRef]

15. Liu, X.; Liu, J.; Wu, Q.; Zhao, F.; Wang, G. Experimental study on pulse radar target probing in RFS based on interrupted transmitting and receiving. *Chin. J. Aeronaut.* **2018**, *31*, 575–583. [CrossRef]

16. Liu, X.; Liu, J.; Zhao, F.; Ai, X.; Wang, G. A novel strategy for pulse radar HRRP reconstruction based on randomly interrupted transmitting and receiving in radio frequency simulation. *IEEE Trans. Antennas Propag.* **2018**, *66*, 2569–2580. [CrossRef]

17. Liu, L.; Wang, S.; Zhao, Z. Radar waveform recognition based on time-frequency analysis and artificial bee colony-support vector machine. *Electronics* **2018**, *7*, 59. [CrossRef]

18. Guo, K.; Qu, Q.; Feng, A.; Sheng, X. Miss distance estimation based on scattering center model using time-frequency analysis. *IEEE Antennas Wirel. Propag. Lett.* **2016**, *15*, 1012–1015. [CrossRef]

19. Guo, K.; Qu, Q.; Sheng, X. Geometry Reconstruction Based on Attributes of Scattering Centers by Using Time-Frequency Representations. *IEEE Trans. Antennas Propag.* **2016**, *64*, 708–720. [CrossRef]

20. Xu, Z.; Ai, X.; Wu, Q.; Zhao, F.; Xiao, S. Coupling Scattering Characteristic Analysis of Dihedral Corner Reflectors in SAR Images. *IEEE Access* **2018**, *6*, 78918–78930. [CrossRef]

21. Barbaross, S.; Lemoine, O. Analysis of nonlinear FM signals by pattern recognition of their time-frequency representation. *IEEE Signal Process. Lett.* **1996**, *3*, 112–115. [CrossRef]

22. Ai, X.; Zou, X.; Yang, J.; Liu, J.; Li, Y. Feature extraction of rotation target based on bistatic micro-Doppler analysis. In Proceedings of the 2011 IEEE CIE International Conference on Radar, Chengdu, China, 24–27 October 2011; pp. 609–612.

MDPI

St. Alban-Anlage 66

4052 Basel

Switzerland

Tel. +41 61 683 77 34

Fax +41 61 302 89 18

www.mdpi.com

Electronics Editorial Office

E-mail: electronics@mdpi.com

www.mdpi.com/journal/electronics

9 783039 361427